Hybridization of Crop Plants

Hybridization of Crop Plants

EDITORS:

WALTER R. FEHR AND HENRY H. HADLEY

Editorial Committee

R. A. FORSBERG A. R. HALLAUER A. W. HOVIN

Managing Editor

D. A. FUCCILLO

Coordinating Editor

MATTHIAS STELLY

American Society of Agronomy and
Crop Science Society of America, Publishers
Madison, Wisconsin, USA
1980

Library of Congress Cataloging in Publication Data

Hybridization of crop plants.

Bibliography: 766 p. + xxvi
Includes index.
1. Plant-breeding. 2. Hybridization, Vegetable. 3. Fertilization of plants. I. Fehr, Water R., 1939- II. Hadley, Henry Hultman, 1917- III. American Society of Agronomy. IV. Crop Science Society of America.
SB123.H9 631.5′23 80-12306
ISBN 0-89118-034-6

Revised in 1982.

The American Society of Agronomy, Inc.
677 S. Segoe Road, Madison, Wisconsin, USA 53711

Printed in the United States of America

FOREWORD

The publications of a scientific society are the chief tangible evidence of its vigor and vitality; they set a standard by which the society can be evaluated. The American Society of Agronomy and the Crop Science Society of America are pleased to have responded to the needs of our members and other agricultural workers and have shared in the development of this book dealing with hybridization of crop plants. Each chapter of this book was under the supervision of a member of an editorial committee jointly appointed by the American Society of Agronomy and the Crop Science Society of America.

This publication serves as a reference on the principles and procedures used to obtain hybrid seed of self and cross-pollinated crop plants. For many crops, descriptions of these procedures are passed orally from experienced to inexperienced persons and are not available in the literature. This book is not intended only as a text, but should be suitable as a reference even for persons inexperienced in plant breeding. It brings together the experience of plant breeders and other scientists in a form which can be used by persons unfamiliar with the crops and is intended for scientists, teachers, and students in plant science disciplines. It does not discuss in detail the breeding or genetics research for which the seed may be used.

The American Society of Agronomy and the Crop Science Society of America are pleased to bring you this special publication on a very important topic in agriculture. We express our deepest appreciation to Dr. Walter R. Fehr of Iowa State University and Dr. Henry H. Hadley of the University of Illinois, coeditors of the book and to the many authors.

ROY G. CREECH
Crop Science Society of America
Mississippi State, Mississippi

JOHN PESEK
American Society of Agronomy
Ames, Iowa

February 20, 1979

PREFACE

Hybridization and self-pollination are important aspects of research for the genetic improvement of crop plants. Hybridization is used to develop new genotypes, evaluate their performance, or exploit hybrid vigor. Self-pollination is used to evaluate genotypes and develop pure lines for cultivar development or genetic research.

This book was prepared to meet the need for written information on methods of hybridization and self-pollination of crop plants. A written description of methods used to obtain desired seed has not been available for many crop species. Successful procedures commonly are communicated verbally among scientists, and new techniques often are not considered important enough for publication in a journal. Lack of written information has hindered persons from learning successful methods without investing considerable effort in contacting scientists who have experience with a crop. Students also have not been able to obtain an overview of hybridization and self-pollination methods.

The book is intended to serve as both a text and a reference. The eight introductory chapters provide an overview of principles to be considered in hybridization and self-pollination, and should be of particular value to teachers and students. The remaining chapters provide detailed information on the procedures used for specific crops. Authors of the crop chapters followed identical subject headings to facilitate comparison of methods used in different species.

Preparation of a book of reasonable length and readability restricted the number of references that could be included by the authors. They were asked to include review articles and key references that would serve as a starting point for readers interested in a more exhaustive review of a subject. Personal communication with scientists provided considerable information, and each source of such information is documented.

The successful completion of this book is the result of the interest and participation of many scientists in the American Society of Agronomy and the Crop Science Society of America. Their first contribution was in the form of letters written early in 1976 to the societies in which they requested the preparation of a book that would bring together this important body of knowledge. The executive committees of the societies acknowledged this request, and appointed a feasibility committee to investigate the need, import, and contents of such a book. The committee, consisting of Walter R. Fehr, Chairman, Marvin K. Aycock, Jr., Jon L. Geadleman, Earl C. Gilmore, Darrell A. Miller, and G. Allan Taylor, recommended in December 1976 that the book be written. The Board of Directors of the societies approved the recommendation and agreed to cooperate in the venture. An editorial committee was appointed in January 1977. Over the next 2 years, numerous society members helped to identify qualified authors, provide information to those selected to prepare chapters, and review manuscripts. Their contribution is gratefully acknowledged.

It is with deepest gratitude that we acknowledge the contributions made by the authors. They were recommended to the editorial committee by their fellow scientists and are knowledgeable in a particular subject area or with a crop species. The quality of the book reflects their diligence in contacting colleagues for information, reviewing the literature, obtaining appropriate figures, and sharing their own experiences. We thank them for sharing their time and talents in service to others.

We acknowledge the contribution of the editorial committee: Robert Forsberg, Arnel Hallauer, and Arne Hovin, in preparing the outline for the book, selecting authors, and coordinating the preparation of individual chapters. The counsel of Domenic Fuccillo, managing editor, and the work of the societies' staff in the final editing of the book are gratefully acknowledged. We also thank the societies for financial support and encouragement.

Walter R. Fehr, Ames, Iowa
Henry H. Hadley, Urbana, Illinois
Coeditors
April 1979

CONTRIBUTORS

R. E. ALLAN, USDA-SEA, Department of Agronomy and Soils, Washington State University, Pullman, WA 99164

W. POWELL ANDERSON, Department of Agronomy, New Mexico State University, Las Cruces, NM 88003

A. K. AUCKLAND, International Crops Research Institute for the Semi-Arid Tropics, 1-11-256 Begumpet, Hyderabad 500016, A.P., India

D. K. BARNES, USDA-SEA, Department of Agronomy and Plant Genetics, University of Minnesota, St. Paul, MN 55101

R. D. BARNETT, Department of Agronomy, University of Florida, Agricultural Research and Education Center, Quincy, FL 32351

E. C. BASHAW, USDA-SEA, Department of Soil and Crop Sciences, Texas A&M University, College Station, TX 77843

BENJAMIN H. BEARD, USDA-SEA, Agronomy and Range Science, University of California, Davis, CA 95616

H. T. BLACKHURST, Department of Horticultural Sciences, Texas A&M University, College Station, TX 77843

F. A. BLISS, Department of Horticulture, University of Wisconsin, Madison, WI 53706

D. A. BOND, Plant Breeding Institute, Cambridge, England

RAYMOND D. BRIGHAM, Texas Agricultural Experiment Station, Texas A&M University, Lubbock, TX 79401

C. M. BROWN, Department of Agronomy, University of Illinois, Urbana, IL 61801

BYRON L. BURSON, USDA-SEA, Grassland, Soil, and Water Research Laboratory, P.O. Box 748, Temple, TX 76501

GLENN W. BURTON, USDA-SEA, Georgia Coastal Plain Station, Tifton, GA 31794

W. RONNIE COFFMAN, Plant Breeding Department, International Rice Research Institute, Manila, Philippines

VERNE E. COMSTOCK, Department of Agronomy and Plant Genetics, University of Minnesota, St. Paul, MN 55101

L. G. DALTON, Department of Sorghum Breeding, Pioneer Hi-Bred International, Inc., Plainview, TX 79072

WALTER DEDIO, Agriculture Canada, Research Station, Morden, Manitoba, Canada R0G 1J0

R. K. DOWNEY, Agriculture Canada, Research Station, Saskatoon, Saskatchewan, Canada S7N 0X2

W. ANSON ELLIOTT, Department of Agronomy and Plant Genetics, University of Minnesota, St. Paul, MN 55108

WALTER R. FEHR, Department of Agronomy, Iowa State University, Ames, IA 50011

R. A. FORSBERG, Department of Agronomy, University of Wisconsin, Madison, WI 53706

J. M. GREEN, International Crops Research Institute for the Semi-Arid Tropics, 1-11-256 Begumpet, Hyderabad 500016, A.P., India

EARL T. GRITTON, Department of Agronomy, University of Wisconsin, Madison, WI 53706

J. P. GUSTAFSON, Plant Sciences Department, University of Manitoba, Winnipeg, Manitoba, Canada R3T 2N2

H. H. HADLEY, Department of Agronomy, University of Illinois, Urbana, IL 61801

A. R. HALLAUER, USDA-SEA, Department of Agronomy, Iowa State University, Ames, IA 50011

ALFRED HAUNOLD, USDA-SEA, Department of Crop Science, Oregon State University, Corvallis, OR 97331

R. M. HERRARA, Plant Breeding Department, International Rice Research Institute, Manila, Philippines

A. W. HOVIN, Agricultural Experiment Station, Montana State University, Bozeman, MT 59717

T. HYMOWITZ, Department of Agronomy, University of Illinois, Urbana, IL 61801

NORMAN I. JAMES, USDA-SEA, National Program Staff, Beltsville, MD 20705

A. JONES, USDA-SEA, U.S. Vegetable Laboratory, U.S. Department of Agriculture, Charleston, SC 29407

KAZUO KAWANO, Centro Internacional de Agricultura Tropical, Cali, Colombia

A. J. KLASSEN, Agriculture Canada, Research Station, Saskatoon, Saskatchewan, Canada S7N 0X2

P. F. KNOWLES, Department of Agronomy and Range Science, University of California, Davis, CA 95616

E. N. LARTER, Plant Science Department, University of Manitoba, Winnipeg, Manitoba, Canada R3T 2N2

D. A. LAWES, Welsh Plant Breeding Station, Aberystwyth, Wales

JOSHUA A. LEE, USDA-SEA, Department of Crop Science, North Carolina State University, Raleigh, NC 27650

NELS R. LERSTEN, Department of Botany and Plant Pathology, Iowa State University, Ames, IA 50011

KOERT J. LESSMAN, Department of Agronomy, New Mexico State University, Las Cruces, NM 88003

D. J. MAJOR, Agriculture Canada, Research Station, Lethbridge, Alberta, Canada T1J 4B1

HAROLD G. MARSHALL, USDA-SEA, Department of Agronomy, Pennsylvania State University, University Park, PA 16802

D. F. MATZINGER, Department of Genetics, North Carolina State University, Raleigh, NC 27650

J. CREIGHTON MILLER, JR., Department of Horticultural Sciences, Texas A&M University, College Station, TX 77843

D. D. MOREY, Department of Agronomy, University of Georgia, Coastal Plain Experiment Station, Tifton, GA 31794

F. J. MUEHLBAUER, USDA-SEA, Department of Agronomy and Soils, Washington State University, Pullman, WA 99164

A. J. NORDEN, Department of Agronomy, University of Florida, Gainesville, FL 32611

S. J. OPENSHAW, Department of Agronomy, University of Illinois, Urbana, IL 61801
JOHN F. OSBORNE, Soil Analysis Division, Department of Land Development, Phaholyothin Road, Bangkhen, Bangkok, Thailand
ROBERT L. PLAISTED, Department of Plant Breeding and Biometry, Cornell University, Ithaca, NY 14853
M. H. POULSEN, Royal Veterinary and Agricultural University, Taastrup, Denmark
ERIC D. PUTT, Agriculture Canada, Research Station, Morden, Manitoba, Canada R0G 1J0
CHARLES M. RICK, Department of Vegetable Crops, University of California, Davis, CA 95616
W. A. RUSSELL, Department of Agronomy, Iowa State University, Ames, IA 50011
K. F. SCHERTZ, Department of Soil and Crop Sciences, Texas A&M University, College Station, TX 77843
D. SHARMA, International Crops Research Institute for the Semi-Arid Tropics, 1-11-256 Begumpet, Hyderabad 500016, A.P., India
D. P. SINGH, Jute Agricultural Research Institute, Barrackpore-743101, West Bengal, India
A. E. SLINKARD, Crop Development Centre, University of Saskatchewan, Saskatoon, Saskatchewan, Canada S7N 0X2
GARRY A. SMITH, USDA-SEA, Crops Research Laboratory, Colorado State University, Fort Collins, CO 80523
R. R. SMITH, USDA-SEA and Department of Agronomy, University of Wisconsin, Madison, WI 53706
R. E. STAFFORD, USDA-SEA, Texas A&M University Agricultural Research and Extension Center, Vernon, TX 73684
T. M. STARLING, Department of Agronomy, Virginia Polytechnic Institute and State University, Blacksburg, VA 24601
G. R. STRINGAM, Agriculture Canada, Research Station, Saskatoon, Saskatchewan, Canada S7N 0X2
CHARLES W. STUBER, Department of Genetics, North Carolina State University, Raleigh, NC 27650
NORMAN L. TAYLOR, Department of Agronomy, University of Kentucky, Lexington, KY 40546
C. E. TOWNSEND, USDA-SEA, Crops Research Laboratory, Colorado State University, Fort Collins, CO 80523
L. J. G. VAN DER MAESEN, International Crops Research Institute for the Semi-Arid Tropics, 1-11-256 Begumpet, Hyderabad 500016, A.P., India
E. A. WERNSMAN, Department of Crop Science, North Carolina State University, Raleigh, NC 27650
V. E. WILSON, USDA—SEA, Department of Agronomy and Soils, Washington State University, Pullman, WA 99164
HAROLD WRIGHT, Department of Quality Control, DeKalb Ag-Research, Inc., DeKalb, IL 60115
DEMETRIOS M. YERMANOS, Department of Plant Sciences, University of California, Riverside, CA 92502

CONTENTS

1 Environmental Effects on Flowering

D. J. Major
Agriculture Canada
Lethbridge, Alberta

Flower and seed development in crop plants is determined by the response of the plant to environmental factors. Understanding these relationships permits the plant breeder to obtain hybrid and self-pollinated seed from a broad range of cultivars in the field, growth chamber, or greenhouse. This chapter describes the general response of plants to environmental factors that influence flowering and shows how these factors may be manipulated to facilitate production of seed.

I. RESPONSE OF PLANTS TO ENVIRONMENT

Crop plants differ widely in their response to environmental factors. They differ among species, among cultivars within species, and among stages of plant development within a cultivar. The response reflects differences in area of adaptation and in the biological mechanisms that have evolved for coping with adverse environments.

The main environmental factors influencing flowering are day length, also referred to as photoperiod, and temperature. Species are usually classified as short-day or long-day types. A short-day plant flowers when the day length is less than its critical length and a long-day plant flowers when the day length is longer than its critical length. Some species are considered to be day-neutral, but it is unlikely that all of the cultivars of any species are completely unaffected by day length. Corn has been called a day-neutral species (Bickford and Dunn, 1972) but short-day responses have been found even in early-maturing corn hybrids (Hunter et al., 1974).

Temperature influences the rate of development and the number of flowers that produce mature seeds. Cool-season crops, such as oat and barley produce maximum seed at about 25 C, whereas tropical crops like pigeonpea have optimum production above 30 C.

The genetic variation in response to day length and, to a lesser degree, temperature has allowed many species to adapt to a wide range of latitudes and altitudes. Soybean, a short-day species, is grown from the tropics to 50° N Lat. Soybean cultivars adapted to the tropics will not flower at day lengths of 14 hours, whereas those adapted to the higher latitudes will flower under continuous light. Genotypes from high altitudes tend to grow well at cool temperatures.

The plant is also affected differently by the environment at various phases of plant development. For example, day length may affect the initiation of flowering of a plant, but not its post-flowering development. Some soybean cultivars flower as rapidly under continuous light as on 14-hour days, but seed maturation is delayed by continuous light (Polson, 1972). The number of days of the correct day length required to induce flowering in some species decreases as plants get older (Borthwick and Parker, 1938a; Velasco and Manuel, 1955; Sasamura, 1960).

II. PHASES OF PLANT DEVELOPMENT

Germination of seed begins the life cycle of crop plants. Adequate moisture and temperature are the universal requirements for germination, although many species also exhibit some form of dormancy that must be overcome. Dormancy can be associated with hard seed coats, day-length sensitivity, the need for afterripening, or the need for exposure to low temperatures.

The juvenile phase covers the period from germination to flower bud formation. The rate at which new leaves are formed is primarily a function of temperature, but the time from emergence to flower initiation is affected by day length. Size of the mature plant is often determined by the duration of the juvenile phase because plants flowering late have initiated more leaves and, because they have more internodes, are taller than those flowering early. Rate of development is also influenced by soil moisture, soil type, and fertility.

Formation of the flower bud begins the reproductive phase. For determinate species and cultivars, the initiation of new leaves stops when flowering starts; however, leaves and stems may continue to enlarge. Corn, a determinate plant, does not initiate new leaves after the tassel is initiated. For indeterminate plants, new leaves are initiated after flowering has begun. In most indeterminate plants, the initiation of new leaves stops after the last flower forms. Cassava is an indeterminate plant that initiates flowers, leaves, and branches simultaneously. Soybean has both determinate and indeterminate cultivars.

Some winter annuals and biennials require exposure to low temperatures before reproductive development can be initiated. Vernalization is defined as the acquisition or acceleration of the ability to flower by a chilling treatment (Chouard, 1960). Vernalization does not induce flowering, but makes the plant more receptive to the day-length stimulus. Unvernalized winter rye plants flower in continuous light, but take twice as long as vernalized plants (Gott et al., 1955).

The reproductive phase lasts from initiation of the flower until fertilization of the flower. The length of this phase is determined by day length in some species. For example, the reproductive primordia of rice can revert to vegetative primordia if photoinduction of the plant is not complete (Noguchi et al., 1967); panicle emergence of rice can be delayed or prevented if the day length is too long after floral initiation (Vergara and Lilis, 1966, 1968); or the time from floral initiation to anthesis for sorghum (Quinby, 1972) and soybean (Parker and Borthwick, 1939a; Nagata, 1958) can be delayed by long day lengths. In the long-day cereals—wheat, barley, and oat—this phase is hastened by long day lengths (Guitard, 1960; Griffiths, 1961; Ormrod, 1963; Khatri, 1964).

The time from fertilization of the flower to maturation of the seed is called the ripening phase. It can be divided into three periods, according to the rate of dry matter accumulation in the seed—the lag period, the linear period, and the period when the rate levels off before maturity (Johnson and Tanner, 1972). The lag period can be as short as 1 week in wheat (Rawson and Evans, 1971) or as long as 20 days in soybean (Major et al., 1975a). Its duration also can be influenced by day length and temperature. Short days and warm temperatures shorten the lag period in soybean, a short-day species (Major et al., 1975b).

Rapid seed filling occurs during the linear period, and its rate and duration are influenced by day length and temperature. High temperatures shorten the duration of seed filling and increase the rate of seed filling in sorghum, rice, and wheat. In short-day plants, the delaying effect of cool temperatures in the fall season may be offset by the shortening day length.

III. DAY LENGTH

Day length has no effect on floral induction until the plant has attained a minimum amount of growth, a part of the life cycle often referred to as the basic vegetative phase (BVP). Species and cultivars differ widely in length of BVP. Some wheat cultivars respond to day length immediately after emergence, but most soybeans do not respond until after the unifoliolate leaves have opened. The BVP of rice cultivars varies from 10 to 63 days after germination, and its length is controlled by at least two genes (Vergara et al., 1965). Variation in the BVP also has been observed in corn (Rood and Major, 1981).

After a plant has completed BVP, initiation of flowers frequently depends on the day length. The critical photoperiod is defined as the maximum day length at which a short-day plant will flower and the minimum day length at which a long-day plant will flower. The optimum photoperiod is the day length at which the plant will flower in the shortest time after BVP is completed.

Knowledge of the critical photoperiod is essential when day length is used to delay the flowering of cultivars. The critical photoperiod of 'Biloxi' soybean, a short-day cultivar, is 13.5 hours. A breeder could use a 14-hour day to prevent flowering of this cultivar until another parent was available for hybridization.

The critical photoperiod can differ markedly among cultivars of a species and frequently is associated with the latitude of adaptation. Culti-

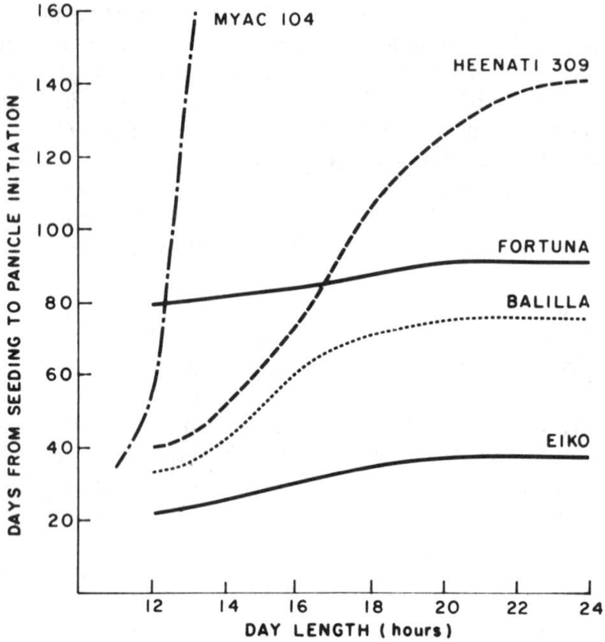

Fig. 1—Day length response curves of five rice cultivars based on data of Best (1961).

vars from low latitudes have a shorter critical photoperiod than those from high latitudes. Soybean cultivars adapted to the southern U.S. have a critical photoperiod of about 13.5 hours; those adapted to northern Minnesota and Western Canada do not have a critical photoperiod. The critical photoperiod of sorghums adapted to the United States is longer than that of tropical sorghums (Miller et al., 1968).

Knowledge of the optimum photoperiod for a species is useful for obtaining flowering and seed development in the shortest possible time. The optimum photoperiod is not known for most crop species; however, breeders commonly use about a 12-hour day to obtain rapid flowering of short-day species and 20 to 24 hours for long-day types. The reduction in number of days to flowering may or may not be linear for each hour of change in day length from the critical to the optimum photoperiod. A day-length response curve is useful in estimating when floral initiation will occur at various day lengths. The response curves in Fig. 1 are adapted from data of Best (1961) and show day-length responses of five rice cultivars. The cultivars 'Eiko' and 'Fortuna' have very low sensitivity to day length, but Eiko has a short BVP and Fortuna has a high BVP. The cultivars 'Balilla', 'Heenati 309', and 'Myac 104' have essentially the same BVP but have moderate, high, and very high day length sensitivities, respectively.

The optimum photoperiod for rapid flowering may not be optimal for obtaining desirable flowers for hybridization, because at the optimum photoperiod there may be fewer flowers, which have less pollen, and which may be more cleistogamous than at day lengths closer to the critical photoperiod. Artificial lighting can be used to extend the day length of short-day plants to obtain suitable flowers for hybridization in winter greenhouses or in latitudes lower than their area of adaptation.

Factors to consider in creating artificial day lengths by supplemental lighting are timing, duration, level, and quality. Day length in the field or greenhouse can be extended before sunrise and after sunset, or night length can be interrupted. Night interruption is effective because the response of plants to day length is actually a function of duration of night rather than day. A short-day plant technically should be referred to as a long-night plant and a long-day plant as a short-night plant. A period of light during the night should delay flowering of short-day plants and hasten that of long-day plants.

Timing of the day length in a growth chamber can facilitate hybridization during a selected time of the day. For hybridization of peanuts, Banks (1976) began a 12-hour day at 1630 hours so that flowers were available for emasculation the next morning between 0800 and 1000 hours.

The duration of artificial lighting is a function of the critical photoperiod of the cultivar. Day length in the field depends on the light level at sunrise and sunset that is adequate for flower control. The effective day length for rice is approximately equal to the astronomical day length. Takimoto and Ikeda (1961) indicated that light levels for measuring day length of Biloxi soybean should start when the light level of daylight increased to 10 lux in the morning and decreased to 200 lux in the evening.

The light level used to extend the day length or interrupt the night is an important consideration. A relatively low level may be required in some species and cultivars. Extending the day length with supplemental illumination of only 1 lux is sufficient to delay flowering of rice (Vergara et al., 1969). As light level increases, the time from seeding to flowering of soybeans has been found to be proportionate to the level of light used to extend the natural day (Major and Johnson, 1974). Differences among cultivars must be considered, however, in establishing the minimum light requirement. Francis et al. (1970) concluded that there were three types of cultivar responses in corn: those that are insensitive to day length and light level, those that require a minimum level of 11 to 22 lux to delay flowering, and those that require a minimum level of 54 lux to delay flowering. Plant age is also a factor influencing the light level requirement. A level of 5.4 lux will prevent flowering of young Biloxi soybeans, but a higher light level is required to prevent flowering of older plants (Hamner, 1940).

Light level during the day can influence flowering and the success of hybridization. In barley, the number of days from sowing to flower initiation is linearly related to the logarithm of light level (Paleg and Aspinall, 1964). Flowers will be initiated in Biloxi soybeans in an 8-hour day provided that the light level is at least 1,076 lux (Borthwick and Parker, 1938b). Peanuts require a high light level of not less than 45% of full sunlight (Chapter 31). Potato should have at least 16 hours of light with a minimum intensity of 20 klux. Light intensity may also modify the vernalization response. Vernalization of winter broad beans is required at 18 klux, but not at 30 klux (Blondon, 1975).

Cloudiness influences light level and temperature, an effect that enhances hybridization of some species and retards hybridization of others. Flowers of sweet potato remain open longer on cloudy days than on hot, sunny days, and hybridization of field pea is more successful when days are cool and bright than when hot, dry, and windy. Cloudy, humid weather can cause flower and pod abortion of pigeonpea. Supplemental illumination is commonly used in greenhouses to overcome low light levels during winter.

The efficiency of artificial lighting can be associated with light quality. For example, red light is at least 30 times more efficient for controlling flowering in soybean than blue light (Parker et al., 1946). Cultivars of soybean that have the E_3 gene are sensitive to incandescent light but insensitive to fluorescent light (Buzzell, 1971), and flowering of rice is delayed slightly by green light (Kondo et al., 1942).

IV. TEMPERATURE

Temperature greatly affects development of all species, but its direct effect on floral initiation and development is highly variable among and within species. While each plant responds simultaneously to temperature and photoperiod, the degree of response to each stimulus is also variable.

The vernalization requirement for flowering of winter annuals and biennials is directly related to temperature. Temperature also is the primary factor that determines floral initiation in species with limited day-length response. Corn, for example, responds to day length for only 15 to 25 days after emergence in temperate regions, so the effect of temperature on flowering is relatively more important than that of day length (Breuer et al., 1976). The flowering and maturation of corn can be predicted by accumulating the temperatures recorded between 10 and 30 C (growing degree-days) during the growing season (Cross and Zuber, 1972).

There are critical minimum and maximum temperatures for floral initiation and development in most plants. The critical minimum for panicle differentiation of rice is 15 to 18 C (Yatsuyanagi and Takeuchi, 1959; Inouye, 1964; Noguchi and Kamata, 1965), temperatures below 21 C or above 32 C will delay floral initiation and reduce seed set of soybeans (van Schaik and Probst, 1958; Hamner, 1969), and the critical minimum and maximum temperatures for peanut flower initiation and development are 21 and 33 C, respectively (Bolhuis and de Groot, 1959).

The critical temperatures for hybridization are more restrictive than those for floral initiation and development. Cool temperatures increase the tendency for flowers to self-pollinate before they are large enough to manipulate, and coolness reduces the amount of pollen shed. On the other hand, high temperatures shorten the duration of stigma receptivity and the duration and viability of pollen. A temperature of 33 C results in low viability of peanut pollen (De Beer, 1963). Soil temperature in the podding zone is critical for development of peanuts (Ono et al., 1974); the critical minimum is 15 to 17 C, critical maximum of 37 to 39 C, and optimum of 31 to 33 C.

The response of many plants to day length is influenced by temperature. The relative importance of day and night temperatures for altering the day length response has not been the same in all studies. Manuel and Velasco (1956) found that low night temperatures hastened flowering of rice in short days, whereas Roberts and Carpenter (1965) did not obtain separate effects of day and night temperatures. Night temperatures in controlled environments are more important than day temperatures for altering the critical photoperiod of cowpea and soybean (Parker and Borthwick, 1939b; Huxley and Summerfield, 1974), but inhibitory night temperatures probably do not occur often enough in the field to seriously affect flowering (Johnson et al., 1960).

The optimum temperature for hybridization is lower for crops adapted to temperate regions than for those adapted to the tropics. The optimum is about 19 C for potato (Chapter 34) and field pea (Chapter 23), which are cool-season crops. Hybridization of pigeonpea is successful up to 40 C (Chapter 33) and flower and seed set of pearl millet will occur between 25 and 45 C (Chapter 32).

V. MOISTURE

Adequate soil moisture is necessary to obtain vigorous plants that will have a minimum amount of flower and seed abortion. Most species and cultivars do best on soils that are well drained and have good capacity for water retention.

Species and cultivars differ in the ability to withstand excess or deficient moisture. A high percentage of flowers abort in tobacco under drought, and plants can be severely damaged or killed in 24 to 48 hours when exposed to saturated soils and high temperatures. Pearl millet does not tolerate flooding, but can be hybridized under considerable moisture stress.

Most species reduce seed production when moisture stress occurs during flowering and early seed filling. It is best to irrigate during the critical period for hybridization. If irrigation is not available, moisture stress can be reduced by providing more land area per plant and by removing excess flowers from the plant.

Soil moisture can influence seed set of species that develop fruit below ground. Optimum soil moisture content in the podding zone of peanuts is about 40% of the total soil volume in the field, regardless of soil moisture content in the rooting zone (Ono et al., 1974). Eighty to 85% of soil capacity is best for greenhouse plantings of peanut (Chapter 31).

Maturity can be hastened by reducing available water toward the end of the growing season. Some crops, such as rape, will even begin to flower again if irrigated after a dry period. If the water supply is reduced, care must be taken to ensure that the seed obtained is sufficiently large to obtain adequate viability.

Relative humidity is most important at the time of hybridization or self-pollination. There is an upper limit of relative humidity for all crops, because eventually the pollen will clump and be difficult to handle, or disease infestations may occur. When relative humidity is too high, flowers of some species can be collected and dried with a dessicent. An excessively low relative humidity can result in failure of the pollen grain to germinate on the stigma. Enclosing the stigma in its floral parts or covering the flower with a bag may increase the relative humidity and improve the percentage of seed set.

Controlling humidity is difficult in most situations. Some growth chambers have humidity controls that can be set to the desired level and humidifying machines are available for greenhouses. Wetting the greenhouses or growth chamber floors will increase humidity.

VI. SOIL FERTILITY

Successful hybridization and seed set occur when soil fertility is maintained at a level that will sustain healthy green plants. Soil tests are useful to determine the amount of nutrients that should be added.

High N fertilization is desirable for some crop species and cultivars, but not others. Buckwheat and sweet potato produce excessive vegetative growth if soil N levels are high (De Jong, 1972). The excess vegetative growth can result in lodging, flower abortion, and delayed maturity. On the other hand, N fertilization at twice the normal rate increases seed production of tobacco.

Some species have special nutrient requirements for seed development. Peanuts require high amounts of Ca to avoid the formation of pods that are empty or contain shrivelled seeds (Chapter 31).

Restrictions on fertility and plant development are beneficial for flowering of some species. Flowering of potato is improved by placing a piece of the seed tuber on a brick and covering it with sand (Thijn, 1954). After roots from the seed tuber have penetrated the soil below the brick, the sand is washed away and tubers are removed as they are formed. Restricting root growth of sweet potato by planting in a pot can increase flowering.

VII. TECHNIQUES FOR SYNCHRONIZING FLOWERING DATES

A. Multiple Planting Dates

The most common method used to ensure the proper timing of male and female flowers is to plant on more than one date. This is particularly true when the time of flowering is not precisely known for the parents. Even when flowering dates are known, multiple planting dates can be used to increase the duration of flowering and permit the breeder to remake crosses that failed in the first attempt. When mating early and late-flowering parents, the late parent may be planted on a single date and the early parent planted on several dates thereafter.

The number of days between planting dates may not cause a corresponding change in flowering dates. A 7-day interval between planting dates may result in a 3-day or a 10-day interval between flowering dates. Reduced or increased flowering intervals are caused primarily by changes in day length and temperature that occur between planting dates. The number of days to flowering of early, midseason, and late cultivars of pigeonpea was evaluated at Hyderabad, India (17°N Lat.). Early cultivars flowered 85 days before late types when planted in July, but only 68 days earlier when both types of cultivars were planted in September (Chapter 33).

Different planting dates in the fall season have little or no effect on date of flowering for crops that require vernalization. Artificial vernalization can be used for multiple planting dates in the greenhouse or for spring planting in the field.

B. Day Length

The delay or hastening of flowering can be accomplished by manipulation of day length, as described in section III. Altering flowering date by artificial day lengths is a common practice in sugarcane, a short-day species. Early flowering is induced by decreasing the natural day length, and flowering is delayed by interrupting the dark period with light or extending the day length after floral initiation occurs naturally (Chilton and Paliatseas, 1956; Coleman, 1962; James and Miller, 1971). Cultivars of pearl millet with different day-length responses will flower simultaneously and can be hybridized under a day length of less than 12 hours (Burton and Stansell, 1971).

C. Temperature

Flowering can be manipulated by control of temperature, as described in section IV. Cool temperatures generally delay and warm temperatures hasten flowering. Excessively high temperatures can slow development of many species. Optimum temperatures for sorghum seem to be between 20 and 35 C, and flowering is delayed by temperatures outside this range. A temperature of 12 C at night results in little sorghum growth, even with optimum day temperatures (Chapter 41).

D. Plant Density

Reduced plant density encourages tillers and branches. These generally flower later than the main stem and prolong the duration of flowering of the plant. Flowers on tillers and branches, however, may be smaller and more difficult to manipulate than those on the main stem.

E. Pruning

Pruning can be used to delay or extend the flowering period. Removal of the growing point on the main stem promotes tillers and branches that flower later. In some forage species, the plants are cut back to the ground, and the secondary growth produces more flowers and seed than the first growth. Such a practice in other species could significantly reduce seed yield.

Pruning of leaves can delay flowering without damaging the main stem. The top leaves of young corn plants are sometimes destroyed by fire to delay the flowering of male plants in fields for hybrid seed production.

Removal of flowers and seed from a plant may cause it to prolong flowering. Removing all flowers each day can cause peanuts to live 240 days or more and insures that flowers will be available over a long period (Bear and Bailey, 1973). The flowering of pigeonpea can also be extended 30 to 60 days by removal of new flowers.

Girdling the stem of sweet potato can increase flowering (Edmond and Ammerman, 1971). A slanted cut is made about half way through the primary stem and a piece of wood or cardboard is wedged into the cut to prevent the stem from healing.

F. Grafting

It is possible to induce flowering of a late-flowering plant by grafting a scion of it to the stock of a plant in bloom. In soybean, a late-flowering plant introduction was grafted to a stock of an adapted cultivar that had just begun to bloom (Kiihl et al., 1977). Mature seed was obtained from the scion of the plant introduction before the nongrafted stock began to flower. The flowers on grafted scions were normal and readily used for hybridization.

In some breeding programs, potato is grafted onto tomato roots to promote flowering of difficult clones (Chapter 34). This technique increases the amount and duration of flowering and reduces the abortion of flowers before and after pollination.

G. Hormones

The flowering of plants is ultimately triggered by a hormone or a balance of hormones. This aspect of flowering is poorly understood, but has great potential in the hybridization of crop species. At present, many long-day plants can be induced to flower sooner by the addition of gibberellic acid (Thimann, 1977). Gibberellic acid may replace the vernalization requirement of *Lolium* spp. (Evans, 1969). Inducing short-day plants to flower is more difficult although some short-day plants can be induced to flower by ethylene-producing chemicals, such as chloroethane phosphonic acid. This chemical is being used commercially by pineapple growers.

VIII. CONDITIONS FOR SUCCESSFUL HYBRIDIZATION AND SELF-POLLINATION

Successful hybridization and self-pollination depend on the availability of healthy plants with an adequate number of flowers. Proper management of plants before flowering includes consideration of soil fertility, soil moisture, day length, temperature, and pest control. Both inadequate and excessive levels of soil fertility and moisture can result in plant development that is less than optimum for flower fertilization and retention. Day lengths with adequate light duration, level, and quality are of particular importance in winter greenhouses and growth chambers. Temperatures near the optimum for the crop species should be provided whenever possible. Pest control includes elimination of weed competition and adequate management of diseases and insects.

Elimination of plant stress during flowering and the early stages of seed development will increase the chances of obtaining a high percentage of successful hybridizations and self-pollinations. An adequate moisture supply is critical, and supplemental irrigation should be available whenever possible. If moisture stress is imminent and no means are available for applying water, interplant and intraplant competition for moisture may be reduced by thinning unnecessary plants and pruning branches or tillers. Moderate temperatures and high relative humidity reduce transpiration, thereby diminishing the effect of inadequate soil moisture.

The duration of stigma receptivity and pollen viability is influenced by temperature and relative humidity. Both excessively cool or hot temperatures are unfavorable for pollination and fertilization. Cool temperatures reduce pollen shed and high temperatures shorten the duration of pollen shed and pollen viability. A reduction in relative humidity can be minimized by enclosing the stigma in the floral covering or by covering it with a bag.

IX. TECHNIQUES FOR RAPID SEED PRODUCTION

Rapid production of hybrid and self-pollinated seed can reduce the length of time required to develop a new cultivar. However, shortening the life cycle by manipulation of the environment will reduce seed yield per plant. It also may reduce flower size and increase cleistogamy, thereby limiting such techniques to the production of self-pollinated seed.

Dormancy and vernalization are survival mechanisms common to some species, and these can be overcome artificially. Seed of species that are difficult to germinate because of a hard seed coat should be scarified before seeding. Other types may require more specialized treatments to overcome dormancy. Wild rice must have the pericarp removed from around the embryo, hop needs a chilling period of 5 to 8 weeks, and some rye cultivars need an afterripening period of about 20 days. In some cases, the afterripening period can be overcome chemically with potassium nitrate or gibberellic acid (Spicer and Dionne, 1961).

Treatment of germinating seeds or seedlings in growth chambers at temperatures of 2 to 5 C will substitute for natural vernalization or overcome the biennial nature of a species. In winter wheat, exposure to 2 to 5 C in short days for 8 weeks is the most effective way of meeting the vernalization requirement (Grant, 1964). Sugarbeet is a biennial and the shortest life cycle occurs when it is exposed to 4 to 7 C for about 10 weeks during floral induction (Gaskill, 1952).

Continuous light is the most suitable method of reducing the time to flowering of long-day plants. Short day lengths promote development of short-day plants, but the actual day length used will depend on the optimum photoperiod. For plants of tropical origin, an 8-hour day may be most suitable. The use of an 8-hour day in Puerto Rico reduced the days from planting to flowering of pigeonpea by 4 months in medium maturity cultivars and 7 weeks in late cultivars (Chapter 33). A 12-hour day may be adequate for crops adapted to more temperate regions.

The temperatures most suitable for rapid seed production generally vary from about 25 to 35 C, depending on whether it is a cool or warm-season crop. Some species, such as pigeonpea or pearl millet, will develop most rapidly at temperatures above 35 C.

Moisture stress may be used to reduce the time taken from flowering to maturity, but stress should not be used until the developing seeds are well established and will not abort.

Soil fertility should be limited to the amount needed to produce a small plant with a few seeds. Nitrogen stress during the seed-filling period will result in smaller, earlier-maturing seeds. Phosphorus tends to have the opposite effect to N, so a liberal application of P may also hasten maturity.

High population densities often appear to hasten maturity, but this is probably the result of fertility and moisture stresses that occur late in the seed-filling period. With excess fertility and moisture, increased population density may actually delay maturity.

Most crops harvested for their seed have been selected for large seed. The embryos of such seed may reach maturity long before maximum seed dry weight is reached. Thus it may be possible to harvest seed 10 to 15 days after the linear phase of seed filling has begun. The seed may be small and shrivelled, but may still have excellent viability. When this procedure is followed, it is advisable to cut the entire plant and dry it in the field or the greenhouse. This method essentially imposes severe moisture stress, and curing the whole plant allows some translocation of sugars and minerals into the seed during the drying period.

REFERENCES

Banks, D. J. 1976. Peanuts: germplasm resources. Crop Sci. 16:499–502.

Bear, J. E., and W. K. Bailey. 1973. Effect of seed treatment and germination on seed dormancy of peanuts, *Arachis hypogea.* J. Am. Peanut Res. Educ. Assoc. 3:122–126.

Best, R. 1961. Some aspects of photoperiodism in rice (*Oryza sativa* L.). Elsevier Publishing Co., N.Y. 87 p.

Bickford, E. D., and S. Dunn. 1972. Lighting for plant growth. The Kent State Univ. Press, Kent, Ohio.

Blondon, F. 1975. Les exigences pour la floraison de trois lignes de *Vicia faba* L. (feverole). C. R. Acad. Agric. 61:1068–1081.

Bolhuis, G. G., and W. de Groot. 1959. Observations on the effect of varying temperatures on the flowering and fruit set in three varieties of groundnut. Neth. J. Agric. Sci. 7:317–326.

Borthwick, H. A., and M. W. Parker. 1938a. Effectiveness of photoperiodic treatments of plants of different age. Bot. Gaz. 100:245–249.

———, and ———. 1938b. Photoperiodic perception in Biloxi soybeans. Bot. Gaz. 100:374–387.

Breuer, C. M., R. B. Hunter, and L. W. Kannenberg. 1976. Effects of 10- and 20-hour photoperiod treatments at 20 and 30 C on rate of development of a single-cross maize (*Zea mays*) hybrid. Can. J. Plant Sci. 56:795–798.

Burton, G. W., and J. R. Stansell. 1971. An automatic darkbox to induce flowering in short-day plants in midsummer. Crop Sci. 11:595–596.

Buzzell, R. I. 1971. Inheritance of a soybean flowering response to fluorescent-daylength conditions. Can. J. Genet. Cytol. 13:703–707.

Chilton, S. J. P., and E. D. Paliastseas. 1956. Studies on the flowering of sugar cane. Proc. Int. Soc. Sugar Cane Technol. 9:652–656.

Chouard, P. 1960. Vernalization and its relations to dormancy. Annu. Rev. Plant Physiol. 11:191–238.

Coleman, R. E. 1962. Control of flowering and the use of pollen storage as techniques in a sugar cane breeding programme. Proc. Int. Soc. Sugar Cane Technol. 11:553–540.

Cross, H. Z., and M. S. Zuber. 1972. Prediction of flowering dates in maize based on different methods of estimating thermal units. Agron. J. 64:351–355.

De Beer, J. F. 1963. Influences of temperature on *Arachis hypogea* L. with special reference to its pollen viability. Versl. Landbouwkd. Onderz. (Wageningen) N.R. 69.2:1–81.

De Jong, H. 1972. Buckwheat. Field Crop Abstr. 25:389–396.

Edmond, J. B., and G. R. Ammerman. 1971. Sweet potatoes: Production, processing, marketing. AVI Publishing Co., Westport, Conn.

Evans, L. T. 1969. The induction of flowering. MacMillan and Co., N.Y.

Francis, C. A., D. Sarria V, D. D. Harpstead, and C. Cassalett D. 1970. Identification of photoperiod insensitive strains of maize (*Zea mays* L.). II. Field tests in the tropics with artificial lights. Crop Sci. 10:465–468.

Gaskill, J. O. 1952. A new sugarbeet breeding tool—two seed generations in one year. Agron. J. 44:338.

Gott, M. B., F. G. Gregory, and O. N. Purvis. 1955. Studies in vernalisation of cereals. XIII. Photoperiodic control of stages in flowering between initiation and ear formation in vernalised and unvernalised Petkus winter rye. Ann. Bot. (Lond.) 19:87–126.

Grant, M. N. 1964. Vernalization and days to anthesis of winter wheat under controlled temperature and light. Can. J. Plant Sci. 44:446–450.

Griffiths, D. J. 1961. The influence of different daylengths on ear emergence and seed setting in oats. J. Agric. Sci. 57:279–288.

Guitard, A. A. 1960. The influence of variety, temperature, and stage of growth on the response of spring barley to photoperiod. Can. J. Plant Sci. 40:65–80.

Hamner, K. C. 1940. Interrelation of light and darkness in photoperiodic induction. Bot. Gaz. 101:658–687.

————. 1969. *Glycine max* (L.) Merrill. p. 62–89. *In* L. T. Evans (ed.) The induction of flowering; some case histories. Cornell Univ. Press, Ithaca, N.Y.

Hunter, R. B., L. A. Hunt, and L. W. Kannenberg. 1974. Photoperiod and temperature effects on corn. Can. J. Plant Sci. 54:71–78.

Huxley, P. A., and R. J. Summerfield. 1974. Effects of night temperature and photoperiod on the reproductive ontogeny of cultivars of cowpea and of soyabean selected for the wet tropics. Plant Sci. Lett. 3:11–17.

Inouye, J. 1964. Effect of temperature on flower bud initiation and fruiting of rice plants grown on artificial culture medium. Proc. Crop Sci. Soc. Jpn. 32:330–332.

James, N. I., and J. D. Miller. 1971. Photoperiod control in the USDA sugar cane crossing program. Proc. Int. Soc. Sugar Cane Technol. 14:341–347.

Johnson, D. R., and J. W. Tanner. 1972. Calculation of the rate and duration of grain filling in corn (*Zea mays* L.). Crop Sci. 12:485–486.

Johnson, H. W., H. A. Borthwick, and R. C. Leffel. 1960. Effects of photoperiod and time of planting on rates of development of the soybean in various stages of the life cycle. Bot. Gaz. 122:77–95.

Khatri, M. I. 1964. Photoperiodic and vernalization studies in New Zealand wheat. Pak. J. Sci. 16:307–311.

Kiihl, R. A. S., E. E. Hartwig, and T. C. Kilen. 1977. Grafting as a tool in soybean breeding. Crop Sci. 17:181–182.

Kondo, M., A. Terasaka, and M. Umino. 1942. Experimental studies on photoperiodism in rice (III). Nogaku Kenkyu 33:1–24.

Major, D. J., and D. R. Johnson. 1974. Effect of light intensity on the development ment of field grown soybeans. Crop Sci. 14:839–841.

———, ———, and V. D. Luedders. 1975a. Evaluation of eleven thermal unit methods for predicting soybean development. Crop Sci. 15:172–174.

———, ———, J. W. Tanner, and I. C. Anderson. 1975b. Effects of daylength and temperature on soybean development. Crop Sci. 15:174–179.

Manuel, F. C., and J. R. Velasco. 1956. Further observations on the photoperiodic response of Elon-elon rice. Philipp. Agric. 40:421–432.

Miller, F. R., D. K. Barnes, and H. J. Cruzado. 1968. Effect of tropical photoperiods on the growth of sorghum when grown in 12 monthly plantings. Crop Sci. 8:499–502.

Nagata, T. 1958. Studies on the flowering and fruiting of summer vs. autumn soybean types. Proc. Crop Sci. Soc. Jpn. 27:87–90.

Noguchi, Y., and E. Kamata. 1965. Studies on the control of flower bud formation by temperature and daylength in rice plants. V. Response of floral induction to temperature. Jpn. J. Breed. 15:86–90.

———, T. Nakajima, and T. Yamaguchi. 1967. Studies on the control of flower bud formation by temperature and daylength in rice plants. VII. Elimination of photoinductive effects by keeping under long-day conditions. Jpn. J. Breed. 17:20–24.

Ono, Y., K. Nakayama, and M. Kubota. 1974. Effects of soil temperature and soil moisture in podding zone on pod development of peanuts. Proc. Crop Sci. Soc. Jpn. 43:247–251. [English summary.]

Ormrod, D. P. 1963. Photoperiodic sensitivity of head differentiation, culm elongation, and heading in some spring wheat and spring barley varieties. Can. J. Plant Sci. 43:323–329.

Paleg, L. G., and D. Aspinall. 1964. Effects of daylength and light intensity on growth of barley. II. Influence of incandescent light on apical development. Bot. Gaz. 125:149–155.

Parker, M. W., and H. A. Borthwick. 1939a. Effect of photoperiod on development and metabolism of the Biloxi soy bean. Bot. Gaz. 100:651–689.

———, and ———. 1939b. Effect of variation in temperature during photoperiodic induction upon initiation of flower primordia in Biloxi soy beans. Bot. Gaz. 101:145–167.

———, S. B. Hendricks, H. A. Borthwick, and N. J. Scully. 1946. Action spectrum for the photoperiodic control of floral initiation of short-day plants. Bot. Gaz. 108:1–26.

Polson, D. E. 1972. Day-neutrality in soybeans. Crop Sci. 12:773–776.

Quinby, J. R. 1972. Influence of maturity genes on plant growth in sorghum. Crop Sci. 12:490–492.

Rawson, H. M., and L. T. Evans. 1971. The contribution of stem reserves to grain development in a range of wheat cultivars of different height. Aust. J. Agric. Res. 22:851-863.

Roberts, E. H., and A. J. Carpenter. 1965. The interaction of photoperiod and temperature on the flowering response of rice. Ann. Bot. (Lond.) 29:359-364.

Rood, S. B., and D. J. Major. 1981. Diallel analysis of the photoperiodic response of maize. Crop Sci. 21:875-878.

Sasamura, S. 1960. Studies on the relation between the plant age and the degree of sensibility to short day length in the late paddy rice variety Zuiho. Proc. Crop Sci. Soc. Jpn. 28:355-358.

Spicer, P. B., and L. A. Dionne. 1961. Use of gibberellin to hasten germination of *Solanum* seed. Nature (Lond.) 189:327-328.

Takimoto, A., and K. Ikeda. 1961. Effect of twilight on photoperiodic induction in some short day plants. Plant Cell Physiol. 2:213-229.

Thijn, G. A. 1954. Observations on flower induction with potatoes. Euphytica 3: 28-34.

Thimann, K. V. 1977. Hormone action in the whole life of plants. Univ. of Massachusetts Press, Amherst.

van Schaik, P. H., and A. H. Probst. 1958. Effects of some environmental factors on flower production and reproductive efficiency in soybeans. Agron. J. 50: 192-197.

Velasco, J. R., and F. C. Manuel. 1955. The photoperiodic response on Elon-elon rice. Philipp. Agric. 39:161-175.

Vergara, B. S., T. T. Chang, and R. Lilis. 1969. The flowering response of the rice plant to photoperiod. A review of the literature. Int. Rice Res. Inst. (Los Banos) Tech. Bull. 8.

———, and R. Lilis. 1966. Studies on the responses of the rice plant to photoperiod. II. Effect of the number of photoinductive cycles on a seasonal rice variety BPI-76. Philipp. Agric. 50:9-14.

———, and ———. 1968. Responses of the rice plant to photoperiod. III. Effect of different photoperiods after panicle initiation on the emergence of the panicle. Philipp. Agric. 52:61-65.

———, S. Puranabhavung, and R. Lilis. 1965. Factors determining growth duration of rice varieties. Phyton Rev. Int. Bot. Exp. 22:177-185.

Yatsuyanagi, S., and T. Takeuchi. 1959. Ecological studies of rice variety. III. The problem of the temperature and the efficiency during the generative growth period. Proc. Crop Sci. Soc. Jpn. 18:164-168.

2 Reproduction and Seed Development

Nels R. Lersten
Iowa State University
Ames

Cultivated species comprise less than 1% of about 300,000 species of flowering plants; nevertheless, they are quite diverse in their reproductive structures and behavior. This chapter deals mostly with the general life cycle of angiosperms and discusses some differences between dicots and monocots and individual species. It provides the background needed to understand the structures, reproductive processes, and problems of hybridization in the individual crop species discussed in later chapters. The account is based on some broad works (Davis, 1966; Frankel and Galun, 1977; Hayward, 1938; Heslop-Harrison, 1972; Maheshwari, 1950; Palser, 1975), as well as more specialized references that will be cited in appropriate sections.

I. THE INFLORESCENCE

Most plants have many flowers clustered in a predictable pattern called an inflorescence. Most crop plants have one of two basic types, the cyme or the raceme.

The cyme is a determinate inflorescence consisting of primary, secondary, tertiary, and often additional floral branches (Fig. 1). The terminal flower on the primary stem develops first, then secondary floral branches develop below it. Each secondary branch has a terminal flower that develops first, then tertiary branches develop below. This branching process continues, thereby expanding the inflorescence and extending the flowering duration of the plant. The cyme is present in potato, for example (Fig. 2).

A more diverse group of inflorescences is based on the raceme, in which flowering begins at the base and proceeds toward the top. The arrows in Fig. 1 do not imply an evolutionary pathway for inflorescence types, but

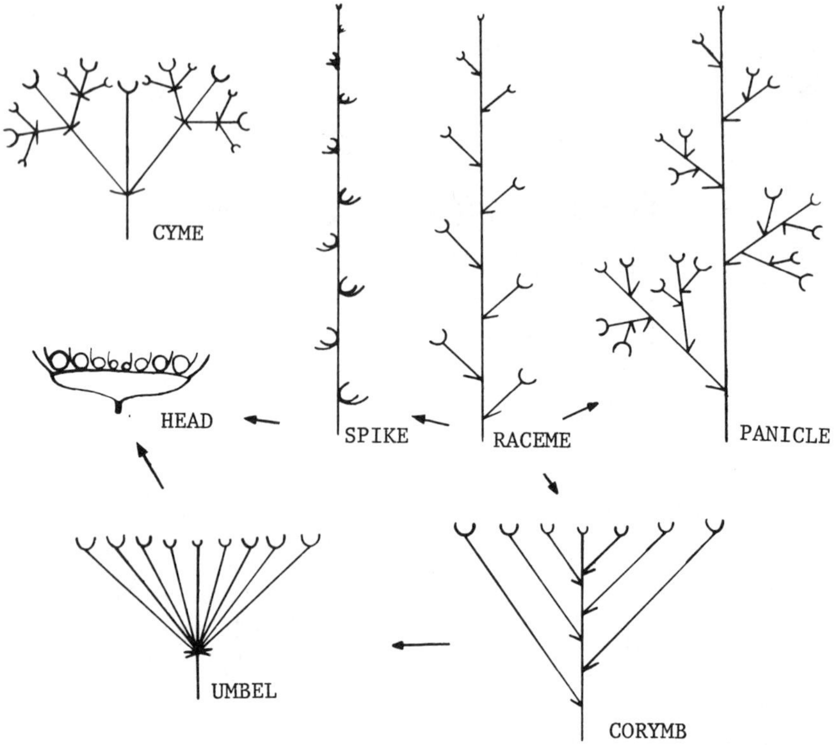

Fig. 1—Common types of inflorescences found in cultivated plants: cyme and various forms of the raceme. Types of racemes include the branched raceme (panicle), raceme with sessile flowers (spike), raceme with pedicels unevenly elongated to produce a flat-topped cluster (corymb), pedicels emerging from a common level (umbel), and umbel with crowded, sessile flowers (head). The arrows are not meant to imply the evolutionary pathways for inflorescence types, but are primarily to show how these types can be derived from a raceme.

show how the various types can be derived from a raceme. Figure 3 shows that a typical raceme consists of a primary stalk (peduncle) with flowers along its length borne on short stalks (pedicels). The soybean has a typical raceme, but various lengths and number of flowers can be found among cultivars and in different environments.

The panicle of oat is a branched raceme (Fig. 4). The spike of barley, wheat, triticale, and many other grasses is a raceme whose flowers are sessile; they are borne on such short pedicels that they seem to be attached directly to the primary stalk (Fig. 5). The spikelet, not the multiple florets that may occur within a spikelet, is the basic unit of the grass inflorescence. The corymb (Fig. 1) is a raceme with pedicels of unequal length that produce a flat-topped cluster of flowers. This type does not occur in cultivated plants, although a corymbose raceme is found in crambe and certain other mustards. The simple umbel has pedicels that elongate from a common point, such as in onion (Fig. 6). An inflorescence of multiple umbels occurs in carrot and other cultivated Umbelliferae. The head (capitulum) may be thought of as an umbel with crowded, sessile flowers like those in sunflower (Fig. 7).

Fig. 2—Inflorescence of potato, an example of a cyme.

Fig. 3—Raceme of sweetpea.

Fig. 4—Panicle of oat.

II. THE FLOWER

A. General Morphology

Flowers of crop plant species differ from each other in size, numbers of appendages, and in other ways. Many species have flowers that consist of a number of leaf-like sepals that compose the calyx, colorful petals that form the corolla, the female pistil (gynoecium), and the male stamens (androecium). The pistil consists of one or more carpels, each carpel having an ovary, style, and stigma. The stamen is composed of a filament and anther. The sepals, petals, carpels, and stamens vary in number and may be separate or fused. In grass species, the calyx and corolla are replaced by glumes, of varying prominence, which embrace the spikelet, and a lemma and palea that enclose the pistil and stamen of each floret within the spikelet.

Flowers of cultivated legumes such as soybean, cowpea, and alfalfa have flowers similar to each other (Fig. 8). The calyx consists of five sepals fused to varying degrees. The corolla includes a large standard petal, two

Fig. 5—Spikes of triticale (left) and hexaploid wheat (right).

small wing petals, and two fused petals collectively called the keel. The pistil has a single carpel whose ovary contains ovules that vary in number among species. The stigma-tipped style varies in shape and length among species. For example, the style of the common bean is spiral and has short hairs near the stigma, whereas the style of soybean is curved and lacks hair. Legume flowers have 10 stamens, 9 in one group with fused filaments, and 1 that is free (Fig. 9). Nectar glands at the base of the ovary make the flowers of many legume species attractive to bees.

A spikelet, the floral unit of a grass inflorescence, consists of a pair of glumes that embrace one or more florets (Fig. 10). Glumes of oat are prominent and enclose several florets, but the glumes of rye are small and at the base of a single floret (Fig. 10). Each floret consists of a lemma that commonly bears an own and a palea (Fig. 10). The grass pistil has an ovary containing one ovule, and a short style with a branched, feathery stigma. Most grasses have three stamens; rice, however, is an example of a grass species with six stamens.

Cross-pollination, also called allogamy, is assured in crops with dioecious plants, such as hemp. A high frequency of cross-pollination also occurs

Fig. 6—Simple umbel of wild onion.

Fig. 7—Head (capitulum) of sunflower.

Fig. 8—Intact flower of soybean, a representative legume.

among monoecious plants (Fig. 11), those with pin and thrum flowers (Fig. 12), and those with protandry or protogyny. Self-pollination, referred to as autogamy, is promoted in perfect flowers that are cleistogamous, such as in oat and soybean.

The stamens (♂) and the pistil (♀) are the structures directly involved in sexual reproduction. They occur in various arrangements and may mature at different times (Lloyd and Webb, 1977). Several sets of terms exist to describe the possibilities (Frankel and Galun, 1977). The more common ones are:

A. ♂ and ♀ expression in individual flowers
   1. ♂ and ♀ in one flower: bisexual, hermaphroditic, monoclinous, perfect
      a. pollen shed before stigma is receptive: protandry
      b. stigma matures and ceases to be receptive before pollen is shed: protogyny
      c. stigma receptive, and pollen shed, after flower opens: chasmogamy
      d. stigma receptive, and pollen shed, in closed flower: cleistogamy
   2. Perfect flowers of two types on same plant.
      a. long styles and short stamens: pin flower
      b. short styles and long stamens: thrum flower
   3. ♂ and ♀ in separate flowers: diclinous, unisexual, imperfect
      a. ♂ flower: male, staminate
      b. ♀ flower: carpellate, female, pistillate

B. Flower distribution on plants
   1. ♂ and ♀ flowers on one plant: monoecious
   2. ♂ and ♀ flowers on separate plants: dioecious
   3. ♂, ♀, and perfect flowers: mixed, polygamous
      a. on same plant: polygamomonoecious
      b. on separate plants: polygamodioecious

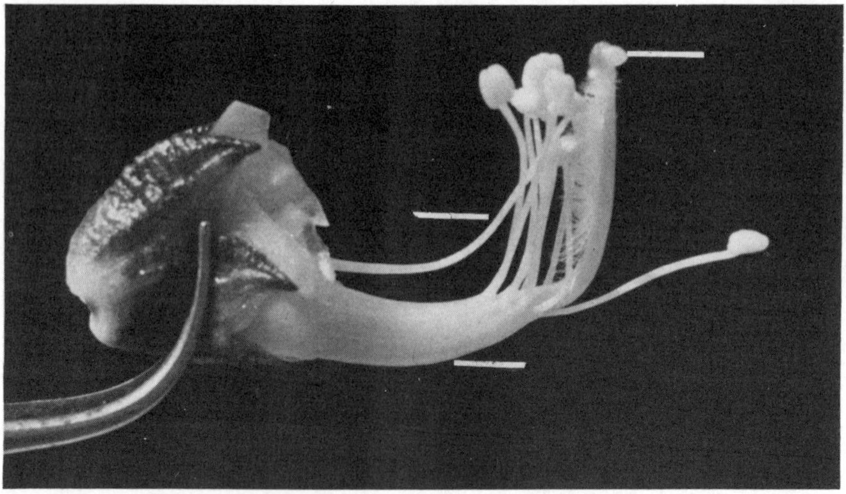

Fig. 9—Cowpea flower with calyx and corolla removed. Upper line indicates stigma at tip of style, middle line touches the solitary stamen, and lower line points to the fused staminal sheath of the other nine stamens.

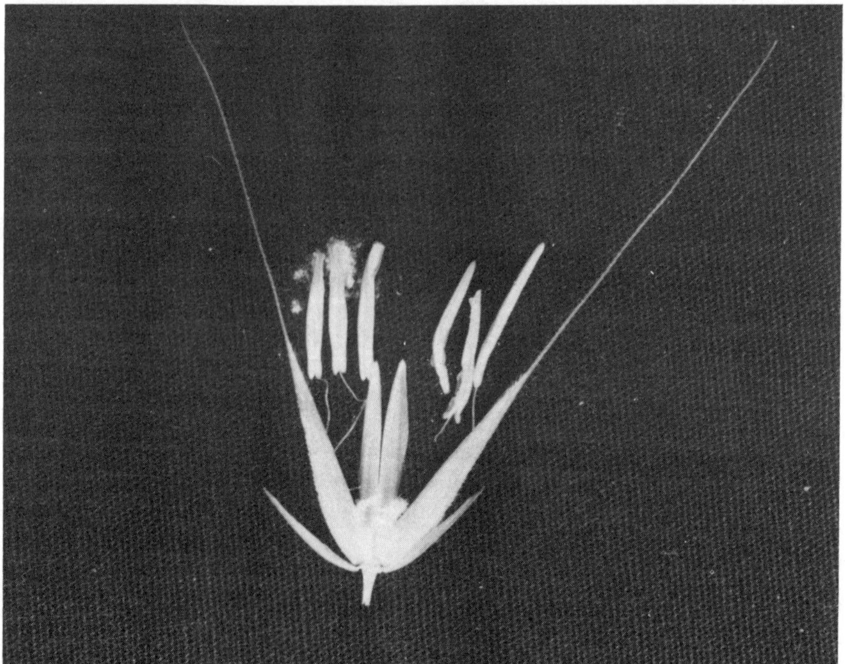

Fig. 10—Spikelet of rye.

Fig. 11—Monoecious inflorescences in corn. Left: terminal male inflorescence. Right: Axillary female inflorescence.

Fig. 12—Pin and thrum flowers of buckwheat. Left: Pin flower with long style and short stamens. Right: Thrum flower with short style and long stamens.

B. Stamen and Pollen Development

A stamen consists of a sterile filament bearing an anther. Four cylindrical pollen sacs (locules) can be first distinguished within the anther as four separate rows of densely cytoplasmic cells (Fig. 13). Cell division continues in each locule until a certain number of microspore mother cells have formed. The microspore mother cells undergo meiosis to form microspores, which develop into pollen. The locules are surrounded by a nurse layer called the tapetum (Fig. 13). There are three to five additional cell layers between the tapetum and epidermis, and a few to many parenchyma cells

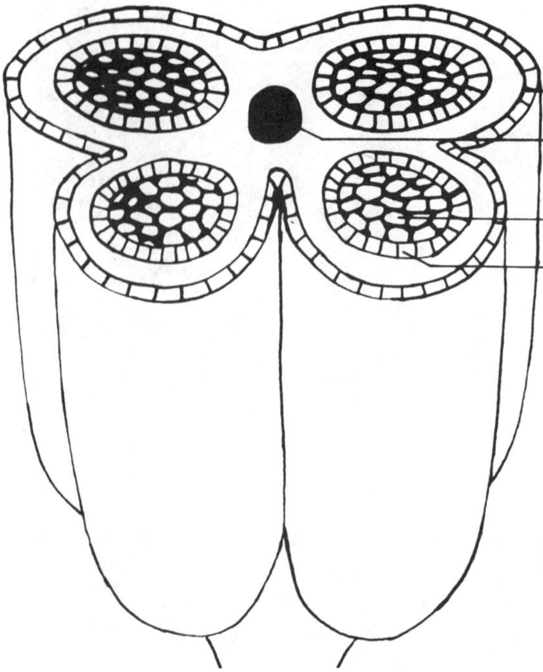

Fig. 13—Three-dimensional view of an anther with upper half removed to show four pollen sacs, each at the microspore mother cell stage.

which serve as connective tissue between the tapetum and the single vascular bundle of the stamen.

Cultivated species differ in stamen length, size of the anther, number of microspore mother cells, and amount of connective tissue. Grasses are unusual because each microspore mother cell remains in contact with the tapetum throughout development, until the mature pollen is shed.

Each microspore mother cell secretes around itself a gelatinous sheath made up of a carbohydrate called callose. This substance isolates each microspore mother cell during meiosis and for some time thereafter. Opinions differ as to why the callose coating is needed.

The tapetal cells of most plants become binucleate before, or during meiosis. In some species, however, tapetal cells become multinucleate through repeated nuclear mitosis. Dandelion may have up to 16 nuclei per tapetal cell. In all legumes, in contrast, the tapetum remains uninucleate.

The tapetum remains as a recognizable layer, even after it loses its cell walls. In most dicots, the tapetum remains in place at the periphery of the locule (parietal tapetum), while in most monocots it grows inward shortly after meiosis and appears to engulf the immature pollen grains (invasive tapetum). There are, however, many exceptions; e.g., sunflower (a dicot) has an invasive tapetum, whereas all grasses (monocots) that have been studied have a parietal tapetum. The tapetum is thought to contribute to pollen growth in one or more ways, depending on the species.

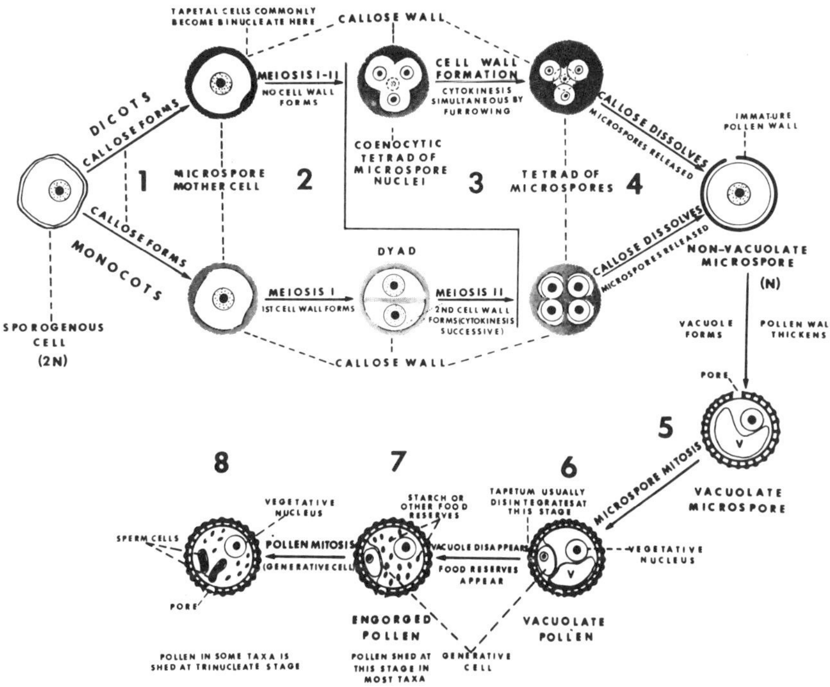

Fig. 14—A diagrammatic scheme showing general features of pollen development of representative monocot and dicot pollen. The numbers delimit successive stages from microspore mother cell (stage 1) to mature pollen (stages 7, 8). From Laser and Lersten (1972).

Each microspore mother cell undergoes the two divisions that constitute meiosis, resulting in four haploid nuclei and, eventually, four pollen grains (Fig. 14). Stages 1 to 4 in Fig. 14 show differences that often occur in dicot and monocot pollen development. Cell walls in dicots do not ordinarily form until after meiosis II. Thus for a short period the four haploid nuclei are in a common cytoplasm. The two walls form simultaneously to delimit the microspores. In monocots, a new wall forms after each meiotic division, so that a cellular dyad stage is followed by a cellular tetrad.

The young microspores, still embedded in callose, begin using food reserves to build the future pollen wall. While this wall is still immature, the callose dissolves and the microspores are released into the fluid of the pollen sac. As the pollen wall thickens, each microspore enlarges and becomes more vacuolate. After the large central vacuole develops, the microspore nucleus divides mitotically (Fig. 14). After one nucleus migrates to the wall, a thin cell wall forms around it and a bit of cytoplasm to delimit the generative cell. The other nucleus remains in the larger cell, now called the vegetative cell. Most pollen is shed in the two-celled condition but, in some plants, most notably grasses, the generative cell divides again and pollen is shed with two sperm cells already formed. Each of these is a true cell with a nucleus, cytoplasm, and cell wall.

The already wall-less tapetum commonly disintegrates completely after the vegetative and generative cells are formed. At the same time, the pollen grains begin to fill with food reserves. Oily, proteinaceous substances from the tapetum may be deposited on the pollen wall and these may be important later when pollen interacts with the stigma.

A mature pollen grain wall has an inner layer (intine), and an outer layer (exine) of considerable complexity in which subdivisions can be seen. The outer part of the exine commonly has a sculptured pattern of ridges, mounds, large or small spines, or other features. These occur singly or in various combinations (Figs. 15 and 16). In addition, the exine fails to form in one or more places, creating pores or furrows which expose the underlying intine (Fig. 16). Grasses and most other monocots have only one pore or furrow (Fig. 15) whereas most dicots have three (Fig. 16). Sugarbeet and spinach, in contrast, are examples of dicot pollen with many pores. These thin places allow the intine to bulge out at the time of germination and become the cell wall of the pollen tube.

Further information on pollen development and mature structure can be found in Heslop-Harrison (1971), Linskens (1974), Mascarenhas (1975), and Vasil (1973).

C. Ovule and Embryo Sac Development

The ovule-bearing or seed-bearing part of a flower is the pistil. The term pistil has lost some of its precision because it is used to describe both a carpel and a gynoecium. A carpel consists of an ovary, style, and stigma. The term gynoecium represents all the carpels in a flower. A gynoecium may have one carpel as in legumes, several separate carpels as in raspberry and strawberry, or several carpels which have varying degrees of fusion, as in potato, cotton, and the majority of cultivated plants. Where there is extreme fusion of carpels, as in buckwheat, sugarbeet, and cultivated grasses, the compound ovary will have a single locule, and will appear superficially to be simple. The terms pistil and gynocoeium will be considered equivalent in this chapter.

Ovules develop as outgrowths from a local area (placenta) on the inner ovary wall. The chamber in which an ovule develops is called the locule. There may be one ovule per locule, as in grasses or buckwheat; two ovules, of which one aborts, as in cherry and related members of the Rosaceae; several ovules per single locule, as in legumes; or many ovules in several fused locules, such as in tomato or cotton.

Whatever the number and arrangement, each ovule at maturity consists of a main body (nucellus) containing the embryo sac, one or two cup-like sheaths (inner and outer integuments), and a basal stalk (funiculus) which connects the ovule to the placenta (Fig. 18J). The integuments have a pore at the apex (micropyle) through which the pollen tube enters. The ovule in the vast majority of species is bent back on itself at least 90°, and in many species, such as legumes, mustard, and potato, it is bent back 180° so that the micropyle is adjacent to the placenta. In a small number of species, e.g., walnut and some grasses, the ovule is straight. In grasses such as corn, the ovule is almost straight (Fig. 17).

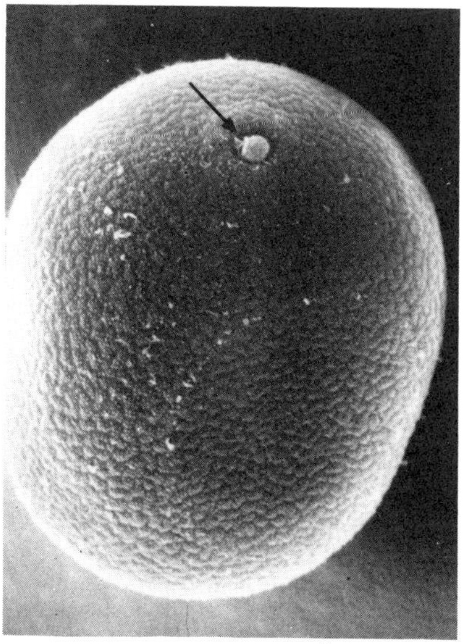

Fig. 15—Scanning electron microscope view of sorghum pollen grain. Grass pollen has inconspicuous exine patterns. The arrow indicates the single pore, characteristic of grasses and monocots in general. The exine plug over the pore becomes a tight cork when the pollen is dehydrated and shrunken. (Photo, courtesy of Dr. H. T. Horner, Jr.)

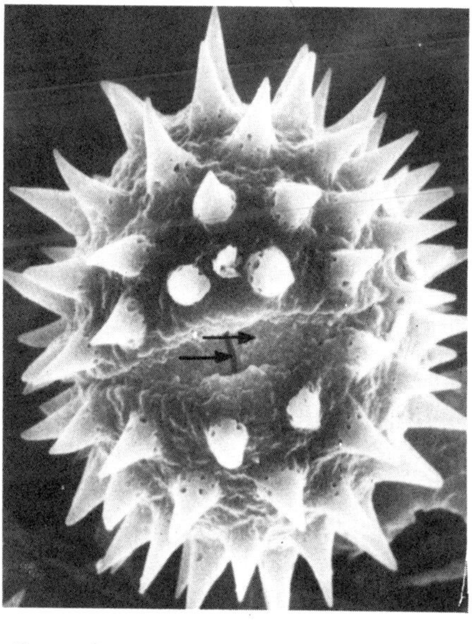

Fig. 16—Sunflower pollen grain, a scanning electron microscope view. Outer exine bears spines with tiny pores. Upper arrow indicates inner portion of exine exposed in a furrow. Lower arrow indicates the pore in the furrow, below which occurs the intine. There are three sets of furrows with pores, two of them obscured on this grain. This is commonest pattern in dicots. Most dicot pollen is not as conspicuously spiny. (Photo, courtesy of Dr. H. T. Horner, Jr.)

The embryo sac is a sac-like structure, commonly composed of seven cells, which develops from one cell of the nucellus. One of the cells in the embryo sac is the egg cell, which becomes the embryo after fertilization. Embryo sacs develop variously and the number and arrangement of cells within the sac vary at maturity, but more than 70% of flowering plants, and nearly all cultivated species (lily and onion are exceptions) are of the so-called normal type.

In the normal type of embryo sac development, one cell in the nucellus enlarges to become the megaspore mother cell (Figs. 18A to E). It undergoes meiosis, resulting in a row of four haploid megaspores (Fig. 18G). Like the microspore mother cell and microspore tetrad, the megaspore mother cell and the megaspore tetrad are coated with callose. The outer three cells almost invariably degenerate (Fig. 18H), while the innermost megaspore enlarges to form the embryo sac. The callose dissolves at about this time.

Three mitotic divisions occur within this cell, resulting in eight nuclei (Fig. 18I). Three nuclei remain at each end and two migrate to the middle to become the polar nuclei. Each of the three nuclei near the micropyle becomes enclosed, along with a small amount of cytoplasm, by a thin wall. These become the egg cell and two accompanying synergid cells. The three nuclei at the other end are also each encased in walls and become antipodal cells. The function of antipodals is unknown; they may remain unchanged, degenerate, or divide and form a mass of cells, as in grasses and crucifers (Fig. 20E). The mature embryo sac consists of the enlarged megaspore, called the central cell, within which are six cells and the two polar nuclei (Fig. 18J).

III. MEIOSIS

Meiosis in angiosperms occurs only in microspore mother cells and megaspore mother cells. The process is virtually identical in both (Bennett, 1977). Figure 19 illustrates the configurations taken by, and the movements of, the chromosomes as they proceed through meiosis I and II.

Premeiosis is recognized as a period during which no visible events occur, but genetic material (DNA) is replicated, so that each chromosome can enter meiosis with double the amount of DNA. During meiosis I, also called reduction division, the two sets of chromosomes in a nucleus pair, exchange portions of their chromosomes by a process called crossing-over, then separate and move into different nuclei. It is divided into four stages, of which the first, prophase I, is the longest and most complex. For this reason prophase I is subdivided into five substages:

a) leptotene. The diffuse chromosomes of premeiosis have coiled and shortened to the extent that they are visible as a jumble of threads.

b) zygotene. As they continue to shorten, chromosomes begin to pair with each other. This pairing is remarkably exact and specific.

c) pachytene. The chromosomes continue to shorten and their subunits, the chromatids, exchange segments by crossing-over.

d) diplotene. Chromosome pairs can be seen separating from each other, except where crossovers have occurred. They continue to shorten.

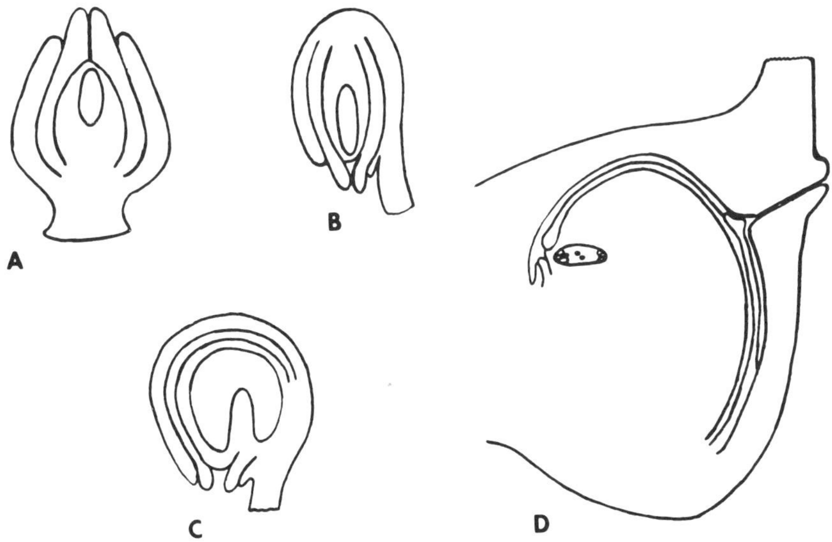

Fig. 17—Ovule shapes shown diagrammatically. A. Straight. B. Bent back 180° with embryo sac also reversed. C. Bent back 180° but embryo sac not completely reversed. D. Ovule of corn showing condition between A and B.

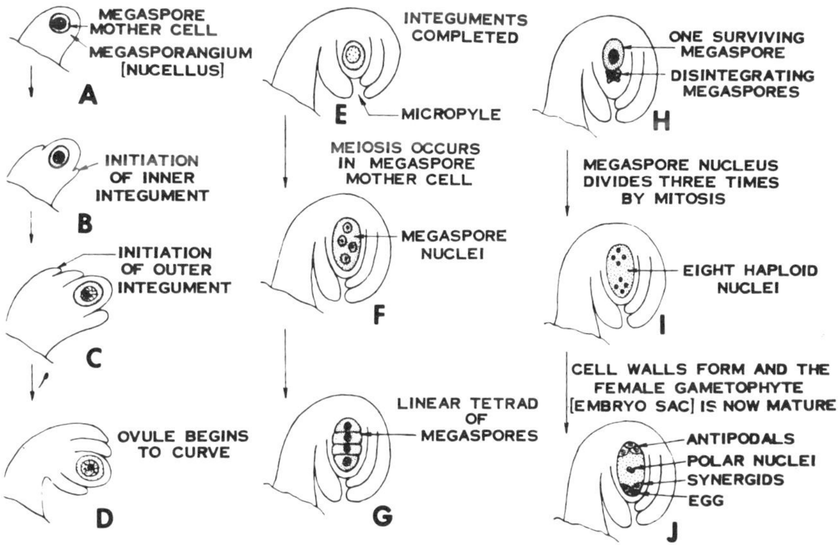

Fig. 18—General features of ovule and embryo sac development. The ovule is bent back on its stalk (funiculus) and embryo sac development is of the normal type. (Reprinted by permission from *Course Book in General Botany* by John D. Dodd © 1977 by Iowa State Univ. Press, Ames.)

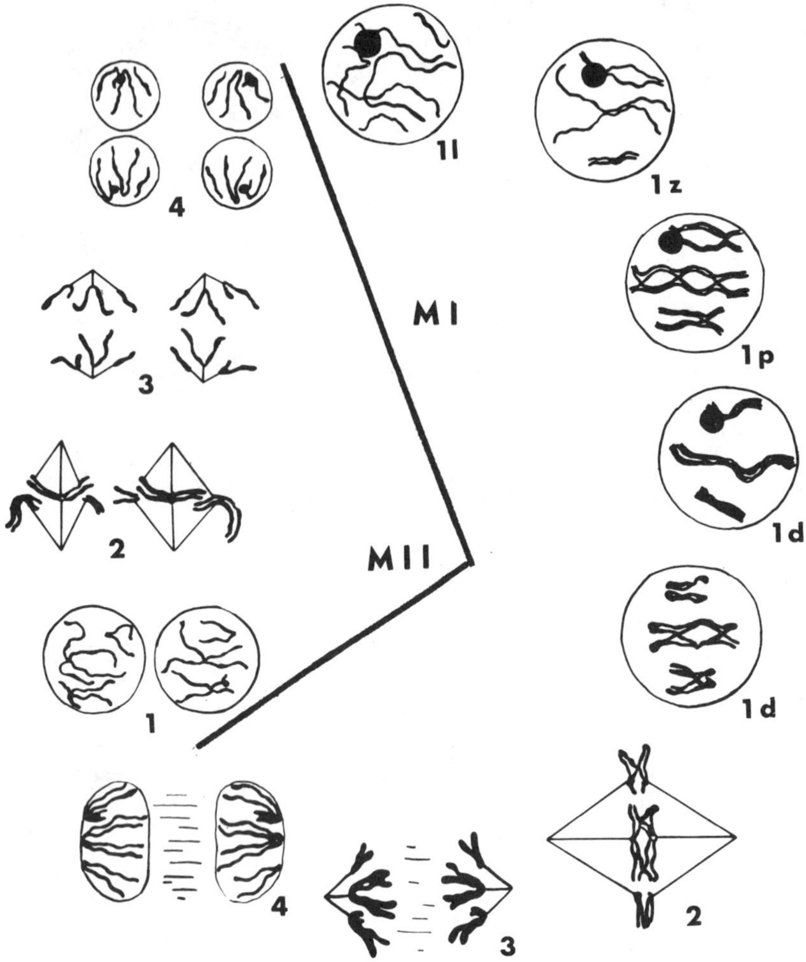

Fig. 19—Stages of meiosis illustrated by a microspore mother cell of a monocot. Meiosis I (MI): Prophase consists of the leptotene (1*l*), zygotene (1z), pachytene (1p), diplotene (1d), and diakinesis (1d) stage. At metaphase I (2), nuclear membrane and nucleolus have disappeared, the chromosomes are lined up on the equatorial plate, and spindle fibers have formed. During anaphase I (3), sets of bivalents move to opposite poles. At telophase I (4), each set of bivalents is surrounded by a newly-formed nuclear membrane. Meiosis II (MII): Prophase II (1), metaphase II (2), anaphase II (3), and telophase II (4) are a part of the second division that results in a tetrad of cells.

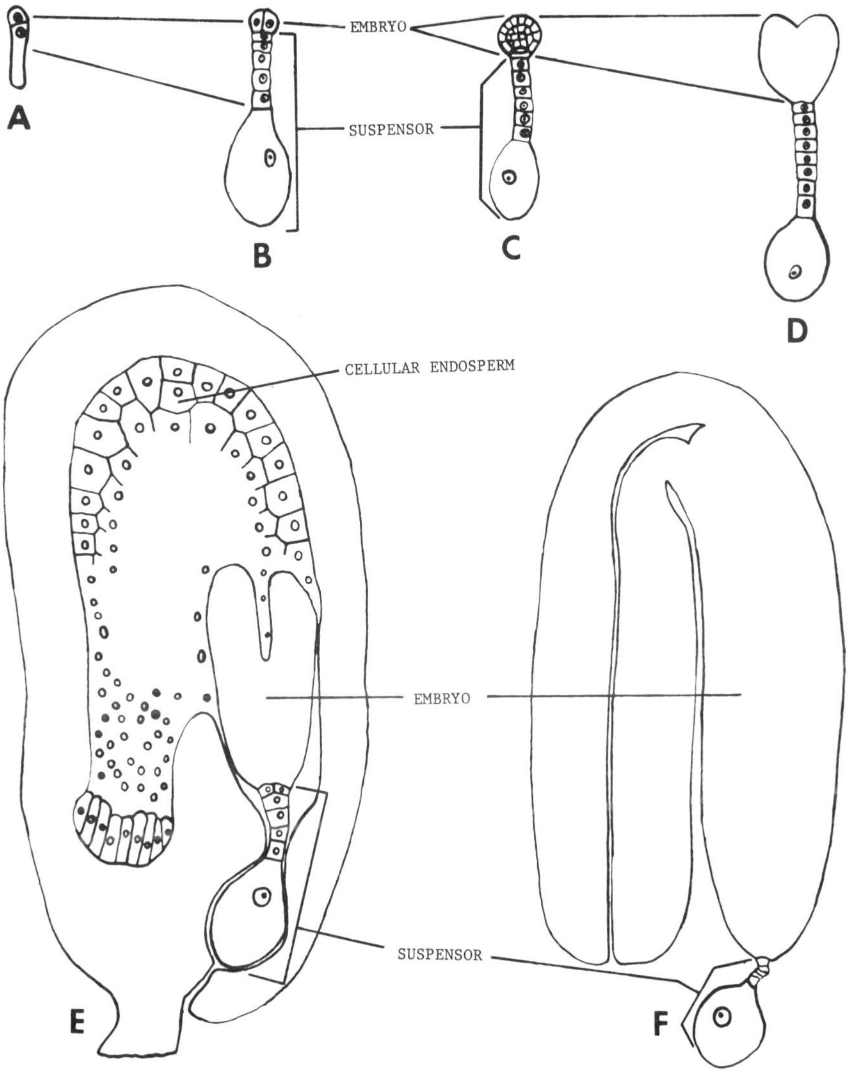

Fig. 20—Stages of embryo development in a dicot, based on *Capsella bursa-pastoris,* a crucifer. All stages are drawn to the same scale. A. First division of zygote to produce first cell of embryo and first cell of suspensor. B. Four-celled embryo with mature suspensor. C. Globe stage of embryo. D. Heart stage of embryo (cellular detail omitted). E. Torpedo stage of embryo set in developing seed. Note acellular milk stage of endosperm and peripheral development of cellular endosperm. F. Mature embryo folded on itself. Note start of collapse of the suspensor, the remnant of which will be destroyed during germination.

e) diakinesis. The chromosomes are at their shortest, and the nuclear wall and nucleolus disappear.

The second stage is metaphase I, during which the shortened chromosomes, still attached at points of crossing-over, line up within the cell in the central plane called the equatorial plate. Spindle fibers are formed.

During anaphase I, the chromosome pairs separate from each other and begin to move toward opposite ends of the cell. Each chromosome appears to be pulled by one or more threadlike spindle fibers attached at various places along its length.

When the two sets of chromosomes reach opposite ends of the cell, telophase occurs and, in most megaspore mother cells, a cell plate forms and two cells result. The interval between meiosis I and II is usually brief, and the chromosomes remain fairly short. In some plants there may be little or no interval between the two.

Meiosis II does not involve crossing-over. The two chromatids of each chromosome separate and the new cells formed are haploid. The stages of meiosis II are prophase (not subdivided), metaphase, anaphase, and telophase. At the end of telophase, cell plates are again deposited.

Determining how long meiosis lasts is technically difficult. Only a small number of plants have been studied, but these are mostly cultivated species (Bennett, 1977). Meiosis lasts about as long in the microspore mother cells of the anther as in megaspore mother cells of the ovules. The range among the approximately 40 species that have been studied is 1 to 17 days. For example, meiosis takes 24 to 51 hours in the cereal grasses (e.g., wheat, barley) that have been studied.

It should be re-emphasized that meiosis occurs only in two special kinds of cells, microspore mother cells in the anther and the megaspore mother cell in the ovule. Meiosis is the process by which matching chromosomes pair, exchange genetic material, and separate. After fertilization, two sets of chromosomes, one from the pollen and one from the egg, will again be together in all cells of the embryo. The straightforward outline presented here does not reflect the many possible complicating factors involved in meiosis following hybridization.

IV. POLLINATION AND POLLEN-STIGMA INTERACTION

A. Pollination Mechanisms in Cultivated Plants

Late in anther development, the four locules become two because of the disintegration of separating cell layers. When pollen is ready to be shed, drying and shrinking causes one or more slits to form on the sides of the anthers. More rarely, such as in potato, pores open at the tip and release pollen in saltshaker fashion. Pollen is commonly carried by insects or wind to cross-pollinate another flower. Some species are self-pollinated. Still other plants have both types of pollination, with self-pollination occurring either as a response to normal seasonal changes or adverse weather.

Many cross-pollinated species are adapted to insect pollen carriers. McGregor (1976) wrote a practical book dealing with insect-pollinated plants. It covers insect pollination in about 150 species of monocots and dicots.

B. Pollen-Stigma Interaction

Stigmas of cross-pollinated flowers may collect pollen from many different plants. Several mechanisms have evolved to screen out unwanted pollen and stimulate growth of desired pollen (Heslop-Harrison, 1975a).

Disintegrating tapetal cells from the parent plant may deposit an oily, proteinaceous coating on pollen. The proteins in this substance may later act as a signal to be recognized by protein on the stigma surface. Mustard is a well-studied example of this mechanism. Proteins also may be incorporated within the intine layer of the pollen wall below the pores or furrows during pollen development. When pollen with such protein lands in stigmatic fluid, hydration of the intine releases them to act in recognition. The difference between exine and intine protein is that the protein coating on the exine is from the parent plant, whereas the intine proteins are produced by the pollen grain.

Stigmas also may have recognition proteins. These can occur as a proteinaceous cover in the so-called dry stigmas (e.g., grasses, mustard, or sunflower), or they may be present in the stigmatic fluid secreted by wet stigmas (e.g., potato). A recent comprehensive review and survey of types of stigmas was made by Y. Heslop-Harrison and Shivanna (1977). They identify additional subtypes under the dry and wet categories.

Interaction between pollen and stigma proteins often governs pollen germination and pollen tube penetration into the style. The details of what happens on the stigma in both compatible and incompatible crosses differ somewhat among the species studied. Incompatible pollen may swell but not germinate; it may form a short tube which does not penetrate the stigma; or, there may be a slight penetration which is halted by a callose barrier formed in the stigma epidermis.

A protein incompatibility reaction occurs between the stigma and foreign pollen, and much is known about it. A still unknown mechanism controls pollen inhibition in species that are self-incompatible, i.e., plants in which pollen will not germinate on the stigma of the same flower, or perhaps not even on stigmas of other flowers on the same plant. A detailed discussion of the genetic systems involved in incompatibility reactions was presented by Heslop-Harrison (1975b).

In some species, pollen-pistil interaction occurs in the style, where pollen tubes secrete enzymes which interact with proteins secreted by the cells lining the style. Incompatible tubes stop growing because of adverse chemical reactions or because they are simply unable to take up nutrients provided by the style. Hybridization in a few such species has been accomplished by removing the style and replacing the stigma on the stub, thereby giving the pollen tube only a short path to the ovule (Maheshwari, 1950).

A third location of pollen-pistil interaction is at the micropyle, where secretion from the ovule has been reported in several plants. The secretion has been reported to attract compatible tubes, while foreign tubes grow randomly in the ovary until their food reserves are gone.

Some plants have no barriers to foreign pollen. Foreign pollen tubes can enter and cause destruction of the embryo sac. This is inefficient because such ovules do not produce seeds. The prevention of this waste has doubtless been a factor in the evolution of more elaborate barriers closer to, or on, the stigmatic surface.

V. COMPATIBLE POLLEN GERMINATION AND TUBE GROWTH

When the miniscule pollen grains are exposed to air, they become extremely dehydrated. When a grain lands on a compatible stigma, it takes up water and begins to swell immediately. In seconds, the intine expands and the wall-borne enzymes and proteins are released. Continued water absorption and renewed metabolic activity in the pollen grain allows the pollen tube to emerge and grow. Some of the pollen enzymes mediate absorption of nutrients, and others digest stigma cell walls, allowing the tube to pass between cells. A pollen tube never enters cells until it reaches the embryo sac.

The style can accommodate pollen tube growth in two ways. In general, dicots tend to have solid styles, with a column of transmitting tissue consisting of cells with a thick, nutrient-rich wall which is digested by the tube as it grows between these cells (Sassen, 1974). In cotton, the pollen tube-style interaction has been studied in detail (Jensen, 1972).

Monocots tend to have a hollow style opening directly onto the stigma, with only the stigma cells as a barrier. The pollen tube grows in a nutrient fluid secreted by cells lining the inner surface of such a style. Lily is the most intensively studied representative of this type (Mascarenhas, 1975). The grasses are a conspicuous exception, having solid styles, each with a strand of well-defined transmitting tissue. The pollen tube in grasses, however, must first grow between cells of the long, feathery stigma for a considerable distance before reaching the style.

The pollen tube has an ordinary cell wall which is thin and rapidly growing at its tip. Protoplasm is concentrated in the tip and is sealed off from older parts of the tube by occasional callose plugs. The tube cell nucleus may lead or lag behind the generative cell. The generative cell, in two-celled pollen, divides to become two sperm cells before the pollen tube reaches the embryo sac.

When the pollen tube enters the ovary, it may be guided mechanically to the micropyle of the ovule by a continuation of the stylar secretory tissue, by a cuplike outgrowth of the ovule which guides the tube, or by secretion of a substance through the micropyle, which acts as a chemical guide. The exact mechanism differs among plants. In some plants, there may be no special mechanism, and tubes appear to grow randomly, striking a micropyle only by accident. In flowers with fused ovaries but separate styles, as in carrot, there may be a horizontal plate of transmitting tissue at the base of the styles to allow random growth into any of the ovaries.

The micropyle may be a distinct opening or it may appear closed by an overlapping of inner and outer integuments (e.g., cotton, some legumes). A carbohydrate-containing fluid secreted by the nucellus may help to open a micropyle with overlapping integuments and attract the pollen tube. Such a secretion has been described in the grass *Paspalum*.

After the tube passes through the micropyle, it must grow through a certain number of nucellar layers to reach the embryo sac. At one extreme is sunflower, in which the synergids protrude through the nucellus and are immediately available at the micropyle. At the other extreme are such plants as cotton, corn, and grape, with an embryo sac deep in the nucellus. Other plants, such as tomato or potato, have fewer nucellar layers. The swelling of the embryo sac during development usually causes destruction of at least some nucellar layers. The tube grows between nucellar cells and digests their cell wall.

Elapsed time from pollen germination to pollen tube entry into the micropyle varies from minutes to months, but 6 to 25 hours is probably most common among cultivated plants. In soybean, it takes 8 to 10 hours. In some cereal grasses, such as wheat, less than 1 hour may elapse (Bennett, 1973). Woody plants tend to be slower, and the process can take weeks or months in some trees.

VI. PENETRATION OF EMBRYO SAC AND FERTILIZATION

After penetrating the micropyle and nucellus, the pollen tube enters the embryo sac by means of enzymatic digestion of the embryo sac wall. The condition of the embryo sac cells at the time of pollen tube entry varies with the species: 1) antipodals may be unchanged, degenerated, or actively proliferating more cells; 2) polar nuclei may have fused or may still be separate; 3) synergids may be unchanged or one may have degenerated.

Pollen tubes have been reported to enter the embryo sac at different places, but they most commonly grow into the conspicuous cell wall proliferation (filiform apparatus) of one of the synergids. All electron microscope studies have reported that the tube enters the degenerating synergid (Mogensen, 1978). In the synergid cell, the tip of the pollen tube ruptures and the cytoplasm and sperm cells flow out. Rupturing of the pollen tube may be caused by osmotic conditions within the synergid cell. The details remain elusive as to how each sperm cell travels to either the egg cell or polar nuclei, gets into the cell and nucleus, and mingles its chromosomes with those of the female. Cotton, barley, and wheat have been studied most carefully (Jensen, 1972; Bennett, 1973). It is known that one sperm and the polar nuclei fuse before union of sperm and egg nuclei. In sunflower, for example, fusion occurs 2 to 3 hours earlier, and in wheat 10 to 15 hours earlier.

The sperm cell has a thin envelope of cytoplasm around the nucleus which is devoid of plastids. Cytoplasmic inheritance could occur if the sperm cell cytoplasm mingled with egg cytoplasm at the time of fertilization. There are reports that cytoplasm does enter the egg cell, but other reports that only the sperm nucleus enters. Even in species where fertilization can be predicted within a few minutes, it is nearly impossible to obtain correctly oriented preparations at just the right instant. Jensen (1972) and Kapil and Bhatnagar (1975) reported recent information on fertilization.

Polyspermy is penetration of an embryo sac by more than one pollen tube, or the occurrence of more than two sperm cells per tube, both resulting in multiple sperms entering an embryo sac. This has been observed in about 55 dicot and monocot species (Vigfusson, 1970), chiefly cultivated plants, and it may be a fairly common event according to some workers. All reports are from cross-pollinated species, however. Penetration of an egg cell and fertilization by more than one sperm has been rarely reported, and Vigfusson feels that this has not yet been shown unambiguously.

VII. ENDOSPERM

Endosperm provides nutrition for the embryo. Endosperm is initiated by the fusion of one of the sperm nuclei with the two polar nuclei, resulting in a triploid primary endosperm nucleus in the cytoplasm of the central cell of the embryo sac. (This requirement of two sperms for sexual reproduction in angiosperms is unique.) The usual triple fusion occurs earlier than fertilization of the egg, and the endosperm begins growing before the embryo. Endosperm may develop in at least four ways, of which three occur in cultivated plants.

The most common way is for the endosperm nuclei to divide initially without cell walls being laid down. This milk stage of the endosperm is most spectacular in the coconut and is well known in corn, but it also occurs in other cereal grasses, legumes, and most other cultivated plants. Walls form later, starting at the periphery of the embryo sac, resulting eventually in a cellular endosperm (Fig. 20E). In later divisions, a cell wall is laid down after each nuclear division.

A few plants have cellular endosperm from the beginning (e.g., tobacco). Another rather small group, including some grasses and the maples, have only a non-cellular liquid endosperm in which cell walls never develop.

The outermost endosperm layer is often distinguished as the aleurone layer because it contains protein bodies called aleurone grains. This layer is developed to various degrees in both dicots and monocots. It is well developed in cereal grasses and in lettuce, and has been intensively studied there with respect to its influence on stimulating germination.

It is not uncommon for the endosperm to form saclike or filamentous outgrowths which extend into adjacent maternal tissue. Such outgrowths are thought to increase nutrient absorption. They are known in over 50 families, including some cultivated legumes.

Endosperm cells are triploid at first because of the fusion of three nuclei. Later, endosperm cells may become even more polyploid. Endosperm in onion may reach 12n and in corn 1,000n. Even higher levels have been reported from certain wild species. High endosperm ploidy often involves unusual variations of mitosis.

One can speculate as to why the triploid condition stimulates endosperm formation. One theory postulates hybrid vigor and states that this mechanism allows the endosperm to compete successfully for available nutrients with the diploid cells surrounding the embryo sac. Another theory is that the male chromosomes in the endosperm act as an intermediary between the embryo and the surrounding diploid maternal tissue. Brink and Cooper (1947) discuss these and other theories in detail.

In sugarbeet, spinach, amaranths, and other members of the Chenopodiaceae and Amaranthaceae, the nucellus persists and becomes filled with food reserves. This tissue, called perisperm, is diploid and occurs in addition to endosperm. The embryo presumably is able to absorb both of these reserves.

VIII. THE EMBRYO

The fertilized egg (zygote) is the first cell of the embryo. Fertilization usually occurs after endosperm initiation, and the zygote remains undivided until at least a small amount of endosperm has been produced. In barley, the first division of the zygote occurs about 9 hours after the first endosperm division, when eight endosperm nuclei are already present.

The first cell wall usually forms a basal cell toward the micropyle and a terminal cell away from the micropyle. The basal cell usually forms a multicellular suspensor which does not become part of the mature embryo. In lettuce, however, part of the suspensor does contribute cells to the primary root. Most commonly, the suspensor is a short filament of cells, with the tip cell sometimes swollen and polyploid, as in mustard and the garden bean (Fig. 20). In soybean, the suspensor is merely a rudiment of three to four cells, whereas in garden pea, chickpea, and some other legumes the suspensor becomes unusually massive.

The suspensor was previously thought to act as a ramrod to push the embryo further into the nutritive endosperm. More recent studies on some of the mustards and beans suggest that in these plants it may be more important in absorbing and transferring nutrients to the embryo during early stages of development; the suspensor may even secrete enzymes that control certain aspects of embryo growth. The suspensor is prominent early, but it soon ceases growth and is inconspicuous when the embryo is mature (Figs. 20 and 21).

The terminal cell produced by the first division of the zygote will become the embryo after passing through several stages. In some plants, daughter cells of the basal cell also contribute to the embryo. Continued mitotic divisions result first in a sphere of up to several hundred cells (globe stage). Toward the end of this stage the first indications of internal differentiation are seen. Up to this time, the pattern of development is similar among dicots and monocots (Figs. 20A to C and 21A).

In dicots, the globe stage is followed by the heart stage, in which two lateral multicellular wings arise which will become the pair of cotyledons (Fig. 20D). The cotyledons are the first or seed leaves of the plant. As the cotyledons grow, the embryo axis (hypocotyl-radicle axis) elongates and a torpedo stage is recognized (Fig. 20E). Further development varies with the species. An epicotyl (rudimentary main shoot) may develop and additional embryonic leaves may form (e.g., pea, soybean), or an epicotyl may be lacking in the mature seed (e.g., cotton, potato, sugarbeet). The mature embryo may be straight (e.g., lettuce) or, more commonly, coiled or folded (e.g. cotton, tomato (Fig. 20F). There may be endosperm around the mature embryo, as in potato and tomato, or all endosperm may have been absorbed and stored in the cotyledons, as in cultivated legumes.

The major distinction between dicot and monocot embryo development is that in monocots only one cotyledon forms. The developing cotyledon assumes a terminal position quite early, so that the embryo shoot tip appears to be lateral (Figs. 21B to D). The single cotyledon in many monocots remains within the seed coat, slowly absorbing endosperm during germination. In onion, only the tip of the cotyledon remains within the seed; the rest of it elongates and arches above ground, where it becomes green and photosynthetic. Except for the cotyledonary difference, embryo and endosperm development are remarkably similar in most dicots and monocots.

The grass embryo, however, is different from that of most other monocots. There is disagreement about what is a cotyledon in grasses, but the scutellum behaves as if it were one, remaining wholly within the seed to absorb endosperm and transfer it to the seedling. The epicotyl of the embryo axis is ensheathed by the tubular coleoptile, often interpreted as the first leaf after the cotyledon. The radicle is ensheathed by the protective coleorhiza (Fig. 21E). Because of several differences in structure and germination behavior, there is controversy about the interpretation of parts of the grass embryo (Rost and Lersten, 1973).

The integuments enclosing the nucellus enlarge and change their cellular features as the embryo develops. Both integuments may contribute to the seed coat, or only one may do so. The nucellus ordinarily becomes crushed and disappears (Fig. 20E), but in some plants it remains as a paper-thin layer just below the seed coat (e.g., peanut). In grasses, the seed coat

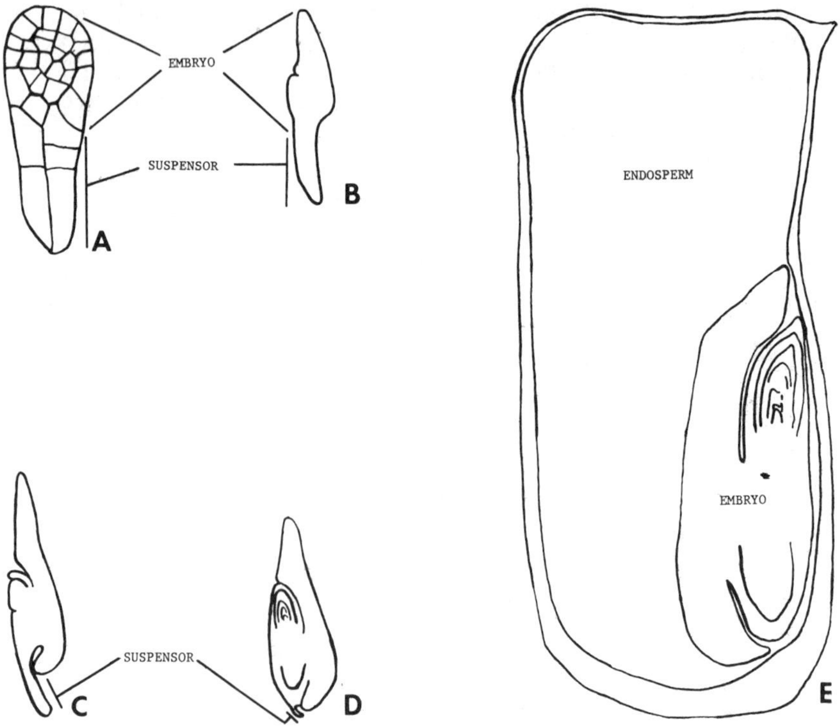

Fig. 21—Stages of embryo development in corn, illustrating general monocot features and special features of grasses. Stages are not all drawn to the same scale. A. Globe stage (6 days after fertilization) approximately equivalent to Fig. 20C. Suspensor not sharply delimited from embryo. B. Young cotyledon stage (12 days old) showing laterally displaced shoot tip and large suspensor. There is no greatly enlarged cell at tip of suspensor. C. Later stage (20 days old) showing early coleoptile development. D. Embryo 35 days old with embryo axis and several young leaves. Note relative size of suspensor. E. Mature (45-day-old) embryo in the caryopsis. No trace remains of suspensor.

becomes tightly appressed to the ovary wall (Fig. 21E) so that a separate seed coat cannot be easily distinguished (Rost and Lersten, 1973). The details of seed coat structure and development vary so much that individual species cannot be considered here.

IX. APOMIXIS AND MALE STERILITY

A. Apomixis

Apomixis, from two Greek words meaning "without mixing," is a general term for several ways of non-sexual reproduction which involve some or most of the apparatus of sexual reproduction. An embryo sac usually forms, sometimes in normal fashion, from the haploid megaspore, in other plants from the unreduced diploid megaspore mother cell or even from another diploid cell of the nucellus. Within this embryo sac, the egg cell and polar nuclei usually form an embryo and endosperm without fertilization. In some types of apomixis, a pollen tube and sperm cells are required to stimulate development, but do not fuse with the nucleus. Apomixis is dealt with in detail in Chapter 3.

B. Male Sterility

It has been known for a long time that in some plants pollen may be aborted, stamens malformed, or stamens may be lacking entirely. This occurs sporadically in wild species, and if it confers an advantage because it ensures cross-pollination, such plants may be selected for and the species may become dioecious. This seems to occur frequently in angiosperms because, while only 5% of the flowering plant genera are dioecious, they are scattered among 75% of the families (Crowe, 1964).

When stamen or pollen failure is noted among plants of cultivated species, such plants can often be selected and propagated for breeding purposes because they have become obligate outbreeders whose progeny cannot be the result of self-pollination. The practical benefits of having such male-sterile plants in otherwise male-fertile species has stimulated much research into both the application and the underlying mechanism of male sterility.

The male sterile condition can be the result of genes which are inherited according to ordinary Mendelian laws. Such genetic male sterility is well-known and is used by breeders in several crop species. In cytoplasmic male sterility, there are factors in the maternal cytoplasm which induce sterility. A combination of cytoplasmic and genetic factors is known to control sterility in still other lines. A knowledge of which category a particular male-sterile fits is necessary for one to know how to restore fertility, and also to indicate how predictable and stable the male sterile condition will be. Practical application of male sterility for plant breeding and commercial hybrid seed production is discussed in Chapters 4 and 8. A detailed review of the genetic aspects is in Frankel and Galun (1977).

Whatever the type of sterility, the visible evidence in the flower is failure of normal stamen or pollen development. In different species, or in various male sterile lines within a species, the stage of development at which stamen or pollen abortion occurs is variable. The stage at which abortion occurs may also be affected by day length, temperature, and other field conditions.

If stamens develop and microspore mother cells are produced, abortion may occur at any stage up to the mature microspores (Frankel and Galun, 1977; Laser and Lersten, 1972). Most abortions occur at about the tetrad stage, a smaller number at or near meiosis or closer to microspore maturation.

There are several hypotheses as to what causes abortion. The tapetum in some dicots and monocots has been observed to swell and expand inwards, seemingly crushing the microspores. Failure of the tapetum to secrete substances at the right time has also been implicated. Amino acid deficiencies and differences in mitochondria have been proposed as possible causes of pollen abortion in some grasses. Finding answers has proved difficult, but much research is in progress. Discovering the reasons for abortion is desirable before we can intelligently attempt to predictably induce male sterility.

REFERENCES

Bennett, M. D. 1973. Cell development in the anther, the ovule and the young seed of *Triticum aestivum* L. var. Chinese Spring. Phil. Trans. (Cambridge Univ.), Ser. B. 266:39–81.

———. 1977. The time and duration of meiosis. Phil. Trans. Roy. Soc. London, Ser. B, 277:201–226.

Brink, R. A., and D. C. Cooper. 1947. The endosperm in seed development. Bot. Rev. 13:423–541.

Crowe, L. K. 1964. The evolution of outbreeding in plants. I. The angiosperms. Heredity 19:435–457.

Davis, G. L. 1966. Systematic embryology of the angiosperms. John Wiley and Sons, New York.

Frankel, R., and E. Galun. 1977. Pollination mechanisms, reproduction and plant breeding. Springer-Verlag, Berlin.

Hayward, H. E. 1938. The structure of economic plants. The Macmillan Co., New York.

Heslop-Harrison, J. (ed.). 1971. Pollen: development and physiology. Butterworths, London.

———. 1972. Sexuality of angiosperms. p. 133–291. *In* F. C. Steward (ed.) Plant physiology, a treatise. Vol. VI C. Academic Press, N.Y.

———. 1975a. Male gametophyte selection and the pollen-stigma interaction. p. 177–190. *In* D. L. Mulcahy (ed.) Gamete competition in plants and animals. North-Holland Publ. Co., Amsterdam.

———. 1975b. Incompatibility and the pollen-stigma interaction. Ann. Rev. Plt. Physiol. 27:403–425.

Heslop-Harrison, Y., and K. R. Shivanna. 1977. The receptive surface of the angiosperm stigma. Ann. Bot. 41:1233–1258.

Jensen, W. A. 1972. The embryo sac and fertilization in angiosperms. University of Hawaii—Harold L. Lyon Arboretum lecture no. 3. Univ. of Hawaii, Honolulu.

Kapil, R. N., and A. K. Bhatnagar. 1975. A fresh look at the process of double fertilization in angiosperms. Phytomorphology 25:334–368.

Laser, K. D., and N. R. Lersten. 1972. Anatomy and cytology of microsporogenesis in cytoplasmic male sterile angiosperms. Bot. Rev. 38:425–454.

Linskens, H. F. (ed.). 1974. Fertilization in higher plants. North-Holland and American Elsevier Publ. Co., Amsterdam.

Lloyd, D. G., and C. J. Webb. 1977. Secondary sex characters in plants. Bot. Rev. 43:177–216.

Maheshwari, P. 1950. An introduction to the embryology of angiosperms. McGraw-Hill Book Co., N.Y.

Mascarenhas, J. P. 1975. The biochemistry of angiosperm pollen development. Bot. Rev. 41:259–314.

McGregor, S. E. 1976. Insect pollination of cultivated crop plants. USDA Handbook No. 496.

Mogensen, H. L. 1978. Pollen tube-synergid interactions in *Proboscidea louisianica* (Martiniaceae). Am. J. Bot. 65:953–964.

Palser, B. F. 1975. The bases of angiosperm phylogeny: embryology. Ann. Mo. Bot. Gard. 62:621–646.

Rost, T. L., and N. R. Lersten. 1973. A synopsis and selected bibliography of grass caryopsis anatomy and fine structure. Iowa State J. Res. 48:47–87.

Sassen, M. M. A. 1974. The stylar transmitting tissue. Acta Bot. Neerl. 23:99–108.

Vasil, I. K. 1973. The new biology of pollen. Naturwissenschaften 60:247–253.

Vigfusson, E. 1970. On polyspermy in the sunflower. Hereditas 64:1–52.

3 Apomixis and its Application in Crop Improvement

E. C. Bashaw

USDA-SEA and Texas A&M University College Station

Seed is formed in some species of higher plants by an asexual method of reproduction called apomixis. The apomictic seed develops in the ovary, as in sexually reproducing plants, but the embryo is formed without union of the sperm and egg. As a rule, the embryo originates by mitotic division of a nucleus in a somatic cell of the ovule and the male gamete does not fuse with the nucleus. Except in special types of apomixis, the zygote has exactly the same chromosome complement and genetic constitution as the mother plant and the offspring is a true clone, just as if it had developed from a vegetative organ. Hybridization between completely apomictic plants, also called obligate apomicts, is impossible except in rare instances when the sperm may unite with the unreduced nucleus of the female. Unless a mutation occurs in an obligate apomict, there is no natural variation among the progeny from which to select new types. Were it not for the fact that sexual or partially sexual plants can be found in most apomictic species, there would be little prospect for improvement through breeding.

For many years, apomixis was considered a total barrier to plant breeding and most geneticists were not anxious to find apomixis in the species they sought to improve. Darlington (1939) expressed the thinking of scientists of his time in a statement that, "apomixis is an escape from sterility, but an escape into a blind alley of evolution," inferring that all apomicts are doomed for extinction. As basic research on reproductive systems and breeding problems increased, it became apparent that apomixis is not a rare phenomenon in the plant kingdom and certainly is not an evolutionary "blind alley." Plant scientists have encountered apomixis again and again in diverse, unrelated, plant families, indicating that it is widespread, but often unnoticed in many genera. Some form of apomixis has been reported in at least 30 families of higher plants, including more than 300 species (Gustafsson, 1947; Nygren, 1954, 1967).

We now recognize that apomixis is the most common method of reproduction in many species of *Citrus* and berries, as well as guayule and numerous perennial forage grasses. Apomixis recently has been reported in grain sorghum and pearl millet and is believed to be present in several other important crops. Breeders of certain forage grasses have long been challenged to overcome the problems of reproductive isolation created by apomixis. Plant scientists have devoted considerable effort to investigations of the cytological and genetic basis of apomixis. These studies and the discovery of sexual plants in species previously thought to be completely apomictic eventually led to a new concept of apomixis in plant improvement—the possibility of controlling and manipulating apomixis as a plant breeding tool. Complete control of apomixis would afford the breeder and producer a system having the efficiency of reproduction through vegetative organs, but with the convenience of propagation through asexually developed seed. The most significant achievement, however, would be the ability to combine the desirable characteristics of two parents and simultaneously produce true-breeding apomictic F_1 hybrids with permanently fixed heterosis. Successful breeding programs recently have been developed with a few apomictic forage grasses by crossing a sexual plant with obligate apomicts and selecting and testing their apomictic F_1 hybrids.

To determine the potential for breeding an apomictic species, it is necessary to understand the apomictic mechanisms involved and to know if sexual reproduction occurs in the species. Individual plants of some species (Kentucky bluegrass, sorghum, guayule), called facultative apomicts, are capable of both sexual and apomictic reproduction in the same ovule. Some offspring from these plants are identical to the mother plant (apomictic origin) while others are different and variable (sexual origin). Since it is impossible to control method of reproduction in the ovules of facultative apomicts, the breeding of these species poses special problems because of unpredictable progeny variation (Section VI B).

I. MECHANISMS OF APOMIXIS

The mechanism of apomixis refers to the cytological basis for embryo development which determines the genetic constitution of the offspring. Many authors favor a broad definition of apomixis that includes reproduction through purely vegetative organs such as rhizomes, stolons, and buds; however, plant breeders tend to restrict the term to asexual seed formation. This is logical because the role of apomixis in the hybridization of higher plants concerns only the gametophytic generation. Apomixis will be used in this chapter to designate only those mechanisms for asexual seed development, collectively called agamospermy. There are four major mechanisms for agamospermy: apospory, diplospory, adventitious embryony, and parthenogenesis. The distinction among mechanisms is based on the site of origin and subsequent developmental pattern of the cell which gives rise to the embryo. Origin and development can be determined by cytological studies of young ovaries at the stage when megasporogenesis and embryo sac development would be expected to occur.

In the following discussion, photomicrographs will be used to compare sexual and apomictic development in ovules at critical stages required for identification of apomictic mechanisms. Megasporogenesis and development of the embryo sac in a typical sexual ovule are shown in Figs. 1A to F. The megaspore mother cell differentiates from a cell in the second (hypodermal) cell layer of the nucellus near the micropyle before the integuments are fully developed (Fig. 1A). At this stage, the orientation of the micropyle of the young ovule is at a right angle to the vertical axis of the ovary. Meiosis results in a linear tetrad of megaspores and soon after the second meiotic division the three megaspores nearest the micropyle disintegrate (Fig. 1B). The nucleus of the remaining megaspore divides and the two daughter nuclei migrate to opposite ends of the cell (Fig. 1C). Each nucleus undergoes two additional divisions producing an eight-nucleate embryo sac, which differentiates to form a mature gametophyte with two synergids, an egg, two polar nuclei, and three antipodals (Fig. 1D). Often the synergids disintegrate very early and in some species, especially of the Gramineae, the antipodals may continue to divide and form a cluster of cells (Fig. 1E) at the tip or side of the embryo sac. In many sexual species, remnants of the antipodals persist after embryo and endosperm begin to develop (Fig. 1F).

A. Apospory

In apospory, the embryo and endosperm develop in unreduced embryo sacs derived from somatic cells at various locations in the ovule. A normal megaspore mother cell develops in the same manner described for the sexual ovule and begins to undergo meiosis, but in apospory it may or may not complete both divisions to form a linear tetrad of megaspores. During this stage one or more somatic cells in the ovule and their nuclei begin to enlarge and usually resemble megaspore mother cells (Fig. 2A). The meristematic appearance of these cells contrasts sharply with that of the surrounding somatic cells, which remain inactive or senescent. At some stage, usually before a mature sexual embryo sac is formed, the megaspore or young sexual embryo sac aborts and is replaced by the developing aposporous sacs. The appearance of an ovule with developing aposporous cells (Figs. 2B and C) is quite different from that of a sexual ovule which will have a single embryo sac. For example, compare Fig. 2B and C with the two-nucleate sexual sac in Fig. 1C. In grain sorghum and some other species, initiation and development of aposporous embryo sacs may occur much later.

The nuclei in some aposporous cells undergo one or more mitotic divisions and the cells may enlarge considerably, but nuclear division and cell enlargement vary even in the same ovule. At the mature embryo sac stage, the ovule contains one or more enlarged aposporous cells with varying numbers of unreduced nuclei (Figs. 2D and E). These cells tend to resemble embryo sacs, but differentiation within the sacs is variable. Differentiation is always very limited in some species (buffelgrass and dallisgrass), and their aposporous sacs never contain the antipodals present in sexual embryo sacs. In these species, apospory is easy to distinguish from sexual development in mature ovules by the absence of antipodals and lack of organization within the aposporous sac. The contrast between undifferentiated aposporous sacs and sexual embryo sacs may be observed by comparing Figs. 2D and E with

Figs. 1D and E. However, in some species (Kentucky bluegrass and grain sorghum) one or more of the aposporous sacs may differentiate fully and look exactly like a sexual embryo sac at anthesis. In these species, it is necessary to follow the early development of embryo sacs to accurately identify apospory and to determine if the sexual embryo sac also develops.

Apospory is by far the most common mechanism of apomixis in higher plants. In most species it is characterized by the presence of multiple "embryo sacs," however, some aposporous species characteristically form a single embryo sac, which may or may not resemble a sexual sac.

B. Diplospory

In diplospory, the embryo and endosperm develop in an unreduced embryo sac derived from the megaspore mother cell. The megaspore mother cell (Fig. 2F) differentiates like that of sexual ovules, but its nucleus does

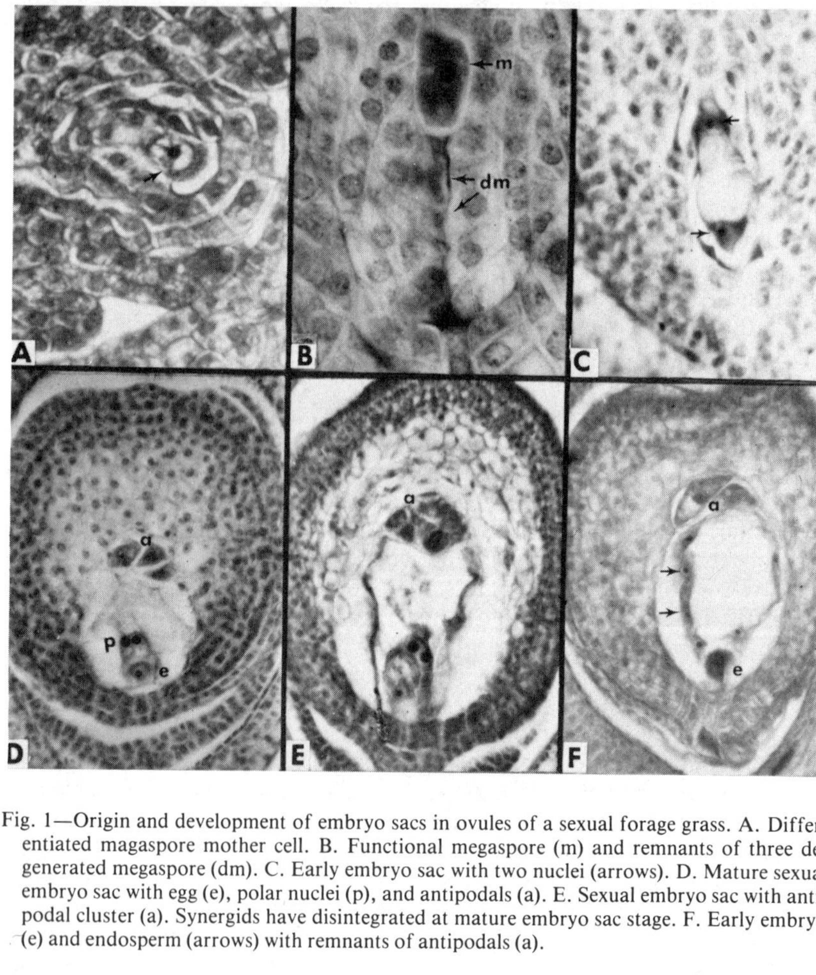

Fig. 1—Origin and development of embryo sacs in ovules of a sexual forage grass. A. Differentiated magaspore mother cell. B. Functional megaspore (m) and remnants of three degenerated megaspore (dm). C. Early embryo sac with two nuclei (arrows). D. Mature sexual embryo sac with egg (e), polar nuclei (p), and antipodals (a). E. Sexual embryo sac with antipodal cluster (a). Synergids have disintegrated at mature embryo sac stage. F. Early embryo (e) and endosperm (arrows) with remnants of antipodals (a).

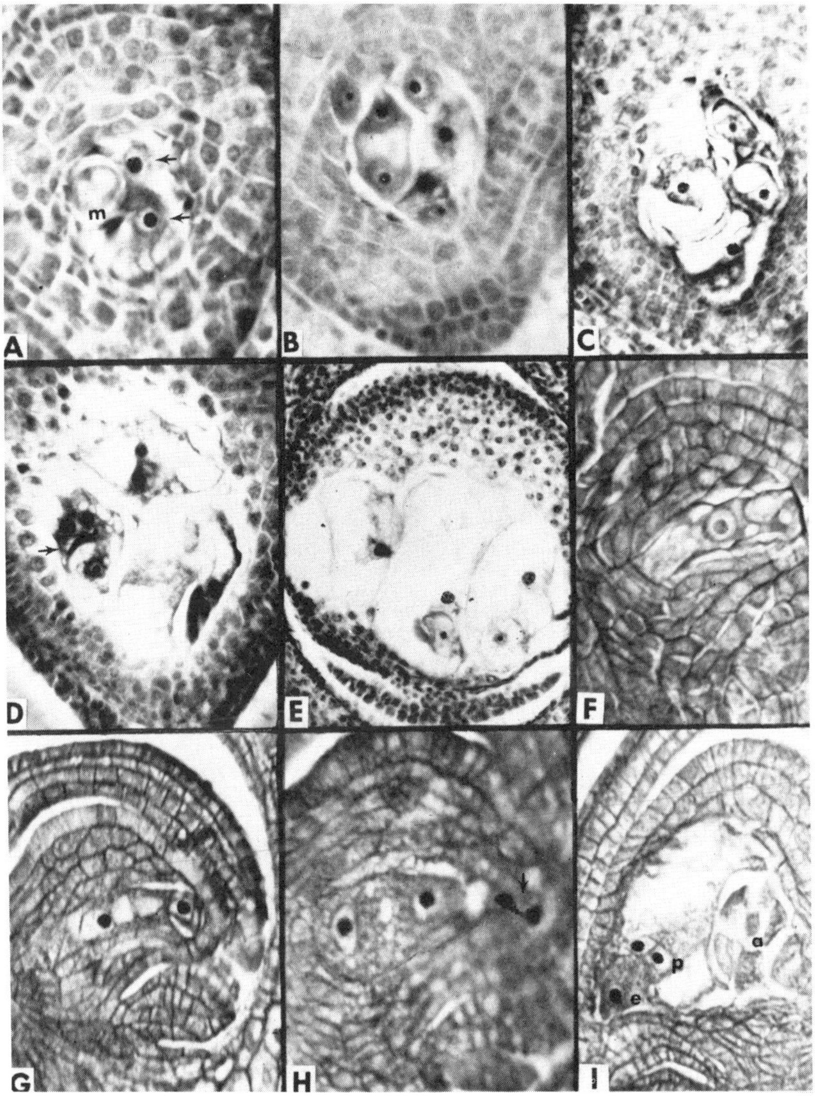

Fig. 2—Origin and development of embryo sacs in ovules of aposporous and diplosporous forage grasses. *Apospory:* A. Aposporous cells (arrows) developing adjacent to degenerated megaspore (m). B to D. Aposporous cells in various stages of development (note resemblance to sexual egg and polar nuclei in Fig. 1D arrow). E. Multiple aposporous embryo sacs in mature ovule (note absence of antipodal cells). *Diplospory:* F. Megaspore mother cell. G. Unreduced embryo sac with two nuclei. H. Two-nucleate embryo sac and remnants of degenerated megaspores (arrow) in a sexual ovule. I. Fully differentiated diplosporous embryo sac with egg (e), polar nuclei (p), and antipodals (a).

not undergo meiosis. Instead, the nucleus divides mitotically and the cell enlarges considerably, but does not divide. Eventually this cell fills the area that the linear tetrad would occupy in sexual ovules. After the first division of the nucleus, the nuclei migrate to opposite ends of the cell (Fig. 2G) and may closely resemble the two-nucleate stage of a sexual embryo sac (Fig. 2H). However, the two-nucleate sexual sac usually develops nearer the center of the ovule and remnants of degenerated megaspores may be visible toward the micropyle. Further division of the nuclei and differentiation of diplosporous embryo sacs varies with different species.

In the few known diplosporous species of agronomic crops, the embryo sac differentiates normally and, when mature, it looks exactly like the sexual embryo sac of the species (Fig. 2I). Cytological identification of diplospory is usually impossible at or beyond the two-nucleate embryo sac stage. The area of the ovule, other than the megaspores and embryo sac, is completely normal and shows no evidence of apomixis. The cytologist must carefully follow the early developmental stages to be certain of the origin of the embryo sac. The absence of meiosis and a linear tetrad of megaspores is the best evidence of diplospory.

There are other forms of diplospory in which meiosis is initiated, but the embryo sac receives the unreduced chromosome number either due to failure of reduction division or spontaneous doubling of the chromosome number of the functional megaspore after meiosis. When chromosome doubling occurs after meiosis, the progeny of heterozygous plants will not be uniform. This type of diplospory has not been reported in species of agronomic importance and is of little concern to plant breeders at this time.

C. Adventitious Embryony

In adventitious embryony, the embryo develops in somatic cells of the ovule, integuments, or ovary wall and is initiated as a bud-like structure through mitotic division of the cell nucleus. No embryo sac is formed, and in the early stages of development these embryonic structures resemble bud tissue or the globe stage of a developing proembryo. Eventually this mass of rapidly dividing cells differentiates and forms a typical embryo. No polar nuclei are associated with these embryos because there is no embryo sac. Little is known about the later stages of development, but it is assumed that in species in which endosperm develops, it must arise from polar nuclei in normal embryo sacs.

Adventitious embryony is the usual mechanism of apomixis in *Citrus* species. It is rare in other higher plants, but this mechanism has been observed occasionally in some grasses, usually along with apospory.

D. Parthenogenesis

A parthenogenic embryo develops directly from the reduced egg nucleus in a sexual embryo sac without fertilization by the sperm (haploidy). This phenomenon probably occurs occasionally in all sexual species, and it cannot be identified cytologically. Parthenogenesis is usually detected by the presence of haploid offspring. Normally parthenogenesis is infrequent, random, and spontaneous. However, the tendency for parthenogenesis in lines of some species (corn, cotton) is genetically controlled.

Use of the term parthenogenesis in plant literature is often confusing and varied. Some define it as development of an embryo from any presumed "egg cell" without pollination, regardless of the origin of the embryo sac. In aposporous apomicts, there is often no basis for designating a particular nucleus as the egg. Initiation of embryonic development without pollination occurs frequently in some species and the role of pollination is problematical. For the plant breeder, restriction of the term parthenogenesis to the familiar phenomenon of haploidy appears to be a logical solution to the confusion encountered in the literature.

E. Special Cases

There are some unusual reproductive phenomena that might properly be classified as apomixis. Androgenesis refers to a rather rare form of haploidy in which the embryo develops directly from the sperm after it enters the embryo sac. Apparently this is a non-recurrent event in most species and poses little or no problem or opportunity for plant breeding. However, genetic control of androgenesis has been reported in some lines of corn (Kermicle, 1969).

A rare form of parthenogenesis, called semigamy, has been reported in some species, including Pima cotton (Turcotte and Feaster, 1969). A sperm nucleus penetrates the egg cell but does not fuse with the egg nucleus. Subsequently, the egg and sperm nuclei divide independently, a process resulting in a heterogeneous haploid embryo that contains sectors of both male and female origin. Turcotte and Feaster (1969) were able to produce homozygous lines of Pima cotton by using colchicine to double the chromosomes in cells of the different sectors in haploid plants derived from the haploid embryos. Results of breeding studies showed that semigamy was controlled genetically in some lines.

Sometimes an unreduced egg nucleus in an apomictic embryo sac is fertilized by the sperm. There is no information available on the frequency of fertilization of an unreduced egg, but it has been reported in numerous species. While this is not a true form of apomixis, according to the classical definition of the term, its occurrence has considerable significance. For example, this represents one important method for natural increase in ploidy level in apomictic species because a whole genome from the male parent is added to the chromosome complement of the mother plant. Implications of this process in the breeding of apomicts are discussed in Section VI.

II. EMBRYO DEVELOPMENT IN APOMICTS

Although the embryo is formed without union of the male and female gametes, pollination is necessary in many apomicts for full development of the embryo and endosperm. The term pseudogamy denotes that pollination is an absolute requirement for development of a viable seed. The role of pollination in apomicts is not fully understood, but it is assumed to function as a stimulus for initiation of embryo and endosperm development, for fertilization of the polar nuclei, or as a requirement for full maturation of the embryo.

The apomictic grasses are generally pseudogamous and, in most species, it seems that the polar nuclei must be fertilized before the endosperm can develop. Chromosome counts of dividing endosperm cells have confirmed fertilization of the polar nucleus or nuclei in a few species. However, it is impossible to obtain clear and conclusive results in most cases because of the dense cytoplasm present in endosperm cells. Initiation of embryo development before pollination has been observed in ovules of apomictic species of *Poa, Paspalum, Cenchrus,* and many other grass genera, but the endosperm did not develop and the embryo did not differentiate until after pollination. In most ovules, neither the embryo or endosperm develops until after pollination.

Development of the embryo and endosperm of apomicts after pollination usually follows the normal pattern of sexual plants in the species. There is no apparent difference in the embryology of sexual and apomictic plants of *Cenchrus, Paspalum,* and *Poa*, except when the embryo starts to develop before pollination. In these cases, cell division of the embryo usually stops before or at the globe stage and there is no further development until pollination. After pollination, the endosperm cells usually divide several times before any perceptible change is seen in the proembryo. In the absence of embryo development before pollination, first division of the unreduced egg nucleus usually occurs after the endosperm starts to develop. Subsequent growth and differentiation is generally not different from normal development described in Chapter 2.

In aposporous apomicts, development of two or more embryos is quite common. Usually one embryo will be situated in the normal position and others may appear at random in the ovule. Initially, endosperm may develop in more than one aposporous embryo sac, but apparently these fuse during development and form a uniform endosperm in the mature seed.

III. INDICATORS OF APOMIXIS

There is nothing about the appearance of a single apomictic plant that suggests apomixis, but its progeny will have certain characteristics that may serve as useful indicators of apomictic reproduction. Detection of apomixis on the basis of morphological indicators requires careful observation of the progenies of individual plants by a person who is thoroughly familiar with mode of pollination of the parent and the variability expected among its sexually produced offspring. For example, lack of variation among the progeny of a heterozygous or cross-pollinated plant and production of offspring that closely resemble the maternal parent suggest apomixis. These same characteristics are expected in the progenies of homozygous, self-pollinated, sexual plants and would not cause one to suspect apomixis in them.

Obligate apomixis can usually be recognized easily in a heterozygous or cross-pollinated species. The offspring of obligate apomicts are completely uniform and all are identical to the maternal parent. The striking uniformity of an obligate apomict will usually be apparent when new accessions are first evaluated, unless the seed has been mixed. A progeny test of variant plants should be conducted using open-pollinated seed to determine if these are sexual, partially sexual (facultative), or simply plants of a different apomictic strain. It is unlikely that obligate apomictic plants will appear suddenly in a plant breeding program of sexual species. If, however, the progeny of a hybrid or selection fails to segregate as expected, it should be suspected as a possible apomict and subjected to further evaluation. Such a plant could be a valuable asset to the breeder for reasons discussed in Section VI.

Detecting facultative apomixis may be considerably more difficult, but the same indicators and principles apply. Facultative apomicts are all capable of some sexual reproduction and, unless the parent is homozygous and self-pollinated, there will be some variant plants among its progeny. The relative proportion of progeny that originate sexually or by apomixis may vary considerably even in the same plant. Therefore, the primary indicator of facultative apomixis is the presence among the progeny of a disproportionately high number of identical offspring that resemble the maternal parent along with some plants that are distinctly different in appearance and presumably of hybrid origin. Generally, facultative apomicts do not produce obligate apomictic offspring and those plants that resemble the maternal parent will continue to produce variable progenies. It is known that offspring of some facultative apomicts frequently show a high level (up to 98%) apomictic reproduction.

Detection of apomixis in relatively homozygous or self-pollinated plants by morphological indicators would probably occur only in a plant breeding or genetics program. The most obvious evidence of apomixis would be expected in a hybridization program. Assuming that the breeder's normal emasculation technique is reliable, or that the female parent is male-sterile, the appearance of maternal type plants among or instead of F_1 hybrids would suggest possible apomixis. Other indications of apomixis would be failure of F_1 hybrids to segregate normally, or the sudden appearance of true-breeding lines in early generation progenies.

Other reliable indications of apomixis exist when chromosome number, meiotic behavior, and fertility of a biotype are known. For example, an aneuploid chromosome number that is constant for the biotype or good seed production in plants with great meiotic irregularity provide good evidence of apomixis (Muntzing, 1933). Fertility in triploids, hybrids from wide crosses, and other plants normally expected to exhibit sterility also suggest apomixis. A high percentage of twin seedlings is frequently encountered in aposporous apomicts. I have observed identical twins (up to 22%) and occasional triplets in buffelgrass as a result of apospory. Facultative aposporous apomicts often produce one embryo in the sexual embryo sac and one or more embryos in apomictic sacs. Uniformity in protogynous species, where early stigma exsertion may enforce crossing, might also indicate apomixis.

The breeder should be alert for any unusual breeding behavior that might suggest deviation from sexual reproduction. Spontaneous increase in the ploidy level of an offspring may suggest unreduced gametes and a tendency toward apomixis. At best, these are only indicators of apomixis that may serve to encourage further studies to confirm the method of reproduction.

IV. TECHNIQUES FOR CONFIRMING APOMIXIS

A progeny test combined with cytological study of megasporogenesis and embryo sac development are usually required to confirm the presence of apomixis and to identify the mechanism. Neither study alone is adequate for conclusive results. As noted previously, the mechanism of apomixis is determined by observing the source and subsequent development of the cell that gives rise to the apomictic embryo. This can only be determined by cytological studies. Unfortunately, cytological observation does not indicate how frequently an asexual seed is formed in facultative apomicts. This information is gained by observing the percentage of maternal-like plants among the progeny. Often the potential for apomictic reproduction is far greater than the percentage of apomictic embryos developed. Some strains of Kentucky bluegrass regularly produce aposporous embryo sacs in over 98% of the ovules, but the percentage of their offspring produced by apomixis may be far less. With highly heterozygous plants, the progeny test provides a reliable estimate of the frequency of apomixis, but it tells nothing about the mechanism involved except for parthenogenesis. Results of studies of selfed progeny of relatively homozygous plants are usually inconclusive. Here one must attempt to cross the suspected apomictic plant with a male parent carrying a dominant marker.

Cytological study of method of reproduction is critical and deserves special consideration. In apospory, replacement of the functional megaspore by an aposporous cell may occur quickly and unobserved. This is particularly critical in species in which the aposporous sac differentiates fully and closely resembles a sexual embryo sac. Where differentiation is limited or numerous aposporous sacs are formed, identification is relatively easy. Identification of displospory is even more difficult because of the small size and early age of the ovule when meiosis occurs. To detect diplospory, one must determine whether first division of a megaspore mother cell is meiotic or mitotic. The meiotic stage can easily be missed in ovules collected for study because three members of the linear tetrad of megaspores nearest the micropyle usually abort and disppear early in a sexual plant. To identify both apospory and diplospory, one must observe individual ovules representing all stages of development from initiation of the megaspore mother cell to formation of the mature embryo sac.

The Paraffin sectioning is a reliable method for systematically examining anatomical features in the ovule because it preserves spatial relationships essential for correct interpretation. For accurate determination of mechanism of apomixis, it is necessary that the ovule be aligned to give true longitudinal sections through gametophytic tissue. Events that occur during megasporogenesis, especially presence or absence of a linear tetrad of megaspores, abortion during or following meiosis, and the appearance of adjacent nucellar cells are critical factors in the identification of the apomictic mechanism. The safranin-fast green stain (Johansen, 1940) is usually satisfactory for overall morphological features and resting nuclei. A good chromatic stain, such as hematoxylin, is essential if chromosome details are desired.

Embryo sac squash techniques (Bradley, 1948) are often useful for study of mature ovules, especially if one desires to see the intact contents of the embryo sac. I have encountered problems with this technique in certain apomictic species that have multiple embryo sacs. It is extremely difficult to distinguish between nucellar embryo sacs and antipodal cells in species in which the antipodals divide profusely. The paraffin section technique is definitely preferred over ovule squashes for study of early megasporogenesis.

Several excellent clearing techniques are available for observation of internal details of ovaries using the phase contrast microscope. These preparations permit viewing the ovule as a unit and show all cellular detail in proper context. A methyl salicylate technique suggested by Crane (1978) has given good results with small ovaries of grass. It is helpful for the cytologist to first become familiar with typical apomictic mechanisms using the paraffin section technique before using squashes or clearing methods.

V. GENETICS OF APOMIXIS

Apomictic species have been poor subjects for genetic study in the past. The data for most experiments have not been conclusive enough to establish the exact mode of inheritance or to permit identification of specific genotypes that condition apomixis. Results of the successful crosses between sexual and apomictic plants that have been reported indicate that the basic processes responsible for apomictic reproduction are genetically controlled. Most of the earlier research was conducted with facultative apomicts, which are capable of both apomictic and sexual reproduction at the same time. Inheritance data from facultative apomicts generally have been confusing and difficult to interpret because there is no certain way to determine whether individual offspring originated from sexual or apomictic reproduction.

Fairly recent discovery of completely sexual plants in species that are predominantly obligate apomicts opened up new opportunities for genetic studies. These plants are capable of only one method of reproduction, apomixis or sexual, and the origin of each offspring is known. Hybrids and progenies from crosses of purely sexual plants with obligate apomicts provide data that can be accurately analyzed statistically for fit to hypothetical ratios. There are a few reports available from these crosses that present reliable inheritance data.

Results of studies conducted with buffelgrass, an aposporous apomict, showed that method of reproduction is controlled by two genes and epistasis (Taliaferro and Bashaw, 1966). These experiments were conducted with selfed progenies of a rare sexual plant that is heterozygous for method of reproduction, and with hybrids produced by emasculating the sexual plant and pollinating it with pollen from obligate apomictic plants. Data on method of reproduction of 614 selfed progeny gave a sexual to apomictic ratio of 13:3. When the sexual plant was crossed with two different obligate apomictic biotypes the frequency of sexual to apomictic F_1 hybrids fit a 5:3 ratio. On the basis of these data, the genotype of the sexual parent was hypothesized to be *AaBb,* where dominant gene *B* conditions sexual reproduction and is epistatic to dominant gene *A,* which conditions apospory. Under this hypothesis, the genotype of the two apomictic male parents would be *Aabb.* Assuming that the double recessive *aabb* would reproduce sexually because of the absence of the dominant *A* gene, one would expect the 13:3 and 5:3 ratios observed in the experiments.

When the sexual buffelgrass plant was crossed with birdwoodgrass, a closely related apomictic (obligate apospory) species, the same ratios were obtained in the F_1 generation (Read and Bashaw, 1969). Subsequently, the sexual plant was crossed with a plant having the only other possible genotype for apomixis, *AAbb.* This is comparable to a test cross for mode of reproduction, and the resultant hybrids comprised the expected 1:1 ratio of sexual to apomictic plants. These results confirmed the hypothesized genetic model and showed that obligate apospory can be manipulated just as other genetic characters in a breeding program.

Somewhat similar results were obtained from an experiment with bahiagrass (Burton and Forbes, 1960). The common tetraploid biotype of this species is an obligate aposporous apomict, but a diploid strain ('Pensacola') is completely sexual. The chromosome number of the diploid was doubled and the resultant sexual autotetraploid was successfully crossed with the common apomictic biotype. F_2 ratios obtained from this cross suggested that apomixis was recessive to sexuality and controlled by a few recessive genes. The data were close enough to expected autotetraploid ratios to permit postulation of an *aaaa* genotype for the apomictic parent.

Little is known about the inheritance of diplospory. Here the mechanism for development of the apomictic embryo sac differs entirely from apospory. Meiosis is somehow suppressed and the unreduced megaspore mother cell is stimulated to form a functional embryo sac by mitotic division. The few experiments that have been conducted with diplosporous apomicts indicate that this mechanism is also genetically controlled. Both sexual and diplosporous apomictic plants were recovered among the progeny of a cross of sexual and apomictic weeping lovegrass (Brix, 1974; Voigt and Bashaw, 1972, 1976). There is not enough information at this time to explain the genetics of diplospory, but the data suggest simple inheritance.

Transmission of apomixis to hybrid progenies has been demonstrated in some cross-pollinated facultative apomicts by using genetic markers to identify the offspring of sexual origin. Experiments in numerous unrelated genera clearly indicate a genetic basis for apospory in facultative apomicts, but mode of inheritance seems to vary in different species. In guayule, conditions necessary for expression of apomixis appeared to be conditioned by three recessive genes (Powers, 1945). Results of studies with the *Bothriochloa-Dichanthium* complex indicated that apomixis was dominant to, but independent of, sexuality and controlled by no more than one gene per genome (Harlan et al., 1964). Apomictic tetraploid plants were assigned the genotype *AAaa,* and the sexual ones the genotype *aaaa*. Experiments with guineagrass also indicated that sexuality in that species is dominant to apomixis and that method of reproduction is probably controlled by two loci (Smith, 1972; Hanna et al., 1973).

There is yet no biochemical or physiological information on the factors responsible for apospory. This information is needed as a basis for understanding the nature of gene action in apomixis and the fundamental difference between mechanisms. A comparison of the genetic data from aposporous apomicts with observed cytological behavior suggests that one genetic factor serves to activate nucellar cells, causing otherwise senescent tissue to become meristematic and produce unreduced aposporous embryo sacs while simultaneously suppressing development of the sexual tissue. Apparently a second factor, when dominant, completely nullifies the activating factor and restores sexuality. It would seem that a primary difference between obligate and facultative aposporous apomicts is lack of suppression of sexual development in the facultative type.

VI. HYBRIDIZATION AND BREEDING OF APOMICTS

Apomixis can either prevent breeding progress or be a useful tool of the plant breeder. The first requirement for successful improvement of apomictic species is the availability of sexual or partially sexual plants for use as female parents in a hybridization program. Except for occasional fertilization of an unreduced egg, an obligate apomictic plant cannot be used as the female parent. Most apomicts produce viable pollen and can serve as male parents just as effectively as sexual plants. Once sexual or partially sexual plants are discovered, the success and efficiency of the breeding program depends largely on whether the species is an obligate or facultative apomict. Apomictic plants generally are highly heterozygous and when the apomictic barrier is broken, there is abundant variability from which to select new types. Apomicts do not usually require special crossing techniques. Hybridization methods described in later chapters for sexual plants should be effective for sexual × apomictic crosses in those crops. Since apomictic plants tend to be highly heterozygous and hybrids may be difficult to identify, reliable methods of emasculation and genetic markers are important in plants subject to self pollination.

A. Obligate Apomicts

Obligate apomixis can be used as an effective plant breeding tool once sexual plants have been identified in the species. The principle involves control and manipulation of method of reproduction through hybridization with obligate apomicts in the species. Depending upon the inheritance of apomixis, a portion of the offspring in the F_1 and succeeding generations will reproduce by obligate apomixis and breed true. Heterosis, when present in apomictic progeny, will be permanently fixed in succeeding generations. For maximum efficiency, the breeder must determine the genetic constitution of the parents and obtain inheritance data on important characteristics as in any improvement program. When this information is available and the appropriate apomictic male parents are selected and crossed with the sexual plant, one simply selects and tests the apomictic offspring that appear to have the desired combination of agronomic traits. Apomictic plants also may be selected in succeeding generations because some sexual progeny will continue to segregate for mode of reproduction.

The procedure used in the breeding of apomictic buffelgrass and birdwoodgrass is illustrated in Fig. 3. With these species, we use a heterozygous, sexual, buffelgrass plant as the female parent and cross it to various apomictic male parents. The resultant apomictic F_1 hybrids are obligate apomicts and breed true while sexual hybrids from the same cross produce segregating progenies (Fig. 4). The sexual plant is heterozygous for method of reproduction, therefore its selfed progeny also provide a source of apomictic plants for selection. The first apomictic cultivar, 'Higgins', released from a buffelgrass breeding program, was selected from selfed progeny of a sexual plant. Primary emphasis in the program is placed on the selection of superior apomictic F_1 hybrids to take advantage of maximum hybrid vigor. Two cultivars, 'Neuces' and 'Llano', were developed in this manner. Selection of apomictic progeny is possible in succeeding generations following hybridization, but there is a risk of some reduction in hybrid vigor. We have noted considerable loss in vigor in F_2 and F_3 progenies of sexual hybrids. A similar situation was observed in bahiagrass in which high forage yield was maintained in apomictic hybrids, but was rapidly lost in generations of the sexual progenies (Burton and Forbes, 1960).

There is no reason to expect that all sexual plants discovered in obligate apomictic species will be heterozygous for method of reproduction. In many cases, one or two generations following hybridization may be required for the expression of apomixis. Early identification of sexual parents with good combining ability and heterozygous for method of reproduction would be desirable. As with any breeding program, not all sexual plants will necessarily be desirable parents. Cytological studies are particularly useful for early determination of mode of reproduction in hybrids because it eliminates the need for conventional progeny tests. Backcrossing sexual F_1 hybrids to their apomictic parents may speed up development of apomictic lines.

If sexual plants are not available in the species, other approaches should be considered. It may be possible to break apomixis through hybridization with sexual plants of closely related species. If the species relationship is close enough, it should be possible to introduce sexuality into the species through interspecific hybridization. Because apomixis tends to preserve fertility by eliminating the sexual mechanism, one may be able to produce useful obligate apomictic, interspecific hybrids with good seed set. A word of caution is in order here because apomixis does not always insure fertility. In pseudogamous species, viable pollen is necessary for seed set, and apomixis would not necessarily overcome sterility in incompatible plants. Apomictic common dallisgrass seldom produces over 40% seed set in spite of an apparently adequate quantity of viable pollen. In this species, the problem seems to be associated with competition for nutrition among the vast number of nucellar embryo sacs formed in the ovule (Bashaw and Holt, 1958). Pollen quality of fountaingrass is extremely poor and few seed are formed from self-pollination. However, when this species is pollinated with buffelgrass pollen, over 20% seed is formed pseudogamously (Simp-

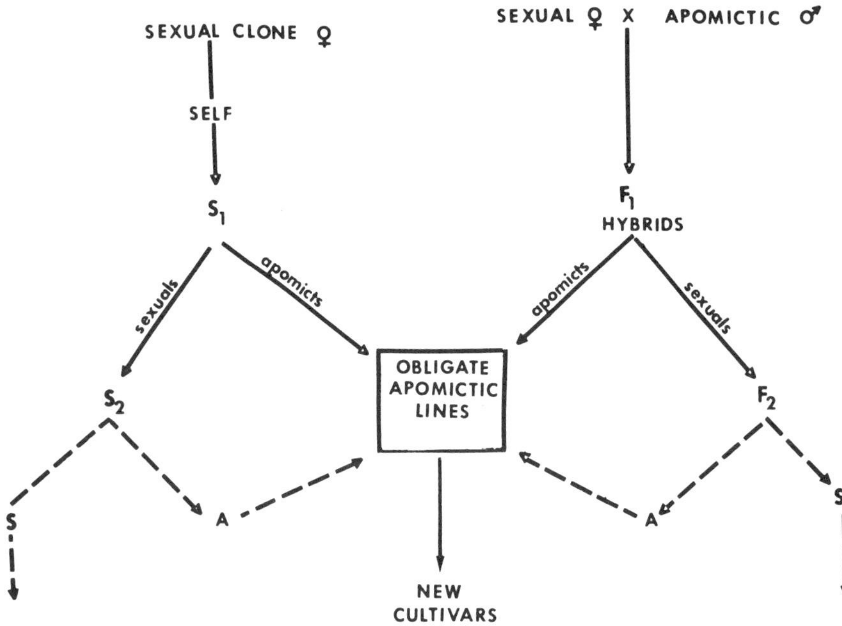

Fig. 3—Diagram of methods used to produce apomictic buffelgrass cultivars. Sexual parent is heterozygous for method of reproduction (Taliaferro and Bashaw, 1966).

Fig. 4—Heterogenous progenies from one sexual (S) F_1 plant and uniform progenies from two obligate apomictic (A) F_1 plants of buffelgrass.

son and Bashaw, 1969). The breeder also should be alert to the possibility afforded by fertilization of an unreduced egg in the apomict. Although this phenomenon is relatively rare, it provides yet another way to introduce new characters. In obligate apomicts, these offspring breed true and may prove immediately useful in the breeding program.

B. Facultative Apomicts

Breeding of facultative apomicts is much more difficult than breeding obligate types because of inability to control variation in their progenies. The presence of some sexuality in each plant allows for hybridization, but the progenies continue to segregate for method of reproduction through succeeding generations. In spite of this fact, some useful cultivars have been developed, especially in forage crops in which some degree of variability may not be a serious problem (Kentucky bluegrass, guineagrass, bluestem).

There are several important factors to consider in planning a breeding program with facultative apomicts. The fact that highly apomictic offspring are often produced by these species provides an opportunity to reduce variation. It would be advantageous to select highly apomictic parents for hybridization because their progeny tend to retain a high level of apomixis. Choice of parents that are relatively similar in important characteristics, such as height and maturity, also should be considered. As I mentioned earlier, the use of parents with distinctive marker characteristics will greatly facilitate identification of hybrids and detection of variation in succeeding generations. Perhaps the most important factor in breeding of facultative apomicts is to determine stability of each line through several generations before releasing a cultivar. Problems in commercial seed production are discussed in Section VIII.

C. Possibilities for Inducing Apomixis

Breeders of sexual species may wish to find sources of apomixis because of its potential. This raises the question of possibilities for inducing apomixis into sexual species. Apparently no research has been designed specifically to induce apomixis, but results of other studies suggest several possibilities, including mutation. One of the first apomictic strains of grain sorghum was derived from irradiated stock (Hanna et al., 1970). Facultative apomixis was observed in pearl millet following treatment of seed with thermal neutrons and diethyl sulfate (Hanna and Powell, 1973). I believe that the sexual buffelgrass plant described earlier originated as a spontaneous mutation from apomixis to sexuality. These results suggest that induction of apomixis by mutation is feasible, but there are no general rules for identifying candidate species for this research. Perhaps the best possibilities would be species with known apomictic wild relatives. Transfer of apomixis from these relatives through hybridization would seem to offer a more promising approach at this time. At present, we lack the necessary information on the physiological basis of apomixis to plan experiments involving growth regulators. When the basic factors responsible for initiation of apomictic embryo sacs are understood, chemical induction of apomixis or sexuality may be possible.

VII. ENVIRONMENTAL EFFECTS ON APOMIXIS

Environmental factors may affect the stability of the reproductive process in facultative apomicts, but such effects have not been observed in obligate apomicts. Apomictic and sexual embryo sacs develop simultaneously in aposporous facultative apomicts, such as Kentucky bluegrass, and apparently there exists a delicate balance in the competition for survival within the ovule. Sometimes environment can upset this balance in favor of one of these gametophytes. For example, cytological studies of several strains of Kentucky bluegrass showed that aposporous embryo sacs developed in at least 98% of the ovules of each strain. However, when these strains were grown under diverse environments, a higher frequency of aberrant and weak plants occurred among the progeny when seed was produced over a wide range of anthesis and short photoperiod as compared with a narrow range of anthesis and long photoperiod (Hovin et al., 1976).

Day length had a significant effect on the relative frequency of apomictic and sexual embryo sacs in ovules of *Dichanthium aristatum* (Poir.) C. E. Hubb. Over 14 hours of daylight during development of inflorescences resulted in a much lower incidence of apomictic sacs than day lengths less than 14 hours (Knox and Heslop-Harrison, 1963; Knox, 1967). Results of cytological and breeding studies with Kentucky bluegrass revealed 15 different ways in which embryos may develop (Grazi et al., 1961). These results show the importance of knowing the potential sources of variation, as well as the characteristic behavior of specific germplasm in succeeding generations and under diverse environments.

Thus far, we have not observed environmental effects on embryo sac development in obligate apomicts or the possible appearance of facultative apomixis in succeeding generations from obligate apomicts. After hybridization of sexual plants with obligate apomicts, it was found that the F_1 offspring and their progenies have always been either completely sexual or obligate apomicts. Environment sometimes does affect pollen quality, an important factor in pseudogamous apomicts. Environment may also affect the total number of aposporous embryo sacs that develop in an ovule.

The possible effects of environment on diplospory are not known. Such effects could be important in these species because only one embryo sac, either sexual or apomictic, is formed and the potential for embryonic development is far less than in most aposporous apomicts.

VIII. SEED PRODUCTION OF APOMICTS

Commercial seed production of obligate apomicts is simple and involves the same general principles that apply to vegetatively reproduced cultivars. There is no danger of outcrossing in these species, and isolation is required only to prevent mechanical mixture during harvest. With buffelgrass, we normally skip two rows between seed production plots. All classes of seed are genetically the same because there are no sexual generations from an obligate apomict. A seed field of perennials can be maintained indefinitely, if it is kept free of contamination by other species and volunteer plants of other strains of the same species. The use of male sterility to facilitate hybridization of sexual plants with pseudogamous apomicts would require that fertility be restored in the F_1 because pollination is required for seed development.

Seed production of facultative apomicts requires special considerations, especially regarding establishment of standards. First, the breeder must determine the degree of variation that is acceptable for a particular crop. One must then determine the nature and extent of variation characteristic of each line. This information can only be determined by growing the line for sufficient generations to obtain a history of variability. Because environment may affect mode of reproduction, all advanced strains should be evaluated for variation in the area and under those conditions expected where commercial seed production is anticipated. Usually all plants in facultative apomictic lines will be capable of some sexual reproduction, and the proportion of sexual to apomictic progeny may remain relatively stable or increase. Isolation to prevent outcrossing of facultative apomicts is just as essential as it is with sexual species. It will be necessary to establish standards for commercial seed production of each individual line being considered for release. Close cooperation between the breeder, seed grower, and farmer will be required to determine whether a line meets the requirements for purity and uniformity necessary for successful farm production.

REFERENCES

Bashaw, E. C. 1962. Apomixis and sexuality in buffelgrass. Crop Sci. 2:412–415.

———, and E. C. Holt. 1958. Megasporogenesis, embryo sac development and embryogenesis in dallisgrass, *Paspalum dilatatum* Poir. Agron. J. 50:753–756.

———, A. W. Hovin, and E. C. Holt. 1970. Apomixis, its evolutionary significance and utilization in plant breeding. p. 245–248. Proc. 14th Int. Grassland Congr.

Bradley, M. V. 1948. An acetocarmine squash technique for mature embryo sacs. Stain Technol. 23:29–40.

Brix, K. 1974. Sexual reproduction in *Eragrostis curvula* (Schrad) Nees. Z. Pflanzenzuecht. 71:25–32.

Burton, G. W., and I. Forbes, Jr. 1960. The genetics and manipulation of obligate apomixis in common bahiagrass (*Paspalum notatum* Flugge). p. 66–71. Proc. 8th Int. grassland Congr.

Crane, C. F. 1978. Apomixis and crossing incompatibilities in some Zephyranthaceae. Ph.D. Thesis. Univ. of Texas, Austin.

Darlington, C. D. 1939. The evolution of genetic systems. p. 149. University Press, Cambridge, England.

de Wet, J. M. J., and J. R. Harlan. 1970. Apomixis, polyploidy and speciation in *Dichanthium.* Evolution 24:270–277.

Grazi, F., M. Umaerus, and E. Akerberg. 1961. Observations on the mode of reproduction and the embryology of *Poa pratensis*. Hereditas 47:489–541.

Gustafsson, A. 1947. Apomixis in higher plants. Lund (Sweden) Haken Ohlson's Boktryekeri.

Hanna, W. W., and J. B. Powell. 1973. Stubby head, an induced facultative apomict in pearl millet. Crop Sci. 13:726–728.

———, ———, J. C. Millot, and G. W. Burton. 1973. Cytology of obligate sexual plants in *Panicum maximum* (Jacq.) and their use in controlled hybrids. Crop Sci. 13:695–697.

———, K. F. Schertz, and E. C. Bashaw. 1970. Apospory in *Sorghum bicolor* (L.) Moench. Science 170:338–339.

Harlan, J. R., M. H. Brooks, D. S. Borgaonkar, and J. M. J. de Wet. 1964. The nature and inheritance of apomixis in *Bothriochloa* and *Dichanthium*. Bot. Gaz. 125:41–46.

Hovin, A. W., C. C. Berg, E. C. Bashaw, R. C. Buckner, D. R. Dewey, G. M. Dunn, C. S. Hoveland, C. M. Rineker, and G. M. Wood. 1976. Effects of geographic origin and seed production environments on apomixis in Kentucky bluegrass. Crop Sci. 16:635–638.

Johansen, D. A. 1940. Plant microtechnique. McGraw-Hill Book Co., Inc., N.Y.

Kermicle, J. L. 1969. Androgenesis conditioned by a mutation in maize. Science 166: 1422–1424.

Knox, R. B. 1967. Apomixis: seasonal and population differences in a grass. Science 157:325–326.

———, and J. Heslop-Harrison. 1963. Experimental control of aposporous apomixis in a grass of the Andropogoneae. Bot. Notiser. 116:127–141.

Muntzing, A. 1933. Apomictic and sexual seed formation in *Poa.* Hereditas 26:115–190.

Nogler, G. A. 1975. Genetics of apospory in *Ranunculus auricomus.* IV. Embryology of F_3 and F_4 backcross offspring. Phytomorphology 25:485–490.

Nygren, A. 1954. Apomixis in angiosperms. Bot. Rev. 20:577–649.

———. 1967. Apomixis in the angiosperms. Handb. Pflanzenphysiol. 18:551–596.

Powers, L. 1945. Fertilization without reduction in guayule (*Parthenium argentatum* Gray) and a hypothesis as to the evolution of apomixis and polyploidy. Genetics 30:323–346.

Rao, N. G. P., and U. R. Murty. 1972. Further studies on obligate apomixis in sorghum. Ind. J. Genet. 32:379–383.

Read, J. C., and E. C. Bashaw. 1969. Cytotaxonomic relationships and the role of apomixis in speciation in buffelgrass and birdwoodgrass. Crop Sci. 9:805–806.

Simpson, C. E., and E. C. Bashaw. 1969. Cytology and reproductive characteristics in *Pennisetum setaceum.* Am. J. Bot. 56:31–36.

Smith, R. L. 1972. Sexual reproduction in *Panicum maximum* Jacq. Crop Sci. 12: 624–627.

Stebbins, G. L. 1950. Variation and evolution in plants. p. 643. Columbia Univ. Press, N.Y.

Taliaferro, C. M., and E. C. Bashaw. 1966. Inheritance and control of obligate apomixis in breeding buffelgrass, *Pennisetum ciliare.* Crop Sci. 6:473–476.

Turcotte, E. L., and C. V. Feaster. 1969. Semigametic production of haploids in Pima cotton. Crop Sci. 9:653–655.

Voigt, P. W., and E. C. Bashaw. 1972. Apomixis and sexuality in *Eragrostis curvula.* Crop Sci. 12:843–847.

———, and ———. 1976. Facultative apomixis in *Eragrostis curvula.* Crop Sci. 16:803–806.

4

Sources, Maintenance, and Utilization of Parental Material

R. A. Forsberg
University of Wisconsin
Madison

R. R. Smith
USDA-SEA and University of Wisconsin
Madison

Plant hybridization by man is a modern technique in the evolutionary time span of crop plants. The demonstration of sex in plants by Camerarius in 1694 undoubtedly led to the realization that plants could be improved through the use of superior parents, already a practice in animal breeding. Fairchild was credited by Bradly with making the first artificial plant hybrid in 1719 (Roberts, 1929), and many crossing experiments followed in the 18th and 19th centuries both to study the laws of inheritance and to develop improved cultivars.

As basic concepts of genetics, cytology, reproduction, and selection were published, verified, and accepted, both before and after the rediscovery of Mendel's laws, hybridization followed by selection gradually replaced selection within established populations as the prime method in crop improvement. Once plant breeders learned that they could develop improved cultivars by crossing parents with complementary traits, the importance of assembling a wide array of breeding stocks for potential use as parents became readily apparent.

The choice of parents to be crossed is one of the most important decisions made by a plant breeder. These decisions are influenced by the short and long range goals of each program and by the breeding method to be used. Modern-day cultivars represent elite combinations of favorable genes which govern a large number of traits. Hence, a large proportion of the progenies from new crosses likely will be inferior to existing cultivars. These progenies, however, may be valuable stepping stones in an improvement program. Factors influencing choice of parents, which in turn create the need for genetic variability among available breeding stocks, include level of agronomic performance, reaction to diseases and insects, and quality standards required to meet processing and nutrition standards. Choice of parents will be discussed in more detail in Section II.A.

I. SOURCES OF GERMPLASM

Plant breeders are continually searching for germplasm which might be useful in attaining the objectives of their breeding programs. A great diversity of germplasm is available for most crop species. Seed or plants of released germplasm lines or cultivars generally are maintained by the originating station, and stocks are furnished upon request. Much informal exchange of nonreleased germplasm also takes place among breeders. For some species, national and international performance nurseries are important vehicles for mutual examination and exchange of germplasm among breeders.

If a breeding program is just being established, local ecotypes and cultivars from areas with similar environments may be good sources of breeding material. These types often possess sufficient adaptation and tolerance to pests necessary to be useful as parental material.

National and international genetic resource centers maintain viable collections of germplasm of economically important crop plants. These international centers have become world collection headquarters for many crop plants:

International Rice Research Institute (IRRA), Manila, Philippines

International Center for Tropical Agriculture (CIAT), Cali, Colombia

International Crops Research Institute for the Semi-Arid Tropics (ICRISAT), Hyderabad, India

International Center for Maize and Wheat Improvement (CIMMYT), El Batan, Mexico

International Center for Agricultural Research in Dry Areas (ICARDA), Iran, Lebanon, and Syria

International Potato Center (CIP), Lima, Peru

International Institute of Tropical Agriculture (IITA), Ibadan, Nigeria.

Another important germplasm repository is the Asian Vegetable Research and Development Center (AVRDC), Taiwan.

Other centers, such as the Germplasm Resources Laboratory, Agriculture Research (AR), Science and Education Administration (SEA), U.S. Department of Agriculture (USDA), Beltsville, Md., and Germplasm Laboratory, Bari, Italy, collect and maintain a wider spectrum of species. Creech and Rietz (1971) outlined the structure of the U.S. germplasm introduction and maintenance system. In addition to the USDA Germplasm Resources Laboratory, other important components of the U.S. system include the National Seed Storage Laboratory, Fort Collins, Colo., and four regional plant introduction stations. The regional facilities are located at Agricultural Experiment Stations at Geneva, N.Y.; Experiment, Ga.; Ames, Iowa; and Pullman, Wash.

A relatively new and promising program is that of the Laboratory for Information Science in Agriculture (LISA), a newly formed nonacademic department in the College of Agricultural Science, Colorado State University, Fort Collins, Colo. (GRIP, 1979). This research group was organized in 1973 by the International Board for Plant Genetic Resources (IBPGR) as the Information Sciences/Genetic Resources (IS/GR) program at the Univ. of Colorado, Boulder, Colo. The initial assignment—to investigate, design, develop, and eventually implement a global system for the cataloging and retrieval of genetic resource data and information (IS/GR, 1977)—is underway. Initial steps have included the activation of LISA's Germplasm Resources Information Project (GRIP), a computer-assisted information system to service the U.S. National Plant Germplasm System (NPGS). It is anticipated that the participatory nature of GRIP, by which input is received from federal, state, private industry, and educational sectors, also will facilitate germplasm and information exchanges among countries.

Under the auspices of the IBPGR, Hawkes et al. (1976) compiled a bibliography of plant genetic resources. This bibliography and a supplement prepared by Williams (1976) provide a comprehensive listing of genetic resources.

Primary and secondary centers of origin are prime geographic locations for the direct collection of genetically diverse plant and seed materials. In earlier explorations, one or two scientists collected many species at one time and collected only the plant types which appeared useful at that time. The current practice is to concentrate on one or a few species, and usually the collector is a scientist who is knowledgeable about those species. The regions are sampled systematically to obtain an adequate representation of the genetic diversity. Some collections, however, are made for the purpose of locating specific desirable characteristics, such as resistance to diseases. Current U.S. explorations are coordinated through the USDA Germplasm Resources Laboratory.

Many countries require a quarantine isolation, a phytosanitation certificate, inspection, or some combination of these for all seed or live plant materials that are introduced. In the United States, materials must be introduced through the Plant Germplasm Quarantine Station, Beltsville, Md., before they can be included in the National Plant Germplasm System. In specific cases, material is forwarded to isolated (quarantine) locations for evaluation. For example, small grains are evaluated in a nursery at Mesa, Ariz., where the plants are observed for seedborne diseases. Individual states or provinces also may have rules and regulations governing the shipment or introduction of seeds or plant material. Scientists should take special precautions as they exchange plant and seed materials directly with their colleagues in other countries.

II. EVALUATION AND MAINTENANCE OF PARENTAL MATERIAL

A. Evaluation

A knowledge of performance capabilities of parental breeding stocks is essential if the choice of parents is to be a judicious one. Such knowledge is accumulated by the plant breeder either through study of performance data or through constant screening of, and familiarity with, breeding lines in his nurseries. In addition to local and statewide tests, regional nurseries allow evaluation of elite lines of specific crops over a wide geographical area. New introductions also must be classified for specific traits before they can be wisely used as parents.

It is important to evaluate the agronomic and disease responses of potential parents under the soil and climatic conditions of the geographical region in which the germplasm will be used. Thus, most screening and classification are performed in field nurseries, with supplemental tests often conducted in greenhouses or growth chambers.

Tests with potentially dangerous races of pathogens or insects not widely dispersed in a specific geographic region are best limited to the indoors, to special facilities, such as the USDA Puerto Rico rust nurseries, or to areas in which the dangerous pest is already prevalent. Also, the response of adult plants to pests may differ from the response of seedlings, and both light and temperature may critically influence level of disease infection.

Plant introductions, either as pure lines or heterogeneous populations, may need to be purified or rogued to establish the population desired for use as parental stock. Maturation time, plant height, response to diseases, and seed color are traits frequently used in the purification of pure lines. It may be necessary to determine chromosome numbers if ploidy-level mixtures are suspected.

B. Maintenance and Development

Both the mode of reproduction and type of product marketed for commercial use influence how parent material is maintained and in what form. While parent materials of self-pollinated crops are generally maintained as pure lines, parent materials of cross-pollinated crops are maintained as inbred lines or heterogeneous populations. In a few crops (potato, bermudagrass, sugarcane), parent material is maintained vegetatively. Maintenance and developmental procedures in relation to mode of reproduction will be discussed in this section.

1. Self-Pollinating Species

Maintaining parent material as pure lines in self-pollinated crops is relatively simple provided that routine care to prevent plant and seed mixtures is taken during harvesting, threshing, seed storage, handling, and planting. Familiarity with the characteristics of lines, families, or populations is a prime requisite for detection of off-types arising from outcrossing or seed mixtures. Fortunately, F_1 hybrids resulting from outcrossing often demonstrate sufficient hybrid vigor that they can be detected among the other uniform members of a population. The appearance of a dominant trait in a recessive line, such as a beardless wheat in a bearded line, strongly indicates that the beardless off-type is an F_1 resulting from an outcross or is a direct seed mixture. A recessive off-type, if not a seed mixture, would indicate that an outcross occurred two or more generations earlier. Off-type plants are generally discarded and not used as parents because only half or none of their pedigree is known. If the off-type possesses some desirable trait, further selection may be necessary to assure homozygosity. The amount of outcrossing in self-pollinating species varies with environments and species, but bagging is rarely required to maintain purity.

Some deciduous wild species shatter at maturity, and seeds must either be harvested promptly or the inflorescence enclosed in a bag before maturity. The panicles of deciduous species of oats can be enclosed in cloth, pollen-tector (tassel), or glassine bags which are tied to a stake or cane pole for support. A more tedious and less efficient system is to remove the nearly mature spikelets by hand from each panicle each day.

2. *Cross-Pollinating Species*

The maintenance of heterogeneous populations and inbred lines is more difficult in cross-pollinated species, such as corn, rye, sorghum, alfalfa, red clover, and most turf and pasture grasses, than in self-pollinated species.

a. Heterogeneous Populations. Species such as alfalfa and red clover rely on insects for pollination. The bumblebee (*Bumbus* spp), honeybee (*Apis mellifera* Linneus), leafcutter bee (*Megachile* spp), and alkali bee (*Nomia meladeri* Cocherell) are the most common insects used. By maintaining proper isolation, wild or native colonies of these insects can be used for pollination within parental populations. However, when a large number of parent populations are to be individually pollinated, small colonies of honeybees or boards of leafcutter bees are enclosed in screened cages with the flowering plants. A small colony of honeybees or 200 cells of leafcutter bees are sufficient to pollinate 150 to 200 plants of alfalfa or red clover.

Species which rely on wind for the dispersal of pollen, such as corn, rye, and most grasses, generally can be effectively isolated by means of distance or border rows. Seeds from the border-row plants are not included in the germplasm pool.

Populations of cross-pollinated species can also be maintained by controlled or hand pollination. This procedure works well for corn, sorghum, or sugarbeet because one pollination produces many seeds. Production of large amounts of seed by hand pollination in alfalfa, rye, red clover, or turfgrasses is tedious and inefficient. In addition to plant-by-plant pollination, a bulk of pollen from many plants can be dispersed manually on bagged ears of corn or panicles of sorghum. This procedure may lead to some inbreeding but, unless genetic drift is severe, it is an effective procedure for germplasm maintenance.

The size of the population and the number of pollinations required to maintain gene frequencies are dependent on the genetic structure and intended use of the population. Broad-based germplasm pools require numerous pollinations among a large number of individuals to retain existing genetic variability. If the intent is to maintain specific qualitative characters, then population size and number of pollinations can be minimal. Procedures to estimate the number of individuals required for retaining qualitatively and quantitatively inherited characters in the population have been described (Hanson, 1959; Graybill and Kneebone, 1959).

b. Inbred Lines. The amount of homozygosity that can be tolerated in a cross-pollinated crop varies among species. In spite of a reduction in plant vigor, inbred lines are relatively easy to develop and maintain by selfing in corn. The same is possible in rye but reduction in plant vigor may be more pronounced. However, the vigor of alfalfa and red clover plants declines markedly following two or more generations of selfing. In these latter species, relaxation of selfing by sib mating allows retention of more vigor, with levels of heterozygosity intermediate between inbred-line homozygosity and open-pollinated heterozygosity.

Conventional or normal inbreds used either as components in synthetic cultivars or as parents of hybrids produced without the use of male sterility are maintained by conventional selfing or sibbing procedures. Maintenance in breeding nurseries requires protection of the female from foreign pollen followed by self-pollination until the desired level of homozygosity is reached after which sib pollination by hand is satisfactory. Open pollination (sibbing) in isolation blocks is more efficient for production of larger amounts of seed.

Three types of inbred lines must be developed and maintained when a cytoplasmic-genetic system is used to produce hybrids as in the case of hybrid corn, sorghum, pearl millet, sunflower, and other species. Where this biological capability exists, the female parent has factors in the cytoplasm of cells which render it male-sterile in the absence of appropriate restorer genes in the nucleus. The cytoplasm of the female gamete, but not that of the male, is transmitted to the offspring making it possible to produce female populations which are free of any male-fertile individuals.

The female parent in the cytoplasmic-genetic system must have two different, but phenotypically identical, forms. One form is male-sterile, with sterile cytoplasm (S) and with nonrestoring genes (rf rf) in the nucleus. The other form must be male-fertile, with normal, fertile cytoplasm (N) and also with nonrestorer genes in the nucleus. The male-sterile version is referred to as the A line, and the male-fertile or normal version is called the maintainer line, B line, or 0 line. The A line generally is developed by crossing a B line as male to a male-sterile female with sterile cytoplasm and nonrestoring genes, followed by repeated backcrossing to the B line. After the A line has been developed, A-line seed is produced for line maintenance by crossing the sterile A line (S rf rf) with the B line (N rf rf). All offspring from this cross will be male-sterile. Maintainer or B lines are maintained by routine selfing, sibbing, or open pollination in isolation.

The third type of inbred line in the cytoplasmic-genetic system is that used as the restorer male parent. Accordingly, it must have the homozygous dominant genotype for fertility restoration (Rf Rf). These restorer or R lines can be developed by using a line with restorer genes as the donor parent, and the genotype (line) desired for use as a restorer line as the recurrent parent in a series of backcrosses. It is very convenient if a donor parent also has sterile cytoplasm (S Rf Rf) and is used as the female because the presence of the restorer allele in segregating populations after each backcross is phenotypically apparent, i.e., S Rf rf × N rf rf → S Rf rf (fertile) and S rf rf (sterile), and an additional test cross or selfing generation is not required for genotype identification. The R lines are fertile and are maintained by routine selfing, sibbing, or open pollination in isolation.

c. Incompatibility. Most crop species that have some form of incompatibility (gametophytic, sporophytic, or a combination) are naturally cross-pollinated species, although some exceptions occur. If the incompatibility system is used for the development of commercial cultivars (hybrids, lines, synthetics), the genetic types have to be maintained individually as parental stocks (Frankel and Galun, 1977). This can be done by selfing (overcoming incompatibility by bud pollination, heat, chemicals), backcrossing, or cloning.

3. Seed Storage

Maintenance of seed stocks with adequate germination is basic to crop improvement. Seed viability is prolonged by decreased moisture content of seeds, lower storage-room temperature, and lower air-oxygen percentages (Roberts, 1972). Relative humidity also must be considered because it influences the moisture-content equilibrium attained by seeds in storage. Although special containers and elaborately controlled environments are required to maintain viability of certain crop seeds, timely harvest, drying if necessary to a moisture content of 10.0 to 12.5%, and storage at temperatures less than 21 C will ensure adequate levels of seed longevity for most crops. Storage requirements and decreases in viability over time vary among genera, among species within a genus, and among cultivars within a species.

According to Harrington's (1960) rule of thumb for safe storage of seeds, the sum of the percent relative humidity plus the temperature in degrees Fahrenheit should not exceed 100. This rule appears to be useful for tobacco (Chapter 47), sorghum (Chapter 41), and warm-season grasses (Chapter 50). The rule was compatible with 3-year storage data for reed canarygrass, a crop whose seed viability decreases rapidly during processing and storage (Bass, 1967). For this species, temperature plus humidity combinations up to 120 were satisfactory for storage periods of 3 years or less, with lower temperatures more important than lower humidity. However, for soybeans, humidity is more important than temperature (Fehr, 1979). A temperature-humidity summation of less than 60 is desired for long-term storage of corn (Chapter 19). Wild rice is uniquely different from most crops in that seeds must be stored under very high relative humidity (Chapter 52). Recent studies suggest that seed viability of small-seeded legumes (alfalfa, red clover) can be maintained for 20 years by storage in common food freezers at −5 to −15 C (Rincker, 1974).

Seeds are most often dried with forced heated air, but small seed lots may be sun-dried, vacuum-dried, or infrared-dried. Chemical desiccants may be used. The higher the seed moisture content, the lower the initial drying temperature must be. Safe drying temperatures for many crops are listed in Table A4.2 of Roberts and Roberts (1972). Data in their Table A4.1 indicate that the equilibrium moisture content of most field and vegetable crop seeds is below 12.5% at 50% relative humidity with temperatures ranging from 12 to 30 C. Strangely enough, temperature has a rather small effect on this equilibrium relationship. For example, data from Hubbard et al. (1957) show that a decrease in temperature from 35 to 25 C increases the moisture content of wheat by only about 1% at 20% relative humidity and by only about 1.5% at 75% relative humidity.

Insect activity in stored seeds is minimized if seed-moisture content and storage temperature are kept low. Insect activity is usually absent below 8% seed-moisture content, but it increases up to about 15% moisture. The optimum temperature for greatest stored-grain insect activity ranges from 29 to 38 C. Thus, conditions desired for maintenance of seed viability (less than 12.5% seed moisture and an air temperature of 21 C) are compatible with conditions desired to limit insect activity. Paradichlorobenzene or napthaline crystals and insecticide vapor strips in seed storage containers or spaces will reduce or eliminate most insect activity.

Seed viability is decreased by the presence of oxygen (Roberts, 1972). Thus, reduction of the oxygen pressure provides another means of extending seed viability of species with storage problems. Oxygen pressure may be reduced by storage in carbon dioxide, nitrogen, argon, or in a vacuum (Peterson et al., 1956; Roberts, 1961; Harrison, 1966).

4. *Vegetative Propagation*

a. Cloning. Vegetatively-propagated germplasm is the most difficult to maintain for either a short or long time. Examples of such germplasm are potato, sugarcane, and certain forage species (*Cynodon, Digitaria, Andropogon*). These can be maintained on a short-term basis as tubers, rhizomes, cuttings, or bulbs. Long-term maintenance in this manner is expensive and generally requires considerable space. A few species can be maintained in the seed form through selfing, but often the desired type is a heterozygote and heterozygosity would be lost by selfing.

b. Tissue Culture. Recent advances in tissue culture may allow vegetative stock to be grown on artificial media and stored at low temperatures for extended periods. The term "tissue culture" is generally applied to growth systems for organ, callus, suspension, cell, protoplast, or meristem culture. Meristem tissue culture consists of growing the excised apex of the plant on a nutrient medium. This is not a simple procedure, because each species has specific nutrient requirements for growth on artificial media. Morel (1975) lists three main features of plantlets grown in vitro which make them valuable for long-term storage: 1) they are small and slow to develop; 2) they facilitate high rates of clonal multiplication; and 3) they can easily be kept free of viruses, fungi, and bacteria. D'Amato (1975) pointed out the potential genetic variability of tissue culture materials and suggested that they are great reservoirs of genetic information which can be exploited when needed. If long-term storage is desired, then this procedure should be employed only when genetic stability in vitro is assured. The storage of plant tissue cultures at extremely low temperatures (0 to −193 C) appears possible; however, sufficient data are not yet available to completely assess the potential of this technique.

III. UTILIZATION

A. Techniques to Facilitate Hybridization

Numerous techniques are used to achieve simultaneous flowering among genotypes which differ in maturity (Chapter 1). Delayed planting of early-flowering lines or use of differential temperatures in controlled environments frequently results in concurrent flowering times. Cutting back of silks on the female inflorescence (ear) of corn is a commonly used delay tactic when the inflorescence (tassel) of the designated male sheds pollen at a later time. In many grasses, cutting back of primary tillers in the early tiller-elongation stage of growth stimulates development of secondary tillers that head later. It is important to note that any special treatment which results in less-than-optimum plant growth may also result in lower seed set in crosses: flower parts may be smaller and more difficult to manipulate; smaller anthers produce less pollen; and, in field nurseries, delays in flowering may shift the time of crossing from cooler into warmer (undesirable) seasonal periods.

Artificial adjustments of day length may be necessary to initiate flowering in photoperiod-sensitive lines. In regions where day length varies considerably from summer to winter seasons lines which have delayed flowering in summer field nurseries due to short nights may flower concurrently with other lines in winter greenhouses under the normal, long-night conditions. If extensive crossing is planned between diverse types, e.g., between tropical and northern U.S. corn, the entire nursery may have to be planted in a more day-length-neutral region in the southern U.S. or Central America. South American maize lines and populations have notoriously late silking and tasseling dates under the short nights of more northernly regions.

Biennial plants that have telescoped stems in the first year of growth and elongate after winter chilling can be induced to elongate at any time, even in the first year of growth, if gibberellin is applied (Villiers, 1975).

Seed dormancy may delay germination and this, in turn, may cause unwanted differences in flowering time among lines. If dormancy is known or suspected to occur, seed drying, scarification, cold treatment of dry seeds, or cold and/or nitrate treatment of imbibed seeds are techniques commonly used to hasten germination. Exposure of imbibed seeds to temperatures between 5 and 10 C for as long as 7 days is effective for many species of agricultural crops and vegetables (MacKay, 1972). In many species, potassium nitrate can replace the requirement for, or reinforce the effect of, other dormancy-breaking agents such as light and temperature (Schwendiman and Shands, 1943; Roberts, 1969; MacKay, 1972). The germination substrate of sand or filter paper is initially moistened with a 0.2% solution of KNO_3, but only water is used for further moistening.

Seeds of tobacco and certain varieties of lettuce require light to induce germination, but only imbibed seeds respond to light (Villiers, 1975). The wavelengths promoting seed germination are in the red region of the spectrum (660 nm).

Pollen storage and pollen shipment may also be used to facilitate hybridization in special instances. Pollen viability, like seed viability, is influenced by relative humidity, temperature, gas atmosphere, and oxygen pressure. Species vary considerably in their pollen-storage requirements, but the optimum relative humidity usually is between 6 and 60% (Stanley and Linskens, 1974). Pollen grains of many species lose viability at either very high or very low humidity. Gramineae pollen especially is noted for its very short survival time. It loses viability within 1 day under dry conditions. Stanley and Linskens (1974) summarized the effect of storage conditions on pollen germination (in vitro) for many species. Freeze-drying of pollen is an effective storage procedure, but careful rehydration before use is critical.

Species which rely on insects for pollination as a routine maintenance procedure also rely on insects for hybridization. The techniques involved are similar in each case and they are discussed in Section II.B.2. of this chapter.

Knowledge of parental characteristics may influence the use of specific lines as the male or female parent. Lines with large anthers or lines whose floral parts are unusually stiff and difficult to manipulate are more frequently used as males. Lines which undergo natural anthesis earlier each crossing day are often used as males while late-flowering lines frequently are used as females. These and other convenience techniques merely improve the efficiency of a crossing program. Male sterility and dioeciousness influence which lines must be used as male and female parents. In interploidy crosses, the line with the highest chromosome number is usually used as the female parent although results vary from genus to genus. See Chapters 5 and 7 for additional discussions of these techniques.

Genetic markers assist in ascertaining hybridity, but usually their use is one of convenience and not necessity. The most frequent challenge is to detect a self-pollination that may have occurred before or during emasculation. If an F_1 plant cannot be positively identified as phenotypically intermediate between its parents, or if a dominant marker was unavailable for use on the male side, then uniformity within the F_2 population (i.e., the female parent's phenotype) may indicate that selfing occurred. Such a determination in the F_2 will not be possible if seeds from different F_1 plants of the same cross are bulked before planting the F_2 population. This problem emphasizes the need to be familiar with parental phenotypes, to carefully evaluate the F_1 plants, and to discard suspected off-types. The appearance of an off-type dominant trait or of one or more plants taller and more vigorous than most plants within a homogeneous line indicate that either outcrossing or a mechanical mixture has occurred.

Hybridization between species in the same or different ploidy level may require special techniques, and these are discussed in detail in Chapter 7. Embryo culture may be necessary to prevent abortion following fertilization, or synthesis and utilization of bridging types, such as autotetraploids or amphiploids, may be necessary to achieve interploidy gene transfer.

B. Utilization

The maintenance of breeding stocks based mainly on botanical principles of reproduction was discussed in Section II.B. of this chapter. This section will deal with the utilization of parental material to develop the types of seed or plant materials used for commercial production. A species can be propagated commercially by seeds of pure line, hybrid, open-pollinated synthetic, or apomictic cultivars or by vegetative propagation of superior genotypes. The use of parental material to develop each type of product will be discussed.

1. Pure-Line Cultivars

The commercial form of a highly self-pollinated species is generally a pure-line cultivar. Development of new cultivars by hybridization consists of 1) selection of parents for attributes desired in the progeny, 2) hybridization of the parents to form a breeding population, 3) natural self-pollination of progenies to develop homozygous (pure) lines, 4) conducting performance tests to identify superior lines, and 5) increase of seed of new cultivars for eventual distribution to farmers. Cultivars developed by this process usually serve as parents for further crosses in breeding programs.

Lines in a germplasm collection commonly are evaluated for characters of interest, particularly pest resistance. A line with a desired character may be used as the donor parent in a backcrossing program to further improve an otherwise superior genotype.

Choice of parents may reflect the skill or art of the plant breeder or, in addition, it may be based on a computer analysis of detailed line measurements and a prediction of progeny performance (Barnard and Grafius, 1971). Either method requires thorough knowledge of parental performance.

2. Hybrid Cultivars

The use of hybrid seed for commercial production is restricted to species in which the increased cost of hybrid seed production is offset by superior yields due to hybrid vigor, in comparison to seed costs and yields of nonhybrid forms. Requirements for hybrid seed production include the presence of hybrid vigor (heterosis), the elimination of fertile pollen in the female parent, and adequate pollination and fertility restoration by the male parent. If these requirements can be met utilizing available germplasm and technology, the actual steps in development of a hybrid cultivar usually include 1) the generation of superior germplasm pools through selection and recombination phases of cyclic selection programs, 2) development of inbred lines by enforced natural or manual self-pollination and by backcrossing, and 3) testing selected inbreds in hybrid combinations to identify superior crosses. These requirements and steps will be discussed in the following paragraphs. Commercial production of hybrid seed is a specialized operation that is described in Chapter 8.

a. Hybrid Vigor. This characteristic represents an increase in performance of hybrid individuals over that of the parents. Although the amount of hybrid vigor varies considerably among and within species, it is generally greater in open-pollinated species where dominance gene action is important. Hybrid vigor usually is less in self-pollinated species where additive gene action predominates. For example, the seed yield of hybrid corn can be more than 100% greater than yields of the inbred parents. However, hybrid vigor in solid-seeded fields of barley or wheat has been low enough that increases in seed yields have not yet offset the higher costs of using hybrid seed compared with using pure-line cultivars.

b. Elimination of Fertile Pollen from the Female Parent. Elimination of fertile pollen from the female parent may be achieved by hand emasculation, by genetic factors, or by cytoplasmic-genetic interactions. Hand emasculation is used in corn, a monoecious species, by pulling or cutting the male inflorescence (tassel) from plants in the female parent population (Chapter 8).

Male sterility in the female parent can be controlled by genes or by a cytoplasmic-genetic system. Genetic male sterility is controlled by nuclear genes in which the alleles for sterility generally are recessive to the alleles for fertility. The genetic system has not been used for commercial hybrid seed production because it has not been possible to eliminate fertile plants from the desired male-sterile plants in the female population. The reason follows. Genetic male-sterile plants (ms ms) must be maintained by mating them with male-fertile (Ms ms or Ms Ms) individuals. Half of the F_1 plants from an ms ms × Ms ms cross would be sterile (ms ms), but the remaining plants would be fertile (Ms ms). One problem is that the unwanted male-fertile plants in the female population might disseminate pollen and reduce the effectiveness of the desired male parent. A second problem is that, except in unique situations, it has been impossible to identify and remove male plants from within the female population before harvest, or to separate after harvest the hybrid seeds produced on the male-sterile females from the selfed-seeds produced by the undesired male-fertile sister plants. Gene-controlled sorting characters such as seed color, size, and shape, plant height, growth type, chemical roguants, and seedling lethals like albinos were discussed by Wiebe and Ramage (1969) as being potentially useful in the balanced tertiary trisomic system of producing hybrid barley. Eslick (1969) discussed specific "selective" genes and Hockett and Eslick (1969) listed seven approaches suggested by several workers for the production of hybrid barley. So far, none has proven effective. However, Ramage (1977) described the use of genetic male sterility to develop improved combinations of genes in barley and wheat populations. Each cycle involves a crossing generation, the F_1 generation, and selection of desired male-sterile genotypes in F_2. The difference in size of seed from monogerm female and multigerm male parents is used as a separation technique in the production of hybrid sugarbeet seed (Chapter 43).

c. Hybrid Combinations. If a cytoplasmic-genetic system exists in the species, combinations of A, B, or R lines from different inbreds can be used to produce single-cross, three-way, or double-cross hybrids without hand emasculation (Fig. 1). All single-cross and three-way cross offspring will be male-fertile. Only 50% of the plants from the double cross will be male-fertile, but pollen shed may be sufficient to completely pollinate all plants (Chapter 8). This cytoplasmic-genetic system is currently used to produce large quantities of hybrid corn, sorghum, sugarbeet, pearl millet, and sunflower seed.

Other procedures for generating male-sterile female parents have been studied, but problems in developing or isolating highly male-sterile female populations have occurred (Lafever and Patterson, 1964; Ramage, 1965). Investigations concerning the use of chemical gametocides to create male sterility are continuing. None of the chemicals is in commercial use at the present time due to inconsistent results among different genotypes and environments.

d. Pollen Transport. Pollen can be transported from the male to female parent by wind, insects, or by hand. Hybrid seed production is most economically and reliably obtained in cross-pollinated species whose natural mode of pollen transport is by wind, such as in corn, sorghum, and pearl millet.

Insect pollination is less reliable than wind for large-scale production of hybrid seed. Bees may prefer the flowers of male-fertile plants more than male-sterile flowers, resulting in inadequate pollination of the female parent. Such insect preference has been a major obstacle for hybrid seed production of alfalfa and broadbean. The destruction of pollen vectors during chemical insect control can be a serious problem and has been an obstacle for production of hybrid cotton seed. Despite these difficulties, hybrid seed production of sunflower by insect pollination has been successful.

Hand pollination has been used to produce hybrid tomatoes. Bulk pollen collected from the male is brushed onto the stigma of female flowers. The cost of hybrid seed is high when it is produced in this manner.

e. Parent Selection. The choice of parents is an important consideration in obtaining high quality seed of a superior hybrid at a reasonable price. The breeder must consider both hybrid performance and seed production during the development of parent lines.

A desirable female parent must have adequate yield potential. When inbred lines cannot produce adequate seed, a modified single cross, a three-way cross, or a double cross may be used. A modified single cross is similar to a three-way cross, except that parents 1 and 2 are closely related (Fig. 1). The cross of parents 1 and 2 produces a near-hybrid that is used as the female parent for large-scale hybrid seed production in combination with parent 3, a distinctly different parent. The female parent for the modified single, three-way, or double cross is a near-hybrid or true hybrid that will produce more seed than an inbred line.

The female parent is selected for its ability to produce high-quality seeds of proper size and shape. Seed should be high in germination, produce strong and vigorous seedlings, and have the ability to retain germination and vigor under prolonged storage. When the cytoplasmic-genetic system is used, the female must consistently display male sterility over a range of environments.

The male parent must produce sufficient pollen for an adequate duration. Inbred lines of corn are scored for pollen production during the final phase of their evaluation, and lines with inadequate pollen are either used as females, modified genetically to increase pollen production, or dis-

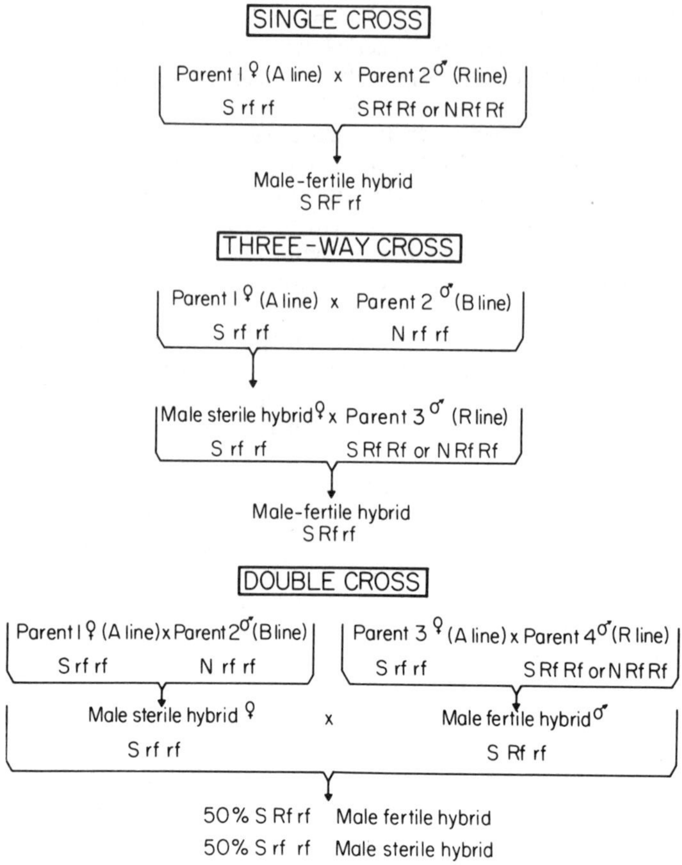

Fig. 1—Use of male-sterile A lines, maintainer B lines, and restorer R lines in the development of hybrids.

carded. For the cytoplasmic-genetic system, the male parent must provide the hybrid with the restorer genes needed to assure adequate pollen production in a commercial field (Fig. 1).

3. Open-Pollinated Cultivars

Seeds of open-pollinated cultivars or synthetics are used for commercial plantings of many cross-pollinated species, including rye, forage legumes and grasses, and corn in developing countries. The cultivar may be 1) a synthetic developed by planned matings of selected genotypes or 2) an improved population developed either by recurrent selection and recombination (Chapter 5) or by bulking seed of several superior inbred lines.

Self-incompatibility is common in species propagated as open-pollinated cultivars. Their flowers are perfect, and the male produces viable pollen; however, either the pollen cannot fertilize the ovule of the flower or it is less competitive than pollen from other genotypes. The frequency of self-pollination tends to be highly variable among genotypes. Individual plants of the synthetic population are highly heterozygous and therefore, self-pollination cannot be used to maintain the population. Component plants, lines, or families are maintained by selfing, sibbing, or by vegetative propagation.

The development of a new synthetic includes 1) development of breeding populations with improved frequencies of favorable genes, 2) the extraction of improved inbred lines from the populations or testing and identification of superior individual clones within the population, 3) formation of potential synthetics by intermating superior lines or clones, 4) identification of superior combinations in performance tests, and 5) seed increase of superior synthetics for release and distribution to the farmer.

An improved breeding population may itself be released as an open-pollinated cultivar following continued selection and recombination phases of recurrent selection programs. Recurrent selection schemes involve 1) intermatings among population members, 2) evaluation of full-sib or half-sib families in performance trials, 3) selection of parents or matings with superior performance, and 4) genetic recombination among selected materials to generate an improved population. In reciprocal recurrent selection programs, step 1 intermatings are made between individuals from two different but concurrent improvement programs.

4. Apomictic Cultivars

Many perennial forage grasses reproduce by apomixis, the embryo of the seed developing without union of sperm and egg cells. The genotype of the seed is the same as that of the mother plant, just as cuttings from the clone of a plant are identical. Apomictic cultivars pose unique problems and possibilities for improvement, and the reader is referred to Chapter 3 for a thorough discussion of this topic.

5. Vegetatively Propagated Cultivars

A few species of crop plants are commercially produced by vegetative propagation of a single genotype. Stem cuttings are used to propagate cassava and sugarcane, and tubers of potato are cut into sections for planting. Plants of a cultivar are highly heterozygous, but they are all identical genetically (homogeneous).

Development of a new cultivar is accomplished by 1) crossing selected clones to develop a breeding population, 2) sowing the hybrid seed to obtain plants that can be vegetatively propagated, 3) evaluating the individual genotypes in performance tests, and 4) releasing superior individuals for commercial production. Problems of outcrossing and seed mixtures are not encountered during commercial propagation of such cultivars.

REFERENCES

Barnard, John, and J. E. Grafius. 1971. Parental selection by computer. Res. Rep. No. 139, Michigan State Univ. Agric. Expt. Stn., East Lansing.

Bass, L. N. 1967. Response of reed canarygrass (*Phalaris arundinacea* L.) seed to storage conditions. Proc. AOSA 57:124–129.

Creech, J. L., and L. P. Reitz. 1971. Plant germplasm now and for tomorrow. Adv. Agron. 23:1–49.

D'Amato, F. 1975. The problem of genetic stability in plant tissue and cell cultures. p. 333–348. *In* O. H. Frankel and J. G. Hawkes (eds.) Crop genetic resources for today and tomorrow. Cambridge Univ. Press, Cambridge, England.

Eslick, R. F. 1969. Balanced male steriles and dominant pre-flowering selective genes for use in hybrid seed production. p. 292–297. *In* R. A. Nilan (ed.) Barley genetics II. Proc. 2nd Int. Barley Genetics Symp., Pullman, Washington.

Fehr, W. R. 1979. Personal communication. Dep. of Agronomy, Iowa State Univ., Ames.

Frankel, R. 1977. Pollination mechanisms, reproduction and plant breeding. Springer-Verlag, N.Y.

Graybill, F. A., and W. R. Kneebone. 1959. Determining minimum populations for initial evaluation of breeding material. Agron. J. 51:4–6.

GRIP. 1979. Monthly newsletter of the germplasm resources information project of Laboratory for Information Science in Agriculture (LISA), Colorado State Univ., Fort Collins. Vol. 1(6).

Hanson, W. D. 1959. Minimum family size for the planning of genetic experiments. Agron. J. 51:711–715.

Harrington, J. F. 1960. Drying, storing, and packaging seed to maintain germination and vigor. Seedsmen's Dig. 11:16, 56.

Harrison, B. J. 1966. Seed deterioration in relation to storage conditions and its influence upon germination, chromosomal damage and plant performance. J. Natl. Inst. Agric. Bot. 10:644–663.

Hawkes, J. G., J. T. Williams, and Jean Hanson. 1976. A bibliography of plant genetic resources. International Board for Plant Genetic Resources. AGPE: IBPGR/76/4. Rome, Italy.

Hockett, E. A., and R. F. Eslick. 1969. Genetic male-sterile genes useful in hybrid barley production. p. 298–307. *In* R. A. Nilan (ed.) Barley genetics II. Proc. 2nd Int. Barley Genetics Symp., Pullman, Washington.

Hubbard, J. E., F. R. Earle, and F. R. Senti. 1957. Moisture relations in wheat and corn. Cereal Chem. 34:422–433.

IS/GR. 1977. Information Sciences/Genetic Resources Program Annual Report for 1977. Univ. of Colorado, Boulder.

Lafever, H. N., and F. L. Patterson. 1964. Cytological and breeding behavior of an aneuploid line of *Avena sativa* L. Crop Sci. 4:535–539.

MacKay, D. B. 1972. The measurement of viability. p. 172–208. *In* E. H. Roberts (ed.) Viability of seeds. Chapman and Hall Ltd., London.

Morel, G. 1975. Meristem culture techniques for the long-term storage of cultivated plants. p. 327–332. *In* O. H. Frankel and J. G. Hawkes (eds.) Crop genetic resources for today and tomorrow. Cambridge Univ. Press, Cambridge, England.

Peterson, A., V. Schlegel, B. Hummel, L. S. Cuendet, W. F. Geddes, and C. M. Christensen. 1956. Grain storage studies. XXII. Influence of oxygen and carbon dioxide concentrations on mold growth and grain deterioration. Cereal Chem. 33:53–66.

Ramage, R. T. 1965. Balanced tertiary trisomics for use in hybrid seed production. Crop Sci. 5:177–178.

———. 1977. Varietal improvement of wheat through male sterile facilitated recurrent selection. ASPAC Food and Fertilizer Tech. Center (Taiwan) Tech. Bull. No. 37.

Rincker, C. M. 1974. Effect of frequent thawing on viability of red clover seed in cold storage. Crop Sci. 14:749–750.

Roberts, E. H. 1961. The viability of rice seed in relation to temperature, moisture content, and gaseous environment. Ann. Bot. 25:381–390.

———. 1969. Seed dormancy and oxidation processes. p. 161–192. *In* H. W. Woolhouse (ed.) Dormancy and survival. Symp. Soc. Exptl. Biol. No. XXII.

———. 1972. Storage environment and the control of viability. p. 14–58. *In* E. H. Roberts (ed.) Viability of seeds. Chapman and Hall Ltd., London.

———, and D. L. Roberts. 1972. Moisture content of seeds. p. 424–437. *In* E. H. Roberts (ed.) Viability of seeds. Chapman and Hall Ltd., London.

Roberts, H. F. 1929. Plant hybridization before Mendel. Princeton Univ. Press, Princeton, N.J.

Schwendiman, A., and H. L. Shands. 1943. Delayed germination or seed dormancy in Vicland oats. J. Am. Soc. Agron. 35:681–688.

Stanley, R. G., and H. F. Linskens. 1974. Pollen. Springer-Verlag, N.Y.

Villiers, T. A. 1975. Dormancy and the survival of plants. Inst. Biol. Studies in Biol. No. 57. Edward Arnold Ltd., London.

Wiebe, G. A., and R. T. Ramage. 1969. Hybrid barley. p. 287–291. *In* R. A. Nilan (ed.) Barley genetics II. Proc. 2nd Int. Barley Genetics Symp., Pullman, Washington.

Williams, J. T. 1976. A bibliography of plant genetic resources (Suppl.) AGPE: IBPGR/76/36. Rome, Italy.

5 Mating Designs, Field Nursery Layouts, and Breeding Records

Charles W. Stuber

USDA-SEA and North Carolina State University
Raleigh

Selecting the kinds of mating techniques and arrangements depends upon 1) predominate type of pollination—self or cross, 2) type of crossing used—artificial or natural, 3) type of pollen dissemination—wind or insect, 4) unique features, such as cytoplasmic or genetic sterility, 5) purpose of project—breeding or genetic, and 6) size of population required. In this chapter, mating designs, field planting arrangements, and general procedures for maintenance of records are described and suggested for breeding or genetic research.

I. MATING DESIGNS

Breeders and geneticists use simple and complex designs. A simple design, the top cross, is frequently used to evaluate a series of breeding lines, selections, or clones. Several designs of intermediate complexity are available for either generating synthetic populations or recombining selected entries in recurrent selection programs. Estimation of genetic variances or effects often requires a rather complex mating scheme. Some designs (such as Design I, Section I F) might be considered multipurpose. Such designs can be used both to estimate genetic variances and to generate families for use in either full-sib or half-sib recurrent selection schemes. All designs involve some type of hybrid crosses. Some also may include self-pollinations.

Before choosing a specific design, the researcher should carefully evaluate the objectives of the project and determine which, if any, design will effectively satisfy those objectives with available resources. Familiarity with the analyses of data that will be generated is also critical to the choice and effective use of the design.

A. Basic Types of Hybrid Crosses

Although the term hybrid is defined as the cross between two genetically dissimilar parents, it is probably more commonly used to denote the F_1 from a cross of two inbred lines. Crosses among clones, open-pollinated cultivars, or other genetically dissimilar populations are included in the F_1 hybrid population category.

Four types of hybrid crosses used most frequently are the single (two-way) cross, three-way cross, backcross, and double cross. The single cross involves two parents:

$$A \times B \rightarrow F_1 = \text{single cross.}$$

As the name implies, the three-way cross involves three different parents:

$$A \times B \rightarrow F_1 \times C \rightarrow \text{three-way cross.}$$

The backcross involves the mating of a single cross to one of its parents:

$$A \times B \rightarrow F_1 \times A \rightarrow \text{backcross.}$$

The double cross involves the mating of two different single crosses as follows:

$$\begin{array}{ccc} A \times B & & C \times D \\ \downarrow & & \downarrow \\ F_1 & \times & F_1 \\ & \downarrow & \\ & \text{Double cross.} & \end{array}$$

B. Backcross

The backcross breeding method is a form of recurrent hybridization widely and effectively used to improve lines or cultivars that excel in most attributes, but lack one or a few desirable characteristics (Briggs, 1938). Characteristics transferred by backcrossing are usually highly heritable traits, such as resistance to disease or insect pests. However, quantitative traits that are moderately heritable, such as maturity, plant height, seed size and shape, also may be adjusted by backcrossing. The technique is used in both cross and self-pollinated species, but has been used infrequently in most perennial forage crops.

Backcrossing is normally initiated by making an F_1 hybrid between the nonrecurrent (donor) parent and the recurrent parent. This F_1 is then backcrossed to the recurrent parent to produce the BC_1. A number of BC_1 individuals are grown and those with the desired trait from the donor parent are selected and crossed to the recurrent parent to produce the BC_2. BC_2 individuals selected for the desired trait are crossed to the recurrent parent. This procedure, shown below, is continued until the desired number of backcrosses is completed:

$$A \times B \rightarrow F_1 \times A \rightarrow BC_1 \times A \rightarrow BC_2 \ldots \rightarrow BC_n.$$

Hayes et al. (1955) presented tables showing the number of plants needed in a backcross to have a reasonable chance that at least one plant will have the desired combination of genes.

For simply inherited traits, usually five to eight backcrosses are considered sufficient. If the desired trait is genetically recessive, a selfed generation will be required after each backcross to identify the desired genotypes which will be crossed to the recurrent parent. Simultaneous selfing and backcrossing is often practiced to identify heterozygotes for crossing to the recurrent parent. Some breeders routinely produce F_2 and F_3 generations after alternate backcrosses. This mating procedure is particularly important when the desired trait is controlled by more than one or two dominant genes. It also allows the breeder to select more intensively for the type of the recurrent parent.

In some crops, such as sugarcane and potato, backcrossing to the same recurrent parent is undesirable because of the resulting effects of inbreeding. Therefore, different clones are used in successive generations to avoid extreme inbreeding.

Commercial cultivars of some crops may be multiclone or multiline synthetics. Cultivars such as these are often improved by using each individual clone in a backcrossing scheme and again intercrossing to produce the improved synthetic. For example, mildew resistance was added to 'Kenstar', a 10-clone red clover synthetic, by crossing each clone to a resistant cultivar (Taylor, 1977). After five backcrosses, an F_2 was produced from each improved clone to allow segregation for detection and selection of the homozygous-dominant, resistant types. These selections were intercrossed to produce the new disease-resistant synthetic.

C. Topcross

The topcross mating scheme involves the crossing of a number of selections, lines, or clones to a common parent (tester) which may be a cultivar, an inbred line, single cross, etc.

$$A \times \text{Tester} \rightarrow TC_A$$
$$B \times \text{Tester} \rightarrow TC_B$$
$$\vdots \qquad\qquad\qquad \vdots$$
$$i \times \text{Tester} \rightarrow TC_i$$
$$\vdots \qquad\qquad\qquad \vdots$$
$$n \times \text{Tester} \rightarrow TC_n$$

where the tester is the same for each mating. Progeny in topcross TC_A are half-sibs of progeny in TC_B. In cross-pollinated crops, like corn, this type of mating is commonly an inbred-cultivar cross (Jenkins and Brunson, 1932). It is frequently used for evaluation of general combining abilities for a group of lines, clones, or selections. The topcross method also has been used for initial evaluations of breeding potentials in exotic corn races (Stuber, 1978).

If an isolated crossing block is available, natural hybridization can be used for topcrossing in many species. The tester is used as the male parent and the females (lines or clones) to be tested are either male-sterile or self-incompatible, or are emasculated before pollen shed. The ratio of number of male rows to female rows varies with species. Alternating rows may be required in crops like sugarbeet, whereas two male rows may be adequate to pollinate 10 to 12 female rows in castor (Fig. 1). It may be necessary to ad-

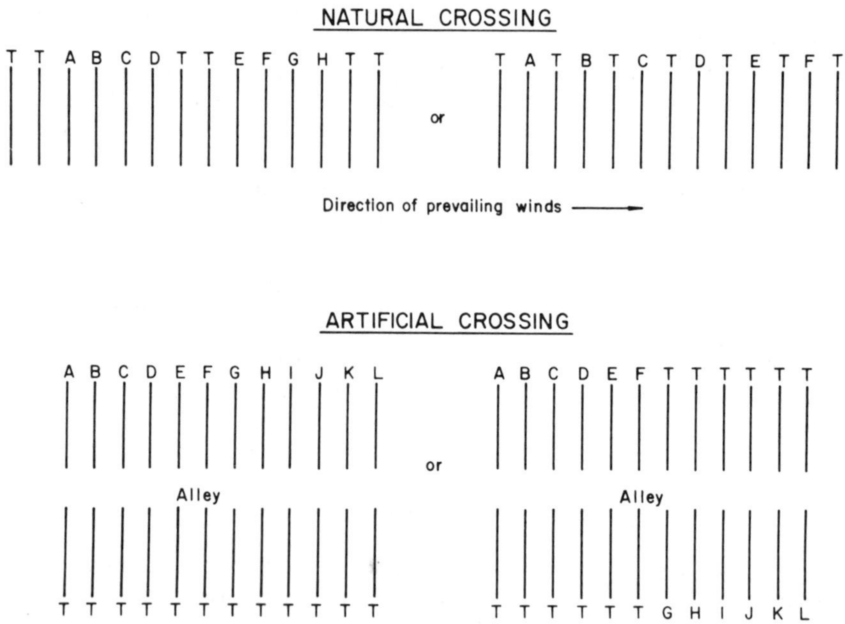

Fig. 1—Topcross nursery arrangements for natural and artificial crossing. The alternating arrangements of testers (T) with entries to be tested (A, B, C, etc.) shown for natural hybridization also may be appropriate for artificial hybridization.

just or delay planting dates if the male and female parents differ in days to flowering. In species that cannot be easily emasculated or where male sterility is not available, it may be possible to utilize visual genetic markers so that crosses can be easily detected. Also, selfing can be prevented in species having self-incompatibility systems. Intercrossing between females can be reduced by planting females a greater distance from other females than from testers. If wind is a factor in pollen dissemination, tester rows should be planted perpendicular to prevailing winds. Replication and randomization of the female genotypes may help to minimize the effects of intercrossing among females.

If isolation is not possible, the tester may be planted in paired rows with the materials to be evaluated (Fig. 1). Hand pollinations are made between rows and the tester may be used either as the male or the female parent. If there is a wide range of maturity among the materials to be evaluated, synchronization of flowering with the tester may be difficult to accomplish in paired rows. In this case, matings may be more efficiently accomplished by planting the materials to be evaluated in separate blocks in order of increasing number of days to flower. The tester is planted in adjacent blocks with specified rows delayed appropriately to effect the desired pollinations.

D. Polycross

The polycross is a mating arrangement for interpollinating a group of cultivars or clones using natural hybridization in an isolated crossing block (Tysdal, 1948). Other designs normally are used for interpollinations if isolation is not possible and hand-crossing is required. The polycross is used frequently for forage grasses and legumes, sweet potato, and sugarcane.

Because the purpose of the polycross is to provide an equal opportunity for each entry to be crossed with every other entry, the field layout is the critical feature of the design. If the number of entries is 10 or less, the Latin square is often used for randomizing. This places each entry in every row and column of the layout (Fig. 2). The number of replications in one Latin square equals the number of entries. Because it is often desirable to have 20 to 30 replications, two or more Latin squares may be necessary. A completely balanced polycross design is outlined by Olesen (1976). If more than 10 cultivars or clones are to be interpollinated, a completely randomized block, with adequate replications, is normally preferred for the polycross (Fig. 2). When the seed is harvested, equal quantities from each replication of each entry are bulked.

Although the polycross requires a minimum of effort for intermating a group of entries, deviations from random mating may occur unless all entries flower simultaneously. Planting of early entries can be delayed appropriately to avoid this problem in annual crops if days to flowering are known in advance. Progeny from each entry have a common parent in the polycross design. Thus, half-sib families are generated which are frequently used for evaluating general combining abilities. The polycross also is often used for generating synthetic cultivars and may be used for recombining selected entries or families in recurrent selection programs.

LATIN SQUARES

C	A	D	B		A	B	C	D		B	C	D	A
D	C	B	A		B	A	D	C		A	B	C	D
A	B	C	D		C	D	B	A		C	D	A	B
B	D	A	C		D	C	A	B		D	A	B	C

RANDOMIZED COMPLETE BLOCK

N	I	P	J		D	G	K	H		O	F	N	A
C	G	A	H		B	O	E	F		B	G	J	P
K	B	O	E		C	L	A	J		E	M	K	H
F	M	D	L		N	M	I	P		D	C	I	L
Rep 1					Rep 2					Rep 3			

Fig. 2—Polycross nursery arrangements showing three Latin squares for 12 replications of four genotypes and a randomized, complete-block design for three replications of 16 genotypes.

E. Diallel and Partial Diallel

The diallel is a mating design to cross three or more parents in all possible combinations. This mating design has been used extensively for developing breeding populations for recurrent selection (Jensen, 1970, 1978). It probably has been used more frequently than any other design in crop plants to estimate general and specific combining ability variances and effects. Genetic interpretations and analyses are provided in numerous papers, such as those by Hayman (1954) and Griffing (1956).

The complete diallel includes all possible matings, including selfs and reciprocals, among a set of entries. These entries are usually homozygous inbred lines, pure lines, or clones, although Gardner and Eberhart (1966) also proposed the use of cultivar cross diallels. For the complete diallel, there are p^2 crosses among p parents. With no selfs or reciprocals, $p(p-1)/2$ crosses are included in the design. In crops like corn, the parental lines are planted in paired rows for each mating (Fig. 3). This requires $p-1$ rows for each parent and a total of $p(p-1)$ rows. Reciprocals may or may not be made, as desired, with this arrangement. Unless the two rows in a pair flower simultaneously, reciprocal matings may not be possible. Selfs can be made on individual plants in the crossing block or in a separate selfing block. For crops such as tobacco, which are multiflowered and set numerous seeds per pollination, single plants may be adequate to represent each line.

For crops with multiple tillers, such as forage grasses, a circular design may be used for diallel matings. One parent of the diallel is planted in the center of a ring of the other parents. The arrangement is repeated with a

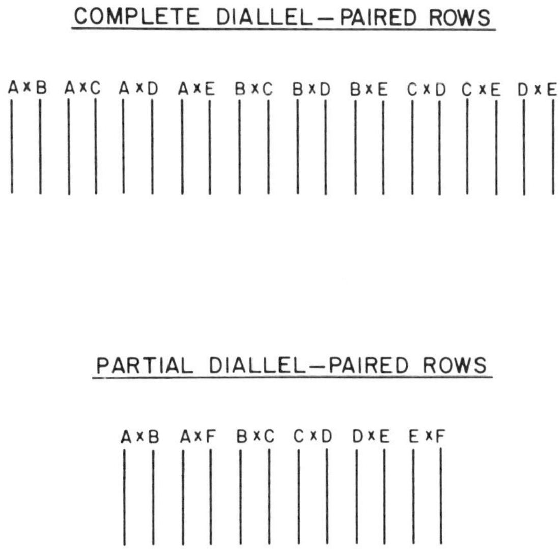

Fig. 3—Paired row nursery arrangements for a complete diallel with five parents and for a partial diallel with six parents.

different parent in the center so that the number of rings equals the number of parents. Panicles of the center parent are isolated under bags with each parent in the ring.

A major disadvantage of the diallel is the large number of crosses generated in the mating scheme. Thus, requirements for space, seed, and labor involved in the crossing block and in the experimental evaluations usually limit the number of parents to no more than 8 or 10. Use of one-half of a symmetric Latin square for pairing the parents for crossing greatly reduces space requirements in the crossing area (Fehr and Ortiz, 1975). With one-half of the symmetric Latin square, the number of rows for each parent in a diallel mating is reduced to $p/2$ and the total number of rows is $p^2/2$. However, the number of parents, p, must be even.

Steps used to construct a symmetric Latin square for an even number of entries are as follows: 1) write the entries in a ring (Fig. 4); 2) select an entry in the ring to be in the upper left-hand corner of the Latin square (starting entry); 3) for the sequence of entries in the first row and column, take the entry immediately clockwise from the starting entry, then the entry immediately counterclockwise from the starting entry, then the next available entry clockwise, then the next available entry counterclockwise, and so on until the row and column are completed; 4) complete the remaining columns by using the first entry in each column as the starting entry and repeat step 3.

All possible pairs of parents are contained in one-half of the symmetric Latin square. For the diallel mating scheme, each column (or row) is considered as a mating unit, and crosses are not made between parents in different units. In each unit, pollen of the first parent is transferred to the second parent, pollen from the second parent is transferred to the third, etc. Thus,

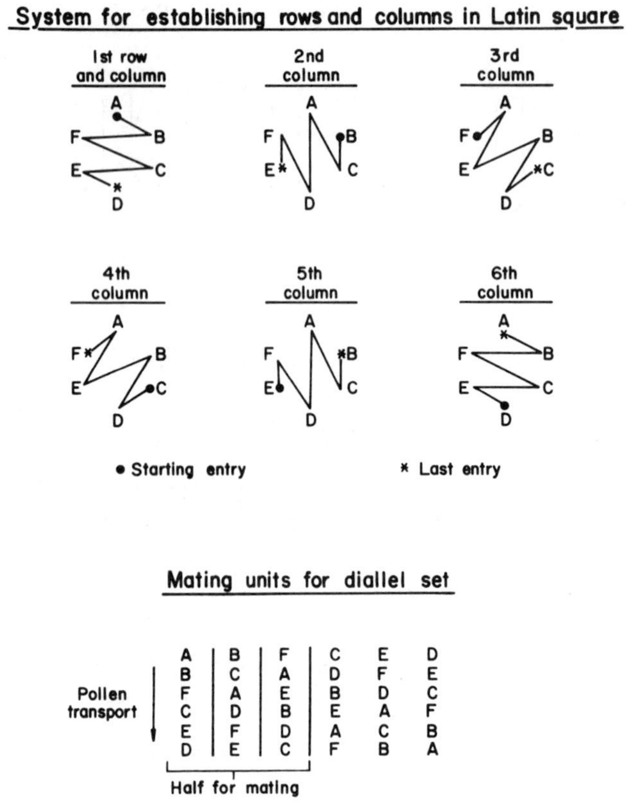

Fig. 4—Procedure for constructing a symmetric Latin square and determining mating units for a six-parent diallel mating.

the first parent in each mating unit is used only as a male and the last parent is used only as a female, unless reciprocal crosses are desired. Reversal of male and female parents may sometimes be necessary to synchronize pollen shedding with receptiveness of stigmas.

Another disadvantage of the diallel concerns the inferences that may be made when used in genetic studies. Unless the parental lines can be assumed to be a random sample from some population, the genetic inferences from the diallel analysis will relate only to that set of lines or to the population that might be generated from mating the lines. Some form of a partial diallel may help to overcome this disadvantage as well as the problems associated with the large number of crosses required for the complete diallel. In the partial diallel, each parent is mated an equal number of times, but not to all other parents as in the complete diallel (Fig. 3).

Several types of partial diallels have been evaluated and reviewed by Dhillon (1975). He suggests that an evaluation of five to seven crosses per parent in a partial diallel should give reliable estimates of general combining ability. Therefore, a large number of parents can be screened for their general combining ability in a partial diallel scheme and the selected parents then might be evaluated in a full-diallel fashion.

Another deviation from full-diallel mating involves simultaneous selfing with a partial diallel-test-crossing design (Matzinger and Cockerham, 1963). Actually, their partial diallel is similar to the Design II mating scheme (Section I G), except that the parents are also self-pollinated. This type of design provides information for estimating certain types of epistatic variances in addition to additive and dominance variances (or general and specific combining ability).

F. Design I

Design I may be considered as multipurpose (Comstock and Robinson, 1948, 1952). It is often used to estimate additive and dominance variances and also can be used to generate families for evaluation in full-sib or half-sib recurrent selection. Frequently, both functions are performed simultaneously.

The design probably has been used more frequently than any other scheme in corn for estimating genetic variances. It is well-adapted for use in most cross-pollinated species and in a few self-pollinated species, such as tobacco, in which an adequate amount of seed is easily produced from each mating. The costly, time-consuming effort required to produce sufficient seed for replicated evaluation trials, utilizing drill-row plots, has essentially precluded the use of this and similar designs in self-pollinating species, such as small grains and soybean. However, Stuber (1970) has outlined a procedure for estimation of additive, dominance, and additive × additive genetic variances using inbred progeny evaluation. The inbred relatives are generated by bulk selfing of the F_1 progenies in mating designs such as Design I and II. These selfed progenies are grown and evaluated in field designs similar to those used for non-inbred progenies. Although the progenies evaluated are inbred to varying levels, genetic variances defined for the non-inbred reference population can be estimated with no restrictions on gene frequencies or number of alleles per locus.

Design I is a nested type of mating design in which each member of a group of parents used as males is mated to a different group of parents used as females (Fig. 5). No female parent is involved in more than one mating. Normally, each male is crossed to four female parents, a cross that generates four full-sib families nested within a half-sib family. Males could be interchanged with females in the design if the matings are more convenient. To avoid assortative mating, matings are arranged in the nursery at planting (Fig. 5). For each group of four randomly chosen females, an alternative random male may be included in case the primary male fails to provide pollen when the females are receptive. The alternative male may be delayed by planting a few days later than the primary male. One or two extra female plants may be grown to assure a complete group of four. If adequate seed is set on more than four females, the extra females are randomly eliminated. In the evaluations of the progenies, blocking of the matings is normally done because of soil variability. For example, 16 families (four sets of four full-sib families) are frequently assigned to a block. In the analysis of variance, each block is treated as a separate experiment with sums of squares and degrees of freedom pooled for a combined analysis.

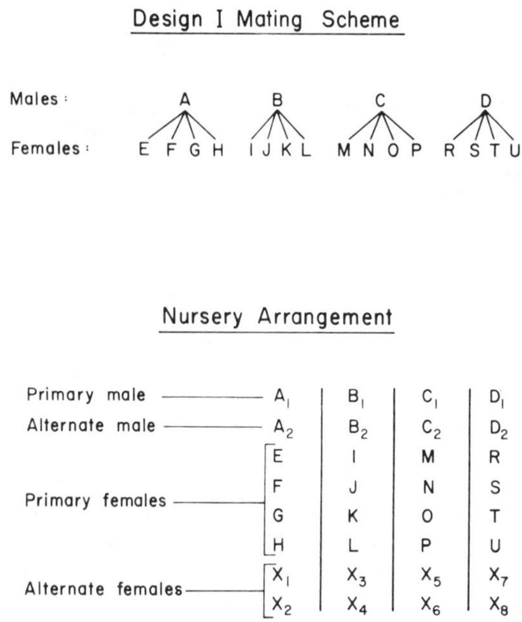

Fig. 5—Design I mating scheme and nursery arrangement showing primary and alternative males and females for each mating group.

G. Design II

Design II, a factorial design, is essentially a modification of Design I (Comstock and Robinson, 1948, 1952). It is used to estimate genetic variances and to evaluate inbred lines for combining ability. Each member of a group of parents used as males is mated to each member of another group of parents used as females (Fig. 6). Frequently four males and four females in each mating group are used to generate a total of 16 full-sib families. This design is well suited to multiflowered plants because each plant can be used repeatedly as both male and female. For single-eared corn, the design is suitable only if the parents are inbred lines from some specified population. Thus, the female is repeated by using several plants from each female line. The mating arrangement in corn usually involves paired-row planting of "male" and "female" lines (Fig. 6). In most cases, reciprocal crosses are assumed to be identical genetically; therefore, seed of reciprocals is bulked for planting progeny evaluation tests. In the analyses of variance, male and female designations are purely arbitrary to facilitate the analyses.

For this design, progenies are normally blocked in the experimental design so that all of the families from the mating of a single group of males to a single group of females remain as an intact unit. For example, a mating group of four males and four females would constitute a block of 16 full-sib families. Analyses of variance would be pooled over blocks for estimation of genetic variance components.

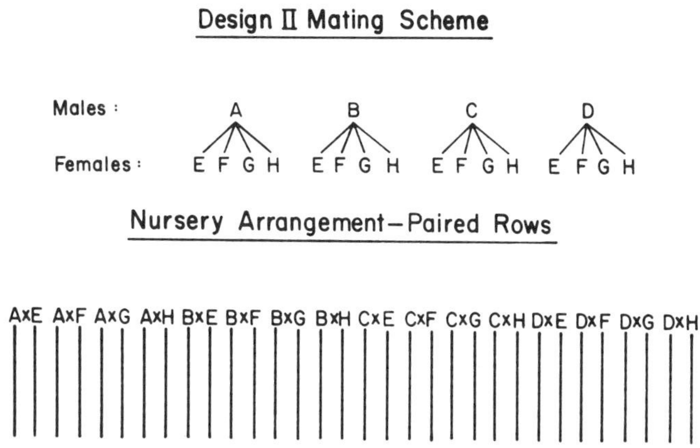

Fig. 6—Design II mating scheme and paired row nursery arrangement.

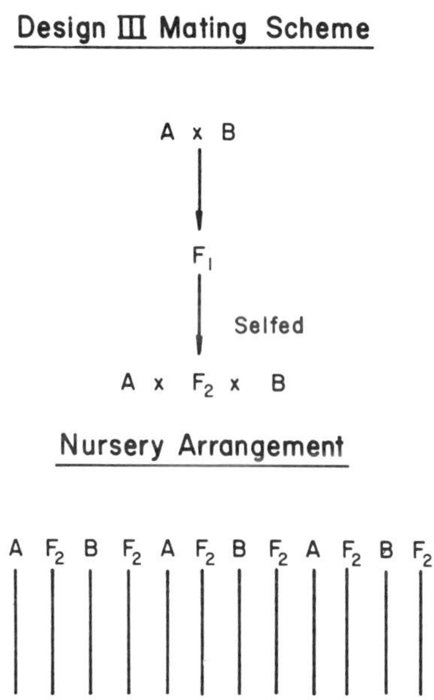

Fig. 7—Design III mating scheme and paired row nursery arrangement.

H. Design III

Design III involves backcrossing of F_2 plants to the two inbred lines from which the F_2 was derived (Comstock and Robinson, 1948, 1952). The F_2 plants are used as male parents (Fig. 7). The number of inbred plants crossed to each F_2 should be large enough to insure sufficient seed for evaluations. As with Designs I and II, progenies are blocked into groups (for example, 16 per group) because of soil variability. Design III is used rather infrequently and primarily to estimate the average dominance of genes. Estimates of genetic parameters tend to be more biased from epistasis in this design than in Designs I and II.

I. Reciprocal Selection Designs

Design I is often used to generate families for recurrent selection programs and is particularly well adapted to the reciprocal recurrent selection scheme (Comstock et al., 1949). When Design I is used for reciprocal recurrent selection, males are selfed as well as crossed to females from the opposite population. After evaluation of half-sib progenies from each male, selfed seed from each selected male is used to produce plants for the recombination cycle which generates the population for the next selection cycle.

A procedure that has become popular in corn for concurrent population improvement and development of single-cross hybrids was developed by Hallauer and Eberhart (1970). It is called reciprocal full-sib selection. The design begins with concurrent interpopulation mating and selfing of S_0 plants from two base populations A(C0) and B(C0). To produce both crosses and selfs on the same plant, the plants must be multieared. Selection of the superior $S_0^A \times S_0^B$ hybrids is based on performance in evaluation tests. Remnant selfed seed produced on the parents of selected hybrids is used to generate populations for the recombination cycle, which, in turn, provides the two new S_0 populations, A(C1) and B(C1), for the next selection cycle. In this respect, it is similar to the reciprocal recurrent selection scheme. Selfed seed from the parents of the superior $S_0 \times S_0$ crosses is also planted ear-to-row for continued selfing and crossing for rapid development of new single crosses. This selfing and crossing for development of single crosses can proceed in each succeeding cycle of recurrent selection to take advantage of the improved genetic constituents in the improved populations. Although the procedure has been specifically designed for two-eared corn, it can be used in any multiflowered crop that can be both self and cross-pollinated. However, its use would be limited to those species in which the primary objective is the production of F_1 hybrids or improved population intercrosses. The procedure also can be modified for use in single-eared corn by selfing in one generation and then crossing the S_1 progeny in the following generation. This modification lengthens the time per cycle.

J. Other Intermating Designs

Many different mating designs can be used to intermate (recombine) groups of genotypes, such as individual plants, inbred lines, clones, or selected families. Objectives for intermating include synthesis of populations, recombination of selected plants or families, and maintenance or increase of populations. Matings generally are designed to insure thorough recombination and an equal genetic contribution of each unit in a random fashion. Although designs like polycross and diallel are frequently used for this purpose, several other mating arrangements may be more suitable.

For recombination of selected entries in a recurrent selection program, the diallel mating design is often used if the number of entries is small (probably ≤ 10). Larger numbers of entries are frequently mated with some type of partial diallel scheme. For example, entry A can be crossed with entries B, C, D, and E; entry B with C, D, E, and F; entry C with D, E, F, and G; and so on. If isolation is possible, the polycross might also be used if adequate replications are provided to assure randomness of pollinations.

Chain (cyclic) crossing involves the use of each plant once as a male parent and once as a female parent, as follows:

$$A \rightarrow B \rightarrow C \rightarrow D \rightarrow E \rightarrow F$$

where the arrows indicate the direction of pollen transport. This procedure insures that each entry is equally represented, if equal amounts of seed are bulked from each mating. However, the procedure provides only a limited amount of recombination. It is useful, however, as a sibbing technique for population or synthetic cultivar increases or for population maintenance if a large number of plants are sampled (>200).

Eberhart et al. (1967) proposed a bulk entry method for population development that provides for thorough recombination when a large number of parents is composited. The method also is called the Irish method. It was specifically designed for corn, but might be modified for use in other plant species. In this method, replicated plantings of individual entries are made in isolation and plants are detasseled before pollen shed. Rows of bulked seed of all entries are planted between rows or ranges of the individual entries to provide pollen as in a top cross nursery (Fig. 8). Ears of each entry are saved and bulked over replications at harvest to represent that entry in the following season. If desired, selection might be imposed, and only ears from the best plants of each entry would be bulked. Individual entries are handled in a similar fashion for several generations with the variation among entries decreasing as recombination progresses. The total number of ears saved in each generation should be quite large (possibly 800 to 1,000) to minimize loss of favorable genes. The bulk entry method may be used with hand pollinations although it was designed to use natural crossing efficiently for recombination.

Another technique for systematically crossing a number of entries to develop a population is often called the multiple cross. For eight parents, it can be diagrammed as follows:

A × B		C × D		E × F		G × H
↓		↓		↓		↓
F_1	×	F_1		F_1	×	F_1
	↓				↓	
	Double cross		×		Double cross	
			↓			
			Multiple cross.			

When only eight parental entries are included, the procedure may be referred to as a double double cross. Additional generations of intermating can proceed by using some type of plant-to-plant or chain crossing technique.

Other procedures for intermating a number of entries for population development may include combinations of diallel, partial diallel, and chain crossing. For example, the procedure may include: 1) first intermating—diallel of the parental entries, 2) second intermating—partial diallel of the F_1 diallel crosses, 3) third intermating—chain (plant-to-plant) crossing among plants from bulked seed from the partial diallel, 4) additional intermatings—chain crossing among plants from bulked seed of the previous intermating. The above procedure can be modified by 1) replacing the diallel with a partial diallel for the first intermating followed by chain crossing, or 2) replacing the diallel and partial diallel by a polycross followed by either plant-to-plant or natural crossing of plants from bulked seed from the previous intermating cycle.

Recombination and intermating in many self-pollinated species is difficult because of problems associated with artificial pollinations and poor seed set. Genetic male sterility has been used quite successfully in crops like barley and soybeans in obtaining natural crosses for synthesizing and intermating composite populations. Using the male-sterile technique, basic populations can be synthesized in several ways such as from single crosses, backcrosses, or diallel crosses. Contribution of the cytoplasm from the male-sterile source can be minimized by using heterozygous fertiles from a maintainer line as male parents in the synthesis of the initial population. This would require hand pollinations. The genetic contribution of the male-sterile genotype can be minimized by backcrossing for one or more generations. The male-sterile factor generally is recessive. Thus, by harvesting seed from only male-sterile plants, crossing is assured and the male-sterile types are maintained automatically in the population. In many species, insects are the pollen vectors. Because of the nonrandomness often associated with insect pollinations, a grid system can be used at harvest to insure sampling of all parts of the intermating block. Details for using male-sterile techniques in selection and intermating schemes have been outlined (Brim and Stuber, 1973; Jensen, 1970, 1978).

K. Inbred Line Development

Self-pollination is an important mating procedure, particularly for the production of inbred lines. A useful characteristic of inbred lines in both self and cross-pollinating species is that they provide repeatable parental genotypes for creating hybrids such as single, three-way, and double crosses. This feature is also important in many genetic studies, both in classical studies and for estimating genetic effects and variances in quantitative traits.

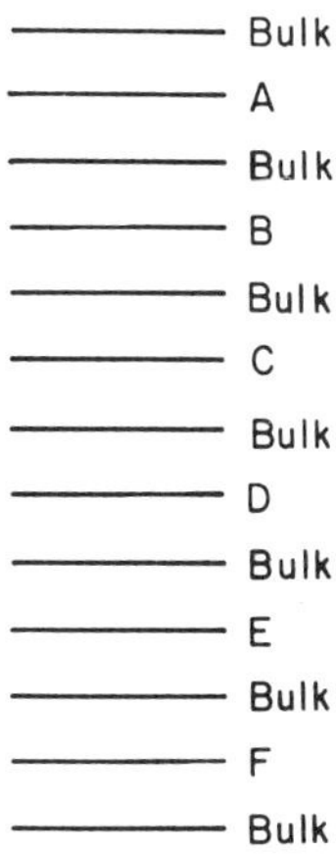

Fig. 8—Bulk entry method for intermating. The bulk normally contains a mixture of all parents, but could exclude parents in adjacent rows to decrease inbreeding.

In species that are normally self-pollinating, no action is required by the breeder to effect self-fertilization. However, it may be necessary to cover the floral parts to prevent contamination from foreign pollen if 100% self-fertilization is required. In monoecious plants like corn, pollen must be manually transported to protected stigmas to assure self-fertilization.

In corn and other cross-pollinated crops, development of inbred lines usually begins with hand self-pollination of individual S_0 plants in a highly heterozygous and heterogeneous population such as an open-pollinated cultivar, a synthetic cultivar, or a cultivar resulting from several cycles of recurrent selection in a population improvement program. A progeny row, usually five or more S_1 plants, is planted from the seed of each selfed S_0 plant. Plants in each row are self-pollinated by hand. In a breeding nursery, plants with undesirable agronomic traits are discarded before pollination. Additional undesirable plants will be discarded before harvest. Selection among progeny rows also may be made but is usually delayed until later selfing generations. Succeeding selfing generations are handled similarly until the desired level of homozygosity is attained.

In a genetics nursery in which random inbred lines are being developed, intentional selection is avoided. In the S_1 generation, a random plant that sets seed is chosen from each progeny row. For the next generation, a progeny row is then planted from seed of each of the randomly chosen S_1 plants. Succeeding generations are handled similarly by single seed descent until the desired level of inbreeding is attained with intentional selection avoided at each generation.

In self-pollinated species such as wheat, oat, barley, tobacco, and soybean, development of inbred lines frequently begins in the F_2 of a cross between two lines or in some synthetic population of heterozygous individuals. Outcrossing is so infrequent in these species that pollination control is unnecessary. However, breeders of self-pollinated crops frequently cover inflorescences with bags in early generations of selfing as a precautionary measure. In other normally self-pollinated species, such as sorghum or cotton, some type of hand pollination control may be required because some outcrossing is usually present.

The double haploid technique for developing homozygous lines is quite useful in some crops. In certain species of cotton, such as *Gossypium barbadense,* haploids can occur as frequently as 1 in 500 seeds. The use of anther culture in tobacco to produce haploid plants is a very efficient method for producing "double haploid" homozygous lines.

II. FIELD NURSERY LAYOUTS

A. General Considerations

Careful planning of the layout of a crossing nursery can save much time by eliminating unnecessary activity from planting through harvest; however, some compromises normally are necessary. Layouts that strive for high efficiency during the pollination period may require additional effort in land preparation and planting. For most crops, the largest portion of time spent in the nursery involves taking notes and making pollinations; therefore, these operations should be given highest priorities for planning. Provisions for cultivation, irrigation, and applications of pesticides also should be considered.

It is usually desirable to locate materials in blocks according to the type of study or the intended pollination procedures. For example, the nursery might be divided into hybridization blocks, selfing blocks, and recurrent selection blocks. Blocks can be subdivided into plots, which are usually single rows.

Procedures for numbering of plots within the nursery depend on individual preferences of the breeder. In general, it is desirable to design the system as simply as possible. A simple system begins with the number 1 each year and a year prefix is carried in record books and on packets of harvested seed (e.g., 79-1, 79-2, 79-3, etc.). Numbering can proceed within blocks in a serpentine fashion. Plot numbers should not be repeated at a location. It is also preferable not to repeat numbers within a year unless a location designation is carried in the records.

Some breeders prefer to use a prefix number or letter to designate specific blocks of material with plots within blocks numbered in a serpentine fashion. For example, 5028 represents plot 28 in block 5. Block, range, and plot numbers may be used, if desired. For example, B1106 represents block B, range 11, and plot 6, in which a range is a series of plots perpendicular to the alley separating the plots. Irrespective of the numbering system used, a field map should be prepared for ease in locating material in the nursery.

Plots in the nursery may be identified with garden stakes placed at the front of the plot or with tags attached to the first plant in the plot. For some blocks, it may be necessary to identify each plot with plot number, entry designations, and possibly, crossing instructions. In other blocks, it may be necessary to identify only every 5th or every 10th plot with its number. Tags or stakes with different colors may be used to identify special materials, such as male-steriles, or different testers if more than one tester is used in a top-cross nursery block.

The size of the plot depends on the crop and mating design and may vary from single hills to rows with 25 or more plants. Planting density also depends on several factors. If the breeder is imposing some type of selection criterion on the material in the nursery, the density will normally be similar to that used in farmers' fields. If, however, maximum seed set on each plant is desired, plants should be more widely spaced. Spacing considerations also may include ease of movement among plots or blocks.

B. Artificial Hybridization

In crossing blocks for artificial hybridization, efficiency of pollination can be maximized if parents for specific crosses can be designated before planting. In many designs, such as the diallel or Design II, parents for specific crosses may be planted in paired rows. Even though parents can be designated before planting, either limited seed supplies or a shortage of land resources may require transporting pollen from a male parent to females in several locations within a crossing block. Also, if some type of selection (e.g., for disease resistance or days to flower) is imposed before pollination, parents cannot be determined before planting. Thus, it may be necessary to transport pollen over some distance to effect the desired pollinations.

Planting arrangements that might be used when it is not practical to plant parents to be mated in adjacent (paired) rows are shown in Fig. 9. Planting of whole blocks on a specific date may make land preparation more convenient, but time may be wasted locating specific parents as they flower. By planting on different dates within parental rows, pollinations may be more convenient but land preparation for later dates of planting may be more difficult.

If the design and available resources allow for planting the parents to be mated in paired rows, pollinations can normally be made more efficiently, thus permitting more hybridizations per unit of effort expended. With paired rows, little or no identification of matings is needed because each row is mated only with the other member of its pair. Also, number of attempted and successful hybridizations is easily determined without extensive record-keeping in field books. Another advantage of paired rows is that the field layouts may permit the use of inexperienced persons for making pollinations. Time of planting of the parents requires careful planning so that flowering is concurrent. Multiple planting dates within or among rows may be required to assure coordination of flowering.

A number of techniques can be utilized to minimize errors in matings. If tillage equipment will accommodate unequal row spacing, parents to be mated can be planted closer to each other than to adjacent pairs, or they could be planted farther apart than adjacent pairs (Fig. 9). Strings or stakes might be used to separate rows that are or are not to be crossed. Stakes or tags of the same color also can be used to identify pairs of parents.

C. Natural Hybridization

Isolation from foreign pollen is a primary requirement in crossing blocks used for natural hybridization. Isolation distances may vary from a few to several hundred meters. Crops with short flowering periods can be effectively isolated in adjacent areas by planting at different dates. Strips of buffer crops also can prevent movement of pollen between isolations in some species. For example, a 1.5-m strip (five rows) of a sorghum-sudan hybrid can provide an effective isolation for a crop like buckwheat. Various types of cages also can be used in a number of crops for isolations.

For natural forces to effect hybridizations, reproductive features must be utilized or modified so that cross-pollination among parental plots is favored and pollination within parental plots is eliminated or minimized. Dioecious species, which have plants with only staminate or only pistillate flowers, are particularly well-adapted for natural crossing. Monoecious species, such as corn, have staminate and pistillate parts separated on the plant and can be easily emasculated for natural hybridization. Even in species with perfect flowers, protandry or protogyny can assure a high proportion of cross-pollination.

Male-sterility factors are used frequently in both self and cross-pollinators to facilitate natural hybridization. Genetic male sterility allows the use of a convenient technique for hybridization and intermating in crops like soybean and barley. Cytoplasmic-genetic male sterility has been used extensively to effect natural hybridization, particularly in commercial hybrid seed production (Chapter 8).

Self-incompatibility systems, controlled by one or more genetic factors, provide a convenient means for natural crossing in several cross-pollinated species. In buckwheat, for example, this system is utilized by eliminating all of the pin flowers in one parent cultivar and all of the thrum flowers in the other cultivar. Many of the cool-season forage grasses and some clovers also can be easily hybridized by using gametophytic or sporophytic self-incompatibility systems.

Genetic markers are often used, particularly in self-pollinators that may be frequently outcrossed, for identification of hybrid seeds or plants. In guar, pubescent lines (*GG*) are used as male parents and crossed to glabrous (*gg*) lines (Stafford and Lewis, 1975). The dominant pubescent trait easily identifies hybrids. A visual genetic marker is also a prerequisite for the Dual Population System proposed by Rachie and Gardner (1975). This system is designed to effect natural outcrossing required for large-scale hybridization in population intermating and improvement programs. The system involves alternate selection of recessive and dominant markers in succeeding generations of intermating, a procedure that assures a high proportion of cross-pollination.

Arrangements of parents in natural crossing blocks should provide for adequate pollen distribution. If the pollen is windborne, direction of prevailing winds should be considered, and long, narrow crossing areas should be avoided. If insects are the primary vectors, planting on a square grid pattern or alternating genotypes within rows can increase the amount of cross-pollination over that achieved when planting in alternate rows.

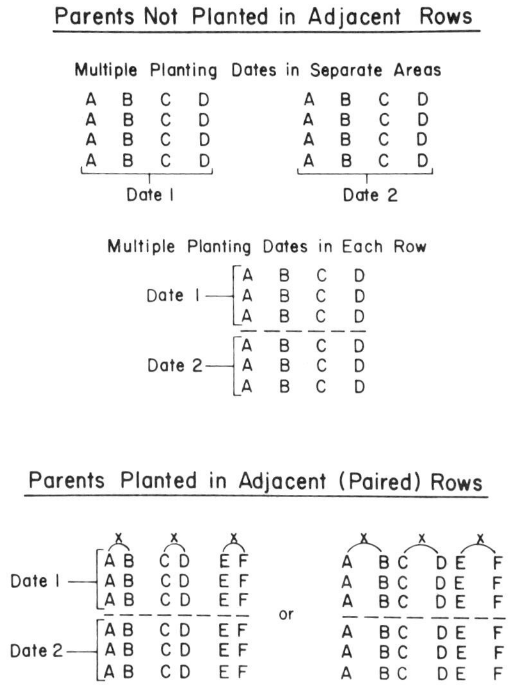

Fig. 9—Arrangements for multiple planting dates when parents cannot be planted in adjacent rows and when parents are planted in adjacent (paired) rows.

III. BREEDING RECORDS

The maintenance of adequate and accurate breeding records is imperative in any successful crop breeding or genetics project. The kinds of records maintained differ so much among individuals that it is difficult to identify any commonly used system. Breeders or geneticists must choose systems that meet their individual requirements.

Any system of record-keeping should be as simple as possible, otherwise the details can become so time-consuming that the breeder may fail to maintain up-to-date records. Because a large proportion of the crosses generated in a breeding program will be discarded, detailed pedigrees need not be recorded each time a cross is made. However, adequate information should be provided in some type of permanent record so that ancestral pedigrees can be ascertained if desired. Pedigrees can be filed in book form, on index cards, or in some type of computerized system. Computerized records require a high degree of organization but can be designed to include considerably more information than those maintained by hand, and retrieval systems can provide complete pedigrees and other pertinent information almost instantaneously.

Some breeders use simple accession numbers in a breeding nursery beginning with 1 each year. In the record book, the number for the current year can be listed in the first column and the number of the previous year in the second column. Other pertinent information follows in later columns, such as amount of selfing and backcrossing. Crosses recorded on seed packets would include accession numbers such as 31 × 25. Most breeders list the female parent first. The year should also be included on the seed packet or can be included as part of the cross designation, e.g., 78-31 × 25. This simple accession number system, which includes the year grown, enables the breeder to check pedigrees when desired and reduces time and errors that might result from listing long pedigrees each time a cross is recorded.

Many plant breeders use a two-digit number denoting the year, followed by a three-digit cross number. This may be preceded by the letter(s) denoting the state or a specific location. For example, NC78056(Red × White) indicates cross number 56 made in North Carolina in 1978 with Red as the female parent and White as the male parent. As generations are advanced in this pedigree system, a selection number (the plant order in the nursery row) is added for each generation. Thus, NC78056-4-9-6 denotes selfed seed from an F_3 plant (number 6) which resulted from selfed progeny of plant number 9 in an F_2 row grown the previous year, which, in turn, resulted from selfed progeny of plant number 4 in an F_1 row. If open-pollinated seed is saved, a zero precedes the selection number. Thus, NC78056-4-9-06 denotes open-pollinated seed harvested from plant number 6. If selfed seeds in a row are bulked, NC78056-4-9-B would so designate and open-pollinated bulked seed would be listed as NC78056-4-9-0B. This system can easily be adapted for greenhouse crosses. For example, 78GH096-5-8 would indicate that the cross (or self) had occurred in the greenhouse in pot number 96.

In some programs involving self-pollinated crops, dash numbers are not added to the basic cross number (e.g., NC78056) until a relatively homozygous line is selected and advanced into a preliminary yield trial. Within a cross, dash numbers are assigned in sequence to those lines entering the performance tests (e.g., NC78056-1). Reselections from a line with one dash number receive a second dash number if one of them is again advanced into a performance test (e.g., NC78056-1-1). This procedure simplifies the line-identification and record-keeping systems.

Breeding records for estimation of genetic parameters, when using mating schemes such as the diallel, Design I, or Design II, can usually be quite simple. For mating designs that utilize randomly chosen parents from some type of random mating population, crosses usually need be identified only by a row number, a number of the plant within the row, and possibly a year or an experiment number. For example, in the Design I mating scheme commonly used in corn, seed harvested from the female plants in row 101 would be identified as 101-3, 101-4, 101-5, and 101-6. A year number (e.g., 77-101-3) might also be included. The two plants, 101-1 and 101-2, would be primary and alternative males. In some modifications of this design the male might be selfed and also harvested. In designs such as the diallel and Design II, both parents of the F_1 crosses will be recorded either by row number or by line designations of the respective parents (e.g., 15 × 38).

REFERENCES

Briggs, F. N. 1938. The use of the backcross in crop improvement. Am. Nat. 72: 285–292.

Brim, C. A., and C. W. Stuber. 1973. Application of genetic male sterility to recurrent selection schemes in soybeans. Crop Sci. 13:528–530.

Comstock, R. E., and H. F. Robinson. 1948. The components of genetic variance in populations of biparental progenies and their use in estimating the average degree of dominance. Biometrics 4:254–266.

———, and ———. 1952. Estimation of average dominance of genes. p. 494–516. *In* Heterosis. Iowa State College Press, Ames.

———, ———, and P. H. Harvey. 1949. A breeding procedure designed to make maximum use of both general and specific combining ability. Agron. J. 41:360–367.

Dhillon, B. S. 1975. The application of partial-diallel crosses in plant breeding—a review. Crop Improv. 2(1–2):1–8.

Eberhart, S. A., M. N. Harrison, and F. Ogada. 1967. A comprehensive breeding system. Der Zuchter 37:169–174.

Fehr, W. R., and L. B. Ortiz. 1975. Recurrent selection for yield in soybeans. Univ. of Puerto Rico J. Agric. 59:222–232.

Gardner, C. O., and S. A. Eberhart. 1966. Analysis and interpretation of the variety cross diallel and related populations. Biometrics 22:439–452.

Griffing, B. 1956. Concept of general and specific combining ability in relation to diallel crossing systems. Aust. J. Biol. Sci. 9:463–493.

Hallauer, A. R., and S. A. Eberhart. 1970. Reciprocal full-sib selection. Crop Sci. 10:315–316.

Hayes, H. K., F. R. Immer, and D. C. Smith. 1955. Methods of plant breeding. McGraw-Hill Book Co., Inc., N.Y.

Hayman, B. I. 1954. The theory and analysis of diallel crosses. Genetics 39:789–809.

Jenkins, M. T., and A. M. Brunson. 1932. Methods of testing inbred lines of maize in crossbred combinations. J. Am. Soc. Agron. 24:523–530.

Jensen, N. F. 1970. A diallel selective mating system for cereal breeding. Crop Sci. 10:629–635.

———. 1978. Composite breeding methods and the DSM system in cereals. Crop Sci. 18:622–626.

Matzinger, D. F., and C. C. Cockerham. 1963. Simultaneous selfing and partial diallel test crossing. I. Estimation of genetic and environmental parameters. Crop Sci. 3:309–314.

Olesen, K. 1976. A completely balanced polycross design. Euphytica 25:485–488.

Rachie, K., and C. O. Gardner. 1975. Increasing efficiency in breeding partially outcrossing grain legumes. p. 285–300. Int. Grain Legumes Workshop Proc., ICRISAT, Hyderbad, India.

Stafford, R. E., and C. R. Lewis. 1975. Natural crossing in guar, *Cyamopsis tetragonoloba* (L.) Taub. Crop Sci. 15:876–877.

Stuber, C. W. 1970. Estimation of genetic variances using inbred relatives. Crop Sci. 10:129–135.

———. 1978. Exotic sources for broadening genetic diversity in corn breeding programs. Proc. 33rd Annu. Corn Sorghum Res. Conf. Am. Seed Trade Assoc. 33:(In press).

Taylor, N. L. 1977. Personal communication. Dep. of Agronomy, Univ. of Kentucky, Lexington, KY 40506.

Tysdal, H. M. 1948. History and development of the polycross technique in alfalfa breeding. Alfalfa Improve. Conf. Rep. 11:36–39.

6 Artificial Hybridization and Self-Pollination

Walter R. Fehr
Iowa State University
Ames

The wide diversity of floral structure and development among crop species has led to an array of methods for artificial hybridization and self-pollination. Despite this diversity, some principles common to all methods must be followed to be consistently successful. The purpose of this chapter is to summarize these general principles. Examples and illustrations are taken from other chapters that describe methods for individual crops. The reader is referred to those chapters for additional detail and references.

I. SELECTION OF PARENTS

Selection of parental material is the first step in a plant breeding program. The choice depends on the objective of the program and the relative merits of available germplasm (Chapter 4). It also may depend on the feasibility of hybridization when genotypes from different species or genera are involved (Chapter 7).

When the choice of parents has been made, hybridization may be facilitated by carefully determining the genotype to be used as either female or male. Male-sterile genotypes automatically would be used as females. The choice is not so obvious in many other situations; nevertheless, proper choice of the female can determine success of the cross. In sweet potato, reciprocal crosses routinely are made between clones, and the relative rates of pollen tube development are measured in the two parents. Compatible sweet potato crosses are those in which pollen grains adhere to the stigma, and pollen tubes penetrate the style. Selection of the female also can be important for interspecific and intergeneric hybridization, for which the breeder must consider chromosome number, seed size, and other characteristics of the parents (Chapter 7).

A common technique in the choice of female parents is the use of genetic markers to identify accidental self-pollinations. Genetic markers preferably are qualitative characters controlled by one or a few genes. The parent chosen as female would have the recessive allele, and the male would have the dominant allele. Seeds or plants from self-pollination would have the recessive character of the female, and hybrids would have the dominant character of the male.

Some genetic markers can be used to differentiate selfed and hybrid seed on the female plant. Waxy endosperm of sorghum is controlled by a single recessive allele and normal endosperm by the dominant allele. In a cross between a female parent with waxy endosperm and a male parent with normal endosperm, waxy seed would result from self-pollination and normal seed from hybridization. Seed markers frequently are used for genetic studies or for determining the amount of outcrossing in a species. They seldom are used as genetic markers in a conventional breeding program because usually only one form of the character is desired and present in breeding lines. Shrunken endosperm, a genetic marker of corn, is an undesirable character in commercial hybrids; therefore, genotypes commonly used by corn breeders have only the dominant allele for normal endosperm.

Genetic markers of seedlings or mature plants are used routinely in some species. To use a seedling or mature plant marker, seeds are harvested from the female parent with the recessive allele. The seeds are sown and the plants examined for the recessive character of selfed plants or the dominant character of hybrid plants. Hypocotyl color of sugarbeet, awnedness of wheat, and pubescence color of soybean are examples of commonly used markers.

Selection of the female parent determines the cytoplasm present in progeny from the cross; therefore, many commercial cultivars may have the same cytoplasm if the same genetic marker is consistently used. Dwarfness of rice recently has become a desirable agronomic character. The dwarf cultivar 'Cina' was used as female in crosses with conventional tall cultivars because dwarfness is recessive. Continued use of dwarfness as a genetic marker has caused many cultivars to have the same cytoplasm from Cina. Lack of cytoplasmic diversity may have undesirable consequences for improvement of the species or susceptibility to plant pests.

II. PREPARATION OF THE FEMALE PARENT

A. Coordination with Time of Pollination

Preparation of the female parent must coincide with availability of pollen from the male. Synchronization of flowering may require use of special techniques at planting or during the growing season, particularly when early and late flowering parents are to be mated (Chapter 1 and 4).

The timing of female preparation during the day depends on the time of pollen availability and the need for immediate pollination of the female after it is prepared. Pollen of lentil is available in the morning, and it is most efficient to emasculate the tiny flower and pollinate immediately while the flower is in hand and readily identified.

Preparation of the female during one period of the day and pollination during another may be required to minimize selfing, to obtain maximum success, or to efficiently spread the work load throughout the day. In northern India and Mexico, hybridization of chickpea is most successful with evening emasculation and morning pollination, even though immediate pollination is possible. Females of corn can be prepared any time of the day with equal success; however, the optimum time of pollen availability will vary with weather conditions. Corn breeders adjust their schedule to pollinate whenever pollen is most available and prepare females the remainder of the day.

A delay of 1 or more days between emasculation and pollination of perfect flowers can reduce the possibility of accidental self-pollination. Flowers of wheat, oat, and barley commonly are emasculated before the anthers are mature and are pollinated 2 or 3 days later when the stigmas become receptive.

Duration of stigma receptivity may influence the delay between emasculation and pollination. The duration is short for species like cowpea, in which receptivity occurs from the day before anthesis until noon of the day of anthesis. In contrast, pollinations of sugarbeet can be made immediately after emasculation or up to 12 days later.

B. Selection of Female Flowers

An important step in artificial hybridization or self-pollination is identification of female flowers at the optimum stage for preparation. Flowers that are too immature may be difficult to manipulate, and the percentage of successful pollinations may be reduced. Flowers that are too fully developed may be accidentally selfed or may be pollinated by an unselected male.

The optimum stage for the female flower commonly is determined by its outward appearance. Knowledge of the relationship between floral appearance and development of the sex organs improves the efficiency of selecting appropriate flowers. The broadbean has a perfect flower; therefore, a flower may serve as either a female or male for hybridization. A broadbean flower used as a female usually is emasculated when the petals are about twice the length of the calyx (Fig. 1). A broadbean flower used as male would be at a more advanced stage of development. In chickpea and flax, flower buds that have a small amount of the petals protruding beyond the calyx are selected for emasculation. Panicles of wild rice to be used as females are prepared during the boot stage.

Differences among genotypes and environments can alter the relationship between outward appearance and development of the sex organs; therefore, precisely determining the suitability of a flower for preparation often requires that the flower be opened. The female flowers of cassava, a monoecious species, are bagged the morning of the day they open (Fig. 2). A drop of nectar at the base of the corolla is the most certain way to tell that the female cassava flower will open shortly (Fig. 3). Florets of wheat appropriate for emasculation have light-green, well-developed anthers and feathery stigmas that extend about one-fourth the length of the florets. If the anthers are yellow or cream-colored, mature pollen may be shed if the anthers are ruptured during emasculation.

Success of hybridization can be influenced by age of the flowers relative to others on the same inflorescence and among other inflorescences on the same plant. The first inflorescences of broadbean generally are considered the best choice, and success is more likely with flowers from the base and middle of the inflorescence than from the top part. A vigorous plant of lentil in the initial stages of flowering is best suited for hybridization. The flowers that first appear in a spike of jute have the highest percentage of successful pollinations.

C. Preparation Without Emasculation

Emasculation is not a universal requirement for artificial hybridization. A few self-pollinated species can be prepared and pollinated immediately without emasculation. Genotypes with male sterility, self-incompatibility, or protogyny, and monoecious and dioecious species do not require

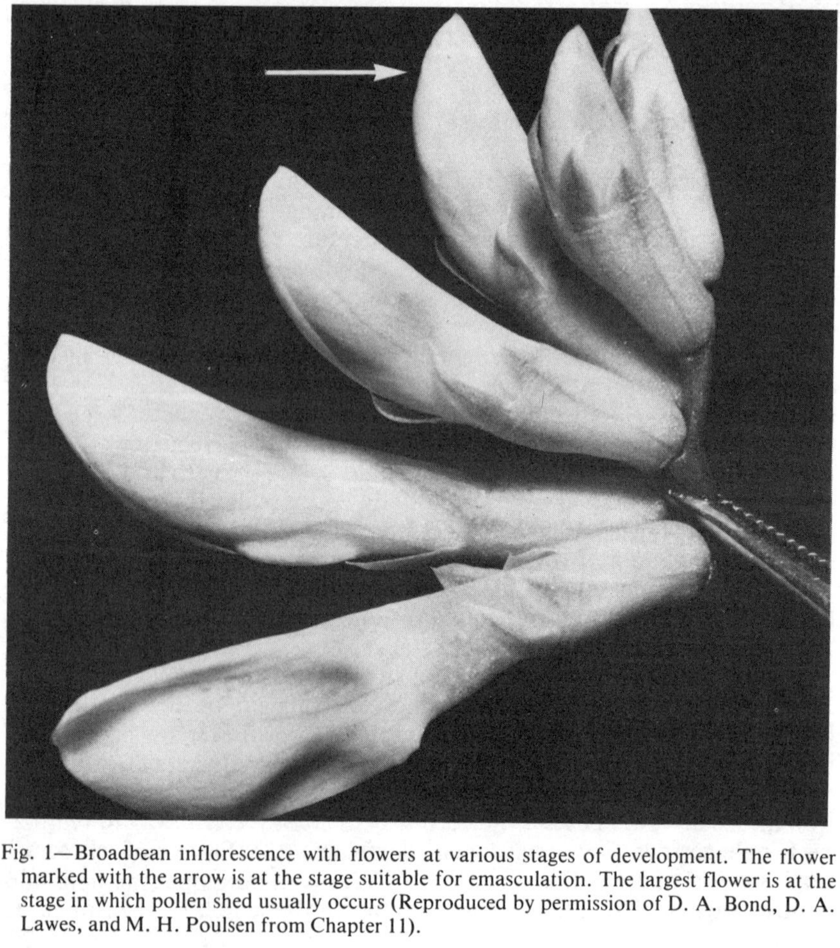

Fig. 1—Broadbean inflorescence with flowers at various stages of development. The flower marked with the arrow is at the stage suitable for emasculation. The largest flower is at the stage in which pollen shed usually occurs (Reproduced by permission of D. A. Bond, D. A. Lawes, and M. H. Poulsen from Chapter 11).

Fig. 2—Branches of cassava with female flowers are covered with a cloth bag in the morning and pollinated later in the day (Reproduced by permission of K. Kawano from Chapter 13).

Fig. 3—A drop of nectar at the base of the corolla of a female cassava flower is the best indication that the flower will open shortly (Reproduced by permission of K. Kawano from Chapter 13).

emasculation.

When self-pollinated species with male-fertile flowers are not emasculated, the stigma must be receptive at least 1 day before anthers of the flower can shed pollen. In soybean, the calyx and corolla are removed, and the stigma is pollinated immediately. Lack of self-pollination in such cases suggests that when cross-pollination is unsuccessful, the flower dries up before its anthers can dehisce.

The most common procedure for preparing a female flower that will be hand pollinated without emasculation is to cover it to avoid accidental pollination. Female flowers of cassava are enclosed in cloth bags (Fig. 2). The female ear of corn is covered with a glassine bag before the silks emerge.

D. Emasculation

Several procedures of emasculation have been developed. Some of the procedures are used in many different species, and others are used in a restricted few.

1. Direct Anther Emasculation

Direct removal of anthers is the most widely used procedure for emasculation. Techniques and instruments vary among species and breeders, but they all have the common objective of manually removing the anthers.

A common first step in direct anther emasculation is removing underdeveloped or overdeveloped flowers from the inflorescence. This reduces competition for nutrients, eliminates accidental self-pollination of emasculated flowers, and prevents accidental harvest of selfed seed. In barley and other grass species, underdeveloped spikelets at the base and tip of the spike and unwanted florets in the remaining spikelets are removed (Fig. 4B). Usually only one flower of a peanut inflorescence is emasculated, and all immature buds and previously pollinated flowers are removed.

The floral parts that cover the sex organs may be entirely or partly removed, or may be left intact. The petals of guar are completely removed to facilitate removal of the anthers. The lemma and palea of barley florets are clipped off just above the anthers (Fig. 4C). There is some evidence that clipping the floret of barley and wheat may slightly reduce seed set and seed weight, but it is much easier and faster than spreading the lemma and palea apart to remove the anthers. Clipping of the florets usually is not practiced if desiccation is a problem. Seed set of jute is poor if the calyx and corolla are removed during emasculation.

The anthers may be extracted during removal of the floral parts covering the sex organs. Sesame breeders routinely pull the corolla from the flower and simultaneously remove the stamens, which are attached to the corolla. The anthers and petals of cotton flowers can be removed with fingernails in one operation.

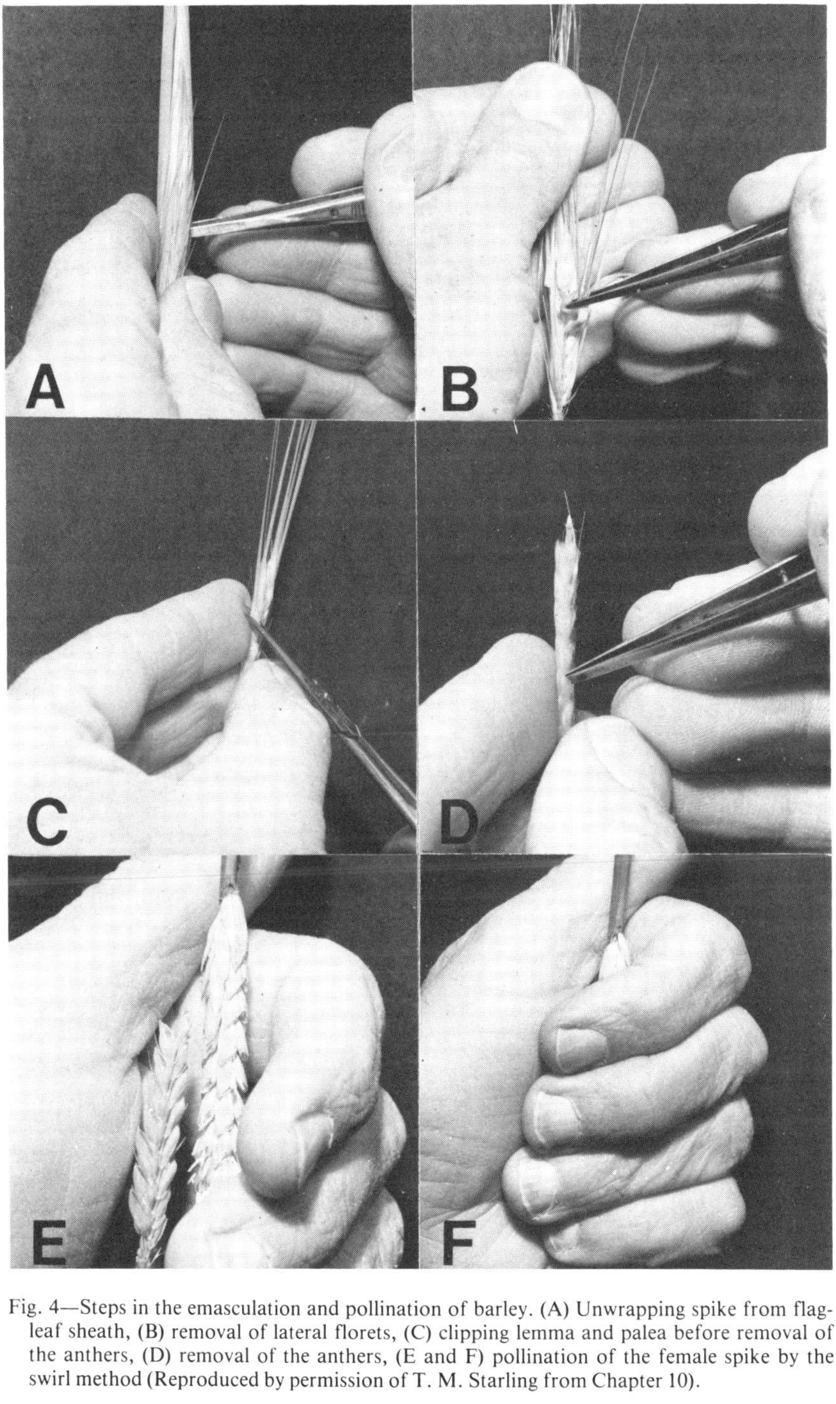

Fig. 4—Steps in the emasculation and pollination of barley. (A) Unwrapping spike from flag-leaf sheath, (B) removal of lateral florets, (C) clipping lemma and palea before removal of the anthers, (D) removal of the anthers, (E and F) pollination of the female spike by the swirl method (Reproduced by permission of T. M. Starling from Chapter 10).

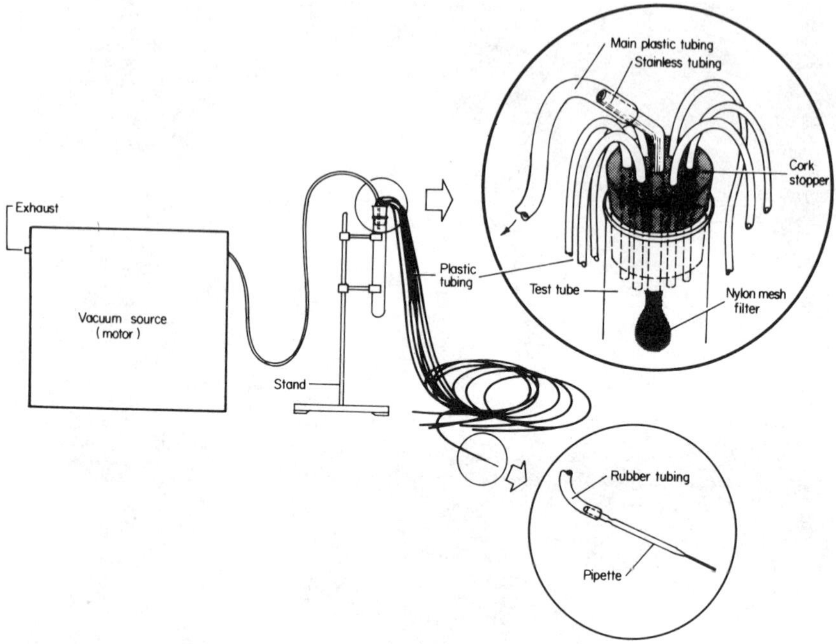

Fig. 5—Schematic drawing of a vacuum system used for suction emasculation of rice at the International Rice Research Institute (Courtesy of W. R. Coffman, International Rice Research Institute, Manila, Philippines).

Anthers commonly are removed with forceps that are appropriate for the size and structure of the flower. Some oat breeders prefer sharp, pointed forceps with curves ends, while others prefer straight ends. Sometimes they file the tip of the forceps to create a flat edge that is less likely to damage the glumes, lemma, and palea. Some barley breeders prefer forceps that can be easily closed to prevent fatigue of the fingers. They may grind the tips to a thin, square shape with a blunt point and use them as a cutting tool for removing the tip of the spike and other unwanted florets. Forceps for emasculation of field pea usually have a sharp point.

A spear-pointed needle, pencil, stick, or scalpel is used instead of a forceps by breeders of a few species. A needle is used to make an incision in the keel of the lentil flower, and the 10 anthers are teased out. The procedure is best suited to conditions of low relative humidity when it is desirable to keep the stigma from drying. A pencil or stick of similar size is preferred by some sorghum breeders to tease the anthers from the spikelet. A scalpel reduced to about half its normal blade width can be used to remove the anthers of potato flowers.

Special tools have been developed to aid emasculation of a few species. A Spinlab device is used for emasculation of cotton. It consists of a circular opening with a cutting surface for removing the petals and a smaller orifice for stripping away the anthers.

Removal of anthers by suction is preferred by some breeders. The basic components of a suction system are an electric aspirator with adjustable vacuum pressure, a filter to prevent anthers from entering the aspirator, and a glass tube inserted in a rubber tube. The system used by the International Rice Research Institute permits several persons to use a single power source (Fig. 5). The glass tube is drawn to a fine point with an aperture appropriate for the size of the anthers. A 1-mm aperture is used for emasculation of alfalfa flowers. The amount of suction is adjusted to permit removal of the anthers without injuring the stigma. A suction of 500 mm Hg is used to emasculate rice flowers.

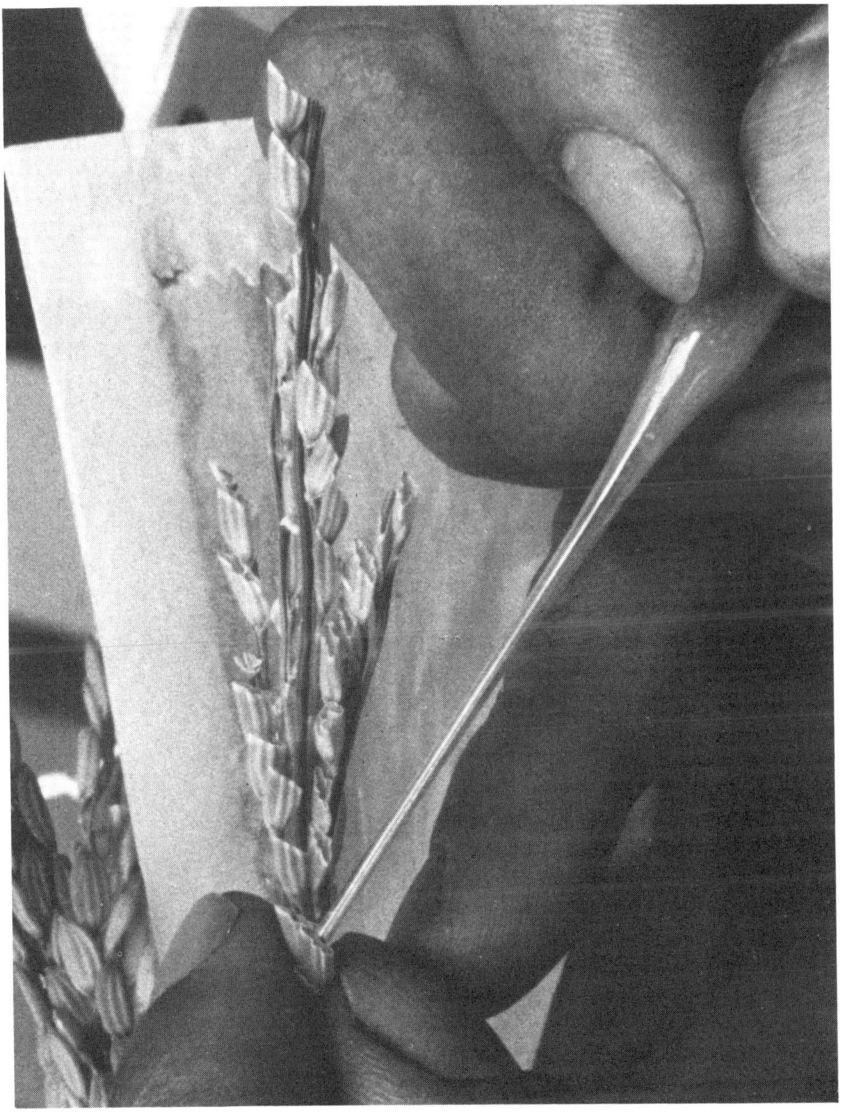

Fig. 6—Pipette placed in the floret of rice to remove the anthers by suction emasculation (Reproduced by permission of W. R. Coffman and R. M. Herrera from Chapter 36).

Flowers for suction emasculation are prepared in the same manner as those in which forceps are used. The glass tube is inserted into the flower and gently moved until all anthers are removed (Fig. 6).

No tools are required to emasculate a few species. Sainfoin, a forage legume, is emasculated by applying pressure to the base of the keel, which causes the staminal column to protrude. The stamens are removed by rubbing them between the thumb and forefinger.

2. *Scissor Emasculation*

Emasculation of rye, barley, and other grass species can be accomplished by cutting the young anthers, but not removing the fragments that remain. A spike is chosen about 1 week before any of its florets will have anthers with mature pollen, a stage several days earlier than for direct anther removal (Fig. 7). The anther tips are removed by cutting off the

Fig. 7—Scissor emasculation of rye, Left to right: Rye spike at the proper stage for emasculation, and emasculation of the spike (Reproduced by permission of D. D. Morey and R. D. Barnett from Chapter 37).

upper one-third of the floret with small, sharp scissors. Care must be taken to assure that all anthers are cut. The cut cannot be made too low because the stigma may be injured excessively. Parts of the anthers below the cut are not removed because they dry up and do not produce viable pollen. The florets are pollinated at the same stage as those prepared by direct anther removal.

Scissor emasculation is preferred by some breeders because it is faster than direct anther removal. The frequency of self-pollination seems to be slightly higher than with direct anther removal, and the use of genetic markers to identify hybrid plants is desirable. The stigmas of florets prepared by scissor emasculation are vulnerable to desiccation and must be properly covered until they are receptive to pollination.

3. *Thermal Inactivation of Pollen*

Viable pollen can be destroyed at temperatures that do not injure the pistil. This principle is used for hot water emasculation of some grass species. Temperatures of 46 to 47 C for 3 min or 48 C for 1 min kill the pollen of smooth bromegrass. Sorghum pollen is killed when exposed to about 47 to 48 C for 10 min. Pollen of rice is inactivated by soaking the panicles for 5 min at 43 C.

An inflorescence is chosen that has just begun to flower and the open florets are removed. The inflorescence is immersed in hot water or is exposed to moist, heated air of precisely the correct temperature. The water usually is contained in a vacuum bottle (Thermos) of adequate size to accept the inflorescence. It is difficult in the field, however, to bend a sorghum plant sufficiently to place the panicle in a bottle. The panicle is enclosed in a sleeve of rubber or plastic tied tightly around the peduncle and open at the top. Water of the proper temperature is poured into the top of the sleeve.

Panicles are allowed to dry before they are enclosed in a bag to await pollination. Florets of rice are pollinated 30 to 60 min after hot water treatment, and any unopened florets are removed.

The principle of thermal inactivation of pollen has been used to achieve temporary male sterility by regulation of greenhouse temperatures. Bluegrass plants become male-sterile, but retain female fertility when maintained at 29 C day and 20 C night temperatures for 11 to 19 days. Low temperatures of 0 to 5 C can be used to achieve temporary male sterility of timothy.

4. *Alcohol Emasculation*

Sterilization of hybridization equipment is routinely accomplished by dipping the instruments into alcohol. Exposing flowers of forage legumes to alcohol for a brief period will inactivate pollen within their anthers.

Alcohol emasculation of alfalfa is accomplished by immersing the whole raceme in a beaker of 57% ethanol for 10 sec and then washing it in water for a few seconds. Exposure of red clover to concentrations of ethanol as low as 66.5% for 10 to 20 sec, followed by a water rinse, prevents self-pollination of self-compatible genotypes.

Species differ in susceptibility of the stigma to injury by exposure to alcohol. A lower percentage of alfalfa flowers set seed when exposed to alcohol than do flowers emasculated by suction. In contrast, concentrations of ethanol up to 88.5% have no effect on the pistil of red clover.

5. Dehiscence Control

High humidity delays anther dehiscence and can be used to facilitate or eliminate emasculation, primarily of grass species. Flowers of some grass species open at the time of pollen shed. To reduce the labor required for direct anther removal from immature buds, the florets are held under high humidity during floral opening so that the anthers do not dehisce. Anthers are removed from open flowers much more easily than from immature ones. In sorghum, open florets are pollinated without anther removal.

A panicle can be kept in high humidity for several days until the desired number of florets have opened and been pollinated. Compared with direct anther removal from immature florets, this method increases the number of florets on a plant that can be pollinated.

High humidity can be achieved by 1) covering the panicles with a paper bag under very humid field conditions, 2) covering panicles with a polyethylene bag, 3) using a humidifier in a greenhouse or growth chamber, or 4) spraying the greenhouse floor with water when natural humidity is high. Temperatures may become excessive under a polyethylene bag; therefore, the bag sometimes is put on the plant in the evening and removed after hand pollination the following morning.

Dehiscence control is considered effective when many seeds are desired and some selfs are not objectionable.

E. Prevention of Contamination

An important consideration during preparation of the female is to prevent undesired pollen from reaching the stigma. One of the most important means of avoiding contamination is to select female flowers at the proper stage of development. Use of flowers that are excessively mature increases the risk of accidental self-pollination. When emasculation is used, inspection of flowers for complete anther removal and accidental anther dehiscence can minimize self-pollination. A magnifier may be necessary to adequately examine flowers that are extremely small (Fig. 8).

Undesired pollen can be present on plant parts, on hybridization equipment, and on the hands of the breeder. When changing parents, the most common practice for disinfection is to rinse the instruments used for emasculation and pollination in alcohol or water. A similar rinse is needed when anthers accidentally dehisce during emasculation. Some breeders prefer to wash their hands when changing parents, particularly during pollination.

Rinsing the flowers is not a common practice for disinfection; however, it is used in a few situations. Sorghum panicles are rinsed in water before emasculation under greenhouse conditions in which free pollen can remain viable for extended periods. A similar rinse is seldom used in the field, where free pollen loses its viability quickly. The stigma of cotton sometimes is disinfected with 30% alcohol to destroy pollen grains when an anther is accidentally crushed during emasculation.

Female flowers of some species must be covered to prevent undesired pollinations, even if emasculation is not necessary. Types of covering include cloth bags (Fig. 2), soda straws (Fig. 9), paper bags (Fig. 10), dialysis tubing, and tissue paper. Covering emasculated flowers prevents desiccation, as well as excluding pollen.

The need for covering the female flowers is not universal among species and may vary among environments. Flowers of tomato, potato, chickpea, and peanut are not contaminated if left uncovered. Heads of safflower are covered with a paper bag in the field, but not in a greenhouse that is free of insects.

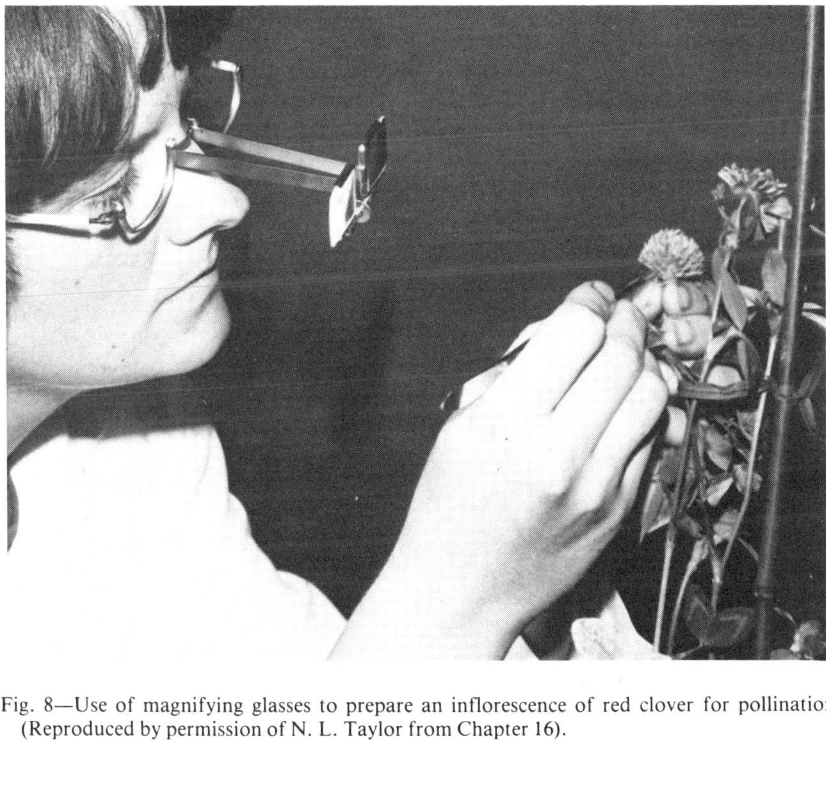

Fig. 8—Use of magnifying glasses to prepare an inflorescence of red clover for pollination (Reproduced by permission of N. L. Taylor from Chapter 16).

F. Identification of the Female Flower

Identification of the female flower may be necessary to locate it for pollination. The identification may be simply the paper bag protecting the flower. Stamping on the bag the date on which the flower was covered aids in determining when the flower is ready for pollination.

A tag may be hung on the plant to record the date of preparation and other necessary information. The same tag may be used at pollination to record additional information (Fig. 11). Tags of different colors may be used to permit quick identification of flowers prepared on different days. Wood stakes or plastic flags may be used to identify plants on which female flowers have been prepared, particularly when the prepared flowers are obvious.

III. POLLINATION

A. Selection of Male Flowers

Pollen collection is most efficient when anther development can be related to outward appearance of the flower or inflorescence. A spike of barley with suitable pollen generally is near emergence from the flag leaf

Fig. 9—Emasculated flowers of Upland cotton. Left, anthers and petals removed showing elongated style and stigma; right, a section of soda straw folded at the end to protect the stigma (Reproduced by permission of J. A. Lee from Chapter 20).

sheath. Pollen of soybean generally is obtained from an open flower with a fresh appearance (Fig. 12).

The relationship between anther development and flower appearance may differ among genotypes within a species, and among locations, days, and time of the day. Availability of pollen can be determined precisely by stimulating the anthers to dehisce. This may involve striking or stroking the inflorescence and watching for pollen, or removing an anther from the flower with a forceps and brushing it on the thumbnail to check for pollen shed.

B. Pollen Collection

Collection of pollen is simplest when it is used immediately for pollination; therefore, preparation of the female usually is timed to correspond with natural pollen shed (Section IIA). Immediate use of pollen assures

Fig. 10—Bag placed over the inflorescence of sesame to protect it from foreign pollen (Reproduced by permission of D. M. Yermanos from Chapter 39).

maximum viability, minimizes identification of the male flower, and minimizes the time required for a pollination. Pollen may be removed from the male flower without removing it from the plant, or flowers may be collected and carried to the female. Tomato pollen can be collected by stripping the anther cone with a flattened or spearpoint needle, and the pollen-laden needle is used to pollinate the stigma of the female parent. Flowers of pigeonpea are removed from the plant, placed on moist filter paper in a covered petri dish, and used for pollination throughout the day.

Pollen carried by wind or insects can contaminate the desired pollen source. To prevent contamination, male flowers may be prevented from opening or covered with a bag. Flowers of sweet potato that are to be used as pollen sources are prevented from opening by placing a paper clip or piece of soda straw over the corolla the day before they open (Fig. 13).

Covering male flowers to prevent contamination usually coincides with pollen collection. Free pollen of grass species loses its viability in a few hours; therefore, male flowers of crops like wild rice, corn, and sorghum

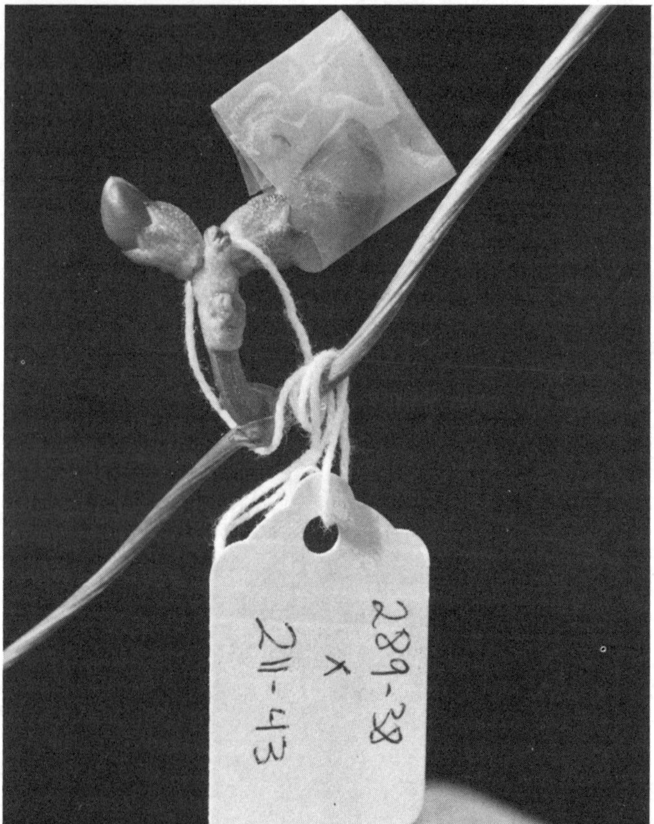

Fig. 11—Wings and standard of cowpea returned to their original position after pollination and secured with tape to prevent desiccation. The tag can be used to identify emasculated female flowers and to record parentage of the cross and other information after pollination (Reproduced by permission of H. T. Blackhurst and J. C. Miller, Jr. from Chapter 21).

are bagged 12 to 24 hours before they are used for pollination. During the period, foreign pollen grains on the flowers lose their viability, additional foreign pollen is excluded, and pollen is collected for use the next day. Before the bag is removed, the male inflorescence is bent horizontally, and the bag is struck sharply to cause additional anther dehiscence.

Pollen of some species is collected from detached branches or flowers. Lateral branches of hop, a dioecious species, are cut from male plants the day before anthers will shed pollen. The branches are taken to the laboratory and placed in water in a glass beaker or test tube that is taped to the inside of a Plexiglas cylinder (Fig. 14). The anthers shed their pollen onto wax paper below the cylinder, from which it is collected and used for pollination. Flowers of tobacco are collected about 24 hours before they are to be used for pollination, placed in a 50-cc beaker containing 10 cc of water, and stored in an insect-free room at room temperature and ambient relative humidity. Although the flowers generally are used within 1 day, they are useful for up to 10 days, and seed has been obtained with 3-week-old flowers if stigmas were heavily covered with pollen.

Special tools have been developed for collection of bulk pollen from some species. A mechanical vibrator is used to collect pollen of tomato and potato (Fig. 15). Vibration of the flower causes the anthers to shed their pollen into a receptacle. A person can use the vibrator to collect enough pollen in 30 min to use for about 1 day of pollination. More elaborate vibrators are available for collection of tomato pollen for hybrid seed production. A small vacuum pump and a spore collector are used to collect pollen of some grass species.

Fig. 12—Soybean flower at the stage when pollen shed usually occurs (Reproduced from Chapter 42).

C. Pollen Storage

Maintenance of viable pollen may be necessary when male and female flowers are not available at the same time. Procedures for pollen storage are well understood for some species, but not for others.

Pollen of some species remains viable for extended periods without special treatment. Pollen of castor will remain viable several days after the flower is picked, and flowers have been sent through the mail when delivery could be made in 3 days or less. Tomato pollen retains good viability under dry room conditions for several weeks.

Low temperature is a common requirement for storage of pollen. Pollen of pearl millet collected in glassine bags and sealed in glass jars at 4 C will retain its viability for over a week, but its viability approaches zero after

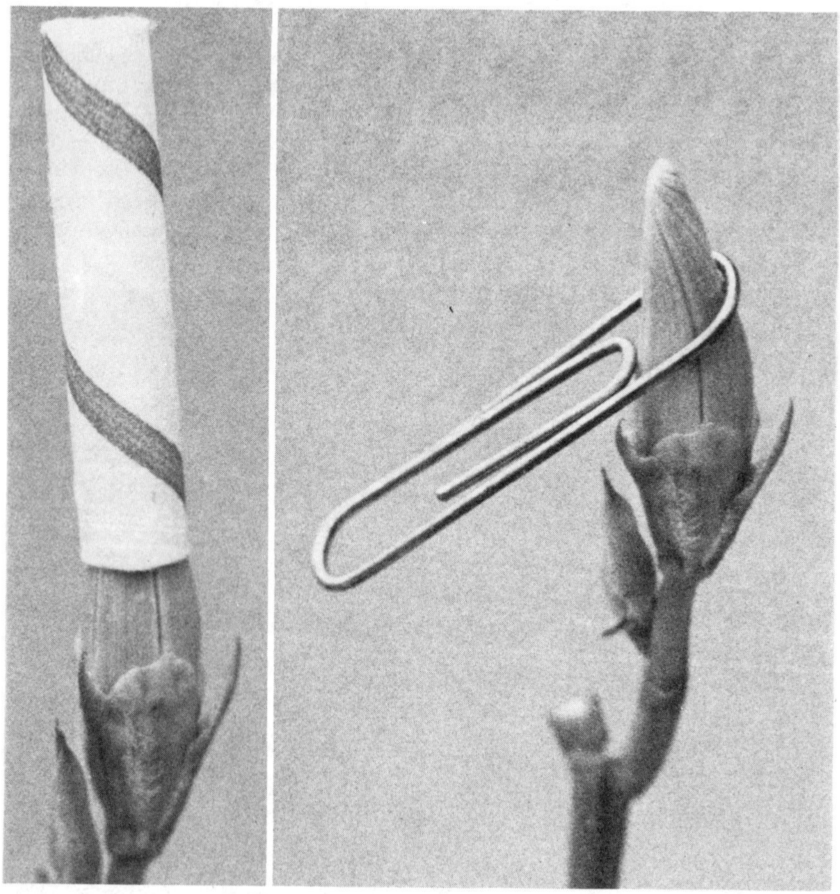

Fig. 13—Male flower of sweet potato closed with a soda straw or paper clip to prevent insects from contaminating the flower before it is used for pollination (Reproduced by permission of A. Jones from Chapter 46).

only 3 days at 27 C. Pollen of potato can be stored for a month at 2.5 C or a year at −24 C.

Low relative humidity is considered important for pollen storage of some species. Potato pollen stored in a desiccator has longer viability than pollen stored at higher humidity. Pollen of sugarbeet stored in closed vials in a desiccator for 5 years at 0 to −10 C has been used to achieve normal seed set.

High relative humidity is preferred for pollen storage of some species of Gramineae. Storage of rye pollen requires relative humidity of 80 to 100%.

D. Pollen Application

1. Direct Pollen Application

The most common method of hand pollination is the direct application of pollen to the stigma using one of several different techniques and instruments. The anthers of tobacco are brushed onto the stigma without any

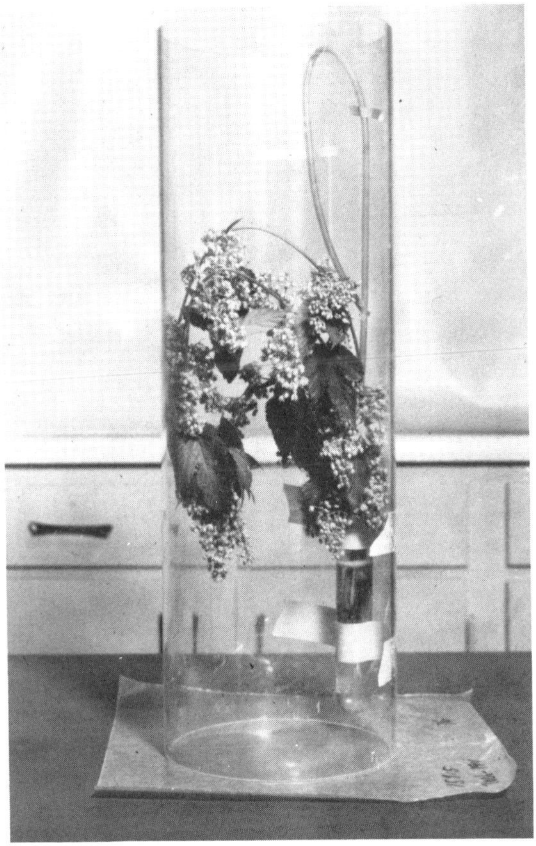

Fig. 14—A detached branch of hop containing male flowers from which pollen will be collected in the laboratory (Reproduced by permission of A. Haunold from Chapter 27).

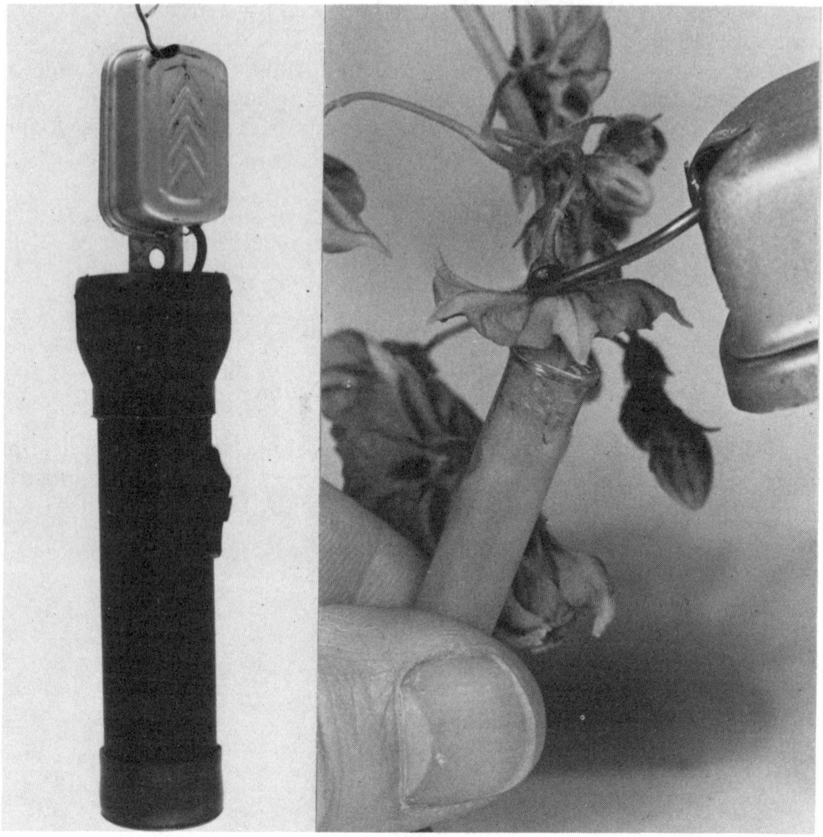

Fig. 15—Collection of bulk pollen of potato with a battery-powered vibrator (Reproduced by permission of R. L. Plaisted from Chapter 34).

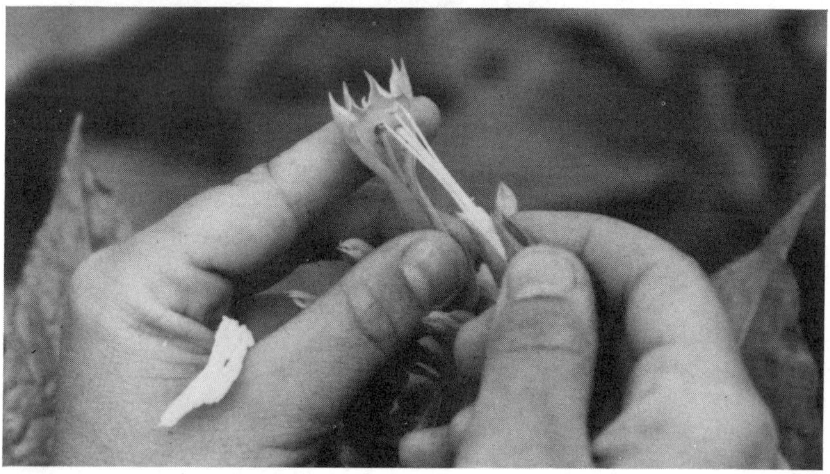

Fig. 16—Application of pollen to the stigma of a tobacco flower (Reproduced by permission of E. A. Wernsman and D. F. Matzinger from Chapter 47).

special instruments (Fig. 16). A pollen-laden stigma of common bean is removed from the male flower with a forceps and brushed onto the stigma of the female (Fig. 17). The anther column of soybean is removed by forceps from the male flower and brushed against the stigma. Pollen of sunflower is applied with a piece of cotton or an artist's bush, or the heads of the two parents may be rubbed together. The alfalfa flower is tripped and its stigma strikes pollen on the tip of a piece of cardboard (Fig. 18). Pollen collected in a bag from the tassel of corn is poured on the silk.

Direct pollen application is the only practical way to pollinate some species. For species in which indirect pollination is possible, the direct method may be used only when pollen is limited.

2. *Indirect Pollen Application*

It is possible in some species to pollinate without placing pollen by hand onto the stigma of each flower. Indirect pollen application involves dissemination of enough pollen close to the female flower that some will

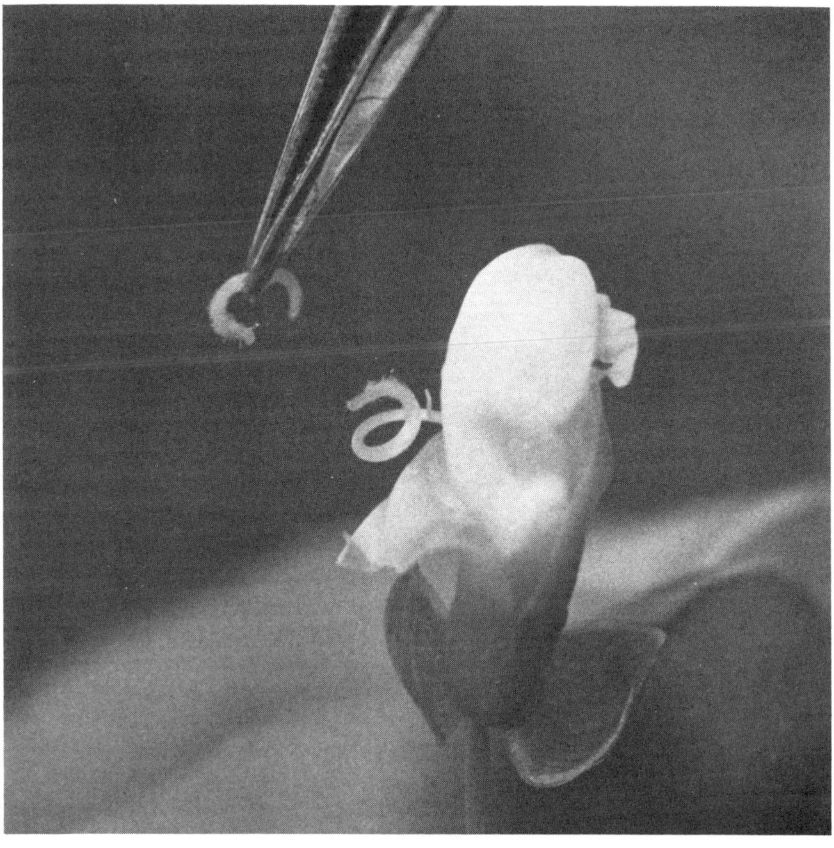

Fig. 17—Pollination of common bean with a pollen-laden stigma from a male flower (Reproduced by permission of F. A. Bliss from Chapter 17).

land on the stigma. Direct pollination of hop, a dioecious species, requires that the bag covering female flowers be removed and pollen brushed against the stigmas. A more convenient procedure of indirect pollination involves inserting a hypodermic syringe into the bag and injecting pollen. The hole is covered with tape or glue, and the bag is agitated to disperse the pollen.

The swirl method of pollination, also referred to as the go-go or twirl method, is used for wheat, triticale, and similar species when the pollen source is plentiful. For this technique, a male inflorescence is chosen that is shedding pollen, and the upper one-third of each floret is removed to facilitate pollen dispersal. The top of the bag protecting the emasculated flowers is cut off, the male inflorescence is inserted and rotated vigorously to stimulate anther dehiscence, then the bag is resealed (Fig. 19). Some breeders prefer to remove the bag protecting the female, use one hand to enclose the female and male inflorescence, and roll the male around the female (Figs. 4E and F). Other breeders prefer to remove the bag protecting the female, insert the female in a paper cone, and shake the male inside it.

The approach method is a form of indirect pollen application used in grass species. Female inflorescences are prepared in the same manner as for direct pollen application. One or more male inflorescences are chosen that have just begun to shed pollen, and a portion of each floret may be removed to expose the anthers. One or more females are positioned slightly below a male inflorescence, and both are enclosed in a bag (Fig. 20). Pollen from the male drops onto the female below to effect pollination. Some breeders strike the bag sharply each day to stimulate pollen shed.

Fig. 18—Tripping an alfalfa flower with a folded piece of cardboard (Reproduced by permission of D. K. Barnes from Chapter 9).

Detached male inflorescences can be used for the approach method (Fig. 20). The detached male inflorescence is placed above the female flowers, and the cut stem is placed in a container of water that has been tied to a pole or taped to the stem of the female plant. Detached inflorescences usually will supply pollen for several days if provided with adequate water.

The approach method has the advantage of providing pollen over an extended period of time and usually results in a higher percentage of seed set than direct pollen application. The method is considered more rapid than direct pollination, but slower than the swirl method.

Mutual pollination is a method of indirect application used to hybridize genotypes with a high degree of self-incompatability, such as forage grasses. Inflorescences of similar maturity from each genotype are placed together at the same height in a bag or isolated enclosure. The bag or stalks are agitated several times a day to stimulate pollen dispersal. Seed from the mutual pollination may be bulked when only two genotypes are used, or kept separate when several male and female parents are in the same enclosure.

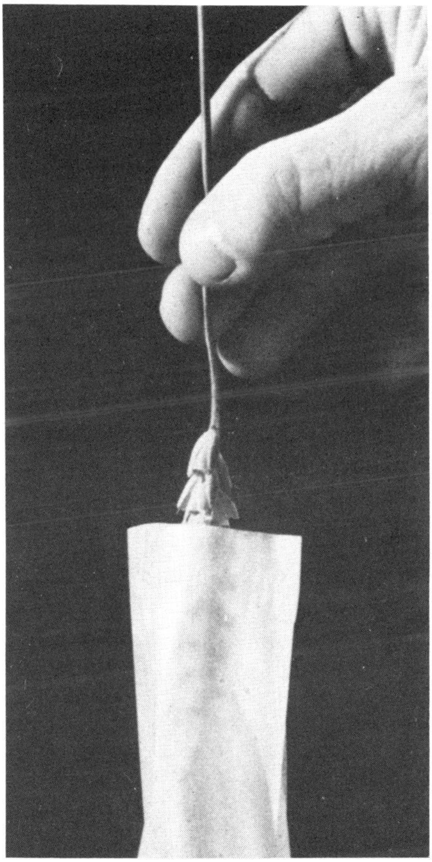

Fig. 19—Pollination of an emasculated spike of triticale by the swirl method (Reproduced by permission of E. N. Larter and J. P. Gustafson from Chapter 49).

E. Prevention of Contamination

Cleaning pollination equipment when changing male parents is essential for avoiding contamination with foreign pollen. This includes fingers and hands when they are in contact with the pollen, such as for the swirl method of pollination (Fig. 4E and F). Ethanol commonly is used to clean equipment. Water also is used, and some persons prefer to place forceps or brushes in their mouth to clean them.

Covering the female after emasculation prevents contamination and minimizes desiccation of the pistil. The covering is removed for pollination and usually is replaced immediately (Figs. 9 and 10). The bag used to collect pollen also may be used to cover the female. The plant parts that enclose the pistil may be opened for pollination, then returned to their original position. The flower also may be held closed to minimize desiccation (Fig. 11).

Fig. 20—A spike of wheat from the male parent is positioned above the female parent for the approach method of pollination. The detached male spike is placed in a reservoir of water that is taped to the stem of the female parent (Courtesy of R. E. Allan, USDA-SEA, Washington State Univ., Pullman, WA 99164).

Bagging the female flower after pollination can reduce seed set of some species. Pod set of pigeonpea is reduced under glassine bags.

Foreign pollen carried by wind or insects can contaminate the source of pollen. Such contamination can be avoided by covering the male flower for the time necessary for the foreign pollen to lose its viability (Section IIIB).

F. Identification of a Pollinated Flower

Pollinated flowers usually are identified by a tag or by the bag used to cover the female. Information recorded for identification may include the parentage of the cross, date of pollination, and number and position of pollinated flowers (Fig. 11). When parents are planted in pairs and only one cross is possible, parentage may be recorded at the time of seed harvest, thereby reducing the time spent recording information during pollination. Pollinated flowers may be marked only with string or plastic ties that are color-coded or stamped to identify different male parents. Such a practice is used for large-scale hybridization of tomato. It also is used for pigeonpea flowers, which are too small to be identified with a tag.

Identification of peanut flowers is unique because the seed develops underground. The portion of the flower that penetrates the soil must be marked with foil or wire, so that it can be identified at harvest.

New buds may develop on an inflorescence after hand pollination is complete and set seed by self- or cross-pollination. Such flowers can be confused with hand-pollinated ones, particularly when the identification tag is placed below the inflorescence and not on an individual flower. Removal of part of the calyx, glume, lemma, or other floral covering can differentiate hand pollinated from naturally pollinated flowers. Systematic removal of new flowers may be necessary to eliminate errors at harvest.

IV. SELF-POLLINATION

Natural self-pollination occurs in species with perfect flowers that are self-compatible and do not open until pollen shed has taken place. A small percentage of cross-pollination occurs in almost all species, but the amount often is so low that manual exclusion of foreign pollen is not practiced. Oat, wheat, and lentil are a few of the species that do not require any outside interference to effect self-pollination.

Some species with perfect flowers must be protected to assure complete self-pollination. Sesame often is described as a self-pollinated species, but cross-pollination up to 65% can occur. Inflorescences of sesame are placed in bags to exclude insects (Fig. 10). Bagging also is used to self forage grasses that are wind pollinated. Selfing of cotton and sweet potato involves closing the corolla with a paper clip, wire, straw, or similar device to exclude insects (Fig. 21). Shaking the enclosed flowers aids pollen dispersal and increases seed set.

Self-pollination of some insect-pollinated legumes requires tripping of the flowers. Four ways in which flowers can be tripped are: inserting a toothpick in the throat of the flower, inserting a toothpick that has the tip covered with emery paper to rupture the stigmatic surface, using a folded cardboard that retains pollen into which the stigma falls (Fig. 18), and gently rolling or squeezing the racemes between the fingers. Rolling the racemes is the most efficient method for alfalfa, but care must be taken to clean hands between pollinations of genotypes. Self-pollination of red clover is most effective by inserting the closed inflorescence in a chamber at 40 C, removing it when the flowers are open, and tripping with a needle or toothpick. Self-pollination of insect-pollinated species also can be accomplished by placing plants in cages with the appropriate insect.

Some insect-pollinated species must be self-pollinated by manually transporting pollen from the anther to the stigma of the same flower. Potato flowers can be selfed by removing pollen from an anther with a scalpel and applying it to the stigma. When large quantities of selfed seed are needed, a vibrator can be used to collect pollen for application to stigmas of the clone (Fig. 15). Seed set of sunflower is increased considerably when heads are brushed with a piece of cotton.

Selfing of monoecious species usually involves the same procedures for female preparation and pollination used for hybridization. The only difference is that pollen for self-pollination is obtained from the same plant on which the female flower occurs. For species like wild rice, which have pistillate and staminate flowers on the same inflorescence, selfing is accomplished by bagging the inflorescence before the pistillate flowers open, and assuring that pollen is dispersed over the stigmas during anther dehiscence.

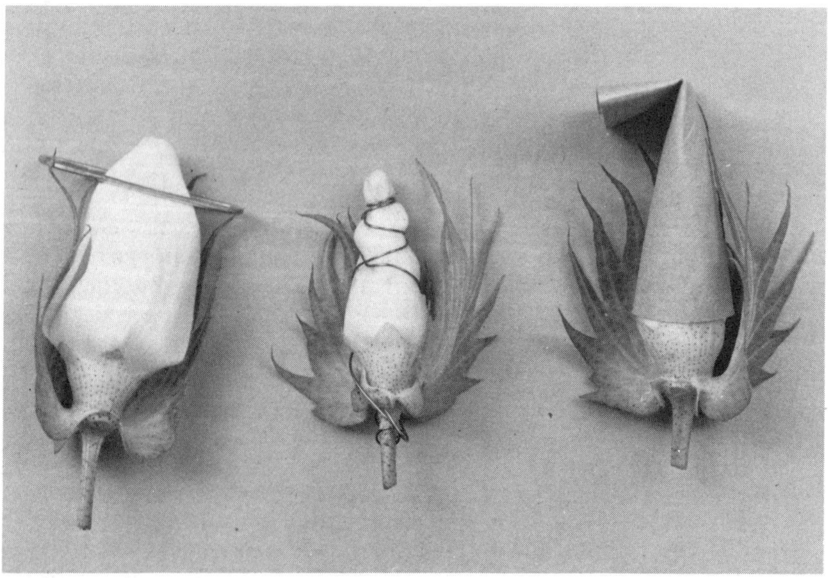

Fig. 21—Three methods of selfing Upland cotton. Left to right: paper clip attached to the end of the corolla; corolla wrapped with copper wire; and cardboard cylinder over the flower (Reproduced by permission of J. A. Lee from Chapter 20).

V. SEED DEVELOPMENT

A. Determination of Successful Pollinations

A visual indication of successful fertilization and seed development is desirable soon after pollination, so that additional pollinations can be made if necessary before flowering ceases. Positive identification of success may be possible in less than a day or may not be possible for a week or more after pollination. Fertilization of sunflower can be determined the morning after pollination by observing withering and receding of the stigmas. Success of oat pollination can be determined 2 to 3 days after pollination by noting a wilted stigma and an enlarged ovary. Success in tobacco can be seen within a week of pollination because unfertilized flowers drop from the inflorescence, while fertilized ones have enlarged seed capsules.

Abortion of developing seed can cause difficulty in determining the success of pollination. Flowers of the common bean that have not been fertilized usually drop 1 to 2 days after pollination. Some pods, however, reach up to 7 cm in length and remain on the plant, but contain only aborted seeds. Broadbean pods frequently abort until they are about 3.5 cm long. Enlargement of the potato ovary can be observed a week after pollination and usually signifies seed development, but the ovary may enlarge without fertilization and later abort.

B. Percentage of Success

Large differences exist among species for the average percentage of successful pollination expected under favorable conditions. Less than 10% of buckwheat flowers set seed after hybridization. Success approaches 100% for pearl millet and sesame. The number of seed obtained from successful pollination of a flower ranges from one in species like sesame and barley to several hundred in tobacco.

Variability in success is affected by environment, health of the plant, the parents involved, and skill of the breeder (Chapter 1). The percentage of success in field pea exceeds 80% when pollinations are made on cool, sunny days with vigorous plants in the early stages of blooming. Only 20% success may be obtained, however, when using older plants under hot, dry, windy conditions. Crosses of field pea involving genotypes with small or fragile flowers also have a low percentage of success.

Skill of the breeder can affect the percentage of successful crosses. Careful preparation of a female flower at the proper stage and adequate pollination with viable pollen can require much practice in some species. The technique used to hold a flower with the fingers can influence the speed, ease, and success of hybridization.

Hormones have been used to a limited extent to improve seed set. Application of 0.005% *p*-chlorophenoxyacetic acid in lanolin on the pedicel or calyx of pollinated tomato flowers may double the percentage of fruit set under unfavorable environmental conditions, when partially sterile parents are used, or for interspecific crosses. A drop of a 100 ppm gibberellic acid solution applied at the base of the anther column of cotton at emasculation promotes fruit set of genotypes that are difficult to hybridize.

7 Interspecific and Intergeneric Hybridization

H. H. Hadley

S. J. Openshaw

University of Illinois
Urbana

Most plant breeders correctly assume that choice of parental materials is a most important step in their breeding programs. They choose parents of a cross only after careful consideration of program objectives and available parental materials. At least one parent likely will be a relatively successful cultivar of the crop (cultigen) to be improved, particularly if much plant breeding effort already has been invested in the cultigen.

A majority of plant breeders prefer that the other parent be from the same biological species, because representatives of the same species usually cross readily to produce fertile hybrids and present little or no hindrance to genetic recombination. Circumstances may indicate, however, that a certain plant breeding problem can be solved only by using a wide cross, one involving representatives of different species or genera. Often such a cross is difficult to make and, even if accomplished, produces a hybrid that is inviable or sterile. Either situation could hinder or even block the transfer of genes from one parental population to the other. The breeder contemplating use of wide crosses, therefore, should make them for reasons that warrant the extra effort they require.

Briggs and Knowles (1967) have listed four reasons for making interspecific hybrids: (i) to transfer one or a few genes from one species to another, (ii) to achieve new character expression not found in either parent, (iii) to produce new alloploid species, and (iv) to determine the relationship of one species to another. The latter two reasons obviously require interspecific hybridization. The first and, perhaps to a lesser degree, the second, correspond to reasons for making intraspecific hybridizations. Breeders making crosses for these reasons, therefore, should thoroughly explore the easier intraspecific path to their goals before committing themselves to the more difficult interspecific route.

Many breeders apparently have concluded that the interspecific path is necessary for achieving their goals, because the literature has become voluminous on the subject of wide crosses in plant breeding programs. A common justification for this choice is that wild relatives and progenitors of our crops can be tapped for "genes carrying special attributes not apparently in the cultivated forms" (Bates and Deyoe, 1973) or that, ". . .an exchange of genes between cultivated and feral or semi-feral species. . .would open a vast potential resource of variability for exploitation by breeders..." (Deakin et al., 1971).

Plant evolutionists recently have urged breeders to collect wild relatives of crop plants and to exploit useful genes they might find in the collections. Zohary et al. (1969) have pointed to the existence of wild diploid progenitors of wheat that constitute large gene pools thus far unexplored and untapped by wheat breeders. The authors deplored the fact that the world wheat collections were extremely deficient in these forms and hoped that wheat breeders would soon collect, screen, and exploit them in their breeding programs.

In a recent discussion of genetic resources in wild relatives of crops, Harlan (1976) suggested that the extent of use of wild relatives in breeding programs depends on: (i) how wild the crop is, (ii) how desperate the situation is, (iii) how great are the pressures to turn out new cultivars, (iv) how available are the materials, (v) how difficult are the materials to use, and (vi) how are the characters of plant breeders or the institutions in which they work. In Harlan's opinion, the use of wild relatives is accelerating rapidly. As more useful results of incorporating genes from these forms into cultivars are reported, more workers will be inclined to use wild genetic resources. "In the future, the need for genetic variability and sources of resistance shall drive us to a much fuller exploitation of *all* the genetic resources that we can assemble" (Harlan, 1976).

The opinions of plant breeders vary concerning the value of interspecific hybridization, but undoubtedly this procedure has taken and will probably maintain a significant position among the plant breeder's tools. The purpose of this chapter is to consider the difficulties associated with the generation of mature fertile or partially fertile interspecific and intergeneric hybrids and the techniques now available to generate them. We shall assume that the main difference between attempts to cross two species from the same genus and two from different genera is degree of difficulty; therefore, the term interspecific will be used to encompass both types of crosses. In presenting our discussion, we shall not attempt to cover exhaustively the extensive literature on the subject. We will cite examples taken mostly from recent publications to illustrate the points made in our discussion. Chapters in the book dealing with specific crops contain additional examples.

I. REPRODUCTIVE ISOLATION BARRIERS

The essential problem in using interspecific hybridization in plant breeding programs is the low probability of obtaining in one individual the desired combination of genes from the parental species. Though one might protest that obtaining a desired genetic combination is the one basic and universal problem of all plant breeding programs, its severity is increased significantly when wide crosses are used. The process of speciation leads to the development of reproductive isolation barriers that maintain the integrity of species by restricting the flow of genes from one to another. The breeder who wishes to temporarily break down those barriers should at least become aware of what they are to improve the chances of success.

Stebbins (1950) proposed a classification of the several types of barriers to genetic interchange between related taxa. His outline will be the basis of our present discussion (Fig. 1).

Sections I and II-A of Fig. 1 correspond to what some persons call prefertilization reproductive barriers, whereas section II-B corresponds to postfertilization barriers. Sections I and II-A1-3 have also been referred to as external barriers; II-A4 and all of II-B as internal barriers. The latter barriers reside in the plant tissues themselves rather than in the outside environment.

A. External Barriers

External barriers to genetic interchange between related populations prevent pollen of plants in one population from falling on stigmas of plants in the other. These barriers are "physical separations in time, space, environment, and specific ecological niches" (Bates and DeYoe, 1973). Combinations of barriers, such as geographical and ecological or ecological and seasonal (flowering time), are more common than individual barriers and tend to reinforce discontinuity between populations.

I. Spatial isolation
II. Physiological isolation
  A. Barriers between the parental species
    1. Ecological isolation
    2. Temporal and seasonal isolation
    3. Mechanical isolation
    4. Prevention of fertilization
  B. Barriers in the hybrids
    1. Hybrid inviability or weakness
    2. Failure of flowering in the hybrids
    3. Hybrid sterility
    4. Inviability and weakness of F_2 and later generation segregates

Fig. 1—Types of reproductive isolation mechanisms (Adapted from Stebbins, 1950).

The barrier of spatial isolation is frequently an easy one to overcome. If it is the sole barrier, individuals from the separated populations can be brought together and crossed to produce fertile hybrids which will allow the free flow of genes from one population to the other. Stebbins (1958) has described this type of behavior and Wright (1976) has summarized numerous examples of such interspecific hybridizations that have succeeded in forest tree improvement programs. When this type of behavior occurs, the two populations should not be considered distinct species according to the precise definition of biological species. The breeder may continue to refer to the populations as species for the sake of convenience, but would be more accurate to label them as subspecies, races, or some other intraspecific category.

Another form of external barrier is ecological isolation. Two populations, even in the same geographical area, may become separated by adaptation to different habitats or ecological niches. Such separation may be incomplete because the habitats may meet in certain areas within the ranges of both populations, thus affording them the opportunity to cross. At these points of contact, F_1 hybrid and later generation hybrid swarms may arise and allow the introgression of genes from one population to the other. The presence of such swarms would indicate to the plant breeder that he could make artificial crosses between plants from the two populations. The absence of such swarms, on the other hand, does not mean he could not obtain hybrids between plants from the different populations. Interpopulation F_1 hybrids and their F_2 and later generation derivatives might not be adapted to either habitat which favors one or the other parent and thus would not persist.

If the related populations we have been discussing are both wild species, their relationships may be of little interest to a plant breeder unless both are closely related to a cultigen. However, if one is a crop species and the other a wild relative, the situation could be of considerable and direct importance. An example of this situation involves sorghum, a cultigen which has many wild relatives. In Africa, a wild race is widely distributed in the tropical forests of the Guinea coast and the Congo, occupying stream banks, alluvial soil, or the edges of paths in the jungle (de Wet et al., 1976). But a wild race is also found around villages and cultivated fields where it hybridizes with cultivated sorghum to produce troublesome weeds. Other wild races in similar ways cross with cultivated races even though adapted to different habitats. Hybrid swarms are not often found, probably because nature is selecting for genotypes adapted to wild habitats and man is selecting, perhaps indirectly and unconsciously, for genotypes adapted to cultivated fields. His selections may become weed companions of the crop and facilitate the transfer of genes from the wild to the cultivated races.

Data on controlled crosses among sorghums suggest that all races (cultivated and wild) hybridize freely in any combination to produce fertile F_1's. Wild races of *Sorghum bicolor* make up an extensive gene pool from which the sorghum breeder should be able to transfer genes to the cultivated races. One can transfer genes with little difficulty because the isolation mechanisms between taxa in this species are generally external and easy to circumvent. The senior author, on the basis of knowledge about crossability between wild and cultivated sorghums, has succeeded without difficulty in transferring two genes for seed shedding from a wild race (*virgatum*) to a cultivated sorghum (broomcorn).

Two populations occupying the same geographical area may be genetically isolated by differences in flowering times. This is another external barrier that may be incomplete, with enough overlapping in flowering periods to allow some crossing. At least the breeder, with an acceptable amount of inconvenience, should be able to modify the flowering periods of plants from one or both of the populations sufficiently to obtain functional flowers from each at the same time.

Stebbins (1958) mentioned an external barrier which he referred to as mechanical isolation. Examples of this barrier are plant species whose flowers have become adapted to insect pollination. Some plant species have evolved jointly with bee species so that one species of plant is pollinated by only one species of bee which, in turn, is attracted only to flowers of that species of plant. Thus, two plant species, each with its own exclusive insect vector, will not hybridize in nature because no insect will visit flowers of both species. Some insects, e.g., honeybees, may even work the flowers of two or more plant species, but never visit flowers of more than one species during the same foraging flight. Interspecific hybridization will not occur in nature, but the plant breeder might make hybrids if no internal isolation mechanism has developed. Possible dependence upon highly discriminating insect vectors potentially interests the breeder who hopes to use such insects to make crosses in a breeding program or in commercial fields for hybrid seed production. Insect vectors with a too highly developed sense of discrimination will probably be useless to the breeder in these situations.

B. Internal Barriers

Internal barriers to genetic interchange between related populations operate through disharmonies between physiological or cytological systems of plants from the different populations. They may: (i) prevent the production of F_1 zygotes, even if pollen from flowers in one population falls on stigmas of flowers in the other, (ii) produce F_1 hybrids that are inviable, weak, or sterile, or (iii) cause hybrid breakdown in F_2 or later generations.

1. Prevention of Fertilization

The prevention of F_1 zygote formation, referred to as cross incompatibility, is caused by disharmony between reproductive tissues of plants from different parental populations. Pollen does not germinate on the stigma; the pollen tube does not completely traverse the style; or the male gamete does not combine with the egg, even though the pollen tube reaches the ovary. These expressions of failure to achieve fertilization are similar to those found within self-incompatible species. One hypothesis to explain cross-incompatibility between species assumes that genes at the S-locus have two functions, preventing (i) self-fertilization within a species and (ii) cross-fertilization between two species. How these two functions can be accomplished apparently is still unexplained, but when one species is self-compatible and the other is self-incompatible, the cross (self-compatible × self-incompatible) often results in fertilization, whereas the reciprocal cross does not. This phenomenon is known as unilateral incompatibility. Other evidence, however, has shown that unilateral incompatibility occurs also in some crosses where both parents are self-incompatible species or both are self-compatible.

Elaborate modifications of the hypothesis based on the S-locus have included other loci to explain the incompatibility reactions in all types of crosses (Abdalla and Hermsen, 1972). Hogenboom (1972), working with interspecific crosses between *Lycopersicon peruvianium* (L.) Mill. and *L. esculentum* Mill., concluded that, in his material, self-incompatibility and interspecific incompatibility were two distinct phenomena based on different genetic systems. He indicated that the inhibition of *L. esculentum* pollen in styles of *L. peruvianum* was governed by a number of independent dominant genes, each governing a distinct process, in the unilateral inhibition of pollen tubes. De Nettancourt et al. (1974) observed the same interspecific cross and concluded that unilateral incompatibility in *L. esculentum* pollen is governed by one gametophyte factor which is either linked or allelic to the S-locus.

Although incompatibility itself appears to be a formidable barrier to interspecific hybridization, Stebbins (1958) observed that failure of pollen growth and fertilization is rarely the primary cause of reproductive isolation between closely related species. Disharmonies within the hybrid zygote and developing plant or in F_2 segregates from fertile F_1 plants are more likely to be the major barriers. Stebbins has discussed these in detail and Allard (1960) has already considered them from a plant breeding standpoint. The reader who wishes to pursue the subject further may consult these references, which are largely the basis for our following brief comments.

2. Hybrid Weakness or Inviability

Some species can be crossed to produce hybrid zygotes, but the F_1's are either inviable or too weak to be of use to the plant breeder. The causes of this hybrid weakness or inviability can be grouped into three categories: (i) disharmonies between genomes of the parental species, (ii) disharmonies between the genome of one species and the cytoplasm of the other, and (iii) disharmonies between the genotype of the F_1 zygote and the genotypes of endosperm or the maternal tissue with which the developing F_1 embryo is associated.

Interactions between the parental genomes probably have a polygenic basis and are difficult to analyze. In a few cases, severe disability or lethality of the F_1 has been determined by one or only a few pairs of genes, but no such case appears to account for an effective reproductive barrier between species (Stebbins, 1958). Other barriers always seem to be involved. A wide sampling of possible parents within two species usually reveals combinations that do not give the lethal or deleterious effect.

Hybrid weakness probably results from the evolution of differences between species in timing of critical processes, such as cell division, organization and differentiation of meristems, and germination of seeds. The combination of genetic messages from the two genomes in the F_1 zygote may be too incoherent to direct a viable pattern of development in the hybrid embryo. Such a situation would constitute a barrier that is difficult or impossible to overcome.

Disharmonies between the genes of one species and the cytoplasm of another are not so difficult to avoid and are much less effective as reproductive barriers. Unless other barriers are present, when the transfer of genes from one species to the other cannot be accomplished through the cross A × B, it may sometimes be accomplished through the cross B × A. A common type of cytoplasmic genome interaction that is deleterious or even lethal involves development of abnormal chloroplasts. This phenomenon has been well illustrated in genetics textbooks by description of Renner's classical work with interspecific hybrids in *Oenothera.* Other deleterious interactions between the cytoplasm of one species and the genome of another have been discussed by Grun (1976).

Successful development of an embryo depends upon the accompanying development of endosperm tissue capable of nourishing the embryo and of a harmonious interaction of related embryo, endosperm, and maternal tissues. To illustrate the effect of different genomes on these tissues, we can assume the genomic contents in species I to be AA for embryo, AAA for endosperm, and AA for maternal tissue, and in species II the embryo is BB, endosperm BBB and maternal tissue BB. A proper balance in dosage effect of genes as they act in these three tissues will have been achieved in both species I and II, although the particular loci involved and individual allelic effects probably differ in the two species. In the interspecific hybrid I × II, genomic contents for the embryo will be AB, endosperm AAB, and maternal tissue AA, whereas in the hybrid II × I they will be AB, ABB, and BB. Possible difference in dosage effects are easy to imagine in these genomic combinations and could account for the observed fact that some interspecific crosses are more successful when made in one direction than if made in the other, while in other cases reciprocal crosses are equally successful or unsuccessful.

In some crosses, serious disharmony does not occur in the embryo (AB), but does in the endosperm (AAB or ABB). Possibly the combination of two genomes from the female parent and one very different genome from the male may produce unfavorable dosage effects. Where the parental species differ in chromosome number, disharmony may result from different ratios of chromosome numbers in endosperm and embryo. To illustrate, if genome A from species I has 10 chromosomes and B from species II has five, the ratios of chromosome numbers in endosperm and embryo tissues would be 30 to 20 (1.5:1) in species I, 15 to 10 (1.5:1) in species II, but 25 to 15 (1.67:1) in the cross I × II and 20 to 15 (1.33:1) in II × I. Less disharmony seems to occur with a ratio of 1.67:1 than with 1.33:1, because, in general, crosses between species differing in chromosome number are more successful when the species with the greater number is used as the female parent. Abnormal endosperm development may be characterized by a low but steady rate of cell division or by an early rapid rate associated with lack of cell wall formation followed by a sudden cessation of mitosis and degeneration of existing endosperm tissue. The former situation may make embryo development difficult, but the latter causes the embryo to starve to death.

Another serious disharmony, termed somatoplastic sterility by Cooper and Brink (1940), sometimes exists in certain combinations of endosperm and maternal tissue. Excessive growth of maternal tissue impairs the capacity for endosperm development and leads to starvation and collapse of the embryo. The term somatoplastic sterility has been applied also to deleterious interactions between endosperm and embryo (Rieger et al., 1968).

3. Hybrid Sterility

Attempts to cross two species may succeed in producing a viable, even vigorous, F_1, but disharmonies between the parental genomes or between the genome of one parent and the cytoplasm of the other may cause the F_1 to be sterile. Such an expression has been called chromosomal sterility (chromosomal hybrid sterility) if caused by structural differences between the parental chromosomes that interfere with their pairing and disjunction at meiosis (Rieger et al., 1968). If the sterility is caused by specific gene complexes, it is known as genic sterility (genic hybrid sterility). Genic sterility usually is due to the genotype of the organism as expressed in the sporophytic phase, but may also include disharmonious genetic combinations in the gametophytic phase of the life cycle.

In wide crosses, sterility of the F_1 often is associated with failure of chromosome pairing (asynapsis) during late prophase and the first metaphase. Asynapsis may indicate lack of homology between chromosomes from the different parents or a lack of synchronization of metabolic processes taking place in early meiosis. One way of distinguishing between the two is to double the chromosome number of the sterile diploid hybrid. In the resulting tetraploid hybrid, each chromosome should have a complete homologue and should associate with it as a pair (bivalent) from late prophase through the first metaphase. If such pairing does not occur in the tetraploid F_1, sterility is probably genic. If pairing is prevalent in the tetraploid F_1, sterility of the diploid F_1 is probably chromosomal.

Whether sterility of an interspecific hybrid is genic or chromosomal may seem of little consequence to the breeder because both lead to reproductive isolation. However, the difference is important to the breeder who wishes to produce a useful alloploid from the hybrid. An alloploid is an individual with two or more different chromosome sets (genomes). An alloploid obtained by interspecific hybridization has genomes from each species in the cross. If sterility is genic, the alloploid likely will be just as sterile as the original F_1; if it is chromosomal, the alloploid will likely be at least partially fertile and of some usefulness to the breeder.

When chromosome pairing is absent in interspecific hybrids, sterility results from abnormal distribution to the gametes of different numbers and combinations of chromosomes from the parental genomes. Only non-reduction gametes or gametes which, by chance, include all or almost all of the chromosomes from one parental genome and none or almost none from the other genome will function. In some interspecific hybrids, however, the chromosomes seem to pair and disjoin normally, yet the hybrid is sterile. Chromosomes of the genomes involved in these cases may still differ structurally by very small inversions, transpositions, etc. Recombination of chromosome segments during meiosis in the F_1 hybrid involving small structural differences can lead to gametes with small, but significant deficiencies and duplications which may render the gametes inviable. Recombination could also produce abnormal F_2 or later generation sporophytes, if the abnormal gametes escape elimination in the gametophytic phase of the life cycle. The former situation would be observed as F_1 hybrid sterility; the latter as F_2 hybrid breakdown.

Intergenomic interaction can result in the elimination of chromosomes from one of the parental genomes in the interspecific F_1 zygote during early mitoses in the developing embryo. The resulting mature plant is highly sterile, but is a haploid rather than a hybrid. Kasha (1974) has used this phenomenon to obtain haploids on a large scale by crossing *Hordeum vulgare* L. × *H. bulbosum* L. In both this cross and its reciprocal, the *bulbosum* chromosomes are eliminated. This behavior indicates that chromosome elimination is not caused by alien cytoplasm. Kasha, however, suggests that some other reported cases of chromosome elimination in interspecific crosses could be caused by incompatibility between a genome and an alien cytoplasm.

Sterility is another expression induced by some combinations of cytoplasm of one species and genomes of another. The cytoplasmic-genetic type of male sterility is a phenomenon that has been recorded for numerous interspecific crosses by plant breeders attempting to exploit hybrid vigor in many crops, including cotton (Meyer, 1969) and wheat (Sage, 1976). Differences in reciprocal interspecific crosses might not always reveal simply inherited male-sterile systems, but nevertheless show a cytoplasmic effect. Palmer and Hadley (1968) found that F_1 hybrids of *Glycine tomentosa* (tomentella Hayata) × *G. tabacina* (Labill.) Beuth. produced pollen with lower stainability and anthers that usually failed to dehisce compared to the reciprocal cross which produced more highly stainable pollen which was shed abundantly. The average number of seed per plant on the former F_1 was 12 while that of the latter was 29.

4. *Hybrid Breakdown*

Some interspecific F_1 hybrids are both vigorous and fertile, yet give rise to F_2 plants that are weak or sterile. Such a situation has been referred to as hybrid breakdown or genetic disability. An illustration of this phenomenon occurs in the F_2 generation from crosses between the cultivated cotton tetraploids, American Upland (*Gossypium hirsutum*) and Sea Island (*G. barbadense*) (Stephens, 1950). The F_1 hybrids between these cottons are easy to make, quite vigorous, and apparently normal in fertility. Some breeders, in fact, are interested in the possible exploitation of the F_1's as commercial hybrids because of the considerable heterosis they exhibit (Marani, 1967). In the F_2 generation, however, most individuals are weak, sterile, or generally unfit agronomically. This result would not interfere with commercial use of F_1 hybrids, but could cause inconvenience in breeding programs with the goal of obtaining inbred lines with combinations of genes from both species. Two explanations have been offered for this type of behavior. One suggests small structural differences between chromosomes of the genomes in the two species (Section IB3). The other assumes complementary genetic systems, such that genotype AABB has been selected for in the one species and aabb in the other. In the F_2 generation, genetic recombination will produce unfavorable genotypes. If many loci are involved in such systems, it is easy to see how few F_2 segregates would be phenotypically successful.

Further discussion of this important barrier to the flow of genes from one species to another would describe the use of interspecific hybrids in breeding programs, a topic that is beyond the scope of this book. Information regarding the severity of hybrid breakdown in F_2 or later hybrid generations from particular crosses, however, might influence the breeder's choice of parental combinations.

II. OVERCOMING REPRODUCTIVE BARRIERS

Man has developed ways of overcoming or circumventing barriers that restrict or prevent genetic interchange between related plant populations. We shall now consider some of the techniques available to the breeder who wishes to cross representatives from populations that have become spatially or genetically separated from one another.

A. Prefertilization Barriers

Plant breeders may increase their chances of obtaining interspecific hybrid zygotes by making judicious decisions and by exploiting special techniques in the following four steps of a crossing program: (i) choice of parental populations, (ii) choice of lines or even individual plants within each parental population, (iii) manipulation or modification of parental plants during or before crossing, and (iv) choice of emasculation and pollination procedures.

1. Choice of Parental Populations

The first opportunity for a breeder to affect the probability of success with wide crosses lies in the choice of parental populations (taxa). The extent of freedom in this choice will vary according to the plant breeding objectives and the phenotypic characteristics of potential parental materials.

Breeders make wide crosses for different reasons ranging from a search for haploids to commercial production of F_1 hybrids. Some reproductive isolation barriers may be tolerable, even favorable, for certain objectives, while others are not. For purposes of discussion, we will assume an objective of transferring a desirable characteristic, like disease resistance, to a cultigen from a wild relative. We will be concerned with the whole range of possible barriers.

The breeder first needs to screen potential parental populations for expression of the desired character. If a choice is possible, individuals should be selected from populations that are most closely related to the cultigen.

Zohary (1973) recently emphasized this point while describing different types of germplasm resources (gene pools) available to the plant breeder. He classified these pools as:

1. cultivated—cultivars of the crop concerned
2. wild—representatives of the wild progenitor stock and companion weeds
3. alien—other species in the genus represented by the crop.

He advised that alien gene pools should be exploited only when potentialities of the cultivated and wild pools apparently have been exhausted.

Harlan and de Wet (1971) proposed three categories of genetic resources that differ somewhat from Zohary's in respect to placement of wild races in the same species that includes the cultigen. Their categories were primary, secondary, and tertiary gene pools. The primary gene pool corresponds to the traditional concept of a biological species, and almost always includes spontaneous (wild and weedy), as well as cultivated races. Crossing is easy among forms in this gene pool, hybrids are fertile, and gene transfer unhindered. The secondary gene pool includes all biological species that will cross with the crop. Gene transfer is possible between the cultigen and members of this pool, but is hindered by weakness, sterility, or inviability of the hybrids. Still the pool is available to the breeder willing to invest some hard work. The tertiary gene pool is rather ill-defined and perhaps most useful in describing the outermost limits of the gene pool of the crop. It includes relatives that will cross with the crop, but which produce anomalous, sterile, or lethal hybrids that require special techniques (embryo culture, bridging crosses, etc.) to obtain or to use (production of alloploids, etc.). The authors did not rank the pools in order of preference, but comparisons of difficulties in gene transfer associated with each emphasizes an order of primary, secondary, and tertiary.

The breeder cannot always determine closeness of relationship between a cultigen and its relatives. Taxonomic classifications based on morphological characteristics may be the best available help, but often do not accurately reflect genetic relationships. Data from crossing experiments are more useful. Biosystematic studies made by cytotaxonomists with an appreciation of plant breeding problems probably serve best. Many data of this type are available for some crops, but meager for others. Breeders working with the latter will have to invest time in attempts to obtain different hybrid combinations. In doing so, they should view taxonomic classifications with a somewhat skeptical eye.

2. Sampling the Parental Populations

After deciding which wild relative or relatives to cross with the cultigen, the breeder should sample adequately both the cultigen and the wild relatives. Choice of specific individual plants or even subgroups within the parental populations can influence: (i) the probability of obtaining a hybrid, (ii) the degree of sterility in hybrids that are obtained, and (iii) the worth of backcross, F_2, or later generations. Considerable genetic variation exists in some cultigens, as well as in their wild relatives. Restriction of crossing attempts to a combination of only a few individuals from the wild species and a few from the cultigen may provide too small a sample for making sound conclusions regarding crossability of the two populations. The results of several recent experiments with wide crosses will illustrate this fact.

Pittarelli and Stavely (1975) crossed autotetraploid *Nicotiana repanda* Willd. ex. Lehm (2n = 4X = 96) with three cultivars of *N. tabacum* L. (2n = 2X = 48). All three crosses produced capsules and seeds, but only the one with cultivar 'SC72' produced F_1 hybrid plants. Hoven (1962) used two accessions from Italy and three from Turkey to represent *Trifolium nigrescens* Viv. (ball clover) as a male parent in crosses with *T. repens* L. (ladino white clover). The accessions from Italy were more cross compatible than those from Turkey. Lambert and Leng (1965) crossed corn inbred HY2 by five strains of teosinte (*Zea mexicana* Schrad, ex. Reeves and Mangelsdorf) and found that the strains differed markedly in their response to backcrosses to HY2. Harlan and de Wet (1977) tested dozens of clones of *Tripsacum dactyloides* (L.) L. in crosses with corn for effectiveness in transferring genetic material, and found that only one clone could be used effectively. Other illustrations of the importance of adequately sampling potential parental populations have been presented in plant breeding texts by Allard (1960) and by Briggs and Knowles (1967).

3. Manipulation and Modification of Parental Plants

Any breeder attempting to make interspecific hybrids should consider making reciprocal crosses, particularly if no knowledge is available for the parental combination. The breeder might prefer crossing attempts in only one direction, like A × B, because A has flowers that are easier to emasculate or because flowers of B produce more pollen. Or perhaps A has a greater chromosome number and the breeder is aware of the general rule that when two species differ in chromosome number, it is better to use the species with the greater number as the female parent. Nevertheless, breeders should try the cross in both directions, even though one is more difficult than the other. For example, Crowder (1953) crossed *Lolium perenne* L. (2n = 14) with *Festuca arundinacea* Schreb. (2n = 42) and found that matings with *L. perenne* as the female parent were much more successful than those with the reciprocal, even though *L. perenne* had the lower chromosome number.

If one obtains seed from both reciprocal crosses, indicating no serious prefertilization barriers, incompatibilities between endosperm and embryo, or somatoplastic sterility, one still should grow and observe the reciprocal F_1 plants for possible post-fertilization barriers caused by cytoplasmic-genomic disharmonies. The cross A × B may be inviable or sterile, but the reciprocal, B × A, may not be. It also may be useful to observe at least a small sample of F_2 individuals from each of the reciprocal crosses. The F_2 segregates derived from cross A × B might be of greater plant breeding value than those from cross B × A, even though the reciprocal F_1 plants seemed to be similar. One example of this type of behavior is male sterility that is based on interactions between the cytoplasm of one species and nuclear genes contributed by the other species. In one cytoplasm the F_2 could show segregation; in the reciprocal, all F_2 plants could be male fertile.

Increasing chromosome numbers (ploidy levels) can affect crossability between two species. Olsson (1963) reported that attempts to cross *Brassica campestris* L. (2n = 2X = 20) × *B. oleracea* L. (2n = 2X = 18) resulted in only 16 F_1 plants from 10,395 pollinations, whereas the same cross on the tetraploid level yielded 133 F_1 plants from 22,884 pollinations. The reciprocal cross (*B. oleracea* × *B. campestris*) failed completely on the diploid level, but on the tetraploid level gave 130 F_1 plants from 18,874 pollinations. Greater success of crosses on the tetraploid level may result from greater gene dosages that break down incompatibility systems.

Chromosome doubling has been particularly successful in crosses between species that differ in number of genomes, as well as total chromosome numbers. Wernesman et al. (1965) obtained no seed of the cross *Lotus tenuis* Wald. et Kit. (2n = 2X = 12) by *L. corniculatus* L. (2n = 4X = 24), however, seed set in reciprocal crosses between an advanced generation tetraploid *L. tenuis* (2n = 4X = 24) and *L. corniculatus* averaged more than eight seed per pod. Crosses between species that differ in number of genomes may fail because of incompatibility between pollen tubes and stylar tissue. Perhaps 2X pollen tubes from 4X sporophytes are too thick to grow down 2X stylar tissue, whereas pollen tubes from 2X sporophytes have no such difficulty in 4X stylar tissue. This could account for cases where reciprocal crosses between untreated parents give different degrees of success.

The ability to double chromosome numbers is an especially useful tool for the breeder working with wide crosses. Numerous techniques that can be used to induce chromosome doubling include temperature shock, wounding, exploitation of certain mutant genes, such as *el* for elongate chromosomes in corn (Alexander, 1957), and the application of various chemicals, such as nitrous oxide and colchicine. The most commonly used technique involves the application of colchicine. This alkaloid can be applied in many ways to produce its chromosome doubling effect.

Colchicine is used most often in an aqueous solution. Glycerine is sometimes added to the solution or the solution is mixed with lanolin to form an emulsion (paste) when a more prolonged treatment is desired and immersion of plant parts in an aqueous solution is not practicable. Strength of solution, duration of treatment, or both usually must be great enough to kill some of the treated tissue in order for treatment to be effective. Treatment must be applied to meristematic tissue. Tissue that develops after treatment will usually be mixaploid, i.e., containing some cells with the original chromosome number and others with an increased number. Selection will occur naturally within the plant, and cells with the original chromosome number usually will have an advantage. The breeder counteracts this advantage by selecting for tissue that appears to be polyploid.

Many variations have been reported in strength of solution, manner of application, kinds of accompanying chemicals or agents, and environmental regimes under which treated plants or seeds are grown. Those interested in pursuing the subject of colchicine and its uses should consult Eigsti and Dustin (1955) for many phases of colchicine research done prior to 1954 and Elliott (1958) who describes the use of this agent and polyploidy in plant breeding. Other leads into literature on the subject can be found in references from this chapter and others on specific crops. We shall present a few examples of successful treatment procedures.

Yermanos and Gill (1967) tested six concentrations of colchicine (0.0125 to 0.4% in water) for induction of polyploidy in *Linum* species. Length of treatment was 3, 6, 12, and 24 hours. Two days after germination, seedlings were immersed in the solution from the crown upward. The roots were not dipped into the solution, but were kept moist with water. The most effective treatment usually was 0.1% colchicine for 6 hours. Treatment of 4-week-old plants was about half as effective as treatment of seedlings. Stems of these plants were trimmed at the top and capped for 24 hours with 3 × 35 mm vials filled with the colchicine solution. Effectiveness of treatment varied with different species.

Sebastiampillai and Jones (1976) compared four colchicine treatments applied to species of *Fragaria.* The most successful, which they designated as the dropper method, consisted of the application of a 0.01 cm^3 drop of colchicine solution (0.5 to 3.0% in water) between the cotyledons to cover the shoot apex. The seedlings were grown in 3.7 cm diam pots, and had emerged from the soil long enough for the apex and cotyledons to turn green. The treated seedlings were placed in a desiccator containing water for 24 hours. The relative humidity was 98 ± 2% and the cotyledons remained moist throughout the treatment period.

A method of possible general use for grasses has been described by Morgan (1976). Young tillers are isolated from a plant to be treated. The leaves of each tiller are trimmed and the roots are removed completely. A sharp scalpel is used to make a triangular incision above the stem apex and the section is removed, leaving a cavity through the leaf bases enclosing the apex. Groups of tillers are placed on a Perspex tray in which holes have been drilled to accommodate 30 bunches of tillers. The tray is placed in a dish containing 150 cm^3 of freshly prepared 0.2% colchicine + 2% dimethyl-sulfoxide for 7 to 8 hours. Following treatment, the tray is transferred to a washing tank and the tillers thoroughly washed in running tap water for at least 4 hours. The tillers are transplanted in shallow boxes containing potting compost. The treated tillers do not survive, but new ones emerge from their bases. When the treated tillers each have produced five new tillers, the latter are subdivided into individual tillers for chromosome counting. Each tiller, with roots removed, is placed for 2 to 4 days in a culture tank containing aerated culture solution until new roots are formed. These roots are used for chromosome counts and tillers with polyploid sectors are isolated and transplanted into pots containing potting compost.

4. Bridging Crosses

The breeder who cannot cross two species, in spite of skillful use of available techniques for overcoming reproductive isolation barriers, may try to circumvent the situation by an indirect approach. Such an approach may consist of a series of crosses involving the cultigen and two or three related species. Although the cultigen and a particular cross-incompatible species are never directly crossed, chromosomes of their genomes are brought together in the same organism.

An experiment by Burk (1967) with tobacco will illustrate. Commercial tobacco, *Nicotiana tabacum* (TT), is susceptible to tobacco mosaic virus, but a wild relative, *N. repanda* (RR), is resistant. These species apparently will not cross, but a third species, *N. sylvestris* Speg. and Comes. (SS), will cross with both. Burk crossed *N. repanda* with *N. sylvestris* to produce a highly sterile F_1 (RS) which he treated with colchicine. The resulting alloploid (RRSS) was highly stable and crossed readily with *N. sylvestris* to produce a sesquidiploid RSS. Subsequent backcrosses to *N. sylvestris* resulted in the loss of R chromosomes, except for the one carrying the gene for resistance to tobacco mosaic virus. It was maintained by selecting virus-free segregates as parents of succeeding backcrosses. After two backcrosses, some plants were enough like *N. sylvestris* to cross with tobacco. Resistant F_1 hybrids from this cross were genomically ST, plus one or more R chromosomes. Although these hybrids were highly sterile, they did set a few seed when pollinated by tobacco. Subsequent backcrosses to tobacco resulted in elimination of S chromosomes, but selection of resistant plants maintained the presence of the desired R chromosome. Burk's original plan was to first produce an alloploid containing some RR pairs of chromosomes before backcrossing to tobacco. Occasional production of nonreduced eggs by the F_1 with ST + R chromosomes, however, accomplished the same purpose as that of an alloploid.

Although the F_1 alloploid represents success in so far as the scope of this chapter is concerned, the breeder's problem is not yet solved at this stage. Complete elimination of S chromosomes and incorporation of a small R chromosomal segment with the gene for tobacco mosaic virus resistance into the T genome by naturally occurring or induced translocation are still required. Use of resistant genes in a T genome or one closely enough related to T to allow pairing of chromosomes from the two genomes would greatly facilitate transfer of the gene for resistance. Consideration of this factor should be made in the choice of the donor parent, if a choice is possible. The gene for resistance, however, may be found only in a genome not closely related enough to T to permit pairing. Radiation of the chromosome addition line (T genomes plus the R chromosome with the gene for resistance) could induce translocation of a short segment of the chromosome with the gene for resistance to a chromosome of the T genome. The classical example of this procedure was reported by Sears (1956) who transferred a short segment of an *Aegilops umbellulata* Zhuk. chromosome with a gene for rust resistance into a chromosome of hexaploid wheat.

An example in which similar genomes have been brought together involves the transfer of genes from the A genome of Asiatic commercial diploid cottons *Gossypium herbacum* L. or *G. arboreum* L. (both 2A) to the A subgenome of American Upland cotton 2 (AD). The 2A species do not cross readily with American Upland, but can be crossed with American wild relatives (2D) to produce a highly sterile AD F_1. Colchicine treatment of this combination yields alloploid synthetic Upland cotton 2(AD) that can be crossed with normal Upland cotton to produce what has become known as a trispecies hybrid. In this hybrid, gene transfer from genome A to subgenome A of the (AD) *hirsutum* genome is relatively easy. The American wild relative (2D) could be considered a bridging species, although the situation differs somewhat from that described by Burk (1967).

Numerous other combinations of crosses and induced ploidy may aid in obtaining forms that include genomes the breeder wishes to bring together. For example, another form of trispecies hybrid might be described by the formula AC (AD). In cotton, Meyer (1974) used a trispecies hybrid, [*G. anomalum* Wawra and Peyr. (2−) × *G. thurberi* Tod. (2D)] × *G. hirsutum* 2 (AD), to obtain Upland types with cytoplasmic male sterility. Sometimes crosses between a diploid species and a tetraploid produce triploid hybrids that are fertile enough to cross back with the tetraploid. Such a triploid F_1 also has been referred to as a bridge. In some cases, triploid F_1 plants are highly sterile, but can be treated with colchicine to produce hexaploids that are fertile enough to cross back to the tetraploid or even to diploid forms.

The use of bridging species in a series of crosses may not be easy. It may require many crossing attempts and the screening of large segregating populations to maintain the desired genes under transfer. Our last example of such a procedure illustrates how complicated and difficult it can be. Hermsen and Ramanna (1973) attempted to transfer disease resistance from *Solanum bulbocastanum* Dun., a diploid wild species, to the cultivated potato, *S. tuberosum* L., a tetraploid. These species would not cross on either the diploid or the tetraploid level. They managed to get the genomes of the two in the same organism by using two additional wild species, *S. acaule* Bitt. (a tetraploid) and *S. phureja* Juz. and Buk. (a diploid). They crossed *S. bulbocastanum* and *S. acaule* to produce a triploid F_1. The chromosome number of the sterile 3X F_1 was doubled to give a fertile hexaploid that was crossed with *S. phureja*. The resulting triple-cross hybrid (tetraploid) was crossed with potato cultivars to give quadruple hybrids, most of which could be backcrossed to the cultivars. As involved as this scheme may seem on paper, it was even more complex to carry out in practice. For instance, 20,000 pollinations were attempted to cross the triple-cross hybrid with potato cultivars, but only 40 quadruple hybrids {[(*S. acaule* × *S. bulbocastanum*) × 2] × *S. phureja*} × *S. tuberosum* were obtained.

5. Choice of Emasculation and Pollination Procedures

A necessity in making any cross is to have receptive stigmas of one parent at the same time pollen is available from the other. This procedure may be more inconvenient for interspecific than for intraspecific hybridization programs. The crop and its relative may have been separated by adaptation to different habitats and may require different environmental regimes to flower or even to grow with vegetative vigor. Techniques used to match flowering dates are discussed in Chapter 1.

Internal prefertilization barriers might be overcome in some cases by attempting a great number of crosses and in other cases by modifying procedures of preparing flowers of the female parent, by improving pollinating techniques, or both. Harlan et al. (1970) took the former approach in making a biosystematic study of the genus *Cynodon* L. C. Rich. They made "tens of thousands" of emasculations and pollinations over a 3-year span to obtain some interspecific or intervarietal combinations. Out of 56 combinations attempted, 18 failed, but 20 yielded from one to five hybrids per 100 attempts. The remaining 18 were somewhat easier to make. Any method of bulk emasculation or pollination that eliminates working with individual florets or flowers would greatly reduce the tedious labor, such as these workers experienced. A hot water emasculation technique, for example, was used successfully by Hadley (1958) to emasculate grain sorghum (*Sorghum bicolor*) heads which were then bagged together with detached heads of johnsongrass to produce several interspecific hybrids.

Methods for eliminating emasculation and hand pollination increase the number of crosses that can be attempted. If marker genes are available in the parental populations, for example, the breeder can cross-pollinate without emasculation, harvest seed from the female seed parent in bulk, and screen the progeny to separate hybrids from offspring resulting from sib crosses or selfs. The wild relative may differ so much from the crop that no marker gene is required because hybrids can be distinguished by their general morphological characteristics.

Examples of such procedures for obtaining interspecific crosses are numerous. Starling (1961) obtained hybrids between *Phalarus arundinacea* L. and *P. tuberosa* var. *stenoptera* (Hack.) Hitch. by bagging together unemasculated panicles of each species. Dewey and Pendse (1968) used a similar technique in obtaining hybrids between *Agropyron desertorum* (Fisch. ex Link) Schult. and *A. cristatum* (L.) Gaertn., except that panicles from the male parent were detached and supported in a carton of water inside the bag. In both examples, seeds from the female parent were planted and the seedlings were screened to identify hybrids. This procedure of producing hybrids is particularly effective if selfing is absent, as in self-incompatible cultigens or wild relatives. In such material, all offspring should be hybrids.

Some form of male sterility is now available in many crops and can be used in much the same way as self-incompatibility for crossing on a large scale. Either genetic or cytoplasmic-genetic male sterility, or even cytogenetically abnormal forms such as aneuploids, may suffice as long as they are female fertile. Both cytoplasmic-genetic male sterility and an unknown type of sterility were used by Bingham (1968) to obtain hybrids between tetraploid *Medicago sativa* L. and diploid *M. falcata* L. Hadley (1958) obtained 15 hybrids between *Sorghum bicolor (vulgare)* and *S. halepense* by bagging together heads of a genetic male sterile (*ms_2 ms_2*) of Texas Blackbull kafir and detached heads of Johnsongrass. He also obtained 13 hybrids by using heads of a cytoplasmic-genetic male sterile of Combine Kafir 60 and heads of johnsongrass in a similar manner.

If plant material is naturally cross-fertilized or is self-fertilized, but modified by male sterility to enforce cross-fertilization, controlled pollination on an individual flower or inflorescence basis may be eliminated. The cultigen and its wild relative may be interplanted in an isolation block where insects or wind can function as pollen vectors. The basic principles for making open-pollinated crosses are discussed in Chapter 5. We shall mention only a few examples of success in making interspecific hybrids in this way. Blowflies were used successfully by Mackay (1973) as pollen vectors in cages to obtain crosses between *Brassica compestris* (turnip) cultivars and *B. napus* L. (rape). Ashri and Rudich (1965) were able to obtain crosses between *Carthamus panaestinus* Eig. and *C. tinctorius* L. (safflower) by growing them in pots near an apiary. Buckner et al. (1976) crossed tall and giant fescue in the greenhouse by using fans to move pollen. Natural wind pollination was used by Porter and Tuleen (1972) to obtain crosses between male sterile wheat and rye, an intergeneric cross with a low percentage of success.

Techniques for preparing female flowers and pollinations are for the most part the same for wide crosses as for intraspecific crosses. These techniques are summarized in Chapter 6 and are illustrated in chapters on specific crops. We shall consider here only a few techniques particularly useful in making wide crosses. They relate primarily to incompatibilities that prevent fertilization.

One technique used to overcome cross-incompatibilities between species is to graft the embryos from germinating seeds of one species onto the endosperms of germinating seeds of the other species. Hall (1954) found that mature wheat plants that developed from wheat embryos grafted to rye endosperms and pollinated with rye pollen produced five times the number of hybrids as ungrafted controls.

Mixing incompatible pollen with compatible pollen has been successful in overcoming certain incompatible reactions. Working with poplars, in which certain interspecific matings are characterized by failure of germinated pollen to penetrate the stigma, Knox et al. (1972a, b) mixed inviable, compatible pollen with viable, incompatible pollen. They killed the compatible pollen by gamma irradiation, by freezing and thawing, or by exposure to methanol. The treated pollen, which they termed recognition pollen, was used to obtain interspecific hybrids. Recognizing that pollen grains release proteins when moistened, they hypothesized that non-enzymatic proteins may be concerned with compatibility reactions. A mixture of incompatible pollen with a protein extract obtained from compatible pollen enabled them to obtain hybrids from normally incompatible matings.

Willing and Pryor (1976) found that in poplar, incompatibility between pollen grain and foreign stigmatic surfaces could be broken by application of organic solvents to the stigma, or treatment of incompatible pollen with organic solvents would render the pollen compatible. The lipoidal mixture removed from the pollen by this procedure was designated tryphine. Pollen coated with tryphine from another poplar species carried the incompatibility response of the coated tryphine. They hypothesized that incompatibility was controlled by a two-factor system, one factor in the stigma and one in the pollen. The absence of either would disrupt the incompatibility process.

The biological interactions occurring between pollen and pistil are discussed in Chapter 2. The failure of pollen from one species to germinate on stigmas of another is common. Martin (1970) attempted 254 different interspecific combinations out of a possible 552 among 24 species from 16 families. Pollen germination occurred in only 11 cases. Some workers have circumvented this reproductive barrier by physically removing the stigma or disrupting its surface. Swaminathan (1955) overcame failures of certain *Solanum* interspecific crosses by removing the stigma and replacing it with a drop of an agar-sucrose gelatin medium known to support pollen tube growth in many *Solanum* species.

If pollen does germinate on a foreign stigma, the resulting pollen tubes may grow so slowly through the style that the egg degenerates or the flower drops before fertilization can take place. This problem may be solved by application of growth regulators that promote pollen tube growth or fruit development. Larter and Chaubey (1965) found that gibberellic acid applied to culms of autotetraploid barley increased pollen tube growth when pollinated by rye. The application of growth regulators could overcome a prefertilization barrier by promoting more rapid growth of pollen tubes or by increasing longevity of the pistil so that pollen tubes have time to grow farther at the same rate. Such regulators also can be used to increase embryo survival after fertilization has occurred.

Surgical procedures have been used to shorten styles of one species, so that pollen tubes from the other could reach the egg. Mangelsdorf and Reeves (1939) severely cut back the silks of corn before dusting them with pollen from the short-styled tripsacum parent. Kanta and Maheshwari (1963) overcame incompatibility by injecting pollen suspensions into surgically opened ovaries in four different species. They also succeeded in culturing excised pistils and pollen on the same medium and obtained crossed seed. Although we are not aware of such techniques being used to make interspecific hybrids, it would seem to be useful for species in which the surgery is feasible.

Protoplast fusion may be a potential method of achieving hybrid combinations that are otherwise unobtainable. This technique has been used to produce interspecific hybrids in *Nicotiana* (Carlson et al., 1972; Smith et al., 1976) and in *Petunia* (Power et al., 1976). In both cases, however, hybrids also can be obtained by sexual fertilization. Prerequisites that must be met before protoplast fusion can become a useful hybridization technique include: (i) culture of parental and hybrid protoplasts, (ii) fusions of nuclei and of cytoplasms, (iii) selection of hybrid cells, and (iv) regeneration of hybrid plants (Bhojwani et al., 1977). Protoplast fusion has interesting possibilities and may become potentially useful to plant breeders. At present, however, we know of no examples where it has made possible a wide cross that could not be obtained by more conventional methods (see Snowcroft, 1977).

B. Postfertilization Barriers

Thus far we have considered primarily methods for overcoming prefertilization barriers to wide crossing. The remaining techniques we shall consider are related chiefly to postfertilization barriers and include (i) application of flower and fruit setting hormones, (ii) use of mixed pollinations, (iii) embryo culture, (iv) grafting, and (v) production of alloploid F_1's.

1. Growth Promoting Substances

One difficulty in obtaining interspecific hybrids in plants that have multilocular ovaries may be that seed set is too low to prevent abscission of flowers and young fruits. Thus the few F_1 hybrid zygotes or young developing embryos that do result from a successful pollination are lost. Some breeders have succeeded in retaining flowers and fruits by applying growth regulators immediately after pollination. In some cases, even though fruit set is achieved by such application, the embryos fail to develop (Deakin et al., 1971). In other cases, fruit set is maintained long enough for embryos to grow sufficiently large to be cultured. Al-Yasiri and Coyne (1964) were able to maintain a high percentage of developing pods for 30 days from interspecific crosses in *Phaseolus* compared to abortion in the control 15 days after pollination. Thirty days was long enough for embryos to become large enough for culturing. Payan and Martin (1975) tested gibberellic acid, alphanaphthalene acetamide, and indole butyric acid for their effectiveness in preventing abscission of flowers and fruit in *Passiflora* (passion fruit). Each growth regulator was mixed with lanolin in 1% and 0.1% concentrations; the resulting pastes were applied directly to ovaries following pollination. An application of 1% gibberellic acid was judged most effective in stimulating fruit set.

2. Mixed Pollinations

Mixed pollinations have been used to accomplish essentially the same objective as the application of growth regulators. Beasley (1940) obtained interspecific crosses in *Gossypium* (cotton) by mixing 6 to 12 pollen grains of the female parent with an excess of pollen from the male parent. Bolls resulting from pollinations with the mixed pollen contained some large seeds of the female parent (selfs), plus some small seeds that were hybrids. Payan and Martin (1975) made interspecific hybrids in *Passiflora,* a genus with three stigmas per flower, by pollinating two of the stigmas with pollen from a different species and one with pollen from a compatible plant in the same species.

3. Embryo Culture

Where a hybrid embryo is formed but fails to develop normally due to the inability of the endosperm to nourish it, the embryo can sometimes be cultured and germinated on an artificial medium. Laibach (1929) used such a technique to obtain hybrids from *Linum austriacum* × *L. perenne* L., a cross which would otherwise have produced shrivelled, inviable seeds.

Although various methods and media have been employed for embryo culture, the general features of the techniques are quite similar. The embryo should be excised at the time it has reached its maximum development on the female plant and before it begins to disintegrate. Dissection and transfer operations should be performed under sterile conditions to avoid contamination. The particulars of the culture medium may vary depending upon the species of concern, but in general, media are liquid or partly solidified with agar and contain mineral salts, a carbohydrate source, vitamins, hormones, and sometimes an organic nitrogen source. Immature embryos may also require addition of coconut milk (itself a liquid endosperm) or malt extract (Rappaport, 1954). Nutritional requirements of embryos and osmotic potential of the medium may depend upon their stage of development (Mauney, 1961; Pecket and Selim, 1965; Wilmar and Hellendoorn, 1968). Younger embryos generally require higher osmotic values to continue development, with lower osmotic values promoting germination.

Culturing of barley embryos will be used to illustrate some of the general principles. Barley inflorescences are collected, wiped with a cloth dampened in 10% Chlorox solution, placed in polyethylene bags, and stored in a refrigerator until use (Norstog, 1965). Embryos should be excised and cultured on the day of collection for best results.

Hands and working area are cleansed with ethanol. The lemma and palea are stripped from a floret and the exposed caryopsis is placed on a flamed slide under a dissection microscope at 25 to 50× magnification. Watchmaker's forceps and bacteriological inoculating needles that have been ground to a fine point are used to excise embryos and transfer them to the medium. For embryos smaller than 0.2 mm, the portion of the ovule containing the embryo can be excised and transferred to a small drop of sterile paraffin oil where the embryo can be teased away. This procedure prevents drying of the smaller embryos. A transferred embryo should be positioned with the scutellum in contact with the medium.

Cultures can be kept in petri dishes or small vials containing a suitable medium. Norstog (1973) reported that 0.2 mm embryos cultured on Barley Medium II produced plantlets in 2 to 3 weeks. Cultures may be maintained at 20 to 30 C in darkness or light, although light tends to inhibit germination. Embryos that fail to germinate after cessation of growth can be induced to do so by transferring them to a new spot on the medium (Norstog, 1965). If necessary, embryos can be transferred to new cultures. After germination, hybrid plants with one or more roots and two to three leaves can be transplanted to a sterile soil consisting of 1/2 sand:1/2 sphagnum to which liquid inorganic fertilizer is applied. Plants should be protected from strong sunlight for the first few days (Kruse, 1974).

Various modifications of the embryo culture technique have been successfully employed. Ziebur and Brink (1951) reported that entire barley endosperms when placed on the agar culture medium promoted the growth of nearby barley, *Raphanus,* and *Capsella* embryos. Kruse (1973, 1974) transplanted excised hybrid embryos onto barley endosperms that had been previously cultured on the agar medium. He used this method to recover hybrids from intergeneric crosses involving *Hordeum vulgare* as a female parent crossed with *Secale, Triticum,* and *Agropyrum.* Harberd (1969) successfully cultured *Brassica* embryos by placing halved ovules in a liquid medium under constant agitation, where many embryos were washed out and subsequently grew. This method eliminated the technical skill normally required in dissecting the embryo away from the endosperm, as well as the need for stringent sterile conditions. McLean (1946) used embryo culture in obtaining interspecific *Datura* hybrids. She found that seedlings that developed only shoots could be grafted onto young *D. stramonium* plants. In addition to the culture of immature embryos, artificial media have been used to induce mature embryos or entire seeds to germinate in instances where the hybrid embryo survived to maturity, but failed to germinate when sown (Lammerts, 1942; North and Wills, 1969; Skirm, 1942; Skovsted, 1935).

4. Grafting

One of the most obvious expressions of hybrid inviability is abnormal chlorophyll development. If this expression results from unfavorable interaction between the cytoplasm of one parent and the genome of the other, use of the reciprocal cross may solve the problem. If the reciprocal cross fails because of other barriers, such as unilateral incompatibility or deleterious interactions between genomes, the breeder might resort to grafting the F_1 seedlings onto normal plants of either parent. Smith (1943) succeeded in maintaining F_1 tissue from crosses between *Melilotus alba* and *M. dentata* in this manner. Another expression of hybrid inviability is the failure of F_1 seedlings to develop a root system adequate for survival. Coe (1954) solved this problem with interspecific hybrids in sugarbeet by grafting the F_1 seedlings onto sugarbeet seedlings. The good root system of the sugarbeet stock allowed the F_1 scions to produce flowers for crossing.

Examples of grafting are rather difficult to find for field crops, perhaps because workers do not usually think such crops are capable of this type of propagation. While some crops such as the monocots are difficult or impossible to graft, other crops can be grafted, at least on the small scale required for providing functional flowers of interspecific hybrids.

5. Producing Alloploids

Numerous interspecific hybrids are highly sterile. If sterility is genic, the breeder may be able to do little except to try backcrossing the F_1 as the female parent to the cultigen which is to be the recipient of the desirable genes. Since it is difficult to determine whether sterility is complete or just very high, one should make as many attempts as practicable before abandoning the project. Sterility may be chromosomal and doubling the chromosome number of the F_1 may be used to produce an alloploid. If the F_1 is a cross between two distantly related diploid species, the alloploid form will be an amphidiploid where each chromosome has one complete homologue. Pairing should be normal and the plant should be at least partially fertile.

If the interspecific F_1 came from a cross between a diploid and a tetraploid species, the F_1 would be a triploid either of the form ABC or AAB depending upon whether the parents had one genome in common or not. In either case, however, the F_1 would be highly sterile. Alloploids can be produced from the sterile F_1's by colchicine treatment. Alloploid sectors in inflorescences of treated plants are often quite easy to detect by their fertility which stands out sharply against the contrasting background of sterility in the rest of the inflorescence. Doubled F_1's from the two types of triploids would differ somewhat in cytological behavior in that one would be 2(ABC), an allohexaploid, while the other would be AAAABB, an autoallohexaploid. Both types are likely to be more fertile than the F_1's from which they came, and both have been synthesized and used in plant breeding programs featuring interspecific crosses.

In summary, one or more of the following procedures may be used by the breeder to obtain interspecific hybrids of potential value in a breeding program:

1. Assemble a collection of cultivars, related cultigens, and wild relatives for use as potential parents.
2. Provide an adequate environmental regime to induce flowering of potential parents.
3. Try many parental combinations of cultivars and wild relatives.
4. Make reciprocal crosses.
5. Change ploidy levels of one or both parents before crossing.
6. Use combinations of three or more parents in bridging crosses.
7. Increase the number of possible crossing attempts by use of efficient emasculation and pollination procedures.
8. Apply growth regulators to the female parent to allow or promote pollen germination and pollen tube growth, and to prevent flower or fruit abscission.
9. Make mixed pollinations.
10. Make grafts to obtain a hybrid zygote, to save an inviable F_1, or both.
11. Perform surgery on the pistil.
12. Use protoplast fusion.
13. Culture F_1 embryos.
14. Backcross partially sterile or male sterile F_1 plants as the female parent to the cultigen.
15. Produce alloploid F_1 plants.

REFERENCES

Abdalla, M. M. F., and J. G. Th. Hermsen. 1972. Unilateral incompatibility: hypotheses, debate and its implications for plant breeding. Euphytica 21:32–47.

Alexander, D. E. 1957. The genetic induction of autotetraploidy: A proposal for its use in corn breeding. Agron. J. 49:40–43.

Allard, R. W. 1960. Principles of plant breeding. John Wiley and Sons, Inc., N.Y.

Al-Yasiri, A., and D. P. Coyne. 1964. Effects of growth regulators in delaying pod abscission and embryo abortion in the interspecific cross *Phaseolis vulgaris* × *P. acutifolius.* Crop Sci. 4:433–435.

Ashri, A., and J. Rudich. 1965. Unequal reciprocal natural hybridization rates between two *Carthamus* L. species. Crop Sci. 5:190–191.

Bates, L. S., and C. W. Deyoe. 1973. Wide hybridization and cereal improvement. Econ. Bot. 27:401–412.

Beasley, J. O. 1940. Hybridization of American 26-chromosome and Asiatic 13-chromosome species of *Gossypium.* J. Agric. Res. 60:175–181.

Bhojwani, S. S., P. K. Evans, and E. C. Cocking. 1977. Protoplast technology in relation to crop plants: progress and problems. Euphytica 26:343–360.

Bingham, E. T. 1968. Transfer of diploid *Medicago* ssp. germplasm to tetraploid *M. sativa* L. in 4x-2x crosses. Crop Sci. 8:760–762.

Briggs, F. N., and P. F. Knowles. 1967. Introduction to plant breeding. Reinhold Publishing Corp., N.Y.

Buckner, R. C., P. B. Burrus II, and L. P. Bush. 1976. Cytological, morphological and agronomic characteristics of tall × giant fescue hybrids and their amphiploid progenies. Crop Sci. 16:811–816.

Burk, L. G. 1967. An interspecific bridge-cross-*Nicotiana repanda* through *N. sylvestris* to *N. tabacum.* J. Hered. 58:215–218.

Carlson, P. S., H. H. Smith, and R. D. Dearing. 1972. Parasexual interspecific plant hybridization. Proc. Natl. Acad. Sci. USA 69:2292–2294.

Coe, G. E. 1954. A grafting technique enabling an unthrifty interspecific hybrid of *Beta* to survive. Am. Soc. Sugarbeet Technol. 8(Pt. 2):157–160.

Cooper, D. C., and R. A. Brink. 1940. Somatoplastic sterility as a cause of seed failure after interspecific hybridization. Genetics 25:593–617.

Crowder, L. V. 1953. Interspecific and intergeneric hybrids of *Festuca* and *Lolium.* J. Hered. 44:195–203.

de Nettancourt, D., M. Devreux, and U. Laneri. 1974. Genetical and ultrastructural aspects of self and cross incompatibility in interspecific hybrids between self-compatible *Lycopersicon esculentum* and self-incompatible *L. peruvianum.* T.A.G. 44:278–288.

deWet, J. M. J., J. R. Harlan, and E. G. Price. 1976. Variability in *Sorghum bicolor.* p. 453–463. *In* J. R. Harlan, J. M. J. deWet, and A. B. L. Stemler (eds.) Origins of African plant domestication. Mouton Publishers, The Hague.

Deakin, J. R., G. W. Bohn, and T. W. Whitaker. 1971. Interspecific hybridization in *Cucumis.* Econ. Bot. 25:195–210.

Dewey, D. R., and P. C. Pendse. 1968. Hybrids between *Agropyron desertorium* and induced-tetraploid *Agropyron cristatum.* Crop Sci. 8:607–611.

Eigsti, O. J., and P. Dustin, Jr. 1955. Colchicine in agriculture, medicine, biology and chemistry. Iowa State College Press, Ames.

Elliott, F. C. 1958. Plant breeding and cytogenetics. McGraw-Hill Book Co., Inc., N.Y.

Grun, P. 1976. Cytoplasmic genetics and evolution. Columbia Univ. Press, N.Y.

Hadley, H. H. 1958. Chromosome numbers, fertility, and rhizome expression of hybrids between grain sorghum and Johnsongrass. Agron. J. 50:278–282.

Hall, O. L. 1954. Hybridization of wheat and rye after embryo transplantation. Hereditas 40:453–458.

Harberd, D. J. 1969. A simple effective embryo culture technique for *Brassica.* Euphytica 18:425–429.

Harlan, J. R. 1976. Genetic resources in wild relatives of crops. Crop Sci. 16:329–333.

———, and J. M. J. deWet. 1971. Toward a rational classification of cultivated plants. Taxon 20(4):509–517.

———, and ———. 1977. Pathways of genetic transfer from *Tripsacum* to *Zea mays.* Proc. Natl. Acad. Sci. USA 74(8):3494–3497.

———, ———, K. M. Rawal, M. R. Felder, and W. L. Richardson. 1970. Cytogenetic studies in *Cynodon* L. C. Rich. (*graminae*). Crop Sci. 10:288–291.

Hermsen, J. G. Th., and M. S. Ramanna. 1973. Double-bridge hybrids of *Solanum bulbocastanum* and cultivars of *Solanum tuberosum.* Euphytica 22:457–466.

Hogenboom, N. G. 1972. Breaking breeding barriers in *Lycopersicon.* 5. The inheritance of the unilateral incompatibility between *L. peruvianum* (L.) Mill. and *L. esculentum* Mill. and the genetics of its breakdown. Euphytica 21:405–414.

Hoven, A. W. 1962. Interspecific hybridization between *Trifolium repens* L. and *T. nigrescens* Viv. and analysis of hybrid meiosis. Crop Sci. 2:251–254.

Kanta, K., and P. Maheshwari. 1963. Intraovarian pollination in some *Papaveraceae.* Phytomorphology. 13:215–229.

Kasha, K. J. 1974. Haploids from somatic cells. p. 67–87. *In* K. J. Kasha (ed.) Haploids in higher plants—advances and potential. Univ. of Guelph, Guelph, Canada.

Knox, R. B., R. R. Willing, and A. E. Asford. 1972a. Role of pollen-wall proteins as recognition substances in interspecific incompatibility in poplars. Nature 237: 381–383.

———, ———, and L. D. Pryor. 1972b. Interspecific hybridization in poplars using recognition pollen. Silvae Genet. 21:65–69.

Kruse, A. 1973. *Hordeum* × *Triticum* hybrids. Hereditas 73:157–161.

———. 1974. An *in vivo/in vitro* embryo culture technique. Hereditas 77:219–224.

Laibach, F. 1929. Ectogenesis in plants. J. Hered. 20:200–208.

Lambert, R. J., and E. R. Leng. 1965. Backcross response of two mature plant traits for certain corn-teosinte hybrids. Crop Sci. 5:239–241.

Lammerts, W. E. 1942. Embryo culture an effective technique for shortening the breeding cycle of deciduous trees and increasing germination of hybrid seed. Am. J. Bot. 29:166–171.

Larter, E., and C. Chaubey. 1965. Use of exogenous growth substances in promoting pollen tube growth and fertilization in barley-rye crosses. Can. J. Genet. Cytol. 7:511–518.

Mackay, G. R. 1973. Interspecific hybrids between forage rape (*Bassica napus* L.) and turnip (*Brassica campestris* L. ssp. *rapifera*) as alternatives to forage rape. 1. An exploratory study with single pair crosses. Euphytica 22:495–499.

Mangelsdorf, P. C., and R. G. Reeves. 1939. The origin of Indian corn and its relatives. Texas Agric. Exp. Stn. Bull. 574.

Marani, A. 1967. Heterosis and combining ability in intraspecific and interspecific crosses of cotton. Crop Sci. 7:519–522.

Martin, F. W. 1970. Pollen germination on foreign stigmas. Bull. Torrey Bot. Club 97:1–6.

Mauney, J. R. 1961. The culture *in vitro* of immature cotton embryos. Bot. Gaz. 122:205–209.

McLean, S. W. 1946. Interspecific crosses involving *Datura ceratocaula* obtained by embryo dissection. Am. J. Bot. 33:630–638.

Meyer, V. G. 1969. Some effects of genes, cytoplasm, and environment on male sterility of cotton (*Gossypium*). Crop Sci. 9:237–242.

———. 1974. Interspecific cotton breeding. Econ. Bot. 28:56–60.

Morgan, W. G. 1976. A technique for the production of polyploids in grasses. Euphytica 25:443–446.

North, C., and A. B. Wills. 1969. Interspecific hybrids of *Lilium lankongense* Franchet produced by embryo-culture. Euphytica 18:430–434.

Norstog, K. 1965. Development of cultured barley embryos. I. Growth of .1–.4 mm embryos. Am. J. Bot. 52:538–546.

———. 1973. New synthetic medium for the culture of premature barley embryos. In Vitro 8:307–308.

Olsson, G. 1963. Induced polyploids in *Brassica.* p. 179–192. *In* E. Akerberg, A. Hagberg, G. Olsson, and O. Tedin (eds.) Recent plant breeding research—Svalof 1946–1961. John Wiley and Sons, N.Y.

Palmer, R. G., and H. H. Hadley. 1968. Interspecific hybridization in Glycine subgenus *Leptocyamus.* Crop Sci. 8:557–563.

Payan, F. R., and F. W. Martin. 1975. Barriers to the hybridization of *Passiflora* species. Euphytica 24:709–716.

Pecket, R. C., and A. R. A. A. Selim. 1965. Embryo-culture of *Lathyrus.* J. Exp. Bot. 16:325–328.

Pittarelli, G. W., and J. R. Stavely. 1975. Direct hybridization of *Nicotiana repanda* × *N. tabacum*. J. Hered. 66:281–284.

Porter, K. B., and N. A. Tuleen. 1972. Cross breeding male sterile wheat (*Triticum aestivum* L.) and rye (*Secale* spp.). Crop Sci. 12:706–708.

Power, J. B., E. M. Frearson, C. Hayward, D. George, P. K. Evans, S. F. Berry, and E. C. Cocking. 1976. Somatic hybridization of *Petunia hybrida* and *P. parodii.* Nature 263:500–502.

Rappaport, J. 1954. *In vitro* culture of plant embryos and factors controlling their growth. Bot. Rev. 20:201–225.

Rieger, R., A. Michaelis, and M. M. Green. 1968. A glossary of genetics and cytogenetics. Springer-Verlag, N.Y.

Sage, G. C. M. 1976. Nucleo-cytoplasmic relationships in wheat. Adv. Agron. 28: 267–300.

Sears, E. R. 1956. The transfer of leaf rust resistance from *Aegilops umbellulata* to wheat. p. 1–22. *In* Genetics and plant breeding. Brookhaven Symposia Biol. No. 9.

Sesbastiampillai, A. R., and J. K. Jones. 1976. Improved techniques for the induction and isolation of polyploids in the genus *Fragaria.* Euphytica 25:725–732.

Skirm, G. W. 1942. Embryo culturing as an aid to plant breeding. J. Hered. 33:210–215.

Skovsted, A. 1935. Cytological studies in cotton. III. A hybrid between *Gossypium davidsonnii* Kell. and *G. sturtii* F. Muell. J. Genet. 30:397–404.

Smith, H. H., K. N. Kao, and N. C. Combatti. 1976. Interspecific hybridization by protoplast fusion in *Nicotiana.* J. Hered. 67:123–128.

Smith, W. K. 1943. Propagation of chlorophyll-deficient sweet clover hybrids as grafts. J. Hered. 34:135–140.

Snowcroft, W. R. 1977. Somatic cell genetics and plant improvement. Adv. Agron. 29:39–81.

Starling, J. L. 1961. Cytogenetic study of interspecific hybrids between *Phalaris arundinacea* and *P. tuberosa.* Crop Sci. 1:107–111.

Stebbins, G. L., Jr. 1950. Variation and evolution in plants. Columbia Univ. Press, N.Y.

————. 1958. The inviability, weakness, and sterility of interspecific hybrids. Adv. Genet. 9:147–215.

Stephens, S. G. 1950. The internal mechanisms of speciation in *Gossypium.* Bot. Rev. 16:115–149.

Swaminathan, M. S. 1955. Overcoming cross-incompatibility among some Mexican diploid species of *Solanum.* Nature 176:887–888.

Wernesman, E. A., R. L. Davis, and W. F. Keim. 1965. Interspecific fertility of two *Lotus* species and their F_1 hybrids. Crop Sci. 5:452–454.

Willing, R. R., and L. D. Pryor. 1976. Interspecific hybridization in poplar. T.A.G. 47:141–151.

Wilmar, J. C., and M. Hellendoorn. 1968. Embryo culture of brussels sprouts for breeding. Euphytica 17:28–37.

Wright, J. W. 1976. Introduction to forest genetics. Academic Press, N.Y.

Yermanos, D. M., and K. S. Gill. 1967. Induction of polyploidy in *Linum* species. Crop Sci. 7:423–427.

Zieber, N. K., and R. A. Brink. 1951. The stimulative effect of *Hordeum* endosperms on the growth of immature plant embryos in vitro. Am. J. Bot. 38:253–256.

Zohary, D. 1973. Gene-pools for plant breeding. p. 177–183. *In* R. Moav (ed.) Agricultural genetics. John Wiley and Sons, N.Y.

————, J. R. Harlan, and A. Vardi. 1969. The wild diploid progenitors of wheat and their breeding value. Euphytica 18:58–65.

8 Commercial Hybrid Seed Production

Harold Wright
DeKalb AgResearch, Inc.
DeKalb, Illinois

Hybrid vigor has been recognized as a widespread phenomenon in plants and animals for many years. Corn breeders first recognized its potential for increasing yields and developed procedures for commercial production of hybrid seed. Commercial hybrids now are used extensively in many crops, including corn, sorghum, sugarbeet, and sunflower. Research is being conducted on many other crops that may permit the widespread use of commercial hybrids in the future.

Commercial hybrids have the greatest potential for crops in which the hybrid seed can be produced reliably and economically. Three biological requirements for successful hybrid seed production include the presence of hybrid vigor, elimination of fertile pollen in the female parent, and adequate pollination by the male parent (Chapter 4). If the biological requirements have been met in a species, a practical program of seed production on a large scale must be developed before hybrids can be used by farmers.

The purpose of this chapter is to describe practical aspects of commercial production of hybrid seed. The principles will be illustrated with four crop species whose biological characteristics necessitate the use of different techniques for some aspects of seed production. Corn is a monoecious species that is naturally cross-pollinated by wind. The flowers of sunflower are perfect, but the stamens develop before the stigma is receptive and it is cross-pollinated by insects. Sorghum is a self-pollinated species that has been successfully adapted to commercial hybrid seed production. Wheat is a self-pollinated species that has been studied extensively for hybrid production, but has not yet been adopted as a commercial hybrid on a large scale.

I. TYPES OF HYBRIDS

The most common type of commercial hybrid is a single cross between two unrelated inbred lines (A × B). Single crosses provide the greatest opportunity for expression of hybrid vigor and usually have higher yields than other types of hybrids. They also provide maximum uniformity for seed characteristics that are important in use of the crop and for height and maturity that can facilitate harvest. The main disadvantage of a single cross for corn and sunflower is that the inbred line used as the female parent usually produces lower seed yields than the types of females used for producing other hybrids (Fick, 1978). Inbred lines of sorghum and wheat have good seed yields and single crosses are used almost exclusively for those crops.

The primary purpose of modified single cross, three-way cross, modified three-way cross, and double cross hybrids is to reduce seed cost by use of a more productive noninbred female parent in the seed field. A modified single cross hybrid (A′ × A) × B is produced by crossing two closely related inbred lines (A′ and A), and using the related single cross as female parent for hybridization with an unrelated inbred (B). The closely related inbreds are progeny from the same breeding population or they may be developed by backcrossing (Chapter 5). A three-way cross (A × B) × C is produced by crossing two unrelated inbred lines (A × B) and using the single cross as female parent for hybridization with an unrelated inbred (C). A modified three-way cross (A × B)(C′ × C) and double cross (A × B)(C × D) involve crossing two single cross hybrids in the seed field. One hybrid of the modified three-way cross is a single cross between two related inbred lines (C′ × C), whereas all four inbreds of a double cross are considered unrelated.

II. OBTAINING SEED OF PARENT LINES

A basic requirement for production of hybrid seed is an adequate supply of foundation seed of the parents. Foundation seed is referred to as basic seed in many countries. A specialized department for foundation seed production is maintained by major seed companies. This department obtains a small quantity of seed of parent lines that have been developed by plant breeders and is responsible for increasing seed of the parent lines to the quantities required for commercial hybrid production. Seed companies that do not maintain complete research and foundation seed departments obtain their foundation seed from seed producers who increase and maintain seed stocks of publicly and, in some cases, privately developed lines.

The key to good foundation seed production is to secure a pure source of the parent from the plant breeder and to maintain a high degree of purity during its increase. Production of pure seed depends on careful selection and isolation of a field, roguing off-type plants, and careful harvesting and seed conditioning.

A. Types of Parents

The types of parents that must be increased as foundation seed depend on the system used to eliminate fertile pollen in the female parents. The cytoplasmic-genetic system of male sterility requires three lines to produce a single cross hybrid; the A line (male-sterile), B line (male-fertile maintainer), and R line (male-fertile with restorer genes) (Chapter 4). Foundation seed production of the A line resembles that of a hybrid, with the A line being the female and the B line the male. Factors considered for optimizing seed set in a hybrid seed field must be considered for production of the A line (Section IIC). Low ratios of A line (female) to B line (male) often are used. For example, a 1:1 ratio is used in sorghum to assure adequate pollination and minimum contamination from undesired pollen (Quinby and Schertz, 1970). Synchronization of flowering dates usually is not a problem because the A and B lines are nearly identical, except for their cytoplasm. Foundation seeds of the B and R lines are produced by open-pollination in isolated fields.

Three-way crosses produced with cytoplasmic-genetic male sterility involve maintenance and production of four lines, an A and B line of one inbred and male-fertile inbreds of the other two. Double crosses involve six lines, A and B lines of two inbreds and male-fertile inbreds of the other two (Chapter 4).

Artificial emasculation (detasseling) can be used for hybrid seed production of corn (Section VI). In such cases, all parents are male-fertile and foundation seed of each is produced by open-pollination in isolated fields.

Both cytoplasmic-genetic male sterility and detasseling are used to produce certain hybrids of corn because hybrid plants produced with the male-sterile method may not shed enough pollen in some environments to obtain maximum seed set (Craig, 1977). To avoid risk for the farmer, about 25 to 50% of hybrid seed produced by detasseling can be blended with 50 to 75% of the identical hybrid produced by the male-sterile method. Hybrid plants from the detasseled fraction will shed enough pollen in the farmer's field to supplement any deficiency in pollen shed of the other fraction.

B. Field Selection and Isolation

Certification standards for purity and isolation are closely followed to assure production of foundation seed that is acceptable in commercial seed production areas, including countries that require seed certification labels. Genetic purity of parent lines is critical because it reduces the cost of roguing in large hybrid seed fields.

Contamination from volunteer plants of the same or different species is avoided by the use of fields that produced a crop in the preceding season that is easily identified or of no problem in the foundation field. For example, volunteer plants of soybean do not have to be removed from a seed field of corn.

Foundation fields require isolation from sources of pollen from other lines, hybrids, or weeds that can cause undesired cross-pollination. Proper isolation can be a major problem in areas where a crop is grown commercially because wind and insect-borne pollen can be carried considerable distances. The isolation distance used for foundation seed fields of corn, sorghum, and wheat is a minimum of 200 m (Craig, 1977; Quinby and Schertz, 1970). Isolation distances used for sunflower seed production vary among producers, and range from 0.8 km to 6.4 km (Smith, 1978).

Different planting dates can be used for some species and locations to achieve isolation by differential time of flowering. A 30-day difference in planting date has been used to obtain isolation between sunflower fields that do not have adequate distance between them.

Some weed species can produce contaminating pollen and can be a problem in achieving proper isolation. Wild sunflower can be a problem for sunflower seed production (Smith, 1978). Sudangrass, shattercane, and Johnsongrass can be a source of contaminating pollen for grain sorghum (Quinby and Schertz, 1977).

C. Roguing

Off-type plants must be rogued from the field, preferably before pollination. Off-types can result from volunteer plants of a previous crop grown in the field; undesired cross-pollination in the previous seed crop; mixing of seed from different genotypes during planting, harvest, or conditioning; or mutation.

Roguing in large foundation wheat fields is not practical because they are seeded in narrow rows with high plant populations. The problem is minimized by preparing a pure source of seed each year. Individual heads representative of the line are selected each year and seed from each is sown in an individual row the following year. Off-type rows are discarded, individual heads are selected from uniform rows, and the remainder is harvested in bulk. The bulk seed is sown on 1 to 2 ha at about one-third normal plant density in rows spaced 0.3 m apart. This small field is the last phase in the increase program in which roguing is practical; therefore, foundation and hybrid production fields seeded at normal plant densities do not have the benefit of roguing.

Roguing for male-fertile plants in a male-sterile line requires special care. For example, wheat must be rogued daily during the 7 to 10 days of blooming. The most effective roguing is done in the early morning or late evening when wind velocity usually is low and the sun is at a lower angle. Winds greater than 15 km/hour cause the plants to move excessively and reduce roguing effectiveness significantly. Heavy rain may cause delays in roguing of up to 24 hours by washing away extruded anthers from male-fertile plants.

D. Harvest and Seed Conditioning

Harvest and seed conditioning must be completed without mixing seed of different lines. Foundation seed is harvested and processed in the same manner as commercial hybrid seed (Sections VII, VIII and IX). Small, specialized threshers that can be cleaned readily are used to harvest small increases of pure seed of wheat, but conventional combines are used for large foundation seed fields.

III. FIELD SELECTION

Desirable fields are important contributors to purity, yield, and quality of hybrid seed. Hybrid seed fields are selected that are best suited for a particular crop, and generally are located in areas of commercial production. Fields are avoided that produced the same crop in the preceding season. It is important that the farmer-cooperator be concerned about proper soil fertility, tillage operations, control of plant pests, and harvest procedures.

The field must be isolated from sources of undesired pollen using distance or time isolation similar to that of foundation seed (Section IIB). Isolation distances may be reduced by planting a barrier of the male parent between the source of contamination and the seed production area (Fig. 1).

Fig. 1—A hybrid seed field of corn in which the female parent has been detasseled. A barrier of rows of the male parent on the edges of the field produces an abundance of pollen that has a greater chance of pollinating the female parent than does contaminating pollen from adjacent fields.

The purpose of the barrier is to produce an abundance of pollen that has a greater chance of pollinating the female parent than does pollen from the source of contamination. Reduction in the isolation distance and width of the barrier varies with production conditions and the hybrid being produced. In corn, isolation distance may be reduced at least 7 to 8 m for each 1 m width of the barrier.

IV. OPTIMIZING SEED SET

Proper selection of fields and farmer-cooperators is a key factor in obtaining high seed yields. Maximum seed set also requires management of the crop to obtain an adequate amount of pollen when the female flowers are receptive. Adequacy of pollen production is a primary consideration in determining the ratio of the female to male parent (female:male ratio) planted in the field. The female:male ratio depends largely on the abundance and duration of pollen production by the male parent. Ratios for sorghum vary from 3:1 to 6:1, corn from 2:1 to 4:1, sunflower from 2:1 to 7:1, and wheat from 1:1 to 3:1 (Craig, 1977; Fick, 1978; Miller et al., 1974).

The width of each strip of the female and male parent is strongly influenced by the size of the available equipment for planting and harvest. For example, in corn and sorghum, 6:2 or 12:4 ratios of female to male rows are less complicated and less costly to handle with conventional planters and harvesters than are narrower strips in a 3:1 ratio (Fig. 1 and 2).

Interplanting male rows between evenly spaced rows of the female has become increasingly popular for hybrid corn production, especially with crosses involving a vigorous male parent with abundant pollen. For example, female rows would be planted in rows spaced 1 m apart and the male rows would be planted between them at a 50 cm spacing. The number of interplanted male rows would be a function of the female:male ratio. The male rows are destroyed after pollen is produced to avoid the accidental harvesting of seed from the male parent.

Crops that are insect-pollinated must have adequate pollen vectors present at flowering to optimize seed set. Hives of honeybees are placed in sunflower fields during flowering at a rate of 0.5 to 2.5 hives/ha (Smith, 1978).

Synchronization of flowering of the parents is essential for optimum seed set, and may require special techniques at planting or during the growing season (Chapter 1). Parents that differ widely in flowering date should be avoided in hybrid combinations whenever possible because hybrid seed production becomes more complicated, costly, and unreliable.

When parents differ in flowering date, one parent must be hastened or delayed to obtain pollen shed when the female is receptive. Planting parents on different dates, commonly referred to as delayed or split planting, is the most common procedure for altering flowering dates (Fig. 3). Two planting dates for the male parent also may be used to extend the pollen shedding period of some crops. The number of days in delay for planting the earlier parent of corn and sorghum can be estimated from the number of growing degree-days required for each parent before flowering begins. The equation used by the U.S. National Oceanic and Atmospheric Administration to compute growing degree-days for a 24-hour period is [(Daily maximum temperature + daily minimum temperature) ÷ 2] − 10 C. The maximum temperature used is 30 C and the minimum is 10 C. For example, if the maximum temperature for a 24-hour period is 28 C and the minimum is 15 C, there are [(28 + 15) ÷ 2] − 10 = 11.5 growing degree-days. If the maximum is 32 C and the minimum 7 C, there are [(30 + 10) ÷ 2] − 10 = 10 growing degree-days. Number of growing degree-days is a useful indicator of flowering dates, but flowering also may be altered by soil temperature, soil moisture, fertilizer, hours of sunlight, and other growing conditions.

Fig. 2—A hybrid seed field of grain sorghum with a ratio of four rows (light colored) of the male parent to 12 rows of the female parent (dark colored).

Flowering of sunflowers is strongly influenced by photoperiod. Planting date studies in which the parent lines of sunflower are planted at regular intervals during the growing season are helpful in establishing the need for and timing of split plantings. Drought and unusual temperatures do not affect time of flowering of sunflower as much as they do corn and sorghum.

Alteration of flowering date by split planting is not reliable in species that require vernalization (Chapter 1). A 10-day delay in fall planting of winter wheat results in approximately a 1-day delay in blooming. Results can be inconsistent among years because of variation in the length and severity of the cold period.

Clipping or burning leaves of young plants has been used to delay flowering of corn, sorghum, and wheat. Additional techniques used to cause small changes in flowering date are fertilizer application, irrigation, and variable planting depths.

More than one technique can be used to synchronize flowering in a field. The choice depends on what is practical as well as effective. When treatments are needed, injury to the female parent that will reduce its seed production should be avoided. Delayed planting, clipping, burning, and moisture stress should involve only the male parent, whenever possible (Fig. 3).

Fig. 3—Rows of the male parent in a hybrid seed field of corn were planted later than the female parent to synchronize their time of flowering.

V. ROGUING

The removal of off-type plants from both the female and male parents is necessary to maintain the genetic purity of hybrid seed. Female parents with cytoplasmic-genetic male sterility must be rogued for male-fertile plants before pollen is produced. The frequency of roguing is determined by the number of off-types and rate of development of flowering. Early in the growth of the crop, removal of the entire plant is advisable. Cutting below the flower primordia may be adequate when plants are more developed. Sunflower heads, however, continue to shed pollen when removed from the stalk; therefore, the detached head is placed face down on the soil to prevent contact with pollinating insects (Smith, 1978).

Unique characteristics of a crop can influence the roguing procedure. Fields of sunflower planted north-south are more easily rogued than those planted east-west. The heads of sunflower usually face the east after the ray flowers are fully developed (Smith, 1978). Planting rows north-south permits the roguers to look westward at the heads, whereas rows planted east-west are difficult to rogue when walking eastward. Roguing large fields in hybrid wheat production is hampered by use of solid seeding (Section IIC), thus the only plants removed are the occasional off-type wheat plants or those of other crops, such as rye.

VI. ARTIFICIAL EMASCULATION

Hand emasculation of the female parent is not economically feasible for commercial production of hybrid seed of most crop species. Corn is an exception because of its moneocious reproductive structure. Removal of the tassel (detasseling) by hand or machine eliminates pollen from the female parent of a corn hybrid (Fig. 4 and 5).

The time of detasseling is based on the percentage of tassels shedding pollen relative to the percentage of receptive female silks. Standards used by some companies are illustrated in Table 1. Another common standard calls

Table 1—Standards used for detasseling hybrid seed fields of corn to insure genetic purity.†

	Percent of silks in the female parent that are receptive				
Percent of tassels in the female parent that are shedding pollen	Up to 0.25	0.26 to 4.0	4.1 to 10.0	10.1 to 25.0	Over 25
0.0 to 0.24	Blue	Blue	Blue	Blue	Blue
0.25	Blue	Blue	Blue	Blue	Green
0.26 to 0.50	Blue	Blue	Green	Red	Red
0.51 to 1.0	Blue	Green	Red	Red	Red
Over 1.0	Blue	Red	Red*	Red*	Red*

† Blue = Hybrid seed from the field will have minimal contamination from self-pollination of the female parent. Green = Hybrid seed from the field may be blended with seed from a field designated as blue to reduce the percenage of contamination from self-pollinated seed. Red = Hybrid seed from the field may be discarded or blended with seed from a field designated as blue to reduce the percentage of contamination from self-pollinated seed. Red* = Detasseling operations are suspended in the field and no seed is harvested because of excessive self-pollination of the female parent. Seed from fields are candidates of grow-out plots.

for removing the tassels so that the percentage of female plants shedding pollen at any one time is never in excess of 1%, or that the total percentage of females shedding pollen on any 3 successive days is never in excess of 2%. A tassel is considered to be shedding when 2.5 cm of the center spike has exposed anthers.

Hand detasseling is performed by persons who walk through the field or who ride on a tractor-mounted platform (Fig. 4). Hundreds of workers, primarily young people, are required to cover the extensive hectarage of hybrid seed corn production. Hand detasseling causes little or no damage to the plant, is unaffected by nonuniform height of the female parent, and can be done when fields are too wet for mechanical equipment. Its disadvantages are the lack of adequate labor in some production areas and high labor costs.

Fig. 4—Hand detasseling in a hybrid seed field of corn.

Mechanical detasselers have been developed to reduce the need for hand labor (Fig. 5). Two common types of mechanical detasselers are cutters and pullers. One type of cutter removes tassels with a rotating blade operated on either a horizontal plane or tilted in the direction of operation. A second type of cutter removes tassels with two revolving small-toothed circular saw blades placed in a V mounting with the cutting edges offset. One blade revolves five to six times faster than its counterpart. The puller types use either rubber rollers or pneumatic rubber tires that are mounted in a V position and tilted in the direction of operation. All of these machines have guides to direct the tassel into the cutting or pulling mechanism.

A primary disadvantage of mechanical detasselers is that they remove the top leaves of the plant with the tassel (Fig. 5). The average number of leaves removed ranges from 1.5 to 2.5, and is dependent on uniformity in height of the female parent, amount of tassel extrusion from the plant whorl before pollen is shed, and skill of the operator in adjusting height of the cutting or pulling mechanisms.

Female parents respond differently to removal of leaves. Seed yields of some parents decrease significantly as the number of leaves removed increases, and the percentage of the desirable flat size kernels of these parents usually decreases and percentage of less desirable round size kernels increases. Growing conditions influence the amount of reduction in seed yield and quality caused by leaf removal. Greater loss occurs during periods of high temperatures and limited soil moisture.

Machine detasseling currently is supplemented by hand detasseling to remove missed or late maturing tassels.

Fig. 5-Mechanical detasseling of corn with a rotary cutter. There is a ratio of six female rows (detasseled) to two male rows (with tassels).

VII. HARVEST

Harvesting of foundation or hybrid seed requires proper timing, adjustment of equipment, and control of moisture to avoid yield loss, seed mixtures, and damage to the seed. Moisture content of the harvested seed varies among crops.

Corn seed is harvested as soon as the kernels reach physiological maturity, which occurs when kernel moisture is approximately 30 to 38%. Craig (1977) indicated several advantages of harvesting corn seed at that moisture level. Harvest of the seed crop as soon as it is mature 1) reduces risk of freeze injury, 2) avoids excessive field losses from mechanical pickers, 3) reduces risk of delays due to adverse weather conditions, 4) reduces losses from insect damage, and 5) reduces losses from ear and stalk rots and other diseases. All of these factors contribute to the quality of the seed crop not only in appearance but in physical damage.

Ears of corn usually are harvested with pickers that leave a part of the husks intact, instead of with combines that shell the grain. Machines (huskers) in the field or seed conditioning plants are used to remove husks before ear sorting and drying. The advantages of ear harvest are that (a) seed can be harvested at a high moisture level without excessive seed damage, (b) damaged, diseased, and off-type seed can be removed by sorting ears before drying, (c) ears can be dried more uniformly than shelled seed, and (d) harvest of ears results in higher seed quality.

Sorghum, sunflower, and wheat seeds usually are harvested at moisture levels that are safe for storage without drying. Sorghum and wheat seed are safely stored at 13% moisture or less, and sunflower seed is stored at about 9.5% in the northern U.S. and 11% in the western U.S. (Smith, 1978). Seed of the three crops can be harvested at higher moisture and dried artificially, but such a practice is cumbersome and expensive, and can result in reduced seed quality. The seed is harvested with commercial combines equipped with rasp-bar cylinders that are set at the lowest threshing speed possible to avoid mechanical injury to the seed. Seed damage also can be minimized by avoiding the use of augers for moving seed within the combine and during transportation to the storage bin. Combines must be thoroughly cleaned with brushes, air pressure, and vacuum cleaners between the harvest of different parents. It is difficult to gain access to all parts of the combine for cleaning; therefore, the first several bushels of seed harvested of a new parent usually are discarded.

Seed of all crops must be handled gently to avoid seed damage during harvest, drying, storage, and conditioning. Dropping seed excessive distances to hard surfaces can cause considerable reduction in germination percentage. The use of seed ladders, rubber pads, slides, conveyor belts, and similar equipment often determines the difference between high and low quality seed.

VIII. SEED DRYING

Drying of seeds is necessary when they are harvested at moisture levels that are too high for direct storage. The length of time seed may be held between harvest and drying depends on the percentage of moisture in the seed and the air temperature. For example, ear corn at 25% moisture or higher should be placed in a drier within 6 to 8 hours after harvest when the air temperature is above 21 C. Reduction of seed moisture within the shortest time after harvest will avoid deterioration of germination and seedling vigor.

Drying temperatures should not exceed 44 C for most seed. Unheated outside air can be used for limited drying, if it is warm and dry. There are three prerequisites for effective and efficient drying of seed.

a) The drier should be filled to a capacity that will allow seed to be dried in 4 days or less.
b) Air should move uniformly through all sections of the drier. Excessive foreign material such as husks, silks, straw, and hulls can impede drying. The top layer of seed should be flat and parallel to the floor of the drier, not in a cone-shaped pile.
c) Air movement should be reversed periodically to avoid layers of extremely dry or high moisture seed.

IX. SEED CONDITIONING

Preparation of seed for marketing may include cleaning, sizing, treating, and bagging. The number of these operations used and the equipment involved varies among crops.

Corn seed that was shelled by harvesting with a combine may require cleaning before drying. Ear corn is shelled after drying and cleaned for immediate sizing and treating or for storage. Cleaning involves air removal of light-weight foreign material. It also may include the removal of large foreign material by passing seed through an upper round hole screen of 10.4 mm (26/64 inch) or 11.2 mm (28/64 inch) and removal of small seed by passing it over a lower round hole screen of 6.0 mm (15/64 inch).

Corn seed is separated into different sizes to improve the accuracy of seed spacing with plate-type planters and to a lesser extent with plateless planters. The screens used to obtain various sizes of seed are presented in Table 2. The seed is passed over a suction cleaner (aspirator) after drying to remove broken and light weight seed that was not removed during cleaning. Heavy cob particles occasionally present in the large seed sizes are removed with a gravity separator. Seed usually is treated with a fungicide to protect it from various soil organisms. An insecticide treatment may also be applied for protection against storage insects.

Table 2—Sizes of hybrid corn seed as determined by kernel thickness and width.

		Kernel width†—Separation by round hole screens				
		6.4–6.8	6.8–7.2	7.2–8.0	8.0–8.8	8.8–10.4
Kernel thickness†	Thru 4.8	Flat size	Flat size	Flat size	--	--
Separation by	Thru 5.2	--	--	Flat size	Flat size	Flat size
slotted screens	4.8–5.2	Round size	Round size	--	--	--
	Over 5.2	Round size	Round size	Round size	Round size	Round size

† Size dimensions in mm.

Corn seed may be packaged by weight in bags containing 22.7 kg (50 lb) or 25.5 kg (56 lb), or by number of kernels, usually 80,000 kernels/bag. Sale of seed on the basis of kernels per bag has become common because farmers can determine readily the number of bags needed based on their intended planting rate and hectarage.

Conditioning of sunflower seed is similar to that of corn. Seed is cleaned with round hole screens ranging in diameter from 4.0 to 9.5 mm, then it is sized based on thickness and width (Smith, 1978). Four common sizes are 11,000 seeds/kg, 13,250 seeds/kg, 15,500 seeds/kg, and 17,600 seeds/kg. Seed usually is not treated with a fungicide before bagging. Bags of seed may weigh 11.4 or 22.7 kg or may be based on 80,000 or 120,000 seeds per bag.

Seed of sorghum and wheat is cleaned with air and screens, but is not sized. Sorghum seed usually is treated with a fungicide and insecticide and bagged in units of 22.7 kg. Hybrid wheat seed usually is treated with a fungicide, and an insecticide treatment also is used by some seed producers. Hybrid wheat seed is sold in bags weighing 22.7 kg.

X. QUALITY CONTROL

Inadequate genetic purity or seed quality can influence the performance of a hybrid in commercial crop production. Proper quality control involves inspection and records for production and handling of both foundation and hybrid seed. Major areas of inspection and records are the source, quality, and purity of foundation seed; field isolation; field diagrams and planting procedures; plant populations of female and male parents; roguing; detasseling corn fields; harvesting; sorting or cleaning immediately following harvest; drying; storage; sizing; treating; bagging; analysis and germination of seed; and approved labeling for state and federal regulations.

Samples of seed harvested from foundation or hybrid fields may be planted in grow-out plots to verify genetic purity. Grow-out plots of corn, sorghum, and sunflower may be grown in a winter nursery for rapid verification. Winter wheat is sown only 2 to 3 months after harvest; therefore, grow-out plots usually cannot be used to eliminate inferior seed lots.

A special type of wheat grow-out plot includes only male-sterile lines grown in isolation from self-pollinated wheat. The plots provide information on sterility in addition to readings made by the field inspector at pollination time. The plots also allow the seedsman to make accurate readings of self-fertility at maturity. Bagging of individual heads before blooming can be an additional aid, especially if some partial self-fertility is suspected.

Germination tests are made on seed from all fields. This information is used to eliminate seed of inferior quality and for packaging seed on the basis of viable kernels per bag. Seed analyses are conducted to determine the amount of weed seeds and foreign material.

XI. STORAGE

Storage of large volumes of hybrid seed requires systematic records. Warehouse records include location of each hybrid and the quantity of various sizes. Warehouse equipment must be adequate to handle seed without loss or damage.

Seed that is used for planting the following year does not require special storage facilities, except in tropical climates with high temperature and humidity. It is a common practice, however, to store foundation seed of parents for more than 1 year, and some companies prefer to store surplus quantities of hybrid seed. A common rule for safe storage is to maintain the sum of relative humidity and temperature (C) below 60. Relative humidity of 50% at 10 C frequently is used for long-term storage.

XII. MARKETING

Companies use several methods for merchandizing their hybrid seed. A common method in the northern U.S. is to utilize farmers as salespersons. In the southern U.S., various agricultural stores include hybrid seeds as part of their merchandize. A very small proportion is sold by mail.

Educational programs and demonstration plots are used to provide information to the farmer. Agronomists are available in many companies to assist the farmers in planning and in remedying special production problems.

ACKNOWLEDGEMENTS

The author wishes to acknowledge the assistance and cooperation of Leo B. Olson, Dwight E. Glenn, Charles F. Krull, and Elmer E. Williams, DeKalb AgResearch, Inc., in the preparation of the manuscript.

REFERENCES

Craig, W. F. 1977. Production of hybrid corn seed. p. 671–719. *In* G. F. Sprague (ed.) Corn and corn improvement. Am. Soc. of Agron., Inc., Madison, Wis.

Fick, G. N. 1978. Breeding and genetics. p. 279–338. *In* Jack F. Carter (ed.) Sunflower science and technology. Am. Soc. of Agron., Inc., Madison, Wis.

Miller, J. F., K. J. Rogers, and K. A. Lucken. 1974. Male-sterile wheat seed production in North Dakota. Crop Sci. 14:702–705.

Smith, D. L. 1978. Planting seed production. p. 371–386. *In* Jack F. Carter (ed.) Sunflower science and technology. Am. Soc. of Agron., Inc., Madison, Wis.

Quinby, J. R., and K. F. Schertz. 1970. Sorghum genetics, breeding, and hybrid seed production. p. 73–117. *In* Joseph S. Wall and William M. Ross (ed.) Sorghum production and utilization. The AVI Publishing Co., Inc., Westport, Conn.

9 Alfalfa

D. K. Barnes
USDA-SEA and University of Minnesota
St. Paul

Alfalfa (*Medicago sativa* L.) is an important forage species for hay and pasture. Although it originated in southwestern Asia, it is well adapted to a wide range of climates and soils in the United States, where about 11 million ha are grown annually. Alfalfa has been referred to as "Queen of the Forages" because of its high yields and feeding value. Between 1900 and 1975 more than 160 cultivars were developed for production in North America (Barnes et al., 1977). Most of the newer cultivars were selected for improved adaptation and multiple pest resistance.

I. PARENTAL MATERIAL

The genus *Medicago* is widely distributed and comprises an array of diverse species that are either annual or perennial. The most recent taxonomic studies of the perennial species concluded that *M. sativa* is polymorphic. Lesins and Gillies (1972) defined the complex as *M. sativa-falcata-glutinosa*, and Gunn et al. (1978) designated it as the *M. sativa sensu lato* complex. Crosses can be madc among subspecies in the *M. sativa* complexes. Crosses can be made with the cultivated tetraploids and wild diploids without special preparation of the parents (Bingham, 1968).

II. PLANT CULTURE

A. Field

Successful alfalfa seed production under field conditions requires proper insect control, adequate pollination, and the adjustment of cultural and management practices to local conditions (Pedersen et al., 1972). High seed yields require well-drained soils which are low in alkali and soluble salts, and a rooting depth of 1.2 m or more. Gravelly and shallow soils should be avoided. Irrigation practices should prevent severe plant stress and promote slow, continuous flowering through the entire production period without excessive stimulation of vegetative growth. Areas with low relative humidities and moderate to high temperatures are preferred for alfalfa seed production because of a low incidence of leaf diseases, long periods of pollinator activity, and favorable harvest conditions. Applications of fertilizers to established stands in the United States and Canada, and foliar applications of major and minor elements to seed fields in California, usually have failed to increase seed yields.

A critical aspect of seed production is crop management so that the flowering period coincides with the period of least competition from other pollen sources, the greatest activity of pollinators, and the most favorable weather conditions (Pedersen et al., 1972). In some areas, these circumstances require removing the first and possibly the second growth for forage and then producing seed on the subsequent regrowth.

Rincker (1976) reported that seed increases during the year of establishment can be maximized by starting seedlings in the greenhouse and transplanting them to the field at a rate of 9,000 to 10,000 plants/ha. He found that transplants yielded more than twice as much seed as direct-seeded plants during the year of establishment; however, seed yields were similar during subsequent production years.

B. Growth Chamber and Greenhouse

Successful alfalfa seed production in the growth chamber and greenhouse requires photoperiod and temperature adjustment plus insect and mite control. Most alfalfa cultivars are long-day plants (Bula and Massengale, 1972), but cultivars differ in their quantitative response to photoperiod. Flowering is usually sparse at photoperiods of 12 hours or less, but at 14 to 16-hour photoperiods most plants will flower. Nittler and Kenney (1964) reported that the most profuse flowering was under continuous light. They also reported that flowers increased when light level was increased from 12 to 24 μ E sec^{-1} m^{-2} under photoperiods long enough to promote flowering.

Elgin and McMurtrey (1977) compared alfalfa seed production with four types of supplemental lighting (high-pressure sodium, metal halide, mercury vapor, and incandescent) at an 18-hour photoperiod. The amount of flowering and subsequent seed production was directly related to the quantity of light supplied. The most effective light source was the high-pressure sodium lamp.

Temperature is another factor that can influence flowering in alfalfa (Massengale et al., 1971). Greenhouse and growth chamber conditions most commonly used for alfalfa seed production include temperatures of 24 to 30 C and at least 16 hours of light.

III. FLORAL CHARACTERISTICS

Essentially all annual species are cleistogamous and are exclusively self-pollinated. Generally, the perennial species require tripping, and will set seed from either self or cross-pollination.

The flowers of perennial species are morphologically complex and have a unique tripping mechanism. The alfalfa flower is a classic example of an evolutionary adaptation for cross-pollination by insects (McGregor, 1976).

A detailed description of floral development and floral morphology was prepared by Barnes et al. (1972). Alfalfa flowers are borne on simple racemes, with about 10 flowers per raceme. At the base of each flower the receptacle is enclosed by the calyx tube consisting of five undiverged sepals terminated by five lobes. The receptacle serves as the base for the corolla, pistil, stamens, and nectary. Alfalfa has a papilionaceous corolla that consists of five petals: a large standard, two lateral wing petals, and two fused keel petals (Fig. 1a). The 10 stamens form a tube in which 9 filaments are fused. The 10th stamen is nearest the standard and is free (Fig. 1b). Filaments alternate long and short so that, during development, anthers fit tightly around the stigma in a double ring (Fig. 2a). The pistil consists of a single carpel that develops a superior ovary, a smooth awl-shaped hollow style, and a well-defined stigma (Fig. 2c). The ovary contains an average of 10 to 12 ovules. The nectary is located in the base of the receptacle (Fig. 1c) and covers the area from the base of the pistil back to the base of the keel petals (Teuber, Albertsen, and Barnes, 1977). The nectary is from 2 to 8 cells thick and contains 7 to 25 stomata from which nectar is exuded. A nectar reservoir is located above the nectary.

Before pollination can be successful in alfalfa, the sexual column (ovary and staminal column) must be tripped (released). This is usually done by bees foraging for nectar and pollen. In some instances, the flower can be tripped by environmental factors such as hot, dry winds, and rain. Larkin and Graumann (1954) indicated that the two major forces involved in the tripping mechanism are pressure exerted by the sexual column from cells under tension at the juncture of the staminal tube and the keel, and the restraining mechanism of the keel petals that cohere due to interlocking projections of cutinized tissue in the appressed petal surfaces. Tripping takes place whenever the restraint of the appressed keel petal is reduced or becomes less than the pressure of the staminal column. Plants differ in ease of tripping.

Four bud stages are recognized in alfalfa: straight bud, pointed bud, hooded bud, and erect standard (Coffman, 1922). Anther dehiscence usually occurs in the pointed bud stage (Fig. 2b). A cuticular membrane forms a continuous film over the stigma, thereby preventing pollination before tripping. When tripping occurs, the stigmatic membrane usually ruptures as the stigma strikes either the bee or the standard petal. Alfalfa pollen is sticky and readily adheres to pollinating insects. If the stigma fails to come into contact with foreign pollen, its own pollen germinates and produces pollen tubes. Self-pollinated alfalfa has fewer flowers that form seed pods and fewer seed per pod than cross-pollinated alfalfa.

The alfalfa flower gives sufficient protection to keep the pollen viable and the stigma receptive from the late bud stages until the flower begins to wilt. The time flowers remain open varies among genotypes and environments from about 5 to 16 days. Alfalfa pollen has been successfully collected and stored from 2 to 6 months in sealed vials (Hanson, 1961; Lehman and Puri, 1967). Pollen storage has rarely been used because alfalfa is perennial, has an indeterminate flowering habit, and can be easily propagated vegetatively. With a combination of field and greenhouse facilities, flowering plants can be produced throughout the year.

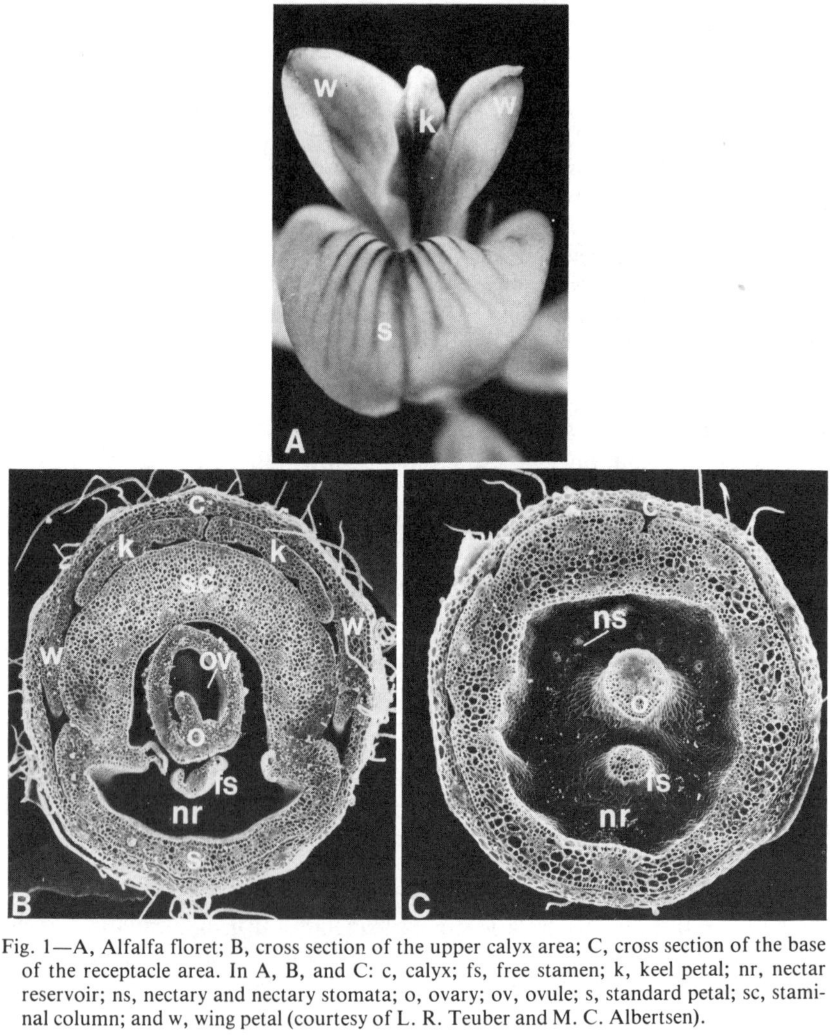

Fig. 1—A, Alfalfa floret; B, cross section of the upper calyx area; C, cross section of the base of the receptacle area. In A, B, and C: c, calyx; fs, free stamen; k, keel petal; nr, nectar reservoir; ns, nectary and nectary stomata; o, ovary; ov, ovule; s, standard petal; sc, staminal column; and w, wing petal (courtesy of L. R. Teuber and M. C. Albertsen).

IV. ARTIFICIAL HYBRIDIZATION AND SELF-POLLINATION

A. Equipment

The equipment needed for pollination is similar to that described for clover (Chapter 16).

B. Preparation of the Female

Successful self and cross-pollinations can be made with any flowers that are not wilted and that have fertile ovules and pollen. Cross-pollination by hand without emasculation has often been used to produce seed for plant breeding studies. For critical studies in which self-pollinated seed cannot be tolerated, either the female parent can be emasculated before pollination or a male-sterile plant can be used as the female parent. Anthers and pollen can be removed from a flower during emasculation by either suction or alcohol (Tysdal and Garl, 1940). With both methods the standard petal is clipped and the flower is gently tripped. With vacuum emasculation, a glass tube, drawn to a 1-mm tip and attached to a vacuum source, is used to re-

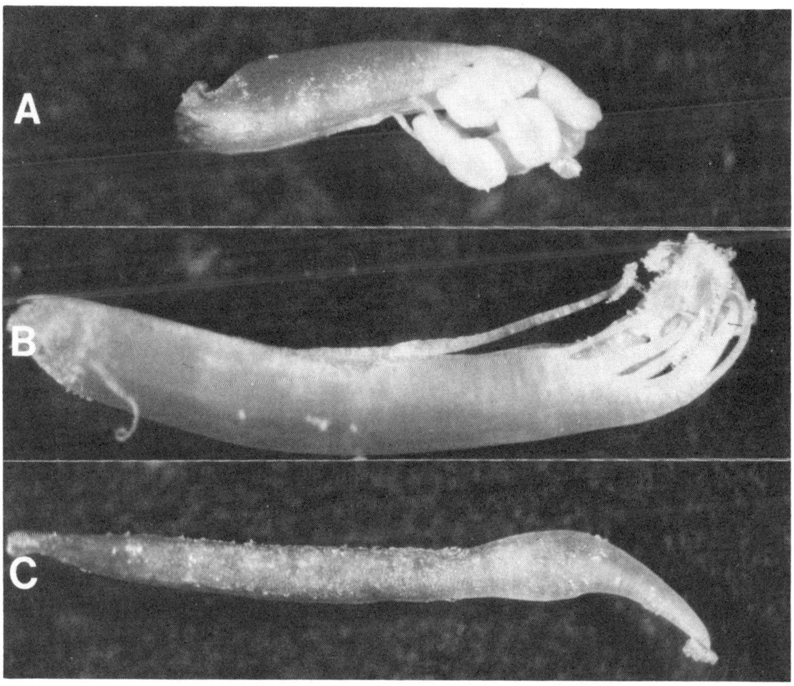

Fig. 2—Alfalfa reproductive development. A, alternating long and short stamens at straight bud stage of development; B, dehisced pollen surrounding stigma at pointed bud stage of development; C, mature ovary, style, and stigma.

move the anthers. A low-powered magnifier can be used to determine thoroughness of pollen removal. A skilled operator can expect normal seed set (three to five seeds/pod) of 99% + F_1 seed from vacuum-emasculated flowers. With alcohol emasculation, the whole raceme is immersed in a small beaker of 57% ethyl alcohol for 10 sec, then rinsed in water for a few seconds. Alcohol is more effective than vacuum for emasculation. However, alcohol causes more injury to the stigma, so that only about 26% of the flowers produced pods compared to 60% for vacuum-emasculated flowers and 76% for nonemasculated controls (Tysdal and Garl, 1940). It is recommended that an emasculated, unpollinated control be included when emasculations are made to determine the effectiveness of procedures and operators.

C. Pollination

Pollinations can be made at any time of the day or night. Cross-pollination by hand without emasculation is accomplished by alternately tripping flowers of the two parents into a folded cardboard pollen reservoir (Fig. 3). When parents have normal amounts of pollen and average self-fertility, about 85% of the seed produced will be hybrid (Pedersen and Barnes, 1973). By randomly tripping flowers from many plants into the same pollen reservoir, bulk seed on a population of plants can be produced.

Fig. 3—Tripping alfalfa floret with folded cardboard.

The bench position of plants being pollinated should be changed periodically to reduce patterns of nonrandom pollination. Viands and Barnes (1977) estimated that about 3.5 g of seed can be produced on a population of plants in the greenhouse per person-hour of pollination. The average number of seed/flower pollinated is about 1.1 from self-pollination and 4.4 from cross-pollination (Pedersen and Barnes, 1973).

Self-pollination requires tripping the flowers without introducing foreign pollen. The four usual methods of tripping flowers include inserting a toothpick in the throat of the flower, inserting a toothpick that has the tip covered with emery paper to rupture the stigmatic membrane, depositing the pollen into a folded cardboard which retains the pollen for the stigma to fall into (Fig. 3), and gently rolling or squeezing racemes between the fingers. Barnes and Stephenson (1971) found that all four methods produced the same average number of seed per flower tripped. However, rolling racemes was about three times more efficient (pollinations per hour) than the other three methods. Alcohol should be used to clean hands between treating plants.

D. Factors Affecting Efficiency

Genetic markers are often used to identify hybrids in pollination studies. The most useful genetic markers are distinctive ones that can be identified at the seedling stage, and are not detrimental to plant growth. About 75 simply inherited traits have been studied in alfafa. The characteristics of most of these traits have been summarized by Barnes and Hanson (1967). The best dominant marker available is the red root trait (Clement and Stanford, 1966). Anthracnose resistance can also be used as a genetic marker and identified in the seedling stage (Campbell et al., 1974). The best recessive markers available are the golden tip trait (Elgin, 1972) and the green hypocotyl (white flower) trait (Stanford, 1951). Yellow and purple flower color have been useful genetic markers in many studies; however, their primary disadvantage is that progenies must be grown to flowering for classification (Barnes, 1966). All gene markers described in tetraploid alfalfa are apparently controlled by tetrasomic inheritance.

V. NATURAL HYBRIDIZATION

Field and cage-planting designs can be important in alfalfa seed production because alfalfa, which depends on insects for pollination, varies in attractiveness to pollinators (Boren et al., 1962), and produces both selfed and crossed seed. Other factors, such as degree of self-sterility and differential time and amount of flowering, also influence the extent of crossing and selfing among parent plants. Seed often is produced by enclosing plants in a cage with honeybees, leafcutter bees, or bumblebees. Excessive amounts of selfing (in some cases as much as 90%) may occur in field cages because individual honeybees work a single clone instead of randomly visiting all clones (Hanson et al., 1964). The amount of natural crossing in the field between two noninbred alfalfa populations can vary significantly. Pedersen (1968) reported that 53% of the seed was hybrid. Kehr (1973) observed differences in crossing from 32 to 96%, with an average of 50%. Kehr reported that in some situations crossing percentages were significantly higher with an alternate-plant design than with an alternate-row design. He suggested that alternate plants reduced the possibility of pollinators being orientated to large blocks of similar plants.

Information on incompatibility and sterility systems in alfalfa was summarized by Barnes et al. (1972). The gametophytic incompatibility system only partially prevents self-fertilization. Some plants show virtually no reduction in fertility after self-fertilization, whereas other plants produce no selfed seed. There is some evidence for inhibition of pollen-tube growth within the stigma and style after selfing. The principal expression of self-incompatibility appears to be in pollen tube and ovule interactions. Busbice (1968) proposed that reduced seed yield during inbreeding results primarily from a loss of heterozygosity in the zygotes which causes lethality.

Both genetic and cytoplasmic-genetic male-sterile systems occur in alfalfa. The genetic system is characterized by complete degeneration of the microspores (Childers and McLennan, 1960). The first cytoplasmic male-sterile system (NS) was described in North America by Davis and Greenblatt (1967), Bradner and Childers (1968), and Pedersen and Stucker (1969). This system shows various degrees of viable pollen production (Barnes and Garboucheva, 1973; Barnes et al., 1974). A second cytoplasmic male-sterile system (ES) has been described in Europe by Guy (1978) and Staszewski and Jakubowska (1978). Genetic steriles have not been used to produce hybrid alfalfa cultivars. The most efficient pollen control method for hybrid seed production is cytoplasmic male sterility. A few three-way hybrid cultivars have been produced utilizing cytoplasmic-genetic male sterility (Barnes et al., 1972). Even though the system is workable, the seed production costs for F_1 hybrids are higher than those for synthetic cultivars. Seed production on male-sterile plants must be improved before significant quantities of F_1 hybrids can be produced commercially.

VI. SEED DEVELOPMENT, HARVEST, AND STORAGE

Pod formation in alfalfa first becomes evident 2 to 3 days after pollination. In successful pollinations, the ovary lengthens and extends several millimeters above the staminal column. In unsuccessful pollinations, the flowers abscise. Seeds are mature when the pods begin to turn brown. Time from pollination to seed maturity is 28 days at a constant temperature of 25 C and 40 days at 20 C.

The two methods for harvesting field-grown seed are windrow-curing and threshing with a pickup combine, or spraying desiccant chemicals and direct combining (Pedersen et al., 1972). Greenhouse-produced seed is usually hand-harvested, air-dried at 35 C, and threshed in some type of laboratory thresher. Most threshers for small-seeded legumes are custom-made and are usually designed to rub pods and seeds between two rough rubber belts or surfaces. One type of thresher for small seed lots was described by Norwood (1967).

Hard seeds, those impermeable to water, are common in alfalfa. According to Gunn (1972) the percentage of hard seeds is governed by edaphic and climatic factors during and after seed maturation and by genetic factors. Hard-seed percentages of 40 to 50 are common in northwestern-grown alfalfa, whereas seed lots produced in the southwestern U.S. usually have less than 20% hard seed. Greenhouse-grown, hand-harvested seed is often nearly 100% hard and should always be scarified. Hard seeds usually are scarified with custom-made equipment that rubs or polishes the seed against a rough surface. The seed coat needs to be scratched without injuring the embryo. Infrared heat treatment (Works and Erickson, 1963) and radiofrequency electrical treatment (Nelson et al., 1977) have also successfully reduced hard seed content.

Alfalfa seed can be long-lived under certain conditions (Gunn, 1972). Reports of viable seed after 25 years of storage in dry conditions are not uncommon. When stored under warm, moist conditions, longevity of the seed will be significantly reduced. Gunn (1972) has suggested the following optimum conditions for long-term storage: fresh, mature seeds with high initial viability, stored in an environment with less than 10% moisture (preferably about 5%), at temperatures near or below 0 C. Low moisture is more important than low temperature.

VII. TECHNIQUES FOR SPECIAL SITUATIONS

Hormones have not been used in either hybridization research or seed production. Grafting techniques have been described (Bernardo et al., 1970), but have not been used in alfalfa seed research. Embryo culture has not been highly developed in alfalfa; but it should improve as emphasis on interspecific hybridization is increased. Recent advances in research with alfalfa callus tissue culture (Bingham et al., 1975) indicates a likelihood of future success in embryo culture research.

REFERENCES

Barnes, D. K. 1966. Flower color inheritance in diploid and tetraploid alfalfa; a re-evaluation. USDA Tech. Bull. no. 1353.

———, E. T. Bingham, J. D. Axtell, and W. H. Davis. 1972. The flower, sterility mechanisms, and pollination control. p. 123–141. *In* C. H. Hanson (ed.) Alfalfa science & technology. Am. Soc. of Agron., Madison, Wis.

———, ———, R. P. Murphy, O. J. Hunt, D. F. Beard, W. H. Skrdla, and L. R. Teuber. 1977. Alfalfa germplasm in the United States: genetic vulnerability, use, improvement, and maintenance. USDA Tech. Bull. no. 1571.

———, and R. A. Garboucheva. 1973. Intra-plant variation for pollen production in male-sterile and fertile alfalfa. Crop Sci. 13:456–459.

———, and C. H. Hanson. 1967. An illustrated summary of genetic traits in tetraploid and diploid alfalfa. USDA Tech. Bull. no. 1370.

———, M. W. Pedersen, J. H. Elgin, Jr., J. D. Axtell, R. A. Garboucheva, and D. M. Smith. 1974. Pollen production classification in male-sterile alfalfa:location and observer effects. Crop Sci. 14:308–310.

———, and M. G. Stephenson. 1971. Relative efficiencies of four self-pollination techniques in alfalfa. Crop Sci. 11:131–132.

Bernardo, E. N., E. L. Sorensen, and R. H. Painter. 1970. Cleft grafting alfalfa. Crop Sci. 10:102–103.

Bingham, E. T. 1968. Transfer of diploid *Medicago* spp. germplasm to tetraploid *M. sativa* L. in 4x-2x crosses. Crop Sci. 8:760–762.

———, L. V. Hurley, D. M. Kaatz, and J. W. Saunders. 1975. Breeding alfalfa which regenerates from callus tissue in culture. Crop Sci. 15:719–721.

Boren, R. B., R. L. Parker, and E. L. Sorensen. 1962. Foraging behavior of honey-bees on selected alfalfa clones. Crop Sci. 2:185–188.

Bradner, N. R., and W. R. Childers. 1968. Cytoplasmic male sterility in alfalfa. Can. J. Plant Sci. 48:111–112.

Bula, R. J., and M. A. Massengale. 1972. Environmental physiology. p. 167–184. *In* C. H. Hanson (ed.) Alfalfa science and technology. Am. Soc. of Agron., Madison, Wis.

Busbice, T. H. 1968. Effects of inbreeding on fertility in *Medicago sativa* L. Crop Sci. 8:231–234.

Campbell, T. A., J. A. Schillinger, and C. H. Hanson. 1974. Inheritance of resistance to anthracnose in alfalfa, *Colletotrichum trifolii.* Crop Sci. 14:667–668.

Childers, W. R., and H. A. McLennan. 1960. Inheritance studies of a completely male sterile character in *Medicago sativa* L. Can. J. Genet. Cytol. 2:57–65.

Clement, W. M., Jr., and E. H. Stanford. 1966. Red-root in alfalfa: Inheritance and relationship with flower color. Crop Sci. 6:569–570.

Coffman, F. A. 1922. Pollination in alfalfa. Bot. Gaz. 74:197–203.

Davis, W. H., and I. M. Greenblatt. 1967. Cytoplasmic male sterility in alfalfa. J. Hered. 58:301–305.

Elgin, J. H., Jr. 1972. Inheritance of the golden tip trait in alfalfa. Crop Sci. 12: 693–694.

———, and J. E. McMurtrey, III. 1977. Effect of four greenhouse supplemental light sources on alfalfa flowering and seed production. Can. J. Plant Sci. 57: 1213–1216.

Gunn, C. R. 1972. Seed characteristics. p. 677–687. *In* C. H. Hanson (ed.) Alfalfa science and technology. Am. Soc. of Agron., Madison, Wis.

———, W. H. Skrdla, and H. C. Spencer. 1978. Classification of *Medicago sativa* L. using legume characters and flower colors. USDA Tech. Bull. no. 1574.

Guy, P. 1978. Unpublished data INRA station d'Amelioration Fourragere, 86600 Lusignan, France.

Hanson, C. H. 1961. Longevity of pollen and ovaries of alfalfa. Crop Sci. 1:114–116.

———, H. O. Graumann, L. J. Elling, J. W. Dudley, H. L. Carnahan, W. R. Kehr, R. L. Davis, F. I. Frosheiser, and A. W. Hovin. 1964. Performance of two-clone crosses in alfalfa and an unanticipated self-pollination problem. USDA Tech. Bull. no. 1300.

Kehr, W. R. 1973. Cross-fertilization of alfalfa as affected by genetic markers, planting methods, locations, and pollinator species. Crop Sci. 13:296–298.

Larkin, R. A., and H. O. Graumann. 1954. Anatomical structure of the alfalfa flower and an explanation of the tripping mechanism. Bot. Gaz. 116:40–52.

Lehman, W.F., and Y. P. Puri. 1967. Rates of germination and tube growth of stored and fresh alfalfa (*Medicago sativa* L.) pollen on agar medium. Crop Sci. 7:272–273.

Lesins, K., and C. B.Gillies. 1972. Taxonomy and cytogenetics of *Medicago*. p. 353-386. *In* C. H. Hanson (ed.) Alfalfa science and technology. Am. Soc. of Agron., Madison, Wis.

Massengale, M. A., A. K. Dobrenz, H. A. Brubaker, and A. E. Bard, Jr. 1971. Response of alfalfa (*Medicago* spp.) to light interruption of the dark period. Crop Sci. 11:9-12.

McGregor, S. E. 1976. Insect pollination of cultivated crop plants. USDA Agric. Handbook no. 496.

Nelson, S. O., W. R. Kehr, L. E. Stetson, and W. W. Wolf. 1977. Laboratory germination and sand emergence responses of alfalfa seed to radiofrequency electrical treatment. Crop Sci. 17:534-538.

Nittler, L. W., and T. J. Kenney. 1964. Induction of flowering in alfalfa, birdsfoot trefoil, and red clover as an aid in testing for varietal purity. Crop Sci. 4:187-190.

Norwood, B. L. 1967. A laboratory thresher for small seeded legumes. Crop Sci. 7: 162-163.

Pedersen, M. W. 1968. Seed number and position in the pod in relation to crossing in alfalfa, *Medicago sativa* L. Crop Sci. 8:263-264.

———, and D. K. Barnes. 1973. Alfalfa pollen production in relation to percentage of hybrid seed produced. Crop Sci. 13:652-656.

———, G. E. Bohart, V. L. Marble, and E. C. Klostermeyer. 1972. Seed production practices. p. 698-720. *In* C. H. Hanson (ed.) Alfalfa science and technology. Am. Soc. of Agron., Madison, Wis.

———, and R. E. Stucker. 1969. Evidence of cytoplasmic male sterility in alfalfa. Crop Sci. 9:767-770.

Rincker, C. M. 1976. Alfalfa seed yields from seeded rows vs. spaced transplants. Crop Sci. 16:268-270.

Stanford, E. H. 1951. Tetrasomic inheritance in alfalfa. Agron. J. 51:274-277.

Staszewski, Z., and B. Jakubowska. 1978. Unpublished data. Inst. of Plant Breeding, Radzikow, Poland.

Teuber, L. R., M. C. Albertsen, and D. K. Barnes. 1977. Floral morphology associated with differences in nectar production in alfalfa (*Medicago sativa* L.). Am. Soc. Agron. Abstr. p. 73.

Tysdal, H. M., and J. R. Garl. 1940. A new method for alfalfa emasculation. J. Am. Soc. Agron. 32:405-407.

Viands, D. R., and D. K. Barnes. 1977. Unpublished data. Dep. of Agron. and Plant Genet., Univ. of Minnesota, St. Paul.

Works, D. W., and L. C. Erickson. 1963. Infrared irradiation—an effective treatment for hard seeds in small seeded legumes. Idaho Agric. Exp. Stn. Res. Bull. 57.

10 Barley

T. M. Starling
Virginia Polytechnic Institute and State University
Blacksburg

Barley (*Hordeum vulgare* L.) is an important feed and food grain grown in temperate parts of the world. It grows well where the ripening season is cool, rainfall moderate, and soil medium-well to well-drained. World production has increased by 60% during the past 15 years. Major producing countries are the USSR, China, Canada, France, United Kingdom, and United States. The most important uses are animal feed, malt for beverage and food products, and human food. Spring barley is predominantly grown, with winter types being grown in regions with mild winters.

I. PARENTAL MATERIAL

Hordeum species are classified into four sections—BULBOHORDEUM, CEREALIA, HORDEASTRUM, and STENOSTACHYS (Price, 1968). Grain-producing forms are in the section CEREALIA. The basic chromosome number is seven with species in CEREALIA being diploid, and those in other sections being either diploid, tetraploid, or hexaploid. Included within CEREALIA are three cultivated species (*H. vulgare* L., *H. distichum* L., and *H. irregulare* E. Aberg and Wiebe) and two wild species (*H. spontaneum* C. Koch and *H. agriocrithon* E. Aberg). Rajhathy et al. (1963) consider all species within CEREALIA to be *H. vulgare* since there is almost complete fertility in crosses among them.

While gene exchange among species within CEREALIA is achieved easily, exchange between this section and other *Hordeum* species and other genera has rarely been reported. Natural hybrids are not known and, although artificial hybrids have been made with species from other sections, their sterility restricts transfer of genes into species of CEREALIA.

A few successful crosses between barley and other genera have been reported. Among these are *H. vulgare* × *Secale cereale* L. (Thompson and

Johnston, 1945) and *H. vulgare* × *Triticum aestivum* L. em. Thell. (Thomas et al., 1977). These intergeneric crosses are sterile and can be maintained only as vegetative clones. The cross with *T. aestivum* was achieved by using barley as the female parent, which is contrary to general expectations.

One of the largest barley collections in the world, consisting of more than 20,000 cultivars, lines, introductions, and wild species, is maintained by the USDA. Germplasm samples and information on entries are available from Curator, Small Grains Collection, Bldg. 046, Beltsville Agricultural Research Center, Beltsville, MD 20705.

II. PLANT CULTURE

Hybridization is facilitated by having vigorous, healthy plants. Soil fertility, moisture, temperature, and lighting are important considerations for production of plants, regardless of whether grown in the field or under artificial conditions. Each location for growing plants has advantages and disadvantages. The choice will be determined by facilities available, population size, crossing techniques to be used, and other considerations.

A. Field

Space limitations are rare and cost of producing plants is less in field nurseries than in greenhouses or growth chambers. Field-grown plants generally tiller well, are hardy, and shed abundant pollen. However, weather conditions in the field can be quite adverse at flowering, making plants difficult to hybridize.

Requirements for growing plants in the field are similar to those for the commercial crop or for breeding and yield nurseries. Best soils are well-drained loams and clay loams with a pH of at least 6.0. Where lime is needed, it should be applied before or at the time of seedbed preparation. Complete fertilizers, depending on soil test results, should be applied before or at planting. Nitrogen applications must be adequate for good growth, but excess quantities cause lodging. With fall-sown barley, part of the nitrogen should be applied at planting and an additional amount applied as top-dressing in early spring. Where supplemental moisture is available, adequate amounts should be supplied to keep plants growing vigorously, especially up to and through flowering.

Low temperatures and short days favor vegetative growth, tillering, and the development of good root systems, while higher temperatures and longer days stimulate seed production. Under field conditions, little can be done to modify photoperiod, temperature, and relative humidity; however, one can plant at an appropriate time to obtain flowering under the most favorable conditions. If working entirely with spring types, planting should be done early in the spring in order that plants will tiller and produce good root systems under short, cool days and will flower before weather gets too hot. With winter types, one should plant early in the fall in order that plants may become well established and tillered prior to cold weather. When working with both spring and winter types, selecting an appropriate planting time becomes difficult. Spring types likely will winterkill if fall planted. If all are spring planted, planting must be sufficiently early to provide adequate cold for vernalizing winter types. Even if winter types are vernalized, they likely will flower later than spring types. One could plant winter types in the fall and spring types in the spring, but spring types usually will flower later than winter types. Problems of this sort favor use of greenhouses or growth chambers to grow plants for hybridizing.

Plants or rows in the field should be arranged so they are accessible for emasculating and pollinating. The design in the nursery will depend on whether one is working with individual plants or rows and the number of crosses to be made. Breeders sometimes use plants from yield or breeding nurseries for crossing purposes. If special plantings are made, plants are spaced 10 to 20 cm apart in short rows or they are solid-planted in short rows with 0.6 to 1 m between rows and rows arranged in tiers with alleyways 0.6 to 1 m wide. This arrangement allows easy access to all rows.

B. Growth Chamber and Greenhouse

Advantages of growing plants in greenhouses or growth chambers rather than in the field include better control of temperature, lighting, water, and relative humidity; less chance of outcrossing and of infection by seed-borne diseases; more opportunity to manipulate flowering to match maturities; more planning of work for several days ahead since weather generally is not a factor; and better scheduling of crossing when it does not conflict with field work. In addition, plants generally contact fewer diseases, pots can be moved and adjusted for effective emasculation and pollination of plants, and seed quality is good. Disadvantages include space limitations, usually a smaller number of tillers per plant, large capital investment, and high cost of energy for lighting, heating, and cooling.

Plants can be grown in pots of various sizes, but a 15-cm-diam clay pot works well. Smaller pots limit tillering, require more frequent watering, and do not provide an adequate base for support of tall plants. Larger pots require more bench space and growing medium, and are difficult to move when needed to facilitate emasculation and pollination. One plant per pot is usually adequate, with two to four being the limit, depending on pot size. Gravel or sand benches are more satisfactory than solid tables because roots grow through holes in pot bottoms into sand and fine gravel, which hold moisture and allow less frequent watering.

Several media are satisfactory for growing plants, and most soil mixes should be acceptable if they have adequate fertility. Because barley grows best in a medium having a pH of 6.0 to 6.5 it may be necessary to add limestone. Even though starting with adequate fertility, it often is necessary to add a water-soluble fertilizer once or twice during the growing period. Plants should have adequate moisture, but the growing medium should not be kept wet continually.

One should attempt to simulate in growth chambers or greenhouses conditions which favor good growth in the field. This means providing cool, short days until the plants have tillered well and then lengthening days and increasing temperatures to initiate flowering. Reid (1977) suggested planting in the greenhouse in late November and maintaining a temperature of 13 C during the day and 7 C at night. If these conditions are maintained for 6 to 7 weeks, both winter and spring types will tiller well and most winter types will be vernalized. At the end of this period, supplemental light should be provided for 2 to 3 hours in the middle of the night and the temperature should be increased. Interrupting the dark period is better than lengthening natural daylight hours and requires a briefer period of supplemental light. A clear, 100-W incandescent bulb placed every 2 m^2 of bench space is adequate to provide the needed far-red spectrum. Day temperature should be increased to 16 C and night temperature to 10 C. Three to 4 weeks later, day and night temperatures should be increased another 3 C, while maintaining the same light schedule. This temperature and light schedule should be maintained until plants mature. Under these conditions, spring types head in early March and winter types from late March through April. When using growth chambers where temperature and light can be fully controlled, date of planting is not critical. In regions where temperatures are cool during early fall, plantings in the greenhouse may be made earlier than November.

Every effort should be made to grow plants under conditions which delay flowering until the spike is well developed, because such spikes are much easier to work with, set a higher percentage of seed, and produce larger seed than poorly developed ones. This primarily is a matter of keeping temperatures and day lengths within the ranges suggested above. Temperatures above 21 C often cause pollination to occur while the spike is deep within the boot. Flower development and pollen shedding are reduced during periods of cool, cloudy weather. This can be partially overcome in the greenhouse by keeping lights on during such periods.

Greenhouse or growth chamber plants require some type of support when tillers start elongating. Plants generally are attached to bamboo stakes or metal rods by means of string, wire, or plastic ties. Sometimes plants are surrounded by a frame covered with wire mesh. Adequately supported plants should require no additional support when glassine bags are placed on them at the time of emasculation.

III. FLORAL CHARACTERISTICS

Barley has an incomplete flower because it lacks sepals and petals. Florets are perfect, with stamens and pistil being enclosed within the same floral structures. The inflorescence is a spike with alternating rachis nodes on each side and with three spikelets at each node. Each spikelet normally consists of a pair of narrow glumes and one floret. The central floret is largest, with laterals ranging from fully fertile to vestigial in different species and cultivars. In six-rowed cultivars, lateral florets are largely fertile, while in two-rowed types they are largely sterile. The outer lemma and inner palea (closest to the rachis) of each floret enclose the stamens and pistil. When awns are present, they are an extension of the lemma. The pistil consists of a two-lobed ovule and a two-branched feathery stigma. Featheriness tends to be reduced in smooth-awned types. There are three stamens, each composed of a two-lobed (four locules) anther and a slender filament which originates at the base of the ovule. Filaments are capable of extensive elongation at flowering and often push anthers outside their enclosures. Two anthers are located toward the palea and the third is toward the lemma. At the base of the ovule toward the lemma are two lodicules which swell and force the lemma and palea apart, exposing the stigma, if pollination does not occur at normal flowering time (Pope, 1944).

Barley is normally self-pollinated, with anthers often dehiscing before spikes emerge from the leaf sheaths (boots) and usually without flowers opening (Pope, 1944). The degree of spike emergence prior to anthesis varies widely with environment and cultivar. Hot weather prior to flowering favors early anthesis (Harlan et al., 1943). Six-rowed types tend to emerge farther before anthesis than two-rowed types. Under field conditions, pollination often is practically complete before awn tips appear, especially in two-rowed types. In the greenhouse, flowering can be delayed until spikes are almost out of the boot if temperatures prior to flowering are kept below 21 C. Under high temperatures or toward the end of the flowering season, spikes flower while deep in the boot and peduncle and floral parts are extremely tender and fragile.

The first florets to mature on a spike are in the central row in the mid to upper part of the spike. This is a good position to check for stage of maturity. Florets mature in both directions from this section, with tip and base florets maturing up to 2 days later. Lateral florets mature later than central ones, but in the same order.

Anthony and Harlan (1920) studied the period of stigma receptivity and pollen viability under field conditions. They emasculated at a stage when pollination might naturally have occurred the following day. Seed set was 100% when pollination occurred 2 days following emasculation and decreased to 0% at 6 days. In their opinion, hybridization failures were more often due to faulty pollen than to any lack of stigma receptivity. They tested pollen viability from anthers in various stages of development by pollinating flowers emasculated 48 hours earlier. The period of pollen viability was short. There was no seed set from immature pollen, 30% when pollen was from anthers 2 to 3 hours away from normal dehiscence, 60% from dehiscing anthers, 40% from anthers which had dehisced 2 to 3 hours earlier, and less than 1% when pollen was from anthers that dehisced 24 hours earlier. Mature pollen grains exposed to free air were shrunken but still germinable after 2 min of exposure. After 10 min, the pollen lost its germinative properties completely. Pollen could be kept for several days by picking spikes just before pollen matured and putting them in an ice-box. When pollen was needed, they placed stems in water in a warm room and the spikes completed development. Reid (1977) held immature spikes in a refrigerator with stems in water to delay flowering.

Pope (1939) studied viability of pollen and ovules of barley following storage at three temperatures: 2.2, 4.4, and 10 C. The storage chamber was dark and maintained at 82% relative humidity. Culms of barley with some spikes emasculated and others not emasculated were cut from plants above the third node below the spike. Their cut ends were placed in vessels of water and they were placed in cold storage within 3 days after emasculation. The extreme limit of storage found for the production of seed from stored, emasculated spikes pollinated with fresh pollen was 42 days at 2.2 C, 29 days at 4.4 C, and 14 days at 10 C. Seeds produced on these spikes germinated, but were reduced in size and sometimes required special treatment, such as germinating on filter paper and transferring to soil. The extreme limits of pollen storage were 26 days at 2.2 C, 19 days at 4.4 C, and less than 14 days at 10 C. Viable pollen was scarce at all temperatures after 1 week. Cold storage of both pollen and emasculated spikes gives a degree of flexibility in matching parents for making crosses.

Harlan et al. (1943) studied effects of temperature on seed set in barley crosses under field conditions at Aberdeen, Idaho. They found that temperatures prevailing 5 days preceding emasculation had a decided effect on seed set. With temperatures in the 15 to 21 C range, pollination up to 4 or 5 days following emasculation gave excellent seed set. At 32 to 38 C, the flowers were ready to pollinate 1 day after emasculation and seed set was reduced with subsequent pollinations. They attributed their failures in hybridization to periods of hot weather during which spikes aborted after emasculation and when it was practically impossible to use certain cultivars because pollen ripened when peduncles and glumes were very immature. Anthony and Harlan (1920) observed little development of anthers in the field during periods of cold, wet weather. Anthers rupture as they lose moisture and as the temperature rises; therefore, the most active period of pollen shedding usually is mid-morning.

Pope (1937) found that pollen germinated in the greenhouse within 5 min after reaching the stigma. Male gametes entered the pollen tube within 10 min, reached the level of the micropyle in 40 min, and within 45 min the two male gametes had entered the embryo sac.

IV. ARTIFICIAL HYBRIDIZATION OR SELF-POLLINATION

A. Equipment

Equipment for emasculating and pollinating barley consists of dissecting forceps, dissecting scissors, tags for recording parents and dates, glassine bags for covering spikes, paper clips for closing glassine bags, and pencils. The scissors should have fine points, a screw joint, and measure about 115 mm in length. Forceps should have fine or medium-fine points with guidepins and a length of 115 mm. They should have a delicate spring to prevent tiring of the fingers while working. The points of the forceps may need modifying to suit one's needs. I grind the points to make them thin and square-edged, and medium-fine rather than fine. Thin, square edges provide a cutting tool for removing spike tips and accessory florets; medium-fine points are less likely to rupture anthers while emasculating. It is convenient to have the forceps and scissors strung around the neck.

Tags used in identifying parents and dates should be sufficiently large to record necessary information, but not large enough to cause bending of the peduncle. A size of 45 by 28 mm with string attached is quite appropriate. Bags for covering heads generally are made from glassine paper, about 38 mm wide by 153 mm long. Dialysis tubing also can be used for covering and can be obtained with a diameter of 25 mm and in rolls of various lengths. Pieces have to be cut to desired lengths and one end stapled or taped. Pope (1944) suggested that, if glassine bags are not available, spikes may be wrapped in ordinary tissue paper or toweling and tied with string.

A stool of comfortable height with storage compartment for equipment and supplies is necessary for field crossing because most of the work will be done in a sitting position. In the greenhouse, one may work standing or sitting and may adjust the height of plants by placing them on inverted pots, benches, or other items of varying heights.

B. Preparation of the Female

Biffen (1907) and Pope (1944) described in detail the mechanics of emasculating and pollinating barley. In general, a spike ready for emasculation will have anthers that are plump and light green to yellowish in color. The spike usually will be in the boot, but will have enlarged until the edges of the flag-leaf sheath have separated to partially expose it. Awns may have emerged 30 to 50 mm depending on the awn length characteristics of the cultivar. A check of central florets in the middle of the spike should suffice to indicate appropriate development. A spike generally premature for emasculating is shown in Fig. 1A, and Fig. 1B shows one that generally would be at the appropriate stage.

Emasculations may be made at any time of day; however, many prefer to emasculate during afternoons because mornings generally are better for pollinating. When emasculating during mornings, one frequently encounters pollen shed and has to remove florets in advanced stages of development or discard entire spikes.

The flag-leaf sheath (boot) is opened either by forcing the leaf edges apart with both thumbs or by splitting the sheath lengthwise with the point of the forceps (Fig. 2A). A gentle rotation and forcing apart of the sheath exposes the spike. The sheath and its attached flag leaf are cut off at the level of the first node of the rachis (Fig. 1C). If cut lower, the weak peduncle may not support the weight of the spike. Lateral florets and all undeveloped florets at the tip and base of the spike are removed using either the forceps (Fig. 1D and 2B) or by grasping the awns between the forefinger and thumb and pulling gently out and downward. Lateral florets of two-rowed barley sometimes bear viable pollen and should be removed from the spike.

From six to eight florets on each side of the rachis are sufficient for a simple cross. When many seeds are desired, the lateral florets also may be emasculated. The most common practice followed for anther removal is to clip the lemma and palea at a level just above the anthers with scissors as described by Biffen (1907) (Fig. 1E and 2C). Proper height for cutting can be determined by holding the spike to light and seeing the position of anthers. Pope (1944) suggests a slanting cut with the lower end toward the operator. The cut generally is initiated near the middle of the lemma and palea at an angle of 30 to 45° from horizontal. The three anthers are removed with forceps (Fig. 2D). If florets are not well developed, two anthers may be hidden in the edge folds of the lemma and palea. In addition to facilitating anther removal, clipping also causes flowers to open widely in a day or two, exposing stigmas for pollination (Fig. 1F). To assure that anthers are removed from all florets, it is best to start with the top floret on one side of the rachis, remove all anthers as you go down the spike, then move to the top floret on the other side and proceed likewise.

Wells and Caffey (1956) described scissor emasculation of wheat and barley. Clipping florets just below anther tips several days before normal dehiscence proved to be an effective and efficient emasculation procedure. The green anther bases were not removed. Clipping in this manner severely pruned the stigma, but had little effect on seed set. Because of smallness of the barley anther, considerable care was necessary to assure that all anthers were cut, especially in the lateral florets where anther tips were at a lower level than in central ones. They postulated that failure of the lower anther portion to form viable pollen after removal of the top portion was probably due to desiccation and death of immature pollen grains. Wells (1962) later

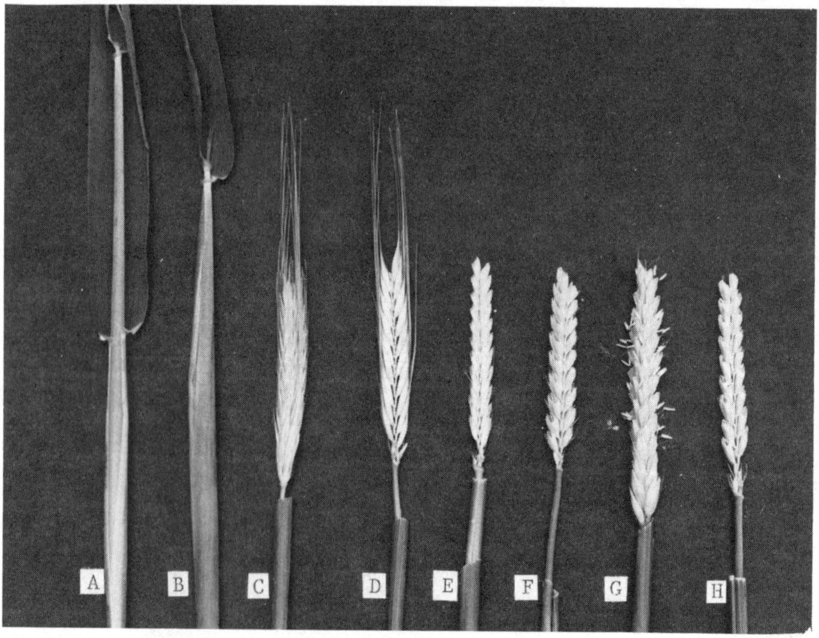

Fig. 1—Barley spikes at various stages of emasculation and pollination. A, Premature for emasculation; B, approximate stage for emasculating as shown by separation of flag-leaf sheath edges and emergence of awns; C, spike unwrapped from flag-leaf sheath with sheath cut off at spike base; D, with lateral florets removed, leaving two central rows; E, clipped florets with anthers removed; F, 2 days after emasculation; G, extrusion of anthers from florets which have been clipped prior to use of spike in pollination; H, a spike 4 days after pollination.

reported on more extensive studies with this technique and indicated a range of 0.7 to 3% self-pollination from scissor emasculation. The use of marker genes to identify hybrid plants would be helpful when using this technique.

Following emasculation, spikes should be covered immediately with glassine bags, sections of dialysis tubing, or tissue paper to prevent outcrossing. If the covering is heavy, it should be attached to a stake to prevent breaking of the peduncle. In the field, coverings should be attached securely at the base of the spike with a paper clip to prevent blowing off. A tag should be hung on the culm showing the plant number or parentage and date emasculated.

C. Pollination

Emasculated flowers generally are ready to pollinate 2 days following emasculation, but may be ready after 1 day during periods of high temperatures. Spikes ready for pollination will have the lemmas and paleas separated and the feathery stigma branches exposed. Pope (1944) suggested that closed flowers found between widely opened ones should be removed prior to pollination because selfing probably has occurred.

From early to mid-morning is the most effective period for collecting pollen and pollinating. Some pollen generally may be found throughout the day, becoming quite scarce past mid-afternoon, especially during periods of high temperature. The stigma appears receptive at all times of day if pollen is available.

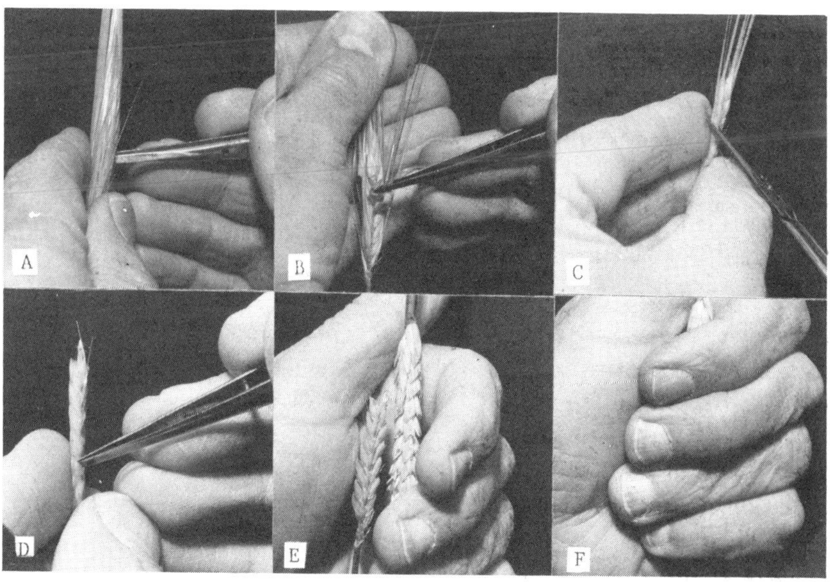

Fig. 2—Steps in emasculating and pollinating barley. A, unwrapping spike from flag-leaf sheath; B, removal of lateral florets; C, clipping across lemma and palea of florets prior to removal of anthers; D, position of fingers while holding spike and forceps during anther removal; E and F, position of female and male spikes in the hand when pollinating by twirl method.

Pollen collecting and pollinating generally are done concurrently rather than having periods of time devoted to each. A spike with suitable pollen is one near emergence from the flag leaf sheath and in which anthesis normally would occur that day. It would be slightly more advanced than the stage described for emasculation. Anthers ready to dehisce will be plump and yellow.

Various methods are used for collecting and applying pollen, depending on individual preference, amount of pollen available, and number of florets to be pollinated. A slow, but reliable, method is collecting individual anthers from spikes cut from plants of the male parent. Florets in the proper stage, usually main florets and some laterals, are cut just above the anther tips. In a matter of minutes, anthers ready to dehisce are pushed out of the flower by elongating filaments (Fig. 1G). These anthers may be used directly or stored in the palm of the hand, on the wrist at the base of the thumb, or in a suitable clothing fold or small container. After removing the cover from an emasculated spike, single anthers are broken against the lemma and palea edges of each floret, allowing pollen to fall on the stigma. A single good anther may be used to pollinate several florets if pollen is abundant.

Pope (1944) described a more rapid method of pollinating which consists of shaking a spike that has dehiscing anthers inside a paper cone which encloses the emasculated spike. Clipping the florets of the male parent facilitates pollen shed. Several female spikes may be pollinated with one good spike of the male parent by enclosing them within the same cone. As new male parents are used, fresh cones are required because pollen may stick to the inside surface of the cone. Wells (1962) described a technique for cone pollinating that was similar to Pope's (1944). He made cones from medium-sized milk shake cups by removing the bottoms, cutting a vertical strip to reduce the diameter, and restapling to form cones approxiately 6.5 cm in diameter at the top and 4 cm at the bottom.

Reid (1977) suggested using one hand to form an enclosure around the emasculated spike, and then twirling the clipped pollen parent spike around the emasculated spike by rolling the peduncle between the thumb and index finger of the other hand as the two spikes are encased loosely (Fig. 2E, F). Hands should be washed when pollen parents are changed.

Anther extrusion and dehiscence can be hastened by placing clipped male spikes in the sun. In the greenhouse, Reid (1977) places the clipped spike on white paper about 15 to 20 cm below the bulb in a crook-necked lamp. Heat, not light, seems to be responsible for speeding anther extrusion and dehiscence.

Rosenquest (1927) described a pollination technique designated as the *approach method.* On spikes used as female parents, all outer glumes, lemmas, and stamens were removed. A male spike near maturity was selected and the outer glumes and lemmas were removed. The two spikes were then fastened together with clips, the male spike against and slightly above the female. Both spikes were covered with a paper bag which was shaken three to four times on succeeding days. This system was adapted for field use by placing the male culm in a bottle of water.

Spikes of female parents must be covered following pollination, generally with the same material used to protect them following emasculation, to prevent outcrossing. Wells (1962) studied the effect of several types of coverings on seed set and seed size. When using kraft paper, wax paper, and glassine paper, there was no difference in percentage seed set. However, kernels covered with kraft paper were 23.8% heavier than those produced under the other coverings.

Following pollination, the male parent and date of pollination are added to the tag attached at emasculation time. The tag may then be folded, clipped to the glassine bag, or cut with scissors to indicate that pollination has been completed.

Success in hybridizing will vary with individuals, general vigor of plants, and prevailing climatic conditions. Inexperienced persons often have little success because they multilate emasculated florets or choose poor pollen sources. An experienced worker expects better than 90% seed set if adequate pollen is available and climatic conditions are favorable.

D. Factors Affecting Efficiency

Several simple techniques may be used to synchronize flowering of parents, a necessity for efficient hybridization. Parents may be planted at two or more dates. Variations in temperature, day length, or both may be used to alter flowering if two or more rooms are available in greenhouses or growth chambers. Flowering is accelerated by growing plants under high temperatures and long days and is delayed with low temperatures and short days. Johnson and Taylor (1958) found temperature effects on heading to be almost negligible at 13-hour photoperiods, but increased temperatures greatly accelerated development with 17-hour photoperiods.

When a cultivar heads too early, the flowering culms may be removed, causing secondary tillers to develop. Early removal of primary culms stimulates development of secondary tillers better than if removed at an advanced stage of development. Spikes of secondary tillers may be small, but generally will be adequate for use as female or male parents. Cold storage of excised culms as suggested by Pope (1939) is useful when flowering of parents is not simultaneous.

The inheritance of numerous traits in barley has been thoroughly studied (Smith, 1951; Nilan, 1964). Many are conditioned by single dominant genes which can be used as marker genes to verify hybridization among F_1 plants.

V. NATURAL HYBRIDIZATION

The extent of natural crossing generally is less than 0.2% (Stevenson, 1928; Robertson and Deming, 1931). Cultivar and seasonal differences exist, with cultivars being a greater source of variation than seasons. Six-rowed types tend to have more natural crossing than two-rowed types, and natural crossing tends to be higher in awnless, awnleted, and naked types.

It is possible to increase natural hybridization through the use of genetic male-steriles. Suneson (1940) was the first to report a genetic male-sterile, conditioned by a single recessive gene. This trait made it possible to produce hybrid seeds on a scale heretofore impossible in barley. Ways in which male sterility can be used in barley improvement were suggested by Suneson (1945) and Suneson and Wiebe (1962). Because male sterile plants fail to produce seed unless fertilized by foreign pollen, its use provided new possibilities for genetic recombination on a continuing basis in populations with this trait, especially when seed is saved only from male sterile plants. Through the use of male sterility, a large number of composite cross populations have been produced and made available to barley breeders throughout the world (Wesenberg and Craddock, 1975). The availability of male sterility has greatly expanded possibilities for recombination and recurrent selection.

When using conventional breeding procedures, and particularly in backcrossing, male sterility can be used effectively to eliminate emasculation. This trait has been backcrossed into a large number of spring and winter cultivars (Hocket et al., 1968). When using male sterility, female spikes should be prepared for pollination by clipping florets and bagging prior to stigma receptivity. Pollination is done in the same ways as previously described for emasculated spikes.

VI. SEED DEVELOPMENT, HARVEST, AND STORAGE

Within 2 to 4 days after pollination, one can tell if fertilization has occurred (Fig. 1H). Kernels reach maximum length in 7 days and mature in approximately 26 days (Harlan, 1920). Seed development in the greenhouse may be enhanced if glassine bags are removed from spikes about 4 days after pollination. If this is done, florets that have not set seed should be removed to eliminate possible outcrossing. Spikes may be harvested and placed in coin envelopes when peduncles turn yellow or seeds at the tip are mature, which usually is approximately 30 days after pollination.

Following harvest, seeds should be kept in a dry place at temperatures below 38 C until fully dried. Threshing generally is by hand to avoid seed damage and loss. Dried seeds may be stored for short periods at room temperature, but should be stored in a freezer or refrigerated seed-storage room for long periods.

Freshly harvested seeds of some cultivars exhibit dormancy. Brown et al. (1948) found that most cultivars germinated at 10 C, whereas a considerable number were dormant at 30 C. Removal of seed coats over embryos gave prompt germination even in the most dormant cultivars. An association of dormancy with winter growth was noted. Reid (1977) suggested avoiding dormancy by harvesting 10 to 14 days after pollination and planting directly. Seedlings from seed harvested this early tend to be weak until fully established.

VII. TECHNIQUES FOR SPECIAL SITUATIONS

Pope (1935, 1942) produced barley seeds with detached spikes. Culms harvested prior to flowering or following emasculation, placed in water, and subsequently pollinated produced seeds which were slightly more than one-third as heavy as normal kernels produced under similar conditions. These small kernels required special handling, such as germinating on blotters, to produce viable plants. This technique could be used to bring excised culms from one location for pollination at another, to hold spikes under refrigeration until a pollen source is ready, to save a pollinated spike which may have broken from the plant, or to bring a large number of spikes together for pollination by the approach method.

REFERENCES

Anthony, Stephen, and H. V. Harlan. 1920. Germination of barley pollen. J. Agric. Res. 18:525–536.

Biffen, R. H. 1907. The hybridization of barleys. J. Agric. Sci. 11:183–206.

Brown, Edgar, T. R. Stanton, G. A. Wiebe, and J. H. Martin. 1948. Dormancy and the effect of storage on oats, barley and sorghum. USDA Tech. Bull. 953.

Harlan, H. V. 1920. Daily development of kernels of Hannchen barley from flowering to maturity at Aberdeen, Idaho. J. Agric. Res. 19:393–430.

———, M. L. Martini, and Harland Stevens. 1943. The effect of temperature on seed set in barley crosses. J. Am. Soc. Agron. 35:316–320.

Hockett, E. A., R. F. Eslick, D. A. Reid, and G. A. Wiebe. 1968. Genetic male sterility in barley. II. Available spring and winter stocks. Crop Sci. 8:754–755.

Johnson, L. P. V., and A. R. Taylor. 1958. Note on the effect of photoperiod and temperature on the development of spike primordia in barley. Can. J. Plant Sci. 38:122–123.

Nilan, R. A. 1964. The cytology and genetics of barley, 1951–1962. Washington State Univ. Res. Studies Monogr. Suppl. no. 3. Vol. 32, no. 1.

Pope, M. N. 1935. The production of barley seed through post harvest pollination. J. Hered. 26:411–413.

———. 1937. The time factor in pollen-tube growth and fertilization in barley. J. Agric. Res. 54:525–529.

———. 1939. Viability of pollen and ovules of barley after cold storage. J. Agric. Res. 59:453–463.

———. 1942. Rate of growth of the embryo of young barley seeds. J. Am. Soc. Agron. 34:200–201.

———. 1944. Some notes on technique in barley breeding. J. Hered. 35:99–111.

Price, P. B. 1968. Interspecific and intergeneric crosses of barley. Barley: origin, botany, culture, winterhardiness, genetics, utilization, pests. USDA Agric. Handbook 338:85–95.

Rajhathy, T., J. W. Morrison, and S. Symko. 1963. Interspecific and intergeneric hybrids in *Hordeum. In* S. Broekhuizen et al. (ed.) Barley Genetics I. Proc. 1st Int. Barley Genet. Symp., Wageningen, Netherlands.

Reid, D. A. 1977. Personal communication to T. M. Starling. USDA, ARS, Western region. Univ. of Arizona, Tucson.

Robertson, D. W., and G. W. Deming. 1931. Natural crossing in barley at Fort Collins, Colorado. J. Am. Soc. Agron. 23:402–406.

Rosenquest, C. E. 1927. An improved method of producing F_1 hybrid seeds of wheat and barley. J. Am. Soc. Agron. 19:968–971.

Smith, Luther. 1951. Cytology and genetics of barley. Bot. Rev. 17:1–51, 133–202, 285–355.

Stevenson, F. J. 1928. Natural crossing in barley. J. Am. Soc. Agron. 20:1193–1196.

Suneson, C. A. 1940. A male sterile character in barley. J. Hered. 31:213–214.

———. 1945. The use of male sterile in barley improvement. J. Am. Soc. Agron. 37:72–73.

———, and G. A. Wiebe. 1962. A "Paul Bunyan" plant breeding enterprise with barley. Crop Sci. 2:347–348.

Thomas, J. B., K. A. Mujeeb, R. Rodriquez, and L. S. Bates. 1977. Barley × wheat hybrids. Cereal Res. Comm. 5:181–188.

Thompson, W. P., and D. Johnston. 1945. The cause of incompatibility between barley and rye. Can. J. Res. 23(C):1–15.

Wells, D. G. 1962. Notes on the hybridization of barley and wheat. Crop Sci. 2:177–178.

———, and H. R. Caffey. 1956. Scissor emasculation of wheat and barley. Agron. J. 48:496–499.

Wesenberg, D. M., and J. C. Craddock. 1975. Personal communication through the 1974 Barley Newsletter. USDA, ARS, Aberdeen, Idaho, and Beltsville, Md., respectively.

11 Broadbean (Faba Bean)

D. A. Bond
Plant Breeding Institute
Cambridge, England

D. A. Lawes
Welsh Plant Breeding Station
Aberystwyth, Wales

M. H. Poulsen
Royal Veterinary and Agricultural University
Taastrup, Denmark

The faba bean (*Vicia faba*) is an erect, annual legume thought to have originated in either Afghanistan (Ladizinsky, 1975) or the Near East (Zohary, 1977). It has been cultivated since Neolithic times (Schultze-Motel, 1972). No wild ancestor has been found. The faba bean is now grown as a cool-season crop in subtropic and Mediterranean climates, and mainly as a spring-sown crop in temperate climates. Some winter sowing of relatively hardy cultivars takes place in England and France. In 1971, the Food and Agriculture Organization (FAO) statistics showed China to have had 3 million out of the world's estimated 4.7 million ha; other important faba bean-producing countries include Italy, Spain, the United Kingdom, Egypt, Ethiopia, Morocco, and Mexico.

Protein content of the grain varies from 25 to 35% of the dry matter. The smaller-seeded *V. faba equina* and *minor* may be handled with the same machinery as used for drilling and harvesting cereals. These types are mainly used as the high-protein fraction of animal feed, whereas the large-seeded *V. faba major* is mainly used as human food in fresh, conserved, or dry form. Faba bean is occasionally used in mixtures with other species for forage or for green manure.

I. PARENTAL MATERIAL

Taxonomically, *Vicia faba* is a member of the section *Faba* of the genus *Vicia,* and several wild species, including *V. narbonensis, V. galilaea,* and *V. hyaeniscyamus,* have been assigned to this section (Ladizinsky, 1975). No successful crosses have been made between *Vicia faba* and these or any other species.

Muratova's (1931) intraspecific classification based mainly on seed size has been widely used. She recognized subsp. *faba* (vars. *major, equina,* and *minor*) and subsp. *paucijuga.* However, Hanelt (1972) recognized subsp. *faba* (vars. *faba* and *equina*) and subsp. *minor* var. *minor,* including *paucijuga,* as a geographical race. Cubero (1974) distinguished only the four varieties *faba, equina, minor* and *paucijuga.* The botanical variety *faba* is still commonly known as *V. faba major,* the largest seeded type within the species. All varieties are diploid with $2n = 12$. Tetraploidy and polyploidy may be induced artificially by colchicine. However, only one verified case of a reproductive tetraploid *V. faba* is known (Poulsen and Martin, 1977).

Crosses between subspecies and between all botanical varieties of *V. faba* are possible, though Abdalla (1977) succeeded with *equina* × *major* and *paucijuga* × *major* only when *major* was the male parent. The breeder also has available variability in locally adapted populations from many parts of the world. Due to the partially allogamous breeding system of the species, a relatively wide variation is maintained within natural populations as well as populations and cultivars created by farmers or plant breeders. A series of naturally occurring and induced mutants has been described (Sjodin, 1971). Genetic collections are maintained at the Zentralinstitut für Genetik und Kulturpflanzenforschung, Gatersleben, East Germany; at the Swedish Seed Association, Svalov, Sweden; at the International Centre for Agricultural Research in the Dry Areas, Aleppo. Syria; at the Germplasm Institute, Bari, Italy; and at the U.S. Department of Agriculture, Beltsville, Md.

II. PLANT CULTURE

A. Field

Soil should be fertile and mineral content, particularly potassium, should be high. Soil pH is important and should be at least 6.5. An efficient *Rhizobium* strain should be present or mineral nitrogen supplied.

Soil moisture must be adequate to ensure good early plant establishment, successful seed set, and prevention of shedding of immature pods. Most faba bean crops are grown on fertile soils with high moisture-holding capacity.

Seed yield per area is relatively unaffected by plant spacing. The small-seeded types are usually grown at a density of around 40 plants per m^2, whereas the large-seeded types are spaced at about 15 plants per m^2.

Seeds should be sown deep, at least 8 cm, to ensure good establishment and to avoid damage by preemergence weed killers and birds. Seeds are usually sown at 5 to 25 cm spacing in rows 25 to 80 cm apart. At least 25 by 50 cm is necessary for access to single plants to be hand-pollinated.

The most common pest is *Aphis fabae,* but severe attacks may occur by weevil (*Sitona lineatus*), pollen beetles (*Meligethes* spp.), thrips (*Kakothrips robustus*), nematodes (*Ditylenchus dipsaci*), and bean beetles (*Bruchus* spp.). Common diseases are chocolate spot (*Botrytis fabae*), leaf spot (*Ascochyta fabae*), powdery mildew (*Erysiphe polygoni*), root rots (*Fusarium* spp. and *Rhizoctonia* spp.), and several viruses. *Ascochyta fabae,* nematodes, and two of the viruses (Broad Bean Stain and Echtes Ackerbohnenmosaik) can be seed transmitted. In the Mediterranean region, the parasitic broomrapes (*Orobanche* spp.) may severely limit production. Control measures for pests and diseases have been described (Litzenberger, 1974).

B. Growth Chamber and Greenhouse

Plants are best grown singly in pots that are no less than 15 cm in diameter. The growing medium should be compost which contains sufficient nutrients to promote growth and flowering and to sustain plants until seeds reach maturity. Because of hard seed coats, it may be necessary to scarify some seeds before sowing to improve evenness of germination. Growing plants should not be allowed to dry out, and automatic watering is preferred.

Hybridization is usually successful, provided the environmental requirements for full development of flowers are met. Except for very early flowering cultivars, flower initiation shows a quantitative long-day response, and long days are required for full development of the initiated inflorescences (Evans, 1959). Sixteen to 18 hours per day are commonly given in controlled environments. The red end of the spectrum is necessary for flowering. Mercury vapor lamps, supplemented by tungsten or by tungsten/halogen bulbs, and high-pressure sodium lamps (400 W/m^2) have been used successfully.

Blondon (1975) reported that, with 18,000 lux of light, winter beans require either cool temperatures (12 to 17 C) or vernalization followed by long days. Use of 30,000 lux and long days eliminate the need for vernalization. Flowering of spring beans in his experiments was independent of temperature, photoperiod, and light level treatment.

Seeds will set under temperatures ranging from 10 C to 30 C, but tolerance of these extremes could depend on genotype. When the temperature is above 23 C in the dark period, a reaction inhibitory to flower initiation occurs.

Common pests in the greenhouse are aphids (*Aphis fabae, Acyrthosiphon pisum, Myzus persicae*), red spider (*Tetraychus urticae*), and whitefly (*Trialeurodes vaporariosum*). These must be kept out of the greenhouse or controlled chemically.

III. FLORAL CHARACTERISTICS

Bees visiting the front of the flower for nectar, pollen, or both can cause cross-pollination. Pollen vectors other than bees have not been reported. Under field conditions, outcrossing may vary considerably, but figures between 25 and 50% have been most commonly reported (Fyfe and Bailey, 1951; Picard, 1953, 1960; Hanna and Lawes, 1967; Bond and Pope, 1974; Poulsen, 1975). Tripping of the flower (releasing the stigma and stamens from the keel petal) is necessary for self-pollination of some genotypes; thus, a bee's visit to the front of a flower can result in self-, cross-, or, what is most likely, mixed pollination.

The complete flowers are borne in axillary racemes and have a typical papilionaceous structure (Fig. 1). Sepals combine into a single five-toothed calyx. The corolla is irregular and made up of five petals—the standard, two wings, and two lower petals that are united along their outer edge to form a keel.

The flower has 10 stamens, with the upper one being physically free. Filaments of the other nine are united in a sheath which encloses the ovary. There is a single ovary with two to four ovules or, in the large-seeded type, up to nine ovules arranged along the inner, upper suture. The fruit is a pod which opens in two valves. Seeds have two large cotyledons and may weigh from 200 to 2,000 mg each.

Fig. 1—Broadbean inflorescence with buds and flowers at various stages of development. The bud marked with the arrow is at the hooded stage and suitable for emasculation. The lowermost flower is at the stage which is usual for pollination, although this can be done at several stages after emasculation.

In most genotypes, flowers develop and mature starting from the 5th to 10th node progressing up the stem. Pollen is shed and forms a plug before the flower opens. The stigma is receptive from a few days before flower opening until the standard petal collapses, which may be a period of from 3 to 5 days.

Pollen-grain numbers and viability differ with genotype, as well as with environment. At high temperatures of 30 C, pollen life is limited to about 1 day, but at lower temperatures of 15 C pollen may survive for several days. Fertilization has been found to occur 24 hours after pollination (van Cruchten, 1977).

IV. ARTIFICIAL HYBRIDIZATION AND SELF-POLLINATION

A. Equipment

Emasculation is usually carried out with small- to medium-sized forceps with fine, but not sharp points. Another technique which has been used successfully and with greater speed is to remove the anthers by means of suction. For this method, a glass tube drawn to a point of 1.5 mm aperture is fitted by a long flexible tube to a suction pump with 0.7 kg/cm^2 of negative pressure operating through a filter to collect the anthers. Pollen is usually transferred to the stigma with the points of forceps, a small paint brush, a piece of filter paper, or the pistil of the pollen donor.

B. Preparation of the Female

When selecting flower buds for emasculation, one should consider their stage of development and the physiological capabilities of the plant to supply the selected flower and the developing pod with nutrients until maturity.

Emasculation must be carried out before the anthers dehisce. As a rule this means at the hooded stage of bud development, when the petals are about twice the length of the calyx (Fig. 1). If, in specific material, pollen has already been released by the hooded stage, younger buds with intact anthers must be chosen for emasculation.

For physiological reasons, the early inflorescences are believed to be the best choice. Good seed sets, however, have been obtained with late flowers under conditions where early seed set has failed or was deliberately prevented. Success is more likely with flowers on those parts of the inflorescence that are nearest the stem. The number of flowers per inflorescence is usually reduced to two or three; especially if many inflorescences on the same plant are to be used.

Selected flower buds may be emasculated at any time of the day. The flower is held between thumb and forefinger and two or three of the sepals covering the keel are torn off with forceps. By a sidewards movement, the points of the forceps are inserted under the standard and wing petals. The pressure on the forceps is gently released and the standard and wing petals folded back and held (Fig. 2A and B). With forceps, the keel is slit open along the suture of the two keel petals. For convenience, removal of part or all of the keel has been practiced, but this is not necessary and damage to the keel should be kept to a minimum to avoid exposure and desiccation of the style and stigma. When the keel is opened, the 10 anthers are removed, preferably by the filaments so that pollen is not squeezed out of the anthers (Fig. 2C). As an extra precaution, forceps and anthers should not be moved higher than the hair surrounding the stigma. Great care must be taken not

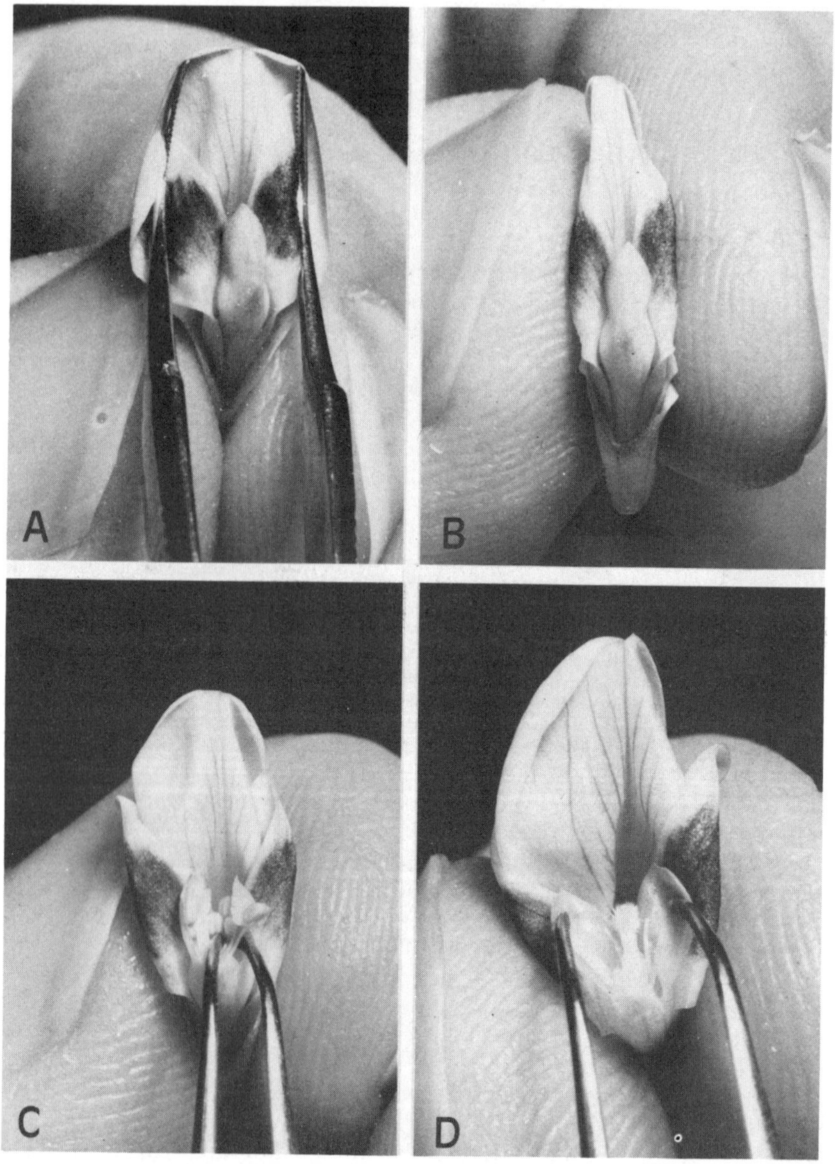

Fig. 2—Emasculation procedure. A, insertion of forceps to part wing petals; B, folding back wing petals between thumb and forefinger; and C, removal of anthers by forceps around the filaments. Photo shows D emasculated bud with all 10 anthers removed.

to cut or tear the style, which can easily break at the bend. When all the anthers have been removed (Fig. 2D), the bud is released and the standard and wing petals rearranged to cover the keel. If the suction pump method is used, the point of the glass tube is used to slit open the keel petals and the anthers are removed by suction. In other respects, the procedures are the same.

C. Pollination

Emasculated flowers may be pollinated immediately after emasculation, but for some genotypes it may be better to wait until the flower is fully developed, which at 25 C takes 2 to 3 days. Pollen may be obtained at any time of the day by tripping a flower, or by slitting the keel and collecting the plug of sticky whitish-grey pollen. Little is known about storage of viable pollen of *Vicia faba*, but experience has suggested that if detached flowers are kept at 5 C, pollen will stay viable for at least 1 week. The pollination technique is to apply ample pollen to the surface of the stigma with forceps or other instrument. The flower is entered either through the slit in the keel petal made for emasculation or by pressing down the wing and keel petals to expose the stigma. The keel petals should preferably appear closed after pollination. Dry conditions favor pollen ripening and flower opening, but petals should be left intact for hybridization to maintain humidity within the flower.

It is easiest to pollinate the flower when it is fully open in the late afternoon, but pollination is effective at any time of day. The stigma remains receptive until the standard petal collapses. It is general experience that application of a relatively large amount of pollen is necessary to ensure fertilization of all ovules in the ovary. When this is required, the pollen plug taken from one flower should not be used to pollinate more than three emasculated flowers. The forceps or paint brush used for pollinating must be cleaned of pollen by dipping into ethyl alcohol and drying before changing pollen donor.

Occasionally, unemasculated flowers are pollinated if genetic markers are being used to identify crosses in the next generation. Pollination of unemasculated young flowers or even buds with undehisced anthers does not prevent selfing.

If a bee-proof greenhouse or cage is not available, flowers may be protected against bee visitation by bagging the plant or stem in a net of about 2 mm mesh. Bags should be in position before the flowers open, and may be removed after emasculation and pollination have been completed and pods have set.

Often a single tagging of an inflorescence is sufficient to identify the emasculated flower and also later to add details of the pollination. Where many different crosses are made on the same stem, it is an advantage to keep a record of the number of the plant or stem, the inflorescence, and the flower. Flowers are easily numbered upwards from the base of the inflorescence because a scar remains if a flower drops or is removed. A record of the female parent, the dates of emasculation and pollination, the identity of the pollen donor, and the numbers of pods and seeds obtained is useful for reference, and for following the progress of the hybridization program.

Under good conditions in a greenhouse, a success rate of 50 to 70% is common for experienced personnel. The number of seeds per pod is usually three or four, although some large-seeded types may produce up to nine seeds per pod.

D. Factors Affecting Efficiency

It is important to have good access to the plants, especially to the females. Where plants can be grown in pots in a bee-proofed greenhouse, they can be moved to a convenient position for emasculation and pollination. It is general practice to encourage seed setting by reducing vegetative development. This is done by removing the stem apex and unwanted lateral tillers when a sufficient number of inflorescences have been developed or initiated.

Success of a crossing program also depends on the matching of flowering dates of the intended parents. The most common method is to plant at about 10-day intervals, so that the range of flowering dates will be covered. A difference of 10 days between sowings may result in only a few days between flowering dates. Early-flowering parents can be held at lower temperatures or their flowers removed when there are no flowers on the parent with which they are to be crossed.

Genetic markers are available for checking crosses or for determining the proportion of selfing when pollination is made without emasculation. Hilum color and testa color are maternally controlled and are frequently used to identify hybrid plants. No markers are known which can be used to identify hybrid seed; however, electrophoretic detection of isoenzymes may provide such a tool in the future. Simply inherited characteristics which have often been used to estimate the degree of outcrossing include black hilum (dominant to colorless), black testa (dominant to the common buff), green testa (recessive to buff), and all-white flower (recessive to normal-colored flower). Seed size and shape, though not simply inherited, are useful when examining progenies of *minor* × *major* or reciprocal crosses, and for roguing contaminated stocks.

V. NATURAL HYBRIDIZATION

In the production of synthetic cultivars, natural hybridization is under only partial control because of the amount of self-fertilization which occurs in the open field (Section III). Experimental attempts have, however, been made for complete control of hybridization by use of male sterility in the production of F_1 hybrid cultivars.

The components of synthetic cultivars may be inbred lines or populations. In some cases they number from 3 to 10, but more commonly 4 or 5 components are used. When first constituting the synthetic, seed of the components can be mixed in equal quantities and the crop pollinated in isolation. This method is common where the components are populations. An alternative method, which is often used when the components are inbred, is to plant the components so they can be identified in the plot or cage.

The seed they produce is then mixed in equal quantities, thereby controlling the proportions of the offspring of the components into the second generation. A cage of about 50 m^2 is sufficient with pollination by honey bees or bumble bees for production of the syn-1 generation, if commercialization on 3,000 ha is expected at about syn-5. The amount of selfing and sibbing relative to crossing in the syn-1 depends on the matching of the flowering periods of the components; their ability to self pollinate, which may be related to their degree of inbreeding; and the efficiency of the pollinating insects. If maximum crossing is not achieved in syn-1, relatively more is expected in later generations and equilibrium of yielding ability may be expected at syn-3 (Bond, 1971; Wright, 1977).

A genetic male-sterile, *ms ms*, has been used to make crosses on a small scale (Bond et al., 1964). Two sterility-inducing cytoplasms have been found together with maintainer and restorer lines (Bond et al., 1966; Berthelem, 1976; Berthelem and Le Guen, 1976), but both are unstable and revert to too high a proportion of fertile plants during large-scale multiplication. Male sterility has, however, allowed the testing of hybrids and the combining ability of parents in more environments than would have been possible by hand crossing.

For small-scale production of hybrids, alternate rows of male-sterile and male-fertile parents, 60 cm apart, have been used. Where commercial hybrid production was experimentally simulated, alternating blocks of 1.5 m width to suit drilling and harvesting machines were employed. Male and female parents were in a 1:1 ratio. One problem is that pollen-collecting bees visit fewer flowers on male-sterile than male-fertile blocks (Bond and Hawkins, 1967; Tasei, 1976).

The isolation requirement for certified seed of commercial *Vicia faba* is in most countries a distance of 200 m. Small plots of breeding material have shown 0.6% contamination at 180 m and even 0.4% when separated by 1 km (Pope and Bond, 1975). In areas where *V. faba major* is frequently grown in gardens, the only way of achieving complete isolation is by caging.

For caging, a net of 3-mm-mesh size to keep out pollinating bees is the main requirement. A very small mesh will keep out more light and should be avoided. Zips at corners of the cages facilitate entry and exit. Supports can be of any material which can be joined to make a frame for the nets. Cages should be about 2 m in height and anchored against wind. The size and shape usually vary according to locally available materials which are adapted according to the needs of breeders. Cages which can be dismantled and transported and which are erected only at flowering time are preferred to permanent structures. The costs of cages will depend on their size, the type of insect screens, and framing material available.

VI. SEED DEVELOPMENT, HARVEST, AND STORAGE

Once pods have developed to a length exceeding 3.0 to 3.5 cm, it is probable that they will develop further to maturity. Before this stage, abortion frequently occurs. Some genotypes may occasionally develop unfertilized ovaries.

In the field in northern and central Europe, seed is usually harvested during August and September, and in the Mediterranean area from April to June. In the United Kingdom, harvest is about 40 weeks after sowing for winter beans and 25 weeks after sowing for spring beans. In the Mediterranean region, the growing period is 20 to 30 weeks and in Scandinavia and Canada about 15 to 20 weeks. In a greenhouse following sowing in February, seed may be harvested within 20 weeks.

For hybridized flowers, seed is shelled from pods by hand. For harvesting field multiplication plots, a conventional cereal combine harvester fitted with the appropriate seed sieves, reduced cylinder speed, and a more open concave setting is used. The modern laboratory threshers are suitable for single plants.

Some mechanical admixtures can be identified by seed size and shape, hilum color, and testa color. This is easier with inbreds than for populations. A check on correctly hybridized seed has to wait until plants are grown in the next generation (Section IV D.).

Seed viability can deteriorate rapidly if the seeds are not either harvested dry or stored at a low temperature. Seed harvested at less than 14% moisture can be stored for 1 to 2 years at a temperature of +5 C or even in a normal laboratory without too much humidity. For storage of up to 5 years, the temperature should be in the range +2 C to −2 C, and it is important that airtight containers be used, as seed moisture content is more critical than temperature. For storage over 5 years, seed moisture content should be reduced to 7% and the temperature to −18 to −20 C.

REFERENCES

Abdalla, M. M. F. 1977. Intraspecific unilateral incompatibility in *Vicia faba* L. Theoret. Appl. Genet. 50:227–233.

Berthelem, P. 1976. Amelioration genetique de la feverole. Cultivar 89:48–53.

———, and J. Le Guen. 1976. Feverole. Rapport d'Activite 1971–77. Station d'amelioration des plantes, Le Rheu, France. p. 119–175.

Blondon, F. 1975. Les exigences pour la floraison de trois lignees de *Vicia faba* L. (feverole). C. R. Acad. Agric. 61:1068–1081.

Bond, D. A. 1971. Breeding methods in field beans (*Vicia faba* L.). p. 119–126. *In* Rep. of Eucarpia Fodder Crops Section Meet., Lusignan, France, September 15 to 17, 1970. CNRA, Versailles 1971.

———, J. L. Fyfe, and Gillian Toynbee-Clarke. 1964. Male-sterility in field beans (*Vicia faba* L.). II. Yield trials of F_1 hybrid winter beans produced with the aid of male sterility. J. Agric. Sci. (England) 63:235–243.

———, J. L. Fyfe, and Gillian Toynbee-Clarke. 1966. Male-sterility in field beans (*Vicia faba* L.). III. Male-sterility with cytoplasmic type of inheritance. J. Agric. Sci. (England) 66:359–367.

———, and R. P. Hawkins. 1967. Behaviour of bees visiting male-sterile field beans (*Vicia faba*). J. Agric. Sci. (England) 68:243–247.

———, and M. Pope. 1974. Factors affecting the proportions of cross-bred and selfed seed obtained from field bean (*Vicia faba* L.) crops. J. Agric. Sci. (England) 83:343–351.

Cubero, J. I. 1974. On the evolution of *Vicia faba.* Theoret. Appl. Genet. 45:47–51.

Evans, L. T. 1959. Environmental control of flowering in *Vicia faba* L. Annu. Bot. (London) 23:521–546.

Fyfe, J. L., and N. T. J. Bailey. 1951. Plant breeding studies in leguminous forage crops. 1. Natural crossbreeding in winter beans. J. Agric. Sci. (England) 45: 141–147.

Hanelt, P. 1972. Die infraspezifische Variabilitat von *Vicia faba* und ihre Gliedering. Kulturpfl. 20:75–128.

Hanna, A. S., and D. A. Lawes. 1967. Studies on pollination and fertilization in the field bean (*Vicia faba* L.). Annu. Appl. Biol. 59:289–295.

Ladizinsky, G. 1975. On the origin of the broad bean *Vicia faba* L. Isr. J. Bot. 24: 80–83.

Litzenberger, S. C. 1974. Guide for field crops in the tropics and sub-tropics. Agency for Int. Dev., Washington, D.C. p. 129–137.

Muratova, V. 1931. Common beans (*Vicia faba*). Bull. Appl. Bot. Genet. Plant Breed. Suppl. 50:285.

Picard, J. 1953. Recherches sur la feverole. Annu. Amelior. Plantes 3:57–106.

————. 1960. Donnees sur l'amelioration de la feverole de printemps (*Vicia faba* L.). Annu. Amelior. Plantes 10:121–153.

Pope, M., and D. A. Bond. 1975. Influence of isolation distance on genetic contamination of field bean (*Vicia faba* L.) seed produced in small plots. J. Agric. Sci. (England) 85:509–513.

Poulsen, M. H. 1975. Pollination, seed setting, cross-fertilization and inbreeding in *Vicia faba* L. Z. Pflanzenzuchtg. 74:97–118.

————, and A. Martin. 1977. A reproductive tetraploid *Vicia faba* L. Hereditas 87:123–126.

Schultze-Motel, J. von. 1972. Die archaologische Reste der Ackerbohne *Vicia faba* und die Genese der Art. Kulturpfl. 19:321–358.

Sjodin, J. 1971. Induced morphological variation in *Vicia faba* L. Hereditas 67: 155–180.

Tasei, J. N. 1976. Les insectes pollinisateurs de la feverole d'hiver (*Vicia faba* var. equina L.) et la pollinisation des plantes male steriles en production de semence hybride. Apidologie 7(1):1–38.

van Cruchten, C. 1977. Personal communication. Station d'Amelioration des Plantes, 21034 Dijon, France.

Wright, A. J. 1977. Inbreeding in synthetic varieties of field beans (*Vicia faba* L.). J. Agric. Sci. (England) 89:495–501.

Zohary, D. 1977. Comments on the origin of cultivated broadbean, *Vicia faba* L. Israel J. Bot. 26:39–40.

12 Buckwheat

Harold G. Marshall
USDA-SEA and Pennsylvania State University
University Park

Buckwheat (*Fagopyrum* spp.) once was common on farms in the north eastern and north central U.S., but today, the crop is of minor importance. Buckwheat is useful as an emergency crop, in double-cropping systems, or as a cash grain crop in northern areas where the growing season is short. Buckwheat grain is used for feeding livestock and poultry and as food for humans. The groats contain 12 to 15% crude protein with a 5 to 7% lysine content (Pomeranz and Robbins, 1972), and are used as a thickening agent and breakfast cereal. Buckwheat flour is used in pancake mixes and flaked breakfast foods.

There has been relatively little effort to improve buckwheat through hybridization, partly because of the difficulties of working with a species which possesses a high degree of self-incompatibility.

I. PARENTAL MATERIAL

Buckwheat belongs to the Polygonaceae family. Species that have been commercially grown are *F. esculentum* Moench (*F. sagittatum* Gilib.), *F. emarginatum* Moench, and *F. tartaricum* (L.) Gaertn. Crosses can be made within these species and between the first two species. There have been no reports of interspecific hybrids with *F. tartaricum*. *F. esculentum* is the major crop species and will be of primary concern in this chapter. Hybridization techniques for *F. esculentum* and *F. emarginatum* are the same.

II. PLANT CULTURE

A. Field

Buckwheat has a reputation for growing on poor soil and thriving in a wide variety of soil types. Fertilizer applications, however, generally will increase grain yields when the crop is grown on soils of low fertility (Marshall,

1969). In most soils, applications of phosphorous are likely to give the most consistent increase in yields. Under certain conditions, buckwheat grain yields also will be increased by applications of potassium and nitrogen. Nitrogen fertilization generally is not advisable on fertile soils because it tends to cause excessive vegetative growth at the expense of seed development. Soil tests are the best guide for establishing fertilizer requirements, but in the absence of a soil test 168 to 224 kg/ha of 10–20–20 fertilizer should be applied.

Buckwheat thrives under relatively moist, cool conditions. Temperature and moisture stresses are the factors most likely to reduce seed set. In temperate zones, buckwheat should be planted 10 to 12 weeks before the first killing frost in the fall. This increases the probability that peak blossoming will occur after the highest summer temperatures. High temperatures increase water absorption and shorten the life span of the plant (Hall, 1950). Air temperatures of 17 to 19 C (day) are most satisfactory during blossoming (B. O. Ruszkowska, 1965; M. Ruszkowska, 1965). At higher temperatures, pollen tube growth is likely to stop before sufficient extension can occur for fertilization (Sugawara, 1956). Satisfactory crossing results can be achieved under higher daily maximum temperatures than those cited, but pollinations should be made during early morning. Temperatures can be reduced by moving portable shades over the plants for a couple of hours during and after pollination.

Buckwheat needs about 70 mm of rainfall from germination to full bloom, and another 20 mm from full bloom to maturity (M. Ruszkowska, 1965). The leaves of buckwheat tend to wilt when most other crop plants are showing no stress, and a water deficit may inhibit fertilization or cause abortion of the zygote. Irrigation is advisable during periods of drought and high-temperature stresses.

Where crossing is done in the field, the plants should be spaced at least 30 cm apart in rows 90 cm apart to provide space for working. The plants should be tied to stakes to prevent lodging.

B. Growth Chamber and Greenhouse

Standard potting mixes, fertilization, and watering schedules that provide good growth of other crops work well for buckwheat. Greenhouse and growth chamber temperatures are best if they do not exceed 19 C. In most areas, summer greenhouses must have artificial cooling equipment to provide optimum temperatures for buckwheat hybridization.

Relative humidities during and after pollination in buckwheat should not fall below 50 to 60% for best results (M. Ruszkowska, 1965). Humidification equipment in growth chambers is helpful.

Buckwheat is considered indeterminant, but photoperiod may influence growth habit and the time and rate of flowering (Shok and Scully, 1955; Sugawara, 1958; M. Ruszkowska, 1965). Special photoperiods generally are not necessary for successful buckwheat hybridization, but more desirable plants may result when buckwheat is grown under intermediate day lengths of 12 to 15 hours. Zebrak (1959) found that day lengths of 9 hours or less retarded the onset of flowering, and B. O. Ruszkowska (1965) reported that fewer inflorescences were produced under diminishing day length. Satisfactory results can be obtained in growth chambers at light levels of 18 klux or more. Light level in the greenhouse normally is not a limiting factor, but plants may become spindly when grown during relatively dark winter seasons.

Because of the plant heights attained by buckwheat, growth chambers should have a minimum of 1.5 m distance between the lights and the top of the shelf or bench.

III. FLORAL CHARACTERISTICS

Buckwheat plants are herbaceous, erect annuals with single main stems that usually bear several branches. The plants produce blossoms daily for about 4 weeks. Buds and flowers are densely clustered in racemes at the ends of branches or on short pedicels that arise from the axils of leaves. Within a raceme, the general blossoming sequence is from the base to the apex and the open flowers hang at various angles. Flowers are open for only 1 day, and the stigmas normally are receptive as long as the flower is open and not withered.

The flowers of *F. esculentum* are perfect, but incomplete. There are no petals, but the calyx is composed of five petal-like sepals. This species has plants bearing one of two floral types. The pin type has flowers with long pistils and the short stamens (Fig. 1), and the thrum type has flowers with

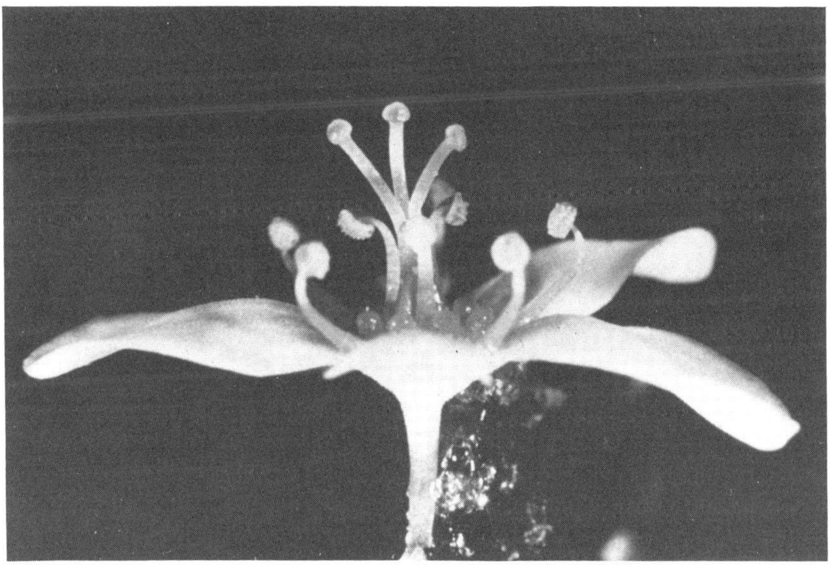

Fig. 1—Typical pin flower of buckwheat showing the long, three-parted style with knoblike stigmas and the short stamens. The anthers have dehisced and the pollen grains are exposed.

short pistils and long stamens (Fig. 2). Flowers with styles and stamens of the same length (Shock-Bodmer, 1934; Esser, 1953; Marshall, 1969) and lines with only one floral type (Marshall, 1969; Fesenko and Antonov, 1973), have been reported. The pistil consists of a one-celled superior ovary, three-parted style with knoblike stigmas, and eight stamens. Three of the stamens closely surround the pistil and open outwards, while the other five are inserted more outwards, alternating with the sepals, and open inwards (Fig. 1). Glands which secrete nectar are located at the base of the ovary.

The flowers of *F. tataricum* differ from those of the two other domestic species. They are smaller, have inconspicuous greenish-white sepals, are unattractive to insects, and have flowers with anthers and stigmas on the same level.

Fagopyrum esculentum is naturally cross-pollinated, primarily by insects. Although some selfing occurs, it usually is prevented by a heteromorphic, sporophytic incompatibility system. Seed is produced when cross-pollinations occur between the pin and thrum flowers. Pollen tube growth from incompatible pollen stops soon after stigma penetration in short styles, but will eventually extend about two-thirds the length of typical long styles. In addition to differences in morphology and incompatibility reactions, the pin flowers have larger pollen grains than thrum flowers.

Inheritance of flower type is reported to be monogenic (Garber and Quisenberry, 1927). A 1:1 ratio of plants with the two flower types can be expected in a population because of the incompatibility system. Sharma and Boyes (1961) indicated that the locus probably is a complex one similar to

Fig. 2—Typical thrum flower of buckwheat showing the short, three-parted style and the long stamens (two were removed). The anthers have dehisced and the pollen grains are exposed.

the model proposed by Lewis (1954) for *Primula*. The complexity of the genetic control is supported by my development of self-fertile lines that breed true for reduced style length within the pin flower complex (Marshall, 1969). Pollen tubes apparently are inhibited in their growth, but reach the embryo sac because of the reduced style length.

Although the incompatibility system in buckwheat is not absolute, efficient hybridization generally is dependent upon cross-pollination between pin and thrum flowers. Controlled cross-pollinations between similar flower types are not likely to yield hybrid seeds, except those between genotypes with pin flowers and reduced style length.

The F_1 plants normally cannot be selfed to produce an F_2 generation. One must produce enough hybrid seed to have both pin and thrum plants of the desired type for sib-mating in the F_1 generation. The choice of breeding methods for buckwheat is restricted.

Pollen viability and longevity undoubtedly are affected by temperature and humidity, but specific data are limited. Buckwheat anthers dehisce early in the morning, and essentially turn inside out during the process. The pollen is sticky and is arranged in rows on the walls of the inverted anthers. This stickiness facilitates the accumulation of pollen on the bodies and extremities of insects visiting the flowers. The amount of time between the opening of the flower and anther dehiscence can be increased by raising the relative humidity, but specific data are not available.

After compatible pollinations, pollen tubes reach the base of short styles in 5 to 15 min and of long styles in 15 to 20 min (Shock-Bodmer, 1930; Morris, 1951; Tatebe, 1956). Information on time to fertilization is inconsistent. Stevens (1912) reported that a three-celled proembryo was present 18 hours after compatible pollinations. He found no evidence of fertilization 72 hours after incompatible pollinations, but some pistils had embryos in the 8 or 16-celled stage 96 hours after such pollinations. Mahony (1935) reported that fertilization occurred between 48 and 60 hours after compatible cross-pollinations. Modilevskjii (1951) found that the length of time to fertilization varied with conditions. On the average, pollen tubes reached the embryo sac in 6 to 8 hours, but under favorable conditions fertilization had occurred by 3 hours after compatible pollinations.

IV. ARTIFICIAL HYBRIDIZATION AND SELF-POLLINATION

A. Equipment

Buckwheat flowers can be emasculated by using either a forceps or a low-capacity electric aspirator. The electric aspirator should be equipped with about 1.4 m of flexible rubber tubing with 5 mm inside diameter, and a glass tubing tip drawn to an inside diameter that will allow the anthers to pass through. A vacuum of about 45 cm Hg is adequate to remove the anthers. The use of a binocular-loupe or magnifying glass (5× or more) is essential to be sure the anthers have not dehisced and that the stigmas are free of pollen. A small pair of scissors also is helpful to remove unwanted racemes during plant preparation.

B. Preparation of the Female

Emasculation is not necessary if a small amount of selfing can be tolerated. Because of the self-incompatibility, mixed-pollinations should result in fertilizations by sperm nuclei from compatible pollen. Emasculation, however, is the only sure way of preventing self-fertilization.

Emasculations can be made either in the afternoon the day before pollination or early in the morning the day of pollination. The latter technique is easiest because the flowers are open and the anthers are more accessible. In either case, a young raceme should be selected and all old flowers should be pinched off with the tweezers. If one parent is used to pollinate all emasculated flowers on a plant, flowers on a raceme may be emasculated for several days in succession. All remaining buds should be removed when cross-pollinations on a raceme are completed.

The objective during emasculation is to remove all eight anthers before dehiscence. If emasculations are to be made on open flowers early in the morning, some procedure to delay dehiscence is necessary, especially when more than a few flowers are to be done. Morris (1951) set her plants in a moist chamber on the previous evening to keep the pollen from shedding before and during emasculation. For my hybridization work in the winter greenhouse, I have been able to delay dehiscence by using a night temperature of about 13 C before the day of emasculation. The temperature should not be raised until emasculations are completed. If the flower is open or partially open, the anthers are easily removed by suction or with a forceps.

Buds that will bloom the following day can be emasculated without danger of self-pollination, but the task is tedious and best done with suction. The axis of the raceme and the receptacle area of the flower are grasped between the thumb and finger, one or two sepals are removed with the forceps or held down between thumb and finger, and the eight anthers are removed with the suction tip of the aspirator. One must open the calyx sufficiently to see all filaments and ascertain that all anthers have been removed. This procedure is difficult, and the hybridizer will prefer to work with open flowers.

Protection of flowers probably is not necessary in greenhouses that exclude insects. If winged insects or strong drafts are likely to be present, a raceme with emasculated flowers should be covered with a light-weight glassine paper bag for pollen control. The open end can be closed with a paper clip. If support of the bag is necessary, one method is to put a metal eyelet in one corner of the bag to tie it to a stake or branch. If growth chambers have vigorous forced-air movement, bagging of the emasculated flowers is advisable. Bagging is necessary in the field, and care is required to prevent breakage of the axis of the raceme.

C. Pollination

A brush of camel hair can be used to transfer pollen between flowers for cross-pollination. Pollen mixture can be avoided by using a different brush for each pollen parent or by cleaning brushes in alcohol prior to re-use. A more tedious method of cross-pollination is to use a forceps to remove flowers from the pollen parent and brush them across the female flowers to transfer pollen from anthers to stigmas.

Pollen should be collected and transferred as soon as possible after anther dehiscence in the early morning. The presence of pollen generally is determined either by touching an anther with the thumbnail or by viewing the anthers with a binocular-loupe or magnifying glass. I know of no information concerning storage of pollen.

The number of flowers setting seed following controlled hybridization is usually less than 10%, even under ideal conditions. Low seed set is characteristic of buckwheat, even after compatible natural cross-pollination. Some improvement may result from removing all but a few racemes on the female plant (M. Ruszkowska, 1965).

Identification of individual flowers used in crossing generally is not practical or necessary when only one pollen parent is used for a raceme. A good procedure is to write the male and female identification on a small tag and hang it at the base of the axil of the raceme.

No useful genetic markers for identifying hybrid seed of plants have been described in buckwheat. The chances of obtaining selfed seeds, however, are remote because of the self-incompatibility.

Self-pollination can be accomplished with a fine brush of camel hair by brushing the open flowers gently while moving from raceme to raceme. The best way to avoid mixed pollination is to use a different brush for each plant. Marshall (1969) developed inbred lines using this technique on individual plants isolated in different greenhouse sections. Individual plants also may be self-pollinated by placing them in screened cages with house flies (*Musca domestica* L.) or honeybees (*Apis mellifica* Linne).

V. NATURAL HYBRIDIZATION

The self-incompatibility mechanism in buckwheat can be used to make crosses between cultivars or strains—generally on a multiplant basis (Elagin, 1968; Alyeksyeyeva, 1971). The scheme is to eliminate all the pin flower plants in one cultivar and all the thrum flower plants in the other cultivar. Several field layouts can be used to facilitate checking the plants for flower type and to provide a satisfactory plant density in the remaining stand. One approach is to plant the two parents in adjacent, single-row plots with 30 cm between rows and 60 cm between pairs of plots to provide space for passage while checking and eliminating plants. Because the pin and thrum flower plants occur in about a 1:1 ratio, a 10 cm planting space between seeds should provide enough remnant plants of the desired flower type. A thicker stand causes difficulty in checking individual plants. Because of the branching of buckwheat, one plant every 30 cm of row will give good results. Another technique is to plant hill plots on about 60-cm centers, four to six seeds per hill, and subsequently eliminate all but the desired plant type from a hill.

The isolation requirements for buckwheat have not been precisely established. Nettevic and Fesenko (1964) reported that interpollination takes place only between closely adjacent rows. They used 8 to 10-m-wide strips of tetraploid buckwheat to isolate diploid strains because tetraploid and diploid strains will not intercross. I have separated diploid strains with 1.5-m-wide strips (five rows) of sorghum × sudan hybrids planted at the same time as the buckwheat.

In a single-cross experiment with inbred lines, almost complete pollen control was achieved by enclosing each crossing block (4 m^2) on three sides with 1.2-m-high muslin (Marshall, 1978). The open side was away from prevailing winds. The partially enclosed blocks were spaced 2 m apart. Based on bee foraging studies by Butler et al. (1943), little interblock bee visitation is likely to occur.

Single plants or small groups of plants also may be completely isolated by covering them with muslin stretched over wire or wooden frames. Gauze is pervious to pollen and is not satisfactory. Caging can cause problems because the plants must be covered for several weeks during the blossoming period. Certain insects may increase under the cages and injure the plants or blossoms. In wet weather, fungus problems may develop under such cages and, in hot weather, high temperatures may be a problem. The fabric covering should be light in color and sheer.

VI. SEED DEVELOPMENT, HARVEST, AND STORAGE

The seed of buckwheat is a fruit which is an achene. Seeds develop rapidly and successful fertilizations can be detected in 3 to 5 days after fertilization. Because of the seed abortion that occurs in buckwheat, a variable percentage of these will not develop viable seeds.

The seeds of buckwheat are green in color until they are physiologically mature, then the hull (pericarp) turns black, brown, or gray in either a solid or mottled pattern. The crossed seeds should be harvested when physiologically mature because mature seeds may sprout on the plants under prolonged wet conditions.

Where relatively few crossed seeds per female plant are to be harvested, they can be hand-picked and placed in paper packets for storage. These packets should be identified with the male and female plant, pot, and plot numbers, or cross number. This information should be recorded elsewhere with parental identifications.

If all the seeds on individual plants are to be harvested, the simplest procedure is to strip off all the seeds by hand and place them in appropriate-sized packets or bags. Trash can be picked out by hand if only small amounts of seed are involved. For larger quantities of seed, trash and aborted seeds can be removed by screening or by forced air in a seed cleaner, or both. When several plants or small plots are to be harvested following natural hybridization, the plants can be cut with a sickle or a small gasoline-powered mower equipped with a catch pan. The plants should be carefully bundled and placed in a kraft paper or sheer cloth bag to prevent seed loss because of shattering. The bundles should not be large and some method of drying may be necessary because of the high moisture typical of buckwheat plant material at harvest time. The maximum temperature for drying is 43 C. After the plants are dry, threshing can be accomplished by beating the bags against a surface and removing the plant material. The remaining seeds can be cleaned as described above.

Small-grain plot threshers can be used with buckwheat, but the cylinder speed must be slowed down sufficiently to avoid damaging or dehulling the seeds. For the small amounts of seed involved in hybridization work, hand threshing is the best choice.

Buckwheat seed from hybridization should be stored where it will be safe from insects and rodents. Moisture content should be 12% or less. The longevity of buckwheat seeds is shorter than that of seeds of the true cereals. If seeds are to be retained for more than 1 or 2 years, storage at a temperature of 5 C or less and 40% relative humidity or less will prolong longevity. Any reduction in these two environmental components will help improve longevity.

VII. TECHNIQUES FOR SPECIAL SITUATIONS

A number of Russian scientists have used grafting, generally referred to as vegetative hybridization, in their breeding work with buckwheat. Much of this work was influenced by Michurin's ideas regarding heritable changes resulting from the influence of the rootstock on the scion. Petelina (1953) obtained so-called diploid × tetraploid hybrids by grafting, and reported F_1 heterosis for plant height and amount of grain per plant. Nettevic (1959) reported that the pollen preference of female parents was changed by grafting them on other cultivar, and that grafting could be used to overcome incompatibility in certain wide crosses. Based on modern Russian literature, it seems that vegetative hybridization is no longer in use. Grafting has not been used in buckwheat hybridization work in countries outside the Soviet block.

REFERENCES

Alyeksyeyeva, O. S. 1971. The characteristics of heterosis in buckwheat. [In Russian]. Nauk. pratsi. Nauk.-dosl. in-t zemlerobstva i tvarinnitstva zakhidn. r-niv USSR 1971 (17):231–237. (Plant Breed. Abstr. 45, 8291).

Butler, C. G., E. P. Jeffree, and H. Kalmus. 1943. The behavior of a population of honeybees on an artificial and on a natural crop. J. Exp. Biol. 20:65–73.

Elagin, I. N. 1968. The use of heterosis in buckwheat. p. 125–132. *In* Heterosis in plant breeding. Leningrad: 'Kolos'. [In Russian]. (Plant Breed. Abstr. 41, 4506).

Esser, K. 1953. Genome doubling and pollen tube growth in heterostyle plants. [In German]. Z. indukt. abstamm.-u. VererbLehre 1953 (85):28–50.

Fesenko, N. V., and V. V. Antonov. 1973. New homostylus form of buckwheat. [In Russian]. Byul. nauch.-tekhn. inform. VNII zernobob. kul'tur 1973 (5):12–14. (Plant Breed. Abstr. 46, 10172).

Garber, R. J., and K. S. Quisenberry. 1927. The inheritance of length of style in buckwheat. J. Agric. Res. 34:181–183.

Hall, W. C. 1950. Growth and development of buckwheat under differential temperature gradients. Bot. Gaz. 111:331–343.

Lewis, D. 1954. Comparative incompatibility in angiosperms and fungi. Adv. Genet. 6:235–285.

Mahony, K. L. 1935. Morphological and cytological studies on *Fagopyrum esculentum.* Am. J. Bot. 22:460–475.

Marshall, H. G. 1969. Isolation of self-fertile, homomorphic forms in buckwheat, *Fagopyrum sagittatum* Gilib. Crop Sci. 9:651–653.

———. 1978. Studies of inbreeding depression, breeding behavior, and heterosis with inbred lines of buckwheat. Crop Sci. 19:110–114.

Modilevskjii, Ia. S. 1951. A cytological and embryological description of the grain types of a cultivated buckwheat showing premature cessation of development. Bot. Zh. SSSR 1951. (8):3–14. [Translated from Russian by R. Ericson, 1959].

Morris, M. R. 1951. Cytogenetic studies on buckwheat. J. Hered. 42:85–91.

Nettevic, E. D. 1959. Selectivity of fertilization in vegetative-sexual hybridization of buckwheat. [In Russian]. Agrobiologija 1959. (4):533–539. (Plant Breed. Abstr. 31, 1781).

———, and N. V. Fesenko. 1964. A biological method of isolating common buckwheat. [In Russian]. Selekcija i Semenovodstvo 1964 (29):41–44. (Plant Breed. Abstr. 34, 5583).

Petelina, N. N. 1953. Vegetative hybridization of buckwheat. [In Russian]. Selekcija i Semenovodstvo 1953 (5):37–40. (Plant Breed. Abstr. 24, 1043).

Pomeranz, Y., and G. S. Robbins. 1972. Amino acid composition of buckwheat. J. Agric. Food Chem. 20:270–274.

Ruszkowska, B. O. 1965. Studies on buckwheat. XIII. Analysis of the fruitification structure of buckwheat on the background of environmental conditions. [In Polish with English summary]. Zaklad Roslin Zbozowych IUNG, 1965:77–97.

Ruszkowska, M. 1965. Studies on buckwheat. Effect of growth and development on the yield. [In Polish with English summary]. Zaklad Roslin Zbozowych IUNG, 1965:3–76.

Schock-Bodmer, Helen. 1930. Heterostyly of *Fagopyrum esculentum.* Investigation of pollen tube growth and the suction strength of the style and pollen grain. [In German]. Schweizerische Bot. Gez. Berichte 39:4–15.

———. 1934. The heterostyly problem: pistil characteristics and pollen tube growth in *Fagopyrum esculentum.* [In German]. Planta 1934 (22):149–152.

Sharma, K. D., and J. W. Boyes. 1961. Modified incompatibility of buckwheat following irradiation. Can. J. Bot. 39:1241–1246.

Shok, Y., and N. J. Scully. 1955. Nature of the photoperiodic responses of buckwheat. Bot. Gaz. 117:134–141.

Stevens, N. E. 1912. Observations on heterostylous plants. Bot. Gaz. 53:277–308.

Sugawara, K. 1956. On buckwheat pollen. III. The relation between pollen germination and temperature. [In Japanese]. Proc. Crop Sci. Japan 28:264–265. (Plant Breed. Abstr. 27, 1648).

———. 1958. On the injury of pistil growth. Retardation of pistil growth as influenced by day-length. [In Japanese]. Proc. Crop Sci. Soc. Japan 26:269–270. (Biol. Abstr. 33, 19329).

Tatebe, T. 1956. Physiological studies on fertility in buckwheat. V. Studies on the behavior of the pollen and pollen tube. [In Japanese with English summary]. Jap. J. Breed. 6:156–162.

Zebrak, E. Z. 1959. Photoperiodic reaction of tetraploid and diploid buckwheat. [In Russian]. Doklady Akad. Nauk. SSSR 1959 (128):176–178. (Field Crop Abstr. 14, 50).

13 Cassava

Kazuo Kawano
Centro Internacional de Agricultura Tropical
Cali, Colombia

Cassava (*Manihot esculenta* Crantz) is one of the most important calorie-producing crops in the tropics. It is called *mandioca* (Brazil, Paraguay, and Argentina), *yuca* (other Spanish-speaking countries), *tapioca* (Asia), and *manioc* (French-speaking Africa). Cassava originated in Latin America and has probably completed the major part of its diversification there (Leon, 1976).

The crop accounts for 59% of tropical root and tuber production (FAO, 1976). Its production is fairly well distributed among the tropical zones of America, Africa, and Asia. Approximately 300 million people in the tropics depend on cassava as a major source of their calorie intake (Nestel, 1973). Aside from human consumption, cassava has important uses as animal feed and for starch and alcohol production.

I. PARENTAL MATERIAL

All cultivated cassava belong to the species *Manihot esculenta* in the family *Euphorbiaceae.* The species is considered allotetraploid with 36 chromosomes that form 18 bivalents at meiosis (Umanah and Hartmann, 1973). The genus *Manihot* contains more than 100 species (Viegas, 1937).

Limited research has been done on the genetic and evolutionary interrelationship among these species. Only few wild species within the genus, for example, *M. glaziovii, M. saxicola, M. melanobasis, M. catingae,* and *M. dichotoma,* have been used in hybridization with cultivated cassava. Among these, *M. glaziovii* has produced outstanding results in introducing resistance to cassava mosaic disease (Jennings, 1970).

There is great germplasm variability within *M. esculenta* for several agronomically important characters such as root yield, harvest index, root dry matter content, HCN content, and root perishability after harvest

(CIAT, 1973, 1974, 1975, 1976). Resistances to some of the major diseases and insects such as cassava bacterial blight (*Xanthomonas manihotis*), *Cercospora* leaf spot, *Phoma* leaf spot, and thrips have been identified (CIAT, 1975). The total germplasm variation within the species is far from being fully exploited. Presently, the major part of cassava hybridization is being done within the species.

Small germplasm collections are scattered among many tropical countries, but the collection at the Centro Internacional de Agricultura Tropical (CIAT) is the only one on an international scale. It has approximately 2,400 accessions collected throughout Latin America. The CIAT germplasm unit is helping to establish a large collection in Brasilia, Brazil. A good duplication of the CIAT germplasm is being established at the International Institute of Tropical Agriculture (IITA), Ibadan, Nigeria, in addition to native African germplasm.

II. PLANT CULTURE

A. Field

Cassava is successfully grown between 30° N Lat and 30° S Lat, from sea level to about 2,000 m, under annual precipitation regimes from 600 to 6,000 mm, and with a soil pH between 3.8 and 8.0. The species tolerates hot climates, but does not do well when daily average temperatures drop below 18 C. Commercial cassava production uses stem cuttings, and the roots are harvested between 8 and 18 mo after planting.

No systematic study has been conducted on the effect of environmental factors on flowering and seed set. Many genotypes flower reasonably well under a wide range of environmental conditions. The plants seldom bear receptive flowers during the first 6 mo. For tropical conditions, the majority of cassava cultivars flower from 8 to 16 mo after planting. Several cultivars have never flowered in 2 years of being in the field.

Flowering is rare during a long dry season, so that irrigation of the pollination field during the dry season is important. Large spacing between plants seems to give the plants a better chance to flower. Hand pollination is extremely difficult when cassava is planted at a commercial production spacing of 1 by 1 m; therefore, a planting distance of more than 1 by 2 m, is recommended.

B. Growth Chamber and Greenhouse

There is no information on the production of cassava in a growth chamber or greenhouse. A cassava plant, ready for pollinations, usually occupies more than 2 m^2 of area and is 2 m tall. Thus, considerable greenhouse space would be required.

III. FLORAL CHARACTERISTICS

A flower bud is formed every time the plant branches, however, most of the flower buds formed during early growth stages are abortive. Flowers cannot be obtained from nonbranching types. Cassava is monoecious. On a given branch, female flowers always open first and the male flowers follow after 1 or 2 weeks.

Both self and cross-pollinations occur naturally in cassava. There seems to be no physiological or genetic mechanism to prevent self-fertilization in normally flowering types. No cross-incompatibility has been found so far. The pollen are relatively large in size and sticky, therefore, natural pollination by wind is unlikely. Several species of wasps (mainly *Polistes* spp) and bees are the main pollinators in Colombia and West Africa, respectively (Hahn, personal communication).

Self-pollination varies between 0 and 100%, depending upon the genotypes and planting distance (CIAT, 1974). A distance of 30 m between pure stands of different genotypes was sufficient to prevent cross-pollination between two populations (CIAT, 1974). However, 500 m is suggested for a perfect isolation of two populations in genetic studies.

IV. ARTIFICIAL HYBRIDIZATION AND SELF-POLLINATION

A. Equipment

Large cloth bags measuring 20 by 25 cm are used to protect flowers from undesired pollen. Small cloth bags measuring 10 by 18 cm are used to catch matured seeds. The bags can be made of any cotton cloth of 60 to 80 mesh, and should have a string attached to close the mouth of the bag on the plant. Tags of 3 by 4 cm are used to identify pollinated flowers. Bottles of 3 to 5 cm in diameter and 5 to 8 cm high are used to transport male flowers during pollination.

B. Preparation of the Female

After a little experience, the breeder can determine with relative ease in the morning which particular female flower will open that day (Fig. 1). The surest way to distinguish them is to open one petal of an unopened female flower in the morning. If a drop of nectar is seen on the basal part of the pistil, the flower will open in the afternoon of the same day (Fig. 2).

Emasculation is not usually needed because female and male flowers are separate. Male flowers open 1 to 2 weeks after the female flowers opened within the same inflorescence. By the time male flowers open, the female flowers of the same inflorescence have developed into fruits or have died.

Fig. 1—Female flowers of cassava that will open in the afternoon of the same day. The pistil is enclosed within the corolla.

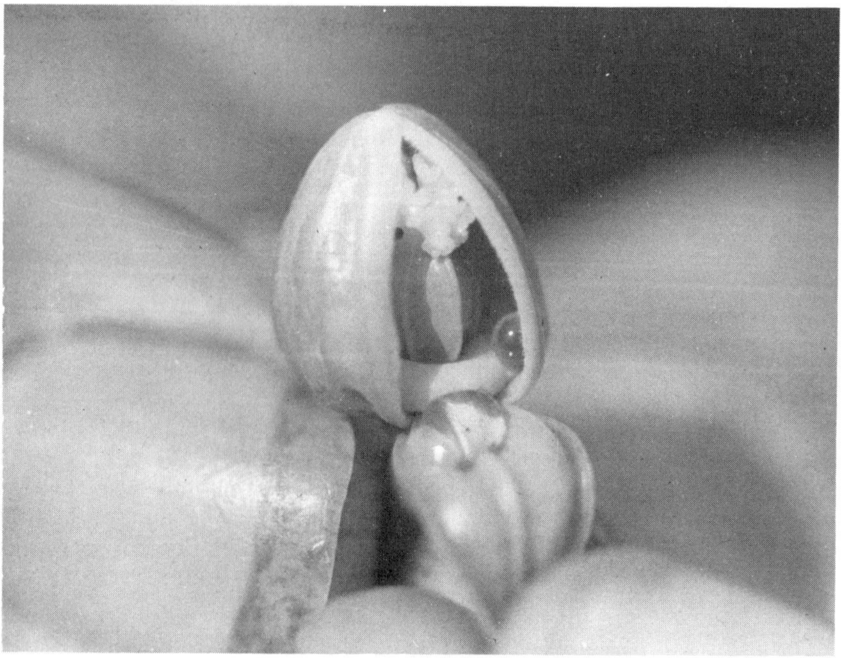

Fig. 2—A drop of nectar in the female flower is the surest indication that the flower will open shortly.

Female and male flowers usually begin to open from 1200 to 1400 hours and remain open for about 1 day (Fig. 3 and 4). To prevent stray pollinations, the flower branches are covered by large cloth bags which also identify the female flowers to be pollinated during the day (Fig. 5).

C. Pollination

Recently opened male flowers are picked off the branch during the first hours of the afternoon and carried in small bottles. Pollination can be done most efficiently between 1300 and 1700 hours. Because both female and male flowers are large and the pollen is sticky, pollination is easy and requires no special tool (Fig. 6). One male flower can be used for the pollination of three female flowers.

After pollination, the female flowers can be covered with a large bag or left exposed. Success rate is higher if the female flowers are left exposed after pollination in some genotypes. The possibility of hybridization by stray pollen seems to be very low once the stigma is pollinated, even if the female flowers are left exposed after pollination. Each flower branch on which the pollination is made is marked with a tag indicating the cross combination, date of pollination, and number of female flowers pollinated.

If the female flowers are exposed after pollination, they must be covered with a small bag 1 or 2 weeks after pollination. By this time the pollinated female flowers will have developed into young fruits. The bag protects young fruits from fruit fly attack and catches mature seeds which fall off naturally about 3 mo after pollination.

One female flower can produce up to three seeds, however, it is difficult to obtain an average of 2.0 seeds per female flower. In 1973 when the cassava hybridization was started at CIAT, an average of 0.60 seeds per female flower was obtained. In 1977, it was improved to 1.39 seeds per female flower (32,000 hybrid seeds obtained from 23,000 female flowers pollinated). Selection of a good female parent is one of the most significant reasons for this improvement. About 150 female flowers can be hand-pollinated daily by one person.

Cassava is a highly heterozygous species and has extremely high inbreeding depression (CIAT, 1974). After one cycle of selfing, some plants are so weak that they cannot produce enough male and female flowers for further hybridization. Thus, selfing is not a major part of hybridization programs in cassava.

D. Factors Affecting Efficiency

Matching flowering dates of genotypes to be hybridized may be a problem. Flowering on a single plant usually lasts for more than 2 mo, therefore, planting parental genotypes every 2 or 3 mo is a practical solution. Male flowers usually outnumber female flowers, which frequently make the number of female flowers available for pollination a limiting factor for mass production of hybrid seeds. In practice, genotypes to be used as female parents are planted in larger numbers and at more planting dates.

Fig. 3—Open female flowers at the stage for pollination. The multibranched stigma is in the center of the flower.

Fig. 4—Male flower with pollen on the anthers.

Fig. 5—Branches with female flowers to be pollinated on the same day are covered with a cloth bag in the morning.

Fig. 6—Pollination is easy and requires no special tool.

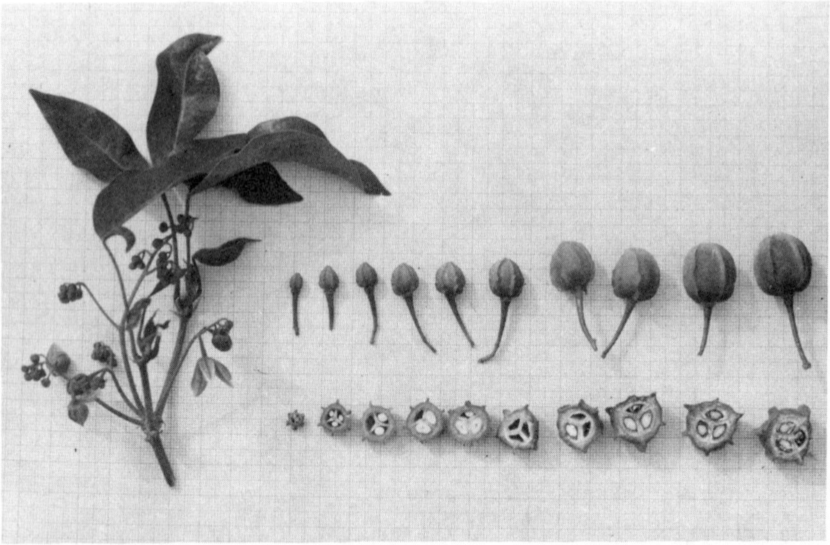

Fig. 7—Within a week after pollination, the female flower develops into a round-shaped fruit which contains up to three seeds and reaches full size in another 2 weeks.

The stigma remains receptive for up to 24 hours and dried pollen remains viable for 6 days. Eight to 19 hours are required for fertilization to occur (Chandratna and Nanayakkara, 1948). It is easiest and probably most efficient to pollinate female flowers that have just opened with pollen taken from male flowers that have recently opened. Pollen taken from a normally developed male flower is highly effective, regardless of genotype. On the other hand, genotypes differ significantly in their ability as female parents in terms of number of seeds set per female flower. A good genotype can give an average of up to two seeds per female flower pollinated, while some poor ones never give more than 0.4 seeds per female flower. Selection of highly fertile genotypes as female parents is a critical factor.

V. NATURAL HYBRIDIZATION

When a cassava breeder chooses natural hybridization as a major tool of his hybridization work, male-sterile genotypes should be used. Two types of male-sterility are common in cassava. In one type, male flowers drop before they reach maturity. In the other, male flowers fully develop, but the anthers do not contain pollen. Populations derived from natural pollinations of male-fertile genotypes include a higher proportion of inferior plants (presumably the result of selfing) than populations derived from natural pollination of male-sterile genotypes.

VI. SEED DEVELOPMENT, HARVEST, AND STORAGE

Within a week after pollination, the female flower develops into a round-shaped fruit containing up to three seeds and reaches full size in 2 weeks (Fig. 7). About 3 months after pollination, the fruits crack and the mature seeds can be harvested.

Newly harvested seeds demonstrate a tendency for dormancy. When seeds are stored at about 24 C, acceptably high germination is obtained 3 to 12 months after harvest. When seeds are kept at room temperature for more than 2 years, germination percentage drops drastically. When seeds are stored at 5 C and 60% relative humidity, no germination loss occurs in 3 years (IITA, 1976).

REFERENCES

Chandratna, M. F., and K. D. S. S. Nanayakkara. 1948. Studies in cassava. II. The production of hybrids. Trop. Agric. 104:59–70.

Centro Internacional de Agricultura Tropical. 1973. Annual Report. CIAT, Cali, Colombia. p. 59–118.

———. 1974. Annual Report. CIAT, Cali, Colombia. p. 53–109.

———. 1975. Annual Report. CIAT, Cali, Colombia. p. B1–57.

———. 1976. Annual Report. CIAT, Cali, Colombia. p. B1–76.

Food and Agriculture Organization. 1976. Production year book. Vol. 30. FAO, Rome, Italy.

International Center for Tropical Agriculture. 1976. Annual Report. 1976. IITA, Ibadan, Nigeria.

Jennings, D. L. 1970. Cassava in Africa. Field Crop Abstr. 23(3):271–275.

Leon, Jorge. 1976. Origen, evolution, and early dispersal of root and tuber crops. p. 20–40. *In* Proc. 4th Symp. Int. Soc. for Tropical Root Crops. IDRC-080e. Int. Dev. Res. Center, Ottawa, Canada.

Nestel, Barry. 1973. Current utilization and future potential for cassava. Chronic cassava toxicity. p. 11–26. IDRC-010e, Int. Dev. Res. Center, Ottawa, Canada.

Umanah, E. E., and R. W. Hartmann. 1973. Chromosome numbers and Karyotypes of some Manihot species. J. Am. Soc. Hortic. Sci. 98(3):272–274.

Viegas, A. P. 1937. Estudos sobre a Mandioca. Instituto Agronomico do Estado de Sao Paulo, Brazil.

14 Castor

Raymond D. Brigham
Texas A&M University
Lubbock

The castor plant (*Ricinus communis* L.) is a perennial of indeterminate growth habit that is believed to have originated in Abyssinia. Castor is now distributed throughout the tropics and subtropics, and is well adapted to the temperate regions of the world. Major producers and exporters of castor oil are Brazil and India. The United States is the largest importer and user, with over 55 million kg of oil imported annually. During the 1960's, up to 30,000 ha of castor were grown in the United States, but its production diminished to zero in the early 1970's. Castor oil is a versatile chemurgic raw product, and is used in hundreds of industrial applications. It is the chief raw material for production of sebacic acid, which is used in producing synthetic fibers and resins. The paint and varnish industry is a large user of dehydrated castor oil. Other users include manufacturers of plastics, artificial leather, brake fluid, soap, perfume, cosmetics, textiles, and lubricants (Zimmerman, 1958). Seed of open-pollinated cultivars and hybrids developed in the United States contain 48 to 50% oil on a dry weight basis.

I. PARENTAL MATERIAL

Ricinus is considered to be a monotypic genus and belongs to the Euphorbiaceae or Spurge family. *Ricinus communis* L. encompasses the many polymorphic types known in the world. Several of these were designated early as subspecies, but they are cross-compatible and are not true subspecies as usually defined in other plants.

The USDA National Seed Storage Laboratory, Fort Collins, Colo., maintains over 900 accessions of castor in the United States. The USDA Southern Regional Plant Introduction Station, Experiment, Ga., has over 270 accessions in a collection. Important centers are also in Campinas, Brazil, and Hyderabad, India. In the Northern Caucasus region of USSR, the Kuban Experiment Station of the All-Union Institute of Plant Industry (VIR) has an extensive collection, as does the All-Union Research Institute of Oil Crops (VNIIMK) in Krasnodar. Israel, Yugoslavia, Hungary, and several other countries also have centers that maintain limited germplasm collections.

II. PLANT CULTURE

A. Field

Castor plants are adapted to a wide range of well-drained soils free of compacted or impervious layers. Sandy loams and loams usually produce optimum yields. Castor is not suited to saline soils, those subject to erosion, or those with low fertility. A moderate level of nitrogen is recommended for nursery plantings. Excessive nutrients and moisture can lead to a highly vegetative plant which is low in seed production. Standing water can cause the death of plants from insufficient oxygen in the root zone. Thirty-eight to 50 cm of rainfall are required during the growing season for satisfactory plant growth and seed production. In arid lands, 60 to 100 cm of irrigation water may be needed for best yields.

To allow normal branching, plants should be spaced 50 cm or more apart in the row when rows are 1 m apart. Plants can also be established in hills in a grid system using various spacings.

Most castor selections bloom satisfactorily under the photoperiod available during the summer growing season in the United States. However, duration of photoperiod and light level may influence formation of flower primordia in certain genotypes (Scully and Domingo, 1947). Light level and quality are usually adequate for field culture of castor. Temperatures in excess of 37 C can cause abortion of seed in certain environments, and unusually low temperatures contribute to poor seed set.

Excessive relative humidity (60 to 70% and above) and frequent rain showers encourage a complex of fungus diseases on the developing capsules, their pedicels, and, in some cases, the raceme axis. Damage may vary from a slight reduction in seed weight to complete destruction of the raceme.

Castor plants do well in winter nurseries, such as in southern Florida, and can withstand light frosts without appreciable damage.

B. Growth Chamber and Greenhouse

Plants can be grown satisfactorily under greenhouse conditions in soil benches, large crocks, pots, and buckets, if a suitable soil mix and adequate bottom drainage is provided. A 2:1:1 mixture of soil, sand, and peat moss over a 3- to 5-cm layer of gravel has been used with good results. Plants can also be grown in ground beds. Starting seedlings in peat cups and transplanting to the ground bed lessens the chance of skips in the greenhouse planting. Spacings should be similar to those in field nursery plantings.

The need for additional nitrogen fertilizer can usually be determined by appearance of the plant, and small amounts should be added periodically as needed. Plants should only be watered enough to promote vigorous growth since excess moisture can lead to excessively high relative humidity and encourage diseases.

Lights may be used over castor plants when overcast winter conditions prevent the reception of normal sunlight, or when day length is shorter than desired. Plants are tolerant of varying photoperiods, and light level and quality are not critical.

Growth chambers can be used, but plants should be grown in small containers to limit their size. Shifriss (1964) found a continuous light level of 15 klux, a constant temperatue of 21 to 23 C, and relative humidity of 40 to 50% to be favorable for growth, flowering, and seed production. In both greenhouse and growth chambers, frequent observations for pests should be made. White flies [*Trialeurodes vaporariorum* (Westwood)] and two-spotted spider mites (*Tetranychus urticae* Koch) are often found on castor plants grown in greenhouses and growth chambers. Inadequate control measures can lead to leaf drop, reduced plant growth, and possible loss of plants.

III. FLORAL CHARACTERISTICS

Inflorescences of castor are borne terminally on the main and lateral branches. The main stem ends in a raceme, usually the largest on the plant, which is designated the *first* or *primary raceme* (Fig. 1). After the first raceme appears, two or three branches arise at the nodes immediately below it. Each of these branches terminates in a raceme after four or more nodes have formed. These racemes are commonly called the *second set* or *secondary racemes.* Branches arise from the nodes just beneath secondary racemes, ultimately terminating in a *third set* or *tertiary racemes.* This sequence of development continues as long as the plant remains alive and healthy and has a proper environment for growth, thus producing the indeterminate growth habit of the species. The sequential development of racemes along any one axis makes it possible for a plant to have racemes in all stages of development from the bud stage to complete maturity.

The racemes of castor are usually monoecious, with the pistillate flowers on the upper 30 to 50% and staminate flowers on the lower part of the inflorescence. Raceme size, number of flowers, and proportion of male to female flowers vary greatly. Both male and female flowers are apetalous, each being enclosed in a fused perianth, and are borne on pedicels of varying lengths. The bud of the male flower is flattened conically, and the five segments of the perianth surround a cluster of numerous stamens bearing branched filaments. The male flowers are yellow at anthesis, and become pale yellow with time. The female flowers are conical and more pointed than the male flowers. The perianth encloses a pistil with a three-carpellate ovary and sessile stigmas. Usually each of the three carpels has a single ovule, and each ovule produces one seed. The three stigmas of each ovary are reddish or green, branched near the apex, and have a papillate surface.

Plants with two, four, or more locules per capsule have been observed, in contrast to the normal three. Hermaphroditic flowers are also found occasionally.

Under field conditions, the fine, light-weight castor pollen is carried to the stigmas mainly by wind, although bees and other pollinating insects are usually found in flowering racemes of castor, and may carry pollen from one plant to another. The anthers of the male flowers burst explosively upon drying, which tends to scatter the pollen grains. Due to a lack of any barriers such as self-sterility within the plant, selfing also occurs under natural conditions, and the tendency is most pronounced in the central racemes. Natural outcrossing of dwarf-internode castor can be in the range of 70 to 90% for the Texas High Plains (Brigham, 1967).

Pollen is normally shed from 2 to 3 hours after sunrise until late afternoon, but usually peaks near mid-morning. Meinders and Jones (1950) observed nearly all pollen was shed between 0730 and 1130 hours in June at Stillwater, Okla. Cool weather can delay the shedding of pollen until late morning.

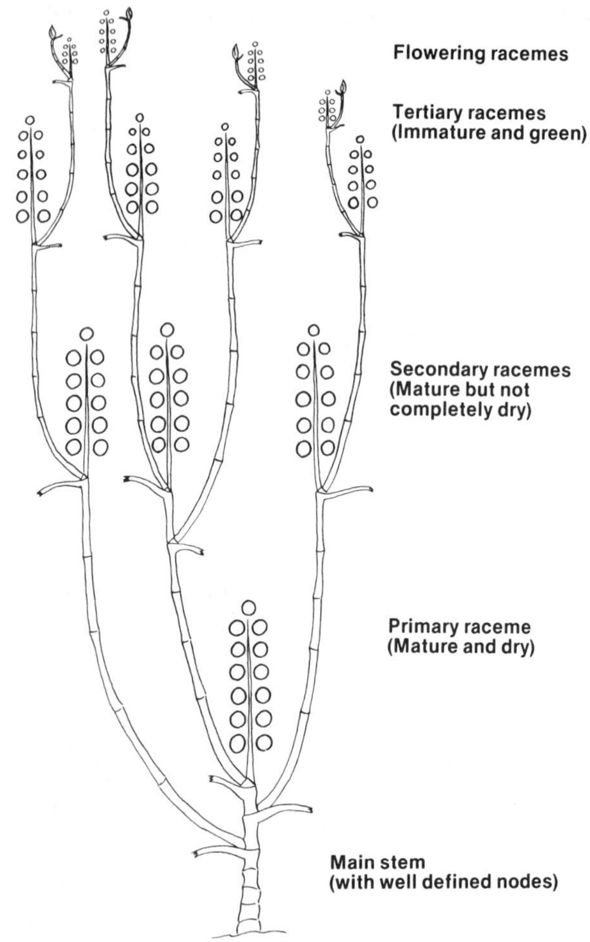

Fig. 1—Diagram of mature castor plant showing sequential raceme development.

The staminate flowers on the lowest portion of the raceme axis reach maturity first, and anthesis usually occurs a short time before the earliest pistillate flowers open on the upper portion of the axis. This timing provides an ample source of pollen for the first pistillate flowers that open and become receptive. Pistillate flowers may bloom over a 21-day period, and staminate over a 35-day period (Pustovoit et al., 1967). The staminate flowers dry within a few days after anthesis and usually drop off of the raceme axis. The stigma is fully receptive a few hours after the flower opens, but is often difficult to pollinate immediately after the flower opens. The branches of the stigma cease to grow shortly after fertilization takes place. If pollen is not available or if it is excluded from the stigmas by artificial means, such as covering with a paper bag, the stigmas continue to elongate, reaching 10 to 20 mm in length in some cases. Depending upon climatic conditions, stigmas may be receptive up to 7 days during favorable weather.

The length of time between pollination and fertilization is not documented, but should be several hours to a day, depending upon length of the stigma and temperature during pollen tube growth.

IV. ARTIFICIAL HYBRIDIZATION OR SELF-POLLINATION

A. Equipment

An apron with several pockets is commonly used to carry equipment for crossing and selfing castor. The equipment consists of paper bags, a plier-type stapler with chisel-point staples, marking pencil, forceps, camel-hair artist's brush, field books, and record books. Usually a plastic bottle of denatured ethanol is used for removal of pollen from forceps or camel-hair artist's brush and fingers when changing male parents. Bags for covering racemes are usually prepared beforehand and by cutting a V-shaped portion from the open end of the bag with large shears or a paper trimmer. Kraft paper, sorghum-pollinating bags are preferred for general use, but kraft grocery bags can be used under arid conditions where waterproof glue to hold the bag together is not a requirement. Parchment paper bags with waterproof glue have also been used with good results.

B. Preparation of the Female

Racemes of monoecious plants should be selected immediately before pistillate flowers open. Primary racemes are preferred because they are larger and produce more seed than secondary racemes. Staminate flowers, opened or unopened, are removed from the lower portion of the raceme axis with the fingers or forceps. If staminate flowers have already opened and anthesis has occurred, the perianth of the pistillate flowers should be inspected closely to be sure they are free of stray pollen grains. Care must be taken to remove any staminate flowers and hermaphroditic flowers which may be interspersed among the pistillate flowers. Removal of staminate flowers and covering with a paper bag may be done at any time of day, but small racemes are more turgid and tend to break off more easily in the mornings; therefore, afternoons are favored for emasculating and bagging racemes.

The main stem is usually sufficiently strong to support a paper bag covering the primary raceme. The small vegetative bud or buds at the base of the raceme, plus one or two adjacent leaf petioles, should be removed to insure a firm attachment of the bag when covering secondary racemes (Fig. 2). The bag is placed carefully over the raceme, the notch in one side is placed against the next available leaf petiole, and the ends of the bag are

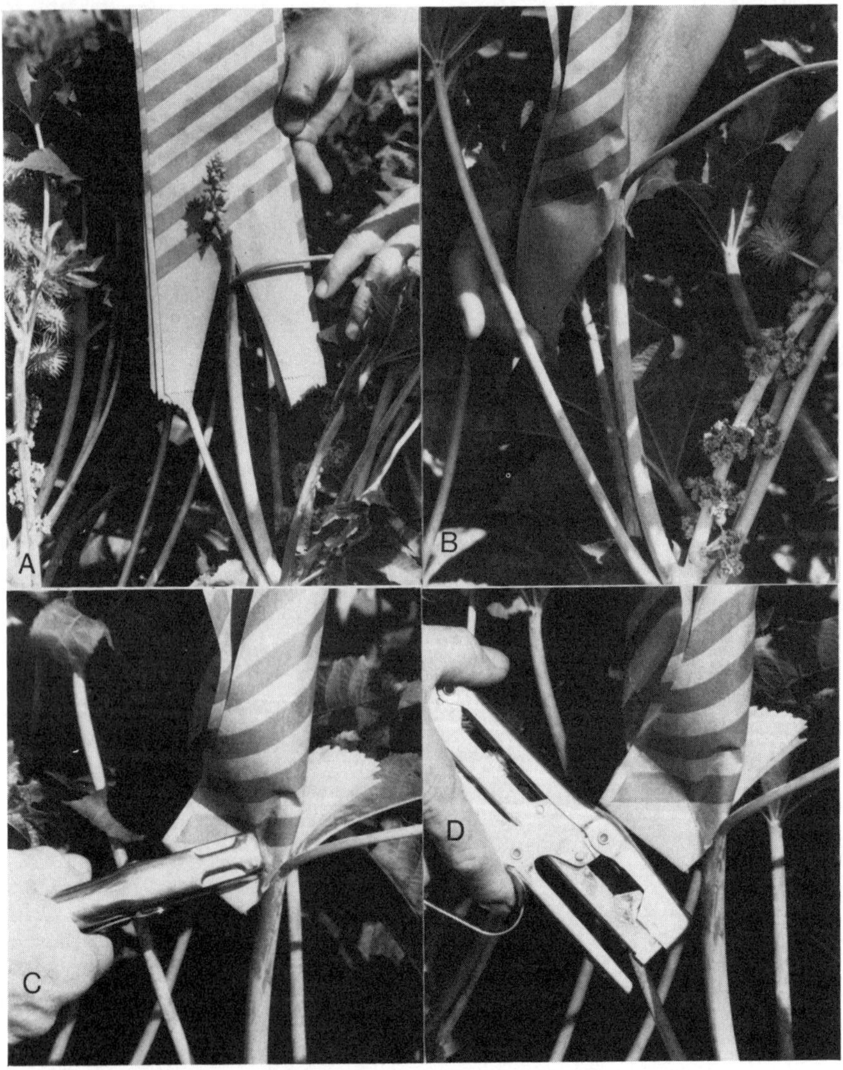

Fig. 2—Bagging operation. A, secondary raceme properly prepared for bagging; B, bag is placed over raceme until notch is against petiole; C, tails of bag are folded snugly against stem and stapled; D, plier-type stapler and properly installed bag.

brought together and folded snugly against the stem. A staple is placed through the four layers of paper. Care must be taken to position the staple as close as possible to the stem so that a slight pull on the bottom of the bag does not allow the bag to slip toward the raceme. If the bag slips, it should be repositioned and restapled. The bags are normally stored flat and must be opened and fully expanded before they are placed over a raceme, otherwise the flowers tend to be compressed and misshapen. Paper clips have been used to secure bags in areas where wind is not a factor, but a chisel-pointed staple applied with a plier-type stapler is more satisfactory in most areas. When bagging pistillate racemes, a careful check should be made for any interspersed staminate or hermaphroditic flowers.

At the time of bagging, the date and time can be marked on the bag with a waterproof marker. Small, white parchment bags are often used temporarily to cover the raceme after emasculating, as the raceme axis is quite short. Because larger parchment or kraft bags are used for selfing or following pollination of racemes, one can visually pick out the small bags, indicating that pollen is yet to be applied.

C. Pollination

Depending upon weather conditions, many of the pistillate flowers will normally be open and receptive 5 to 7 days after bagging. The proper interval can easily be ascertained by removing a bag from an emasculated raceme and inspecting it for receptive pistillate flowers. If staminate flowers are used for direct application to the stigmas, pollination can be carried out at any time of day when pollen is shedding freely from the anthers, but should be accomplished when wind movement is at a minimum.

The staminate flowers are usually collected in the morning shortly before or after anthesis and stored in a small paper bag, plastic petri dish, or other container that will not crush or compress the flower. To obtain bulk pollen, the bag containing staminate flowers collected earlier is shaken vigorously after the anthers have dehisced to encourage additional pollen shed. The pollen is shaken to one corner of the bag and the corner is torn off to gain ready access to the pollen.

Pollen will remain viable several days after the flower is picked, and will still be usable the second or third week if stored under room conditions (Pustovoit, 1967). Packets of pollen have been sent through the U.S. mail with good success when 2 to 3 days delivery time was possible. Pollen can also be frozen or freeze-dried successfully and will remain viable up to 1 year (Haarring et al., 1969).

Before application of pollen, the raceme on which the female flower is located should be carefully checked and any interspersed staminate or hermaphroditic flowers removed.

When the stigmas are well exposed, the pistillate flowers can easily be pollinated by dusting pollen on the stigmatic branches. This can be done by holding the dehisced staminate flower with forceps and lightly touching it to the stigmas. One staminate flower can be used to pollinate several pistillate flowers, but should be discarded when pollen is sparsely evident upon touching to the thumbnail. When pollen has been collected in bulk from staminate flowers, a camel-hair artist's brush may be used to apply the pollen. Fingers, forceps, and brushes should be cleaned with alcohol when changing from one male parent to another.

The raceme is covered with a bag immediately after pollen is applied and the source of pollen and date of pollination is recorded on the bag. If all pistillate flowers are not receptive at first pollination, a second pollination can be made after a few days.

Unless exceptionally hot or wet weather follows pollination, seed set should be nearly 100%. From the pollination of 40 pistillate flowers, 120 seeds are possible, since three seeds are normally produced in each capsule.

In routine selfing, monoecious racemes are selected and prepared as above, except the staminate flowers are not removed. The date of bagging is usually marked on the bag. The initials or identifying mark of the person applying the bag may be added to facilitate a check on proper preparation of the plant and fastening of the bag by individuals working in the nursery. It is helpful to face the bags in one direction as much as possible (usually facing plant no. 1) so workers performing follow-up operations can proceed along the row in one direction and see the date the bags were placed on the racemes. Depending upon weather conditions, many of the pistillate flowers usually will be open and receptive 5 to 7 days after bagging, and the bags should be agitated for best pollination and seed set. This is customarily done by holding opposite edges of the bag about halfway from the top, and making several short, quick pumping motions, similar to pumping a bellows. This causes the pollen to move upward in a cloud, where it falls on receptive stigmas. Care must be taken not to loosen or tear the bag and bags should be dry; therefore, this operation is often carried out the latter part of the day. Pumping of bags should be continued until all pistillate flowers have opened and all stigmas have received pollen. This may require several trips over to the nursery at 4 to 5-day intervals after bagging commences. Workers should be alert for loosened, torn, or split bags; vegetative shoots or leaves growing in the bag with the raceme; and racemes which have elongated until the apex of the raceme is touching the bottom of the bag. If left unattended, the apexes of extremely long racemes of certain genotypes may force their way through the bag, exposing receptive stigmas to foreign pollen, and resulting in seeds of unknown male parentage.

When capsules have all set seed, the selfing bags may be removed to allow drying, and to avoid growth of fungi on capsules and their pedicels. The paper bag is removed, folded in half lengthwise, folded around the base of the raceme, and stapled. If the bag is not torn, it can be used to harvest the capsules from the raceme. The bag is slipped over the raceme that bears the dry capsules, gripped near the base of the raceme, and pulled upward, stripping the capsules into the bag, which is then folded and stapled shut. Row and selection number plus other information should be recorded on the bag.

D. Factors Affecting Efficiency

Breeding nurseries are usually planted in a series of short rows with parental lines to be frequently used in crossing grouped together. If large amounts of hybrid seed from a cross are required, the male and female parents are often planted adjacent to each other to facilitate hybridization. If sufficient land is available, the two parental lines can be planted in isolation and plants of the female line can be emasculated or rogued before natural pollination. Spatial requirement for isolation will vary with wind movement, but should be at least 800 m and not downwind from other castor plantings.

Because of the indeterminate nature of the castor plant and the progressive flowering of numerous racemes, it is usually not necessary to alter flowering dates unless the plants are extremely late flowering, are planted extremely late, or are grown in a short growing season. When one parent is much earlier than the other, racemes in the bud stage can be removed to delay onset of flowering.

Genetic markers for identifying hybrid seed are not easily interpreted, and are usually not relied upon in a breeding program. The markers most used for identifying hybrid plants are red or mahogany stems dominant to green stem color; waxy bloom dominant to absence of bloom; and spined dominant to spineless capsules. The gene for dwarf-internode is also recessive, and plants with this characteristic can easily be differentiated from normal-internode plants that may be involved in crosses.

V. NATURAL HYBRIDIZATION

Use of hybrid seed of castor has been feasible since the development of suitable female lines beginning in the 1950's. True male sterility has not been found in castor, but the absence of staminate flowers causes racemes to be completely pistillage. A simply inherited recessive gene, now designated *ff*, for absence of staminate flowers (Katayama, 1948; Claassen and Hoffman, 1950) permitted production of the first commercial F_1 hybrid by roguing the normal monoecious plants from the backcross F_1 line to obtain a population of 100% pistillate plants in the female rows of hybrid seed fields. This procedure required several trips over the field, and was laborious and expensive. Identification of a type of femaleness that is genetically unstable and may undergo sex reversal (Shifriss, 1960; Brigham, 1961) was next reported. Before this system was utilized in hybrid production, a unique sexual mechanism combining the gene for pistillate flowers and environmentally sensitive genes for interspersed staminate flowers was developed by Zimmerman and Smith (1966). This led to the release of a female line, designated CNES-1, which was widely used in the production of commercial F_1 hybrids (Fig. 3). This line is normally pistillate under moderate temperatures, but produces interspersed staminate flowers under high temperatures. In crossing fields, usually only one or two roguings of the female line are necessary to insure that all flowering plants are pistillate, and to remove off-type plants that appear.

In practice, female lines are obtained by selecting all-female plants in backcross populations, growing the progenies under a warmer and longer than usual growing season, roguing before flowering plants that are monoecious, and harvesting bulked seed from both those plants that remain pistillate and those that produce pollen from interspersed staminate flowers. Under proper environmental conditions, this bulked seed will produce plants in seed rows that are 100% pistillate on the first and second sets of racemes. Constant selection pressure must be maintained in the foundation seed fields to prevent excessive expression of the genes for interspersed staminate flowers which can result in selfing of the female parent.

A genetic type with a fused perianth of the male flowers has been reported (Subramanian and Sivasubramanian, 1975). This anomaly prevents normal pollen shedding, and may find use later in developing female lines for use in commercial hybrid seed production.

The layout of isolated crossing blocks or fields is similar to that used for other wind-pollinated crops. Usually two rows of male parent are planted for every 10 to 12 female rows. Plantings should be oriented so that prevailing winds blow across the rows for best pollen distribution. The male parent should be earlier flowering than the female to avoid the necessity of planting at different times. The male and female parents should be rogued for off-type plants just prior to blooming and the female parent should be rogued for any pollen-shedding plants at the early bloom stage of the primary raceme. Isolation distance should be at least 800 m and not downwind from other castor plantings. All wild castor plants growing along roadsides or in fence rows within 800 to 1,000 m of the isolated planting should be removed.

VI. SEED DEVELOPMENT, HARVEST, AND STORAGE

After pollination, capsules on the racemes begin to expand if fertilization has occurred and normal seed development is in progress. When the seeds have matured, capsules begin to dry, and the three locules separate or

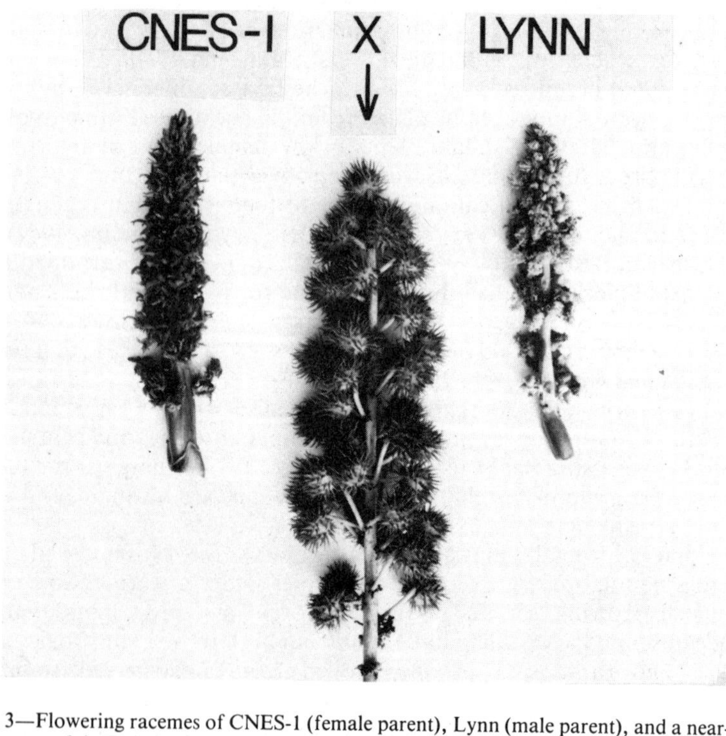

Fig. 3—Flowering racemes of CNES-1 (female parent), Lynn (male parent), and a near-mature raceme of the F_1 hybrid used for commercial castor seed production.

partially separate at the sutures (Fig. 4). Capsules are then removed from the plant either by placing a paper bag over the raceme and stripping the capsules from the raceme axis, or hand stripping with the use of leather gloves or a special stripping cup (Weiss, 1971).

Dry capsules may be stored for long periods prior to hulling the seed from the capsules. Small hullers used in breeding programs in the United States usually are equipped with rubber-covered vertical discs approximately 30 cm in diameter (Schoenleber and Taylor, 1954). One disc is stationary, while the other is rotated by an electric motor, which also powers a cleaning fan to separate hulls from seed. The distance between the face of the discs is adjustable, permitting efficient hulling without excessive cracking of the seed coat. Small, hand-powered hullers are described by Weiss (1971). Although the seeds are large, the seed coat is fragile and easily broken if mechanical pressure is applied.

Seed can be stored in paper envelopes, paper bags, or cloth bags after harvest. Preferably, storage should be in a refrigerated seed storage facility,

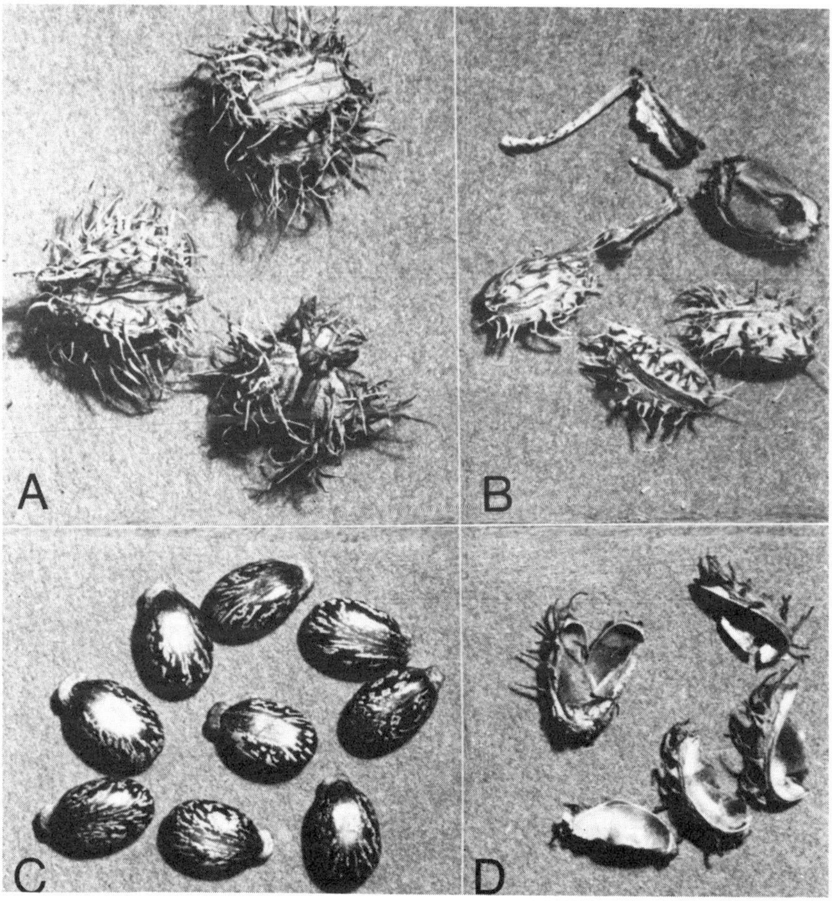

Fig. 4—Capsules and seed. A, mature dry capsules; B, segments (carpels) of a capsule that contain the seed; C, clean seed; D, hulls removed from seed.

but germination remains fairly high for several years when seed are stored at room temperature. If individual packets or bags can be stored in tight-fitting metal or wood storage containers, paradichlorobenzene crystals or nuggets can be added to repel insects.

VII. TECHNIQUES FOR SPECIAL SITUATIONS

Spraying of castor seedlings in the two to four-leaf stage with gibberellin A_3 has been reported to increase the female tendency on the primary, secondary, and tertiary racemes (Shifriss, 1961). In contrast, male tendency has been induced by use of ethyl hydrogen-1-propyl-phosphonate (Philipos and Narayanaswamy, 1976). Embryo culture and grafting have generally not been used in castor research programs.

Because castor is a true perennial, it is possible to transplant the mature plants near the end of the growing season. The main stem is usually cut 12 to 20 cm above the ground line, and side branches are pruned to within 10 to 12 cm of the main stem. The root system is dug sufficiently deep and pruned to fit inside a large container (nominal capacity of 20 to 25 liters), or is transplanted to a sufficiently large hole dug in soil. Roots can be kept moist by immediately wrapping them in wet cloth after digging. New shoots arise from dormant buds at the nodes of the main stem and the side branches. Care should be taken in cutting the main stem and branches to prevent splitting or crushing the stem. A single blow with a sharp machete has been used successfully for the above-ground pruning operation.

In areas where winter temperatures are subfreezing for only a short time, castor plants can be pruned and covered with soil in the fall of the year, and uncovered in spring after danger of frost is past. The main stem is carefully cut 10 to 14 cm above the ground line, and side branches are removed before covering the stem with soil. After uncovering in spring, buds emerge from the nodes of the main stem, and new branches with racemes follow. Some plants do not survive due to root rot, rodent damage, etc., but over 50% recovery has been reported (Brigham, 1961).

REFERENCES

Brigham, R. D. 1961. TSP 10, a dwarf-internode S-pistillate castorbean line for possible use in producing F_1 hybrids. CR 62-61, USDA-ARS-CRD and Texas Agric. Exp. Stn.

———. 1967. Natural outcrossing in dwarf-internode castor, *Ricinus communis* L. Crop Sci. 7:353–355.

Claassen, C. E., and Albert Hoffman. 1950. The inheritance of the pistillate character in castors and its possible utilization in the production of commercial hybrid seed. Agron. J. 42:79–82.

Haarring, Rainer, L. H. Zimmerman, and J. D. Smith. 1969. Short and long period storage of pollen of *Ricinus communis* L. Crop Sci. 9:17–18.

Katayama, Y. 1948. The progeny of female individuals in castorbeans. (In Japanese) Jap. J. Genet. 23:19.

Meinders, H. C., and M. D. Jones. 1950. Pollen shedding and dispersal in the castor plant *Ricinus communis* L. Agron. J. 42:206–209.

Philipos, H. V., and S. Narayanaswamy. 1976. Induced potential for male sex expression in *Ricinus communis* L. by ethyl hydrogen-1-propylphosphonate. Protoplasma 87:71–77.

Pustovoit, V. S. (ed.). 1967. Handbook of selection and seed growing of oil plants. Moscow. 352 p. [In Russian]. Translated into English in 1973 and available as TT71-50100 through U.S. Dep. of Commerce, NTIS, Springfield, Va.

Schoenleber, L. G., and W. E. Taylor. 1954. Development of a portable castor bean huller. Agric. Eng. 35:327–336.

Scully, N. J., and W. E. Domingo. 1947. Effect of duration and intensity of light upon flowering in several varieties and hybrids of castor bean. Bot. Gaz. 108: 556–570.

Shifriss, Oved. 1960. Conventional and unconventional systems controlling sex variations in *Ricinus*. J. Genet. 57:361–388.

———. 1961. Giberellin as sex regulator in *Ricinus communis*. Science 133:2061–2062.

———. 1964. Growth and sexuality of *Ricinus communis* L. in a constant environment. Am. Nat. 98:187–189.

Subramanian, M., and P. Sivasubramanian. 1975. X-ray-induced fused perianth in *Ricinus communis* (L.). Rad. Bot. 15:307–308.

Weiss, E. A. 1971. Castor, sesame and safflower. Leonard Hill, London.

Zimmerman, L. H. 1958. Castorbeans: A new oil crop for mechanized production. Adv. Agron. 10:257–288.

———, and J. D. Smith. 1966. Production of F_1 seed in castorbeans by use of sex genes sensitive to environment. Crop Sci. 6:406–409.

15 Chickpea

A. K. Auckland
L. J. G. van der Maesen
International Crops Research Institute for the Semi-Arid Tropics
Patancheru nr. Hyderabad, India

Cicer arietinum L., the cultivated chickpea, is an annual species that originated in south eastern Turkey (Ladizinsky, 1975) (Fig. 1). Cubero (1975) identified two groups based on seed size, *macrocarpa* and *microcarpa.* The *macrocarpa* are referred to as kabuli and the *microcarpa* as desi by plant breeders in the Indian subcontinent and the International Crops Research Institute for the Semi-Arid Tropics (ICRISAT). The kabuli and desi are the main groups of immediate interest to genetic improvement within the species.

Kabuli chickpeas have large seed (>25 g/100 seeds) that are ramshead or rounded in shape and of creamy color. There are usually 1 or 2 seeds/pod. The plants are medium to tall with no anthocyanin, and have white flowers and large leaflets (10 to 20 mm). They are characteristic of the Mediterranean area and adapted to spring plantings in Afghanistan, Iran, and countries westward. In contrast, desi chickpeas have small, irregularly shaped seed of various colors. There are usually 2 or 3 seeds/pod. The plants are short and contain anthocyanin, the flowers are usually purplish, and the leaflets are small. They are adapted to winter plantings in Pakistan and countries eastward. Some desi cultivars are grown in Afghanistan and Iran and are adapted to spring plantings there (Auckland, 1977). The kabuli types have been more modified by human selection than the desi types (Ramanujam, 1976).

The total world area of chickpea is about 11 million ha. India produces 74%, Pakistan 10%, and Ethiopia 4% of the total crop. Most chickpeas are consumed locally and there is little export trade.

I. PARENTAL MATERIAL

The genus *Cicer* belongs to the tribe *Cicereae,* family *Leguminosae. Cicer* differs from related genera in *Vicieae* by its glabrous style, inflated pods, and glandular pubescence.

Van der Maesen (1972, 1973) presented taxonomic keys to the 39 species of *Cicer* and an informal intraspecific classification of *C. arietinum* L., based on the "subspecies" distinguished by Popova (1937). Of these species, 31 are perennials and 8 are annuals. Eleven species have been identified as diploids ($2x = 16$). A diploid, annual species, *C. reticulatum,* was recently discovered (Ladizinsky, 1975).

Ladizinsky and Adler (1976) made crosses between *C. echinospermum, C reticulatum,* and cultivated chickpea. These diploid species are closely related both morphologically and by their seed protein profile (Ladizinsky and Adler, 1975). *Cicer echinospermum* differed from the cultivated species by a reciprocal translocation and their hybrid was sterile. *Cicer reticulatum,*

Fig. 1—Branch of chickpea plant with flowers and pods.

however, crossed readily with four red-flowered cultivars of *C. arietinum*. The hybrids were fertile and meiosis was normal. One white-flowered cultivar when crossed with *C. reticulatum* produced a quadrivalent, an Anaphase 1 bridge, and a fragment at meiosis, indicating that this cultivar and *C. reticulatum* differ by a translocation and a paracentric inversion. Pollen fertility of the hybrid was low and no seed was formed. The difference in meiosis of the hybrids involving cultivars of *C. arietinum* and *C. reticulatum* indicate chromosome repatterning within *C. arietinum* and suggests cytogenetic differences between cultivar groups within the cultivated species.

Interspecific hybridization could be of particular importance in the future (Auckland and Singh, 1977a). *Cicer reticulatum* is resistant to *Ascochyta* blight, but susceptible to wilt (Nene, 1977). *C. pinnatifidum* is reported resistant to *Uromyces* rust in Mexico (Lopez, 1977). F1 hybrids have been backcrossed to the cultivated species at ICRISAT.

ICRISAT maintains the world germplasm collection of *Cicer arietinum* L. cultivars (about 11,140 accessions) and long-term storage facilities are being set up (van der Maesen, 1975). A breeding program was initiated in 1973 and new strains or segregating populations are issued to national chickpea improvement programs (Singh and Auckland, 1975; Auckland and Singh, 1977b).

The wild annual species *C. cuneatum* Hochst. ex Rich., *C. bijugum* Rech., *C. chorassanicum* (Bge) M. Pop., *C. echinospermum* P.H. Davis, *C. judaicum* Boiss., *C. pinnatifidum* Jaub. et Sp., *C. reticulatum* G. Ladizinsky, and *C. yamashitae* Kitamura are also available at ICRISAT. They are, however, maintained with difficulty and seed supplies are limited.

II. PLANT CULTURE

A. Field

Chickpea grows well on a range of well-drained soils if the roots can continue to reach the receding moisture. A pH of 6.5 to 8.0 is suitable. Flowering and fruiting are earlier on light, dry soils than on heavy soils. Responses to fertilizers are slight, but inoculation with *Rhizobium* is beneficial in new areas and may induce early flowering. The duration of the crop varies from 100 to 170 days. The receding water table and high temperatures at the end of the season accelerate crop maturity.

In all areas, the crop is grown during the dry season and survives on residual moisture or irrigation. Rainfall during the season can damage the crop physically, promote diseases, reduce the efficaciousness of hybridization, and endanger seed set.

In eastern Asia, chickpeas are sown in October to November. Day length and temperature decrease after planting and only in March, when most of the crop is ripe, does the day length surpass 12 hours. In contrast, the crop in the western Asia and Mediterranean regions is sown in March to April when the photoperiods are short and temperatures are low initially. In Ethiopia the crop is planted in September and there is little variation in day length or temperature during the growing season.

In general, long days induce early flowering and short days delay flowering (Eshel, 1966; van der Maesen, 1972; Pandey et al., 1976), but differential cultivar response has been noted (van der Maesen, 1972; Ladizinsky and Adler, 1975).

Papadakis (1966) classified the chickpea as a winter legume withstanding temperatures between 8 C minimum and 22 C maximum during the coldest month. Frost resistant cultivars have been reported (Popova, 1937). In India, during the flowering period, the average minimum temperatures are 10 to 14 C and average maximum temperatures are 25 to 31 C. These ranges are considered optimum for hybridization and seed set in most cultivars. During ripening, night temperatures rarely fall below 14 C, and maximum temperatures may rise to 35 C.

The interaction between light and temperature effects on chickpeas are not yet fully understood. Nanda and Chinoy (1960) suggested an inverse relationship of light and temperature for normal growth during the vegetative period. Eshel (1966) considered the influence of day length more important than temperature.

Low seed setting is associated with high humidity and cloudiness in chickpeas (Howard et al., 1915; Kadam et al., 1938; Sen and Mukherjee, 1961). Aziz et al. (1960) considered both high humidity and low temperatures, as well as low humidity and high temperatures to adversely affect seed setting. Eshel (1968) and van der Maesen (1972) did not find pollination or fertilization to be affected by relative humidity per se and suggested that low seed setting is due rather to suboptimum soil humidity which affects flowering.

Van der Maesen (1972), working in greenhouses in the Netherlands, found that shading (25% and 75% of full sunlight) reduced the number of flowers per plant. Saxena and Sheldrake (1976) reported that the interception of 80% full sunlight did not reduce the yield of field grown chickpea under the high light level conditions of Hyderabad, India. They suggested that water was the primary factor limiting yield.

For all crossing block nursieries grown in the field, a minimum spacing of 75 cm between rows and 30 cm between plants in the row is recommended to ensure minimum physical damage to plants. Cutworms can cause loss of stand at the seedling stage and it may be advisable to use a persistant soil insecticide like aldrin. Later in the season, *Heliothis armigera* is the major pest and endosulfan at 0.7 kg active ingredient/ha can be used for control (Reed, 1977). Protection from *Fusarium* and *Rhizoctonia* wilt is difficult and it is best to select a nursery site which has no history of these diseases. If blight diseases (*Ascochyta, Botrytis,* or *Alternaria*) occur in the area, spraying with 0.3% Dithane M-45 (Mancozeb) at 7 to 10 day intervals after flowering is completed will reduce the disease incidence (Nene, 1977).

The establishment of field nurseries for hybridization during a normal growing season in natural chickpea environments presents few problems, provided residual soil moisture is adequate and rainfall during the growing season is minimal. Off-season nurseries have, however, presented major problems. Chickpeas sown in June (average daily temperature 25 C) in Israel did not flower (Eshel, 1966). If chickpeas are planted out of season in peninsular India, they fail to flower due to high (over 40 C) temperatures or rainfall. Nurseries grown by ICRISAT during the four monsoon seasons of

1974 to 1977 in the high altitude and relatively dry, north Indian valleys of Lahaul and Kashmir were unsatisfactory. Excessively low or high temperature, rainfall, and disease caused either outright losses, the harvest of limited seed of low viability, or both. Differential cultivar response to these atypical conditions was extreme. Some southern India desi types and related segregates never produced seed. Representative generation advancement of lines and populations was impossible.

During the 1973 and 1974 summer season, ICRISAT (India) used Lebanon, a natural environment for spring-planted kabuli chickpeas, as a site for off-season advancement. It was possible to successfully advance and select in Lebanon segregating material that was adapted to western Asia from kabuli × desi or kabuli × kabuli crosses. Crosses of Indian desi parentage were dwarfed in this environment and the lack of obvious phenotypic variability precluded selection within them (Auckland and Singh, 1977a). Both kabuli and desi types produced good quality viable seed in the Lebanon and it was possible to successfully advance generations towards homozygosity from all the cross combinations.

B. Growth Chamber and Greenhouse

Summerfield et al. (1977), were able to grow chickpeas in controlled environments which approximate the field phenotype. The growth cabinets were set at the mean values of day length (11.5 hours) and day (28 C) and night (14 C) temperatures of the chickpea growing season in southern India (17°N Lat). Relative humidity was maintained at 65 ± 5%, which corresponds to the mean value of monthly maximum and minimum values over the same period (October to March) in that area. Warm white fluorescent tubes were the most suitable sources of artificial illumination for optimum morphogenetic development of chickpea in growth cabinets.

Van der Maesen (1972) grew chickpeas in greenhouses in the Netherlands. Satisfactory flowering of most cultivars was obtained by controlling temperatures between 17 and 35 C, maintaining relative humidity at 50 to 60%, and growing the plants in river sand or a well-aerated garden soil of pH 6.5 to which nutritive elements were added.

III. FLORAL CHARACTERISTICS

For all practical hybridization and breeding purposes, chickpeas can be regarded as a self-pollinated species (Ayyar and Balasubramanyan, 1935; Eshel, 1968), but a small amount of natural cross pollination by bees has been noted (Howard et al., 1915; Niknejad and Khosh-Khui, 1972).

Self-pollination takes place 1 or 2 days before the flower opens. In India, the flowers open on 2 successive days. On the first day, they open from about 0900 to 1500 hours and close about 1800 hours. The second day they open earlier and close sooner (Ayyar and Balasubramanyan, 1935).

Chickpeas have typically papilionaceous flowers. Solitary flowers are borne in axillary racemes, although cultivars with two or three flowers racemes are known (Fig. 1). The pedicels (6 to 13 mm) are straight when flowering and curved when bearing fruit. The calyx (8 to 10 mm) has five partly joined sepals. The corolla (8 to 11 mm) may be greenish white, purplish pink, red or blue in color. The standard petal is broad and clawed, the wings are free and the keel incurved. There are 10 stamens, 9 fused and 1 free (6 to 8 mm), and an upturned style (3 to 4 mm) with a slightly broadened stigma (van der Maesen, 1972).

Eshel (1968) categorized the developmental stages of bud and flower in chickpea. These stages (Fig. 2) are:

A—Closed bud: in this stage the stigma is immature and the anthers are still at the base of the bud.

B—Hooded bud: the corolla has elongated, and the anthers are about half the height of the style. The stigma is receptive and remains so until stage D. Emasculation is done at this stage.

C—Half-open flower: the anthers have now reached the same height as the stigma and the pollen is mature just before the anthers dehisce. Pollination now takes place and the keel remains closed preventing any foreign pollen from reaching the stigma. Pollen for crossing is collected at this stage.

D—Fully open flower: the anthers are shrivelled, while the standard and wings are fully expanded. Fertilization takes place 24 hours after pollination (Eshel, 1965).

E—Fading flower: the postfertilization stage, in which the ovary starts to elongate.

IV. ARTIFICIAL HYBRIDIZATION AND SELF-POLLINATION

A. Equipment

The equipment needed for preparation of the female parent consists of fine forceps, 95% alcohol for sterilization of forceps, colored cotton thread, and a magnifying lens on a head band. For pollination, forceps, glass tubes for male flower storage, marker tags, an indelible pencil, alcohol for sterilization of forceps, and a magnifying lens are needed.

B. Preparation of the Female

Only flowers with purple peduncles should be selected in desi cultivars because those without anthocyanin pigmentation are often scheduled for natural flower drop (Pandya, 1977). Hybridization in northern India is not begun until after the first pod has formed (Bahl and Gowda, 1975; Pandya, 1977).

A large flower bud where the petals are just showing should be selected (Fig. 2B). The bud for emasculation should be held lightly at its base, between the thumb and first finger. The front sepal may be drawn back or snipped off. The keel petal should be gently pushed downwards with fine forceps. This exposes the anthers which can be removed from the filaments. The peduncle, stigma, and style are fragile and should not be touched during the operation.

Colored cotton thread tied loosely around the stem below the emasculated flower and removed sepal serve for identification. There is usually no need to protect emasculated flowers from cross-pollination. In hot, dry climates, shade and irrigation prevents flower dessication (Khosh-Khui and Niknejad, 1971; Eser, 1976). New buds emerging after the emasculated flower should be removed for a few days afterwards (Eser, 1976).

C. Pollination

Emasculation commonly is carried out in the afternoon of the day previous to pollination (Ayyar and Balasubramanyan, 1935). In northern India, good pod setting is obtained by emasculation between 1400 and 1800 hours and pollination the next morning between 0800 and 1300 hours. As

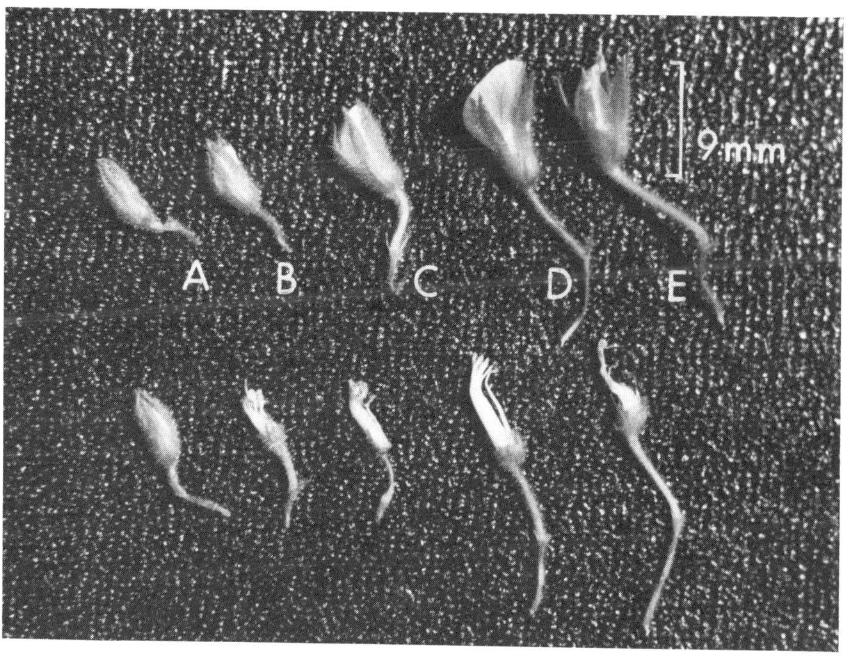

Fig. 2—Flowers of chickpea at five stages of development. Petals and sepals are intact in the upper row and removed in the lower row. For emasculation, only petals are removed. (A) Immature flower at the closed bud stage, (B) hooded bud stage when emasculation is done, (C) half-opened flower at the proper stage for pollen collection, (D) fully opened flower after pollination has occurred, (E) faded flower stage when the ovary begins to expand.

the season advances, days become warmer and morning emasculation followed by immediate pollination is more effective (Bahl, 1977; Pandya, 1977). Simultaneous emasculation and pollination was best in southern India (Singh and Auckland, 1975), Chile (Bascur, 1977), Israel (Khosh-Khui and Niknejad, 1971), Spain (Cubero, 1977), and Turkey (Eser, 1977), but evening emasculation and next day pollination was more effective in Mexico (Lopez, 1977).

Half-open flowers should be selected for pollen collection (Fig. 2C). Pollen at this stage is mature, yellow in color, and slightly sticky. Pollen viability remains high for 24 hours, and then rapidly declines and becomes greyish or cream-colored and powdery. It is easier to collect the entire keel petal and filaments rather than individual anthers. These should be placed in a corked glass tube to avoid dessication. Gently holding the keel petal and filaments with forceps, the pollen should be applied liberally, but lightly, to the stigma of the emasculated female parent, avoiding damage to stigma, style, or pedicels. Large numbers of pollen grains produce longer tubes than single grains. The distance between stigma and ovule is 7 mm and fertilization takes place 24 hours after pollination. Pollen viability remains constant throughout the flowering period (Eshel, 1968).

The pollinated flower can be tagged on the stem below and labelled with an indelible pencil. Under normal circumstances, there is no need to protect the pollinated flower.

The natural rate of pod setting is between 18 and 59% (Aziz et al., 1960; Eshel, 1965). Singh and Auckland (1975) reported 24% pod set with artificial pollination the same day as emasculation and 15% the day after emasculation, but the success rate of individual persons ranged from 5 to 50%. Eser (1977) reported only 10 to 20% success with pollination the day after emasculation, but 70 to 80% when emasculation was immediately followed by pollination. The high success rate was attributed also to early morning hybridization (0700 to 1000 hours), moistening the stigma with a soft brush dipped in sugar solution, and resetting the pollinated flower with moist fingers to its original position.

Retig (1971) pollinated flowers at the hooded bud stage (Fig. 2B) before anther dehiscence and without emasculation of the female parent. He obtained a 35% pod set from hybridized flowers, 67% of which were proven hybrids. This method has been employed in Australia with a success rate of 70% (Knights, 1977).

D. Factors Affecting Efficiency

A cultivar plot of four rows, each 10 m long with 75 cm between rows is used by ICRISAT, but smaller programs may require only two rows.

At Hyderabad, India, flowering continues for about 3 weeks in all cultivars and simultaneous flowering of most cultivars can be arranged by planting the crossing block nursery with three sowing dates at 2-week intervals. This technique for matching flowering dates is widely used.

Purple flowers are dominant to white flowers and have been used to identify F_1 plants in many genetic studies and breeding programs (Eshel,

1968; Khosh-Khui and Niknejad, 1971; Retig, 1971). It is possible that maximum genetic advances for yield will be achieved in the future by the introgression of desi into kabuli types (Auckland, 1977; Auckland and Singh, 1977a) and that flower color will continue as an important genetic marker. Ladizinsky and Adler (1976) have also shown that the purple flower color of *C. echinospermum* and *C. reticulatum* is dominant to white flower color in *C. arietinum.*

Van der Maesen (1972) compiled a reference list of simply inherited characters which can be used for genetic markers. They include purple dominant to blue flower; purple dominant to green stems; single peduncle dominant to double flowered peduncle; erect dominant to prostrate plant habit; rough dominant to smooth seed coat, and resistance to Ascochyta blight dominant to susceptibility.

V. NATURAL HYBRIDIZATION

Farooq (1978) reported that male sterile plants were produced in the cultivar 'BEG-482' after mutagenic treatment by X-rays and EMS. Sethi and Gowda (1977) found two male-sterile F_2 plants in a cross between genetically diverse parents. However, male sterility has not yet been exploited in chickpea breeding.

VI. SEED DEVELOPMENT, HARVEST, AND STORAGE

If crossing was a success, the pedicel will remain fresh and a pod will start forming within 5 or 6 days. Pods may be harvested when dry and parchment-like and the seeds rattle in the pod. Threshing should be done by hand, taking care not to damage the seed coat, and the seeds packeted and identified by the cross name or number. Chickpeas soon lose viability if stored at high temperatures and high humidity, but will normally retain viability under natural conditions from one season to the next. If kept longer, they should be stored at 10 C or cooler.

VII. TECHNIQUES FOR SPECIAL SITUATIONS

Chickpea research is in its infancy and techniques using embryo culture or grafting for special situations in hybridization have not been reported. Bahl and Gowda (1975) tried using a growth regulator, applied to the flower bud, to improve pod setting by retarding the abscission layer formation in the pedicel, but without marked success.

REFERENCES

Auckland, A. K. 1977. Breeding chickpeas at ICRISAT. p. 133-140. *In* Induced mutations for improvement of grain legume production in South East Asia (1975). Technical Document. International Atomic Energy Agency, Vienna, Austria no. IAEA-203.

———, and K. B. Singh. 1977a. The exploitation of natural variability for the improvement of chickpea. p. 83–95. *In* Muhammed Amir, Rustem Aksel, and R. C. van Borstel (ed.) Genetic diversity in plants. Plenum Publishing Corporation, New York.

———, and ———. 1977b. An international approach to chickpea breeding. p. 10–13. *In* Plant Breeding Papers, Vol. 2. Third Int. SABRAO Conference, Canberra, Australia.

Ayyar, V. R., and R. Balasubramanyan. 1935. Anthesis and pollination in Bengal gram (*Cicer arietinum*) Madras Agric. J. 23-5:170–8.

Aziz, M. A., M. A. Khan, and S. Shah. 1960. Causes of low setting of seed in gram (*Cicer arietinum*). Agric. Pakistan 11:37–48.

Bahl, P. N. 1977. Personal communication. Division of Genetics, Indian Agricultural Research Institute, New Delhi-12, India.

———, and C. L. L. Gowda. 1975. Pod setting in crosses of Bengal gram. Ind. J. Genet. 35-1:13–16.

Bascur, G. B. 1977. Personal communication. Instituto de Investigaciones Agropecuarias, Estacion Experimental, La Platina, Chile.

Cubero, J. I. 1975. The research on the chickpea (*Cicer arietinum*) in Spain. p. 117–122. *In* Int. Workshop on Grain Legumes. ICRISAT, Hyderabad, India.

———. 1977. Personal communication. Instituto Nacional de Investigaciones Agronomicas, Cordoba, Spain.

Eser, D. 1976. Heritability of some important plant characters, their relationships with yield and inheritance of blight resistance in chickpea. [In Turkish, with English summary.] Ankara Universitesi, Ziraat Fakultesi Yayinlari 620. p. 40.

———. 1977. Personal communication. Plant Growing and Breeding Dep., Faculty of agriculture, Univ. of Ankara, Turkey.

Eshel, Y. 1965. Studies on flowering, seed setting and ripening of chickpeas (*Cicer arietinum* L.). Ph.D. thesis. Hebrew Univ. of Jerusalem, Israel. [In Hebrew with English summary.]

———. 1966. Phenology of chickpea. Research Report 1965. Sci. Agric. Jerusalem 1:542.

———. 1968. Flower development and pollen viability of chickpea (*Cicer arietinum* L.). Israel J. Agric. Res. 18:31–33.

Farooq, F. 1978. Mutagenic studies in *Cicer arietinum* L. Ph.D. thesis. Dep. of Botany, Osmania Univ., Hyderabad, India.

Howard, A., G. L. C. Howard, and A. R. Khan. 1915. Some varieties of Indian gram. Mem. Dep. Agric. India 7:213–235.

Kadam, B. S., R. K. Kulkarni, and S. M. Patel. 1938. Flowering and pod setting in gram. J. Am. Soc. Agron. 30-7:547–557.

Khosh-Khui, M., and M. Niknejad. 1971. Flower colour inheritance in reciprocal crosses of white and black chickpeas (*Cicer arietinum* L.). Israel J. Agric. Res. 21-2:81–82.

Knights, E. J. 1977. Personal communication. Agricultural Research Institute, Wagga Wagga, N.S.W. 2650, Australia.

Ladizinsky, G. 1975. A new *Cicer* from Turkey. Notes Roy. Bot. Gard. Edinb. 34: 201–202.

———, and A. Adler. 1975. Seed protein electrophoresis of the annual species of *Cicer* L. Israel J. Bot. 15:1–7.

———, and A. Adler. 1976. The origin of chickpea *Cicer arietinum* L. Euphytica 25:211–217.

Lopez, G. H. 1977. Personal communication. Centro de Investigaciones Agricolas de Sinaloa, Apartado Postal 356, Culiacan, Sinaloa, Mexico.

Nanda, K. K., and J. J. Chinoy. 1960. Effect of vernalization and photoperiodic treatments on *Cicer arietinum.* J. Plant Physiol. 3-1:31–44.

Nene, Y. L. 1977. Personal communication. ICRISAT, A. P., India.

Niknejad, M., and M. Khosh-Khui. 1972. Natural cross-pollination in gram (*Cicer arietinum* L.). Indian J. Agric. Sci. 42-4:273–274.

Pandey, R. K., M. C. Saxenz, B. B. Singh, and B. K. Singh. 1976. Physiological studies in gram. Proc. All-India Coordinated Research Project on Pulses. 12th Workshop (Rabi Pulses), Varanasi.

Pandya, B. P. 1977. Personal communication. G. B. Pant Univ. of Agriculture, Dep. of Plant Breeding, Pantnagar, India.

Papadakis, J. 1966. Climates of the world and their agricultural potentialities. Buenos Aires. p. 173:109.

Popova, G. M. 1937. Grain leguminosae 25-71. *In* N. I. Vavilov (General ed.), Wulff, E. V. (ed.) Flora of cultivated plants of the U.S.S.R. Moscow-Leningrad.

Ramanujam, S. 1976. Chickpea: *Cicer arietinum* (Leguminosae-Papilionatae). p. 157–159. *In* N. W. Simmonds (ed.) Evolution of crop plants. Longman, London.

Reed, W. 1977. Personal communication. ICRISAT, A. P., India.

Retig, B. 1971. Hybridization methods in chickpea (*Cicer arietinum* L.). 2. Crossing without emasculation. Israel J. Agric. Res. 21-3:113.

Saxena, N. P., and A. R. Sheldrake. 1976. ICRISAT annual report 1975–1976. Chickpea physiology. p. 176.

Sen, S., and D. Mukherjee. 1961. Preliminary studies of defective seed setting in gram. Sci. Cult. 27-4:185–188.

Sethi, S. C., and C. L. L. Gowda. 1977. Personal communication. ICRISAT, A. P., India.

Singh, K. B., and A. K. Auckland. 1975. Chickpea breeding at ICRISAT. p. 3–17. *In* Int. Workshop on Grain Legumes. ICRISAT, Hyderabad, India.

Summerfield, R. J., F. R. Minchin, and E. H. Roberts. 1977. Growing chickpeas (*Cicer arietinum* L.) in controlled environment growth cabinets. Univ. of Reading Mimeo Publ. p. 20.

van der Maesen, L. J. G. 1972. *Cicer* L., a monograph of the genus, with special reference to the chickpea (*Cicer arietinum* L.), its ecology and cultivation. Veenman and Zonen, Wageningen, Netherlands, and Meded. Landb. Hogesch. 72(10). p. 342.

———. 1973. Genetic resources of chickpea. Plant Genetic Res. Newsl. 30:17–24.

———. 1975. Germplasm collection and evaluation in *Cicer* and *Cajanus.* p. 229-237. *In* Int. Workshop on Grain Legumes, ICRISAT, Hyderabad, India.

16 Clovers

Norman L. Taylor
University of Kentucky
Lexington

The clover species described in this chapter are the true clovers belonging to the genus *Trifolium*. The species of major agricultural importance in the United States are red clover (*T. pratense* L.), white clover (*T. repens* L.), crimson clover (*T. incarnatum* L.), and alsike clover (*T. hybridum* L.). Crimson clover is an annual species, and the others are perennial. These clovers are thought to have originated in Asia minor or southeastern Europe. They are grown extensively for hay, pasture, and soil improvement throughout the eastern half of the United States and under irrigation in the Pacific and adjacent states.

I. PARENTAL MATERIAL

The genus *Trifolium* consists of about 240 species divided by taxonomists into about 16 sections (Hossain, 1961). Little or no gene transfer among sections has been possible. In the largest section of *Trifolium,* hybridization has chiefly centered around improvement of white clover. The species has been crossed successfully with *T. nigrescens* Viv., *T. xerocephalum* Fenzl., *T. isthmocarpum* Brot., *T. occidentale* D. E. Coombe, and *T. uniflorum* L. (Trimble and Hovin, 1960; Kazimierski and Kazimierska, 1968, 1970 ; Gibson and Beinhart, 1969; Chen and Gibson, 1970; Pandey, 1957). The only verified hybrids with red clover were obtained with the diploid annuals *T. diffusum* Ehrh. and *T. pallidum* Waldst. & Kit. (Taylor et al., 1963; Schwer and Cleveland, 1972; Armstrong and Cleveland, 1970). Alsike clover has been hybridized with Kura clover (*T. ambiguum* L.), but the hybrids did not flower (Keim, 1953). No verified interspecific crosses involving crimson clover have been obtained.

Seeds of *Trifolium* are available from the Regional Plant Introduction Stations at Geneva, New York (perennials), and Experiment, Georgia (annuals). A recent survey of germplasm in *Trifolium* indicated that seeds of about 85 species are available in the two locations and exploration for new species is underway (Taylor et al., 1977).

II. PLANT CULTURE

A. Field

The clovers generally flower and produce seeds best under soil conditions of medium fertility, medium pH levels, good drainage, and of moderate to heavy texture. White clover will tolerate relatively infertile conditions and sandy soils, whereas red clover succeeds best on non-sandy soils. Alsike clover is adapted to wet soils and will tolerate flooded conditions for short periods.

Soil treatments with phosphorus, potassium, and calcium to obtain the best vegetative growth of clover generally is thought to be adequate for seed production. On certain soils, the addition of minor elements, such as boron and sulfur, may increase seed yields. Nitrogen applications should be avoided because they intensify vegetative growth and lodging under field conditions. The amount of nitrogen fixed by *Rhizobium* species growing symbiotically with clovers is sufficient for adequate flowering and seed production.

Although adequate moisture is necessary for optimum growth of clovers, too much moisture or high relative humidity during flowering and seed set generally is detrimental. Cloudy, rainy weather reduces the number of heads of red clover per unit area, interferes with ovary development, and retards the pollination activity of bees.

Density of 70 to 80 plants/m^2 generally is adequate for seed production. Seeding rates vary with size of seed and range from 1 kg/ha for white clover to 20 kg/ha for crimson clover.

Nursery populations usually are transplanted or sown from 0.3 to 1 m within and between rows depending upon cultivation equipment available. Plantings may be clean cultivated or overseeded with grasses such as Kentucky bluegrass (*Poa pratensis* L.) or tall fescue (*Festuca arundinacea* Schreb.) to control weeds. Herbicides commonly used to control weeds under clean cultivation include Treflan, CIPC, Balan, and 2,4,D-B applied at rates indicated on labels of the products.

Common nursery designs used for open-pollinated seed production include completely randomized designs, with or without replication. However, for polycross nurseries, randomized complete blocks, Latin squares, or lattice designs are used (Chapter 5). Replicated plantings usually are not necessary for maintenance of seed of ecotypes and mass-selected cultivars.

B. Growth Chamber and Greenhouse

Seeds may be germinated in flats of soil, vermiculite, or in petri dishes 2 to 3 months prior to the time plants are needed for selfing or crossing. Seed scarification is often necessary because many of the clover seeds possess hard coats, particularly during the first year after harvest. Seeds may be scarified in any convenient manner provided the embryo and cotyledons are not damaged. One method is to rub the seeds lightly between two sheets of fine-grit emery paper until scarification is accomplished. If the seeds are then placed on a watered paper disc in a covered petri dish, they usually begin to imbibe water in a few hours. Seeds that remain unswollen may be removed for further scarification. Viable seeds that swell but fail to germinate often may be induced to do so by sprinkling activated charcoal on each seed. Seedlings should be transplanted into disinfested soil or other media to eliminate damping-off and other disease organisms and weed seeds. Methyl bromide has been used as a disinfestant, but often results in incomplete killing of white clover seeds. Steam sterilization of some soils under pressure will cause puddling and release of toxic elements, particularly manganese.

Vegetative increase of genetic materials is often necessary for hybridization. The perennial species and some annual species may be increased vegetatively by rooting stolons, rhizomes, stems, roots, or crowns, or by division of crowns. Rooting media include vermiculite, perlite, peat, and sand. Crown cuttings of red clover have been rooted successfully during the winter under short photoperiods in a mixture of equal parts of sedge peat, soil, and sand, and transplanted directly with the rooting medium. Annual species are rooted most successfully in a vegetative stage inasmuch as senescence occurs after flowering.

Most clover species appear to be long-day plants. However, clovers from different latitudes require different photoperiods, and interactions of temperature and day-length requirements for flowering are common. Many annual and some perennial species will flower if sown in a greenhouse at ambient temperatures and exposed to moderately long day lengths. Crimson clover and other annuals will flower earlier if exposed for 2 to 3 weeks to 10 to 15 C and transferred to a higher temperature (Knight and Hollowell, 1958). Most white clovers will flower under 14 to 16 hour photoperiods without vernalization (Beatty and Gardner, 1961).

A general procedure to obtain flowering in 4 to 6 weeks in a greenhouse is as follows: Plants are brought into a greenhouse in the late fall or early winter depending on the vernalization requirement for the species. They are transplanted into a moderately organic soil of adequate fertility to insure good flowering. Plants are subjected to a continuous low temperature (13 to 16 C) for approximately 4 to 5 weeks at a 10 to 12-hour photoperiod. Temperature is increased to 18 to 24 C and day lengths to 14 to 16 hours.

III. FLORAL CHARACTERISTICS

The clovers have a perfect, leguminous flower consisting of a calyx, a corolla, 10 stamens, and a pistil (Fig. 1). The calyx tube terminates in five lobes or teeth. A standard petal, two wing petals, and two keel petals unite at the base to form the corolla tube which is white or cream colored in white clover, reddish pink in red clover, crimson in crimson clover, purple in *T. purpureum* L., and yellow in hop clover (*T. campestre* L.).

Inside the corolla tube are nine stamens and one stigma united into a sexual column. The tenth stamen is free (Fig. 1). The number of ovules per ovary commonly varies from one to four, but may be as many as 10. Flowers are grouped in heads or short spikes and may be either sessile or stalked. At maturity, petals usually are indehiscent and either reflexed (white clover) or erect (red clover). In some species, the calyx enlarges to a bladder-like structure as seeds develop (strawberry clover, *T. fragiferum* L.). Flowers per head vary from one for *T. uniflorum* L. and only a few for subterranean clover, *T. subterraneum* L) to over 100 per head in many species. The ovules mature within or slightly extruded from the calyx and are indehiscent or dehiscent by ventral seams or hardened lids.

About 30% of the clover species are self-incompatible and cross-pollinated by bees, and 70% are self-pollinated (Taylor et al., 1977). Some species (e.g., crimson clover) normally cross-pollinate, but set considerable

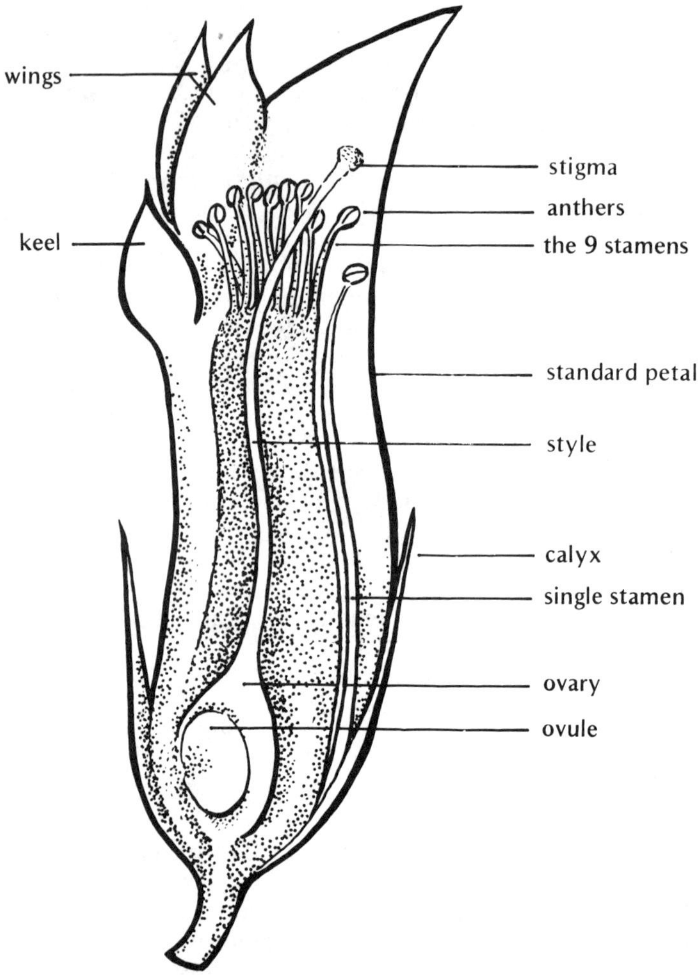

Fig. 1—Idealized diagram of typical clover flower.

seed upon selfing (Knight and Hollowell, 1973). Small flowers are often characteristic of self-pollinated species, and large flowers of cross-pollinated species. The flowers of red and zigzag clover (*T. medium* L.) are so large that honeybees (*Apis mellifera*) may avoid these species if other nectar and pollen sources are available. Bumblebees (*Bombus* spp) are efficient pollinators of the large-flowered clover species.

Among the self-pollinated species, pollination and fertilization may occur before the flowers open. For cross-pollinated species such as red clover, the stage of bloom for optimum seed set appears to be when each flower is about half open. Stigma receptivity and pollen viability continue, but decline throughout a period of 10 days. Red clover flowers first open in the middle of and on the topmost heads of the main stem. The length of time between pollination and fertilization of the egg cell in diploid red clover is between 28 and 32 hours. The first division of the zygote occurs 20 to 33 hours after fertilization (Mackiewicz, 1965).

IV. ARTIFICIAL HYBRIDIZATION AND SELF-POLLINATION

A. Equipment

Equipment used for crossing and selfing includes curved or straight pointed forceps, cuticle scissors, magnifying glasses (usually 10×), small jewelry tags, bamboo stakes, rafia or plastic ties for staking and tying plants, and instruments for transferring pollen (Fig. 2 and 3). Toothpicks to which a small piece of black emery paper has been glued, a sharp pointed lead pencil, or a small folded card are all satisfactory for pollen transfer. This equipment may be obtained from florist supply companies.

B. Preparation of the Female

Prior to flowering, especially in the field, the heads of each plant are protected against pollinating insects with bags of fine muslin, about 9 × 14 cm, which can be closed with a draw string. Bagged heads are supported with wires or stakes appropriate for the height of the plant. For easy manipulation, heads are trimmed to 15 or 20 newly opened flowers in the centers of the heads.

Emasculation is usually not necessary in red and white clover because of self-incompatibility controlled by the gametophytic S-allele system. However, in crimson clover and some self-fertile stocks of red and white clover, emasculation is desirable. Emasculation of self-pollinated species is quite difficult because the flowers may be small and tightly packed in the head. Furthermore, the anthers may dehisce at a very early stage, sometimes even before the petals are extruded beyond the calyx.

White and alsike clover can be emasculated by removal of the corolla. The underside of the corolla is gripped with forceps at a point midway between the tip of the calyx and the tip of the standard. The corolla tube and the attached anthers are removed leaving the pistil intact. This operation has been found effective for emasculation for all florets, except those which are fully opened or immature. For self-compatible genotypes, water is sprayed on stigmas to kill pollen that may have dehisced prior to removal of the corolla.

For species such as red clover in which removal of the corolla also removes the stigma, other emasculation procedures have been devised. One technique is to slit the corolla and calyx longitudinally on the underside, and remove the corolla and staminal column intact without disturbing the stigma. Another method is to remove the wing and keel petals of newly opened flowers in the center of the flowering head by grasping with forceps the uppermost part of the wings and pulling upward. Then, the heads are immersed in 66.5% ethanol for 10 to 20 sec and rinsed with water (Bassiri and Smith, 1972). Each investigator should determine methods of emasculation for a given clover species or particular genetic stocks before using them routinely in genetic and breeding investigations.

C. Pollination

The optimum stage of bloom to pollinate most clover species is shortly after the flowers open. The specific time of day for pollination appears to be unimportant. Cross-pollination of red clover flowers that are half open usually results in the highest seed set. Even 10 days after flower opening, approximately 40% of the flowers produce seeds. Wilted flowers should not be pollinated. The maximum length of time pollen will remain effective is not known for most clover species.

In manual pollinations, the pollen is removed from the plant used as the male by inserting the pollinating instrument (toothpick, etc.) between the standard and the keel and applying downward pressure. This causes the

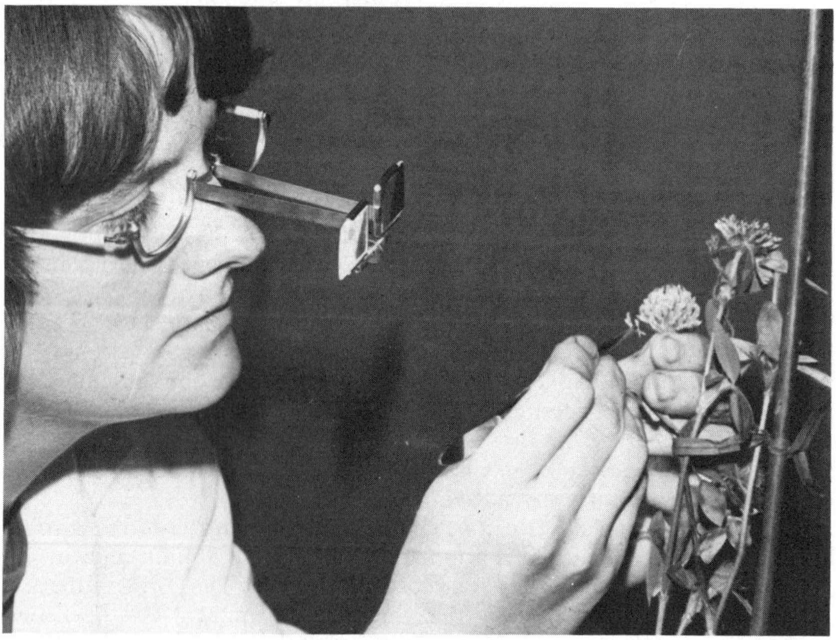

Fig. 2—Removal of flowers from a head of red clover, with the use of magnifying glass.

staminal column to strike the toothpick (Fig. 3). The pollen is checked at this time to see that it is moist and yellow rather than dry and white. Pollen is transferred on the toothpick to stigmas of plants used as females. Pollen from unemasculated female plants will tend to dilute that of the plant used as male and usually only 5 to 10 flowers can be effectively cross pollinated. One collection of pollen will usually pollinate 10 to 15 emasculated flowers. For reciprocal crosses, pollen is collected and applied alternately between paired heads of different plants using the same toothpick or folded card. After all flowers of a particular cross have been pollinated, a small jewelry tag is looped and secured over the stem immediately under the head. Tags are labeled as to parentage using an indelible pencil to prevent loss of the record when the plants are watered. Heads are kept free of water for at least the first 24 hours to prevent abortion of pollen. Before proceeding to the next cross, hands, forceps and other pollinating equipment are washed with alcohol and rinsed with water. If pollinating instruments are to be reused, they are set aside for several days after washing to prevent contamination.

Seeds of red clover can be produced on excised stems as a convenience for crossing at different locations. Stems bearing freshly-opened flowers that have not been crossed by bees are brought to the greenhouse or laboratory from the field. Stems are severed just above the crown and cut ends are immersed in water. In the greenhouse, stems are shortened and inserted in vials containing a 2% sucrose solution prior to making cross pollinations (Kendall and Taylor, 1969).

Self-incompatible clovers may be self-pollinated by three methods: spontaneous (no manipulation); tripping of individual florets by use of a toothpick, card, or lead pencil; or by rubbing the heads between the thumb and fingers. Tripping or rubbing usually produces more seeds than spontaneous selfing.

Red clover plants grown at high temperatures often produce more self seed than those grown at low temperatures. Likewise, exposure to 32 C for 1 to 2 days temporarily changes alsike clover from self-incompatible to self-

Fig. 3—Technique for selfing or crossing red clover using a toothpick to which a small piece of emery paper has been glued.

compatible (Townsend, 1968). A technique to increase self seed set of red clover was developed by Kendall and Taylor (1969). Heads, either excised or on intact plants, may be treated when in bud with some petal color showing. Heads are inserted in a chamber maintained at 40 C and with the lower part of the stems maintained at 25 C. After the flowers open at 40 C, the heads are removed from the chamber and the florets are selfed by tripping with a toothpick. Excised heads or intact plants are maintained at 25 C until seeds mature. Average seed set per head on highly self-incompatible clones may range up to 9.0 with the procedure.

D. Factors Affecting Efficiency

Most of the clovers flower over a rather long period of time and matching flowering dates usually is not a problem. Extremely early flowering genotypes may be maintained under short photoperiods or flowering stems may be removed to delay flowering. To prevent senescence, self-pollinated species should not be allowed to flower until ready for use.

Genetic markers for identifying hybrid seeds or plants have been summarized for white clover by Gibson and Hollowell (1966). Dominant characters include: red leaf, red flecking, red midrib, and white 'v'-marking, cyanidin red corolla, and black seed. Recessive characters are blush-colored corolla, and nonclasping bracts.

Red clover leaf characters include leaf marking that is dominant over nonmarking and yellow leaf-mark that is dominant over white. Red pigment on stems and stipules is conditioned by two dominant factors, but classification is difficult because the intensity of color is influenced by external conditions, especially light (Wexelsen, 1932). White flower-petal is conditioned by three recessive nonallelic genetic factors. Each of the genes in the homozygous recessive condition is epistatic to all other flower colors, except a dominant factor that causes a variegated flower color (Williams, 1935).

Simply inherited recessive characters in crimson clover include white flower color (Sandal, 1955), glabrous leaf, petiolulate leaf attachment, and multifoliolate leaves (Knight, 1969). The inheritance of characters in all clovers may be complicated by different genetic backgrounds and should be investigated in the stocks on hand before attempting their use as genetic markers in controlled crosses.

V. NATURAL HYBRIDIZATION

The following procedure is used for crossing clovers with honeybees. Bees are maintained in 4 to 10-frame hives in an apiary until the clover begins to bloom in the cages. Blooming heads are removed to prevent contamination from pollen carried by bees, and hives are moved into the cages, preferably at night when most bees are in their hives. Bees are fed a 1:1 by volume solution of granulated white sugar and water or commercially prepared bee food until the clover blooms profusely in the cage. Red clover in small cages in Kentucky have yielded up to 560 kg of seed per ha.

Clover may be cross-pollinated with bumblebees by the following technique: Bumblebees are captured by swiftly enclosing a mason jar or other container over the bee. The bees are washed by adding lukewarm water to the jar and shaking to remove and abort pollen. Bees are placed directly into the cage, where after drying, they will begin to cross-pollinate the flowers. Four to six bees generally will pollinate two plants depending on the number of flowers present. In sunny weather, bumblebees usually will live 10 days to 2 weeks, but in wet, cool weather they may die in a few days, and fresh bees should be introduced to complete the crossing. Bumblebees in a glass bee house will be considerably more active and live longer than those outside. Outdoor lights over cages may induce uniform and abundant flowering for pollination of white clover by honey and other bees.

Clover may be cross-pollinated outdoors without cages, but great distances are required particularly to isolate small areas. Volunteer plants of clover are also a hazard, and the isolation area must be observed daily before bees are active. Red clover plants with relatively few blooms usually are heavily contaminated with pollen from the outside when isolated at distances up to 457 m. On the other hand, plot sizes of 274 to 366 m^2 in heavy bloom may be effectively isolated (98 to 99% purity) from other clover at 457 m (Williams and Evans, 1935).

Crossing in the self-incompatible clovers may be controlled by planting propagules of single clones in fields of the species that serve as pollen parents. It is also possible to convert red clover lines to genetic male sterility for crossing. However, bee visitation to male sterile plants is at a minimum, and little seed is produced (Smith, 1971).

Large-scale crossing in the clovers has been most successfully controlled by the gametophytic S-allele system. Many variations have been suggested, but only one scheme has been used to produce a double cross hybrid (Anderson et al., 1972). By this method non-inbred clones of red clover (I_0) are selfed and first generation selfs (I_1) are tested to isolate homozygous S-allele genotypes. Crosses then proceed according to the following diagram:

Material	Clonal identification and S-alleles expected			
	1	2	3	4
Inbred line with homozygous S-alleles	S_1S_1	S_3S_3	S_5S_5	S_7S_7
Single cross: one genotype each	S_1S_3		S_5S_7	
Double cross: four genotypes	S_1S_5	S_1S_7	S_3S_5	S_3S_7

The inbred lines are vegetatively propagated to obtain an isolated field for each single cross. Seeds from the single crosses are blended and sown in an isolated field to obtain double-cross seed. The double-cross is controlled by the S-allele system and the product is true double-cross seed, except for occasional selfs or intra-line crosses. Theoretically, generations beyond the double cross should have reduced forage yields and are not recommended. The principal disadvantage of the method is the high cost of the single-cross seed because of the necessity of establishing large fields of the four inbred clones with low seed yields.

One hybrid produced by this method was tested throughout the United States, but forage yields were not greater than those of the synthetic cultivar 'Kenstar'. Similar single-cross or double-cross hybrids, while theoretically possible in other cross-pollinated diploid clovers, have not been produced to date.

VI. SEED DEVELOPMENT, HARVEST, AND STORAGE

Pollinated clover flowers usually begin to wilt in about 2 days, but non-pollinated heads will remain unwilted and receptive up to 10 days after blooming. Clover seeds may mature as early as 21 days after pollination, but under humid conditions, ripening may take somewhat longer.

Harvested seeds are usually dried for a few days either at room temperature or in a dryer at temperatures not exceeding 45 C. For crosses in which only a few heads are involved, a simple method of threshing is to manually rub seeds from florets between two corrugated rubber mats. For somewhat larger quantities of heads, one useful machine for threshing is the Forsberg seed scarifier in which the emery paper inside the drum is replaced with corrugated rubber. Still larger quantities may be threshed with a 90-cm belt thresher. If very large quantities of heads with stems are to be threshed, a small combine may be used. The Clipper Cleaner (model 2-B) with two screens, and the larger four to six-screen models are available to clean threshed seed. A set of hand-testing screens is often useful for preliminary screening, and a South Dakota seed blower may be used for final cleaning. A Tornado blower-vacuum will clean the machine between lots. Most of this equipment is available from Ferrell-Ross, Oklahoma City, Ok.; Burrows Equipment Co., Evanston, Ill.; or Seedburo Equipment Co., Chicago, Ill. Seed threshing and cleaning equipment are matters of individual choice, and many different methods are available.

Often it is desired to identify clover seeds as to female parent. The seeds of red clover vary from white, yellow, brown, and deep purple. Seeds must be mature for full expression of color, and weathered seeds are likely to be brown. Many other species do not have variable seed colors. Variation within species should be determined prior to threshing and cleaning, if maintenance of genetic identity is likely to be a problem.

Clover seeds may be stored for long periods at 0 C or lower without loss of viability. Red clover seeds lost little viability over a 10-year period when stored in a freezer (Rincker, 1974). Other investigators have stored clover seeds at 40% relative humidity and 4 to 5 C for up to 5 years without great loss of viability.

VII. TECHNIQUES FOR SPECIAL SITUATIONS

Many of the special techniques which have been developed for *Trifolium* are concerned with interspecific hybridization. Chromosome doubling has been shown to increase the success of interspecific hybridization and to increase fertility by formation of amphiploids. Stem cuttings or seedlings are inverted in 0.2% aqueous colchicine for 2 to 6 hours, rinsed in tap water, and transplanted to pots filled with disinfested soil. Doubled branches are identified by the presence of stainable pollen at anthesis for sterile diploid hybrids and by cuboidal pollen (4x) verses rod-shaped pollen (2x) for fertile diploid hybrids. Tip cuttings of doubled branches may be rooted and expected chromosome numbers should be verified.

Nitrous oxide treatment, applicable only to fertile materials, can be applied to red clover and related species for 24 hours at 6 bars atmospheric pressure beginning 24 hours after cross pollination. Up to 49% tetraploids were produced by the treatment in *T. alpestre* L., 71% in *T. pratense,* and 79% in *T. rubens* L. Doubled plants and sectors may be identified by the same methods as used in colchicine-doubled materials.

To overcome post-fertilization barriers, embryos have been cultured on artificial media (Hovin, 1962). Excised embryos are placed in an aqueous Nitsch solution containing 2% sucrose and maintained until trifoliolate leaves form. Seedlings are transferred to a 0.7% agar-solidified Nitsch medium containing 1% sucrose. After 2 to 3 weeks, surviving seedlings are transferred again to an aerated nutrient solution with sucrose. The last transfer is to disinfested soil which is watered with modified Nitsch solution until the plants are well established.

REFERENCES

Anderson, M. K., N. L. Taylor, and R. Kirthavip. 1972. Development and performance of double-cross hybrid red clover. Crop Sci. 12:240–242.

Armstrong, K. C., and R. W. Cleveland. 1970. Hybrids of *Trifolium pratense* × *Trifolium pallidum*. Crop Sci. 10:354–357.

Bassiri, A., and R. R. Smith. 1972. Emasculation of self-compatible red clover clones with ethanol. Can. J. Plant Sci. 52:846–848.

Beatty, D. W., and F. P. Gardner. 1961. Effect of photoperiod and temperature on flowering of white clover, *Trifolium repens* L. Crop Sci. 1:323–326.

Chen, C., and P. B. Gibson. 1970. Meiosis in two species of *Trifolium* and their hybrids. Crop Sci. 10:188–189.

Gibson, P. B., and G. Beinhart. 1969. Hybridization of *Trifolium occidentale* with two other species of clover. J. Hered. 60:93–96.

————, and E. A. Hollowell. 1966. White clover. USDA Agric. Handb. 314. U.S. Government Printing Office, Washington, D.C.

Hossain, M. 1961. A revision of the *Trifolium* in the Nearer East. Notes R. Bot. Gard. Edinb. 23:387–481.

Hovin, A. W. 1962. Interspecific hybridization between *Trifolium repens* L. and *T. nigrescens* Viv. and analysis of hybrid meiosis. Crop Sci. 2:251–254.

Kazimierski, T., and E. M. Kazimierska. 1968. Investigations of hybrids of the genus *Trifolium* L. I. Sterile hybrid *Trifolium repens* L. × *T. xerocephalum* Frenzl. Acta Soc. Bot. Pol. 37:549–560.

———, and ———. 1970. Investigations of the hybrids of the genus *Trifolium* L. III. Morphological traits and cytogenetics of the hybrid *Trifolium repens* L. × *T. nigrescens* Viv. Acta Soc. Bot. Pol. 39:565–592.

Keim, W. F. 1953. An embryo culture technique for forage legumes. Agron. J. 45: 509–510.

Kendall, W. A., and N. L. Taylor. 1969. Effect of temperature on pseudo-self-compatibility in *Trifolium pratense* L. Theor. Appl. Genet. 39:123–126.

Knight, W. E. 1969. Inheritance of multifoliolate leaves, glabrous leaves, and petiolulate leaflet attachment in crimson clover, *Trifolium incarnatum* L. Crop Sci. 9:232–235.

———, and E. A. Hollowell. 1958. The influence of temperature and photoperiod on the growth and flowering of crimson clover (*Trifolium incarnatum* L.). Agron. J. 50:295–298.

———, and ———. 1973. Crimson clover. Adv. Agron. 25:47–76.

Mackiewicz, T. 1965. Low seed setting in tetraploid red clover (*Trifolium pratense* L.) in the light of cytoembryologic analyses. Gen. Pol. 6:5–39.

Pandey, K. K. 1957. A self-compatible hybrid from a cross between two self-incompatible species in *Trifolium.* J. Hered. 48:278–281.

Rincker, C. M. 1974. Effect of frequent thawing on viability of red clover seed in cold storage. Crop Sci. 14:749–750.

Sandal, P. C. 1955. Inheritance of white flower color in crimson clover, *Trifolium incarnatum.* Agron. J. 47:147–148.

Schwer, J. F., and R. W. Cleveland. 1972. Diploid interspecific hybrids of *Trifolium pratense* L., *T. diffusum* Ehrh., and some related species. Crop Sci. 12:321–324.

Smith, R. R. 1971. Inheritance of a male-sterile character in red clover (*Trifolium pratense* L.). Crop Sci. 11:326–327.

Taylor, N. L., P. B. Gibson, and W. E. Knight. 1977. Genetic vulnerability and germplasm resources of the true clovers. Crop Sci. 17:632–634.

———, W. H. Stroube, G. B. Collins, and W. A. Kendall. 1963. Interspecific hybridization of red clover (*Trifolium pratense* L.). Crop Sci. 3:549–552.

Townsend, C. E. 1968. Self-compatibility studies with diploid alsike clover, *Trifolium hybridum* L. III. Response to temperature. Crop Sci. 8:269–272.

Trimble, J. P., and A. W. Hovin. 1960. Interspecific hybridization of certain *Trifolium* species. Agron. J. 52:485.

Wexelsen, H. 1932. Segregations in red clover (*Trifolium pratense* L.). Hereditas 16: 219–240.

Williams, R. D. 1935. Genetics of flower colour in *Trifolium pratense* L. I. Basic white colour (factor c). J. Genet. 31:431–450.

———, and G. Evans. 1935. The efficiency of spatial isolation in maintaining the purity of red clover. Welsh J. Agric. 11:164–171.

17 Common Bean

F. A. Bliss
University of Wisconsin
Madison

The common bean, *Phaseolus vulgaris* L., includes dry or field beans grown for the dry, edible seeds and snap beans (French or string beans) grown for the immature fruits. At least 12 commercial classes of dry beans are produced in the United States each year. The major classes are distinguished primarily by physical properties of the seeds and include the navy or pea bean, small white, small red, pink, red kidney, great northern, pinto, black turtle soup, cranberry, white marrow, flat small white, and yelloweye. Additional genetic variability exists within *P. vulgaris* as landrace cultivars and wild or semidomesticated strains, particularly in South and Central America, where cultivated beans likely originated (Gentry, 1969) and are a major dietary component.

I. PARENTAL MATERIAL

Only four species ($2n = 2x = 22$), *P. vulgaris, P. coccineus* L., *P. acutifolius* Gray var. *latifolius* Freem., and *P. lunatus* L. var. *lunatus,* of the more than 100 within the genus are prominent as cultivated food crops. It has been postulated that the cultivated common bean complex resulted from hybridization among two or three species (Freytag, 1955). These may include *P. coccineus* or its wild ancester, wild *P. vulgaris* such as the forms collected by Gentry, and the South American wild bean *P. vulgaris* var. *aborigineus* Burk. (Berglund-Brucker and Brucher, 1976). The area of the origin of *P. vulgaris* is probably in Mexico and Guatemala (Freytag, 1975), but may extend into South America.

Successful crosses have been obtained between *P. vulgaris* and *P. coccineus,* when the former was used as the female parent (Lamprecht, 1941; Al-Yasari and Coyne, 1966; Smartt, 1970). However, some combinations produce dwarf and misshapen plants with various amounts of sterility and self-incompatibility. Coyne (1964) obtained a complex hybrid of *P. acutifolius* × [*P. vulgaris* × F_1 (*P. vulgaris* × *P. coccineus*)]. Although embryo abortion usually results, successful crosses have been obtained between *P. vulgaris* × *P. acutifolius* using embryo culture (Honma, 1955; Honma, 1956; Coyne, 1964; Smartt, 1970); between *P. vulgaris* × *P. ritensis* with chromosome doubling and embryo culture (Braak and Kooistra, 1975), and *P. vulgaris* × *P. lunatus* using heterozygous parents and embryo culture (Honma and Heeckt, 1959).

The most extensive efforts to systematically collect and maintain germplasm have resulted in the USDA collection at Pullman, Wash.; the *Phaseolus* germplasm bank at the Centro International de Agricultura Tropical, Cali, Colombia; the *Phaseolus* collection at The National Institute of Agricultural Research and the Post-Graduate School, Chapingo, Mexico; and the legume germplasm bank at the University of Cambridge, England (Burns, 1975).

II. PLANT CULTURE

A. Field

Attempts to manually cross-pollinate field-grown plants have met with little success. Plants such as those with a determinate bush growth habit cannot be manipulated easily for crossing. Unfavorable climatic conditions and pests increase damage to the fragile flowers, and a low percentage of successful crosses usually results.

B. Growth Chamber and Greenhouse

Most cross-pollinations are made on plants grown in a growth chamber, greenhouse, or a shelter house with a plastic roof and screened sides to exclude insects. Seeds should be treated with a suitable fungicide prior to sowing. They may be sown directly into clay or plastic pots 15 to 20 cm in diameter, containing fertile, well-drained soil, such as a compost mixture of loam, peat and sand in a 7:3:2 ratio by volume. Autoclaving may be necessary if root rot-inciting pathogens are a problem. One or two plants can be maintained in pots of this size, and any thinning should be done by the time the unifoliolate leaves have expanded. Seeds may also be sown into a medium, such as Perlite, for germination, and the seedlings transplanted into pots containing moist, workable soil when the unifoliolate leaves begin to expand. When germinating in an inert medium, tap water or dilute nutrient solution should be used for watering to avoid nutrient deficiencies.

Seeds and seedlings should be inoculated with an effective strain of *Rhizobium phaseoli* to promote nitrogen fixation. In addition to adequate amounts of N, P and K, the minor elements Ca, Zn, Bo, Mg, and Mo are especially important to beans. Soils high in aluminum (e.g. red tropical soils) should be avoided. Soil pH should be from 6 to 6.5. Two or three applications of a complete nutrient solution can be added to the soil during the growing season, or dilute nutrient solutions applied as foliage sprays prior to or in the early bloom stage.

Bean plants react unfavorably to moisture stress caused by either excess or insufficient water. Waterlogged conditions for 24 to 48 hours will result in leaf and blossom drop and reduced yield. Moisture stress is most critical from shortly before flowering until the seeds are nearly mature. Williams (1962) found that moisture stress affected the fruiting behavior of snap beans almost exclusively by abscission of floral buds and young pods containing small, fertilized ovules. He found no evidence that severe moisture stress upset either pollination or fertilization. High relative humidity reduces the desiccation of flowers after emasculation and is conducive to good seed set, but it also enhances some foliage diseases. Frequent sprinkling of benches and walkways in the greenhouse during pollination on hot days helps to reduce temperature and increase humidity, which favor better seed set. Splashing on the plants during watering should be avoided to reduce possible spread of pathogens.

P. vulgaris is a short-day species and at low latitudes cultivars flower as soon as they are physiologically ready. Many wild and semidomesticated beans and many cultivars which have evolved in the low latitudes fail to flower at higher latitudes until the days become shorter in late summer. Most cultivars bred for temperate, high latitude areas flower over a wide range of photoperiods. Although these types appear to be similar in flowering under long days of 14 to 16 hours, they often perform quite differently when grown under short days of 11 to 12 hours and high temperatures.

Plants which are short-day-sensitive can be induced to flower by growing them in chambers which exclude light for a desired period or during the fall and winter months in temperate areas. Supplemental light should be provided when plants are grown for hybridization during periods of low natural light level. A combination of incandescent and fluorescent lights to provide from 20.5 to 32.0 klux at the plant level is sufficient to promote good plant growth.

The optimum daytime temperature range for plant growth and reproduction is 15 to 27 C, and night temperatures should be 5 to 8 C cooler (Knott, 1957). At temperatures substantially below 15 C, plant growth is delayed, flowers may fail to open normally, and buds may be too small for easy hybridization. Temperatures exceeding 30 C are likely to be deleterious to growth and reproduction. High temperatures often interact with photoperiod to delay flowering (Coyne, 1968), cause spindly plants with low light level, and produce flower drop and pod abortion when occurring with moisture stress.

It is essential that plants be free of diseases and insects. White fly, spider mites, aphids, and leaf miner can be particularly troublesome on plants grown in the greenhouse. Different chemicals should be used periodically to provide control and minimize the opportunity for resistant populations to develop. Systemic insecticides may be added to the soil in the pots at sowing or transplanting, in addition to later application of insecticides to the foliage. High concentrations of pesticides should not be used because bean leaves are quite sensitive. If several generations of plants are to be grown consecutively in a greenhouse, it should be cleaned thoroughly prior to starting new plants.

III. FLORAL CHARACTERISTICS

Phaseolus vulgaris is highly self-pollinated and one can safely assume that no outcrossing will occur among greenhouse-grown plants. The sequence of flowering depends on whether the plant is determinate or indeterminate. In the determinate types, the first flowers to open are those of the inflorescence at the uppermost leaf axil on the mainstem, followed by the lowest flowers in the terminal inflorescence (Wivutvongvana and Mack, 1974; Ojehomon, 1966). In the indeterminate types where there are only axillary inflorescences, the node number at which the first lateral inflorescence differentiates is variable and appears to be a cultivar characteristic (Ojehomon, 1969).

Dehiscence and self-pollination occur in the late bud stage, usually at night, with pollen tube growth beginning almost immediately (Williams, 1962). The flowers usually open in the early morning, probably in response to light and temperature. Freytag (1977) believes that dehiscence depends primarily on maturation of anther sacs and secondarily on lowering of the relative humidity.

The perfect flowers, which are of the typical legume shape, are borne on racemes which arise at a node with a trifoliolate leaf (Fig. 1A). The corolla consists of two petals that are fused to form a spirally twisted, prolonged keel; two wing petals; and a standard. The corolla is enclosed in a small, lobed calyx and two small bracts. Enclosed in the keel are one free and nine fused stamens, each with a bilobed anther sac borne on a long filament (Fig. 1B). The pistil has an ovary containing usually from 5 to 8, and occasionally 9 to 10 ovules. The style is coiled and the stigma is a flattened, slightly depressed pad surrounded by a ridge of pointed hairs (Fig. 2E). The stigmatic surface is covered with a sugary liquid secreted by epidermal cells. The stamen filaments follow the stylar coil such that the anther sacs are appressed to the stigma. Pollen is shed directly onto the stigmatic surface and self-pollination results (Ibrahim and Coyne, 1975; Webster et al., 1977).

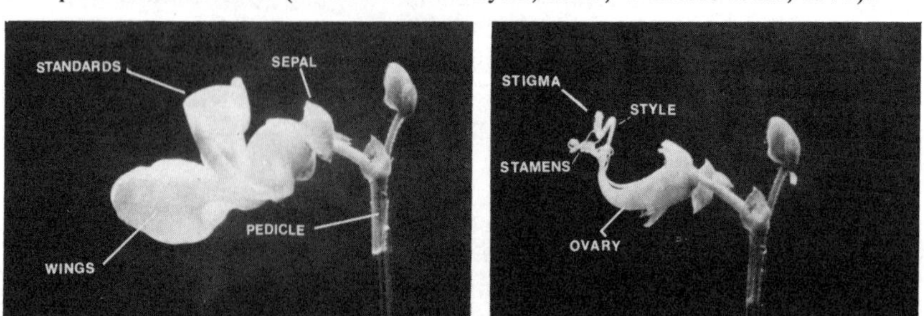

A B

Fig. 1—Flower structure of the common bean at the time of anthesis. A, Side view of open flower; B, flower with calyx and corolla removed.

IV. ARTIFICIAL HYBRIDIZATION OR SELF-POLLINATION

A. Equipment

The equipment necessary for artificial hybridization includes a fine-tipped forceps, either curved or straight depending on the preference of the pollinator; a small bottle of alcohol for rinsing the forceps; cellophane tape

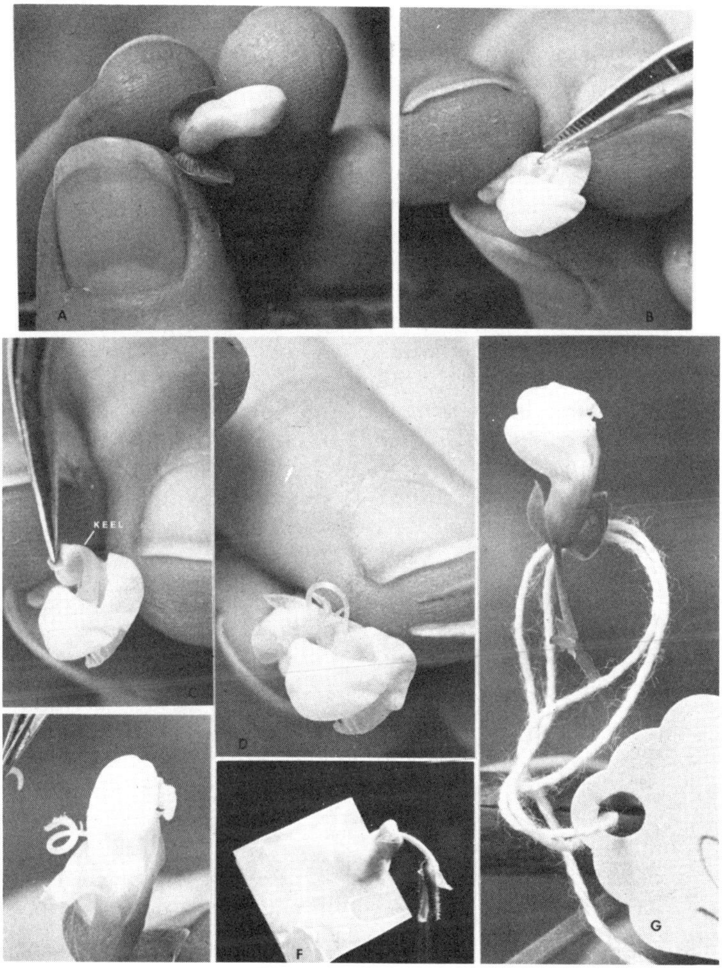

Fig. 2—Steps in artificial hybridization of the bean flower. A, Female bud at proper stage for emasculation; B, opening the standard with the forceps and removal of the wings; C, exposed keel with wings removed; D, coiled style with anthers and wings removed; E, pollination of the female style with a pollen-laden stigma from a male flower; F, enclosure of female bud with cellophane tape after pollination; G, female flower after pollination showing the tag in place around the pedicle. (Photos A, B, C, D, E, G, courtesy of Carol A. Erwin, Dermot P. Coyne, Richard Dodds, Univ. of Nebraska, Lincoln.)

on a dispenser; small string jeweler's tags; coin envelopes about 65 × 110 mm; and a magnifying lens. Plastic, quick-snap tags which are split to the hole at one end to allow the tag to be attached around the flower pedicle, can be substituted for the string tags. A jeweler's headband binocular lens with 3 to 7× magnification can be used if needed. A small box is suitable for carrying and storing the equipment.

B. Preparation of the Female

The stigma is receptive at least 2 days before and 1 day after normal anthesis. Buds which are plump, are showing color, and will open the following day are chosen for the female parent (Fig. 2A). If self-pollination has already occurred, the stigma will be swollen, light green rather than white, dried rather than sticky, and with some remnant pollen attached.

Emasculation and cross-pollination can be made throughout the day, with success depending largely on the skill of the pollinator and weather factors, such as sunlight, temperature, and humidity. An important prerequisite for success is to avoid injuring the stigmas and styles of the emasculated flowers and to prevent their desiccation after pollination. Cross-pollination can be made either with or without emasculation (Buishand, 1956). The bud is held between the thumb and forefinger of one hand and a forceps is held in the other, resting the little finger and ring finger on the hand holding the flower bud. The standard is opened by inserting the point of the forceps into the suture and pushing from side-to-side (Fig. 2B). The wings are carefully removed with the forceps to expose the coiled keel (Fig. 2C). The keel must be removed to allow access to the anthers. A small incision is made near the top of the keel with the point of the forceps. When it has been opened, the upper lip should be grasped with the forceps and peeled up and back. If done properly, the majority of the outer half of the keel can be removed exposing the stigma and anthers (Fig. 2D). Often the keel will break and subsequent efforts must be made to remove enough of it to expose the anthers. After it is pulled off, the 10 stamens are removed carefully so as not to rupture the anther sacs and cause self-pollination. Because all stamens can seldom be removed altogether, it is helpful to count those held by the forceps at each attempt and to dispose of the anthers until all 10 are accounted for. The forceps can be dipped in alcohol periodically to prevent contamination.

If emasculation is not used, the standard is opened with the forceps and bent slightly backward. Both wings are pressed downward causing the stigma to protrude through the opening at the end of the keel. This method is particularly useful in cultivars where four or five buds are grouped closely together and mature simultaneously.

C. Pollination

Most breeders achieve the best success from pollinations made prior to 1100 hours. Where daytime temperatures are extremely high, success has been reported from cross-pollinations made after 1530 hours (Tulmann

Neto, 1977). Flowers are commonly used as a pollen source the day they open. They are removed from the male plant and used immediately in the morning or placed in a plastic bag and kept in a refrigerator for use in the afternoon.

Two procedures commonly used for hand pollination are the rubbing and hooking methods. For the rubbing method, the pollinated stigma of a flower from the male parent is forced from the keel by pressing the wings downward, and is rubbed against the female stigma. For the hooking method, the pollinated stigma is removed from the male flower with a forceps and rubbed against the stigma of the female flower (Fig. 2E). The stigma of the male flower is left in the female flower by hooking it through the style of the latter near the stigmatic surface.

Pollination without emasculation can be completed by rubbing pollen from the pollinated stigma of a male flower onto the protruding stigma of the female flower, then allowing the stigma of the latter to recede into the keel. This does not preclude later self-pollination. A variation of this procedure is the clamping method, in which the pollen-covered stigma of the male is hooked behind the protruding stigma of the female after cross-pollination, preventing it from receding into the keel. This procedure minimizes the chance of later self-pollination and clamps the pollinator stigma in place.

To prevent rapid desiccation of the freshly pollinated stigma, the standard is closed gently after pollination and the bud may be enclosed with cellophane tape (Hikida, 1961). A strip of tape about 40 mm long is formed into a circle by joining the ends, with the sticky side facing inward. After the top of the tape circle is pressed together, it is placed over the standard to keep the bud closed (Fig. 2F). If care is taken to cover only the petals (exclude the sepals) the tape will fall off as the petals wither and the pod develops. To provide more moisture around the bud, a small piece of wet tissue is sometimes placed inside the tape circle. Silbernagel (1968) achieved a 20 to 30% increase in hybridization efficiency by placing unopened anthers near the female stigma, rather than brushing it with a pollen-laden stigma before closing the bud with cellophane tape. After each pollination is completed, a tag is placed around the pedicel of the female flower (Fig. 2G). Information recorded on the tag includes the identification of the female and male parents, date of pollination, and initials of the pollinator.

Ovules nearest the stylar end of the ovary are fertilized first, with fertilization progressing in a linear fashion to the stem end of the pod. The time from pollination to fertilization of the last ovule is about 12 hours. Abscission of flowers and young pods sometimes occurs in response to environmental stress. An indication that plants are growing under stress is the production of club-shaped pods which result when the developing seeds near the stem end (those fertilized last) abort and those near the stylar end continue to develop.

Buishand (1956) reported that success rates of 30 to 40% were common using the rubbing method following emasculation, compared to 70 to 80% success obtained using the hooking method following emasculation. Freytag (1977) reported 70 to 90% success using the hooking method, with each pod containing four to six seeds. The hooking method without emasculation gives good success and is faster, but care must be taken to detect possible self-fertilization.

After the small pod has begun to develop 3 to 4 days after cross-pollination, all subsequent flowers should be removed. It is necessary to prevent all selfed pods from developing. Severe competition results and selfed pods continue to develop while pods containing hybrid seeds wither and drop off. Under good greenhouse conditions each plant should support at least two to three pods, resulting from cross-pollination.

No special techniques are required to obtain selfed seeds if suitable conditions exist. Determinate, bush plants will produce from 3 to 4 pods per plant in the greenhouse, but 20 or more in the field. Indeterminate plants usually produce at least 10 selfed pods in the greenhouse and up to 100 or more in the field. The number of viable seeds per pod varies depending on the genotype of the seed parent and growing conditions surrounding the plants.

D. Factors Affecting Efficiency

Because most genotypes vary in the time from planting to first flower even when environment is favorable, it may be necessary to stagger planting dates to ensure that flowering periods overlap. Genotypes known to be photoperiod and temperature-sensitive should be subjected to conditions that will promote flowering, thus ensuring availability of male and female plants.

Pigmentation and morphological traits are used most often as genetic markers. Purple (or red) pigmentation of plant tissues is dominant to non-pigmentation and can be used to identify hybrid plants from the seedling stage until pod maturation. Purple stippling of the cotyledons of germinating seeds is dominant to green cotyledons, purple hypocotyl to green hypocotyl, purple foliage to green foliage, colored flower to white flower, and purple pod to green pod. Although the pigmentation locus is complex (Prakken, 1970), simply inherited differences in seed coat color have been reported and can be used to detect hybrid plants provided the female parent carries the recessive trait (Smith, 1939).

Because the seed coat is of maternal origin, genes expressed in hybrid cotyledon tissue may be masked and special techniques must be used for detection. Romero et al. (1975) described genetic differences in the electrophoretic banding patterns of polypeptides of the globulin seed protein. Since genes of both female and male parents are expressed in the hybrid cotyledons, a small portion of the cotyledon can be carefully removed for electrophoretic analysis and the remainder of the intact seed used for germination.

Germination habit of *P. vulgaris* (epigeal) and *P. coccineus* (hypogeal) has been used to distinguish interspecific hybrids from selfed progeny. Greatest success has been achieved using *P. vulgaris* as the female parent, and the germination habit of the hybrid is usually intermediate between the two parents. Leaf color, length and width of leaflets, pod shape and flower characteristics were used by Honma and Heeckt (1959) to distinguish F_1 hybrids of *P. vulgaris* × *P. lunatus.* Inflorescence type is generally different enough among species to allow identification of hybrids (Smartt, 1977).

V. NATURAL HYBRIDIZATION

Beans are usually highly self-pollinated, with the amount of outcrossing varying between locations. Several reports have shown outcrossing to be in the range of 0 to 1% (Tucker and Harding, 1975), while others have reported rates as high as 6 to 10% (Antunes et al., 1973). This variability is probably due to the kinds and frequency of insects. Thrips produce low levels of outcrossing, and carpenter bees (*Xylocopa brasilianorum* L.) have been reported to be responsible for levels as high as 15 to 20% in Puerto Rico (Vakili, 1976). The extensive outcrossing is being used there to increase variability in populations prior to selection.

The occurrence of male sterile or partially-male sterile plants in *P. vulgaris* has been reported by Frazier (1967), Ibrahim et al. (1974) and Agbo and Wood (1977), but no genetic analyses were included. Mutschler and Bliss (1980) have reported genetic male sterility due to the expression of two unlinked loci, Ms_1 and $In\text{-}ms_1$, which interact to produce a 13:2:1 F_2 ratio of non-sterile:sterile:lethal individuals.

VI. SEED DEVELOPMENT, HARVEST, AND STORAGE

Most bean cultivars begin flowering 21 to 40 days after sowing unless inhibited by unfavorable photoperiod and temperature. Pods show discernible growth the second day after anthesis. Carr and Skene (1961) found that maximum growth rate of pods occurred from days 2 to 11 and that the pods attained maximum length 12 to 17 days after anthesis. From 8 to 20 days after anthesis, fresh weight of the seeds increases exponentially, then growth rate declines. Seeds begin to dry out and mature about 30 days after anthesis.

Buds that have not been successfully fertilized usually drop within 1 to 2 days after cross-pollination. Pods in which one or more ovules have been fertilized begin to grow rapidly after the second day. Some pods may abscise after a considerable period of growth and others cease to grow after they have reached 50 to 70 mm in length, yet still remain on the plant. They usually contain seeds that have aborted.

Successful hybridizations usually produce one to six seeds per pod depending on the genotype of the seed parent and the conditions that prevail during crossing and seed development. Seeds can be harvested after the pods begin to mature and they are loose inside the pods, about 28 to 30 days after fertilization. Because moisture content of the seeds is still high, they should be dried in the pods with moving air at 40 C for 3 to 4 hours. Seeds are threshed from each pod by hand and placed with the identification tag inside a coin envelope. The pedigree of the cross is recorded on the envelope, which is stored in an insect-proof cabinet. Temperature should not exceed 40 C and relative humidity should be less than 60%. Viable seeds can be maintained in storage for at least 3 years.

VII. TECHNIQUES FOR SPECIAL SITUATIONS

Cuttings can be taken from healthy plants for rooting, particular care being exercised to include a growing point when sampling determinate, bush type plants. Cuttings treated with a rooting compound should be placed in a cool environment having high relative humidity. Adventitious roots develop in 5 to 7 days and cuttings can be transplanted the same as seedlings. Metal or bamboo stakes should be placed near each plant in the pots to provide support, and plastic ties or string can be used to secure the plants to each stake.

ACKNOWLEDGEMENTS

The responses of the following people whose comments and suggestions were extremely helpful are acknowledged: John A. Atkin, D. E. Debouk, David Davis, Desmond D. Dolan, Ken Hill, Shigemi Honma, L. W. Hudson, J. D. Kelly, J. P. Meiners, J. Smartt, Steven Temple, Augusto Tulman Neto, and W. J. Zaumeyer. The reviews, suggestions and additions made by Dermot Coyne, Peter Graham, George Freytag, Joe Sullivan, and Greg Tolla were essential and greatly appreciated. The photos received from Dermot Coyne and those taken by Yu Ma are gratefully acknowledged.

REFERENCES

Agbo, F. M. O., and D. R. Wood. 1977. Observation of male sterility in dry beans. Ann. Rep. Bean Impr. Coop. 20:49–50.

Al-Yasari, S. A., and D. P. Coyne. 1966. Interspecific hybridization in the genus *Phaseolus.* Crop Sci. 6:59–61.

Antunes, I. F., J. G. C. da Costa, and E. H. Oliveira. 1973. Natural hybridization in *Phaseolus vulgaris* in Pelotas, R. S. Brazil. Rep. Bean Improv. Coop. 16:61–62.

Berglund-Brucher, O., and H. Brucher. 1976. The South American Wild bean (*Phaseolus aborigineus* Burk.) as ancestor of the common bean. Econ. Bot. 30: 257–272.

Braak, J. P., and E. Kooistra. 1975. A successful cross between *Phaseolus vulgaris* L. and *Phaseolus ritensis* Jones with the aid of embryo culture. Euphytica 24: 669–679.

Buishand, Tj. 1956. The crossing of beans (*Phaseolus* spp.). Euphytica 5:41–50.

Burns, R. 1975. Genetic resources in beans, *Phaseolus vulgaris,* and CIAT's germplasm bank. Workshop of Genetic Improvement of Dry Beans, *Phaseolus vulgaris* and germplasm resources. CIAT. Cali, Colombia, 14–16 Oct. 1975.

Carr, D. J., and K. G. M. Skene. 1961. Diauxic growth curves of seeds, with special reference to French beans—(*Phaseolus vulgaris* L.). Aust. J. Biol. Sci. 14:1–2.

Coyne, D. P. 1964. Species hybridization in *Phaseolus.* J. Hered. 55:5–6.

———. 1968. Effect of temperature on pod set of varieties of *Phaseolus vulgaris* L. Ann. Rep. Bean Impr. Coop. 11:16–17.

Frazier, W. A. 1967. Sterile bean lines. Ann. Rep. Bean Impr. Coop. 10:15.

Freytag, G. F. 1955. Variation of the common bean in Central America. Ph.D. Thesis, Washington Univ., St. Louis, Mo.

———. 1975. Research related to the origin and improvement of the common bean (*Phaseolus vulgaris* L.). *In* J. Bird and K. Maramorosch (eds.) Tropical diseases of legumes. Academic Press, N.Y.

———. 1977. Personal communication. Mayaguez Institute for Tropical Agriculture, P.O. Box 70, Mayaguez, PR 00708.

Gentry, H. S. 1969. Origin of the common bean, *Phaseolus vulgaris.* Econ. Bot. 23(1):55–69.

Hikida, H. R. 1961. Cellophane tape method for bean crosses. Ann. Rep. Bean Impr. Coop. 4:10.

Honma, S. 1955. A technique for artificial culturing of bean embryos. Proc. Am. Soc. Hort. Sci. 65:405–408.

———. 1956. A bean interspecific hybrid. J. Hered. 47:217–220.

———, and O. Heeckt. 1959. Interspecific hybrid between *Phaseolus vulgaris* and *P. lunatus.* J. Hered. 50:233–237.

Ibrahim, A. M., and D. P. Coyne. 1975. Genetics of stigma shape, cotyledon position and flower color in reciprocal crosses between *Phaseolus vulgaris* L. and *Phaseolus coccineus* (Lam.) and implications in breeding. J. Am. Soc. Hort. Sci. 100:622–626.

———, ———, and G. Emery. 1974. Male sterility and stigma shape alteration in *Phaseolus vulgaris.* Ann. Rep. Bean Impr. Coop. 17:48–49.

Lamprecht, H. 1941. Die ortgreneze zwitches *Phaseolus vulgaris* L. und *P. multiflorus* Lam. Hereditas 27:51–175.

Knott, J. E. 1957. Handbook for vegetable growers. Wiley and Sons, Inc., N.Y.

Mutschler, M. A., and F. A. Bliss. 1980. Genic male sterility in the common bean (*Phaseolus vulgaris* L.). J. Am. Soc. Hort. Sci. 105:202–205.

Ojehomon, O. O. 1966. The development of flower primordia of *Phaseolus vulgaris* (L.) Savi. Ann. Bot. N.S. 30:487–492.

———. 1969. A quantitative study of inflorescence development in *Phaseolus vulgaris.* Ann. Bot. N.S. 30:325–332.

Prakken, R. 1970. Inheritance of color in *Phaseolus vulgaris* L. II. A Critical review. Meded. Landbouwhogesch. Wageningen 70-23:1–38.

Romero, J., S. M. Sun, R. C. McLeester, F. A. Bliss, and T. C. Hall. 1975. Heritable variation in a polypeptide subunit of the major storage proteins of the bean, *Phaseolus vulgaris* L. Plant Physiol. 56:776–779.

Silbernagel, M. J. 1968. Unopened anthers increase hybridization efficiency. Ann. Rep. Bean Impr. Coop. 11:45.

Smartt, J. 1970. Interspecific hybridization between cultivated American species of the genus *Phaseolus.* Euphytica 19:480–489.

———. 1977. Personal communication. Dep. of Biology. The University of Southampton, England.

Smith, F. L. 1939. A genetic analysis of red seed-coat in *Phaseolus vulgaris.* Hilgardia 12:533–621.

Tulmann Neto, A. 1977. Personal communication. Centro De Energia Nuclear Na Agricultura. Caixa Posta 96, Piracicaba, Brazil.

Tucker, C. L., and J. Harding. 1975. Outcrossing in common bean *Phaseolus vulgaris* L. J. Am. Soc. Hort. Sci. 100:283–285.

Vakili, N. 1976. Personal communication. Mayaguez Institute for Tropical Agriculture, P.O. Box 70, Mayaguez, PR 00708.

Webster, B. D., C. L. Tucker, and S. P. Lynch. 1977. A morphological study of the development of reproductive structures of *Phaseolus vulgaris* L. J. Am. Soc. Hort. Sci. 102:640–643.

Williams, D. D. F. 1962. Influence of soil moisture level on flower abscission, ovule abortion and seed development in the snap bean (*Phaseolus vulgaris* L.). Ph.D. thesis, Univ. of Wisconsin, Madison.

Wivutvongvana, M., and H. J. Mack. 1974. Floral anatomy of *Phaseolus vulgaris* L. cvs. Gallatin 50 and Oregon 58. HortScience 9:462–464.

18 Cool-Season Grasses

A. W. Hovin
Montana State University
Bozeman

In this chapter, emphasis is placed on cool-season species of economic or potentially economic importance in the United States. The species belong to the genera *Agropyron, Bromus, Dactylis, Elymus, Festuca, Lolium, Phalaris, Phleum,* and *Poa* (Table 1). Geographic distribution in the United States, adaptation, and primary use of the agronomically important species of these genera are documented elsewhere (Hanson et al., 1969). Although limited information is available for some species, they were included to represent a range of cool-season grasses that a plant breeder may encounter.

I. PARENTAL MATERIAL

The grass genera *Agropyron, Bromus, Dactylis, Elymus, Festuca, Lolium, Phalaris, Phleum,* and *Poa* consist of over 40 species that are of agronomic importance in the United States (Hanson, 1972). Although gene transfer from related species has been achieved in several genera, interspecific hybridization generally is not employed to improve crop cultivars because highly diverse germplasm is available for many agronomically important species. Intergeneric *Lolium* × *Festuca* crosses have been used successfully to improve forage quality of tall fescue (Buckner et al., 1953; Hovin and Buckner, 1966).

Exploration for additional germplasm continues for several cool-season genera to supplement seed accessions available from the USDA Regional Plant Introduction stations. Plant breeders seeking to extend the genetic variability in breeding of grass species need to know latitude, altitude, and chromosome numbers of the original germplasm. Factors such as latitude and previous use of the plant material, including grazing pressure of pastures and range, water supply, and soil type, will affect photoperiodic requirements of germplasm from diverse origins (Sonneveld, 1966). Most perennial cool-season grasses are polyploids and the somatic chromosome numbers of 16 species suggest that some species may have different ploidy series (Table 1).

II. PLANT CULTURE

A. Field

No particular soil type is required for field nurseries; however, a soil is preferred that is well drained, easily tilled, and free of perennial weeds. Grasses tolerate a relatively wide range in soil pH. The fertility level of the soil should be moderately high to stimulate tillering, but excessive amounts of available N may cause lodging (Hanson and Carnahan, 1956). Grasses grown for hybridization purposes generally are seeded directly in rows or transplanted as spaced plants using 1 to 2 m distance within and between rows, depending on species and cultivation practices. Plantings may be cultivated to control weeds. Herbicides used for weed control should be applied at rates indicated on labels of the products. Transplanted or established plants will tolerate higher rates of N than newly seeded areas.

The management of grasses should be aimed at promoting heading and subsequent flowering. Normal light conditions are usually adequate for seed production during the summer months. One needs to keep in mind that heading and flowering of perennial and some annual cool-season grasses are dependent on soil fertility, temperature, and photoperiod. Several grass species must complete a vegetative growth stage (juvenile phase) before plants will respond to floral inductive treatments. The processes necessary for this response remain unknown (Sachs, 1972). The three stages of inflorescence development are: 1) floral induction when apical or axillary meristems respond to the photoperiodic stimulus without showing any morphological alteration of the shoot apex; 2) floral initiation, which is a morphological response to appropriate photoperiod; and 3) floral elongation when subsequent development of the inflorescence takes place (Cooper, 1960). Exposure to a period of low temperature, to a photoperiod less than the critical length, or both, is required for floral induction of most cool-season grasses under field conditions, but the requirements vary with species, ecotypes, and genotypes (Evans, 1964). The vernalization requirement generally is met in the fall or early spring for most cool-season grasses, resulting in one crop of inflorescence-bearing culms during the growing season. For instance, August seeding of reed canarygrass will result in flowering the next year at Ames, Iowa (Carlson, 1977). However, reed canarygrass seeded in August in Minnesota will not complete its juvenile phase before freeze-up; therefore, it remains vegetative the following year (Hovin, 1977).

Recognizable morphological stages of the plant may indicate stage and rate of floral development, although heading date is not a dependable indicator of anthesis date. Because both genetic and environmental factors contribute to the length of time between heading and anthesis, plants with the same heading date may start anthesis at the same time 1 year, but not the following year, even in the same field. The control of this difference in anthesis among environments can be achieved best in the greenhouse or growth chamber.

Table 1—Chromosome number, duration of pollen-dispersal, and time of day of anthesis and pollen shedding for 16 selected cool-season grasses.†

Scientific name	Common name	Chromosome number	Average pollen dispersal period per inflorescence	Time of day of anthesis and pollen shedding
		2N	days	hours
Agropyron cristatum (L.) Gaertn.	Fairway wheatgrass	14	8	1400 to 1800
A. elongatum (Host) Beauv.	Tall wheatgrass	14, 56, 70	7	1400 to 1800
A. intermedium (Host) Beauv.	Intermediate wheatgrass	28, 42	10	1400 to 1800
A. smithii Rydb.	Western wheatgrass	42, 56	7	1400 to 1800
Arrhenatherum elatius (L.) Presl.	Tall oatgrass	28	7	1500 to 1900
Bromus inermis Leyss.	Smooth bromegrass	28, 42, 56, 70	7 to 10	1400 to 1900
Dactylis glomerata L.	Orchardgrass	28	6 to 7	1700 to 2100 (Hovin, 1977)
Elymus junceus Fisch.	Russian wildrye	14	7	1500 to 1900
Festuca arundinacea Schreb.	Tall fescue	42	9	1300 to 1800
F. ovina L.	Sheep fescue	14, 42, 56, 70	6	1600 to 2000
F. pratensis L.	Meadow fescue	14, 28	7	1500 to 2000
Lolium multiflorum Lam.	Annual ryegrass	14	7 to 10	0900 to 1300 (McVeigh, 1978)
L. perenne L.	Perennial ryegrass	14	7 to 10	1000 to 1300 (Funk, 1978)
Phalaris arundinacea L.	Reed canarygrass	14, 28	5	0600 to 1100 (Hovin, 1977)
Phleum pratense L.	Timothy	14, 42	10	0400 to 0900
Poa pratensis L.	Kentucky bluegrass	22–147 (Carnahan and Hill, 1961)	8	0300 to 0800

† By permission of authors, Hanson and Carnahan (1956), except as indicated.

If a seed crop is harvested, the remaining stubble may be burned off in compliance with state or local regulations. Regrowth should not be clipped during the remainder of the growing season to stimulate formation of new primordia for subsequent inflorescences the following year. Stubble and dead regrowth may be burned off in early spring and before nitrogen fertilizer is applied, but this practice will result in lower seed yield than when stubble is burned the previous summer (Elling, 1977).

B. Growth Chamber and Greenhouse

Plants used for hybridization may be started from seed or vegetative propagation of genetic material. Seed may be germinated in petri dishes, flats, or pots filled with a suitable soil mixture, often including sedge or sphagnum peat, well-rotted manure, and sand or vermiculite. Steam-sterilized soil will kill most weeds, but may cause an excess of toxic elements, particularly manganese. Chemical soil sterilants such as methyl bromide are used, but may not effectively kill hard seeds of legumes.

Perennial and some annual grasses may be clonally propagated from stolons, rhizomes, tillers, and axillary shoots from culm nodes. Flowering dates of most genotypes of cool-season grasses can be altered by using different planting dates, temperature regimes, and day length because of the vernalization and day length requirements needed for flowering. A better knowledge of these factors is emerging from recent research, as shown by the following examples.

Rate of inflorescence development of three long-day *Lolium* species was unaffected by the extent of the vernalization period, but was hastened by an increase in temperature, length of photoperiod and number of subsequent long days (Evans, 1960). In Minnesota, we shorten by 50% the time to reach anthesis of reed canarygrass by increasing the greenhouse night temperature from 15 C to 20 C during the long-day treatment (Heichel et al., 1980).

Ketellapper (1960) successfully vernalized seedlings of *Phalaris aquatica* L. (*P. tuberosa* L.) starting at the three-leaf stage with artificial light and short days at 4 C for 8 weeks; however, 6 weeks were too short a period. Reed canarygrass requires a longer juvenile phase before plants will respond to floral induction. Approximately 22-day-old seedlings at the four-leaf stage and with one to two tillers required a vernalization period of 9 to 12 weeks when the treatment was 5 to 7 C and an 8-hour photoperiod (Heichel et al., 1980). Panicles emerged after 15 days at a moderate greenhouse temperature of 25 C day temperature and a 16-hour photoperiod. Age of the plant prior to vernalization had little effect on number of days to panicle emergence following onset of the long photoperiod. The successful treatment sequence for flowering of orchardgrass in the greenhouse was 1 month of 3 C given to seedlings immediately after germination, followed by 1 month at 15 C day/5 C night temperature or 2 months at 20 C with day length less than 13 hours (Fejer, 1966).

In some grass species, short days may be substituted for vernalization, but in Kentucky bluegrass the short-day treatment is effective only at a vernalizing temperature of 0 to 10 C (Sachs, 1972). Short days were found to compensate for the vernalization requirement of orchardgrass, meadow fescue, smooth bromegrass, and reed canarygrass (Hanson and Sprague, 1953). They established grasses from tillers and maintained them under a short photoperiod in the greenhouse at 12 C for 4 to 5 weeks, and followed with a 15 to 16-hour photoperiod at 21 C. Heichel et al., 1980 found that short-day substitution was ineffective before reed canarygrass seedlings had completed their juvenile phase.

The long-day treatment can be achieved in the greenhouse by extending the normal day to 16 hours or to 10 hours, plus 1 or 2 hours of interrupted night-cycle. High light radiance to achieve the long-day effect does not increase inflorescence production or shorten the time for flowering (Sprague, 1948). Within the greenhouse section, light series may be separated by use of black sateen curtains. In areas where the light may be reflected from snow outside the greenhouse, it is helpful to completely cover plants at night to achieve a desired short-day effect.

III. FLORAL CHARACTERISTICS

The spikelet is the floral unit of all grasses and is aggregated in groups or clusters which constitute the inflorescence. The spikelet is in the axil of two leaf-like bracts (glumes) and consists of a short axis (rachilla) on which flowers are located in the axils of two bracts: the palea and the lemma. The lemma, palea, stamen, and pistil are called the floret. During anthesis, the floret opens by changes in turgor of two, rarely three, lodicules at the base of the floret. The duration of the pollen dispersal period and time of day for anthesis and pollen shedding vary among different species (Table 1).

Anthesis occurs early in the morning for some species, including bluegrasses and orchardgrass that respond to stimulation by light, and in the afternoon for other species such as smooth bromegrass and tall fescue that respond to increase in temperature. The grass breeder can delay anthesis in grasses that flower soon after sunrise by keeping them in a dark chamber for an extended night period. Stigmas of self-pollinated and cross-pollinated grasses become receptive to pollination before florets open. The length of time during which stigmas are receptive in the perfect grass flower generally corresponds to the duration of anthesis in the floret. Grasses have trinucleate, short-lived pollen. Without special treatment to maintain pollen viability, grass pollen removed from the plant cannot be stored for more than a few hours. The length of time between pollination and fertilization of an ovule is not known for most forage and turf grasses.

IV. ARTIFICIAL HYBRIDIZATION AND SELF-POLLINATION

A. Equipment

Isolation and pollination bags, paper clips, cotton, scissors, paper or plastic ties, and bamboo or metal stakes are used to prepare plants of the female parent for hybridization. Bags made of clear dialysis tubing is superior to parchment paper or glassine paper bags (Hovin, 1977).

Equipment for emasculation of self-compatible female parents include forceps and a magnifying glass or a dissecting microscope. Fogging machines may be used to raise the humidity in the crossing room or greenhouse and Thermos or similarly insulated bottles are needed for hot water emasculation. Bags of different sizes and quality are needed for isolation and pollination whether the crossing is done outdoors or in a greenhouse. For crossing or self-pollinating plants outdoors, waterproof glue must be specified for construction of the bags. The bags must be sufficiently strong to withstand wind and inclement weather during pollination and seed maturation.

B. Preparation of the Female

Hand emasculation of self-compatible plants is tedious and requires great care to prevent self-pollination. There is no particular time of the day when emasculation should be performed, but it should be completed 2 to 3 days before anthesis.

If the inflorescence is compact, thinning of florets is necessary before anthers are removed. This facilitates emasculation and reduces the possibility of overlooking florets. Working the inflorescence from the tip to the base in a systematic manner reduces chances of injury and reduces the amount of handling of emasculated florets.

The anthers are removed by forceps inserted between the palea and lemma. Artificial fog will retard anther dehiscence and permit the removal of anthers as they extrude (Keller, 1952). In some species, dehiscence is delayed naturally, and anthers may be removed when they are visible outside the floret (Jenkin, 1931).

A magnifying glass or a dissecting microscope may be used to improve the hand emasculation procedure. The latter only can be used when plants are grown in individual pots so they can be placed horizontally on a table and the inflorescence taped in place for viewing during emasculation.

The range of thermal inactivation points for pollen is narrower than that of ovules; therefore, either high or low temperatures may be used for bulk emasculation of self-pollinated grasses. The hot water technique by Stephens and Quinby (1933) has been used effectively for several cool-season grasses (Keller, 1952). For instance, 46 C to 47 C for 3 min or 48 C for 1 min were found satisfactory to induce temporary male-sterility in smooth bromegrass (Tsiang, 1944). The hot water method requires a wide-mouthed insulated jug filled with warm water, a wire loop for stirring, and an accurate thermometer. Containers with thermostatically controlled heating coils also may be used. Preliminary trials should proceed large-scale bulk-emasculation attempts because genotypes may differ in response to treatment. Treated panicles should be allowed to dry before being covered by isolation bags.

A different procedure may be used for potted plants grown in the greenhouse. Temporary male-sterile but female-fertile plants of bluegrass resulted when genotypes with emerging panicles were transferred from a greenhouse section maintained at 20 C to one maintained at 29 C day/20 C night temperatures for 11 to 19 days (Hovin, 1958). Because the minimum inactivation temperature is higher for anthers than for ovules, temporary male sterility also was achieved by placing timothy plants in low-temperature (0 to 5 C) chambers to prevent anther development (Nordenskiold, 1945).

C. Pollination

Most cool-season grasses are cross-pollinated, but some are self-pollinated. The perennial bluegrasses are predominantly facultative apomicts. The mutual pollination procedure is common for cross-pollinated grasses where each parent is used both as female and male parent. When plants are grown in pots in the greenhouse, mutual pollination is accomplished by enclosing inflorescences of both parents in a pollination bag before anthesis. When plants cannot be brought together for mutual pollination, three approaches are available to effect hybridization and seed development: 1) collection and storage of pollen, 2) collection of inflorescences from the male plant, and 3) using detached panicles of both parents. The preparatory work is done usually in the morning, but can be done any time during the day when detached inflorescences are used.

Grass pollen generally is collected at anthesis just before pollination is made. A paper bag is placed over one or several inflorescences and shaken to promote pollen dispersal. The bag is placed over a female inflorescence after the isolation bag has been removed, agitated, and fastened to the stake supporting the isolation bag. In the greenhouse, pollen can be collected on glossy dark paper and transferred to the stigma by a fine brush. The brush must be rinsed in alcohol and dried before attempting a different cross.

The trinucleate grass pollen usually does not store well, even at cold temperatures, unless steps are taken to prevent desiccation. Pollen of ryegrass has been stored successfully for 39 days by mixing it with polyvinylpyrrolidone in a ratio by volume of 1:2 to 2:1 and storing the mixture in gelatin capsules at −20 C (Nitzsche, 1970). He allowed the mixture to reach room temperature before applying it by a brush on stigmas of male-sterile plants. This procedure warrants wider testing with other crops.

When flowering of parents is synchronized, a simple procedure makes use of detached inflorescences as a pollen source (Keller, 1944). The cut culms are placed in a bottle filled with water and fastened to a stake that also supports the isolation bag. The detached inflorescences are placed inside the bag protecting the female parent and should remain higher than those of the female parent during anthesis. I use this procedure to pollinate orchardgrass, reed canarygrass, and tall fescue.

The crossing also can be achieved in a laboratory by mutual pollination of detached inflorescences from both parent clones (Keller, 1943; Burton, 1948). The tagged inflorescences are collected from field or greenhouse-grown plants and placed in tubs of water. The room for crossing may be sparsely lit and have moderate temperature of about 20 C and relative humidity above 50%. The inflorescences in the waterfilled container are kept inside an isolation bag for the duration of pollination and are agitated frequently during pollen shedding. The seed will be of smaller size than that harvested from intact plants. Use of a detached male inflorescence or detached inflorescences from both parents is superior to hand-pollination, even for partially self-compatible species, except for critical genetic studies.

Individual inflorescences should be labelled with a string tag on which parentage and date are recorded. When mutual pollination of two-clone crosses is used, reciprocals may be kept separate when seed is harvested. Reciprocal crosses can be combined following seed cleaning.

D. Factors Affecting Efficiency

Because of the determinate flowering habit of most cool-season grass species, breeders who attempt to cross early-flowering with late-flowering clones face particular challenges. If the crossing is planned to be carried out in the greenhouse during the winter, one may establish several clonal propagules of individual genotypes in separate pots. The propagules may be maintained at different temperature and day length regimes to synchronize anthesis for pollination.

Few genetic markers are available to aid in identifying hybrid seed or seedlings of cool-season perennial grasses. For predominately self-incompatible species, occasional progenies from selfing tend to be less competitive in solid stands and are assumed not to survive and affect yield of plants of hybrid origin. Crosses in spaced-plant nurseries are verified more accurately when compared to clonal propagules of both parents rather than to the female parent.

V. NATURAL HYBRIDIZATION

Plant breeders working with perennial grasses make use of the gametophytic self-incompatibility system which may be controlled by two (Lundquist, 1962) or possibly three or more loci (Spoor, 1976). The two-loci incompatibility reaction occurs when alleles are matched in pollen and style at both loci. Interaction between the two loci has not been demonstrated and the loci act as one physiological unit. The percentage of hybrid seed from crosses between two diploid parents that had been inbred for one generation was estimated to be 83, 71, and 64% for 2, 3, and 4 loci self-incompatibility systems, respectively (England, 1974). Therefore, the two-loci and three-loci self-incompatibility systems are expected to be effective in ensuring a high frequency of hybrids without emasculation or use of male sterility.

Male sterility has been reported for *Agropyron* (Schulz-Schaeffer, 1970), *Lolium* (Nitzsche, 1971; Wit, 1974). and *Dactylis* (Myers, 1946). The genetic male-sterility mechanism in orchardgrass has been used for cultivar development.

In the field, the appropriate planting plan to obtain single-cross, polycross, or topcross seed depends on several factors. Isolation bags may be used successfully to obtain small amounts of seed in areas where strong wind is not a particular problem during pollination and seed maturation. Where wind prevents the use of isolation bags, isolation by distance and time are the preferred alternatives.

Half-sib matings in perennial forage grasses are achieved by use of isolated topcross or polycross nurseries (Chapter 5). Clones to be topcrossed are space-planted among rows of the topcross parent. In the spring, I plant the topcross parent in drilled rows across the gradient of the predominent wind direction and alternate the row spacings of 1 and 2 m. The clones are interplanted either in the spring or fall in the middle of the 2-m spacing between topcross rows on 8 to 10-m centers and are replicated 2 to 3 times or as material permits. Seed is harvested the year following planting. An equal amount of seed from each replicate of individual clones is bulked and identified by clone. After seed harvest, the clones may be dug and the area sprayed with a herbicide to kill remnant tillers. The topcross parent can be used again for a second set of topcrosses.

The polycross nursery must be established with sufficient replications and border rows. As many as 24 to 30 replications of each block are not uncommon. Use of the Latin square will improve random interpollination (Olesen and Olesen, 1973). Equal amounts of seed produced from each replicate of the same clone are bulked for seeding of progeny tests. Seed for the polycross progeny test also can be obtained by compositing seed of individual clones used in diallel crosses.

Field nurseries for diallel crosses should be arranged to facilitate crossing by use of isolation bags or by spatial isolation. The circular crossing procedure requires isolation and pollination bags. One parent clone of, for instance, a seven-clone diallel is planted in the center and the other six clones are planted in a circle around it (Drolsom and Nielsen, 1970). This arrangement is repeated by placing each of the remaining clones in the center of similar crossing blocks. The following year, inflorescences of the center clones are isolated under bags with a similar number of inflorescences from one of each of the outside clones. The bags which are fastened to one of several stakes placed near the center clone must be agitated several times daily to promote pollination during anthesis. This scheme repeats each single-cross from two different circles, and seed from the two replicates may be bulked.

For natural hybridization of self-incompatible genotypes used in two-clone crosses, wirecloth frames that are sufficiently tall to accommodate the fully extended plants may be placed around the clones. Isolation bags made from plastic or unbleached cotton muslin (thread count 35 × 35/cm) are placed over the entire wire frame during pollination. A wooden stake slightly taller than the wire frame is placed inside the frame with one end extending through the isolation bag, which is secured to the stake. The stake is used to agitate the plants to aid in pollination.

Seed increase of experimental strains in the field can be achieved by use of isolation cages, as described by Dunne (1959) and England (1972). Cages may be constructed with creosote-treated fence posts placed in the corners and middle of the long side of an area 1.5 m × 5 m (Berg and Hill, 1975). The height of the posts depends on the height of the grass species. Construction-grade lumber (2 × 14 cm) is nailed to the posts along the bottom, top, and the middle. Plants are allowed to grow uncovered inside the frame until heading when the frame is covered with unbleached cotton muslin (thread count 35 × 35/cm). Seed yields of orchardgrass plants in air-conditioned polyethylene-covered cages were nearly equal to those in muslin-covered cages, but bromegrass yields were greatly reduced (Padmanthan, 1969). Electric dehumidifiers and fans have improved seed set of tall fescue in some years (Berg and Hill, 1975).

A practical degree of pollen isolation was achieved in orchardgrass by construction of a cross-shaped wooden frame (Shenk, 1977). The sides of the structure were covered with unbleached muslin and extended beyond the outside edge of each of four experimental strains planted in separate quadrants. The covered wooden structure, approximately 2.5 m high, prevented most pollen contamination. In this case, absolute pollen control was not the primary objective of seed increase of several experimental strains.

Isolation can be achieved by maturity differences of adjacent strains or cultivars. Some strains, however, have a longer period of anthesis outside the area of origin (Hill and Hovin, 1964). Several late-flowering timothy cultivars of European origin will flower over a 1-month period when grown in Minnesota (Elling, 1977). Consequently, isolation by time must be based on carefully conducted studies in the area of seed production before recommendations are made.

Isolation by distance is the principal means of preventing pollen contamination (Hanson and Carnahan, 1956). The safe distance between strains or cultivars of similar maturity to avoid undesired intercrossing depends on several factors affecting viability and dispersal of pollen. A distance of 400 m is considered by the Association of Official Seed Certifying Agencies to be minimal for foundation seed certification of grasses in the United States. The minimum isolation distance is 160 m for production of registered and 50 m for certified seed.

Natural hybridization in the field for development of synthetic cultivars utilizes clonal propagules of parent material (Syn 0). The parents usually have been selected on the basis of progeny tests and are known to have similar flowering characteristics. Seed of each parent used in the development of the Syn 1 generation is produced in a replicated polycross nursery. After threshing and cleaning, the seed from each replicate is bulked for individual parents. Seed weight and germination records are used in bulking equal quantities of seed from each parent for the Syn 1. The Syn 1 seed, which is often the breeder seed class, is planted in rows that are field-cultivated, rogued, and harvested in bulk for production of Syn 2 seed (foundation seed class). This seed is planted for production of Syn 3 seed which becomes either the registered or certified seed class.

When vegetative propagation of parent clones is possible, large-scale two-clone crosses of perennial grasses may be made in the field by use of isolated crossing blocks. Each block consists of clonal propagules planted alternately in two directions on 0.5 to 1 m centers. To reduce possible pollen contamination, the seed is discarded from the outside plants (Carlson, 1977). This procedure may become economically feasible for production of hybrid grass cultivars when extensive vegetative plantings can be made of parent clones (Hovin et al., 1973).

VI. SEED DEVELOPMENT, HARVEST, AND STORAGE

In species known to have seed shattering problems, seed development must be watched frequently starting 2 weeks after pollination. The seed is generally harvested when the culms have turned brown just below the inflorescence. Although seed harvested at the dough stage will germinate as well as mature seed under favorable conditions (McAlister, 1943), my experience is that immature seed has poor seedling vigor under field conditions and will not store well.

The inflorescence may be cut below the bag to remove it from the stake. Inflorescences and the seed may continue to dry inside the bag. If shattering is a problem, the identity of seed from reciprocal crosses may be maintained by carefully placing inflorescences from reciprocal crosses in separate envelopes.

Whenever inflorescences of the two clones used in mutual pollinations under bag do not remain at the same height at anthesis, only the seed of the lower inflorescences should be harvested to avoid self-pollination. The inflorescences must be identified before they are cut from the plant.

Small seed lots can be threshed by rubbing the panicles with a large rubber stopper against a piece of corrugated rubber mat. Blenders have been used successfully after the blades have been covered with rubber or other tubing materials, and the speed reduced. Several improvised seed cleaning procedures are used by plant breeders. The South Dakota seed blower and other cleaning apparatus are available that will clean small seed lots.

Field-grown seed from replicated clonal nurseries may be threshed with a stationary plot thresher or with a hammermill modified according to procedures described by Kneebone and Brown (1953). Flaming and hammermills have been used successfully on some grass species to remove appendages and cottony filaments of seed.

Grass seed generally needs to be stored at a cool temperature (5 C) and low relative humidity (below 40%) to maintain viability. Properly stored seed of most species will retain viability for many years. Under these conditions, storage insects will not become a problem. Occasionally it becomes necessary to operate a thorough sanitation program in the seed storage area that is subject to fluctuating temperatures and moisture. Fumigation, insecticides, or chemically inert dust may be used. Space treatments with aerosols or mist sprays have limited value as a control measure.

VII. TECHNIQUES FOR SPECIAL SITUATIONS

Hormones have not been widely used to induce flowering or alter flowering of grasses. Gibberellic acid was reported to replace the vernalization requirement in *Lolium* (Evans, 1969). Evans showed that the long-day effect was promoted by gibberellic acid and floral initiation was induced with 2-chloroethyltrimethylammonium chloride when applied with gibberellic acid. This approach warrants more extensive testing with other grasses to determine whether the generation cycle can be shortened.

Ethephon enhances shoot and root growth of axilliary buds in reed canarygrass culms and may be used to increase the rate of clonal propagation (Hovin et al., 1973). Embryo and callus culture have been attempted with some success in cool-season grasses (Conger and Carabia, 1978; Kasperbauer et al., 1979). Similar techniques may be used by plant breeders in the future to manipulate germplasm and to explore new procedures to obtain genetic variability and aid in interspecific and intergenetic crosses.

REFERENCES

Berg, C. C., and R. R. Hill. 1975. Seed production by ryegrass-fescue hybrid derivatives in muslin-covered cages. Crop Sci. 15:52–54.

Buckner, R. C., H. D. Hill, A. W. Hovin, and P. B. Burrus, II. 1963. Cytogenetics and morphological characteristics of progenies of crosses of annual ryegrass × tall fescue hybrids and their amphiploid derivatives. Crop Sci. 3:453–454.

Burton, G. W. 1948. A method for producing chance crosses and polycrosses of Pensacola bahiagrass, *Paspalum notatum.* Agron. J. 40:470–472.

Carlson, I. T. 1977. Unpublished data. Dep. of Agronomy. Iowa State Univ., Ames, IA 50011.

Carnahan, H. L., and H. D. Hill. 1961. Cytology and genetics of forage grasses. Bot. Rev. 27:1–162.

Conger, B. V., and J. V. Carabia. 1978. Plantlet regeneration from callus culture of orchardgrass. Crop Sci. 18:157–159.

Cooper, J. P. 1960. The use of controlled life-cycles in the forage grasses and legumes. Herbage Abstr. 30:71–79.

Drolsom, P. N., and E. L. Nielsen. 1970. Some characteristics of mutual pollination progenies of *Bromus inermis.* Crop Sci. 10:17–19.

Dunne, J. P. 1959. A system for control of pollination in insect and wind pollinated plants. Euphytica 8:76–80.

Elling, L. J. 1977. Unpublished data. Dep. of Agronomy and Plant Genetics, Univ. of Minnesota, St. Paul, MN 55108.

England, F. J. W. 1972. Isolation chambers for controlled pollination in grasses. Euphytica 21:523–526.

———. 1974. The use of incompatibility for the production of F_1 hybrids in forage grasses. Heredity 32:183–188.

Evans, L. T. 1960. The influence of environmental conditions on inflorescence development in some long day grasses. New Phytol. 59:163–174.

———. 1964. Reproduction. p. 126–153. *In* C. Barnard (ed.). Grasses and grasslands. MacMillan and Co. Ltd., London.

———. 1969. Inflorescence initiation in *Lolium temulentum* L. XIII. The role of gibberellins. Aust. J. Biol. Sci. 22:773–786.

Fejer, S. O. 1966. Growth and reproduction of New Zealand, Mediterranean, and hybrid *Dactylis glomerata* after short day and temperature treatments. Can. J. Plant Sci. 46:233–241.

Funk, C. R. 1978. Unpublished data. Dep. of Soils and Crops, Rutgers Univ., New Brunswick, NJ 08903.

Hanson, A. A. 1972. Grass varieties in the United States. Agric. Handb. No. 170. USDA. U.S. Government Printing Office, Washington, D.C.

———, and H. L. Carnahan. 1956. Breeding perennial forage grasses. USDA Tech. Bull. No. 1145. Washington, D.C.

———, F. V. Juska, and G. W. Burton. 1969. Species and varieties. p. 370–409. *In* A. A. Hanson and F. V. Juska (eds.) Turfgrass science. Agronomy 14. Am. Soc. Agron., Madison, Wis.

———, and V. G. Sprague. 1953. Heading of perennial grasses under greenhouse conditions. Agron. J. 45:248–251.

Heichel, G. H., A. W. Hovin, and K. I. Henjum. 1980. Seedling age and cold treatment effects on induction of panicle production in reed canarygrass. Crop Sci. 20:683–687.

Hill, H. D., and A. W. Hovin. 1964. Meiotic irregularities and potential seed production of *Dactylis glomerata* L., *Bromus inermis* Leyss., and *Phleum pratense* L. at different locations. Crop Sci. 4:78–81.

Hovin, A. W. 1958. Reduction of self-pollination by high night temperature in naturally self-fertilized *Poa annua* L. Agron. J. 50:369–371.

———. 1977. Unpublished data. Dep. of Agronomy and Plant Genetics, Univ. of Minnesota, St. Paul, MN 55108.

———, B. E. Beck, and G. C. Marten. 1973. Propagation of reed canarygrass (*Phalaris arundinacea* L.) from culm segments. Crop Sci. 13:747–749.

———, and R. C. Buckner. 1966. Interspecific and intergeneric hybridization in breeding of *Festuca arundinacea*. p. 666–668. *In* Proc. X Int. Grassl. Congr., Helsinki, Finland.

Jenkin, T. J. 1931. The method and technique of selection, breeding and strain building in grasses. Bull. Imp. Bur. Plant Genet. Herbage Plant. Aberystwyth 3:1–34.

Kasperbauer, M. J., R. C. Buckner, and L. P. Bush. 1979. Tissue culture of annual ryegrass × tall fescue F_1 hybrids: Callus establishment and plant regeneration. Crop Sci. 19:457–460.

Keller, W. 1943. Seed production on grass culms detached prior to pollination. Agron. J. 35:617–624.

———. 1944. The bulk hybridization of smooth bromegrass. J. Hered. 35:49–56.

———. 1952. Emasculation and pollination technics. p. 1613–1619. *In* Proc. VI Int. Grassl. Congr. State College, Pa.

Ketellapper, H. J. 1960. Growth and development in Phalaris. I. Vernalization response in geographic strains of *P. tuberosa* L. Ecology 41:298–305.

Kneebone, W. R., and H. W. Brown. 1953. A modified hammermill for processing grass seed. Agron. J. 45:389–391.

Lundquist, A. 1962. The nature of the two-loci incompatibility system in grasses. I. The hypothesis of a duplicative origin. Hereditas 48:153–181.

McAlister, D. F. 1943. The effect of maturity on the viability and longevity of the seeds of western range and pasture grasses. Agron. J. 35:442–453.

McVeigh, K. J. 1978. Unpublished data. Agricultural Service Corp., 5240 Gattin Rd., Salem, OR 97301.

Myers, W. M. 1946. Effects of cytoplasm and gene dosage on expression of male-sterility in *Dactylis glomerata.* Genetic 31:225–226.

Nitzsche, W. 1970. The storage of grass pollen. p. 259–260. *In* Proc. XI. Int. Grassl. Congr., Surfers Paradise, Queensland, Australia.

———. 1971. Cytoplasmatische männliche Sterilität bei Weidelgras (*Lolium* spp). Z. Pflanzenzüct. 65:206–220.

Nordenskiold, H. 1945. Cyto-genetic studies in the genus *Phleum.* Acta Agric. Suecana 1:1–138.

Olesen, K., and O. J. Olesen. 1973. A polycross pattern formula. Euphytica 22: 500–502.

Padmanthan, K. 1969. The effects of environment on pollen development, seed set and seed quality in orchardgrass, *Dactylis glomerata* L. and bromegrass, *Bromus inermis* Leyss. Ph.D. thesis, Penn State Univ. Univ. Microfilms. Ann Arbor, Mich. (Diss. Abstr. Int. 31:534B).

Sachs, R. M. 1972. Inflorescence induction and initiation. p. 348–364. *In* V. B. Youngner and C. M. McKell (eds.). The Biology and utilization of grasses. Academic Press, N.Y.

Schulz-Schaeffer, J. 1970. A possible source of cytoplasmic male sterility in intermediate wheatgrass, *Agropyron intermedium* (Host) Beauv. Crop Sci. 10:204–205.

Shenk, J. S. 1977. Unpublished data. Dep. of Agronomy, The Pennsylvania State Univ., University Park, PA 16802.

Sonneveld, A. 1966. Photoperiodic adaptation of grassland plants. p. 711–715. *In* Proc. X. Int. Grassl. Congr., Helsinki, Finland.

Spoor, W. 1976. Self-incompatibility in *Lolium perenne* L. Heredity 37:417–421.

Sprague, V. G. 1948. The relation of supplementary light and soil fertility to heading in the greenhouse of several perennial forage grasses. Agron. J. 40:144–154.

Stephens, J. C., and J. R. Quinby. 1933. Bulk emasculation of sorghum flowers. Agron. J. 25:233–234.

Tsiang, Y. S. 1944. Variation and inheritance of certain characters of bromegrass, *Bromus inermis* Leyss. Agron. J. 36:508–522.

Wit, F. 1974. Cytoplasmic male sterility in ryegrasses (*Lolium* ssp) detected after intergeneric hybridization. Euphytica 23:31–38.

19 Corn

W. A. Russell
A. R. Hallauer
Iowa State University and USDA-ARS
Ames

Corn (*Zea mays* L.) is an annual crop indigenous to the Western Hemisphere. Columbus encountered corn on his first voyage in 1492, and subsequent explorers found corn being grown by the Indians throughout the Western Hemisphere. It now is grown in every state of the United States and in every important agricultural area of the world. It is grown on more than 120 million ha in the world, and is an important commodity in world trade. Corn is consumed either directly or indirectly by millions of people. About 75% of the corn is fed to animals; thus, indirect consumption is greater than direct consumption. Corn is produced primarily as an energy crop, but specialized versions for protein, oil, waxy, sweet corn, and popcorn are available.

I. PARENTAL MATERIAL

Corn is an extremely variable species that is included in the grass family. Corn and seven other genera are commonly included in the family *Maydeae,* three (*Zea, Euchlaena,* and *Tripsacum*) that are native to the Americas, and five (*Coix, Chionachne, Polytoca, Scherachne* and *Triobachne*) that are native to southeastern Asia and Australia. Some taxonomists include teosinte (*Euchlaena mexicana*) in the same genus as corn (*Zea*). *Euchlaena* and *Tripsacum* are considered the closest relatives of corn. The genera from southeastern Asia and Australia have not been studied extensively and are not considered seriously as relatives of corn.

Intergeneric crosses of *Zea* (corn) have been successful with *Euchlaena* (teosinte), *Tripsacum, Saccharum* (sugarcane), and *Coix* (Goodman, 1965).

Corn and teosinte are easily crossed, have similar chromosomes, and produce fertile offspring. Special techniques usually are required to cross corn and *Tripsacum,* and the hybrids are male-sterile, even after several backcrosses to corn. Corn and sugarcane have been crossed several times, but the hybrid progeny have never attained the reproductive stage. Only one case of a corn-*Coix* hybridization has been reported, and the seeds of the cross were not viable.

Corn is grown from 58° N Lat to 40° S Lat, from below sea level of the Caspian plains up to 3,000 m in the Andes Mountains, and from semiarid regions having 25 cm annual rainfall to tropical regions that receive more than 500 cm annual rainfall. Consequently, collections of corn cultivars have an extraordinary diversity of morphological and physiological traits. In the Western Hemisphere alone, there are some 25,000 collections of corn that make up about 130 distinct races. In the United States, however, more than 90% of the breeding effort is directed to germplasm whose origin traces to not more than three of the 130 existing races (Brown, 1975). Extensive collections of the different cultivars, types, and races are available at Centro Internacional de Mejoramiento de Maiz Y Trigo (CIMMYT), Chapingo, Mexico; the National Seed Storage Laboratory, Fort Collins, Colo.; and the North Central Regional Plant Introduction Center, Ames, Iowa.

II. PLANT CULTURE

A. Field

The relationship between corn culture and climatic conditions was reviewed by Shaw (1977), and an excellent discussion of corn production was given by Larson and Hanway (1977). The purpose of this section will be to consider items that are specific to breeding nurseries.

Good plant stands and vigorous plant growth are necessary in the breeding nursery. A good seedbed is required for inbred materials because of their reduced seedling vigor. Fungicide seed treatment should be used, except in instances in which selection for resistance to soil-borne pathogens is intended. Seed germination requires a soil temperature above 10 C; consequently, the date of planting should be determined by expected temperatures.

Regular fertility practices for the soil type should be used. Plant densities will vary, depending upon specific requirements (Section IV, D). Overplanting and thinning to a certain plant density may be desirable in some instances. Weed control for the whole season is important, particularly in nurseries of inbred lines because these less-vigorous materials do not compete strongly with many weed species. Chemical control of leaf diseases and insects may be necessary in some seasons. Pest control is required in most areas where winter nurseries are grown.

Special environmental conditions for hybridization or self pollination of corn are not required when it is grown in its area of adaptation. Temperature is usually not an important problem, unless it is excessively high when pollen is released. Pollen grains will lose viability very rapidly when the temperature exceeds about 30 C, particularly if the relative humidity is low. Deficient soil moisture at flowering usually causes a delay in silk emergence relative to tassel development, and the silk may cease growth before it has emerged from the ear husk. Moisture stress also may cause tassel firing and, thus, abortion of the male flowers. Even in areas where rainfall usually is adequate for good corn production, supplemental irrigation in breeding nurseries can be advantageous just before or during silking in most years. The effect of a soil-moisture stress varies among genotypes and usually is observed on inbred lines sooner than on single crosses (Tatum and Kehr, 1951).

B. Growth Chamber and Greenhouse

Corn plants do not grow well in the greenhouse during the winter months. The primary problems seem to be day length and light level. Supplemental lighting must be used, but plants will be considerably less vigorous than in the field. Greenhouse culture can be used for special purposes if the number of plants needed may be relatively few; otherwise, winter nurseries in the warmer climates of the low latitudes should be used.

III. FLORAL CHARACTERISTICS

The corn plant is monoecious, with the male flowers in the tassel and the female flowers on the ear shoots (Fig. 1). When the male flower is mature, the anthers are exserted from the spikelet, and pollen is dispersed through a pore at the tip of the anther. Anther exsertion usually begins on the central spike a short distance below the tip. Each spikelet has two flowers, and the anthers are exserted from the upper flower first and then from the lower flower, either later the same day or the following days. When the anther pore opens, the pollen may be completely dispersed in only

Fig. 1—Corn stages. A) Tassel with anthers exserted; B) ear shoot with silks emerged.

a few minutes, or over a longer period as determined by the temperature, humidity, air movement, and genotype. Pollen shed for a tassel may vary from only 1 or 2 days to more than a week.

Pollen is dispersed by wind currents; consequently, extensive cross pollination and little self pollination occurs on an individual-plant basis. Insects cause an insignificant amount of pollen dispersal, but they can cause contamination in controlled self- or cross-pollinations. The amount of pollen dispersed from one tassel will vary among materials. Hybrids will shed more pollen for a longer period than will inbred lines. Kiesselbach (1949) reported estimates that 25,000 pollen grains are produced for each silk on an ordinary ear. With average summer temperature and humidity in the U.S. Corn Belt, dispersal may begin about 3 hours after sunrise and continue for 1 to 3 hours. With cooler temperature and higher humidity, however, dispersal may be delayed until midday and continue most of the afternoon. Once pollen is dispersed into the atmosphere, it will remain viable only a few minutes because of rapid desiccation.

The top ear shoot usually is at the sixth or seventh node below the tassel in U.S. Corn Belt germplasm. There is an axillary bud at each node below the top ear node, but elongation of the cob and silk emergence usually occur at only the upper 2 or 3 nodes. The silks that emerge from the tip of the ear husk are the functional stigmas, and there is one silk for each potential kernel (Fig. 1B). The first silks to emerge usually are from near the basal part of the ear. The rate of silk emergence is controlled by temperature, soil moisture, and soil nutrients; consequently, complete emergence may occur in only 2 or 3 days under favorable growth conditions or may require 5 to 7 days with cool temperatures. Under extreme stress conditions, silk growth may cease completely. Silks usually emerge first from the top ear node and may be followed in a few hours or days by silks from the second ear node and, sometimes, a third.

The silks usually emerge at the top ear node 1 to 3 days after anther dehiscence has begun. Tassel development seems to control development of the ear shoot, and this dominance is greatest for genotypes that produce only one ear per plant in any environment. Prolific genotypes may have no dominance for the tassel, and their silks frequently emerge before the tassel begins to shed pollen.

The pollen grain has a relatively thin outer membrane that gives little environmental protection; consequently, viability may be lost in a few minutes because of desiccation. Everett (1958) found that the best storage temperatures were −7 to +4 C. Storage temperatures near 0 C permitted retention of viability in some pollen grains for as long as 190 to 200 hours after collection. Storage in a beaker at 4 C and 90% relative humidity gave effective pollen up to 6 days (Jones and Newell, 1948). Storage of cut tassels with cut ends in water gave effective pollen up to 9 days. Pfahler (1967) found that a storage period of 1 day at 5 C did not show much change in fertilization ability of pollen grains. Water content of corn pollen had a decisive effect on its storability in liquid nitrogen (Barnabas and Rajki, 1976). The critical moisture content seemed to be about 30% in their study, with a greater content contributing to loss of pollen viability in storage. They observed that pollen with a moisture content of 22.2% and stored in liquid nitrogen at −196 C for 365 days resulted in high seed set.

Silks become receptive as soon as emerging from the ear husk. The length of time for receptive silks on an ear is determined by the time required for all silks to emerge from the husk and the time a silk remains receptive after emergence. Data indicate that it may require up to 5 to 6 days for all silks to emerge from an ear (Everett, 1958; Hallauer and Sears, 1966). Receptivity up to 10 days after emergence has been reported (Walden and Everett, 1961), but it falls off rapidly after that. High temperatures and low humidities probably would decrease this time. Best seed sets occurred with pollinations 3 to 5 days after first silk emergence, but pollinations after 8 days still gave 66% seed set compared with the optimum (Hallauer and Sears, 1966).

IV. ARTIFICIAL HYBRIDIZATION AND SELF-POLLINATION

A. Equipment

Items of equipment required include tassel bags, ear shoot bags, paper clips or a stapler, a knife, a pencil, tags, and an apron (Fig. 2A). The breeder has a choice of several types, sizes, and weights of tassel and ear shoot bags. The bags must be water repellant and made with moisture-proof glue. Paper clips must be slip-proof. Brass-plated clips are best because they rust quickly and will not slip. Small hand staplers may be used instead of paper clips. The knife commonly used is a paring knife of convenient size that must have a sharp blade. Graphite marking pencils are useful for labeling, and the lead should be soft enough to give a good mark on the tassel bag. Tags used for labeling preferably should have metal eyelets. Aprons should have one large pocket to carry up to 100 tassel bags and 3 or 4 smaller pockets to carry ear shoot bags and other pieces of equipment.

B. Preparation of the Female

The ear shoot (husk tip) must be covered by an ear shoot bag before the silks emerge from the husk tip (Fig. 2B). Ear shoots may be covered any time during the day, but it frequently is the first operation of the day because it can be done before pollen shedding begins. The length of ear shoot development from the leaf axil before silks begin to emerge will vary considerable among genotypes and also is affected by environmental conditions. Once ear shoots begin to emerge, the nursery should be checked daily, particularly if temperatures are moderate and soil moisture is adequate to cause rapid growth. The bag should be placed so that it is firmly anchored between the shoot and the auricle of the ear leaf, or sometimes it is expedient to break off the ear leaf and pull the bag down so that it is anchored between the shoot and the stalk (Fig. 2C).

The best ear shoots to use in self or cross-pollination will have had silks emerged under the shoot bag for 2 to 3 days. If a covered shoot has silks emerged more than 2 or 3 days before it can be used, it may be necessary to trim the silks to prevent them growing out of the bag and becoming con-

Fig. 2—Pollination of corn. A) Equipment used for self- and cross pollination; B) bag being placed on ear shoot before silks have emerged; C) ear shoot bag in place before pollination; D) tassel covered by bag the day before hand pollination; E) ear shoot bag lifted to permit pollination; F) tassel bag clipped in place over pollinated ear.

taminated. On the day preceding intended pollination, the extended silks are cut to within 2 cm of the husk tip. This procedure will cause the formation of a brush of silks on which the pollen can be spread the next day (Walden, 1967). If it is necessary to use the ear before all silks have emerged, it may be convenient to cut the ear shoot to within 2 cm of the cob tip so that all silks will be available the next day. Silks trimmed this way usually will grow 2 to 4 cm before being pollinated the next day. The shoot bag must be replaced securely after preparation of the shoot.

C. Pollination

A tassel usually produces its greatest volume of pollen in the second and third days of dehiscence. When the ear is prepared for pollination, the tassel that is the intended source of pollen is covered by a tassel bag the same day (Fig. 2D). The tassel should have at least 1 day of anther exsertion and pollen dehiscence. The tassel bag is placed over the tassel on the day before pollination to eliminate contamination by other pollen that may adhere to the tassel, and it will hold all pollen shed by the tassel while the bag is in place. The bag is held in place by a paper clip or a staple.

Pollen may be collected and used immediately from uncovered tassels that are dehiscing. This procedure may be convenient in a cross-pollination program in which several ears are pollinated by pollen from one tassel or a pollen composite from several tassels. There may be some contamination from tassel to tassel because of wind-blown pollen, but contamination will usually be low because the amount of pollen collected is much greater than the amount of foreign pollen. The technique is satisfactory if a low frequency of contamination can be tolerated; otherwise, it should not be used (Genter, 1976).

For self-pollination, the pollen is taken from the tassel and placed on the silks of the same plant. With reasonable care, contamination will be infrequent because of the excessive amount of pollen grains from the intended source in proportion to foreign pollen.

For cross-pollination, pollen from one tassel is used for one or several ear shoots simply by pouring pollen from the pollen bag directly onto the silks of another plant. Where several pollinations are made and pollen from several tassels is composited, a pollen gun may be used. Clumping of the pollen may occur, particularly if the humidity is high, and it usually is necessary to have a humidity-control compound in the gun to keep the pollen dry.

The pollinating operation must wait until anthers are exserted and pollen is released. This may be about 3 hours after sunrise, but it may be earlier or later in the day, depending upon the temperature and humidity. Pollen dehiscence under the bag will not be complete, however, and shaking the tassel in the bag will give viable pollen most of the day. Controlled pollinations made at 2-hour intervals from 0900 to 1700 hours gave no significant differences in amount of seed set (Hallauer and Sears, 1966). Walden and Everett (1961) found that the optimum time for pollinations ranged from 1530 to 1930 hours.

The silk is exposed for pollination either by lifting the ear shoot bag or by tearing off the closed end of the bag (Fig. 2E). After the pollen has been dusted onto the silk, the tassel bag is quickly placed over the ear shoot, pulled down toward the ear node, and fastened around the stalk either by a nonslip paper clip or staple (Fig. 2F). If the ear shoot bag is left in place after pollination, it may prevent contamination that might be caused by insects, such as grasshoppers (*Melanoplus* spp.) that may chew holes in the tassel bag. Conversely, in some seasons when the ear shoot bag is left in place, it may cause considerable ear-tip rot by harvest time.

When more than one ear on a plant is pollinated, seed set on the lowest ear is most reliable if pollinated 1 day before the ear at a higher node.

Marking on the tassel bag may be necessary to identify parental materials. For this purpose, a moderately soft graphite marking pencil is used. Limited marking is desirable because it is time-consuming. Different-colored, striped tassel bags are available, and in some instances, a color can be used for identification. The pollination date can be marked on the bag the day ahead of pollination by using a date stamp and pad with weather-proof ink. Large numbers or letters should be used when marking on the bag because it may be partly torn before the ear is harvested.

One usually can expect good success from hand pollination with corn. In a drought situation, scarcity of viable pollen for a pollination may reduce seed set and some contamination may occur. Seed set often may be obtained for only half to two-thirds of the ear if pollinations are made 1 to 2 days either too early or too late.

D. Factors Affecting Efficiency

Items that should be considered in planning the field layout of a breeding nursery are hand vs. machine planting, dates of planting, maturities of breeding stocks, breeding procedures, plot sizes, and plant densities. Variable planting dates may be desirable to spread out the work load in the pollination and harvest seasons. Dates of planting should be planned so that physiological maturity will be attained and seed ears can be harvested before damaging frost is expected.

Proper planning of the field layout will enhance the efficiency of a breeding procedure.There usually are selfing nurseries for inbred line maintenance, inbred line increase, and inbred line selection and inbreeding; crossing nurseries for single crosses, three-way crosses, double crosses, and top-crosses; and recurrent selection nurseries. The plot size should be standardized as much as possible.

If one needs a large supply of seed, as in the increase of an inbred line or in the production of a series of single crosses, a relatively low plant density of about 30,000 plants/ha is desirable. One usually obtains the maximum number of seeds per pollination at low plant densities. A higher density, such as 60,000 plants/ha, is used if the purpose is to sample a large number of plants in a heterogeneous cultivar. Higher densities of 60,000 to 70,000 plants/ha should be used in an inbred line selection nursery to facilitate the identification of lines that tolerate stress conditions. If the selection is for prolificacy, a lower density of 30,000 to 40,000 plants/ha may be desirable to permit the individual plant to express its potential for developing more than one ear per plant.

To determine the number of plants to be used per progeny in an inbred selection nursery, one must consider that genetic variation among progenies increases and, within progenies, decreases as inbreeding progresses. Therefore, the opportunity for effective selection within a progeny decreases rapidly as the coefficient of inbreeding increases.

Rows in the breeding nursery are arranged in ranges or tiers, with alleys or walkways separating the ranges. The number of rows per range should be limited to not more than 60 to 70 whenever possible. Duplication of plot numbers in field labeling should be avoided, even in different areas of the nursery, because it may cause problems in seed identification.

For the production of crosses, parents should be planted in paired rows. A string tied loosely at ground level between paired rows is recommended to minimize the risk of crossing between the wrong rows. If several crosses have one parent in common, the common parent is used as male and is planted adjacent to a block of the female parents. These planting arrangements eliminate the necessity of labeling bags after pollination. The flowering dates of two parents frequently do not match, thus necessitating a delayed planting for the earlier-flowering parent. The actual number of days to delay planting for the earlier line may be based on accumulation of growing-degree days after the later parent is planted (Chapter 1). The row for the delayed planting may be obliterated by rain, but it can be marked by planting a few field peas (*Pisum sativum*) on the first date of planting.

An applied breeding program will usually have evaluations of new lines in topcrosses or testcrosses in which there is a common tester parent. It may be most convenient to produce seed of these crosses in an isolated crossing block in which the common tester is the male and the other lines are detasseled and serve as females.

Recurrent selection programs that include crossing selected plants to a tester should have a field design that makes pollination as convenient as possible. The selection and tester populations are planted in adjacent blocks, such as across an alley in two ranges, to minimize walking distance between the two blocks. If flowering of the two populations does not coincide, two dates of planting for one or both will be needed. Even if the two populations do coincide for the flowering date, it is advantageous to plant the female parent a few days earlier than the male so that the silks are fully emerged before pollination begins. Virtually full seed set can be expected if the silks emerge 7 days before pollen shed (Walden and Everett, 1961).

Some mechanical procedures may be used to delay flowering because the growing point has not progressed above the ground even though the seedling has reached the four- to six-leaf stage. Flaming treatments on inbred lines when the plants were 5 cm tall and a second flaming when the plants were again 5 cm tall delayed anthesis 2.2 to 2.6 days and silking 2.3 to 2.8 days (Green, 1949). He observed an average reduction in grain yield of 7%.

Dungan and Gausman (1951), using clipping treatments on eight single crosses, obtained an average pollen-shed delay of 8.1 days when plants 87 cm tall were clipped to 6 cm. They observed an average silk delay of 10.8

days and a yield reduction of 47.6%. Clipping plants 40 cm tall back to ground level caused delays of 5.3 days for pollen shed and 6.1 days for silk emergence, and a 17.7% yield reduction in their study. The most practical treatment for eight inbred lines was clipping to ground level when the plants were 31 cm tall. Such a treatment caused a 4.6-day delay for pollen dehiscence, a 5.6-day delay in silk emergence, and a 23.2% yield reduction.

Cloninger et al. (1974) obtained an average pollen-shed delay of 5.5 days with a clipping treatment at the 4-leaf stage, 7.6 days at the 6-leaf stage, and 6.7 days at the eight-leaf stage for 28 single crosses among eight inbred lines. No data were given for silk emergence. They observed an average yield reduction of 11% for clipping at the four-leaf stage, 38% for the six-leaf stage, and 46% for the eight-leaf stage.

Many corn strains are sensitive to photoperiod; consequently, time to flowering for both male and female flowers may be changed by altering the day length. Low-latitude corn strains, for example, are short-day sensitive and, when grown in a higher latitude, will not develop to anthesis until late summer when the day length decreases. By creating an extension of the night by artificial means, low-latitude strains can be forced to anthesis earlier in the summer.

Genetic markers are not used extensively in corn breeding, except for some special genetic studies or specific traits. One of the first comprehensive cataloging of corn mutants was by Emerson et al. (1935). A detailed list of the corn mutants, the 10 chromosome maps, and colored prints for many of the plant and seed mutants was published by Neuffer et al. (1968).

Seed and seedling markers can be used for the production of homozygous diploids from monoploids (Chase, 1947; Chase, 1952; Goodsell, 1961; Nanda and Chase, 1966; Kermicle, 1969). The most widely used endosperm marker is sugary-1 (su_1), which is responsible for the endosperm type of most commercial and garden sweet corn. Other endosperm mutants that are being used commercially in hybrids or have been studied for possible use in speciality types include waxy, opaque-2, floury-1 and -2, sugary-2, dull, amylose extender, shrunken-2, and brittle-1. Some plant mutants that have been studied for their usefulness in hybrid cultivars include brown midrib, brachytic-2, reduced, compact, gametophyte factors, liguleless, male sterile, and tassel seed.

V. NATURAL HYBRIDIZATION

Natural hybridization of corn by wind pollination is used to advantage in breeding procedures and hybrid seed production (Chapter 8). Seed increases of synthetic cultivars can be made with appropriate isolation from other corn. Certification standards require that a cultivar must be located not less than 200 m from corn of a different color or texture. The isolation may tend to be more stringent in a genetic experiment in which no contamination can be tolerated. Raynor et al. (1972) and Paterniani and Stort (1974) showed less than 1% effective pollination for a relatively short distance from the contaminating source. Hutchcroft (1959) found in 1 year that contamination exceeding the 1% certification requirement was limited to a subplot of 3 × 3 m surrounding the contaminating plant. Sometimes natural barriers, such as trees or dates of planting, can be used to effect isolation.

VI. SEED DEVELOPMENT, HARVEST, AND STORAGE

The mature corn kernel almost invariably contains four roots, the radicle, and three seminal roots. In many hybrids and most inbreds, all these parts have been initiated before the 30th day after pollination (Sass, 1977). Maximum dry weight of the kernel, defined as physiological maturity, is not attained for at least an additional 25 to 30 days (Hillson and Penny, 1965; Hallauer and Russell, 1962). Sprague (1936) found that seed harvested 21 days after pollination and permitted to dry slowly to 16% moisture, germinated 100% in 6 days. Experience has shown, however, that harvest before 50 days after pollination may affect germination, particularly in cold soil, and frequently affects seedling vigor.

If several ears are harvested from the plot or row of a progeny and individual ear identification is not needed, all ears can be placed in a mesh bag, closed with a wire tag, and marked according to plot or row number. If individual ear labeling is necessary, a marked tag can be attached to the ear either by a short nail or rubber band. Tags should be marked as simply as possible to save time and to avoid labeling errors.

Breeding stocks usually will need to be dried before they can be stored. For artificial drying, temperatures are much more critical for immature seed with high moisture content than for mature seed in which the moisture content at harvest is less than 30%. The critical level for immature seed seems to be 40 C (Harrison and Wright, 1929; Wileman and Ullstrup, 1945). Temperatures up to 15 degrees higher can be used safely for seed with less than 30% moisture, provided that air circulation is sufficiently rapid (Wileman and Ullstrup, 1945; McRostie, 1949; Dimmock, 1947). The generally recommended drying temperature is 40 to 43 C (Kiesselbach, 1939; Dimmock, 1947), and seed should be dried to 12 to 13% moisture. Further drying to moisture content as low as 5% is not injurious to the seed.

Shellers, operated either by hand or motor, are available for shelling single ears or lots of ears. Large machines equipped to remove chaff and that can be used to shell single ears or large bulk lots are useful in a large program. All shelling equipment must be self cleaning to avoid mixing among samples. Shelling should be done by hand for studies in which sheller damage cannot be tolerated.

All seed samples should be labeled with the year in which they were produced, and, in some instances, the location will be needed also. For most samples, a field row or plot number will be the only identification needed, and, if the sample is derived from a single-plant selection, a plant number will be used also. For other samples, a pedigree will be needed. Detailed labeling on seed packets should be avoided because it is an easy source of errors.

Two kinds of storage facilities are required, short term and long term. The seed laboratory probably will be satisfactory short-term storage for many items that are grown every year. Bockholt et al. (1969) obtained germination of 80% after 2 years, but only 17% after 3 years, when seed was stored in the laboratory in paper envelopes. When they stored seed dried to 9.5% moisture at room temperature in a sealed container, germination was 66% after 5 years and zero at 11 years.

Long-term storage facilities with controlled temperature and humidity are necessary for seed lots of materials that do not need to be grown every year. Research has indicated that the seed should be dried to 11% or lower before being placed in storage (Sayre, 1948; Goodsell et al., 1955; Bockholt et al., 1969). Storage in sealed containers will increase longevity of the seed. Storage in sealed containers with carbon dioxide or nitrogen showed no advantage in maintaining seed quality (Sayre, 1948; Goodsell et al., 1955).

A rule of thumb for corn seed storage is to maintain a temperature (C) and humidity so that the sum of the two values does not exceed 60. At Iowa State University, seed is stored at approximately 10 C and 45 to 50% relative humidity, and satisfactory germination has been retained up to 20 years on some samples. It must be recognized, however, that seed of some genotypes will not germinate after 15 years. Longer term storage can be obtained by using lower temperature and humidity, but this requires more elaborate equipment.

REFERENCES

Barnabas, Beata, and Erna Rajki. 1976. Storage of maize (*Zea mays* L.) pollen at −196 C in liquid nitrogen. Euphytica 25:747–752.

Bockholt, A. J., J. S. Rogers, and T. R. Richmond. 1969. Effects of various storage conditions on longevity of cotton, corn, and sorghum seeds. Crop Sci. 9:151–153.

Brown, W. L. 1975. A broader germplasm base in corn and sorghum. p. 81–89. *In* Proc. 30th Annu. Corn Sorghum Res. Conf., Am. Seed Trade Assoc.

Chase, S. S. 1947. Techniques for isolating monoploid maize plants. Am. J. Bot. 35: 582.

———. 1952. Monoploids in maize. p. 389–399. *In* J. W. Gowen (ed.) Heterosis. Iowa State Univ. Press, Ames.

Cloninger, F. D., M. S. Zuber, and R. D. Horrocks. 1974. Synchronization of flowering in corn (*Zea mays* L.) by clipping young plants. Agron. J. 66:270–274.

Dimmock, F. 1947. The effects of immaturity and artificial drying upon the quality of seed corn. Can. Dep. Agric. Tech. Bull. 58.

Dungan, G. H., and H. W. Gausman. 1951. Clipping corn plants to delay their development. Agron. J. 43:90–93.

Emerson, R. A., G. W. Beadle, and A. C. Fraser. 1935. A summary of linkage studies in maize. Cornell Univ. Agric. Exp. Stn. Mem. 180.

Everett, H. L. 1958. Increasing the longevity of corn pollen. Annu. Hybrid Corn Ind. Res. Conf. 13:47–53.

Genter, C. F. 1976. Comparison of pollination methods in maize. Crop Sci. 16:442–444.

Goodman, M. M. 1965. The history and origin of maize. North Carolina Agric. Exp. Stn. Tech. Bull. 170.

Goodsell, S. 1961. Male sterility in corn. Crop Sci. 1:227–228.

———, G. Huey, and R. Royce. 1955. The effect of moisture and temperature during storage on cold test reaction of *Zea mays* seed stored in air, carbon dioxide, or nitrogen. Agron. J. 47:61–64.

Green, J. M. 1949. Effect of flaming on the growth of inbred lines of corn. Agron. J. 41:144–146.

Hallauer, A. R., and W. A. Russell. 1962. Estimates of maturity and its inheritance in maize. Crop Sci. 2:289–294.

———, and J. H. Sears. 1966. Influences of time of day and silk treatment on seed set in maize. Crop Sci. 6:216–218.

Harrison, C. M., and A. H. Wright. 1929. Seed corn drying experiments. J. Am. Soc. Agron. 21:994–1000.

Hillson, M. T., and L. H. Penny. 1965. Dry matter accumulation and moisture loss during maturation of corn grain. Agron. J. 57:150–153.

Hutchcroft, C. D. 1959. Contamination in seed fields of corn resulting from incomplete detasseling. Agron. J. 51:267–271.

Jones, M. D., and L. C. Newell. 1948. Longevity of pollen and stigmas of grasses: Buffalograss, *Buchloe dactyloides* (Nutt.) Engelm. and corn, *Zea mays* L. J. Am. Soc. Agron. 40:195–204.

Kermicle, J. L. 1969. Androgenesis conditioned by a mutation in maize. Science 166: 1422–1424.

Kiesselbach, T. A. 1939. Effect of artificial drying upon the germination of seed corn. J. Am. Soc. Agron. 31:489–496.

———. 1949. The structure and reproduction of corn. Nebraska Agric. Exp. Stn. Res. Bull. 161.

Larson, W. E., and J. J. Hanway. 1977. Corn production. *In* G. F. Sprague (ed.) Corn and corn improvement. Agronomy 18:625-669. Am. Soc. Agron., Madison, Wis.

McRostie, G. P. 1949. Some factors influencing the artificial drying of mature grain corn. J. Am. Soc. Agron. 41:425–429.

Nanda, D. K., and S. S. Chase. 1966. An embryo marker for detecting monoploids of maize (*Zea mays* L.). Crop Sci. 6:213–215.

Neuffer, M. G., L. Jones, and M. S. Zuber. 1968. The mutants of maize. Crop Sci. Soc. Am., Madison, Wis.

Paterniani, E., and A. C. Stort. 1974. Effective maize pollen dispersal in the field. Euphytica 23:129–134.

Pfahler, P. L. 1967. Fertilization ability of maize pollen grains. II. Pollen genotype, female sporophyte, and pollen storage interactions. Genetics 57:513–521.

Raynor, G. S., E. C. Ogden, and J. V. Hayes. 1972. Dispersion and deposition of corn pollen from experimental sources. Agron. J. 64:420–427.

Sass, J. E. 1977. Morphology. *In* G. F. Sprague (ed.) Corn and corn improvement. Agronomy 18:89–110. Am. Soc. Agron., Madison, Wis.

Sayre, J. D. 1948. Storage tests with seed corn. Proc. Annu. Hybrid Corn Ind. Res. Conf. 3:57–64.

Shaw, R. H. 1977. Climatic requirement. *In* G. F. Sprague (ed.) Corn and corn improvement. Agronomy 18:591–623. Am. Soc. Agron., Madison, Wis.

Sprague, G. F. 1936. The relation of moisture content and time of harvest to germination of immature corn. J. Am. Soc. Agron. 28:472–478.

Tatum, L. A., and W. R. Kehr. 1951. Observations on factors affecting seed-set with inbred strains of dent corn. Agron. J. 43:270–275.

Walden, D. B. 1967. Male gametophyte of *Zea mays* L. I. Some factors influencing fertilization. Crop Sci. 7:441–444.

————, and H. L. Everett. 1961. A quantitative method for the *In vivo* measurement of the viability of corn pollen. Crop Sci. 1:21–25.

Wileman, R. H., and A. J. Ullstrup. 1945. A study of factors determining safe drying temperatures for seed corn. Indiana Agric. Exp. Stn. Bull. 509.

20 Cotton

Joshua A. Lee
USDA-SEA and North Carolina State University
Raleigh

The cultivated cottons comprise four species in the genus *Gossypium,* tribe *Gossypieae,* family *Malvaceae.* Two species were domesticated in Africa and/or India, one in South America, and one in Mesoamerica.

Cotton is cultivated virtually around the world in tropical and subtropical regions, both under irrigation and as a rainfed crop. Important areas of production are the southern United States westward to California, Mesoamerica and South America, Egypt, the Sudan, the Soviet Union, India, and the People's Republic of China. Lesser amounts are grown in the savannas of Africa, southern Europe, the Near East, and Australia.

Cotton is grown mainly for its lint (seed floss) which is used in spinning yarns for textiles, industrial cloths, and toweling. Smaller amounts are used in the production of sewing thread and for stuffing. The seeds are an important by-product. The refined oil is used in cooking, and the oil-cake residue as a protein feed for ruminant livestock. Linters, the short fibers removed from seed before crushing, are an important source of industrial cellulose.

I. PARENTAL MATERIAL

The Old World cultivated cottons are classified as *Gossypium arboreum* L. and *Gossypium herbaceum* L. Both are diploids with a haploid chromosome number of 13, the basic number for the genus. *Gossypium herbaceum* grows wild in Africa where the plant might have been domesticated. *Gossypium arboreum* is known only in cultivation, or as a ruderal plant, and might have been derived from *G. herbaceum* after that species was domesticated (Hutchinson et al., 1947). *Gossypium arboreum* remains an important crop in India today, whereas *G herbaceum* is today a minor crop grown for local use in the drier areas of Africa and Asia.

Table 1—Species of *Gossypium* listed according to genomic formula and characters contributed for actual, or potential, improvement of the cultivated tetraploid species.

Gossypium species	Genomic formula	Character and authority
tomentosum Nutt. ex. Seem.	$(AD)_3$	Loss of extrafloral nectaries (Meyer and Meyer, 1961).
herbaceum	A_1	Resistance to leafhopper (Knight, 1954).
arboreum	A_2	Resistance to bacterial blight (Knight, 1953).
thurberi Tod.	D_1	Increased fiber strength in *G. hirsutum* (Lewis, 1957).
armourianum Kearney	$D_{2(1)}$	Glabrous leaf (Meyer, 1957).
harknessii Brandg.	$D_{2(2)}$	Cytoplasmic male sterility (Meyer, 1975).
klotzschianum Anderss.	D_3	Gossypol in foliage (Bell and Stipanovic, 1977)
raimondii Ulb.	D_5	Unusual terpenoid aldehydes in foliage (Bell and Stipanovic, 1977).

Gossypium hirsutum L. is native to Mesoamerica and *Gossypium barbadense* to South America. The species are closely related tetraploids with a diploid-like meiosis, the haploid chromosome number being 26. *Gossypium hirsutum* accounts for about 90% of the current world's production of cotton. In addition to the cultivated forms, wild and ruderal races of the two species exist in tropical regions.

Cultivated *G. hirsutum* is commonly called short-stapled cotton, and is most often grown as an annual, the so-called Upland form. *Gossypium barbadense* is classified as a long-stapled cotton. Its strong, fine fibers are favored for the production of sewing thread and luxury fabrics.

In addition to the cultivated cottons, there are at least 31 wild taxa distributed in dry tropical and subtropical latitudes around the world (Fryxell, 1969). These, along with wild stocks and obsolete cultivars of the cultivated species, represent germ plasm sources of actual or potential use in improving cultivated cottons. Beasley (1942) classified *Gossypium* species according to chromosomal size and affinity at meiosis, assigning capital letters to various genomic groups, A, B, C, D, etc. Beasley's classification is convenient when discussing interrelationships within the genus.

Although at least one Soviet worker (Vysotski, 1958) has reported introgression of germplasm from a *Malva* species into *G. hirsutum,* workers elsewhere have not reported crosses between *Gossypium* and related genera. Many diploid and tetraploid species can be crossed with the cultivated diploid species (A) and the cultivated tetraploid species (AD). Table 1 lists some of the species that have been crossed with the cultivated tetraploids with notations on the characters contributed for actual or potential improvement.

All of the tetraploid species intercross readily, although there is frequently genetic breakdown in ensuing progenies. Crossability among *Gossypium* species does not always follow closely known phylogenetic affinities. The Old World cultivated species, the contributors of genome A to the cultivated tetraploids (Beasley, 1942), are difficult to cross with AD species. Pundir (1972) implicated endosperm degeneration in embryo sacs bearing hybrid embryos as the main cause.

Most American D diploids cross well with each other and with AD species. Crosses between *G. klotzschianum* and *G. davidsonii* (both D_3) and other D and AD species usually fail because of a lethal interaction in hybrid embryos. The D_3 species produce viable hybrid plants with certain selections of AD cottons (Lee and Smith. 1970).

The African B diploid, *Gossypium anomalum* Wawr. & Peyr., crosses readily with Old and New World cultivated cottons, and is one of the few species that produces viable hybrids with D_3. The Australian C diploid, *Gossypium sturtianum* J. H. Willis, crosses readily with *G. hirsutum,* however, little exchange of germplasm occurs among the genomes of the derived hexaploid.

A collection of obsolete cultivars of *G. hirsutum* is maintained at the U.S. Cotton Physiology and Genetics Laboratory, Stoneville, Miss., a collection of *G. barbadense* at the U.S. Cotton Research Laboratory, Phoenix, Ariz., and a collection of diploid cottons and race stocks of *G. hirsutum* at Texas A&M University, College Station. Seeds of various cottons are on deposit at the National Seed Storage Facility, Fort Collins, Colo.

II. PLANT CULTURE

A. Field

Cottons are grown on various soils, including sandy loams, lateritic clays, calcarious soils, and valley alluviums. Soils should be well drained. Cotton performs best on fertile soils, but excess nitrogen can lead to vegetative development at the expense of fruit production (Brown, 1938). The North Carolina Extension Service recommends 110 kg of N, 70 kg of K, and 50 kg of P per ha for satisfactory cropping. On sandy soils subject to leaching, the N and K can be applied in split applications. Boron deficiency inhibits fruit set in some areas of the U.S. cotton belt. Boron can be applied in fertilizers, or as a foliar spray at the rate of 5.5 kg/ha in split applications. Soil pH should not be below 6.2.

R. E. Sneed (personal communication, 1972) found that cotton in North Carolina requires about 2.5 cm of rainfall per week to maintain a good rate of growth and fruitation during July and August. Water requirements fluctuate with temperature and the depths to which roots penetrate. Mauney and Phillips (1963) expressed the opinion that moisture levels optimal for vegetative develop in wild stocks of *Gossypium* inhibit flower retention and development. Although cultivated cottons seem less sensitive to high soil moisture level than such primitive kinds, fruition proceeds more uniformly under adequate than with excessive moisture.

Plant density preferences vary among locations. Some stocks of *G. hirsutum* and *G. barbadense* maintained for home use in tropical America often grow into small trees. Cotton grown for annual production is commonly sown at a seeding rate of 9 to 12 seeds per m of row. Lesser densities tend to extend the fruiting season and denser planting leads to a high proportion of unproductive plants.

Cotton seed germinate in 5 days when soil temperature reaches 22 C, and seed should not be planted in soil colder than 16 C. Optimum temperatures for growth of seedling cotton are in the range of 27 C for a daily maximum and 11 C for a nightly minimum. Mauney (1966) showed that high night temperatures in the range of 27 C delay fruition and promote vegetative growth in cultivated *G. hirsutum*. A warm, dry regime is best for the last stages of fruit maturation and harvest.

Herbicides and tillage usually are used for weed control. A preplant herbicide is incorporated in the upper 5 cm of the soil surface or a postplant herbicide is sprayed on the surface. Later applications of herbicides may be applied as a directed spray. Control of insect pests is usually organized and performed on a community-wide basis, with a strong emphasis on field scouting for monitoring pest populations.

Defoliants are applied when approximately one-half to three-fourths of the mature bolls have opened. A dessicant is frequently applied before the defoliant to promote better leaf drop. Seed cotton is harvested with a spindle picker, or with one of various types of stripping machines. Seeds are separated from lint on commercial gins and the lint packed into bales averaging about 227 kg each.

B. Growth Chamber and Greenhouse

Greenhouses are commonly used to grow cotton for crossing and selfing. Growth chambers are not used for such purposes. For most work in the greenhouse, the intent is to simulate conditions for optimal growth in the field. Greenhouses are used mostly for producing cotton during the cool months. A cultivar of *G. hirsutum* can mature a crop of bolls in the greenhouse at Raleigh in about 120 days. A crop can be grown from seed harvested in September in time to provide mature seeds for field planting the following May.

Culture conditions in ground beds in greenhouses do not differ from those in the field. Most of the fertilizer can be applied as a preplant dressing, and water is usually applied in furrows. As in the field, soil pH should not be below 6.2. Greenhouse soils should be sterilized before planting, and methyl bromide is most commonly used. Potting medium should consist of three parts of soil to one part of vermiculite, or other such soil conditioner. About 25 g of complete fertilizer (equals parts of N, P, and K) should be applied to a 32-cm diam pot before planting. Additional N and K can be applied at biweekly intervals at the rate of 8 to 10 g per application. Watering of pots is best done by a time-clock monitoring system. Frequency of watering can be increased as the plants mature. Greenhouse temperatures are controlled by automated systems of heating and cooling, and turbulating devices are used for distributing air. Heating during cool weather should be programmed at 27 C for a daily maximum, and 19 C for a nightly minimum. On sunny days, ambient greenhouse temperature can approach 36 C before cooling is applied.

Cottons sensitive to short days are frequently grown in greenhouses for crossing. Such cottons require careful management to ensure flowering and fruiting (Mauney and Phillips, 1963). Maturing plants should not be exposed to more than 10 hours of light per day, and soil moisture should be limited to the point of stress. The temperature regime described for cultivated cottons is adequate to promote flowering in primitive kinds, as long as the day-night temperature differential does not fall below 8 to 10 C.

III. FLORAL CHARACTERISTICS

Cultivated cottons are generally self-pollinated (Brown, 1938). The stigmas of most cultivars are not exserted clear of the uppermost anthers, so that selfing usually occurs immediately after anther dehiscence (Fig. 1). Some primitive forms of *G. hirsutum* and *G. barbadense* have the stigma exserted well above the anthers. Such separation of anthers and stigma should promote cross-pollination by insect vectors.

A flower bud (square) of *G. hirsutum* grows from the pinhead stage to anthesis in from 21 to 25 days (Brown, 1938). Once in bloom, a cotton flower is receptive to pollination for 8 hours or less.

Cottons normally have perfect flowers. The developing bud is enclosed in three, rarely four, bractioles called the epicalyx (Fig. 1). A five-lobed calyx cups the base of the flower. The mature, unopened flower is tubular, consisting of five petals united at their bases and continuous with an anther column clasping the ovary and style. Petal length can vary from 9 cm in some stocks of *G. barbadense* to 2.5 cm in some cultivars of *G. arboreum.* Corolla color can be creamy white or various shades of yellow. White is most common in *G. hirsutum* and *G. arboreum,* and yellow predominates in *G. barbadense* and *G. herbaceum.* The diploid species and most cultivars of *G. barbadense* display a red spot at the base of each petal, a phenotype rarely observed in cultivars of *G. hirsutum.* The corollas of most species of *Gossypium* flare widely at anthesis, whereas those of *G. barbadense* and the wild species, *G. raimondii,* retain a tubular form.

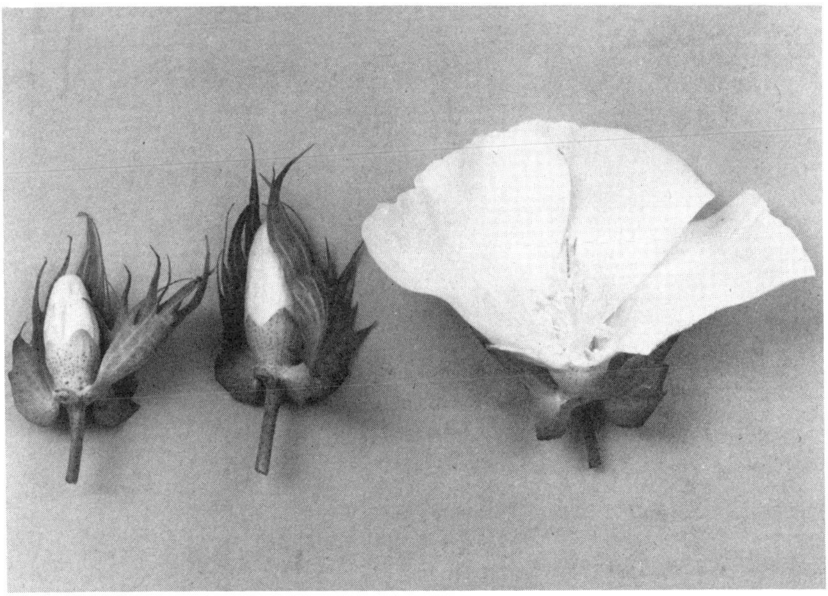

Fig. 1—Cotton (*G. hirsutum* L.) flowers at three stages of development. Left to right: bud 48 hours from anthesis; 24 hours from anthesis (whitebud stage); and a cutaway view of flower at anthesis.

The anther column bears anthers (approximately 100 per flower in cultivated *G. hirsutum*) on filaments of variable length. The tip of the stigma is exserted above the anther column. It is divided into three to five parts, each continuous with a locule of the ovary. The ovules are borne at the central axis of the ovary, and can number as high as 10 per locule in *G. hirsutum,* but fewer in most cultivars of the other domesticated species. The ovary, after fertilization, develops into a valvate capsule (boll); the maturation process requiring from 40 to 80 days. The capsule can be round, ovate, or top-shaped, depending upon the species or cultivar, and can bear from 3 to 10 g of seed and adhering lint.

IV. ARTIFICIAL HYBRIDIZATION AND SELF-POLLINATION

A. Equipment

Equipment needed for hand-crossing and selfing of cotton is neither elaborate nor costly. Items used for self-pollination include paper clips, malleable wire, cardboard cylinders, and shellac and ethyl acetate for sealing the ends of buds (Fig. 3). Selfed or cross-pollinated flowers can be marked in various ways. Varicolored merchandize tags can be adapted by color code for various operations, such as one color for selfed flowers and another for crossing, or various colors can denote different crosses. Miravalle (1965) recommended a rectangular tag with a circular eye large enough to slip over a flower bud. Information concerning the flower can be code-punched into the tag for cross reference with a field book, or informa-

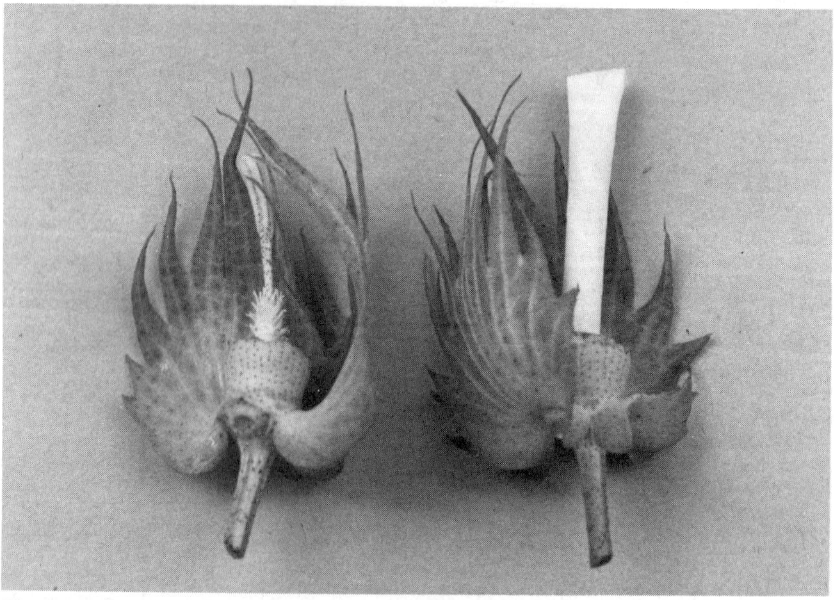

Fig. 2—Emasculated flowers of cotton (*G. hirsutum* L.). Left, anthers and petals removed showing the elongated style and stigma; right, a section of soda straw in place to protect stigma.

tion can be written on the tag in permanent ink. Emasculated flowers can be protected with a length of soda straw (Fig. 2), a cardboard cylinder (Fig. 3), or by placing a paper or a cloth bag over the flower.

Flowers to be pollinated are marked in various ways. Wooden stakes or white flags on heavy wire staffs may be placed in rows near plants bearing emasculated flowers. White cloth strips tied to plants are sometimes used.

B. Preparation of the Female

Progress of floral development in cotton follows a rigid and predictable schedule, provided that the temperature regime is favorable. Flowers approaching maturity are at the whitebud stage, the stage for emasculation, by 1200 hours the day before anthesis (Fig. 1). An experienced observer can predict the onset of anthesis as early as 48 hours before a flower is due to shed pollen. White buds are usually emasculated the afternoon before anthesis.

Some workers emasculate cotton flowers with the fingers, removing the epicalyx, corolla, and anther column in one deft operation. Cutting the petals away with an emasculating device, followed by stripping the anthers, is less injurious to flowers and improves fruit set. Some cottons, notably certain strains of *G. barbadense,* do not tolerate loss of petals before pollination. With such stocks, anthers may be removed through a slit in the corolla, or, if petals are removed, a drop of gibberellic acid solution (100 ppm aqueous) applied at the base of the anther column at emasculation improves fruit set (Brown and Lee, 1976). Treating stigmas of emasculated

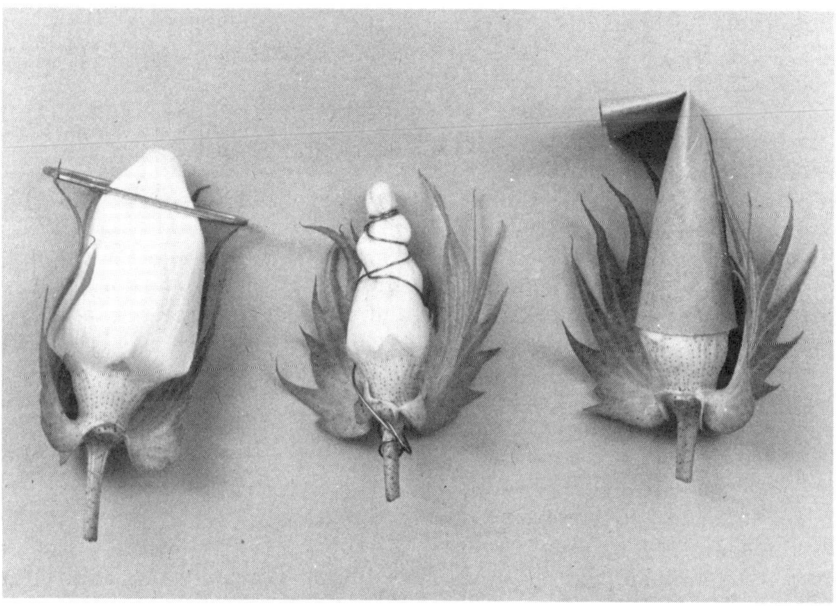

Fig. 3—Three methods of selfing in cotton (*G. hirsutum* L.). Left to right, paper clip attached to end of corolla, corolla wrapped with copper wire, cardboard cylinder over flower. All flowers are at anthesis.

flowers with a 30% ethanol solution destroys extraneous pollen. Stigmas of emasculated flowers are protected by covering with a length of soda straw, a cardboard tube, or a bag.

C. Pollination

Emasculated flowers are pollinated the day following emasculation, usually beginning around 1000 hours. The pollen in whitebuds can effect fertilization if held overnight under refrigeration. Stigmas are not fully receptive to pollination until the morning of anthesis. Pollen does not survive in the field for more than a few hours.

Flowers to be used as pollen sources are prepared the day before anthesis by sealing the corolla with any one of several devices (Fig. 3). With the exception of bags, none of the devices cited forestall the efforts of determined bumblebees; therefore, selfed flowers should be inspected for bee damage before use as pollen sources.

Flowers should be shedding abundant pollen at the time of use. In North Carolina during July, pollen shed in *G. hirsutum* usually commences when shade temperature reaches 27 C, most often around 0900 hours. Some cultivars of *G. barbadense* shed pollen earlier.

For most hand pollinations, the pollen is applied directly from the anthers to the stigma, usually within minutes after the male flower is collected. If pollen shed is tardy, as when temperature is low, heating will promote shed. Stigmas are protected by replacing the shielding device, and the flowers are tagged according to some preplanned system of identification.

Pundir (1972) found that the elapsed time between pollination and fertilization in cultivars of *G. hirsutum* and *G. arboreum* ranged from 24 to 48 hours. Diurnal temperatures in Pundir's experiments ranged between 9 and 38 C.

Research on freeze-drying of cotton pollen for future use is lacking. Small lots can be stored in a refrigerator (2 to 3 C) for at least 24 hours without loss of viability.

Rate of fruit set after pollination of emasculated flowers varies with location, season, and cultivar. R. J. Kohel (personal communication, 1977) reported about 50% success with *G. hirsutum* at College Station, Texas, with the rate falling as the season progressed. At Raleigh, North Carolina, I have recorded as high as 90% success in the field and greenhouse with *G. hirsutum,* but less (70 to 80%) with *G. barbadense* (Brown and Lee, 1976).

D. Factors Affecting Efficiency

Cotton nurseries are rarely elaborate. At Raleigh I commonly use blocks 15 m in length made up of rows 1 m wide. Two-meter alleys are left at the end of the blocks to facilitate access to the rows. Plots for crossing and for selfing are laid out in similar patterns. Often crossing and selfing are attempted in the same row.

Cotton plants commonly bloom over periods ranging from 3 weeks to more than a month. Matching flowering dates for most cultivars is rarely a problem, provided that such is taken into account at planting time. Most cultivars of *G. barbadense* begin flowering 1 to 2 weeks later than *G. hirsutum,* if the two are planted on the same date. There is usually enough overlap in the flowering periods to ensure crossing, particularly when *G. barbadense* is used as the female parent.

Matching the flowering sequences of annual cultivars and cottons sensitive to short days presents a problem requiring careful management. Such cottons must be grown in tropical winter nurseries or in greenhouses under the proper regimes of temperature, moisture, and light duration. Most primitive cottons require the development of 15 or more nodes along the main plant axis before there can be flowering under any environment (Mauney and Phillips, 1963). Two-year-old plants of primitive cottons are commonly used for crossing with cultivated cottons, and the cultivar is most often used as the female parent. Pruning of flower buds delays flowering indefinitely, and can be used to promote synchronous flowering.

Many genetic markers are available for identifying F_1 plants in cotton, but few have been used. The most common is R_1, a dominant allele in *G. hirsutum* that imparts wine-red plant color (Pope et al., 1944). R_1 does not weaken plants, and is readily transferred from one background to another. Alleles with similar effects are present in other cultivated species.

Glandless ($gl_2gl_2gl_3gl_3$) has been proposed as a seed marker (Cross and Richmond, 1959). Normal cultivars of cotton have conspicuous pigment glands in most plant parts, including embryos. The glandless strains of *G. hirsutum* and *G. barbadense* are devoid of these pigment bodies. The glandless stock is used as the female parent and the amount of hybridization is assessed by cutting and inspecting seeds.

Turcotte and Feaster (1967) described a stock of Pima cotton (*G. barbadense*) that exhibited semigamy, a condition wherein the egg and sperm fail to unite after fertilization, each subsequently giving rise to an independent tissue line in the embryo. Such tissues are often haploid. Making a semigamous stock homozygous for a conspicuous marker affords ready identification of sectored F_1 plants, thus providing material for the production of homozygous lines.

V. NATURAL HYBRIDIZATION

Cottons are predominantly self-pollinated with the amount of outcrossing variable among locations. Interplanting of genetically marked stocks is used to assess rate of natural hybridization. Simpson and Duncan (1956) recorded from 29 to 60% outcrossing among 79 cultivars of *G. hirsutum* at Knoxville, Tenn. in an unsprayed field. Although self-incompatibility systems have not been documented for cotton, they found that cultivar rank was highly repeatable over two seasons. In the Mississippi Valley, the San Joaquin Valley, and the High Plains of Texas, outcrossing rarely exceeds 5%.

Bees of various genera serve as pollen vectors in cotton, with the genus *Bombus* being prominent. Hive bees, *Apis mellifera* L., have been used with success in cross-pollinating cotton (McGregor, 1976).

Mass crossing of cotton in the field requires the use of induced or genetic male-sterility. Eaton (1957) induced male sterility in *G. hirsutum* through foliar applications of alpha-dichloroisobutrate (FW-450). Fruits set well on treated plants, but seed yield was sharply reduced because of damage to the growing plant. Chemosterilization remains as a tool of potential worth in the production of hybrid cottons.

Several recessive male-sterile alleles are known in *G. hirsutum,* some imparting complete pollen abortion (Weaver, 1968; Richmond and Kohel, 1961). These have potential use in promoting mass crossing in genetic and breeding programs and have been used in the production of commercial hybrids in India (J. B. Weaver, Jr., personal communication, 1976). Cytoplasmic-genetic male sterility (CMS) is recoverable when the genomes of *G. hirsutum* are transferred to the cytoplasms of various diploid species. The most dependable source of CMS made available to date employs the cytoplasm of *G. harknessii* (Meyer, 1975). This species has also contributed fertility restorer genes.

The chief potential use of CMS is in the production of single-cross hybrids for commercial production. A problem at present is the need to control insect pests in cotton fields while simultaneously preserving pollen vectors. There is evidence that if such vectors are preserved, satisfactory amounts of F_1 seed can be produced. Lee Stith (personal communication, 1977), working in Arizona, used a system of eight rows of CMS to two rows of pollinator and obtained good seed set on the CMS rows. Wild bees served as pollen vectors.

Isolation requirements for intercrossing experiments in subtropical latitudes vary among locations, and perhaps also from year to year. Miller and Rawlings (1967) in an intermating experiment with *G. hirsutum* in North Carolina found a distance of 0.8 km from the nearest cotton to be adequate for isolating their material. Green and Jones (1953) found a buffer strip of 9 rows of cotton effective as a screen for experimental plots in Oklahoma. Pope et al. (1944), working in Tennessee, used strips of corn as barriers to pollen transport.

VI. SEED DEVELOPMENT, HARVEST, AND STORAGE

Cotton plants in the field at Raleigh commence blooming about 65 days after emergence, usually around 4 July, and continue flowering until the plants are loaded with fruit, commonly around 15 August. Bolls set in mid-July are expected to mature in 45 days on a typical *G. hirsutum* cultivar, such as 'Coker 310'. Maturation rates for *G. barbadense* are usually slower than those of *G. hirsutum*, the boll period commonly extended 10 days longer. Late fruits of all cottons take longer to mature, sometimes up to 80 days for *G. barbadense.* Mauney (1961) showed that embryos of *G. hirsutum* ('Coker 100W') from early flowers grew to maximum length in about 30 days.

Hand-crossed and selfed cotton are usually harvested by hand and the lint and seeds separated on a micro-gin. Larger lots are harvested with spindle pickers, or stripping machines, and the lint removed from seeds at commercial ginneries.

The fuzz on cotton planting seed is commonly removed with concentrated sulfuric acid, and the seed dried at 43 C preparatory to storage. Cottonseed with germination of 80% or better can remain viable in well-ventilated storage facilities for at least 2 years. Storage pests, insects and rodents, are rarely a problem with cottonseed. Seed viability can be retained for longer periods under refrigeration. Pate and Duncan (1964) showed that *G. hirsutum* seeds containing 11% moisture remained fully viable for 25 years after storage at 0.5 C, but seeds stored at 13% moisture had only 16% germination. Bockholt et al. (1969) showed that Acala cotton (*G. hirsutum*) seed with 6.2% moisture stored at 10 C retained 51% viability after 26 years. In genetic and breeding programs, small quantities of seed are stored at subfreezing temperatures in cloth bags or in wire-top envelopes.

VII. TECHNIQUES FOR SPECIAL SITUATIONS

Special techniques for facilitating hybridization in cotton include the use of regulatory hormones, and embryo culture using excised embryos or intact ovules. Gibberellic acid (GA) promotes fruit set in cotton. Walhood (1957) reported that GA (100 ppm aqueous) promoted 100% boll set in Acala cotton; however, there was a dramatic increase in the level of seed abortion. Mathur and Mittal (1964) found that GA applied to preblooming cotton at rates ranging from 50 to 200 mg/liter increased flowering, but only the highest rate increased fruit set. Brown and Lee (1976) used GA to promote fruit set after emasculation in *G. barbadense.*

Culture of excised embryos has been used to produce interspecific hybrids in cottons, particularly crosses between Old and New World cultivars. The techniques for such culture as developed by Mauney (1961) have become standard for cotton.

Because yield of normal plants from embryo culture is low, techniques of ovular culture as developed by Beasley (1974) have come into favor for production of difficult hybrids. James McD. Stewart (personal communication, 1977) has used ovular culture to mass produce hybrids between AD (New World cultivar) and A (Old World cultivar) species, thus increasing the availability of diploid germplasm for the improvement of the tetraploid forms.

REFERENCES

Beasley, C. A. 1974. Glasshouse production of cotton flowers, harvesting procedures, methods of ovular transfer, and *in vitro* development of immature seed. Cotton Grow. Rev. 51:293–301.

Beasley, J. O. 1942. Meiotic chromosome behavior in species, species hybrids, haploids, and induced polyploids of *Gossypium* species. Genetics 27:25–54.

Bell, A. A., and R. D. Stipanovic. 1977. The chemical composition, biological activity, and genetics of pigment glands in cotton. p. 244–258. *In* Proc. Beltwide Cotton. Improve. Conf.

Bockholt, A. J., J. S. Rogers, and T. R. Richmond. 1969. Effects of various storage conditions on longivity of cotton, corn and sorghum seeds. Crop Sci. 9:151–153.

Brown, H. B. 1938. Cotton. McGraw-Hill Book Co., Inc., N.Y.

Brown, J., W. G., and J. A. Lee. 1976. Effects of emasculation on boll set in three stocks of cultivated *Gossypium.* Crop Sci. 16:599–601.

Cross, J. E., and T. R.Richmond. 1959. The use of glandless seed to determine the amount of natural crossing in *Gossypium hirsutum.* Agron. J. 51:511–512.

Eaton, F. M. 1957. Selective gametocide opens way for hybrid cotton. Science 126: 1174–1175.

Fryxell, P. A. 1969. A classification of *Gossypium.* Taxon 18:585–591.

Green, J. M., and M. D. Jones. 1953. Isolation of cotton for seed increase. Agron. J. 45:366–368.

Hutchinson, J. B., R. A. Silow, and S. G. Stephens. 1947. The evolution of *Gossypium* and the differentiation of the cultivated species. Oxford University Press, London.

Knight, R. L. 1953. The genetics of blackarm resistance. IX. The gene B_{6m} from *Gossypium arboreum.* J. Genet. 51:270–275.

———. 1954. The genetics of jassid resistance in cotton. IV. Transference of hairiness from *Gossypium herbaceum* to *G. barbadense.* J. Genet. 52:199–207.

Kohel, R. J., and T. R. Richmond. 1962. An evaluation of seed yield potential of completely male-sterile cotton in areas of high and low natural cross-pollination. Agron. J. 54:525–528.

Lee, J. A., and F. H. Smith. 1970. Some chemical properties of hybrid seed from a cross between *Gossypium barbadense* L. and *Gossypium davidsonii* Kell. Crop Sci. 10:379–380.

Lewis, C. F. 1957. Genetic recombinations in a hybrid involving three species of *Gossypium.* Agron. J. 49:455–460.

Mathur, S. N., and S. P. Mittal. 1964. Effects of gibberellin on flowering cotton plants. Physiol. Plant. 17:275–278.

Mauney, J. R. 1961. The culture *in vitro* of immature cotton embryos. Bot. Gaz. 122:206–209.

———. 1966. Floral initiation of Upland cotton (*Gossypium hirsutum* L.) in response to temperatures. J. Exp. Bot. 17:452–459.

———, and L. L. Phillips. 1963. Influence of daylength and night temperatures on flowering in *Gossypium.* Bot. Gaz. 124:278–283.

McGregor, S. E. 1976. Insect pollination of cultivated crop plants. p. 171–189. *In* ARS-USDA Handb. No. 496.

Meyer, J. R. 1957. Origin and inheritance of D_2 smoothness in Upland Cotton. J. Hered. 48:249–250.

———, and V. G. Meyer. 1961. Origin and inheritance of nectariless cotton. Crop Sci. 1:167–169.

Meyer, V. G. 1975. Male sterility from *Gossypium harkenessii.* J. Hered. 66:23–27.

Miller, P. A., and J. O. Rawlings. 1967. Breakup of initial linkage blocks through intermating in cotton breeding populations. Crop Sci. 7:199–204.

Miravalle, R. J. 1965. A tag for use in recognizing and identifying cotton flowers to be pollinated. Crop Sci. 5:74.

Pate, J. B., and E. N. Duncan. 1964. Viability of cottonseed after long periods of storage. Crop Sci. 4:342–343.

Pundir, N. S. 1972. Experimental embryology of *Gossypium arboreum* L. and *Gossypium hirsutum* L. and their reciprocal crosses. Bot. Gaz. 133:7–26.

Pope, O. A., D. M. Simpson, and E. N. Duncan. 1944. Effect of corn barriers on natural crossing in cotton. J. Agric. Res. 68:347–361.

Richmond, T. R., and R. J. Kohel. 1961. Analysis of a completely male sterile character in American Upland cotton. Crop Sci. 16:599–601.

Simpson, D. M., and E. N. Duncan. 1956. Varietal response to natural crossing in cotton. Agron. J. 48:74–75.

Turcotte, E. L., and C. V. Feaster. 1967. Semigamy in Pima cotton. J. Hered. 58: 55–57.

Vysotski, K. A. 1958. Cotton-mallow and cotton-hibiscus hybrids. Bot. Genet. Select. 31:151–161 (Abstr.).

Walhood, V. T. 1957. Effects of gibberellins on boll retention and cut-out in cotton. Proc. Cotton Defoliation Conf. 12:24–30.

Weaver, J. B., Jr. 1968. Analysis of a genetic double recessive completely male-sterile cotton. Crop Sci. 8:597–600.

21 Cowpea

H. T. Blackhurst
J. Creighton Miller, Jr.
Texas A&M University
College Station

The cowpea, *Vigna unguiculata* (L.) Walp, also known as southern pea, blackeye pea, lubia, niebe, coupé, or frijolé is an annual legume. Early evidence favored the Indian subcontinent as its point of origin (Vavilov, 1951). Later investigations indicated an African origin (Faris, 1963). The cowpea was introduced into the United States about 1700. The genus contains some 170 species of herbaceous, erect, semiupright, prostrate-spreading, and twining-climbing forms. This important tropical and subtropical legume is grown for both green and dry seed and for forage. It is an excellent source of protein. It can be successfully grown under a greater diversity of climatic, soil, and cultural conditions than most other legume crops.

I. PARENTAL MATERIAL

The genus *Vigna,* along with *Phaseolus* and *Dolichos,* is included in the tribe *Phaseoleae.* Three subspecies of *Vigna unguiculata* are now recognized: *unguiculata* (cowpea), *cylindrica* (catjang) and *sesquipedalis* (asparagus-bean) (Gunn, 1973). The three subspecies can be crossed, and free gene exchange is possible. Crossing with other cultivated or wild species of *Vigna* has had little success due to failure of either pollen germination or union of gametes. *Vigna unguiculata* is apparently a well-isolated species with limited potential for intergeneric or interspecific crossing.

Partial germplasm collections are maintained in the United States at Experiment, Georgia, and in India, Nigeria, and other countries. The International Institute of Tropical Agriculture, Ibadan, Nigeria has the largest collection, with some 8,000 accessions.

II. PLANT CULTURE

A. Field

Cowpeas can be grown on a wide range of soil types; however, highest yields are usually obtained on sandy loams and lighter clay soils. Cowpeas will perform reasonably well on soils with low fertility. Phosphorus application of up to 100 kg/ha of K_2O is recommended for sandy soil, with lesser amounts, 30 to 60 kg/ha, on loam and clay. High rates of nitrogen are detrimental and can result in excessive vegetative growth, delayed maturity, pod shattering, and reduced nitrogen fixation. Soils low in organic matter should receive no more than 45 kg/ha of nitrogen, as no yield benefits are realized with higher applications. The practical approach to optimum yields would be to provide conditions and cultivars which maximize symbiotic nitrogen fixation as the nitrogen source (Zary et al., 1978).

Cowpeas are known to be relatively drought tolerant, and the amount of supplemental irrigation water needed is not great. Excessive moisture can increase vegetative growth, resulting in reduced yields; however, it is critical that adequate moisture be available at flowering, otherwise pod set will be light and seeds per pod will be reduced.

Bush types are usually planted in rows 75 to 100 cm apart and within-row spacing of 7 to 10 cm at a seeding rate of 17 to 28 kg/ha. Wider spacing of 30 to 60 cm is required for vining types. Seed may be broadcast at rates up to 100 kg/ha when it is grown for forage. In field nurseries, cultivars and breeding lines should be grown in plots separated sufficiently to minimize cross-pollination by bumblebees. Spacing of one plant per 30 cm is recommended for initial evaluation.

Optimal photoperiod for induction of flowering in cowpeas is 8 to 14 hours (Wienk, 1963). The major portion of *Vigna* cultivars are day-neutral, but a few are definitely short-day types. Maturities range from less than 60 days up to 7 to 8 months, depending on genotype and environment (Rachie and Roberts, 1974).

Optimum temperatures for seed germination, pod set, and dry matter production are 27 to 29 C day/22 to 24 C night (Rachie and Roberts, 1974). Relative humidity exerts a very strong influence on pod set, because flower shedding can be significant with high temperatures and a low relative humidity. Sprinkler irrigation is an aid to maintenance of proper humidity during the flowering period. Light level is of little consequence in field production of cowpeas.

B. Growth Chamber and Greenhouse

When growing cowpeas in a growth chamber or greenhouse, every effort should be made to provide cultural and environmental conditions that closely parallel the field optimum. It is especially important that conditions be monitored during germination, and immediately before, after, and during the crossing period. Improper watering is perhaps the single most important factor resulting in poor growth and susceptibility to pests and diseases of plants grown in the greenhouse. It is also advisable to utilize a systemic insecticide, such as Temik, in a soil-less growth media. Photoperiods of 8 to 14 hours with temperatures of 27 C day/22 C night are desirable for satisfactory growth and development of most cultivars. Low light level during the winter, coupled with cool temperatures, can cause etiolation and reduce flowering and subsequent seed set following hybridization. Low light and high humidity can cause mildew problems; however, high humidity, along with moderate temperature, tends to favor seed set of hand-manipulated flowers. It is most desirable to carry out crossing in late spring, summer, and early fall.

III. FLORAL CHARACTERISTICS

The cowpea is self-pollinated and the process is complete before the flower opens. Cross-pollination is usually less than 1%, but will vary somewhat with the cultivar and, more particularly, with the bumblebee or wild honeybee population.

Flowers are born on racemose inflorescences at the ends of peduncles that arise from leaf axils. There are four to six units of flowers per inflorescence. Each unit is a simple raceme with 6 to 12 flower buds (Ojehomon, 1968a). The flower is large, has a straight keel, and is yellowish-white to purple in color. There are 10 stamens, 1 free and 9 fused; the ovary is multiloculate, and the style is bearded along the inner side, ending in an oblique stigma (Fig. 1). There is a high rate of abortion in cowpea, which can shed

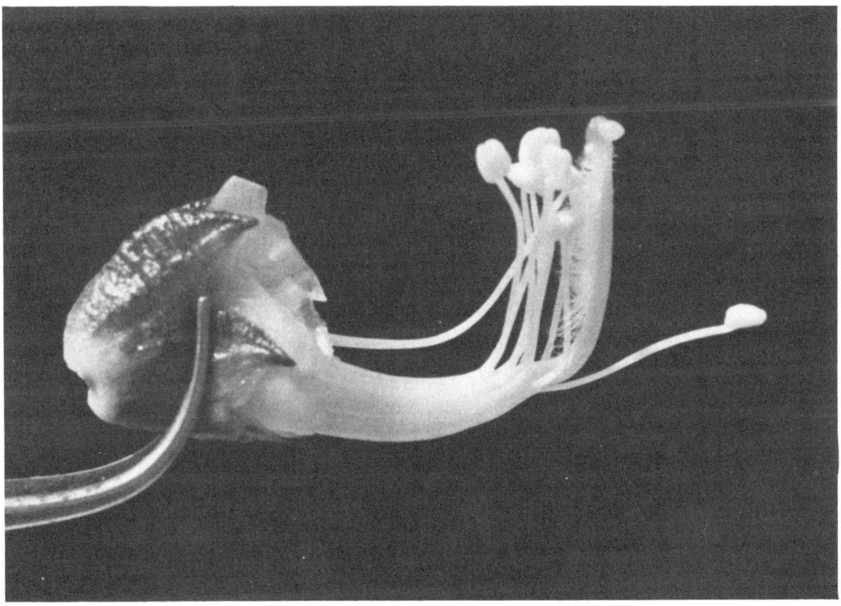

Fig. 1—Mature bud, with corolla removed to expose the 10 stamens, 9 with fused filaments and 1 free. The style has hairs on one surface and a stigma at its tip. The stylar hairs are not receptive to pollen.

70 to 88% of its 100 to 500 flower buds prior to anthesis. Of the remaining buds, about half may abort prematurely under certain environmental conditions, so that only 6 to 16% of the total flower buds produce mature fruits (Ojehomon, 1968b). The fruit is a cylindrical to flat pod, at times septate and dehiscent.

IV. ARTIFICIAL HYBRIDIZATION AND SELF-POLLINATION

A. Equipment

Equipment required for hybridization includes fine forceps, a small vial of alcohol, 2.5 × 3.0-cm marking tags of assorted colors, and soda straws or transparent adhesive tape.

B. Preparation of the Female

Flowers that will open in 1 day should be selected for hybridization. Flowers showing a colored corolla 2 to 2.5 times the length of the calyx will generally be at the proper stage for crossing (Fig. 2). The stigma is receptive

Fig. 2—Flower bud at the proper stage for emasculation.

the day before anthesis and generally to noon on the day of anthesis, depending on temperature and relative humidity. Under very dry conditions, the stigma rapidly loses receptivity.

Emasculation can be accomplished in two ways, depending on pollination methodology. If pollination is to take place in conjunction with emasculation, the standard is separated with forceps along the suture, unfolded, and spread back along with the wings (Fig. 3). The keel is removed with forceps, exposing the stamens and stigma. The stamen should be removed by firmly grasping the filaments and proceeding carefully to insure that all 10 are removed, no anthers burst, and all contact with the stigma is avoided. After emasculation, the stigmatic surface should be checked for the presence of pollen before cross-pollination is attempted. Forceps should be dipped in alcohol between emasculations. Scrupulous care in sterlizing hands, in addition to equipment used in emasculation and pollination, is absolutely essential.

If pollination is to take place on the day following emasculation, an incision can be made with a razor blade at the junction of the corolla and calyx, and the corolla removed with forceps. The stamens are often removed in this process, but generally will need to be removed with forceps. A

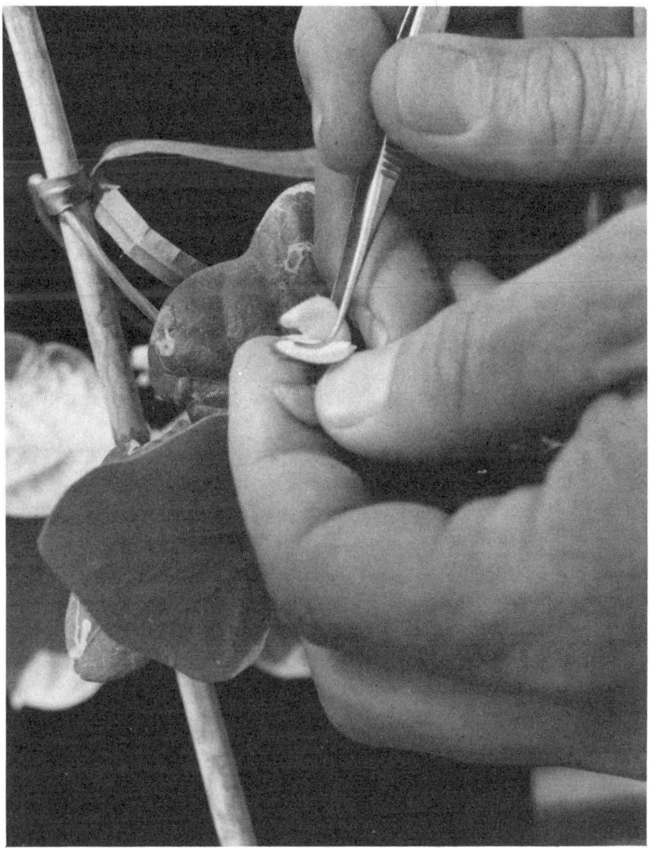

Fig. 3—Pushing back the wings and standard petals to expose the keel during emasculation.

Fig. 4—Placement of a soda straw over an emasculated bud.

Fig. 5—Mature flower from which pollen is available.

short length of soda straw is then pushed over the pistil, the upper end being pinched together with adhesive tape (Fig. 4). This reduces desiccation of the remaining plant tissue. A tag should be placed on the peduncle of the flower with the designation of female parent. The male parent is added to the tag after pollination has been made.

C. Pollination

With the soda straw technique, pollination takes place the morning following emasculation. When the soda straw technique is not used for emasculation, pollination should follow immediately.

Flowers from the male parents should be picked the morning of anthesis (Fig. 5). If pollen from these flowers is to be used for late afternoon pollination, they can be refrigerated in separate bags until needed. The pollen usually will remain viable in the flowers; however, storage for any length of time necessitates placement of anthers in small vials which are then placed in a desiccator over sulfuric acid or calcium chloride.

The pollen will accumulate in the keel, and in most cultivars, it is possible to squeeze pollen out of the keel like toothpaste from a tube, and transfer it directly to the stigmatic surface of the emasculated flower of the desired female parent (Smartt, 1976) (Fig. 6). As soon as the pollen has been applied, the soda straw should be replaced and left until fertilization has taken place (Fig. 4). If pollination is undertaken with emasculation, the

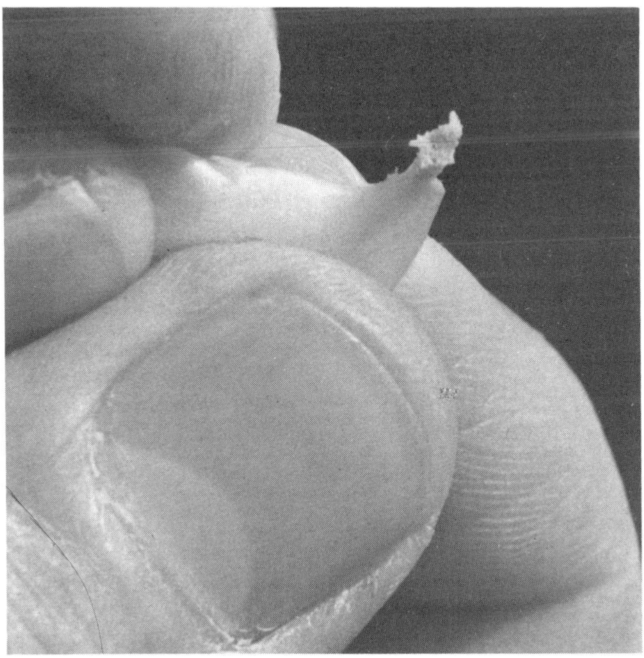

Fig. 6—Squeezing of pollen from the keel for direct transfer to the stigmatic surface of the desired female parent.

wings and standards are returned to the closed position after pollen transfer and secured with a 3 cm strip of adhesive tape to prevent desiccation and contamination by visiting insects (Fig. 7). Regardless of technique used, a tag should be attached at the time of emasculation. After pollination, female parent, male parent, and date should be recorded on the tag. Tags of varying colors may be used to code parents or to identify given crosses, such as F_1, BC_1, BC_2. The tape falls off when the petals separate from the pod. The soda straws will be pushed off by the developing pod. Under favorable conditions, a success rate of 50% can be expected (Fig. 8).

D. Factors Affecting Efficiency

Parents to be crossed should be located relatively close together, with individual plant spacing of 0.5 m within rows. Altering of flowering dates generally is not necessary because most cultivars flower over a sufficiently long period of time to obtain the desired crosses. Unused flowers should be removed daily to extend the flowering period and delay plant senescence. If

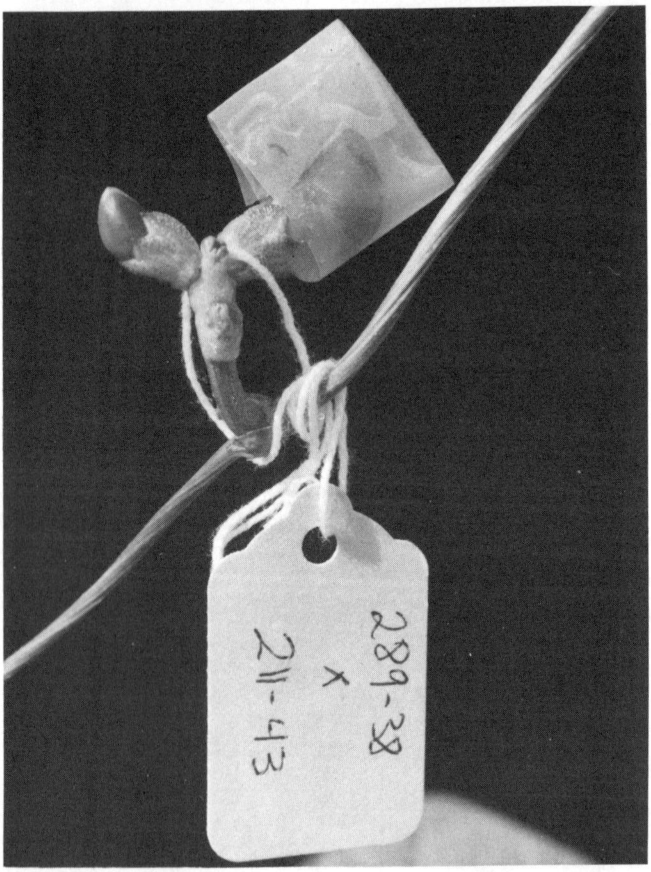

Fig. 7—Wings and standards returned to closed position following pollen transfer, and secured with adhesive tape to prevent desiccation.

flowering time of a parent is uncertain, it might be advisable to make successive plantings so that material of both is available at the critical time. Crosses involving short-day cultivars can be accomplished best during the winter months. Day-neutral types can be grown during the short day, making possible any desired combination.

Where plots are set up for progeny testing and seed increase, pod and seed color patterns have been quite useful for preventing chance mixtures. For roguing purposes, flower color is of value, as the cream types are a golden yellow color, the blackeye a light blue, and the purple hulls show a much more intense anthocyanin accumulation. If plots are rogued by flower color and subsequently by pod color markings, seed mixtures will be minimized. Inheritance patterns of genetic markers, as well as other cowpea characters, have been summarized (Rachie and Roberts, 1974).

A simple classification of the *Vignas* on the basis of pod and seed color has been developed by Brittingham (1946). A system of identifying cultivars through the use of seed shape and color markings has been provided by Ligon (1958).

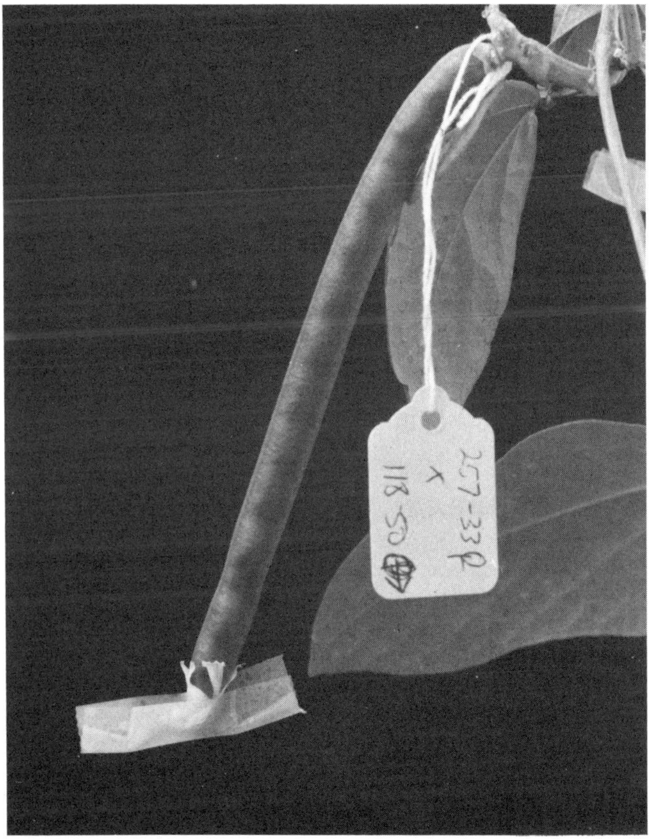

Fig. 8—Successful cross approximately 10 days after pollination.

V. NATURAL HYBRIDIZATION

The cowpea is highly self-pollinated and not suitable for hybrid seed production. Some effort has been expended at the IITA in the identification and development of male-sterile cultivars (Rachie et al., 1975). At this time, hybrid cowpea seed is not produced commercially.

VI. SEED DEVELOPMENT, HARVEST, AND STORAGE

Success of a cross is apparent as early as 2 to 3 days following pollination, and harvest of pods can take place in 3 weeks. The period of development from seed sowing to ripe pods ranges from 60 to 80 days in most cultivars.

The seed is viable and will dry out perfectly if taken from pods that are flacid and beginning to dry. Collection at this time will avoid possible loss from pod shattering, either from natural maturation or accidental striking when pods are dry.

Pods should be placed in individual kraft bags at harvest, and when sufficiently dry they should be crushed and threshed. This process can be accomplished either by hand, for individual pods or small quantities, or with a small mechanical thresher for increase plots. The seeds and debris can be separated either by use of seed sieves, wind, or an electric fan.

Individual seed lots should be placed in seed envelopes or suitable seed bags, perferably cloth with drawstring closure. The most satisfactory storage method is in cloth bags at 5 C. Before the seed is placed in storage, it should be fumigated with carbon bisulfide to prevent weevil infestation. The fumigation can be accomplished in large, airtight metal chambers. For small quantities, fumigation can be carried out in steel or plastic barrels covered with paper or plastic. Small seed lots can be stored at 14 C in sealed containers. Weevils will not hatch at this temperature; however, the eggs remain viable and will hatch soon after the seed is removed from storage. Fumigation with carbon bisulfide will kill both adult weevils and eggs.

VII. TECHNIQUES FOR SPECIAL SITUATIONS

The success of crosses is relatively low. With interspecific crosses, it might be desirable to perform bud pollination or employ artificial auxins to delay flower abscission for a few hours. Naphthalacetic acid in talc dusted into emasculated flowers reduces blossom drop and can result in 30% success in hand crosses (Rachie and Roberts, 1974).

Hormones also can be used when slow growth of pollen tubes results in failure of seed set. A saturated solution of indolebutyric acid and indoleacetic acid in glycerine can be used. The solution should be applied at the time of pollination on the abcission zone of the flower bud.

REFERENCES

Brittingham, W. H. 1946. A key to the horticultural groups of varieties of the Southern pea, *Vigna sinensis*. Proc. Am. Soc. Hort. Sci. 48:478–480.

Faris, D. G. 1963. The origin and evolution of the cultivated forms of *Vigna sinensis*. Can. J. Genet. Cytol. 7:433–452.

Gunn, C. R. 1973. Recent nomenclatural changes in *Phaseolus* L. and *Vigna* Savi. Crop Sci. 13:496.

Ligon, L. L. 1958. Characteristics of cowpea varieties (*Vigna sinensis*). Oklahoma Agric. Exp. Stn. Bull. B-518.

Ojehomon, O. O. 1968a. The development of the inflorescence and extra-floral nectaries of *Vigna unguiculata* (L.) Walp. J. W. Afr. Sci. Assoc. 13:93–114.

————. 1968b. Flowering, fruit production and abscission in cowpea, *Vigna unguiculata* (L.) Walp. J. W. Afr. Sci. Assoc. 13:227–234.

Rachie, K. O., K. M. Rawal, J. D. Franckowiak, and M. A. Akinpelu. 1975. Two outcrossing mechanisms in cowpeas, *Vigna unguiculata* (L.) Walp. Euphytica 24:159–163.

————, and L. M. Roberts. 1974. Grain legumes of the lowland tropics. Advances in Agron. 26:1–132.

Smartt, J. 1976. Tropical pulses. London, Longmans.

Vavilov, N. I. 1951. The origin, variation, immunity and breeding of cultivated plants. (Translated from Russian by K. Starr Chester). Chronica Bot. 13:1–364.

Wienk, J. F. 1963. Photoperiodic effects in *Vigna unguiculata* (L.) Walp. Meded. Landbouwhogesch. Wageningen 63:82.

Zary, K. W., J. C. Miller, Jr., R. W. Weaver, and L. W. Barnes. 1978. Intraspecific variability for nitrogen fixation in southern pea (*Vigna unguiculata* (L.) Walp.). J. Am. Soc. Hort. Sci. 103:806–808.

22 Crambe

Koert J. Lessman
W. Powell Anderson
New Mexico State University
Las Cruces

Crambe (*Crambe abyssinica* Hochst. Ex. R. E. Fries) is an erect annual herb with numerous branches that grows 60 to 90 cm tall and matures in about 90 days (Fig. 1). Interest in crambe lies in the usefulness of its seed oil per se, the erucic acid in this oil, and the residual meal after oil extraction (Downey, 1971). Crambe is one of the richest known sources of erucic acid, and this acid makes up 55 to 60% of the glycerides (Lessman and Berry, 1967; Mikolajczek et al., 1961). The current trend is toward the development of rapeseed cultivars low in erucic acid, but with improved nutritional quality of the oil as a food (Tallent, 1972).

I. PARENTAL MATERIAL

Crambe is believed to be a native of the Mediterranean area. Members of the genus *Crambe,* of which there are about 20 species (Mazzani, 1954), have been grown in Europe, Africa, the Near East, Central and West Asia, and North and South America (Brunn and Matchett, 1963; Mazzani, 1954). The species designation *abyssinica* might imply that the center of origin of this species of *Crambe* was Abyssinia, but this appears unlikely because Vavilo (1946) did not include it among plants considered for the Abyssinia Center.

Seed stocks of *C. abyssinica* were first introduced into the United States from Europe in the 1940's by the Connecticut Agricultural Experiment Station (White and Higgins, 1966). Evaluations of a number of strains of crambe began in earnest in 1958 in Texas and 1962 in Indiana.

Early attempts to evaluate crambe germplasm for genetic variability were not encouraging due to lack of significant differences among introductions of *C. abyssinica* for a large number of characters (Lessman and Meier, 1972; Papathanasiou et al., 1966).

In an attempt to obtain greater genetic variability in *C. abyssinica,* plant selections from PI 247310 (*C. abyssinica*) and PI 279346 (*C. hispanica*) were paired randomly and crossed reciprocally (Meir and Lessman,

1973a). The average seed set from these crosses was about 77%, with no reciprocal differences detected. Significant differences were found among F_3 and F_4 entries for eight characters evaluated, but not for oil percent (Meier and Lessman, 1973b).

Fig. 1—Upper portion of a plant of *Crambe abyssinica* (PI 281734) in flower. (Courtesy of George A. White, SEA-USDA, Beltsville, Md.)

Lessman (1975) evaluated 162 lines of *C. abyssinica* for genetic diversity of six characters. The lines evaluated were selected from a progeny test of 2,000 randomly selected plants out of PI 247310 and PI 279346. The characters evaluated were seed yield, test weight, plant height, oil percent, days to bloom, and percent glucosinolates. Statistical differences were detected among lines for all characters, except percent glucosinolates. These results indicated that progeny evaluation of individual plants was effective in separating genetic variation that exists in the original germplasm. Crossing among diverse germplasm, followed by selection, can be a practical step toward improvement in crambe.

The cultivar 'Prophet' was developed by mass selection for large seed size from PI 247310 (*C. abyssinica*), while 'Indy' was developed similarly for small seed size from PI 249346 (*C. hispanica*). In contrast, 'Meyer' was developed by selection among progenies from crosses of *C. abyssinica* and a *C. hispanica* type (Lessman, 1975). 'Indy' and PI 249346 are now believed to be ecotypes of *C. abyssinica,* rather than different species (Meier and Lessman, 1973a).

White and Solt (1978) reported that the chromosome number of *C. abyssinica* is n = 45, for *C. hispanica* var, hispanica is n = 30, and for *C. hispanica* var. glabrata is n = 15. Three ploidy levels of n = 15, 30, and 45 were found in *C. kralikii.* They also reported that the chromosome number of *Crambella teretifolia* is 11 and that of *Hemicrambe fructiculosa* is 9.

Downey (1971) stated that progress in improving crambe through breeding is hindered by its high level of polyploidy. With the basic chromosome number of crambe reported as 15, the chromosome complement of *C. abyssinica* as 90, and that of *C. hispanica* as 60, the level of ploidy tends to mask the variability that may lie within the species.

II. PLANT CULTURE

A. Field

Christmas et al. (1968) described cultural practices associated with crambe seed production. The following is a resume of their report.

Crambe is a cool-season crop that appears undamaged by temperatures as low as −5 C, but is killed by temperatures of −7 C or lower. Where the growing season is sufficiently long, two crops of crambe may be produced the same year. The first crop is planted in early spring after threat of temperatures below −5 C has passed, and the second around mid-July. Crambe may follow wheat as a second crop.

Seedbed preparation for crambe is similar to that for most other agronomic crops. Seeding rates of 6 to 40 kg/ha of hull-seed have been used with little difference in yields reported. The crop may be seeded broadcast, with a wheat drill or cultipacker seeder, or seeded in rows spaced 0.5 to 0.75 m apart using a corn planter equipped with corn, sorghum, or soybean seed plates. Crambe tends to lodge when planted in rows wider than 0.75 m. Optimum planting depth is about 2 cm, and the seed should not be planted deeper than 2.5 cm.

Well-drained, fertile soils are ideal for crambe production. The crop will not tolerate chronically wet or waterlogged soils and appears to withstand moderate drought conditions.

Data on fertilizer requirements of crambe are inadequate at present, but it appears that crambe's phosphorus and potassium requirements are similar to those of small grains. The crop has responded well to an application of 112 kg/ha of nitrogen.

Weed competition can be a serious problem in crambe production. Preliminary investigations indicate that the herbicides DCPA (Dacthal), propachlor (Ramrod), or a combination of the two will selectively control weeds in crambe (Ross and Williams, 1966). Profluralin (Tolban) provided excellent selective weed control in crambe grown in New Mexico in 1977.

B. Growth Chamber and Greenhouse

No one has reported culturing crambe species in growth chambers or greenhouses per se; however, crambe plants do not appear to be photoperiod-sensitive. Plants grown in continuous light in greenhouse show no

Fig. 2—Flowers and fruits on fruiting branchlets. (Courtesy of George A. White, SEA-USDA, Beltsville, Md.)

obvious modification of number of days from emergence to full-bloom. The observation, however, does not represent results of controlled experimentation.

Greenhouse culturing of crambe plants in pots follows standard practices, with good results obtained using standard greenhouse soil mixtures and two to three standard applications of Hoagland's solution during the growth period. The usual greenhouse pests may be controlled as for other plant species, with one exception. Crambe plants are severely injured or killed by use of standard greenhouse nicofume bombs for control of aphids and other greenhouse insect pests.

III. FLORAL CHARACTERISTICS

The inflorescence of crambe is a long panicle raceme (Fig. 2). Flowering is indeterminate, but early-formed pods usually remain attached until last-formed pods mature. Crambe flowers are typical of *Cruciferae* structure. They have four sepals about 2 mm long, four white petals about 3 mm long, and six stamens. The flowers are primarily self-pollinated, but some natural outcrossing does occur (Beck et al., 1975). The plants flower at or near the termini of their branches, producing a great number of seeds borne singly (Fig. 2).

IV. ARTIFICIAL HYBRIDIZATION AND SELF-POLLINATION

A. Equipment

Standard needle-nosed forceps are used to remove the anthers from each flower.

B. Preparation of the Female

All opened flowers and all pods previously formed on a raceme are removed. One to six (usually two to three) unopened flowers that will open within 24 hours are selected for emasculation. If flowers are in short supply, young flowers that will open within the next 2 to 3 days are retained and more immature flowers are removed.

The selected flowers are emasculated by removing the anthers from each flower with sharp-pointed tweezers.

C. Pollination

Pollination is accomplished by brushing pollen onto the stigmas of the emasculated flowers with an anther from a plant of the male parent.

Each emasculated and pollinated flower is enclosed by a small bag supported by a stake. To avoid wind damage to the bags, the plants are kept in a greenhouse.

D. Factors Affecting Efficiency

Crambe hispanica usually has densely pubescent leaves and stems, while those of *C. abyssinica* are glabrous (Leppik and White, 1975). Beck et al. (1975) segregated a selection of PI 279346 (*C. hispanica*) that produced only glabrous leaves. They found that pubescent-leaf was controlled by a single recessive gene and they suggest that this trait could serve as a suitable genetic marker in crambe studies.

V. NATURAL HYBRIDIZATION

The potential for natural hybridization of crambe appears practical. However, large-scale crossing will be dependent upon the isolation and development of a male-sterile system with identification and management of suitable vectors.

VI. SEED DEVELOPMENT, HARVEST, AND STORAGE

Crambe leaves turn yellow, and the plant defoliates as it approaches maturity. A few days after the last leaves have fallen, the seed pods and small branches turn light golden to straw in color. The seeds should be ready to harvest as soon as this color has progressed down the stem below the last seed-bearing branches.

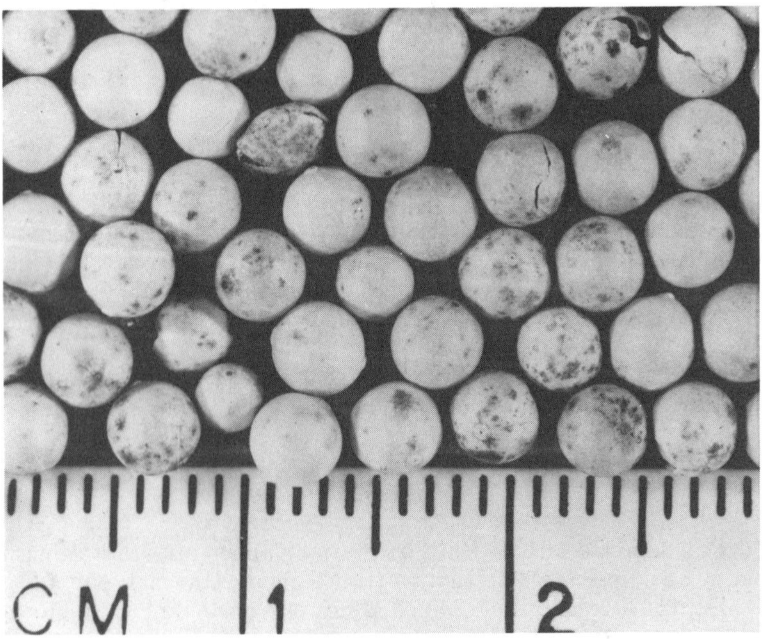

Fig. 3—Mature seeds, showing characteristic spherical fruits enclosing single seeds.

Crambe has been harvested in Indiana by combining directly, while in the Pacific Northwest it was necessary to swath crambe at harvest to facilitate drying.

The one-seeded fruit of crambe is spherical in shape and about 3.5 mm in diam (Fig. 3). The hull remains on the seed at harvest and is considered a part of the harvested product (Papathanasiou and Lessman, 1966). The pods (seeds plus hulls) weigh 7.0 to 7.5 g/1,000 seeds with the hulls accounting for 14 to 20% of the weight of pods grown in Canada (McGregory et al., 1961) and 25 to 40% for those grown in the United States (Earle et al., 1966).

REFERENCES

Beck, L. C., K. J. Lessman, and R. J. Buker. 1975. Inheritance of pubescence and its use in outcrossing measurements between a *Crambe hispanica* type and *C. abyssinica* Hochst. Ex. R. E. Fries. Crop Sci. 15:221-224.

Brunn, J. H., and J. R. Matchett. 1963. Utilization potential of *Crambe abyssinica.* J. Am. Oil Chem. Soc. 40:1-5.

Christmas, E. P., K. J. Lessman, C. B. Southard, and M. W. Phillips. 1968. Crambe: A potential new crop for Indiana. Purdue Univ. Agric. Ext. Serv. Bull. AY-168.

Downey, R. K. 1971. Agricultural and genetic potentials of cruciferous oilseed crops. J. Am. Oil Chem. Soc. 48:718-722.

Earle, F.E., J. E. Peters, I. A. Wolff, and G. A. White. 1966. Compositional differences among crambe samples and between seed components. J. Am. Oil Chem. Soc. 43:330-333.

Leppik, E. E., and G. A. White. 1975. Preliminary assessment of *Crambe* germplasm resources. Euphytica 24:681-689.

Lessman, K. J. 1975. Variation in crambe, *Crambe abyssinica* Hochst. J. Am. Oil Chem. Soc. 52:386-389.

———, and C. Berry. 1967. Crambe and vernonia research results at the forage farm in 1966. Purdue Univ. Agric. Exp. Stn. Res. Progr. Rep. 284. Project 846.

———, and V. D. Meier. 1972. Agronomic evaluation of crambe as a source of oil. Crop Sci. 12:224-227.

Mazzani, B. 1954. Introduction de plantas oleoginosas nuevas para Venzuela, *Crambe abyssinica* Hochst. Agron. Trop. 4:101-104.

McGregory, W. G., A. G. Plessers, and B. M. Graig. 1961. Species trials with oil plants. I. Crambe. Can. J. Plant Sci. 41:716-719.

Meier, V. D., and K. J. Lessman. 1973a. Breeding behavior for crosses of *Crambe abyssinica* and a plant introduction designated *C. hispanica.* Crop Sci. 13:49-51.

———, and ———. 1973b. Heritabilities of some agronomic characters for the interspecific cross of *Crambe abyssinica* and *C. hispanica.* Crop Sci. 13:237-240.

Mikolajczak, K. L., T. K. Miwa, F. R. Earle, I. A. Wolff, and Q. Jones. 1961. Search for new industrial oils. V. Oils of cruciferae. J. Am. oil Chem. Soc. 38: 678-681.

Papathanasiou, G. A., and K. J. Lessman. 1966. Crambe. Purdue Univ. Agric. Exp. Stn. Bull. 819.

Papathanasiou, G. A., K. J. Lessman, and W. E. Nyquist. 1966. Evaluation of eleven introductions of crambe, *Crambe abyssinica* Hochst. Agron. J. 58:587–589.

Ross, M. A., and J. L. Williams, Jr. 1966. Herbicides for weed control in crambe production. Unpublished data. Dep. Botany and Plant Pathology, Purdue Univ., Lafayette, IN 47907.

Tallent, W. H. 1972. Improving high-erucic oilseeds: chemically or genetically? J. Am. Oil Chem. Soc. 49:15–19.

Vavilov, N. I. 1946. The origin, variation, immunity, and breeding of cultivated plants. Chron. Bot. 13:35–36.

White, G. A., and J. J. Higgins. 1966. Culture of crambe: A new industrial oilseed crop. USDA-ARS Production Res. Rep. No. 95.

———, and M. Solt. 1978. Chromosome numbers in *Crambe, Crambella,* and *Hemicrambe.* Crop Sci. 18:160–161.

23 Field Pea

Earl T. Gritton
University of Wisconsin
Madison

The pea (*Pisum arvense* L.) is an annual herbaceous legume adapted to cool, moist climates. The field pea is grown for forage or for its dry, mature seeds which are used as food or feed, while the garden pea is grown for its green, succulent seeds which are used for food. The origin and progenitor of the pea are obscure, but it is one of the oldest of cultivated plants. It was cultivated in Neolithic farming villages of the Near East as early as 7000 to 6000 BC (Zohary and Hopf, 1973). Four centers of origin based on genetic diversity were listed by Vavilov (1926): Central Asia, the near East, Abyssinia, and the Mediterranean. Blixt (1970) indicated that the principal center of genetic diversity is the Mediterranean gene center with secondary centers in the Near East and Ethiopia.

I. PARENTAL MATERIAL

Peas are a legume belonging to the family *Papilionaceae.* One of the subdivisions of this family, *Viciae*, includes the genera *Pisum, Cicer, Vicia, Lens, Lathyrus,* and *Abrus.* Within the genus *Pisum,* Linnaeus distinguished two species, *P. arvense,* the field pea, and *P. sativum,* the garden pea. Species designation has since been given to *P. abyssinicum* (Braun), *P. aucheri* (Jaubert & Spach), *P. elatius* (Stev), *P. formosum* (Alefeld), *P. fulvum* (Sibth & Sm.), *P. humile* (Boiss & Noe), *P. jomardi* (Schrank), and *P. transcaucasicum* (Gov./Stankov). *Pisum aucheri* is synonymous with *P. formosum,* and this species, a perennial, was separated to form the monospecific genus, *Alophotropsis* (Biossier), by Lamprecht (1974). The *Pisum* taxa can be crossed easily and freely exchange all genes so far studied. All have the somatic chromosome number of 2n = 14. Because of the similarities among the species, the genus might best be considered as monospecific, in accordance with the view of Lamprecht (1974), who classified the different forms as ecotypes under *P. arvense.*

Some attempts have been made to cross *Pisum* with other genera of the *Viciae* tribe (McComb, 1975). *Vicia faba* pollen applied to *Pisum* stigmas produced no seeds, though hybrid embryos developed for up to 6 days (Gritton and Wierzbicka, 1975). Other purported hybridizations have not been confirmed.

Pisum germplasm, in the form of seeds, is available from a number of sources. Saint (1973) listed 31 collections in 20 countries, and additional collections are known to exist. The plant introduction station having primary responsibility for *Pisum* in the United States is the Northeast Region Plant Introduction Station, Geneva, N.Y., which maintains about 2,000 accessions. The Weibullsholm-Pisum Genetics Association collection, Landskrona, Sweden, maintains about 2,000 lines. The pea collection of the Institut fur Kulturpflanzenforschung, Gattersleben, DDR, contains about 1,600 lines. The Laboratoro Del Germplasm, Bari, Italy includes a *Pisum* collection.

The accessibility of diverse *Pisum* germplasm should be enhanced with the creation of the Pisum Genetics Association. This international organization is devoted to the study and preservation of *Pisum* material. Information concerning the genetics of *Pisum* and the sources of genetic stocks is published through the informal *Pisum Newsletter.*

II. PLANT CULTURE

A. Field

Peas are normally planted in the field early in the spring after danger of frost has passed, and develop best under cool, moist conditions. The soil fertility requirements of peas are lower than for many other crops, so a moderately fertile soil is adequate for good growth. Nitrogen seldom is applied because peas, in association with the proper *Rhizobia* species, are capable of fixing atmospheric nitrogen. Peas grow best in a soil that is moist, but not wet. Wet soils frequently produce unthrifty plants suffering from root rot.

Some of the plants in the field nursery usually are grown along trellis supports to facilitate observation, classification, hybridization, selection and maximum per plant yields of high quality seeds. Plant populations vary widely depending on the use to be made of the plants, but from one to nine seeds per 30 cm of row are common. Plants may be grown in a row along one or both sides of the trellis.

Single-row, unsupported plots spaced approximately 1 m apart and planted at the rate of about six seeds per 30 cm of row may be used for evaluation of lines and for seed increase. More advanced lines, with more seed available for planting, may be drilled in plots 5 to 20 m long in 10 to 12 rows spaced 18 cm apart.

Because of the labor, expense, and variability of results, limited use is made of winter nurseries in warmer parts of the United States, such as Florida, Arizona, and California. Where such nurseries are employed, they are used primarily for generation advance and seed multiplication, rather than hybridization. Lights are not generally used in the field.

B. Growth Chamber and Greenhouse

Greenhouses and growth chambers are routinely used in pea breeding because they permit two additional generations to be grown in the off-season. In some cases they are cooled for use during the summer, but more commonly they are used only during the winter. Plants may be grown in soil, an inert medium like sand, or in an aerated nutrient solution. Nutrients are supplied as needed. Moisture may be supplied through hand watering or any of various automatic devices, such as trickle irrigation. The medium should not become waterlogged because the plants will not do well.

Common temperatures are a constant 20 C, or 20 C day/13 C night. Photoperiod is about 16 hours, achieved by using any of various types of supplemental lighting in the greenhouse, and usually a combination of fluorescent and incandescent lights in the growth chamber. Low light levels result in unthrifty, etiolated plants. Air filters may be used where atmospheric pollution causes plant injury.

III. FLORAL CHARACTERISTICS

The pea usually is considered to be self-pollinated; however, cross-pollination can be quite extensive with some genotypes and environments. Reported percentages of cross-pollination range from zero (White, 1917) to 60 (Harland, 1948). The proportion generally reported with commercial cultivars is less than 1%. The preponderance of self-fertilization in the pea is due primarily to its cleistogamous nature. Pollination takes place approximately 24 hours before the flower opens (Cooper, 1938). Pollen placed on the stigma germinates in about 8 to 12 hours, and fertilization occurs about 24 to 28 hours after pollination (Gritton and Wierzbicka, 1975).

The inflorescence of the pea is a raceme arising from the axil of a leaf. The peduncle may vary greatly in length and often elongates from 1 cm when first exposed to 6 cm or more when fully elongated. Bracts may or may not develop at the base of the pedicel. There are usually one to three flowers per raceme, though some genotypes have more. The number of flowers per peduncle may vary from one to several on the same plant.

Floral structure of the pea is characteristic of a large number of plants belonging to the family *Papilionaceae* (Fig. 1). The green calyx is five-lobed, with two of the sepals behind the standard, two subtending the wings, and an anterior one subtending the keel. The standard is notched at the top in the middle and keeled at the back of the base. The two wings are smaller than the standard and enclose the two lower petals, which are fused, except at their base, to form a boat-shaped keel covering the pistil and stamens (Fig. 1A).

The filaments of 9 of the 10 stamens are joined for much of their length to form a staminal tube around the ovary. The tenth is free throughout its length. In the very young flower the filaments are shorter than the style, but have elongated by the time of pollen shedding so that the anthers are bent upward and tightly compressed with the style in the tip of the keel.

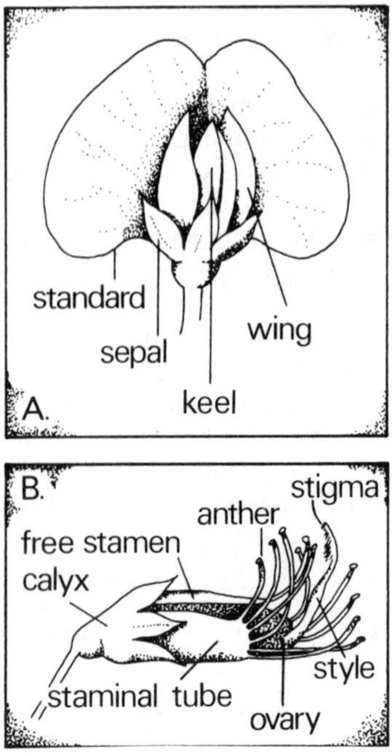

Fig. 1—Pea flower. A, Intact flower showing standard, wings, keel, and sepals; B, petals removed to show pistil and stamens.

The base of the pistil consists of a flattened green ovary containing from 5 to 12 ovules. Extending from the top of the ovary is a slightly flattened, cylindrical style bent at nearly a right angle to the ovary and recurved toward the ovary near its tip. On the inner side of this curve is a brush of stylar hairs. At the tip of the style is an elliptical viscous stigma (Fig. 1B).

The stigma of peas is receptive to pollen from several days before anthesis until about 1 day after the flower wilts (Warnock and Hagedorn, 1954). Pollen can effect fertilization from the time the anthers burst until several days thereafter.

The node at which the first flower is borne is quite regular for a given strain and is one of the characteristics used in describing cultivars. Environment does not alter the lowest node at which floral initiation occurs, node 5, but under a controlled environment of short days and warm nights individual plants have been obtained with 176 flowering nodes (Murfet, 1977). Under field conditions in temperate regions, the number of vegetative nodes before the first inflorescence may vary from 4 for the earliest lines to about 25 for the latest. Early cultivars characteristically produce their first flower at node 5 to 11, while late cultivars generally flower at node 14 or more. Floral formation proceeds sequentially up the stem, and few to many nodes may bear flowers, depending upon the strain and environmental conditions. Where more than one flower is borne per peduncle, the most mature flower is nearest the base of the peduncle and the youngest is at the tip. Thus, at a given node there may be a flower which has already shed pollen and whose wings and standard have already reflexed, a flower shedding pollen but whose wings and standard have not reflexed, and a flower whose anthers have not yet burst. Anthesis proceeds sequentially upward from node to node. Under rapid developmental conditions, only 1 day may separate anthesis between nodes, while with cool, moist, cloudy conditions the separation may be 5 to 6 days.

IV. ARTIFICIAL HYBRIDIZATION AND SELF-POLLINATION

A. Equipment

The equipment needed for artificial hybridization includes a pair of forceps with sharp tips, some alcohol (like 95% ethanol) for killing any pollen that might adhere to them, and tags to identify crosses. The forceps and alcohol are available from medical and scientific catalogs or sales rooms. The tags are available from office supply stores or catalogs.

B. Preparation of the Female

The plant to be used as female is chosen on the basis of vigor and stage of flower development. Greatest success in hybridization is realized from the lower buds on a plant which is just beginning to flower. Some of the last formed flowers at the top of the plant may give little or no seed set.

For best success in hybridization, the flower to be emasculated should have developed to the stage just before anther dehiscence. At this stage, the petals usually will extend beyond the sepals and the anthers can be removed without rupturing (Fig. 2A). Younger flower buds are more difficult to emasculate, are more easily damaged, and may set fewer seeds. Self-fertilization will have already occurred in more mature flowers (Fig. 2B). Where two or more flowers are borne on the peduncle, all but one may be snipped off to facilitate manipulation of the desired bud and to reduce competition for nutrients.

Emasculation may be carried out at any time of the day, but usually is performed immediately before pollination to avoid the necessity of protecting the flower from outcrossing and of identifying emasculated flowers. As the first step in emasculation, the sepal in front of the keel is torn away with the tip of the forceps to permit more open entry to the pistil and stamens. By positioning the forefinger behind the flower and the thumb in front of the flower, a light squeezing pressure can be applied to spread the standard and wings and expose the keel. At this point, some crossers like to bend the standard back and the wings out and hold them with the thumb and finger. The keel can be slit by inserting one of the sharp tips of the forceps about halfway down and slipping upwards or downwards (Fig. 2C). Care must be taken not to insert the tip so far into the keel that it damages the pistil. After the keel is slit, it can be spread apart by inserting both tips of the forceps and allowing them to open. The two halves of the keel can be held down with the thumb and finger for increased exposure of the pistil and stamens, or pressure can be applied with the finger and thumb positioned to spread the keel (Fig. 2C). The 10 anthers are removed by grasping the filaments with the forceps and gently pulling (Fig. 2D). Care must be taken not to overlook the free one. Generally it is not necessary to identify the emasculated flower unless pollination is going to be delayed.

An alternative emasculation procedure is reportedly faster. Instead of splitting the keel and removing the anthers with a forceps, the standard and wings are rolled back simultaneously and held with the thumb and index

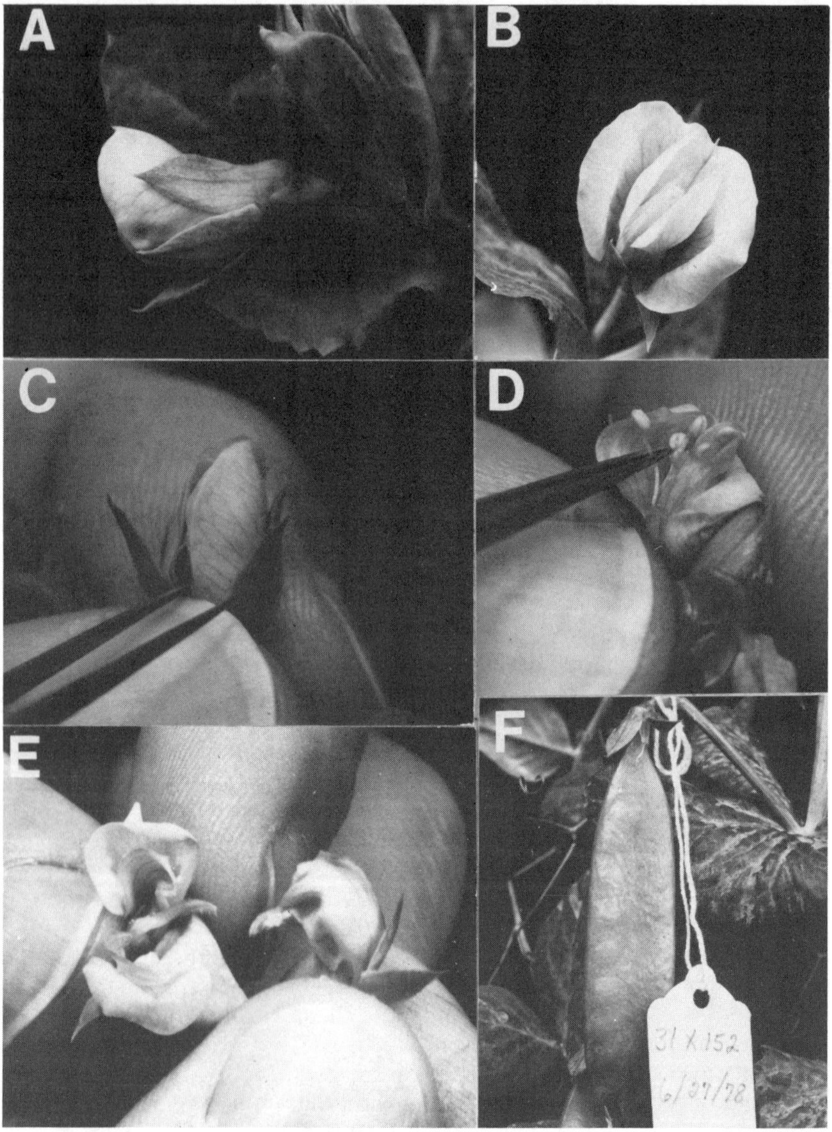

Fig. 2—Pea flower emasculation, pollination, and pod development. A. Bud at appropriate stage for emasculation and pollination; B, flower at appropriate stage for pollen collection; C, slitting the keel; D, removing anthers; E, pollinating emasculated flower; and F, developing pod with identification tag attached.

finger to expose the keel. With the thumb and index finger of the other hand, the upper half of the keel is grasped and torn off along with the anthers. The keel is usually torn near its middle. Care is taken to leave the lower half attached to the base of the flower. Often the single free anther remains in the flower, but it can be removed with the point of a pencil or a fingernail.

C. Pollination

Pollen for crossing can be obtained at any time of the day with the most abundant and viable pollen coming from freshly dehisced anthers. Genotype and environment influence dehiscence, but good pollen usually can be obtained from flowers just prior to their opening or just after. In the field, it is often desirable to emasculate and pollinate in the morning while relative humidity and turgor of the plant are high. In the growth chamber and greenhouse where moisture may be under more control, crossing may be carried on throughout the day.

Pollen can be stored for later use. Layne and Hagedorn (1963) found that untreated pollen could effect fertilization up to 6 days after dehiscence. They found that freeze-drying and vacuum-drying were equally effective as methods of preserving pea pollen, and that pollen longevity and fertilizing capacity were extended for approximately 1 year following vacuum drying and storage at −25 C. Pollen stored under a relative humidity of 35 to 40% at −5 C was still viable after 15 months (Warnock and Hagedorn, 1956).

The heavy, sticky pollen can be collected from a flower at about the stage shown in Fig. 2B and transferred on the tips of forceps, a toothpick, cotton swab, or other instrument. When the male flower can be sacrificed, it is plucked from the plant, the petals removed, and the stigma, style, and anthers used as a brush to transfer pollen to the stigma of the emasculated flower (Fig. 2E). Care must be taken to place the pollen on the stigmatic surface, which is located at the extreme tip of the style, not on the hairs on the concave side of the style which might appear to be the appropriate site (Warnock and Hagedorn, 1954). Pollen must be applied carefully to avoid injury to the style by forcing it too far backward. After pollination, it is important to fold the petals of the pollinated flower back into position to protect the stigma from dessication and to reduce exposure to foreign pollen. Isolation bags or other protective coverings generally are not used.

A tag attached to the pollinated flower serves to identify the cross (Fig. 2F). It should include the identification of the female and male parent, and other information as desired, including certain characteristics of the parents (e.g., disease resistance), date of the cross, and initials of person making the cross. The information can best be written on the tag before it is attached to the flower. The string of the tag should be formed into a loop, carefully slipped over the flower, and snugged around the pedicel. It is especially important that the tag be on the pedicel of the pollinated flower if other flowers on that peduncle are to be allowed to form pods. Absence of filaments and the anterior sepal will help identify the flower which was crossed.

A common error among beginners is to slip the loop over the growing point of the plant, in which case it ends up around the stem rather than on the pedicel of the pollinated flower. The loop of the string can be more easily slipped over the flower and pulled snug if it is first moistened with saliva by drawing it through the mouth. Strain and possible damage to the flower can be reduced by inserting the tips of the forceps into the loop in the string before pulling on it, then using the forceps to help close the loop.

Greatest success from crossing is realized under cool, bright conditions with vigorous plants in the early stages of blooming. Under these conditions, probably 80 to 100% of the pollinated flowers will set seed. Number of seeds per pod will vary with genotype and environment, but usually ranges from four to eight. Poorer results are obtained from crossing older plants under hot, dry, windy conditions or where insects and diseases may attack. These conditions, which are sometimes typical of fields in midseason, may result in only 20 to 50% pod set, with from one to four seeds per pod. Crosses involving small-flowered or fragile types as females also normally result in a low number of successful crosses.

D. Factors Affecting Efficiency

Lines to be crossed in the field should be located near each other to save time in checking flowering and transporting pollen. Lines of similar maturity might likewise be grouped.

The most common method of synchronizing flowering among strains is to use multiple planting dates. Flowering may be prolonged by removing flowers and preventing seed set. Where environmental conditions can be controlled, photoperiod and temperature can be used to adjust time to flower and node of first blossom (Murfet, 1977).

Differences in chromosome structure are common and translocation sterility can be quite high (90%) in the F_1 (Lamprecht, 1974), which should be considered in planning the number of crosses to be made.

A multitude of well-defined genetic markers are available in *Pisum* (Yarnell, 1962; Blixt, 1972). Many of these have been identified as to linkage group, while others await further investigation. The dominant alleles of *r*, *rb* and *i* are expressed in cotyledons of the first filial generation so are especially useful in identifying hybrid seeds following cross pollination. The recessive genes *r* and *rb*, which are inherited independently, both result in the dry seed being wrinkled and the dominant allele causes the seeds to be round. The recessive *i* causes the cotyledons to be green, and *I* results in yellow cotyledons. In some genetic backgrounds, the green of *i* may bleach in light over time and become indistinguishable from the yellow of *I*. Because of the masking effect of the seed coat, it is usually necessary to chip through the testa to expose a portion of the cotyledon when scoring for *I* or *i*.

Color in the flower, seed, or plant is dependent upon a basic anthocyanin gene *A*. Presence or absence of color can be used to identify hybrids where parents differ in this gene and the female is recessive. Plants can be scored easily and quickly for flower color, so it is a popular marker. Many genes affecting the expression of color in plant, flower, or seed are also useful markers.

V. NATURAL HYBRIDIZATION

Natural hybridization has not yet been used in pea breeding programs because of lack of pollination control. Incompatability systems are not known in *Pisum* and adaptable male sterility has not been available. Recently, a simple recessive type of genetic male sterility may have been discovered (Auld, 1978). Although pollen transfer would be dependent upon insects, this will probably not present an insurmountable obstacle to the use of functional male sterility.

VI. SEED DEVELOPMENT, HARVEST, AND STORAGE

The developing pod of successfully pollinated flowers will begin to extend past the tip of the wilted petals in 2 to 4 days. Flowers in which seed development is not initiated will begin to wilt and dry up 2 to 4 days after anthesis, and generally will drop off within a week. The number of ovules developing within the pod can be ascertained by the number of bulges they form in the wall of the pod and by the ovules' silhouettes when the pod is backgrounded by strong light.

A few ovules may be capable of normal germination 18 days after pollination; however, development for 24 to 36 days generally is required for maximum germination. The stage for maximum germination approximates the stage of maximum seed size and dry weight accumulation. Seeds are usually harvested when the pods have turned brown and brittle, and the seeds have dried to a moisture content of 13% or less. Seeds from hand-pollinated pods may be shelled out by hand and the crossing tag placed in the packet along with the seeds. Where large numbers of single plants are selected, a small belt thresher or cylinder thresher may be used. Larger plots are pulled by hand and threshed with a stationary thresher, or a grain combine may be employed to harvest the standing crop.

Pea seeds do not exhibit dormancy or afterripening, though some round-seeded forms may produce seeds with impermeable coats which require scarification for rapid and uniform germination. In temperate climates, sound pea seed may be stored under ambient conditions for 1 year or sometimes 2 years without serious loss in viability; however, high temperature and high humidity contribute to rapid loss in viability. Seed has been successfully stored for 10 years at −18 C and low relative humidity (Blixt, 1974).

VII. TECHNIQUES FOR SPECIAL SITUATIONS

Embryos of pea can be excised and cultured (Bonner and Axtman, 1937). Grafting has been employed (Paton and Barber, 1955). Plants have been vegetatively propagated by rooting stem cuttings (Richter, 1977), and mature plants producing flowers and pods have been obtained from tissue culture of shoot apices (Gamborg et al., 1974).

The assistance of S. Blixt, Landskrona, Sweden, has been especially helpful in preparing this chapter.

REFERENCES

Auld, R. 1978. Personal communication. Dep. of Plant and Soil Sciences, Moscow, ID 83843.

Blixt, S. 1970. *Pisum.* p. 321–326. *In* O. H. Frankel and E. Bennett (eds.) Genetic resources in plants—their exploration and conservation. International Biological Programme. Blackwell Scientific Publications, Oxford.

———. 1972. Mutation genetics in *Pisum.* Agric. Hort. Genet. 30:1–293.

———. 1974. Influence of genotype and storage conditions on germinability of pea seeds. Pisum Newsl. 6:4.

Bonner, J., and G. Axtman. 1937. The growth of plant embryos *in vitro*. Preliminary experiments on the role of accessory substances. Proc. Natl. Acad. Sci., USA 23:453–457.

Cooper, D. C. 1938. Embryology of *Pisum sativum.* Bot. Gaz. 100:123–132.

Gamborg, O. L., F. Constabel, and J. P. Shyluk. 1974. Organogenesis in callus from shoot apices of *Pisum sativum.* Physiol. Plant. 30:125–128.

Gritton, E. T., and B. Wierzbicka. 1975. An embryological study of a *Pisum sativum* × *Vicia faba* cross. Euphytica 24:277–284.

Harland, S. C. 1948. Inheritance of immunity to mildew in Peruvian forms of *Pisum sativum.* Heredity 2:263–269.

Lamprecht, H. 1974. Monographie der gattung *Pisum. In* K. Mecenovic (ed.) Steiermarkische Landesdruckerei, Graz, Austria.

Layne, R. E. C., and D. J. Hagedorn. 1963. Effect of vacuum-drying, freeze-drying, and storage environment on the viability of pea pollen. Crop Sci. 3:433–436.

Murfet, I. C. 1977. The physiological genetics of flowering. p. 385–429. *In* J. F. Sutcliffe and J. S. Pate (eds.) The physiology of the garden pea. Academic Press, N.Y.

McComb, J. A. 1975. Is intergeneric hybridization in the Leguminosae possible? Euphytica 24:497–502.

Paton, D. M., and H. N. Barber. 1955. Physiological genetics of Pisum. I. Grafting experiments between early and late varieties. Aust. J. Biol. Sci. 8:231–240.

Richter, E. 1977. On the vegetative propagation of pea plants. Pisum Newsl. 9:49.

Saint, S. 1973. World list of grain legume collections. p. 28–38. *In* Plant Genetic Resources Newsl. FAO, Rome.

Vavilov, N. I. 1926. Studies on the origin of cultivated plants. Bull. Appl. Bot. Pl. Breed. 16:139–248.

Warnock, S. J., and D. J. Hagedorn. 1954. Stigma receptivity in peas (*Pisum sativum* L.). Agron. J. 46:274–277.

———, and ———. 1956. Germination and storage of pea (*Pisum sativum* L.) pollen. Agron. J. 48:347–352.

White, O. 1917. Studies of inheritance in *Pisum.* II. The present state of knowledge of heredity and variation in peas. Proc. Am. Phil. Soc. 56:487–588.

Yarnell, S. H. 1962. Cytogenetics of the vegetable crops. III. Legumes. A. Garden peas, *Pisum sativum* L. Bot. Rev. 28:465–537.

Zohary, D., and M. Hopf. 1973. Domestication of pulses in the old world. Science 182:887–894.

24 Flax

Benjamin H. Beard
USDA—SEA and University of California
Davis

Verne E. Comstock
University of Minnesota
St. Paul

Remains of a *Linum*-type plant have been found in archaeological digs of Stone Age man in Switzerland, but some believe cultivation of flax (*Linum usitatissimum* L.) originated in Asia. The crop has had a multipurpose role for centuries, although it seems probable that it was first used for food, then as a source of fiber for clothing and other textiles, and finally as a source of oil and protein.

Flax is grown on all continents, but the leading countries in flax seed production in order of importance during recent years are India, Argentina, USSR, Canada, and the United States. Flax is grown for fiber production, but its use for that purpose is limited. The seed oil is an excellent vehicle for paints and varnishes, but it is also used in the manufacture of linoleum, patent leather, soap, and an anti-spalling protectant for concrete. The residue from oil extraction is known as linseed meal, a highly-prized protein supplement for cattle, especially for show animals.

I. PARENTAL MATERIAL

The genus *Linum* of the Linaceae family has over 200 species. *Linum usitatissimum* is the only widely grown species, although *L. angustifolium* Huds. (both n = 15) has been cultivated in some areas. *Linum usitatissimum* is the only species with nondehiscent or semidehiscent capsules suitable for modern commercial harvesting.

Crosses between species with 15 chromosomes have been successful (Tammes, 1928; Gill and Yermanos, 1967a; Bari and Godward, 1970; Seetharam, 1972; Wicks and Hammond, 1978). Crosses between *L. usitatissimum, L. angustifolium* Huds., *L. africanum* L., *L. corymbiferum* Desf., *L. decumbens* Desf., *L. hispanicum* Mill., *L. humile* Mill., *L. nervosum* Waldst., and *L. pallescens* Ledeb. are possible either directly or through bridge crossing. The F_1 progeny are fertile. Gill and Yermanos (1967a) speculate that speciation in the 15-chromosome group has resulted mainly from chromosomal reorganization.

Crosses between *Linum* species with nine chromosomes are also frequently successful, but crosses between nine-chromosome species and *L. usitatissimum* or any other species with 15 chromosomes have not produced seeds (Gill and Yermanos, 1967b). Unfortunately, characteristics of most interest that are not found in *L. usitatissimum* are present only in some nine chromosome species.

Information regarding availability and extent of *Linum* germplasm collections can be obtained from U.S. Department of Agriculture, Science and Education Administration, Agricultural Research, Washington, DC 20250.

II. PLANT CULTURE

A. Field

Flax does not have special soil or fertility requirements, but should be supplied adequate quantities of the essential elements. Flax is not particularly drought-resistant; therefore, irrigation may be necessary to insure adequate soil moisture for optimum seed set. Moisture stress during flowering and early boll development will cause abortion of flowers and young seeds. Low relative humidity will favor pollen shedding, but crosses can be accomplished under high humidity if the temperature is 15 C or above.

A hill or a 1-m row of a parent will provide sufficient flowers for crossing in the field (Dillman and Brinsmade, 1938). If only single plants from a seed source are desired, early thinning to space plants 15 to 30 cm apart is advisable. Adequate space for working is the main consideration for row or hill spacing. Two rows 30 cm apart with 90 to 100 cm between paired rows will provide ample work space.

In areas with summer storms, parent plants should be tied to bamboo stakes to prevent lodging and loss of crossed bolls.

B. Growth Chamber and Greenhouse

Flax will flower and set seed under a wide range of environments and photoperiods. Day lengths exceeding 12 hours and temperatures between 12 and 24 C are considered optimum. At temperatures below 10 C, the anthers will not dehisce.

Soil for greenhouse or growth chamber culture should be sterilized because flax seedlings tend to be susceptible to many soil-borne pathogens. A standard greenhouse mix of soil, sand, and peat moss is satisfactory, but may require fertilizer at monthly intervals. Clean sharp sand, vermiculite, etc., can also be used as the planting medium. Half strength Hoagland's solution should be applied to sterile media at weekly intervals (Hoagland and Arnon, 1950).

Care in watering is very important. Moisture stress from either overwatering or underwatering will cause abortion of flowers and young seeds.

Three to four pots of each parent will provide sufficient flowers for crossing in a growth chamber or greenhouse. Plantings should be repeated at intervals of 3 to 5 weeks if the relative maturities of the parents are unknown. If parental seed is limited, two plants of each parent cultivar will provide an adequate number of flowers for a modest crossing program.

The stage of bloom of the female plants affects the success-to-failure ratio. Early flowers emasculated and hand pollinated will more likely set bolls and have more seeds per boll than later flowers.

III. FLORAL CHARACTERISTICS

Flax flowers open in the early morning with the time dependent on temperature and whether the sky is cloudy or clear (Dillman, 1938). Flowers open about 20 min after sunrise and anthers dehisce 20 min later on a clear day as the temperature gradually increases from a low of 10 C (Dillman, 1938; Howard et al., 1919). Flower opening and pollen shedding are delayed slightly on cloudy days or with cooler temperatures. Petals fall and pollen disappears after 1030 to 1200 hours under clear skies.

The senior author has observed that during foggy, cool weather (10 to 12 C) at La Jolla, Calif., flax flowers do not open until 2 to 3 hours after daylight. Petals remain on the flowers until late afternoon or even until the following day, and pollen remains viable until 1500 to 1600 hours. On sunny days with slightly higher temperatures, flowers open 30 to 40 min after sunrise and petals fall before noon.

Flax is predominately self-pollinated, but outcrossing (less than 2%) occasionally results from insect activity. Dillman (1928) clipped styles just above the ovary at regular intervals after pollination and found that fertilization takes place 3 to 4 hours after pollen shed. The senior author has obtained similar results using emasculated flowers and hand pollination.

Flax flowers are regular, perfect, and borne in terminal, multiflowered panicles (Bailey, 1949). The five sepals, petals, and stamens are attached in whorls below the ovary. The sepals are overlapping, persistent, shorter than the capsule at maturity, and alternate with the petals. The stamens alternate with the petals and are united at the base of the filaments. There are five pistils, each with an ovary, style, and stigma. The ovaries are divided by false septa into two locules each of which contains an ovule. Dillman (1938) described and illustrated four flower types: common or funnel form, tubular, star-shaped or crimped, and large disk-shaped. A common or funnel form is shown in Fig. 1. The seed capsule is commonly called a boll.

IV. ARTIFICIAL HYBRIDIZATION AND SELF-POLLINATION

A. Equipment

Equipment for crossing flax is minimal and inexpensive. Some researchers use a sharp pencil to remove the anthers and colored string or small tags to identify bolls after cross-pollination. Others use small pointed scissors to cut off the sepals, a narrow pointed tweezers for emasculation, and a bent piece of aluminum for attaching the identification tags (Davidson and Yermanos, 1965).

B. Preparation of the Female

Some researchers prepare the female flowers in late afternoon (1500 hours to dark), and pollinate the following morning. Others emasculate and immediately pollinate in a two-step operation during the same morning. With either method, an unopened bud is selected that has the petals protruding about 30 to 60 mm past the sepals (Fig. 2A). The petals are removed to prepare the flower for emasculation. The tip ends of the petals are grasped with the tweezers and removed with a gentle twisting pull. If the tweezers are placed too far down toward the sepals, the anthers and styles will be damaged. If the tweezers are placed too far out on the tip, only small pieces of the petals will be removed. It is possible with practice to remove all of the petals at one time without damage to anthers or styles.

Fig. 1—A fully opened flax flower and one with petals and sepals removed. A centimeter scale is shown.

Another procedure is to peel back the sepals and hold them between the thumb and index finger. The petals overlap in this stage and the tweezers can be inserted between the petals about halfway between the bent back sepals and the tips of the petals. Removal of the first petal exposes anthers

Fig. 2—Flax flowers in various stages of hybridization. A) Flower at proper development for emasculation; B) flower with petals removed, anthers extending beyond the sepals and removed with small tweezers; C) emasculated flower to be used as a female parent; D) pollination by brushing the anthers of the male flower against the stigma of the emasculated female flower. (Photographs provided by W. H. Isom, Univ. of California, Riverside.)

and stigmas. This enables the inexperienced worker to remove the rest of the petals without damage to the stigma. With practice, all of the petals can be removed in one step. A flower with petals removed is shown in Fig. 2B.

The next step is to remove the anthers without damage to the stigma. The filaments usually are separated from the style, and anthers can be removed easily either one at a time or in groups of two or three. Although there are usually five anthers, each flower should be thoroughly examined to be sure all anthers have been removed (Fig. 2C). A few flowers have more than five anthers.

Female flowers should be identified immediately after emasculation. Individual flowers are borne on fragile pedicels that will not support large tags or markings, yet the identification must be placed in such a way that other flowers on the same panicle are not mistaken for the emasculated flower. The identification must be gently and carefully attached to avoid breakage of the pedicel. If tags are used the female identification and the date should be written on one side of the tag before it is put on the pedicel.

Some early workers covered the emasculated flower with a short piece of soda straw, but today most researchers consider this unnecessary. The senior author obtained no seed from 200 emasculated flowers left unprotected and unpollinated. Researchers in a new area should determine if insects transfer pollen to emasculated flowers.

C. Pollination

Stigmas are receptive from the day before the flower opens until the morning of the day after pollen shed. The stigma shrivels the day after pollination. If flowers are emasculated in the late afternoon of the day before the flower opens, the stigma tips will be together. The stigma tips often will be split the following morning. The senior author uses emasculation in the afternoon and feels that split stigmas are receptive and that good seed set follows their pollination. However, Hammond (1978) uses emasculation in the morning and abandons split stigmas because of low seed set.

Under normal conditions, flax flowers open in the early morning and the petals drop in late morning or afternoon, thus any flower with petals has usable pollen. Flax pollen is short-lived, but adequate pollen is generally available from the time the anthers dehisce until the petals start to fall. Pollen viability can be maintained for a few additional hours by storing flowers from the male parent in iced, insulated boxes. Such storage doubles the time available for pollination of emasculated flowers. Pollen availability can be checked by touching an anther to one's thumbnail and observing pollen grains left there.

The petals are removed from the flower to be used for pollen and the anthers rubbed against the stigma of the emasculated flower (Fig. 2D). One male flower can be used to pollinate two or more emasculated flowers, but seed set decreases following the second pollination. If male flowers are limited, a single anther can be removed and used to pollinate an emasculated flower. One male flower can be used to pollinate five emasculated flowers without a noticeable decrease in seed set. If a single anther is used to pollinate two emasculated flowers, the seed set of the second will average about half that of the first. No protection is needed for the pollinated flowers.

There are probably as many methods of identifying crosses as there are persons making them. Two aspects to consider are the identification of the parents of the boll that develops from a cross and records to insure all desired cross combinations are obtained. Short pieces of colored string tied around the pedicel will identify crossed flowers. Different colors of string will identify dates or particular crosses. Small jewelers tags are frequently used and are easily attached to the proper pedicel (Davidson and Yermanos, 1965). These tags do not break the pedical or blow off in the field during high winds. They are large enough to record the parental combination and date of emasculation. The female parent is recorded on one side of the tag at the time of emasculation, and after pollination the identification for the male parent, preceded by 'X' is put on the opposite side. The tag is then folded over the male identification. Folding facilitates determination of flowers that have been pollinated and protects the male identification from sun and rain.

Records of crosses can be kept by listing all female parents on the left side of a sheet of paper and all male parents across the top. Lines are drawn from each parent to form boxes that indicate each desired combination. When a pollination is completed, the date is written in the appropriate square. When the cross is checked for seed set, the date of successful crosses is circled with a red pencil, and the date of unsuccessful crosses is erased. The status of all desired combinations is available at all times on a single page.

Successful hybridization depends on the maintenance of nearly optimum conditions for the plants. Disease, insect, or environmental stresses may cause early abortion. Success rates of experienced researchers may vary among locations and seasons. Beard and Abel (1967) obtained bolls from 84% of their pollinations with an average of 7.7 seeds per boll from crossed flowers compared to 8.8 seeds per boll from flowers allowed to self naturally. Seedling emergence from both crossed and selfed seeds was 70% under field conditions.

D. Factors Affecting Efficiency

The primary consideration for a crossing block is to allow sufficient space to work on one parent without destroying the plants in an adjacent plot. Having the male plants in close proximity to the female parent can save some time, but storage of selected male flowers in an insulated ice chest allows for use of widely separated plots. Many researchers use the end plants of the border rows of yield test plots for crossing.

In the field, flowering of different parents will generally overlap. Plants will bloom for 10 to 14 days and three or four plants per cultivar will provide sufficient flowers for a crossing program. Flowering of flax plants can be prolonged if all flowers are removed. Flowering can be delayed by cutting the main stem and basal branches one or two nodes above the cotyledonary node. This technique will delay flowering for 2 weeks to a month or more.

Flower color or seed color differences are convenient genetic markers for identification of hybrids. Rust genes also can be used. Beard and Comstock (1965) list other genetic markers that may be useful. Complete emasculation of a flax flower is easily accomplished, thus there is a low probability that selfed seed will develop. Emasculated flax flowers are not pollinated with wind or insect-borne pollen.

V. NATURAL HYBRIDIZATION

The first proven cytoplasmic-genetic male sterility was found in flax (Bateson and Gairdner, 1921), however, it has not been used for production of hybrid seed. Commercial hybrid seed production would be a useful development (Hoffmann, 1961; Kenaschuk, 1975). Flax yields are generally low and increased yield from F_1 plants has been reported (Carnahan, 1949; Kasim, 1964; Shehata and Comstock, 1971). Two barriers to hybrid seed production are heavy, sticky pollen that is probably transferred only by insects, and closed or tubular flowers associated with male sterility. Duvick (1959) suggested selection might change the male-sterile flowers from closed or tubular to funnel form, as in *Nicotiana.* Further studies with flax indicated petal configuration could be changed (Comstock, 1970; Dubey and Singh, 1966; Kumar, 1971; Kumar and Singh, 1970, 1972; Thompson and Miller, 1978). Field layouts, isolation requirements, and other procedures for hybrid seed production have not been determined.

VI. SEED DEVELOPMENT, HARVEST, AND STORAGE

Fertilization is completed about 3 hours after pollination (Dillman, 1938). Observations of a single boll 3 days after pollination will usually allow one to discriminate between successful and unsuccessful hybridizations. After 5 days, boll development is a positive indication of success or failure. Seeds set in only a few locules can be determined after 5 or 6 days by the presence of a misshapen boll.

Bolls can be harvested when their color changes from green to yellow, although the boll is not fully mature until it becomes brown. Seed will not germinate when harvested before the boll changes color from green to yellow.

The bolls containing hybrid seed can be crushed with the fingers and the seeds separated from the chaff by hand. Care must be exercised at all times to avoid contamination, because the seeds can accidentally be carried from one lot to another without being noticed. The identification tag should be kept with the boll and later with the cleaned seed. Seeds are usually stored in small coin envelopes.

VII. TECHNIQUES USED FOR SPECIAL SITUATIONS

Laibach (1929) described the use of embryo culture to obtain a species cross, but the technique has not been fully exploited for flax crosses. Flax plants can be grafted, but this technique has not been useful for hybridization.

REFERENCES

Bailey, L. H. 1949. Manual of cultivated plants. The MacMillan Company, N.Y.

Bair, G., and M. B. E. Godward. 1970. Interspecific crosses in *Linum*. Euphytica 19:443–446.

Bateson, W., and A. E. Gairdner. 1921. Male-sterility in flax, subject to two types of segregation. J. Genet. 11(3):269–275.

Beard, B. H., and G. H. Abel. 1967. Labor requirements for manually crossing flax (*Linum usitatissimum* L.). Crop Sci. 7:268–269.

———, and V. E. Comstock. 1965. Flax genetics and gene symbolism. Crop Sci. 5:151–155.

Carnahan, H. L. 1949. The inheritance of oil content and other characteristics in flax and association between characters in the cross of Dakota × Minerva. Ph.D. Thesis. Dep. of Agronomy and Plant Genetics, Univ. of Minnesota, St. Paul.

Comstock, V. E. 1970. Natural crossing onto male-sterile flax and association of male sterility with seed color. Proc. 40th Annu. Flax Institute of the United States.

Davidson, J. M., and D. M. Yermanos. 1965. Flowering pattern of flax (*Linum usitatissimum* L.). Crop Sci. 5:23–28.

Dillman, A. C. 1928. Daily growth and oil content of flax seeds. J. Agric. Res. 37: 357–377.

———. 1938. Natural crossing in flax. Agron. J. 30:279–286.

———, and J. C. Brinsmade, Jr. 1938. Effect of spacing on the development of the flax plant. Agron. J. 30:267–278.

Dubey, D. K., and S. P. Singh. 1966. Use of cytoplasmic male sterility for the production of hybrid seeds in flax (*Linum usitatissimum* L.). Crop Sci. 6:125–126.

Duvick, D. N. 1959. The use of cytoplasmic male-sterility in hybrid seed production. Econ. Bot. 13:167–195.

Gill, K. S., and D. M. Yermanos. 1967a. Cytogenetic studies on the genus *Linum*. I. Hybrids among taxa with 15 as the haploid chromosome number. Crop Sci. 7:623–627.

———, and ———. 1967b. Cytogenetic studies on the genus *Linum*. II. Hybrids among taxa with nine as the haploid chromosome number. Crop Sci. 7: 627–631.

Hammond, J. J. 1978. Personal communication. Dep. of Agronomy, North Dakota State Univ., Fargo, ND 58102.

Hoagland, D. R., and D. I. Arnon. 1950. The water-culture method for growing plants without soil. California Agric. Exp. Stn. Circ. 347.

Hoffman, W. 1961. Lien (*Linum usitatissimum* L.) Handb. Pflanzenzuchtung 5: 264–266.

Howard, A., G. L. C. Howard, and A. R. Khan. 1919.Studies in the pollination of Indian crops. Mem. Dep. Agric. India Bot. Ser. 10:208–211.

Kasim, M. H. 1964. The analysis of yield and its components and seed quality characteristics in diallel crosses among 10 varieties of flax. Ph.D. Thesis. Univ. of Minnesota (Libr. Congr. Card No. NUC 65-94943). Univ. Microfilms. Ann Arbor, Mich. (Diss. Abstr. 64:10,830).

Kenaschuk, E. O. 1975. Flax breeding and genetics. Chap. 7. *In* Oilseed and pulse crops in western Canada. Western Co-operative Fertilizers Limited, Box 2500, Calgary, Alberta T2P 2N1.

Kumar, S. 1971. Male sterility of flax—a multiple allelic interaction. Agra Univ. J. Res. (Science) 20:7–10.

————, and S. P. Singh. 1970. Inheritance of male-sterility in some introduced varieties of linseed (*Linum usitatissimum* L.). Ind. J. Agric. Sci. 40:184–191.

————, and ————. 1972. Inheritance of partial male-sterility in linseed (*Linum usitatissimum* L.). Ind. J. Agric. Sci. 42:34–38.

Laibach, F. 1929. Ectogenesis in plants: methods and genetic possibilities of propagating embryos otherwise dying in the seed. J. Hered. 20:201–208.

Seetharam, A. 1972. Interspecific hybridization in *Linum.* Euphytica 21:489–495.

Shehata, A. H., and V. E. Comstock. 1971. Heterosis and combining ability estimates in F_2 flax populations as influenced by plant density. Crop Sci. 11:534–536.

Tammes, T. 1928. The genetics of the genus *Linum.* Bibliogr. Genet. 4:1–36.

Thompson, T. E., and J. F. Miller. 1978. Cytoplasmic male-sterility in flax: Restorer gene frequency and degree of corolla opening. Crop Sci. 18:983–986.

Wicks, Z. W., III, and J. J. Hammond. 1978. Screening *Linum* species for new sources of genes resistant to *Melampsora lini.* Crop Sci. 18:7–10.

25 Forage Legumes

C. E. Townsend
USDA-SEA and Colorado State University
Fort Collins

This chapter emphasizes those species of forage legumes other than alfalfa, *Medicago sativa* L., and the clovers, *Trifolium* spp., that have an economic or potential economic importance in North America and Europe. Information is considerable for some species, but essentially nonexistent for others. The principal species are identified by both their common and Latin names in Table 1. In general, these species are used for hay or pasture, but some of the winter annuals are used as cover crops.

I. PARENTAL MATERIAL

New cultivars of birdsfoot trefoil (*Lotus corniculatus* L.) have been developed in Canada and the United States with germplasm available within the species (Seaney and Henson, 1970), but interspecific hybridization offers a means of improving certain traits. With the most successful method, the chromosome number of the diploids (Wernsman et al., 1965) or interspecific diploid hybrids (Somaroo and Grant, 1972) is doubled with colchicine, and the resulting tetraploids are crossed directly to tetraploid ($2n=24$) *L. corniculatus.* A less successful approach is to cross the diploids directly with *L. corniculatus.* This procedure requires backcrossing the triploid progenies, which are usually sterile, to the tetraploid parent.

Somaroo and Grant (1971) intercrossed seven diploid species closely related to *L. corniculatus: L. alpinus* (DC.) Schleich. ex. Ramond, *L. japonicus* (Regel) Larsen, *L. filicaulis* Dur., *L. schoelleri* Schweinf., *L. krylovii* Schischk. and Serg., *L. tenuis* Waldst. et Kit. ex Willd., and *L. corniculatus* var. *minor* Baker. In some crosses, interspecific hybrids were grown directly from seeds; in certain crosses, embryo culture was necessary to produce the hybrids; and in other crosses, hybrids could not be produced even with embryo culture. They noted that crosses were easier to make if the self-compatible species were the female parent.

Table 1—Information concerning the breeding and improvement for 10 species of forage legumes.

Species and common name	Growth form and type of inflorescence	Probable origin	Chromosome no. (2n)	Type of inheritance	No. of ovules per ovary		Time from pollination to mature seed
					Mean	Range	
Astragalus cicer L. Cicer milkvetch	Perennial (rhizomatous); Raceme	East-central Europe (Barneby, 1977)	64 (Ledingham, 1960)	Unknown	12 (Townsend, 1971	9 to 15	28 to 35 days (Townsend, unreported)
Coronilla varia L. Crownvetch	Perennial (creeping-rooted); Umbel	Central Europe (Alex, 1971)	24 (Darlington and Wylie, 1956)	Tetrasomic (Berchtold et al., 1973) Probably autotetraploid	16 (Baluch et al., 1973)	14 to 17	46 to 59 days (Al-Tikrity et al., 1974)
Lespedeza cuneata (Dumont) G. Don Sericea lespedeza	Perennial; Compound raceme—1 to 4 flowers, Cleistogamous flowers not in a raceme	N. E. Asia (Henson, 1957)	20 (Hanson and Cope, 1955)	Disomic (Hanson and Cope, 1955)	1 (Cope, 1977)	--	28 to 42 days (Cope, 1977)
Lespedeza stipulacea Maxim. Korean lespedeza	Annual; Flowers borne in leaf axils at the tips of stems	N. E. Asia (Henson, 1957)	20 (Hanson and Cope, 1955)	Disomic (Offutt, 1976)	1 (Hanson, 1943)	--	28 to 42 days (Cope, 1977)
Lotus corniculatus L. Birdsfoot trefoil	Perennial; Umbel	Mediterranean Basin (Seaney and Henson, 1970)	24 (Darlington and Wylie, 1956)	Tetrasomic (Dawson, 1941) (Hart and Wilsie, 1959)	45 (Seaney and Henson, 1970)	20 to 70	24 to 47 days (Seaney and Henson, 1970)
Lupinus albus L. White lupine	Annual; Upright raceme	Mediterranean Basin (Gladstones, 1970)	50 (Gladstones, 1970)	Functional diploid (Lamberts, 1955)	6 (Jaranowski, 1962a)	5 to 7	32 days (Jaranowski, 1962a)
Lupinus angustifolius L. Blue lupine	Annual; Upright raceme	Mediterranean Basin (Gladstones, 1970)	40 (Gladstones, 1970)	Functional diploid (Lamberts, 1955)	5.5 (Jaranowski, 1962a)	5 to 6	25 days (Jaranowski, 1962a)
Lupinus luteus L. Yellow lupine	Annual; Upright raceme	Mediterranean Basin (Gladstones, 1970)	52 (Gladstones, 1970)	Functional diploid (Lamberts, 1955)	5.5 (Jaranowski, 1962a)	5 to 6	25 days (Jaranowski, 1962a)
Onobrychis viciifolia Scop. Sainfoin	Perennial; Raceme	Cultivated for several centuries in the USSR and Europe	28 (Chapman and Yuan, 1968)	Tetrasomic (Chapman and Yuan, 1968; Ditterline, 1977) Probably autotetraploid	1 (Thompson, 1938)	--	42 to 49 days (Ditterline, 1977)
Vicia sativa L. Common vetch	Annual; Raceme	Mediterranean Basin (Hollings and Stace, 1974)	12 (Darlington and Wylie, 1956)	Disomic (Donnelly, 1958)	6 (Donnelly, 1977)	4 to 8	50 to 60 days (Donnelly, 1977)

Amphidiploids produced by doubling the chromosome number of interspecific diploid hybrids were crossed to *L. corniculatus* (Somaroo and Grant, 1972). The crosses were successful when *L. corniculatus* was the male parent, but difficult in the reciprocal crosses. Embryo culture was not used. Because the genetics of diploids is less complicated than that of tetraploids, they suggested that certain desirable traits be improved at the diploid level and then transferred to *L. corniculatus* by way of amphidiploidy.

Attempts to hybridize *Lespedeza* species have met with varying success. Hanson and Cope (1955) cross-pollinated 13 annual and perennial species with emphasis on combinations involving sericea lespedeza [*Lespedeza cuneata* (Dumont) G. Don], Korean lespedeza (*L. stipulacea* Maxim), and common or striate lespedeza [*L. striata* (Thunb.) H. & A.]. The annual species have lower tannin content than the perennials. All successful crosses, however, involved perennial species of Asiatic origin, as follows: *L. cuneata* × *L. latissima* Nakai, *L. latissima* × *L. cuneata, L. cuneata* × *L. inschanica* (Maxim.) Schind., *L. hedysaroides* (Pall.) Ricker × *L. cuneata, L. hedysaroides* × *L. inschanica.* Crosses involving diploid and tetraploid forms of *L. cuneata, L. stipulacea,* and *L. striata* have been unsuccessful (Brinkley et al., 1959). Ovules of the most promising *L. stipulacea* × *L. cuneata* cross began to collapse about 6 days after pollination; the collapse was attributed to endosperm failure. Cope (1957) discussed breeding-related research for the *Lespedeza* species.

Offutt (1962) increased the variability of several characters in *L. stipulacea* by gamma-ray and thermal-neutron irradiations and considered irradiation a useful tool for plant improvement. Donnelly (1971) found many mutants in a mutation breeding program with *L. cuneata* and developed a cultivar from one of the mutants.

The isolating barriers between *Lupinus* species within the Old World and New World groups are different (Jaranowski, 1962a). Hybridization occurs between American species, but not between the Old World species which are the ancestoral forms of the cultivated species white lupine (*Lupinus albus* L.), blue lupine (*L. angustifolius* L.), and yellow lupine (*L. luteus* L.). Jaranowski (1962b) made all possible reciprocal crosses between these three species, except for the cross *L. angustifolius* × *L. albus.* Fertilization occurred in all crosses, but the embryos aborted because of endosperm failure. In certain crosses, embryo abortion was delayed somewhat, and embryo culture was suggested as a possible means of overcoming the sterility barrier. From 5,730 crosses of *Lupinus* species in 81 combinations, Kazimierski (1961) reported the following successful crosses: *L. albus* × *L. jugoslavicus* Kazim. & Now., *L. albus* × *L. termis* Forsk., *L. angustifolius* × *L. linifolius* Roth, and *L. arboreus* Sims × *L. hartwegii* Lindl. Many other interspecific hybridization studies have been made, but they are beyond the scope of this review.

Lamberts (1955) discussed the lack of genetic variability in lupine breeding programs and stated that collections should be made in their center of origin. Forbes et al. (1975) located disease resistance in natural populations of *L. angustifolius* in the western Mediterranean region. Gustafsson and Gadd (1965) suggested that genetic variability might be increased through induced mutations.

Although various *Vicia* species possess certain desirable characters, genetic barriers prevent or hinder their transfer to the more commonly grown forms. To elucidate the problem, Donnelly and Clark (1962) made intraspecific pollinations between strains of common vetch (*Vicia sativa* L.), *V. villosa* Roth, and *V. dasycarpa* Ten. and interspecific pollinations between 30 species pairs for 12 species of *Vicia: angustifolia* L., *articulata* Hornem., *atropurpurea, dasycarpa, ervilia* (L.) Willd., *galeata* Boiss., *grandiflora* W. Koch., *lutea* L., *pannonica* Crantz, *sativa, serratifolia* Jacq., and *villosa.* The *V. angustifolia* parent (2n = 10) was later reclassified as *V. cordata* Wulf. (Donnelly and Hoveland, 1966). Normal pods and seeds were produced from cross-pollinations within *V. sativa* and *V. dasycarpa*. However, crosses between certain strains of *V. villosa* produced only shrivelled seed. *Vicia villosa* apparently had more genetic diversity than did the other two species. The only successful interspecific crosses were *V. villosa* × *V. dasycarpa* and *V. sativa* × *V. cordata.* All F_1 seedlings from the first cross were chlorophyll-deficient. Considerable variability existed among the F_2 plants from the latter cross and more F_2 plants resembled the parental types for vigor than would have been expected if random chromosome segregation and recombination had occurred.

Germplasm possessing several desirable traits has been developed from advanced generations of the *V. sativa* × *V. cordata* hybrid, and three cultivars have been released (Donnelly, 1977). In addition, *V. sativa* and *V. serratifolia* have been successfully crossed and a cultivar has been released from an advanced generation of the hybrid (Donnelly, 1977). Leeuwrik and Elliott (1965) overcame the sterility barriers between *V. sativa* and *V. angustifolia* at the diploid (2n = 12) level by doubling the chromosome number of each species with colchicine and crossing at the tetraploid level.

Interspecific hybridization has not played a role in breeding sainfoin (*Onobrychis viciifolia* Scop.), crownvetch (*Coronilla varia* L.), or cicer milkvetch (*Astragalus cicer* L.). Several *Onobrychis* species have the same chromosome number as *O. viciifolia* (Chapman and Yuan, 1968), but interspecific hybridization studies have not been reported. *Coronilla varia* L. has been crossed with *C. globosa* Lam., *C. glauca* L., and *C. viminalis* Salisb., but the crosses were not successful (Risius, 1977). The prospects of hybridizing *A. cicer* with other related species are not promising because of its high ploidy level.

II. PLANT CULTURE

A. Field

The species determines the type of breeding nursery. The nurseries for the self-pollinated, annual species are typical of those used to develop pureline cultivars. Plants may be spaced in rows or may be bulk-sown. The breeding nurseries for *Lespedeza cuneata* are similar to those of the self-pollinated species. For perennial, cross-pollinated species the desired parental plants or clones are vegetatively propagated and the resulting propagules are placed in crossing blocks. The methods of vegetative propagation for the different species are as follows: *Lotus corniculatus,* by root or stem cuttings (Seaney and Henson, 1970); *O. viciifolia,* by bud cuttings (Ditterline, 1977); *C. varia,* by root or stem cuttings (Cope, 1977); and *A.*

cicer, by rhizomes. Although *L. cuneata* can be vegetatively propagated, the availability of self-pollinated seed negates the need for this practice (Donnelly, 1977). Occasionally, the superior plants in breeding nurseries of perennial species are permitted to flower and set seed while the unselected plants are cut before flowering.

Spacing distance between propagules is determined by whether the species spread by rhizomes or other vegetative means, by expected life of crossing block, and by the type of cultivation equipment. The propagules of rhizomatous and other spreading species should be placed on relatively wide spacings, such as 1.2-m centers, with closer spacings for nonspreading species. The number of clones in a crossing block is variable and may range from 4 to more than 50. The number of replications for each clone is usually three or more with one or more propagules per replication. The randomized complete block is a common nursery design.

Seed usually is produced during the year of establishment, but some clones of certain species, such as *A. cicer,* flower sparsely the first year, and 2 years may be required to produce enough seed for progeny testing. Seed identity is maintained on a maternal plant basis. To overcome some of the nonrandom pollination that occurs with early and late flowering clones, seed should be harvested from the latest as well as the earliest appearing inflorescences on all clones.

B. Growth Chamber and Greenhouse

Several species within this group have flowering requirements that are quite different from those of the more commonly grown forage legumes. The *Lespedeza* species are unique in this respect. Temperature influences the proportion of chasmogamous and cleistogamous flowers in *L. stipulacea* under a 12-hour photoperiod (Hanson, 1943). Chasmogamous flowers predominate at 27 C and above, whereas cleistogamous flowers are the most common at 21 C. Seed set in the chasmogamous flowers is higher in the summer under field conditions than during the winter in the greenhouse. Cleistogamous flowers are highly fertile under both conditions.

Photoperiod influences both the initiation of flowering and flower type in *L. stipulacea, L. striata* (Nakata, 1952), and *L. cuneata* (Bates, 1955). In general, all species remain vegetative at photoperiods of 14 hours or longer and only cleistogamous flowers form at photoperiods of 10 to 12 hours. For *L. stipulacea,* only chasmogamous flowers develop at photoperiods of 12.5 to 13.5 hours (Nakata, 1952). Strains of *L. cuneata* differ in the proportion of flower types produced, but chasmogamous flowers are most frequent under a 13-hour day (Bates, 1955). Plant height and day length also interact to influence the type of flower produced. Vegetative growth ceases when flowering begins.

Initiation of flowering in *Coronilla varia* depends on both photoperiod and temperature (McKee et al., 1972). The species flowers profusely with photoperiods of 15.5 hours or longer with the proper thermo-induction treatment. Exposure to minimum temperatures of 4 to 10 C for various periods of time is enough to induce flowering in most plants. Plant age and genotype influence flowering response.

Flowering in *A. cicer* is enhanced by pretreating the plants at −2 to +5 C and by growing them at 27 C day/21 C night temperatures under a 15-hour photoperiod (Townsend and McGinnies, 1973). A practical method for flower induction is to leave the plants in the field until mid-December, then place them in a greenhouse or growth chamber under a 16-hour or longer photoperiod at temperatures of about 27 C day/21 C night. It is very important to have an adequate light level, such as that provided at a photosynthetic photon flux density of 450 to 500 μE m^{-2} sec^{-1}.

The flowering requirements for the lupine species are not fully understood. Species of Mediterranean origin are long day and respond to vernalization, but temperature appears to play a more dominant role than photoperiod (Gladstones, 1970; Rahman and Gladstones, 1972). The flowering response of *Lupinus albus* to photoperiod, however, is less pronounced than that of *L. angustifolius* or *L. luteus.* Gladstones and Hill (1969) found single dominant genes in *L. angustifolius* and *L. cosentinii* Guss. that removed most of the vernalization requirements.

The flowering requirements for *Lotus corniculatus*, a long-day species, are similar to those of the more commonly grown legumes, such as *M. sativa* L. and *Trifolium* species (McKee, 1963). *Onobrychis viciifolia* is a long-day plant (Ditterline, 1977) and *V. sativa* is a short-day plant (Donnelly, 1977), but their flowering requirements have not been resolved.

III. FLORAL CHARACTERISTICS

The flower structure of the species discussed in this chapter is typical of other legumes (Chapter 2). The flower types and pollinating systems of lespedeza present unusual challenges to the plant breeder. The cleistogamous flowers produce only self-pollinated seed, whereas the chasmogamous flowers produce both self- and cross-pollinated seed in various proportions. Cross-pollination in *Lespedeza cuneata* ranged from 16 to 43% because of a large environmental effect (Cope, 1966).

Meiosis and microgametogenesis appear to be the same for both flower types in *L. stipulacea,* but anthesis and pollen dehiscence occur only in the chasmogamous flowers (Hanson, 1953a). At high temperatures, anthesis generally occurs between 0700 and 1000 hours, but the flowers close before night and generally remain closed. Fertilization takes place 24 to 48 hours after pollination. The stigma appears to be receptive to pollination 1 to 2 days before anthesis. Some pollen grains in cleistogamous flowers may germinate within the anther sacs and their tubes penetrate the anther wall, enter the adjacent stigma, and effect fertilization. The cleistogamous flowers have reduced corolla, stamens, and pistil (Hanson, 1943).

In the cross-pollinated *Lotus corniculatus,* anthers dehisce before the flower is completely open, but the stigmatic membrane must be broken before the pollen will germinate (Tome and Johnson, 1945). Materials released by the ruptured membrane appear to enhance pollen germination. Only about 40% of the ovules develop into mature seed following cross-pollination (Wojciechowska, 1963). The reduced seed set may be due to uneven maturation of ovules within an ovary and to the exhaustion of the supply of stigmatic fluid (Bubar, 1958). The stigmatic fluid dries up within 24 hours and repeated pollinations are ineffective.

Cross-pollination in *O. viciifolia* exceeded 90% under field conditions, but ranged from 8 to 28% for two-clone combinations under isolation with honeybees (*Apis mellifera* L.) as pollinators (Knipe and Carleton, 1972).

Under greenhouse conditions with honeybees as pollinators, *C. varia* averaged 94% cross-pollination (Risius, 1977). *Astragalus cicer* is predominantly cross-pollinated, but no information is available on its percentage of cross-pollination because marker genes have not been discovered. *Vicia sativa* and related species are highly self-pollinated (Donnelly, 1958; Mlyniec, 1962).

Lupines are predominantly self-pollinating with varying amounts of natural cross-pollination. They do not require bee visitation for pollination, but such visitations generally improve seed set. Flowers of *Lupinus angustifolius* are not particularly attractive to pollinating insects, because they do not produce nectar. Natural cross-pollination in *L. angustifolius* ranged from 0 to 12%, depending on genotype, location, year, and honeybee population (Forbes et al., 1971). When grown in alternate rows with other cultivars of the same species, Barbacki and Kapsa (1960) reported no natural cross-pollination for *L. angustifolius,* 4% for *L. albus,* and 10% for *L. luteus.*

IV. ARTIFICIAL HYBRIDIZATION AND SELF-POLLINATION

A. Equipment

The equipment for artificial hybridization is simple and consists of items such as forceps, scissors, toothpicks, and marking tags. The most expensive piece of equipment is the vacuum pump for the suction emasculation technique.

B. Preparation of the Female

Flowers on inflorescences of most species mature from the base to the top; consequently, withered, fully open, and immature flowers may be found on the same inflorescence. Temperature influences the rate of flower development and the length of time the flower remains receptive. In general, flowers remain receptive for 1 or 2 days after maturity.

In any emasculation procedure, the number of newly opened florets on an inflorescence must be reduced to a manageable level, usually 15 to 20. Suction emasculation and alcohol emasculation can be used for both cross- and self-pollinated species (Chapter 6). Seaney (1962) emasculated *Lotus corniculatus* by removing the stamens and fused keel petals with forceps while leaving the standard and two wing petals intact. Knipe (1972) described two methods to emasculate flowers of *O. viciifolia.* The first method consists of removing the standard petal with forceps and forcing the keel away so that the stamens and pistil are exposed. The flowers are immersed in a solution of 47.5% ethyl alcohol for 5 sec and rinsed in water. In the second method, pressure is applied to the base of the keel which causes the staminal column to protrude. The stamens are then rubbed between the thumb and forefinger. Both methods are effective, but the latter is much faster. A common procedure for emasculating the self-pollinated species is to remove the anthers with forceps before pollen shedding (Donnelly and Clark, 1962). Emasculation may not be necessary for either self or cross-pollinated species if 1) genetic markers such as seedling pigmentation or flower color are available or 2) if the female parent is highly self-incompatible.

C. Pollination

Pollinations usually are made within a few minutes or 1 or 2 hours after emasculation; however, *Lotus corniculatus* and *Lupinus* species may be pollinated 2 to 4 days after emasculation (Seaney, 1962; Jaranowski, 1962a). For most species, emasculation and pollination can be made during any time of the day, but the best seed set in *O. viciifolia* follows late afternoon or evening pollinations (Knipe, 1972). In most self and cross-pollinated species, fertilization occurs within 24 to 48 hours after pollination.

Bud pollination is used frequently to cross self-pollinated species (Donnelly and Clark, 1962) and to cross the chasmogamous flowers of *Lespedeza* species. In *L. stipulacea,* Hanson (1953b) recommended that the flower bud be opened by applying gentle pressure with the thumb and forefinger to the ventral and dorsal sides. Sometimes a sharp probe must be used to separate the keel petals. Anthers of the male parent are held by forceps and brushed against the stigma. Self-pollinated species may be bud pollinated in a similar manner. Bud pollinations are successful because the stigma is receptive 1 or 2 days before pollen shedding.

Self-pollination can be accomplished in most cross-pollinated species by rolling the florets between the thumb and fingers, by tripping florets with a plain toothpick or with a toothpick that has a piece of emery cloth or sandpaper glued to one end, or by tripping florets with a folded piece of firm paper, such as heavy brown wrapping paper that has been trimmed to a point on one end. Cross-pollination with or without emasculation can be accomplished with either the toothpick or folded-paper method. Cross-pollinations in *O. viciifolia* are made by collecting pollen on a fingernail and rubbing it against the stigma of an emasculated flower (Knipe, 1972). When the operator's fingers come in direct contact with pollen, they must be washed with alcohol or with soap and water and dried before the next pollination.

Flowers that have been recently emasculated or pollinated do not need to be protected in a greenhouse. In the field, however, small muslin bags or some protective device must be placed over the inflorescence to exclude pollinating insects. Pollinated inflorescences are usually identified with small (1.5×2.5 cm) marking tags. Information, such as parents of the cross and date of pollination, is recorded on the tag. Expected seed set following hand pollination is determined primarily by the number of ovules per ovary. For those species with a single ovule per ovary, it may be difficult to get enough seed for subsequent progeny evaluation.

Little information is available on the storage of pollen. Because all parental plants did not flower at the same time, Grant et al. (1962) stored pollen of *Lotus corniculatus* under refrigeration in small gelatin capsules contained in stoppered shell vials. Pollen of *V. sativa* remained viable for 21 days when stored under low temperature and humidity (Milczak, 1971).

D. Factors Affecting Efficiency

Some clones of perennial species may flower less profusely than others or may be exceptionally slow to flower after they are brought from the field to the greenhouse or growth chamber. For clones that flower less profusely than desired, more propagules must be started. To synchronize flowering, the late-flowering clones may be started earlier than the other clones in the greenhouse or growth chamber.

Simply inherited traits are more common in diploid and functional diploid species than in polyploids. They are easier to use in the diploid than in polyploid species because of the more complex inheritance in the latter. Genetic information is more voluminous on the *Lupinus* species than for any other group in this chapter. Gladstones (1970) lists a number of traits on which genetic information is available; many of these, including flower color, plant color, and seed color, are useful as genetic markers. In *Lespedeza* species, flower color (Hanson, 1953c; Offutt, 1976), tannin content (Cope, 1966), and seed color (Bates and Henson, 1955) serve as genetic markers. Flower color or plant pigmentation traits are available as genetic markers in *V. sativa* (Donnelly, 1958; Clark and Donnelly, 1964), *V. villosa* (Clark and Donnelly, 1964), *V. grandiflora* (Clark and Donnelly, 1964), and *O. viciifolia* (Knipe and Carleton, 1972). The exposed stigma trait may also be used in *O. viciifolia* (Knipe, 1972). In *L. corniculatus,* a flower color trait (Hart and Wilsie, 1959) and cyanogenesis (Dawson, 1941) can serve as markers. Grant et al. (1962) noted that certain characters are dominant in F_1 hybrids of diploid species closely related to *Lotus corniculatus.* Of these characters, red stem coloration, striping on the flower bud, reddish-brown keel tip color, pod and seed stippling, and positive reaction for the presence of hydrocyanic acid in the leaves might be used as markers. A white-flowered trait is available as a marker in *C. varia* (Risius, 1977). Genetic markers have not been reported for *A. cicer.*

V. NATURAL HYBRIDIZATION

Cross-pollinated species are well-adapted to natural hybridization because insect pollinators are required for these crops (McGregor, 1976). In these species, the flower must be tripped before pollination can occur, and under natural conditions the insect pollinator does the tripping. Self-pollinated species are not suitable for natural hybridization because tripping is not required for pollination, however, insect pollination may improve the seed set on some species. Honeybees and bumblebees (*Bombus* species and *Pyrobombus* species) are common pollinators, and certain wild bees may be important pollinators under some conditions. If the population of natural pollinators is low, hives of honeybees should be placed near the crossing block. Commercial beekeepers should be contacted for sources of pollinators. There is little or no information on the distances required for proper isolation of a field planting from others of the same species. Most researchers, however, use the isolation requirements for alfalfa as a guide (Chapter 9).

Conditions favorable for natural hybridization can be provided in greenhouses and in field cages. Desired plants and a hive of honeybees can be placed in a greenhouse that is screened against foreign pollinating insects. When placed in the greenhouse, the pollinators must be free of pollen from the species to be pollinated or contamination will occur. Cages of various sizes, shapes, and construction material may be placed around one or more plants in the greenhouse or field before floret opening (Kazimierski, 1961). The construction material (wire screen, plastic screen, or gauze) must provide for sufficient light to enter the cage and for adequate ventilation. Most cages have supporting structures, but Carleton et al. (1970) described a polyethylene, non-rigid, pneumatic isolation chamber for controlled matings. The cages confine or exclude pollinating insects according to the nature of the study.

Self-incompatibility is an excellent means of enforced outcrossing. The cross-pollinated species are predominantly self-incompatible (gametophytic type), but a few plants in most species set relatively large amounts of seed when self-pollinated (Thompson, 1938; Seaney, 1964; Risius, 1968; Townsend, 1971; Scheetz et al., 1972). Inheritance of self-incompatibility is complex in these species. Most breeders tend to select parents with high levels of self-incompatibiliy to reduce the potential for inbreeding depression that accompanies self-pollination.

To measure maximum cross-pollination among the chasmogamous flowers of *L. cuneata,* Cope (1966) established high and low-tannin plants alternately on about 1-m centers so that either a low or a high-tannin plant had a plant of the opposite type on all four sides. To increase the amount of cross-pollination he suggested that selection be practiced for a high ratio of chasmogamous to cleistogamous flowers (self-pollinated). He also proposed a minimum percentage of chasmogamous seed for breeder and foundation seed. The latter would be accomplished mechanically, because chasmogamous seed is larger than cleistogamous.

VI. SEED DEVELOPMENT, HARVEST, AND STORAGE

Pod development usually indicates seed formation except in interspecific crosses. Number of seed per pod and length of time from pollination to mature seed varies with species and environmental conditions. Mature seed of the *Lespedeza* species and *O. viciifolia* remains in the pod after threshing. Seed resulting from hand pollination is harvested and threshed by hand. When large volumes of seed are involved, some type of small thresher (Harmond and Rampton, 1956) is used. An excellent hulling machine (W. A. Rice Seed Co., Jerseyville, IL 62052) consists of two rubber rollers that revolve in the opposite direction at different speeds. It is especially good for processing some *Astragalus* species with very tough seed pods. The South Dakota seed blower (E. L. Erickson Products, Brookings, SD 57006) is an excellent machine for the final phase of cleaning small seed lots. Seed identity for self-pollinated species is maintained on the basis of individual crosses, while that for the cross-pollinated species may be maintained in the same manner or on the basis of the maternal parent when produced under polycross or open-pollination conditions.

Hard-seed coats are common for *Lotus corniculatus, C. varia, A. cicer, Lespedeza cuneata,* and some strains of *Lupinus* species. Hard-seeded *Vicia* lines, a desirable character for reseeding purposes, have been developed by interspecific hybridization (Donnelly and Hoveland, 1966). Seeds with hard coats are impermeable to water and will not germinate unless scarified or treated with sulfuric acid to permit imbibition of water (Brant et al., 1971). Most mechanical scarifiers process only small quantities of seed. A versatile scarifier has been developed that scarifies a few seeds or several kilograms equally well and is especially suitable for scarifying seeds with exceptionally hard seed coats, such as those of *A. cicer.* (This machine originally was designed by Dr. H. J. Gorz, University of Nebraska, Lincoln, NE 68503, and has been modified by the Giddings Machine Co., Fort Collins, CO 80524.) Seed longevity is favored by storage at low relative humidity and low temperature. Because the hard seed coat also favors longevity, seed intended for storage should not be scarified.

ACKNOWLEDGMENT

The author extends his appreciation to the following people for reviews of this chapter: R. W. Cleveland, C. S. Cooper, W. A. Cope, R. L. Ditterline, E. D. Donnelly, W. F. Keim, and M. L. Risius.

REFERENCES

Alex, J. F. 1971. Botanical aspects of crownvetch. p. 35–42. *In* E. M. Watkin (ed.) The potential for crownvetch in Ontario. Univ. Guelph, Ontario, Canada.

Al-Tikrity, W., W. W. Clarke, G. W. McKee, M. L. Risius, and R. A. Peiffer. 1974. Days from pollination to seed maturity in crownvetch. Crop Sci. 14:527–529.

Baluch, S. J., M. L. Risius, and R. W. Cleveland. 1973. Pollen germination and tube growth after selfing and crossing *Coronilla varia* L. Crop Sci. 13:303–306.

Barbacki, S., and E. Kapsa. 1960. Variability in *Lupinus albus.* Genet. Pol. 1:61–101.

Barneby, R. C. 1977. Personal communication. The New York Botanical Garden. Bronx, NY 10458.

Bates, R. P. 1955. Effects of photoperiods on plant growth, flowering, seed production, and tannin content of *Lespedeza cuneata* Don. Agron. J. 47:564–567.

———, and P. R. Henson. 1955. Studies of inheritance in *Lespedeza cuneata* Don. Agron. J. 47:503–507.

Berchtold, D. J., R. W. Cleveland, and M. L. Risius. 1973. Meiosis of *Coronilla varia* L. Crop Sci. 13:642–645.

Brant, R. E., G. W. McKee, and R. W. Cleveland. 1971. Effect of chemical and physical treatment on hard seed of Penngift crownvetch. Crop Sci. 11:1–6.

Brinkley, R. M., C. H. Hanson, and W. A. Cope. 1959. Growth and morphological characteristics of induced autotetraploids of *Lespedeza cuneata, L. stipulacea,* and *L. striata,* and interfertility of these species at diploid and tetraploid levels. Agron. J. 51:143–147.

Bubar, J. S. 1958. An association between variability in ovule development within ovaries and self-incompatibility in *Lotus (Leguminosae)*. Can. J. Bot. 36:65–72.

Carleton, A. E., D. M. Erb., and R. F. Eslick. 1970. Polyethylene, nonrigid pneumatic isolation chambers for controlled matings. Am. Soc. Agron. Abstr., p. 7.

Chapman, S. R., and M. Yuan. 1968. Cytological and morphological variation in breeding stocks of *Onobrychis.* A preliminary report. p. 93–96. *In* C. S. Cooper and A. E. Carleton (eds.) Sainfoin symposium. Montana Agric. Exp. Stn. Bull. 627.

Clark, E. M., and E. D. Donnelly. 1964. *Vicia:* Inheritance of plant color and recessive epistasis in flower color. Crop Sci. 4:661–662.

Cope, W. A. 1957. The lespedezas. III. Lespedeza breeding and improvement. Adv. Agron. 9:142–157.

———. 1966. Cross-pollination in sericea lespedeza. Crop Sci. 6:469–470.

———. 1977. Personal communication. Department of Crop Science, North Carolina State Univ., Raleigh, NC 27607.

Darlington, C. D., and A. P. Wylie. 1956. Chromosome atlas of flowering plants. The Macmillan Co., N.Y.

Dawson, C. D. R. 1941. Tetrasomic inheritance in *Lotus corniculatus.* J. Genet. 42:49–72.

Ditterline, R. L. 1977. Personal communication. Department of Plant and Soil Science, Montana State Univ., Bozeman, MT 59715.

Donnelly, E. D. 1958. Inheritance of white flower color in common vetch, *Vicia sativa.* Agron. J. 50:763–764.

———. 1971. Registration of Interstate sericea lespedeza. Crop Sci. 11:601–602.

———. 1977. Personal communication. Department of Agronomy and Soils. Auburn Univ., Auburn, AL 36830.

———, and E. M. Clark. 1962. Hybridization in the genus *Vicia.* Crop Sci. 2: 141–145.

———, and C. S. Hoveland. 1966. Interspecific reseeding *Vicia* hybrids for use on summer perennial grass sods in Southeastern U.S.A. p. 679–683. *In* Proc. Tenth Int. Grassland Congress.

Forbes, I., J. S. Gladstones, and H. D. Wells. 1975. Gray leaf spot-resistant *Lupinus angustifolius* L. germplasm from plant exploration in the western Mediterranean Region. Crop Sci. 15:867–868.

———, D. B. Leuck, J. R. Edwardson, and R. E. Burns. 1971. Natural cross-pollination in blue lupine (*Lupinus angustifolius* L.) in Georgia and Florida. Crop Sci. 11:851–854.

Gladstones, J. S. 1970. Lupines as crop plants (review article). Field Crop Abstr. 23: 123–148.

———, and G. D. Hill. 1969. Selection for economic characters in *Lupinus angustifolius* and *L. digitatus.* 2. Time of flowering. Aust. J. Exp. Agric. Animal Husb. 9:213–220.

Grant, W. F., M. R. Bullen, and D. de Nettancourt. 1962. The cytogenetics of *Lotus.* I. Embryo-cultured interspecific diploid hybrids closely related to *L. corniculatus* L. Can. J. Genet. Cytol. 4:105–128.

Gustafsson, A., and I. Gadd. 1965. Mutations and crop improvement. 2. The genus *Lupinus (Leguminosae)*. Hereditas 53:15–39.

Hanson, C. H. 1943. Cleistogamy and the development of the embryo sac in *Lespedeza stipulacea.* J. Agric. Res. 67:265–272.

———. 1953a. *Lespedeza stipulacea:* Stamen morphology, meiosis, microgametogenesis, and fertilization. Agron. J. 45:200–203.

———. 1953b. Controlled hybridization in *Lespedeza stipulacea.* Agron. J. 45: 333–334.

———. 1953c. Inheritance of flower color, resistance to powdery mildew, and other characters in Korean lespedeza. Agron. J. 45:555–558.

———, and W. A. Cope. 1955. Interspecific hybridization in Lespedeza. J. Hered. 46:233–238.

Harmond, J. E., and H. H. Rampton. 1956. A versatile new experimental thresher. Agron. J. 48:424–426.

Hart, R. H., and C. P. Wilsie. 1959. Inheritance of a flower character, brown keel tip, in *Lotus corniculatus* L. Agron. J. 51:379–380.

Henson, P. R. 1957. The lespedezas, I. The origin, history, and development of lespedeza in the United States. Adv. Agron. 9:113–122.

Hollings, E., and C. A. Stace. 1974. Karyotype variation and evolution in the *Vicia sativa* aggregate. New Phytol. 73:195–208.

Jaranowski, J. 1962a. Fertilization and embryo development in the genus *Lupinus.* Part I. Seed development in cases of autogamy. Genet. Pol. 3:209–246.

———. 1962b. Fertilization and embryo development in the genus *Lupinus* Tourn. Part II. Fertilization and embryo development following reciprocal, species hybridization. Genet. Pol. 3:333–368.

Kazimierski, T. 1961. Interspecific hybridization of *Lupinus.* Genet. Pol. 2:97–102.

Knipe, W. J. 1972. Reproduction and genetics of sainfoin (*Onobrychis viciaefolia* Scop.) as they relate to its breeding. Ph.D. Thesis. Montana State Univ., Bozeman, Univ. Microfilms, Ann Arbor, Mich. (Diss. Abstr. Intern. 33:531-B.)

———, and A. E. Carleton. 1972. Estimates of the percentage of self- and cross-pollination in sainfoin (*Onobrychis viciaefolia* Scop.). Crop Sci. 12:520–522.

Lamberts, H. 1955. Broadening the bases for the breeding of yellow sweet lupine. Euphytica 4:97–106.

Ledingham, G. F. 1960. Chromosome numbers in Astragalus and Oxytropis. Can. J. Genet. Cytol. 2:119–128.

Leeuwrik, D. M., and F. C. Elliott. 1965. An allotetraploid from the interspecific cross *Vicia sativa* L. × *Vicia angustifolia* L. Quart. Bull. 47:486–491. Mich. State Univ., East Lansing.

McGregor, S. E. 1976. Insect pollination of cultivated crop plants. USDA Handb. 496. 409 p.

McKee, G. W. 1963. Influence of daylength on flowering and plant distribution in birdsfoot trefoil. Crop Sci. 3:205–208.

———, M. L. Risius, and A. R. Langille. 1972. Flowering of crownvetch as affected by thermo-inductive treatment, photoperiod, plant age, and genotype. Crop Sci. 12:553–557.

Milczak, M. 1971. A study of intervarietal hybrids of common vetch (*Vicia sativa* L.). Hodowla Roslin, Akli Matyzacja i Nasiennictwo 15:101–112. (Plant Breed. Abstr. 43:5982. 1973).

Mlyniec, W. 1962. The mechanism of pollination and generative reproduction in *Vicia villosa* Roth. Genet. Pol. 3:285–299.

Nakata, S. 1952. Photoperiodic response of lespedeza. Plant Physiol. 27:644–647.

Offutt, M. S. 1962. Effect of gamma ray and thermal neutron irradiation on seed of Korean lespedeza, *Lespedeza stipulacea.* Crop Sci. 2:325–330.

———. 1976. An inhibitor gene for flower standard development in Korean lespedeza. Crop Sci. 16:703–706.

Rahman, M. S., and J. S. Gladstones. 1972. Control of lupine flower initiation by vernalization, photoperiod and temperature under controlled environments. Aust. J. Exp. Agric. Animal Husb. 12:638–645.

Risius, M. L. 1968. Crownvetch breeding. p. 64–66. *In* Proc. Second Crownvetch Symposium. The Pennsylvania State Univ. Agron. Mimeo No. 6.

———. 1977. Personal communication. Department of Agronomy. The Pennsylvania State Univ., University Park, PA 16802.

Scheetz, J. G., A. E. Carleton, and J. R. Stroh. 1972. Fertility and inbreeding studies of cicer milkvetch (*Astragalus cicer* L.). Crop Sci. 12:758–759.

Seaney, R. R. 1962. Evaluation of methods for self- and cross-pollinating birdsfoot trefoil, *Lotus corniculatus* L. Crop Sci. 2:81.

———. 1964. Cross and self seed set in birdsfoot trefoil plants selected for self-fertility. Crop Sci. 4:440–441.

———, and P. R. Henson. 1970. Birdsfoot trefoil. Adv. Agron. 22:119–157.

Somaroo, B. H., and W. F. Grant. 1971. Interspecific hybridization between diploid species of *Lotus* (Leguminosae). Genetica 42:353–367.

———, and ———. 1972. Crossing relationships between synthetic *Lotus* amphidiploids and *L. corniculatus.* Crop Sci. 12:103–105.

Thompson, J. R. 1938. Cross and self-fertility in sainfoin. Ann. Appl. Biol. 25: 695–704.

Tome, G. A., and I. J. Johnson. 1945. Self- and cross-fertility relationships in *Lotus corniculatus* L. and *Lotus tenuis* Wald. et Kit. J. Am. Soc. Agron. 37:1011–1023.

Townsend, C. E. 1971. Self-compatibility studies with *Astragalus cicer* L. Crop Sci. 11:769–770.

———, and W. J. McGinnies. 1973. Factors influencing vegetative growth and flowering in *Astragalus cicer* L. Crop Sci. 13:262–264.

Wernsman, E. A., R. L. Davis, and W. F. Keim. 1965. Interspecific fertility of two *Lotus* species and their F_1 hybrids. Crop Sci. 5:452–454.

Wojciechowska, B. 1963. Embryological studies in the genus *Lotus.* Part I. Fertilization and seed development following open and self-pollination of *Lotus corniculatus* L. Genet. Pol. 4:53–63.

26 Guar

R. E. Stafford
USDA-SEA and Texas A&M University
Vernon

T. Hymowitz
University of Illinois
Urbana

Guar, *Cyamopsis tetragonoloba* (L.) Taub., is a summer annual legume introduced from India into the United States in 1903. In India and Pakistan, guar is grown as forage for cattle, as a green manure crop, and as a vegetable for human consumption (Hymowitz and Matlock, 1964).

In the United States, initial emphasis was placed on guar as a green manure crop. It was later determined, that in addition to its soil-building properties, guar produces seed that contain a galactomannan gum for which a multitude of food and non-food applications have been developed. Today, the primary importance of guar in the United States lies in the commercial value of its seed. The center of production is in the sandy land areas of north Texas and southwestern Oklahoma, with smaller production areas in south Texas and southern Arizona.

Guar has not been widely used as a basic research tool. Although numerous advances have been made in guar breeding, improvements in yield have been especially hindered by the lack of information on basic physiological processes and inadequate amounts of research on hybridization procedures. Contributions to the literature by the few researchers working with guar are substantial. The reader is referred to a thorough and comprehensive compilation of the literature by Whistler and Hymowitz (1979).

I. PARENTAL MATERIAL

Cyamopsis has been variously divided by taxonomists into three or four species. Gillett (1958) divided the genus into three species, *C. tetragonoloba* (L.) Taub., *C. senegalensis* Guill. and Perr. and *C. serrata* Schinz. The status of intermediate forms between *C. serrata* and *C. senegalensis* was left unsettled, although he suggested that intermediate forms may be the result of hybridization between the species. Torre (1960)

Table 1—Chromosome number and geographic distribution of species in the genus *Cyamopsis.* †

Species	Diploid chromosome number	Distribution
C. senegalensis Guill. and Perr.	14	Saudi Arabia, Ethiopia, Sudan, Mali, Senegal, Tanzania
C. serrata Schinz	14	S.W. Africa, Botswana, South Africa
C. dentata (N.E. Br.) Torre	?	Rhodesia, Angola, S.W. Africa, South Africa
C. tetragonoloba (L.) Taub.	14	Cultigen

† After Whistler and Hymowitz (1979).

proposed that the intermediate forms be recognized as a distinct species, which he named *C. dentata* (N. E. Br.) Torre. Diploid chromosome number and geographic distribution of the four species are given in Table 1.

From cytological studies, Senn (1938) noted that there was a distinct similarity between the genus *Cyamopsis* and the genus *Indigofera.* He hypothesized that the genus *Cyamopsis* was a derived aneuploid from *Indigofera.* Gillett (1958) preferred to retain *Cyamopsis* as a distinct genus, although it is closely related to *Indigofera.*

The species *C. senegalensis* is a low-growing, annual herb having narrow pentafoliolate leaves and small pods. Seeds are white and average about 1.2 g per 100 seed (Menon, 1973). The species, *C. serrata,* is also a low growing annual, but has narrow trifoliolate leaves. Seed are pink and average about 1.0 g per 100 seed. Both species are reported by Menon to be drought tolerant, disease resistant, and highly dehiscent. Little is known about the annual herb, *C. dentata.* Apparently, it is found in the same semiarid habitats as *C. senegalensis* and *C. serrata.* The chromosome number for *C. dentata* is unknown.

The cultigen *C. tetragonoloba* (formerly *C. psoralioides*) is an erect, annual herb that has never been found growing in the wild (Gillett, 1958). The transdomestication concept has been advanced to explain the origin of the species *C. tetragonoloba* (Hymowitz, 1972). No crossing between species of *Cyamopsis* has been reported in the literature.

In the United States, guar germplasm collections are maintained at the Texas A&M Univ. Agricultural Research and Extension Center, Vernon, Texas; at Oklahoma State Univ., Stillwater, Okla.; and at the Southern Regional Plant Introduction Station, Experiment, Ga. The entire Plant Introduction collection (1050 accessions) and 33 breeding lines and cultivars are stored at the National Seed Storage Laboratory, Fort Collins, Colo.

In India, the most extensive guar germplasm collection is maintained by the National Bureau of Plant Genetic Resources, I.A.R.I. Campus, New Delhi. Other collections are maintained by the Institute of Agriculture, Anand, Gujarat; Millet Breeding Station, Coimbatore, Mysore; Central Arid Zone Research Institute, Jodhpur, Rajesthan; and at the Haryana Agricultural University, Hissar, Haryana. In Pakistan, a guar germplasm collection is maintained by the Agricultural Research Institute, Lyallpur (Whistler and Hymowitz, 1979).

II. PLANT CULTURE

A. Field

Responses of the guar plant to temperature, light, moisture, and mineral nutrition are only partially understood. Research has shown that soil temperatures of 21 C or higher are necessary for guar seed germination (Hymowitz and Matlock, 1964). Guar has an indeterminate growth habit and will remain vegetative and continue to flower and set pods from about 4 to 6 weeks following emergence until terminated by either plant senescence or low temperatures. Mature seed deteriorate rapidly from excessive rainfall and high humidity.

Nutrient requirements for guar have been reported by researchers in India using a sand culture technique. The optimum level of zinc was determined to be 1.0 ppm, boron 0.10 ppm, and copper 0.01 ppm (Nair and Mehta, 1959; Gandhi and Mehta, 1959; Neelakantan and Mehta, 1961). Sastry (1962, 1966) studied the effect of nitrogen, phosphorus, potassium, calcium, magnesium, and sulfur nutrition of guar as it related to local lesion formation by sunnhemp mosaic virus. The optimum nutrient level to avoid the disease was determined to be 420 ppm for nitrogen, phosphorus 31 ppm, potassium 430 ppm, calcium 200 ppm, magnesium 5 ppm, and sulfur 32 ppm.

On fertile soils or where preceding crops have been fertilized, guar usually requires no additional fertilizer. Guar requires high levels of phosphorus. Little critical field research has been conducted to identify optimum fertility levels or soil pH values for guar.

Guar is generally grown under natural rainfall in Texas and Oklahoma. The ability of the guar plant to flower and fruit over an extended period of time partially alleviates yield restrictions caused by droughts of short duration. Prolonged drought stress during the fruiting period is especially yield-limiting. Limited irrigation of breeding nurseries and yield trials has been adopted as a general practice at Chillicothe, Texas.

Guar is generally regarded as a short-day plant. Research in India indicates that guar is photoperiod-sensitive and different light requirements are needed to produce high fodder (vegetable yields) and high seed yields (Paroda et al., 1977). At Vernon, Texas (34°N Lat) guar produces maximum seed yields when planted in June and flowering and pod set occur under 13 to 14 hours of sunlight.

Guar has been successfully grown in winter nurseries at Mayaguez, Puerto Rico. Plant height is reduced, internodes are short with dense clusters of pods on the branches, and fruiting commences near the ground level, indicating sensitivity to short photoperiods. No special requirements appear necessary for advancing breeding lines during the winter season (Stone, 1977).

B. Growth Chamber and Greenhouse

Preliminary data obtained in a greenhouse at Lubbock, Texas, indicated that 13 to 14 hours of light were required to obtain normal plant growth and seed set. Satisfactory flowering and seed set have been obtained in a growth chamber at Vernon, Texas, when plants were grown under 14 hours of light at 27 C and 10 hours darkness at 24 C.

III. FLORAL CHARACTERISTICS

Guar is completely self-fertile and is highly self-pollinated. Floral studies in India indicate that 35 days are required for an initiated bud to develop into a flower (Menon, 1973). Flowers pass through an array of colors from white to deep blue from the bud stage to petal drop, and a mature bud is creamy white in color. Petals develop a pink color just prior to opening. Anthers dehisce about 1.5 to 2 hours before flower opening, and pollen was found to be capable of germination 2 hours before and 11 hours after flower opening. The length of time during which the stigma is receptive has not been reported.

Buds change from a cream color to light pink between 0800 and 1000 hours and petals start opening about 0900 hours (Menon, 1973). Petals continue to open until 1600 hours, with a few flowers opening as late as 0130 hours. The pink petal color changes to deep blue after 1600 hours. Flowers that open early in the morning shed petals the same day, between 1700 and 2000 hours. Late opening flowers shed petals the following morning (Menon, 1973).

The longevity of pollen at room temperature was reported by Menon et al. (1968) to be 11 to 13 hours and maximum pollen germination of 46% was obtained between 0800 and 0900 hours. Each pollen cell normally produced one tube, although two tubes occasionally were observed. The average pollen tube length recorded was 1.6 mm. The viability of pollen stored under refrigeration was not studied.

The inflorescence of guar is a raceme, about 9 to 13 cm long in the branched types and 15 to 20 cm long in the erect or sparsely branched types (Fig. 1). Normally, 40 to 60 flowers are present in an inflorescence of the branched type and 50 to 70 in the erect and sparsely branched types (Menon, 1973). Fruiting pedicels are stiffly erect in the leaf axils (Fig. 1). Flowers are bisexual, about 9 mm long, and almost sessile. The calyx has five sepals and the corolla has five petals (Fig. 2). The standard is circular, the wing petals are oblong, and the keel petals are as long and broad as the wing petals. The keel is blunt and slightly incurved. The pistil consists of one carpel. The ovary is linear, sessile and one-celled containing 6 to 10 ovules. The style is short and incurved with a head-shaped stigma (Menon, 1973; Gillett, 1958; Chanddrasekharan and Ramakrishman, 1928; Jafri, 1966).

Fig. 1—Fruiting raceme of guar at the late bud stage.

There are 10 stamens enclosed in the keel (Fig. 2). The anthers are uniform, and the pollen grains are spherical in shape, averaging 40 to 43 μ in diameter (Menon et al., 1968; Menon, 1973).

Fig. 2—Open guar flower showing petals, stigma, style, and stamen.

IV. ARTIFICIAL HYBRIDIZATION AND SELF-POLLINATION

A. Equipment

Artificial hybridization in guar requires only a pair of forceps, short glassine bags, a hand stapler, and label tags. Since guar is normally self-pollinated, no equipment is needed to produce selfed seed.

B. Preparation of the Female

Hand emasculation and crossing techniques were developed and are being used in India (Chaudhary et al., 1974). Mature buds are selected for emasculation in the late afternoon between 1600 and 1800 hours (Fig. 3). Buds are selected whose stigmas will be receptive to pollen for at least 2 days. All self-pollinated flowers below the selected buds are removed, thus, the lowest buds on the raceme are always the emasculated ones. To avoid damage to the raceme, upper buds are not removed until 3 days after emasculation; however, upper buds should be removed if flowers open during this period. No more than two buds usually are emasculated on a raceme at one time.

The bud should be supported by the fingers during emasculation to prevent damage to the stigma or style. The front sepal is removed with a forceps by gently pulling upward. Petals are removed in like manner. Many of the anthers are removed with the petals. Remaining anthers should be removed and visually checked to insure that they are undehisced. The whole inflorescence, including emasculated buds, is bagged with a glassine bag and stapled at the base of the raceme.

C. Pollination

Buds are pollinated the morning following emasculation, preferably between 0800 and 0900 hours (Chaudhary et al., 1974). To obtain better seed set, buds should be pollinated twice, on succeeding mornings. Newly opened flowers are selected for the pollen source (Fig. 2). Pollination is achieved by gently brushing the anthers across the stigma of the emasculated bud. Mature anthers generally dehisce on contact.

Following pollination, the raceme is labeled with a tag and rebagged. The bag should be removed after pod formation takes place, about 3 to 5 days after pollination. Generally, pods resulting from hybridization contain only two to three seeds and are smaller than those resulting from self-pollination. The shape and size of the pod insure easy identification of pods obtained from hybridization. Pods resulting from self-pollination are shown in Fig. 4. Chaudhary et al. (1974) achieved a 6 to 8% effective pod set using the hand emasculation and crossing technique.

Fig. 3—Guar raceme at the late bud stage. Lower buds are suitable for emasculation.

Fig. 4—Early pod formation about 3 to 5 days after self-pollination.

D. Factors Affecting Efficiency

Guar crossing blocks can be arranged in any convenient design which allows sufficient space between rows for emasculation and crossing. Because guar has an indeterminate flowering habit, special techniques for matching flowering dates of different genotypes do not appear necessary.

The pubescent-glabrous marker is used extensively in making controlled natural crosses in the United States and will be discussed in Section V. Glabrous and pubescent plants in guar can be easily distinguished by visual and tactile examination. The glabrous characteristic is conditioned by a single recessive gene (Kinman, 1962).

Prasada Rao and Patel (1973) found that dark green foliage color is completely dominant over light green foliage color and is conditioned by a single gene. Although the authors did not speculate about the utility of this character, it could conceivably be used as a genetic marker. No genetic markers are available for identifying hybrid seed.

V. NATURAL HYBRIDIZATION

Guar breeding in the United States initially involved the evaluation of Plant Introductions (PI). Most guar PI were heterogeneous and a minor amount of outcrossing could easily pass undetected. It was only after a significant portion of the introductions had been evaluated and single plant selections had been made that the degree of natural crossing became apparent.

A reasonably successful method has been developed for using controlled, natural hybridization for generating new material for selection and evaluation. By utilizing the genetic marker glabrous-pubescent leaf type, hybrid plants can be easily identified and utilized for further evaluation. In this procedure, a pubescent breeding line (male parent) and several glabrous breeding lines (female parents) are planted in a block, well isolated from other guar. Isolation requirements equal to or greater than that for Foundation seed production (200 m) are followed. A common design is to plant every third row with the pubescent parent and plant glabrous lines in the remaining rows. Rows in the crossing blocks are generally 5 m long and spaced 1 m apart. Bees or other insect pollinators are needed to achieve maximum crossing between the pubescent and glabrous lines.

No special treatment of plants is required for natural crossing. Natural crossing between early maturity breeding lines and medium to late maturity lines has been found to be only slightly less than crossing within a given maturity class. Altering flowering dates by date of planting does not seem to be necessary in obtaining natural crosses in guar.

Seed is harvested from the glabrous lines on a single row basis. The following season, several rows 100 m in length are planted from seed of each harvested row. Pubescent plants (hybrids) are tagged during the growing season and harvested on a single plant basis. Detectable natural crossing (pubescent × glabrous) using only wild insect pollinators has ranged from less than 1 to 4.4% at Chillicothe, Texas (Stafford and Lewis, 1975). Other natural crossing (glabrous × glabrous) is less easily detected and has not been utilized.

Genetic male sterility was reported by Mital et al. (1968), but has not been exploited for hybrid seed production. Semisterility caused by reciprocal translocations has been reported in the United States (Kinman et al., 1969) and India (Vig, 1965). It has been of little utility from the standpoint of hybridization techniques.

VI. SEED DEVELOPMENT, HARVEST, AND STORAGE

Successful pollination of hand-emasculated guar flowers can be readily detected after pod formation takes place. Pods are generally small with two to three seed per pod. Guar is commonly harvested either at physiological maturity or shortly after cold temperatures have terminated plant growth. Leaves generally abscise and stems and pods dry to acceptable moisture levels to permit harvest 10 to 14 days following freezing temperatures.

Self-pollinated nursery-row selections are hand-harvested and bagged in burlap sacks. A large sheet-metal funnel, 80 cm across the top and 48 cm at the bottom is used to simplify the bagging process. Bagged samples are air-dried and threshed with a Vogel thresher. Seed is cleaned with a Clipper cleaner. Single-plant selections are cut, bagged in paper bags, air dried, and threshed with a single plant or head thresher. Single-plant selections are commonly cleaned by hand after threshing. Seed is identified by a cross number and a bulk or plant selection number, depending on the selection procedure.

Seed is commonly stored in metal storage trays and boxes at normal room temperatures. Stored grain pests in the United States are generally of little consequence in guar. Naphthalene crystals are routinely placed in the storage boxes as a pest preventive measure. Seed viability can be retained for several years at room temperature.

REFERENCES

Chandrasekharan, S. N., and T. S. Ramakrishman. 1928. Botany of some useful plants. Madras Agric. J. 16:5–11.

Chaudhary, B. S., R. S. Paroda, and K. R. Solanki. 1974. A new crossing technique in clusterbean *Cyamopsis tetragonoloba* (L.) Taub. Curr. Sci. 43:456–459.

Gandhi, S. C., and B. V. Mehta. 1959. Studies on boron deficiency and toxicity symptoms in some common crops of Gujarat. Indian J. Agric. Sci. 29:63–70.

Gillett, J. B. 1958. *Indigofera (Microcharis)* in tropical Africa with the related genera *Cyamopsis* and *Rhynchotropis*. Kew Bull. Add. Ser. 1:1–166.

Hymowitz, T. 1972. The trans-domestication concept as applied to guar. Econ. Bot. 26:49–60.

————, and R. S. Matlock. 1964. Guar: Seed, plant and population studies. Oklahoma Agric. Exp. Stn. Tech. Bull. B-108:1–35.

Jafri, S. M. H. 1966. The flora of Karachi. Karachi Book Corp., Karachi.

Kinman, M. L. 1962. Sesame, sunflower, guar and castorbean investigations. Unpublished Annual Report. USDA-ARS, College Station, TX 77840.

———, E. C. Bashaw, and L. E. Brooks. 1969. Reciprocal translocation in guar, *Cyamopsis tetragonoloba* (L.) Taub. Crop Sci. 9:570–571.

Menon, U. 1973. A comprehensive review on crop improvement and utilization of clusterbean (*Cyamopsis tetragonoloba* (L.) Taub). Department of Agriculture, Rajasthan, Jaipur. Monograph Series. 2:1–51.

———, N. S. Rathore, and P. D. Bhargava. 1968. Pollen studies in guar *Cyamopsis tetragonoloba* (L.) Taub. J. Polynology 4:51–53.

Mital, S. P., B. S. Dabas, and T. A. Thomas. 1968. Male sterility in guar—*Cyamopsis tetragonoloba* (L.) Taub. Curr. Sci. 37:357.

Nair, G. G. K., and B. V. Mehta. 1959. Studies on zinc deficiency symptoms in some common crops of Gujarat (India). J. Agric. Sci. 52:396–401.

Neelakantan, V., and B. V. Mehta. 1961. Studies on copper deficiency and toxicity symptoms in some common crops of Gujarat (India). J. Agric. Sci. 56:293–298.

Paroda, R. S., M. L. Saini, B. S. Jhorar, and K. R. Solanki. 1977. Genetic improvement of guar—problems and perspectives. *In* First guar Research Workshop. 11–12 January. Central Arid Zone Research Institute, Jodhpur, Rajasthan.

Prasada, Rao, K. E., and G. J. Patel. 1973. Inheritance of foliage colour in clusterbeans. Andhra Agric. J. 20(1/2):43–44.

Sastry, K. S. M. 1962. Effect of mineral nutrition on local lesion formation by sunnhemp mosaic virus in guar. Ind. Phytopathol. 15:254–258.

———. 1966. The effect of calcium, magnesium and sulfur nutrition of guar plants on local lesion formation by sunnhemp mosaic virus. Ind. Phytopathol. 19:38–44.

Senn, H. A. 1938. Chromosome number relationships in the Leguminosae. Bibliog. Genet. 12:175–336.

Stafford, R. E., and C. R. Lewis. 1975. Natural crossing in guar, *Cyamopsis tetragonoloba* (L.) Taub. Crop Sci. 15:876–877.

Stone, E. G. 1977. Personal communication. Mayaguez, Institute of Tropical Agriculture, Box 70, Mayaguez, Puerto Rico 00708.

Torre, da A. R. 1960. Taxa Angolensia nova vel minus cognita-l. Men. Junta Invest. Ultram. 2ª Ser. 19:23–66.

Vig, B. K. 1965. Efect of reciprocal translocation on cytomorphology of guar. Sci and Cult. 31:532–533.

Whistler, R. L., and T. Hymowitz. 1979. Guar: Agronomy, Production, Industrial Use, and Nutrition. Purdue University Press, W. Lafayette, Indiana.

27 Hop

Alfred Haunold
USDA-SEA and Oregon State University
Corvallis

The female inflorescence of hop, *Humulus lupulus* L., is used to flavor fermented malt beverages. This practice dates back to the Middle Ages and probably earlier, when large quantities of dried hops were added to malt beverages mainly for preservative purposes. Today, hops are used primarily for flavoring, and the hop dosage is considerably less than it was in ancient times.

Hop is native to Europe, Asia, and North America (Small, 1978). Wild hop plants are found in many parts of the world, but commercial hop production is generally limited to areas between 30° to 52° latitude. The hop plant is sensitive to day length and generally produces poor yields in latitudes too close to the equator. Due to environmental and economic factors, commercial U.S. hop production is presently confined to certain areas in Washington, Oregon, Idaho, and California. The United States presently ranks first in hop production worldwide, closely followed by West Germany. Over 50% of the United States crop is exported annually to more than 70 different countries.

I. PARENTAL MATERIAL

Taxonomists differ in classifying the genus *Humulus* (Burgess, 1964; Davis, 1957; Small, 1978). Most, however, place hop into the *Cannabaceae* (formerly *Cannabinaceae*) family. This family has two genera, *Humulus* and *Cannabis.*

The genus *Humulus* includes two major species, *H. lupulus* L. (common hop) which has $2n = 20$ ($x=10$) chromosomes and *H. japonicus* Sieb. et Zucc. (Japanese hop) which has a chromosome complement of $2n = 16$ ($x=8$) in the female and $2n = 17$ in the male. *Humulus lupulus,* a perennial, is the hop used by the brewing industry. *Humulus japonicus* is an annual that produces only a few resin glands and has no brewing value. Attempts to cross the two species thus far have been unsuccessful.

Common hop is dioecious. Male and female plants have a similar growth habit, but differ markedly in their floral structure (Fig. 1). Occasionally a monoecious plant is found at the diploid and, more frequently, at the polyploid level.

Aneuploids, triploids, and tetraploids are found occasionally in segregating populations, particularly among the seedling progeny of triploids (Haunold, 1970). Polyploids generally resemble diploid hops, except for reduced vigor in some primary trisomics and tetraploids and increased vigor in most triploids.

Research centers that maintain collections of hop germplasm are located at

Dep. of Crop Science, Oregon State Univ., Corvallis, Or.
Dep. of Hop Research, Wye College, Univ. of London, England
German Society for Hop Research, 8069 Hüll, West Germany
Institute for Hop Research, 63310 Žalec, Yugoslavia
Institute for Hop Research, 21470 Bački Petrovac, Yugoslavia
Hop Research Institute, 438-46 Žatec, Czechoslovakia
Institute for Hop Research, Zhitomir, Ukraine, USSR

Other hop research centers are located in Argentina, Australia, East Germany, India, Japan, Mexico, New Zealand, Poland, South Africa, and Spain.

II. PLANT CULTURE

A. Field

The perennial root stock (crown) of *H. lupulus* in early spring produces annual vines from buds below the soil surface. The vines climb a supporting pole or string by twining clockwise to heights of 6 m or more (Fig. 2).

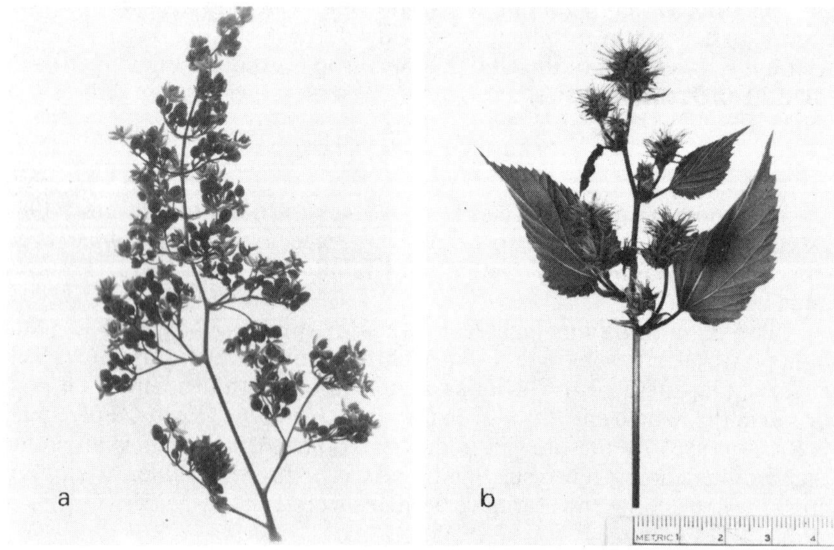

Fig. 1—Hop flowers at maturity. a) Male hop flowers at anthesis shortly before pollen shedding. b) Female hop flowers at the burr stage with receptive stigmas protruding from individual florets. A sidearm of high-yielding cultivars may have over 100 burrs.

Hooked hairs on the leaf petiole and on the main stem aid the climbing process. Depending on the cultivar, flowering occurs in late June to mid-July, at which time male and female plants can be identified.

Hop is adapted to various climates and soil types. Due to the high value of the crop, the choicest agricultural locations are usually selected for hop gardens. They may be well-drained, river bottom soils or alluvial soils of various texture and origin (Brooks et al., 1961; Burgess, 1964).

The hop plant develops an extensive root system. Perennial woody roots frequently penetrate to a depth of 4 m or more. They are supplemented by annual secondary roots near the soil surface, which are formed from perennial roots or from annual vines in late spring or early summer after hilling operations.

The above-ground parts of the plant die during the winter, but crowns that are well covered by soil or snow can tolerate temperatures of −25 C or lower. For optimum spring growth, crown buds require exposure to temperatures below 5 C for at least 4 to 5 weeks.

Soils in most hop growing areas of the United States are neutral or slightly acidic (pH 6 to 7), but occasionally slightly alkaline soils are encountered where hops also grow well (Brooks et al., 1961).

Fig. 2—Hop shoots in early June, about 1 month after training. A yardstick (91 cm) is in the foreground. Sidearms (not yet visible) develop from axillary buds at the stem nodes. Floral initiation takes place at about this stage of growth.

An average American hop crop of about 2,000 kg/ha of dried hop cones, produced on plants that average 5.5 m in height, removes a considerable amount of nutrients from the field. This nutrient loss generally is replaced with commercial fertilizers such as 18:18:18 (NPK) at up to 670 kg/ha before the start of spring operations. Liming with about 6,700 kg/ha of agricultural lime and green manuring with rye, wheat, barley, or ryegrass-vetch mixtures during the off season is used occasionally. Subsequent fertilization may include up to 220 kg/ha of N, applied as either anhydrous ammonia, ammonium sulfate, ammonium nitrate, or urea. Trace elements such as sulfur, boron, zinc, iron, and manganese are applied occasionally.

Growth hormones such as gibberellic acid (GA_3) generally do not stimulate higher hop yields, except for relatively low-yielding cultivars such as 'Fuggle' (Zimmermann et al., 1964; Hartley and Neve, 1966; Thomas, 1970; Thompson and Thomas, 1970). American growers supplement the available soil moisture by artificial irrigation of up to 900 mm during the summer months.

Most commercial hop yards in the United States are planted to permit cross cultivation. Plant spacing may vary from 1.8 × 1.8 m to as wide as 2.4 × 2.4 m, but a 2.1 × 2.1 m or 2.3 × 2.3 m spacing is commonly used, resulting in 2,200 or 1,915 plants/ha, respectively. In early spring, two or three strings are attached to the overhead trellis wires and to the plant below the center of the crown, forming a 'V'-shaped structure for hop shoots to climb (Fig. 2).

Normally two or three shoots are hand-trained on each string in late April or early May. Most remaining basal shoots are removed by hand and later by chemical suckering agents such as dinoseb (2-*sec*-butyl-4,6-dinitrophenol) or paraquat (1,1′-dimethyl-4,4′-bipyridinium ion). In Germany, plant spacing within rows is considerably closer than that used in the United States. Distances between rows may be as close as 1.6 m, and typical plant spacing within rows may vary from 0.9 to 1.4 m, resulting in over 4,000 plants/ha (Kohlmann et al., 1969).

Floral initiation takes place in late May to early June in the northern hemisphere which prompted Thomas and Schwabe (1969) to classify hop as a short-day plant. Flowering of male and female hops, however, occurs during the longest summer days at photoperiods normally exceeding 16 hours per day. Supplemental light for about 5 weeks during the period of floral initiation in the field can dramatically increase cone production (Umeda et al., 1967). The physiological effects of supplemental lighting resemble those caused by gibberellic acid, except that they apparently are not restricted to certain cultivars.

Hop cultivars are usually derived from one seedling plant that was vegetatively propagated; thus, all daughter plants are genetically identical to the mother plant. There is apparently no limit to the number of times a hop plant can be vegetatively propagated. Some commercial hop cultivars still in production today date to the latter part of the 19th century. From a practical standpoint, most hop gardens are removed after 30 to 40 years of production and are replanted after several years of intercropping.

Hops are normally propagated asexually by several methods.

1. *Division of the Woody Rootstock*

The perennial rootstock (crown) is split into a number of pieces, each with at least one bud. One may get from 5 to 50 propagules from a well-developed crown, depending upon its size.

2. *Rhizome Cuttings*

Rhizomes develop from shoots that are layered intentionally or are accidentally covered by soil, or from shoots that grow underground and emerge late or fail to break through the soil surface. They may be from 5 to 20 mm in diameter and from 20 to 100 cm long with several buds at each node. Rhizomes are pruned close to the mother crown and are cut into pieces of about 8 to 15 cm in length for field planting. Smaller pieces may be started in greenhouse pots.

3. *Soft Wood Cuttings*

A vegetative shoot can be cut into small sections composed of at least one pair of leaves and a 5 to 8 cm long stem. The sections may be placed into a mist chamber (Howard, 1965). In 1 to 2 weeks, small roots develop on the stem embedded in the sand and axillary buds start to elongate. Root development can be enhanced considerably by supplementing the carbohydrate resources of the cuttings with additional sugars (Howard and Sykes, 1966) or by treating the basal portion of the cutting with a commercial rooting hormone. Soft wood cuttings may also be placed overnight at 2 to 4 C into a rooting solution composed of 20 ppm indolebutyric acid and 20 ppm boric acid and transferred to the mist bed the next morning, or stems may be dipped briefly in a concentrated solution (about 1,000 ppm) of indolebutyric acid before planting.

B. Growth Chamber and Greenhouse

Except for special purposes such as disease testing, growth chambers are impractical for growing hops. The plants climb to a height of several meters before flowering, which necessitates tall growth chambers.

Hops can be grown successfully from late winter to early fall in greenhouse pots or in a greenhouse soil bed. Extending the day length to about 16 hours by artificial illumination from incandescent or fluorescent lights assures normal development and vigorous growth during early spring. During the winter it is difficult to keep hop plants in an actively growing state even with adequate illumination. Therefore, plants are best moved to cold frames or suitable hardening chambers to experience at least 5 to 6 weeks of near-freezing temperatures. The root stock may also be removed from the pots and be stored at 3 to 5 C.

The temperature requirements for greenhouse-grown hops are not critical. Typical greenhouse temperatures of about 20 to 25 C during the day and 15 to 18 C during the night are ideal, but lower night temperatures or substantially higher day temperatures do not seriously impede plant development.

Due to the dense foliage and luxuriant growth, insect control may be a problem with greenhouse-grown hops. Commercial fumigants or sprays used on other greenhouse plants are generally recommended. The most common pests are two-spotted spider mites (*Tetranychus urticae* Koch), white flies (*Trialeurodes vaporariorum* Westw.), and aphids (*Phorodon humuli* Schrank), listed in order of their difficulty to control. Diseases are rarely a problem in the greenhouse. Powdery mildew (*Sphaerotheca humuli* (DC.) Burr.) is unknown in the western U.S. and propagules infected with downy mildew (*Pseudoperonospora humuli* Miy. et Tak., G. W. Wils.) can be easily recognized and discarded.

III. FLORAL CHARACTERISTICS

Common hop is dioecious. Lateral branches, called sidearms, develop from axillary buds on the main stem of male and female plants in late June. Under field conditions, sidearms may reach a length of 50 to 200 cm each,

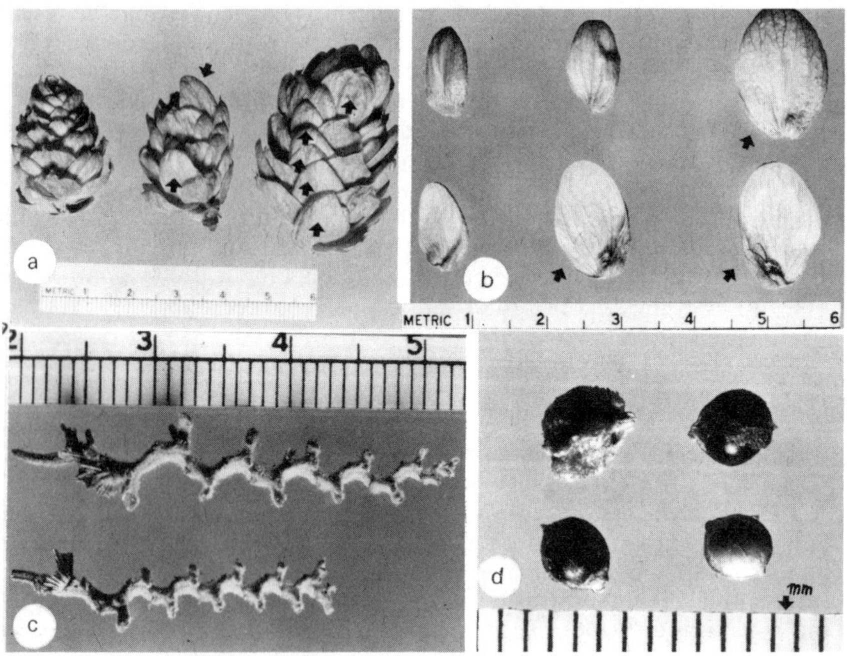

Fig. 3—Seed stage. a) Seeded hop cone (right), seedless cone (left) and partially seeded cone (center) from the same plant (scale in cm). Note larger size of seeded cone and of seed bearing bracteoles (arrows). b) Seed-bearing (arrows) and seedless bracteoles, dorsal view (top) and ventral view (bottom). Note residual stigma attached to the seed, lower right. c) Rachis (strig) of a seeded cone (top) and a seedless cone (bottom) from the same plant. Note differences in size and internode length; both strigs have an identical number of nodes. d) Hop seed with perianth and resin glands (upper left); dark seed coat of well-developed seeds from pollination by different males, with perianth partly removed (upper right and lower left); and tan, non-functional seed, containing an aborted embryo (lower right).

depending on the cultivar. Early development due to loss of apical dominance from mechanical or insect damage to terminal buds is undesirable, because such plants frequently fail to climb properly. On well-developed plants under a 5.5 m trellis, flowering of males or females generally starts on sidearms at nodes about 2.5 to 3 m up the main stem and gradually proceeds toward the top and bottom of the plant.

The mature female inflorescence is a strobile or cone with a zig-zag-shaped rachis known as a strig or core, and two types of small, oval leaf-like structures, the bracts and bracteoles (Fig. 3a,b,c). Two bracts and four bracteoles are found at each strig node (Davis, 1957; Burgess, 1964; Romanko, 1973). They are attached to short pedicels in such a way that in mature cones one bract partly overlaps the base of two bracteoles. The bracteoles produce small, yellow resin glands called lupulin or lupulin glands, primarily at the base where a small pocket is formed and where the seed develop (Fig. 3b,d). Lupulin galnds contain the resins and essential oils values by brewers. The outer seed coat is maternal tissue (perianth), and it, too, is frequently covered with resin glands. Bracts generally carry few, if any, resin glands.

The young receptive inflorescence is 5 to 10 mm in diameter and is composed of 20 or more florets (Fig. 1b). Individual florets consist of a cup-shaped perianth enclosing an ovary and two papillate stigmas that are 3 to 5 mm long. When the stigmas are fully developed, the female inflorescence is called a burr, because of its spiney appearance (Fig. 1b). Female flowers usually develop on secondary and tertiary branches of sidearms and on the upper terminal part of the main shoot, in clusters of 5 to 10 flowers.

Male flowering branches resemble a panicle (Fig. 1a). Male flowers also develop on secondary and tertiary branches of sidearms and on the terminal portion of the main stem. Each flower is 3 to 5 mm in diameter, containing five sepals and five anthers of similar size. All anthers usually are at a similar stage of development and they have comparable pollen-shedding potential. Male flowers produce a small amount of resin glands on the outer side of the sepals and in the furrow of the anthers. Before anthesis, the filament elongates to about 2 to 4 mm. The anther, about 2 mm in length, dehisces in favorable weather by forming a longitudinal slit on each side to release pollen.

Male plants in our germplasm collection frequently start pollen shedding 1 to 2 weeks later than the earliest-blooming females and continue to shed pollen for about 10 to 15 days. Flowering of male or female hop plants occurs from late June through July, depending on the genotype. Pollen, distributed by wind, may travel over distances of several kilometers without loss of viability. If no male plants are present, female flowers remain receptive up to 3 weeks before stigmas begin to drop off. Pollinated flowers develop slightly faster than non-pollinated flowers. However, several days after pollination, freshly pollinated hop flowers appear unchanged from non-pollinated flowers.

Pollen shedding depends on environmental conditions and particularly on air temperature and relative humidity (Brooks and Puri, 1963; Burgess, 1964). Male sidearms frequently have several hundred flowers that may produce from 2 to 5 cm^3 of functional pollen. In Oregon, two stages of pollen shedding are frequently observed, a major one in mid-morning and a second one in late afternoon.

IV. ARTIFICIAL HYBRIDIZATION

A. Equipment

Various types of bags can be used to protect receptive female flowers from open pollination. Bags should be at least 50 cm long and from 15 to 25 cm in diameter to accommodate at least one well-developed sidearm. White, translucent, parchment paper bags or plastic-coated paper bags, similar to those used by bakeries for French-bread wrapping, are ideal. Brown paper bags or special pollinating bags with a clear window are also suitable. Other equipment needed to pollinate females includes a plier-type stapler for securing bags around the sidearm, a 3 to 4 m stepladder, scissors, marking pen with waterproof ink, adhesive tape, an apron to hold paper bags and tools, plastic snap-on marking tags, brightly colored flagging tape, waterproof glue, and cheesecloth for temporary shading.

Pollen can be collected in paper bags directly in the field. When many males have to be visited, the equipment for routine operations includes round clear plastic cylinders, 14 × 45 cm or larger, open on both ends; rimless test tubes, 3 × 20 cm; rubber tubing, about 60 cm long; a bank of fluorescent lights, 4 × 20 W each; a squeeze bottle; a desiccator with saturated calcium chloride ($CaCl_2 \cdot 6\ H_2O$) to maintain 35 to 40% relative humidity at 3 to 5 C; glass shell vials, 15 × 80 mm; cotton; a roll of smooth wax paper; two wooden dowel sticks or metal spatulas, about 45 cm long; 70% alcohol; a refrigerator; adhesive tags; and a syringe, about 1 cm in diameter, with a needle of about 0.5 mm inner diameter.

Fig. 4—Bagging of female sidearm before pollination.

B. Preparation of the Female

Protection of female flowers from outcrossing can be done by planting one or more females in an isolated location. The area should be evaluated for natural seed set and volunteer males must be removed. Bags can also be used to protect females from outcrossing. A vigorous, well-developed female plant of the desired genotype is selected. The plant should have no more than two shoots trained on each string. Before bagging, sidearms should have growth sufficient to distinguish individual flower clusters; however, stigmas should not be visible. One of the opposing sidearms, usually higher than 2.5 m from the ground, is pulled or cut from the main stem. The remaining sidearm is prepared for bagging by removing most leaves. The two large leaves at the main node of the stem are also removed. The sidearm is trimmed to about 50 to 60 cm in length, so that it has two to three nodes and four to six secondary branches. Such a sidearm will provide up to 100 female flowers, each composed of 20 or more florets. The potential seed set per sidearm, therefore, may vary from less than 500 to over 2,000 seeds. In practice, a single pollination produces about 200 to 400 seeds per sidearm. The open end of the bag is folded around the main stem, and is closed with a plier stapler (Fig. 4).

Sidearms should be bagged at least several days before the first florets are receptive. Bagging retards floral development slightly; depending on temperature and weather, it may be up to 1 week or more before flowers become receptive.

Terminal flowers on lateral branches usually are the first to become receptive. They frequently are also the largest burrs. The basal and central florets on each burr are usually receptive first, and late-developing florets near the top of each inflorescence may become receptive up to 1 week later. Additional flowers on lateral branches reach the receptive stage over a 2 to 3-week period and protective bags must remain in place until miniature cones begin to develop and receptive stigmas are no longer visible.

C. Pollination

Paper bags may be placed over pollen-shedding sidearms to collect pollen; however, pollen clings to the inside wall of the bags and is difficult to gather without contamination. A procedure developed at Oregon has been used to collect hop pollen from many genotypes in a relatively short time (Haunold and Zimmermann, 1974). One or two sidearms at anthesis are severed from the selected male plant. The basal stem portion of the sidearm is placed in tap water in a test tube that had been taped inside a plastic cylinder (Fig. 5). A piece of rubber tubing is placed in the test tube extending over the top rim of the cylinder to permit addition of tap water to the tube from a squeeze bottle. The cylinder is placed on a clean piece of wax paper. The top of the cylinder is covered with a paper bag to prevent pollen drift and contamination from adjacent cylinders. A bank of fluorescent lights is placed behind the cylinder for continuous illumination.

Many anthers have dehisced the morning after the sidearm was collected, but lack of wind or vibrations prevents them from shedding most of the pollen. The paper bag on top of the cylinder is carefully removed and the sidearm is gently tapped with a long spatula or a wooden stick. A large amount of pollen is released immediately and rapidly settles on the collection sheet. The portion of the collection sheet outside the cylinder is wiped with a sponge moistened with 70% alcohol. The tapping tool is sterilized in a similar manner. After a few minutes, all pollen has settled and the cylinder is carefully removed. Small insects and plant debris, such as anthers and sepals, are removed with tweezers. Pollen is scraped to one side of the collection sheet with the smooth edge of a piece of plastic or stiff paper and is transferred to a glass vial, which is stoppered with a cotton plug.

One collection from a fresh sidearm usually yields from 1 to 3 cm^3 of pollen. The cylinder is rinsed with hot water and with 70% alcohol. Test tube and rubber tubing are cleaned in a similar manner and set aside to dry. Pollen vials are marked and stored at 2 to 4 C in a desiccator over saturated calcium chloride. Under such conditions, hop pollen remains viable for several months.

Fig. 5—Male sidearm at anthesis, showing setup for pollen collection in the field laboratory.

For controlled pollination, a small amount of collected pollen or a small sidearm with flowers at anthesis is placed into the bag enclosing receptive female flowers. The bag is closed and the pollen is dispersed by agitation.

Pollination can also be accomplished without opening the bag (Haunold and Zimmermann, 1974). About 0.3 to 0.5 cm^3 of pollen is drawn from the storage vial into a small hypodermic syringe. The needle is inserted directly through the wall of the bag to deposit the pollen (Fig. 6). With the transluscent bags described previously, one can see the pollen dropping into the bag. The needle hole is covered with adhesive tape or a bead of glue and the bag is agitated gently to disperse the pollen.

One pollination generally produces from 8 to 15 seeds per cone and up to several hundred seeds on an average sidearm. For maximum seed production, one or more additional pollinations at 3 to 4-day intervals are recommended to pollinate additional florets that may have become receptive. Crossing date and parentage are recorded on a plastic snap-on tag that is slipped over the base of the sidearm or around the main stem near the pollinated sidearm.

For a few days, pollinated stigmas appear unchanged. After about 1 week they wither, turn brown, and begin to drop off. Small bracts and bracteoles begin to form and a miniature cone takes shape. About 3 weeks after pollination, when all florets in the bag are past the receptive stage and fresh stigmas are no longer visible, the bag may be removed.

Fig. 6—Pollinating bagged, female flowers with a hypodermic syringe.

V. NATURAL HYBRIDIZATION

Female plants growing near males are usually wind-pollinated, even with considerable differences in flowering dates. In the Willamette Valley of Oregon, five to seven male plants/ha are routinely planted in commercial hop yards. Open pollination usually results in a seed content of 10 to 20% by weight of the cone. Many brewers prefer seedless hops, but growers obtain significantly higher yields from seeded hops. This yield increase is partly due to the extra weight of the seeds, but also to an increase in cone size and weight, caused by stimulation of cone growth from pollination, regardless of seed set (Haunold, 1975).

Collection of open-pollinated seed from desirable female genotypes is easy, but has certain drawbacks for breeding. The male parent of the cross usually is unknown, and the breeder's choice is limited to evaluation and selection of suitable female parents. The use of open-pollinated seed, however, has resulted in several successful hop cultivars (Burgess, 1964; Romanko et al., 1968; Horner et al., 1972).

Controlled crosses by natural hybridization can be made by either planting a selected pollinator in an isolation garden together with female plants, or by bringing pollen or pollen-shedding sidearms to the isolated location and manually dispersing the pollen.

VI. SEED DEVELOPMENT, HARVEST, AND STORAGE

Large cone size and the appearance of individually stimulated bracteoles indicates successful pollination (Fig. 3a,b). One may also bend back individual bracteoles and examine the developing seed at the base. In Oregon, hop seeds are usually mature 45 to 50 days after pollination. Fully developed seeded cones weigh from 125 to 300 mg each. A single hop seed averages from 4 to 7 mg in weight.

Cones are collected in early September and are dried in a forced-air oven at 40 C for 1 to 2 days. Seeds are gently rubbed out by hand. A set of hop sieves used by hop dealers and hop testing laboratories is ideal for seed cleaning. (Sieves are manufactured by Seedburo Equipment Co., 618 W. Jackson Blvd., Chicago, IL 60606.)

Hop seeds are covered with a layer of perianth and resin glands. This layer is partially removed when seeds are rubbed between the hands and a shiny, blue-black seed coat emerges (Fig. 3d). Empty seeds usually are tan (Haunold, 1975). Some genotypes, however, regularly produce brown or tan seeds that are fully viable.

The seed is a fruit (achene) containing a coiled embryo with two cotyledons. The major food reserves of hop seeds are lipids and proteins. Seeds may be stored for a few months at ambient temperatures or for several years in a refrigerator.

VII. TECHNIQUES FOR SPECIAL SITUATIONS

Hop seeds generally do not germinate well, even when seeded into a well-prepared seed bed under favorable conditions of moisture and temperature. Seeds that overwinter in the field and are exposed to cold temperatures readily germinate the following spring. Various types of pretreatment have been attempted (Williams and Weston, 1957), but cold-temperature pretreatment apparently is the most effective method of enhancing hop seed germination.

Hop seeds can either be seeded directly into pots or greenhouse flats which are placed outdoors during the winter months, if minimum temperatures are near freezing (Keller, 1953). Another procedure that assures good hop seed germination consists of surface sterilization, cold-temperature pretreatment, and germination in a germinator under an 8-hour light period, followed by space planting in the greenhouse (Haunold and Zimmermann, 1974).

REFERENCES

Brooks, S. N., C. E. Horner, and S. T. Likens. 1961. Hop production. ARS-USDA Agricultural Information Bull. 240.

———, and Y. P. Puri. 1963. Atmospheric conditions influencing pollen shedding of hops (*Humulus lupulus* L.). Crop Sci. 3:530–531.

Burgess, A. H. 1964. Hops, botany, cultivation, and utilization. World Crops Books. Leonard Hill, London; and Interscience Publishers Inc., N.Y.

Davis, E. L. 1957. Morphological complexes in hops (*Humulus lupulus* L.) with special reference to the American Race. Ann. Mo. Bot. Gard. 44:271–294.

Hartley, R. D., and R. A. Neve. 1966. The effect of gibberellic acid on development and yield of Fuggle hops. J. Hort. Sci. 41:53–56.

Haunold, A. 1970. Fertility studies and cytological analysis of the progeny of a triploid × diploid cross in hop, *Humulus lupulus* L. Can. J. Genet. Cytol. 12: 582–588.

———. 1975. Use of triploid males for increasing hop yields. Crop Sci. 15:833–840.

———, and C. E. Zimmermann. 1974. Pollen collection, crossing, and seed germination of hop. Crop Sci. 14:774–776.

Horner, C. E., S. T. Likens, C. E. Zimmermann, and A. Haunold. 1972. Cascade, a new continental-type hop variety for the U.S. Brewers Dig. 47:56–62.

Howard, B. H. 1965. Regeneration of the hop plant (*Humulus lupulus* L.) from softwood cuttings. I. The cutting and its rooting environment. J. Hort. Sci. 40: 181–191.

———, and J. T. Sykes. 1966. Regeneration of the hop plant (*Humulus lupulus* L.) from softwood cuttings. II. Modification of the carbohydrate resources within the cutting. J. Hort. Sci. 41:155–163.

Keller, K. R. 1953. Seed germination in hops, *Humulus lupulus* L. Agron. J. 45: 146–150.

Kettner, L. 1976. Hallertauer Hopfenbau, Geschichte und Gegenwart. Pinsker Verlag, Mainburg, West Germany.

Kohlmann, H., A. Kastner, and L. Kamm. 1969. Der Hopfen. Hopfen-Verlag Wolnzach.

Romanko, R. R. 1973. Guide to American hops. p. 1–36. *In* Steiner's Guide to American hops. S. S. Steiner Inc., N.Y.

———, S. T. Likens, and J. C. Shephard. 1968. Talisman, a new American Clusters-tye hop variety. Idaho Agric. Exp. Stn. Bull. 496.

Small, E. 1978. A numerical and nomenclatural analysis of morphogeographic taxa of *Humulus.* Syst. Bot. 3:37–76.

Thomas, G. G. 1970. The effect of gibberellic acid on the growth of the hop, var. Northern Brewer. p. 69–71. *In* Annual Report. Wye College, Univ. of London, England.

———, and W. W. Schwabe. 1969. Apical morphology in the hop (*Humulus lupulus*) during flower initiation. Ann. Bot. 34:849–859.

Thompson, F. C., and G. G. Thomas. 1970. Trials on the effect of gibberellic acid on the variety Northern Brewer. p. 65–68. *In* Annual Report. Wye College, Univ. of London, England.

Williams, I. H., and E. W. Weston. 1957. Hop propagation. I. The germination of hop seeds. p. 108–118. *In* Annual Report. Wye College, Univ. of London, England.

Umeda, U., S. Sasaki, S. Kubo, H. Kitami, and T. Sasahara. 1967. Cultivation of hop plants. U.S. Patent 3,309,815. Offic. Gaz. (U.S. Patent Office) 836:819.

Zimmermann, C. E., S. N. Brooks, and S. T. Likens. 1964. Gibberellin A_3-induced growth responses of Fuggle hops (*Humulus lupulus* L.). Crop Sci. 4:310–313.

28 Jute

D. P. Singh
Jute Agricultural Research Institute
Barrackpore, West Bengal, India

Jute, *Corchorus capsularis* L. and *C. olitorius* L., is a subtropical crop which produces fine fiber of commerce from its bark. *Corchorus capsularis* originated from Indo-Burma and *C. olitorius* from Africa (Singh, 1976).

The crop is cultivated mainly in India, Bangladesh, Brazil, and Taiwan between 20°N Lat and 30°S Lat. It is usually sown between February and June in the northern hemisphere (Finlow, 1939). In Brazil, two crops may be grown in a year, the first in May to June and another in November to December.

The plants normally attain the height of 4 to 5 m. Plant height in this crop is positively correlated with fiber yield (Shukla and Singh, 1967; Singh, 1971).

The crop is harvested for fiber between 120 and 140 days after sowing. The plants are cut at the ground level, tied in 20 to 30 cm diam bundles and left in the field for 2 to 4 days for shedding of leaves. During this period, the tissues of the main stem shrink and rupture, facilitating the entry of microorganisms to rot the stem (Kundu and Mukherjee, 1954; Mukherjee et al., 1954).

Retting is accomplished in slow moving and clean water with the process being completed in 10 to 12 days at 30 to 35 C. The fiber is separated from the main stem by hand, washed, and dried. This fiber serves as a raw material for jute mills.

I. PARENTAL MATERIAL

The species of jute are herbaceous annuals belonging to the family Tiliaceae. Cultivars of *C. capsularis* normally have branches, but some nonbranching types have been collected from Brazil. The *C. olitorius* cultivars collected from African countries, such as Sudan, have helped the Indian plant breeders in developing high-yielding photoperiod-insensitive cultivars. Many cultivars also have been collected from China, Taiwan, Japan, Philippines, USSR,and Pacific regions.

Scientists from Taiwan also have selected some useful cultivars of *capsularis,* like 'Shu-Shang Green Bark' and 'Tainang 1'. Tainang 1 is regarded as being resistant to anthracnose, a high yielder, and insensitive to day length (Chi et al., 1967). Mutation breeding also has helped in widening the source of germplasm in this crop (Singh et al., 1973; Singh, 1970).

Interspecific crosses among the two cultivated jute species are highly incompatible. Initial studies on the reasons for the incompatibility have shown that pollen tubes in interspecific crosses normally grow through the styles (Srinath and Kundu, 1952). Fertilization takes place, and some young embryos and free endosperm nuclei also are formed. However, the embryos degenerate early in development and seed fails to set (Ganesan et al., 1957; Patel and Dutta, 1960). Recently, some interspecific hybrids between *capsularis* and *olitorius* have been reported (Bhaduri and Bairagi, 1968; Islam and Rashid, 1960, 1961).

All the genetic collections of this crop are being maintained at the Jute Agricultural Research Institute (I.C.A.R.), Barrackpore (W.B.), India.

II. PLANT CULTURE

A. Field

The effect of soil fertility on hybridization and seed set in both species of jute has not been studied critically. A sandy loam soil with an application of 2 to 3 q/ha of manure is considered good for growing the crop. The application of 20 kg of N, 30 kg of P, and 30 kg of K per ha 30 days after planting helps in getting good fruit set and seed yield.

The soil should have enough moisture to help the plants grow properly. Low moisture in the soil adversely affects the formation and opening of buds. In such cases, the buds are shrunken and unhealthy and develop into inferior fruits.

Fig. 1—A *Corchorus olitorius* branch with flowers.

Plant populations of 0.2 to 0.3 million per ha are enough to get good seed production. The crop is normally sown in rows 40 to 50 cm apart.

Photoperiod plays a major role in this crop. A short photoperiod induces early flowering, while a longer period delays it. For hybridization and seed set, a photoperiod of 11 to 12 hours is good.

An open sky with good sunshine is considered ideal for hybridization. Cloudy and rainy days are not good for proper pollination and seed set.

A temperature ranging from 28 to 32 C and a relative humidity of 40 to 90% are considered standard for hybridization.

B. Growth Chamber and Greenhouse

To add a generation of crossing each year, jute has been grown in South India below 20°N Lat during winter months (November to March) where the winters are more or less similar to the summers prevailing between 20° and 30°N Lat. Growth chambers and greenhouses have not been used in breeding this crop.

III. FLORAL CHARACTERISTICS

Jute is predominantly self-pollinated, but cross-pollination of 10 to 15% occurs in *olitorius,* mainly by insects. The percentage of cross-pollination in *capsularis* is negligible (up to 5%) and does not require any treatment to enforce selfing.

The jute flower is perfect and small in size. The inflorescences occur on the terminal and lateral branches. The flowers are in groups of two or three in *olitorius* and in groups of two to six in *capsularis* (Fig. 1 and 2). Although the flowers in both species are small, those in *olitorius* are about twice the size of those in *capsularis.* In both species, the number of sepals and petals is usually five. In *olitorius,* there may be six to eight sepals and petals, but in *capsularis* flowers with six sepals and petals rarely occur. Stamens are variable in number from 20 to 60 and surround the stigma. The ovary usually has five locules and often six or more carpels, particularly in *olitorius.* In *capsularis,* there are two rows of ovules in a locule, each row usually consisting of five ovules. In *olitorius,* there is a single row of about 40 ovules.

Fig. 2—A *C. capsularis* branch with a flower and young buds.

IV. ARTIFICIAL HYBRIDIZATION AND SELF-POLLINATION

A. Equipment

It is not necessary to protect *capsularis* flowers against natural crossing, but it is important to control pollination in *olitorius*. Mesh cloth bags of cylindrical shape (60 × 30 cm) generally are used to protect the plants against foreign pollen (Fig. 3). The bags are fitted inside with two bamboo or iron rings, each about 20 cm from the end so that they divide the bags into three more or less equal portions. One end of the bag is closed and tied with strong, 45 cm-long string. The other end is threaded with the same length of another string, so that it can be closed whenever needed. Sometimes cages of the same size made by 16-mesh wire netting are used instead of mesh cloth bags. In wire net bags, both ends of the cage are closed with two cloth cylinders of 30 cm length, and the remaining arrangements are similar to those for mesh cloth bags.

Emasculation, pollination, and bagging require a pair of forceps, a hand lens, cellophane paper bags 16 × 7 cm, and small paper tags. A lens is used most easily if tied to a string and draped around the neck of the worker.

B. Preparation of the Female

In preparing for hybridization or self-pollination, one must be sure that the plant is free of disease or insect injury because unhealthy plants normally do not survive for a long period and the pods do not develop

Fig. 3—Flowers of jute plants enclosed in mesh cloth bags that are tied to a scaffolding.

properly. When a plant is to be selfed, a bamboo cane higher than the plant is placed in the soil. A bag is suspended from the cane over the flowers to be selfed. The drawn string at the open end of the bag is tied to the plant. The rings in the bag prevent the walls from collapsing and prevent injury of the flowers. When a large number of plants are to be selfed, a scaffolding or a framework, with upright and horizontal bamboo canes, has been found to be economical (Fig. 3). The string connecting the bag to the scaffolding should not be too taut because the wind may move bags and cause strain on the plants and damage them.

The flowers which appear first in the spike always set better. For hybridization, all buds are removed from the spike, except the one to be emasculated. Immature and young buds should never be used for emasculation. Emasculation is normally attempted between 1500 to 1800 hours on the day before pollination.

A bud that is ready for emasculation usually has pale yellow anthers and petals. An immature bud has whitish petals, and the anthers are slightly reddish. The filaments in an immature bud are short, and the anthers are below the level of the stigma.

The bud to be emasculated is held at its base between the thumb and index finger of one hand. The sepals and petals are separated with a pair of fine-pointed, straight forceps. Jute flowers are very sensitive to the removal of calyx and corolla, particularly the former. Seed set is poor when the calyx is removed. The anthers are removed with the forceps, then the buds are examined with a hand lens for complete removal of the anthers. The bud is enclosed in a cellophane bag having small perforations that permit the movement of air, but not the entrance of insects (Fig. 4). The bags are tied with a length of light twine to the base of the inflorescence that contains the emas-

Fig. 4—Inflorescences enclosed in the cellophane bags.

culated flowers. On the day of pollination, the emasculated flowers may be identified without difficulty because they usually have their sepals and petals wide open.

C. Pollination

The male flowers are enclosed in cellophane bags on the evening before they are used to avoid contamination from foreign pollen. The male flowers are removed between 0500 and 0700 hours on the day of pollination and placed in covered petri dishes. The buds start opening in the dishes and the anthers dehisce by longitudinal splitting.

Fresh pollen normally is used in jute for pollination. Pollinations attempted under rainy conditions do not result in good seed set. Under such cases, the collected male flowers can be stored in covered petri dishes for 1 or 2 days at 8 to 10 C.

The flowers of *capsularis* close at 1530 hours and those of *olitorius* at 1330 hours. The emasculated flowers usually become receptive the next morning between 0700 and 0800 hours, if the temperature is more than 20 C. Receptivity is delayed by 1 or 2 hours at lower temperatures. The best time for pollination is in the morning up to 0930 hours. In *capsularis,* it may extend to 1100 hours.

The male flower is held with a pair of forceps, and the base of the emasculated flower is held between the thumb and index finger of the other hand. The stigma is lightly touched with the anthers of the male flower until it is covered with pollen, which can easily be observed with the help of a hand lens. A paper tag bearing the cross number is tied onto the pollinated flower before enclosing it in the cellophane bag.

The cellophane paper bags function well in withstanding light showers or wetting by dew during the night. Bags should be removed 24 hours after hand pollination. The pollinated flowers start withering and a small fruit develops about 24 hours after pollination.

If emasculation and pollination have been properly conducted, the expected success may be as high as 90%. Failures may be detected best with the help of a hand lens on the third or fourth day after pollination. Fruits resulting from unsuccessful pollination appear discolored and shrivelled. In some cases, the pruning of the other pods on the branch provides stimulus to the crossed pods, enabling them to develop normally.

D. Factors Affecting Efficiency

Sowing of the cross nursery is in blocks or ranges of rows. The row spacing distance is 45 to 50 cm with plant spacing within the row of 8 to 10 cm. A population of 0.2 to 0.3 million plants per ha is considered sufficient for use in hybridization. The plants are not allowed to grow taller than 120 to 150 cm because it is easier to handle short plants for crossing purpose. Sometimes the tip of the plants are pruned at 40 to 50 days after sowing to induce more branching for production of more flowers.

To secure simultaneous flowering of parents of both *capsularis* and *olitorius,* the early parents may be sown about 30 to 45 days after the late parents.

Cold treatment of seeds at 8 to 10 C before sowing delays germination and results in more vigorous seedling growth in both the species. In such cases, *capsularis* cultivars flower 15 to 18 days earlier than the controls, while flowering is delayed by 10 to 15 days in *olitorius.* Vernalization alters the effects of the photoperiod treatment applied after it (Kar, 1943; Kundu, 1956).

Several genetic markers like pigmentation patterns, leaf shape, size and color, flowering period, branching habit, pod shape and size, seed coat color, and seed size help in identifying hybrid plants and seeds in this crop.

In *capsularis*, the distribution of anthocyanin in an individual plant varies to a great extent. Plants range from full green to full red, through a series of well defined and distinct forms. Plants have been classed into three broad genetic groups: full green, green pigmented, and red (Patel et al., 1944). Three loci control pigmentation: C is required for color production regardless of the genetic make up of the other two loci; A, A^L, and A^R are a multiple allelomorphic series affecting the distribution, as well as the intensity of the pigments; and R causes dilute coloration, and r, intense coloring.

In *olitorius,* the plants are either red or green. The expression of the pigments depends on the sunlight and the age of the plants. It is first observed as a dark red blotch at the base of the stipule. A ring of color slowly is formed around the base of the stem and gradually extends upwards as the plant grows. Genetically the gene for red pigment is dominant over the gene for green.

Cultivars possess different types of leaves, such as narrow lamina, crumpled surface, palmate, undulate, yellow, red, etc. These characters are mostly monogenic and recessive to normal green leaves (Das Gupta and Ghosh, 1954; Singh, 1974).

Some cultivars of jute flower early, while others are late. Early flowering is recessive to late (Ghosh and Sarma, 1965).

Cultivars differ in their branching behavior. Some cultivars are severely branched, while others have no branches. The branching behavior is dominant over nonbranching (Patel et al., 1945).

The pod surface may be smooth or ridged, its shape being oval or round. Such variations exist mostly in *capsularis.* In *olitorius,* only the elongated and stumpy pod shapes are common. Rough pod surface is dominant over smooth, and oval and stumpy pod shapes are recessive to normal (Patel et al., 1945).

In *olitorius,* the cultivars have seeds with different seed coat colors like leek green, steel grey, blackish grey, greyish steel, etc. This character helps in identifying hybrids. Seed coat color is governed by a single factor (Patel et al., 1945).

V. NATURAL HYBRIDIZATION

Jute has a low degree of cross-pollination, and is not good material for natural hybridization.

A monogenic male-sterile line has been identified in an x-ray induced mutant (Rakshit, 1967; Singh et al., 1973). This male-sterile gene has pleiotropic effects and causes several undesirable characters; therefore, it is unsuitable for any hybrid seed production program. Efforts to break this complex have failed, and no recombinant has been isolated that has the male-sterile gene disassociated from such undesirable characters as ribbon leaf and short height of the plant (Mitra, 1976). Efforts are underway to break this pleiotropic complex or to develop a cytoplasmic-genetic male-sterile.

The amount of outcrossing in jute is highly variable and nonrandom (Basak and Gupta, 1972). Because jute is cross-pollinated only by insects, reasons for the variation in outcrossing are attributed to the activity of insects (Bateman, 1950). The insects cross-pollinate only a few plants here and there in the plot, which results in highly variable cross-pollinated population. Under these circumstances, it has not been possible to follow any field layout for natural cross-pollination. An isolation of 10 to 15 m is recommended to minimize cross-pollination.

VI. SEED DEVELOPMENT, HARVEST, AND STORAGE

The time interval between the blooming and maturity of the pods is nearly 6 weeks. The appearance of brown spots, streaks, or patches on healthy pods indicate that the pods are mature and possess healthy seeds. The crossed pods should be collected as soon as they turn brownish black.

In a crop grown for its seed, one should ascertain if the plants have developed a yellow color on their stems. Such coloration indicates a stage when nearly 60 to 80% of the fruits are mature in *olitorius*. In this species, pods shatter and lose seeds at this stage, so it is always advisable to harvest the crop immediately. The problem of shattering does not exist in *capsularis,* and in some photoperiod-insensitive cultivars of *olitorius*. If it rains during this stage of maturity, the quality as well as quantity of seeds is affected adversely. The rain soaks through the pods and some of the seeds germinate in situ or become infected by fungi.

Crossed pods are harvested individually, and the seeds are threshed by hand. Harvesting of naturally pollinated plants or plants from cross-pollination plots is done in bulk. Plants are cut at their base or at mid-stem length and spread on a canvas under the sun. The plants are beaten by sticks or bamboos to shatter the seeds out of the pods. The seeds are winnowed, cleaned, and dried on the canvas in the sun for 4 days. Seeds harvested in bulk are analyzed for germination percentage.

Freshly harvested seeds contain nearly 21% moisture. Seeds stored at this moisture lose viability within a year. After drying, the seeds are treated with organo-mercury preparations against seed-borne fungi. This treatment prolongs viability in storage and controls seedling diseases. After exposure to moisture, treated seeds lose viability more rapidly than untreated seeds (Ghosh et al., 1951). Seeds should be stored in burlap bags with polyethylene under dry room conditions, otherwise they lose viability in less than 1 year (Kundu and Sarma, 1955). Seeds stored properly with 7% moisture retain viability up to 4 years.

VII. TECHNIQUES FOR SPECIAL SITUATIONS

Plant breeders have wished to develop an interspecific hybrid combining the desirable characters of both *capsularis* and *olitorius*. In most cases, the cross *olitorius* × *capsularis* has given very poor fruit set and no viable seeds. The reciprocal cross also gives negative results because the flower aborts shortly after pollination. Sulbha and Swaminathan (1959) used hormones in reciprocally grafted plants and increased the seed set. Islam and Rashid (1960, 1961) applied 300 ppm solution of indole-acetic acid to the pedicels of flowers soon after pollination and reported success in getting the hybrid. Islam (1964) combined hormone application and embryo culture to rear hybrid seedlings of the cross *capsularis* × *olitorius*. A medium supplemented with 0.1% yeast extract and 0.05 ppm each of kinetin and indole-acetic acid proved best for the germination of hybrid embryos which grew into transplantable seedlings.

Moistening the style and ovary with nearly 38% sucrose solution containing 0.01% thiamine hydrochloride or 40 ppm of napthoxy acid just after pollination also has increased fruit set (Roy, 1968).

REFERENCES

Basak, S. L., and S. Gupta. 1972. Quantitative studies on the mating system of Jute (*Corchorus olitorius* L.). TAG 42:319–324.

Bateman, A. J. 1950. Is gene dispersal normal? Hered. 4:353–363.

Bhaduri, P. N., and P. Bairagi. 1968. Interspecific hybridization in jute (*C. capsularis* × *C. olitorius*). Sci. Cult. 34:355–357.

Chi, C. Y., S. T. Hsu, and M. N. Shih. 1967. The breeding of insensitive new jute variety. Tainung 1. J. Agric. Assoc. China 58:59–67 (Ch).

———, S. T. Shiu, and H. T. Chu. 1969. Improvement of jute varieties in Taiwan. Agric. Quart. 11(2):112–119 Ch).

Das Gupta, B., and K. Ghosh. 1954. Genetics of *Corchorus.* VII Inheritance of undulate leaf character. Indian J. Genet. 14:78–81.

Finlow, R. S. 1939. The production of jute. J. Tex. Inst. 30:352–373.

Ganesan, A. T., S. S. Shah, and M. S. Swaminathan. 1957. Cause for the failure of seed setting in the cross *Corchorus olitorius* and *C. capsularis.* Curr. Sci. 25: 292–293.

Ghosh, T., M. Basak, and B. C. Kundu. 1951. Effects of isolation and chemical treatment in relation to storing of jute seeds. Indian Phytopathol. 4:33–34.

———, and M. S. Sarma. 1965. Inheritance of earliness in *Corchorus capsularis* L. Indian Agric. 9(1):1–6.

Islam, A. S., and A. Rashid. 1960. First successful hybrid between the two jute yielding species *C. olitorius* L. (Tossa) × *C. capsularis* L. (white). Nature 105:258–259.

———, and A. Rashid. 1961. A new jute hybrid. J. Hered. 52:287–291.

Iyer, R. D., K. Sulbha, and M. S. Swaminathan. 1961. Fertilization and seed development in crosses between *C. olitorius* and *C. capsularis.* Indian J. Genet. 21: 191–200.

Kar, B. K. 1943. Vernalization in jute. Curr. Sci. 13:130–131.

Kundu, B. C. 1956. A review of research works carried out at the Jute Agric. Res. Inst. (1948–49 to 1955–56). Indian Central Jute Committee, Calcutta, India.

———, and M. K. Mukherjee. 1954. Does jute cultivation exhaust the soil? Indian. Farming 3:14–15.

———, and M. S. Sarma. 1955. Jute seed. I. Nature and characteristics. Agric. Res. Bull. 3. Indian Central Jute Committee, Calcutta.

Mitra, G. C. 1976. Inheritance of male sterile complex mutation in white jute (*C. capsularis* L.). Genetica 47:71–72.

Mukherjee, M. K., A. K. Kundu, S. C. Chakravarty, and G. Halder. 1954. Role of jute (*Corchorus* spp) leaves in the maintenance of soil fertility of jute soils. Sci. Cult. 19:456–457.

Patel, G. I., and R. M. Dutta. 1960. Interspecific hybridization between *Corchorus olitorius* Linn. and *C. capsularis* Linn. and the cytogenetical basis of incompatibility between them. Euphytica 9:89–110.

Patel, J. S., R. L. M. Ghosh, and B. Das Gupta. 1944. The genetics of jute (*Corchorus*) II. Inheritance of anthocyanin pigmentation pattern. Agric. Res. Mem. 3. Indian Central Jute Committee, Calcutta.

———, ———, and A. T. Sanyal. 1945. The genetics of jute (*Corchorus*) II. The inheritance of corolla colour, branching habit, stipule character and seed coat colour. Indian J. Genet. 4:75–79.

Rakshit, S. C. 1967. Induced male sterility in jute (*C. capsularis* L.) Jap. J. Genet. 42:139–143.

Roy, B. 1968. Breeding procedures for jute. I.C.A.R. Tech. Bull. 17:44.

Shukla, G. K., and D. P. Singh. 1967. Studies on heritability, correlation and discriminant function selection in jute. Indian J. Genet. 27:220–225.

Singh, D. P. 1970. A new high yielding *olitorius* hybrid. Sci. Cult. 36:276–278.

———. 1971. Estimates of correlation, heritability and discriminant function in jute (*C. olitorius*). Indian J. Hered. 2(1):65–68.

———. 1974. Genetics of some induced leaf mutations in white jute. Indian J. Genet. 34(2):291–293.

———. 1976. Jute. p. 290–291. *In* N. W. Simmonds (ed.) Evolution of crop plants. Longman Publ., Co., London.

———, B. K. Sharma, and S. C. Banerjee. 1973. X-Ray induced mutations in jute (*C. capsularis* and *C. olitorius*). Genet. Agric. 27:115–147.

Srinath, K. V., and B. C. Kundu. 1952. Cytological studies of pollen tube growth in reciprocal crosses between *Corchorus capsularis* Linn. and *Corchorus olitorius* Linn. Cytol. 17:219–223.

Sulbha, K., and M. S. Swaminathan. 1959. Effect of grafting on fruit set and embryo development in crosses between *Corchorus olitorius* and *C. capsularis.* Curr. Sci. 28:460–461.

29 Lentil

F. J. Muehlbauer
USDA-SEA and Washington State University
Pullman

A. E. Slinkard
University of Saskatchewan
Saskatoon, Saskatchewan, Canada

V. E. Wilson
USDA-SEA and Washington State University
Pullman

Cultivated lentils (*Lens culinaris* Medik.) had their origin in the Near East arc where they were cultivated with other vegetables as early as the seventh millenium B.C. (Zohary, 1972; Williams et al., 1974). Evidence for the Near Eastern origin comes from archaeological remains. The mountainous region between Hindu-Kush and the Himalayas was first suggested as the center of origin (Barulina, 1930), but evidence acquired later supports the Near East (Zohary, 1972).

Lentil was one of the first plants to be cultivated by man, and it has long since been an important food crop in Eurasia and the subcontinent of India (Youngman, 1968; Zohary, 1972). Lentils are a high-protein pulse crop for direct human consumption. They are valuable when grown in rotation with cereals because they have the capability of fixing atmospheric nitrogen in symbiosis with *Rhizobium* spp. In the Western Hemisphere, the Pacific Northwest of the United States, followed by South America and Canada, is the largest producer of lentils.

I. PARENTAL MATERIAL

Lens culinaris Medik. belongs to the Order Rosales, Sub-order *Rosineae,* Family *Leguminosae* and Sub-family *Papilionaceae.* Within *Papilionaceae, Lens* holds an intermediate position between *Vicia* and *Lathyrus,* but it is closer to *Vicia. Lens* contains the cultivated lentil, *L.*

culinaris (synonym *L. esculenta* Moench.), and four wild species; *L. ervoides* Brign., *L. montbretii* Fisch. and Mey., *L. nigricans* Bieb., and *L. orientalis* Boiss. (Williams et al., 1974). Synonyms for the wild species have appeared in the literature, but they are too numerous to be mentioned. Similarities in plant type and pollen-grain morphology suggest that *L. culinaris* was derived from *L. orientalis* (Zohary, 1972; Williams et al., 1974). *Lens culinaris* is diploid, with 2n = 14 chromosomes, as is *L. orientalis* and *L. nigricans.* No chromosome numbers have been reported for *L. ervoides* or *L. montbretii.*

Slinkard (unpublished data, 1977) crossed *L. culinaris* and *L. nigricans* and produced progeny that segregated for seed size and seed coat color. Other attempts to intercross any of the five *Lens* species have not been reported.

Genetic collections of *Lens* for breeding purposes are maintained in the United States, India, and Syria. The U.S. Department of Agriculture Plant Introduction Station, Pullman, Washington, maintains a collection of over 500 accessions that includes material collected in 31 different countries. The International Centre for Agricultural Research in the Dry Areas (ICARDA), located in Aleppo, Syria, has recently established a germplasm collection that now stands at about 2,000 entries (Hawtin, 1977). The Indian Agricultural Research Institute, New Delhi-12, India maintains a collection of over 3,000 lentil lines collected from all lentil growing countries of Asia, Africa, Europe, and America (Sharma and Kant, 1975).

II. PLANT CULTURE

A. Field

Lentil nurseries for field crossing should be located on sites that have not been seeded to lentils for 2 to 3 years. Hard seed dormancy can result in volunteer plants that can cause seed mixtures in parental material or, more seriously, could be used inadvertently as parents in crosses. In the temperate regions, crossing blocks are located on well-drained soil on south-facing slopes to insure rapid emergence and good stands.

In Idaho, Oregon, Washington, and the western provinces of Canada, lentils are planted as early in the spring as possible in a seedbed with a minimum of straw residue on the surface. Soil temperatures rise quickly under cleanly tilled surfaces, and germination, emergence, and growth are enhanced. Delayed seeding usually results in a reduced plant growth, a rapid and brief flowering period, reduced time available for crossing, and greatly reduced seed set. In dry areas, improved germination and emergence can result if the plot area is rolled after seeding to improve contact of the soil with the seed. Under humid or wet soil conditions, this practice is avoided because it may cause soil crusting and reduce seedling emergence.

Plant densities in lentil nurseries are much lower than the density used in commercial production. For crossing purposes, lentils are planted in single-row plots 2 to 5 m long and 60 to 100 cm apart, with a within row spacing of 5 to 15 cm between plants. The additional space given to individual plants promotes branching, increases flower numbers, and prolongs flowering; all of these factors are beneficial to a lentil crossing program. Nurseries for evaluation of bulk populations or selections are planted similarly. The space between rows provided by this system allows individual plant selection within rows and greater seed yields from the selected plants or rows. However, the effects of intergenotypic competition and its influence on selection must be considered in low plant densities.

Lentils apparently are not excessive users of water. Some moisture stress seems to increase flowering and seed set. This effect is often observed on the dryland Palouse hills of eastern Washington and northern Idaho where higher seed yields occur on the slopes and hilltops where moisture is limiting than in the low areas where moisture and nutrients are not limiting. In the low areas, excessive vegetative growth sometimes occurs at the expense of seed yield. Irrigation benefits lentil yields, but must be carefully managed to avoid over-watering.

Lentils respond to nitrogen and phosphorus on soils that have low residual levels of these nutrients. Lentils also respond to potassium, especially on sandy soils. All fertilizer applications should be based on soil tests.

In field nurseries, methods of weed control include use of herbicides before seeding, before seedling emergence, and after emergence. The herbicide used and time of application are determined by the range of weed species present. Although herbicides are available for weed control, hand weeding and cultivation is safer from the standpoint of avoiding injury from herbicides.

The introduction and spread of plant viruses in nurseries by aphids can be controlled by insecticides. At least one virus, pea seedborne mosaic virus (PSbMV), is seedborne in lentils and can drastically affect lentil growth and yield (Hampton and Muehlbauer, 1977). Virus-free seedstocks and good insect control are the only methods currently used to prevent this virus from infecting the nursery and jeopardizing breeding stocks. Resistance to the virus is controlled by a single recessive gene, *sbv*, but the gene has not been incorporated into any known cultivar (Haddad and Muehlbauer, 1977).

B. Growth Chamber and Greenhouse

Lentils can be grown in a standard growth medium of equal parts of soil, sand, and peat moss with the nutrient levels adjusted according to soil tests. During the growth period, adequate amounts of nutrients are maintained in the growth media by addition of nutrient solutions or controlled-release fertilizers applied to the surface. Because lentils fix nitrogen through symbiosis with *Rhizobium leguminosarum,* it is important that rhizobia are either present in the soil or are added with the seed at the time of planting. Overwatering in growth rooms often promotes the growth of damping-off organisms, especially *Pythium ultimum* Trow., which also hastens maturity. Soil fungi can be suppressed by fumigation or steam sterilization of the soil before seeding, seed treatment with a fungicide, careful watering procedures, or drenching the soil with a fungicide. Soil fumigation to control diseases and soilborne insects and weeds seems to give better results than steam sterilization, because root and stem-rotting fungi do not usually recolonize fumigated soil as rapidly as steam pasteurized soil. Maximum seed production on F_1 hybrid plants can be obtained by hydroponic culture (A. Slinkard, unpublished data, 1977).

Flowering and seed set on lentils are enhanced by a photoperiod of 16 hours or longer, with a light level of 10 to 15 klux. Artificial light is used to extend the photoperiod. Lentils are considered to be either long-day or day-neutral plants (Shulka, 1955; Moursi and Gawad, 1963; Saint-Clair, 1972). Regardless of the day-length response, an extended photoperiod up to continuous illumination is conducive to earlier flowering and a reduced life cycle (F. J. Muehlbauer, unpublished data, 1977).

In greenhouses and growth chambers, temperatures of 27 C during the day and 21 C during the night are optimum for lentil growth and seed set (Saint-Clair, 1972); however, night temperatures as low as 15 C are not detrimental (A. Slinkard, unpublished data, 1977). Seed set in greenhouses

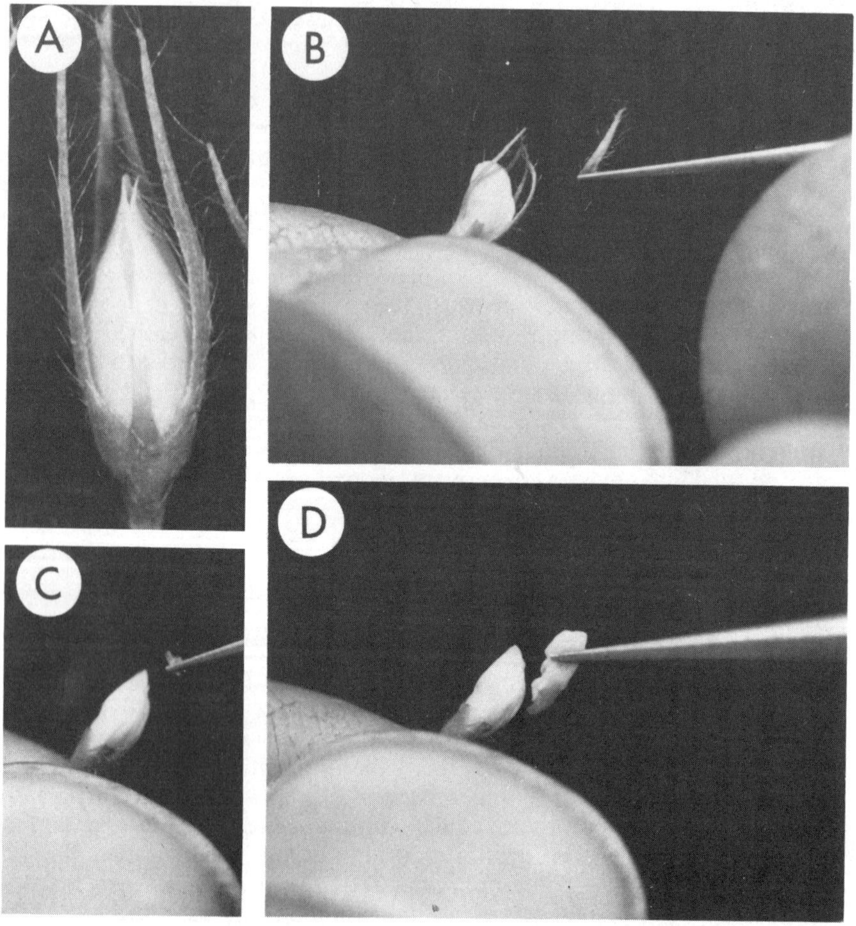

Fig. 1—Lentil flower at the appropriate stage for emasculation, and the preparatory steps in the emasculation procedure. (a) Female flower before emasculation, (b) removal of sepals, (c) removal of a notch from the standard petal, and (d) removal of the wing petals.

or growth rooms is improved by keeping relative humidity at about 50% (Wilson, 1972). Relative humidity can be maintained at this level with a commercially available greenhouse humidifier.

III. FLORAL CHARACTERISTICS

Lentils are naturally self-pollinated and have an extremely low level of natural cross-pollination estimated to be less than 0.8% (Wilson and Law, 1972). This low level of cross-pollination is presumed to be caused by pollen transfer by small insects, such as thrips. Wind and honeybees are ineffective for cross-pollination.

The individual lentil flower is complete and has a typical papilionaceous structure (Fig. 1a). Lentil flowers are borne singly or in multiples on peduncles that originate from the upper nodes of the lentil plant. The flower has a calyx that consists of five equally elongated sepals that equal or exceed the length of the corolla of the unopened flower. The corolla has a standard, two wings, and two lower petals that lie internal to the wings and are united at their lower margins to form the keel. There are 10 stamens, 9 of which are usually joined in a staminal column surrounding the pistil and one stamen remains free (Fig. 2a). The anthers open lengthwise and shed the pollen directly onto the stigma. After the anthers dehisce, the brush of hairs

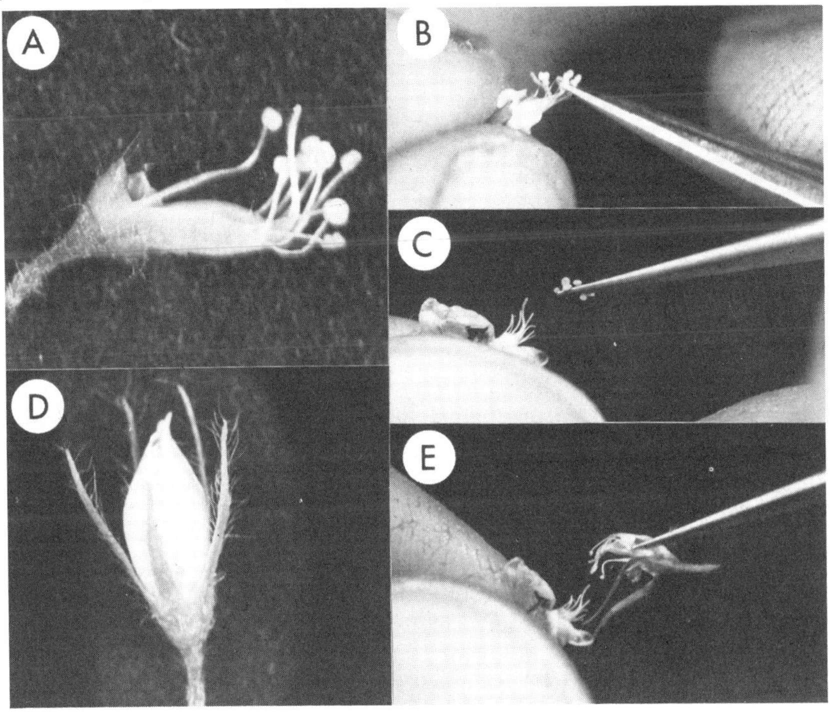

Fig. 2—Completion of emasculation and pollination. (a) Staminal column and stigma before emasculation, (b and c) removal of the stamens with forceps, (d) flower chosen for source of pollen, (e) pollination showing dehisced anthers and pollen-laden stigma of male parent being brushed against the stigma of the female flower.

on the stigma sweep the pollen out of the tip of the keel as the style and ovary elongate. The style usually develops at a right angle to the ovary and is flattened on the outer side. The ovary is flat and glabrous and normally contains one ovule or two that alternate along the margin. The small size and delicate structure of the flower are major factors preventing rapid and readily successful hybridization of lentil.

VI. ARTIFICIAL HYBRIDIZATION OR SELF-POLLINATION

A. Equipment

A dissecting microscope, sharply pointed forceps, alcohol, and small tags are the only equipment needed for lentil hybridization. The dissecting microscope with 5 × to 7 × magnification is used in the laboratory to view parts of the small lentil flower during the steps in emasculation and pollination. In the field, magnifying glasses may be substituted for microscopes. The sharply pointed forceps used for emasculation and pollination are immersed in 95% alcohol between crosses to prevent contamination by foreign pollen. The tags are used to record the parents, the date, and the person making the cross. The tag is affixed to the internode directly below the cross-pollinated flower.

B. Preparation of the Female

Emasculation of lentil flowers is necessary to prevent self-pollination (Wilson, 1972). Female flowers are chosen for emasculation when the tips of the petals reach 50 to 75% of the length of the sepals (Fig. 1a). At that stage of development, almost none of the flowers will have pollinated. Nearly all of the flowers will have pollinated when the petal length is greater than 75% as long as the sepals (Fig. 2c) (Wilson, 1972). Lentil flower petals expand about 25% of the length of the sepals every 24 hours; therefore, pollen is normally shed less than 24 hours after the optimum time for emasculation.

During the emasculation process, the selected flower bud is held between the thumb and forefinger with the suture of the keel facing the operator (Fig. 1b). Care should be taken not to bend or twist the peduncle. Sharply pointed forceps are used to remove the sepals closest to the keel. Several alternative methods of removing the anthers can be used. The easiest method seems to be to remove a notch from the standard (Fig. 1c) followed by removal of the wings (Fig. 1d). Removal of the notch makes it easier to fold back the standard to gain access to the wings. The standard and keel are then folded away to expose the staminal column and stigma (Fig. 2a). The stamens are removed by grasping the filaments with forceps (Fig. 2b) and breaking them free from the staminal column (Fig. 2c). The female flower is now emasculated and ready for pollination.

Some operators do not remove the notch from the standard, but immediately fold it back. They feel that by keeping the standard intact, a higher relative humidity is maintained within the flower after pollination;

thus promoting a higher percentage of seed set. In another procedure, the keel is removed at the same time as the wings to facilitate access to the stamens. However, removal of the keel often results in injury to the ovary and may result in a higher percentage of flower abortion.

C. Pollination

The highest percentage of crossed seed is produced if the lentil stigma is pollinated immediately after emasculation (Malhotra et al., 1978). The stigma retains its receptivity for at least 24 hours under conditions of high relative humidity and low temperature; however, reduced pod set is readily apparent if the stigma is exposed more than 12 hours. Immediate pollination while the emasculated female flower is in hand eliminates the time-consuming task of finding emasculated flowers within the lentil canopy.

Flowers in which the corolla has elongated to three-fourths the length of the sepals are selected as sources of pollen (Fig. 2d). Most flowers at that stage have anthers that have recently dehisced their pollen. Viable pollen suitable for transfer is identified by its bright orange-yellow color and is contained in a flower with a turgid appearance to the keel and wing petals. Pollen retains its viability for several days if flowers are collected immediately after anthesis, placed in petri dishes, and stored in a cool place.

Pollination is accomplished by first removing or folding back the keel and standard of the male flower to expose the 10 dehisced anthers and the pollen-laden stigma. The anthers and pollen-laden stigma are held with the forceps and brushed lightly against the stigma of the emasculated flower (Fig. 2e). The keel and standard of the pollinated female flower are carefully returned to their original position to protect the stigma. In an alternative method of pollen transfer, a spear-pointed needle is inserted through the keel and pollen is collected on the needle point. Pollen is transferred from the pollen-laden needle to the stigma of the emasculated female flower (Malhotra et al., 1978). Another method of pollination used successfully involves removing the standard and wing petals and opening one side of the tip of the keel with sharply pointed forceps to locate the pocket of pollen. The stigma of the emasculated female flower is dipped into the pollen pocket to achieve pollination. Regardless of the method of pollination used, enough pollen should be applied to the stigma to make it visible to the unaided eye, even if this requires the use of several male flowers. Pollinated flowers do not need protection from foreign pollen.

After pollination has been completed, a tag is affixed to the internode directly below the pollinated flower to identify the cross. Tags should not be affixed to the peduncle because of its delicate structure and susceptibility to injury. An alternative method of identifying crosses is to use colored thread looped around the peduncle, with a different color for each parental combination (Malhotra et al., 1978).

The percentage of successful pollinations varies widely by individuals and environments. Wilson (1972) reported 46, 61, and 82% successful manual pollinations by three experienced operators in the greenhouse. Success depends upon an environment with at least 50% relative humidity, high light level, and temperatures between 15 and 25 C. The percentage of successful crosses declines rapidly when the relative humidity drops below 35%, primarily because viable pollen is difficult to gather and transfer. Malhotra et al. (1978) reported 27 to 43% successful pollinations under field conditions in India. Under controlled environmental conditions with 21 day/15 C night temperatures, successful pollinations approaching 90% can be obtained (A. Slinkard, unpublished data, 1977).

D. Factors Affecting Efficiency

Crossing block layouts for lentils are similar to those used for other pulse crops. The layout used most often is to plant 4 to 5 m parental rows 60 to 70 cm from each other with the plants spaced 5 to 15 cm apart within rows. Paired rows of each parental combination facilitate crossing and reduce the movement of personnel within the block. In a variation of this layout, a male parent row is planted with female parent rows on each side, thereby making three-row, two-parent or three-parent crossing blocks. The two female rows in this system can be the same or different parents or can be used for different dates of planting of the same female parent. In another design, a group of parents is planted in adjacent rows across the nursery, with a parent common to all crosses planted perpendicular to those rows. In most crossing-block layouts, provision is made for second and sometimes third plantings close to the first planting to coordinate flowering of the parents or to extend the period of time for crossing.

Cotyledon color is a valuable genetic marker in identifying F_1 hybrids. F_1 hybrid seeds from crosses between yellow cotyledon female parents and red cotyledon male parents can be identified by their red cotyledon color; seeds with yellow cotyledons are discarded as selfs (Wilson, 1972). Yellow cotyledon color also can be used to identify F_1 hybrids in crosses between green cotyledon and yellow cotyledon parents, but care must be taken because green cotyledon color bleaches to yellow with age (A. Slinkard, unpublished data, 1977). Cotyledon color is controlled by the genes *yc* for color and *i* for the recessive color inhibitor so that the genetic constitution of red cotyledons is *II YcYc,* yellow cotyledons *II yc yc*, and green cotyledons *ii Yc Yc* or *ii yc yc* (Slinkard, 1978).

The most reliable genetic marker in use to identify F_1 hybrid seeds is red cotyledon color. The classification of lentil seeds into red and yellow cotyledon color groups is facilitated by the use of ultraviolet irradiation. Under ultraviolet irradiation at a wavelength of 366 mμ, cotyledon colors can be distinguished through the seed coat by the fluorescent properties of red cotyledons. For seeds with black seed coats, a portion of the seed coat must be removed; however, such damage to the seed coat often lowers the germination percentage. Metal shields should be used to protect workers from direct exposure to ultraviolet irradiation.

Genes for flower color can be used as genetic markers. Flower color is presumably controlled by *P* for the major pigmentation gene and *V* for violet flower color. The genetic constitution of different flower colors was suggested as violet, *VV PP*; pink, *vv PP*; white, *VV pp*; and rose, *vv pp* (Lal and Srivastava, 1975).

If cotyledon color is not used, true hybrids can be separated from selfs by the presence of obvious segregation in the F_2 generation. Segregation for seed coat color can be observed and used to identify hybrid populations. This method can be misleading because various seed coat colors and patterns are often borne on the same plant even though the seed coat is maternal and should not segregate. The solution to this problem must await a definitive study on the inheritance of seed coat coloration.

V. NATURAL HYBRIDIZATION

Natural hybridization is not used to make lentil hybrids.

VI. SEED DEVELOPMENT, HARVEST, AND STORAGE

Pod and seed development can be observed within 3 days after pollination. When cross-pollination is successful, the development of the ovary is rapid and seed within the pod can be seen readily within 7 days. When cross-pollination is unsuccessful, the ovary often will enlarge, but abortion will occur at a later date when the developing pod may have reached as much as 50% of its normal size.

Seed losses from pod dehiscence can be avoided by harvesting crosses as soon as the pods take on a yellow-brown color. Harvested pods should be allowed to dry completely in envelopes or bags before the seeds are removed. Harvesting and separating seeds from the pod are always performed by hand.

Lentils, like most large-seeded legumes, lose viability rapidly when stored at high temperature and high relative humidity. Usually hybrid seeds can be kept at room temperature, provided they are sown in a reasonable period of time. For periods of 3 years or more, lentil breeding material should be stored at about 10 C with 30% relative humidity or in a freezer at −20 C.

VII. TECHNIQUES FOR SPECIAL SITUATIONS

Post-harvest dormancy problems with lentil seeds often are encountered when the seeds are not allowed a storage period of at least 3 months. Scarification effectively insures good moisture imbibition of the seeds, but seeds often do not germinate with scarification only. Nearly 100% germination of recently harvested seeds can be obtained by soaking the scarified seeds in an aerated water bath for 4 hours at 23 C (N. Haddad, unpublished data, 1977). When using the aerated soaking technique, individual crosses or lines are identified separately in cloth bags or screened containers. Best results are obtained if the seeds are planted immediately after treatment without drying.

REFERENCES

Barulina, H. 1930. Lentils of the U.S.S.R. and of other countries. Bull. Appl. Bot. Genet. Plant Breed. 40:1–319 (Suppl.).

Haddad, N., and F. J. Muehlbauer. 1978. Inheritance of resistance to pea seedborne mosaic virus in lentils. Crop Sci. 18:613–615.

Hampton, R. O., and F. J. Muehlbauer. 1977. Seed transmission of the pea seedborne mosaic virus in lentils. Plant Dis. Rep. 61:235–238.

Hawtin, G. C. 1977. The ALAD/ICARDA Lentil Program. Lentil Exp. News Serv. 4:1–3.

Lal, S., and R. S. Srivastava. 1975. Inheritance of flower colour in Lentil. Indian J. Genet. Plant Breed. 35:29–30.

Malhotra, R. S., H. S. Balyan, and P. K. Gupta. 1978. Crossing technique in lentils. Lentil Exp. News Serv. 5:7–8.

Moursi, M. A., and A. A. Abd El Gawad. 1963. Lentil photoperiodism. 2. Formative and photoperiodic reaction to light duration. Ann. Agric. Soc. Univ. Ain. Chams. 8:297–304.

Saint-Clair, P. M. 1972. Responses of *Lens esculenta* Moench to controlled environmental factors. Committee on Agriculture, Univ. of Wageningen, the Netherlands. No. 72-12.

Sharma, B., and K. Kant. 1975. Variability for seed characteristics in the world germplasm of lentil. Lentil Exp. News Serv. 2:12–14.

Shukla, T. C. 1955. Photoperiodic induction in *Lens esculenta* Moench. Sci. Cult. 20:389–390.

Slinkard, A. 1978. Inheritance of cotyledon color in lentils. Lentil Exp. News Serv. 5:24.

Williams, J. T., A. M. C. Sanchez, and M. T. Jackson. 1974. Studies on lentils and their variation. I. The taxonomy of the species. Sabras J. 6:133–145.

Wilson, V. E. 1972. Morphology and technique for crossing *Lens esculenta* Moench. Crop Sci. 12:231–232.

————, and A. G. Law. 1972. Natural crossing in *Lens esculenta* Moench. J. Am. Soc. Hort. Sci. 97:142–143.

Youngman, V. E. 1968. Lentils—A pulse of the Palouse. Econ. Bot. 22:135–139.

Zohary, D. 1972. The wild progenitor and place of origin of the cultivated lentil, *Lens culinaris.* Econ. Bot. 26:236–332.

30 Oat

C. M. Brown
University of Illinois
Urbana

The cultivated oat (*Avena sativa* L. and *Avena byzantina* C. Koch) is an annual grass that ranks fifth among cereal grains in world production. Oats are grown on every continent, but their production is of far greater importance in the Northern Hemisphere. The areas of greatest production are in the North Central U.S., the Prairie Provinces in Canada, Western Russia, countries of Northwest Europe, and the British Isles. Of the cereal grains, oats contain the highest quality and quantity of protein and although approximately 90% of the oat crop is used for livestock feed, oats also provide an important source of human food. Prepared foods include rolled oats and dry breakfast cereals.

I. PARENTAL MATERIAL

The origin of cultivated oats is uncertain. Evidence indicates that various species of oats persisted as a mixture in wheat and barley used by several civilizations in areas surrounding and to the north of the Mediterranean Sea before being cultivated as a crop.

The genus *Avena* forms a polyploid series of diploids, tetraploids, and hexaploids with a basic chromosome number of seven. Rajhathy and Thomas (1974) have classified *Avena* into 19 taxonomic species of which 10 are diploid (2n = 14), five are tetraploid (2n = 28), and four are hexaploid (2n = 42). They have further divided the 19 taxa into 11 genome groups each representing a biological species, of which seven are diploid, three are tetraploid, and one is hexaploid. Taxa of the same biological species (same genomes) are cross compatible and produce fertile F_1 hybrids. Crosses between different biological species may exhibit varying amounts of cross-incompatibility and sterility in the F_1 hybrids.

Wild and cultivated species occur in each of the three ploidy levels, but most cultivated oats belong to the hexaploid group. Wild species possess certain desirable traits, such as disease resistance, and gene transfers between wild and cultivated species within the same biological species are usually accomplished with little difficulty. For example, all hexaploid wild and cultivated species are cross-compatible and produce fertile F_1 hybrids. Consequently, gene transfers between cultivated hexaploid species *A. sativa* and *A. byzantina* and the wild hexaploid species *A. fatua* and *A. sterilis* are common in most oat breeding programs. Gene transfers from the diploid and tetraploid wild or cultivated species into the gene pool of cultivated hexaploid oats are complicated by varying degrees of cross-incompatibility and sterility. However, considerable progress has been made in recent years and some transfers have been made, e.g., Aung and Thomas (1976) and Sharma and Forsberg (1977). The oat breeder needs to have a clear understanding of species relationships and the genomic structure of the polyploids of *Avena* to identify suitable pathways for accomplishing gene transfers from the lower ploidy species to the hexaploid species. For a more complete review of the current state of knowledge of species relationships and genomic structure of *Avena,* the reader is referred to Rajhathy and Thomas (1974) and Baum (1977).

Comprehensive germplasm collections of wild and cultivated species of *Avena* are maintained and distributed by the Plant Genetics and Germplasm Institute, Science and Education Administration, U.S. Department of Agriculture, Beltsville, Md., and by the Canada Department of Agriculture, Ottawa, Ontario.

II. PLANT CULTURE

A. Field

Environmental conditions that provide healthy and vigorous oat plants also provide optimum material for making crosses. Most soil types provide a suitable medium for growing oats, provided adequate nutrient, pH, and moisture levels exist. Although the requirements for added nitrogen depend especially upon soil type and previous cropping, some supplemental nitrogen is usually needed. Light-colored, low organic matter soils have a low capacity to supply nitrogen; the deep, dark-colored, high organic matter soils have a high capacity to supply nitrogen. Nitrogen fertilization rates of 32 to 45 kg/ha on dark-colored and 67 to 90 kg/ha on light-colored soils are usually adequate for optimum growth of oats.

Lime, phosphorus, and potassium needs should be determined by soil tests. Most soil fertility specialists consider a pH of 6 or above, Bray P_1 of 56 kg/ha, and an exchangeable K test of 336 kg/ha to be adequate for oats. The need for supplemental application of secondary and micronutrients varies greatly from region to region, and one should be alert for deficiency symptoms.

Oats respond favorably to cool, moist conditions and bright, sunny days. Field-grown oats intended for crossing should be watered when moisture stress is evident. Supplemental water will be beneficial even in the relatively humid oat growing regions, particularly during any extended dry period. Conditions that are favorable for growth of oats are also quite favorable for disease and insect development on the crop, and it is often desirable to use fungicides and insecticides to protect susceptible parents that are to be used for crossing. Birds, which can cause severe damage to field-grown oats, must also be controlled when present in damaging numbers.

B. Growth Chamber and Greenhouse

Optimum conditions for greenhouse and growth chamber hybridization are similar to those for field hybridization. Fertility, moisture, light, and temperature should be maintained so that vigorous plants are available. Adequate light level is an essential requirement for successful crossing in greenhouses and growth chambers. Supplemental light must be available for extending the photoperiod and increasing light level. This is particularly important in the northern latitudes where many cloudy days occur during the short-day winter months.

In Illinois, excellent seed set has been obtained in greenhouses using metal halide lights and a 12 to 13-hour photoperiod. The lights are usually left on during the entire day even on clear days. Temperature is maintained at 25 to 30 C during the day and 20 to 25 C during the night. Light level at bench level under the halide lights approximates 900 $\mu E\ sec^{-1}\ m^{-2}$.

Material for greenhouse crossing can be grown in pots, in soil beds, in nutrient solutions, or using any other method that provides vigorous plants. Plants grown in pots can be moved about for convenience when crossing. Four to five plants can be grown in a 15 cm pot provided adequate light, moisture, and fertility are provided.

III. FLORAL CHARACTERISTICS

All cultivated and wild species of oats are predominately self-pollinating and outcrossing seldom exceeds 0.5%.

The inflorescence of oats is an open panicle consisting of a main axis, which is a continuation of the stem, and lateral axillary branches that arise from alternate sides of nodes of the main axis (Fig. 1; Bonnett, 1966). Branches arising from the main axis are designated first order branches while those arising from the lateral branches are designated second or third order branches. The main axis and each of the lateral branches terminate in a single spikelet. The number of spikelets per panicle may vary from approximately 20 to 50 depending on genotype and on growing conditions.

The oat spikelet consists of several florets enclosed in two empty glumes with the tip of one glume extending slightly above the other (Fig. 2). Usually only the two basal florets of each spikelet are fertile (Fig. 3), but occasionally the lower three will be fertile. The florets are arranged alternately upon a central axis, the rachilla. Oat flowers are perfect and each is enclosed by two chaff-like bracts, the lemma and palea. The palea is located on the side of the floral axis next to the rachilla. Each flower contains three stamens, a pistil, and two lodicules. The lodicules, located between the base of the pistil and the lemma, swell at anthesis causing the flower to open. After anthesis, the lodicules collapse.

Natural anthesis usually occurs in the field in the afternoon, but the exact time of opening varies with environmental conditions and locations. Misonoo (1936) in Japan found that temperature variation was the primary factor that determined time of day for natural anthesis. On some days, natural anthesis did not occur, while on most days it occurred in the afternoon after the temperature had reached its maximum and had begun to decline. Coffman (1937) reported that natural anthesis usually reached its maximum in the field between 1500 and 1600 hours in Virginia, Iowa, and Idaho. Hadden (1952) observed that anthesis usually began at approximately 1530 hours in South Carolina and lasted for about 1 hour. On warm days (32 to 38 C) in Idaho, Coffman and Stevens (1951) found that natural anthesis was delayed until late afternoon. Natural anthesis usually occurs under field conditions in Illinois between 1400 and 1600 hours, but the author has seen it delayed to as late as 2000 hours on hot days.

Misonoo (1936) observed that anthesis occurs in oat flowers in the order of emergence from the sheath. Primary florets open first, but at times the primary and secondary florets open on the same day. Florets on an individual panicle bloom over a period of 8 to 9 days (Brown and Shands,

Fig. 1—Oat panicle.

Fig. 2—Oat spikelet showing ventral (underlapping and longer) and dorsal empty glumes.

Fig. 3—Side view of a spikelet showing dorsal (shorter, overlapping) glume, primary floret, secondary floret, and ventral (longer, underlapping) glume.

1956; Misonoo, 1936). Usually only one or two florets of a panicle bloom on the first day, but on large panicles as many as 10 to 13 florets near the middle of the panicle may bloom on one particular day.

Individual stigmas are usually receptive 1 day before natural anthesis and remain receptive for as long as 3 to 5 days. Length of receptivity is shorter under conditions of high temperature. Excessive mutilation of flower parts during emasculation contributes to floral dessication and more rapid loss of receptivity (Brown and Shands, 1956).

Pollen of oats is most abundant and of highest quality just prior to and during natural anthesis. Suitable pollen for effective crossing can usually be found in individual florets 1 to 2 hours before the flower opens during natural anthesis. Anthers dehisce most of their pollen just before or during opening of the floret, and one can seldom find usable pollen in a floret that has opened. Brown and Shands (1956) found that some oat pollen viability was maintained for several hours up to several days by storage at temperatures of 4 or 11 C. However, their results were somewhat variable and anthers tended to dehisce prior to use, thereby limiting the availability of usable pollen.

IV. ARTIFICIAL HYBRIDIZATION AND SELF-POLLINATION

a. Equipment

The equipment for making oat hybridizations consists of a small pair of sharp pointed scissors, narrow or sharp-pointed forceps, thin glassine bags, paper clips or staples, and small tags with attached strings (Fig. 4). Forceps and scissors are readily available in hospital supply houses or department stores. Forceps are the most important item of equipment. Some hybridists use sharp pointed forceps with curved ends while others prefer

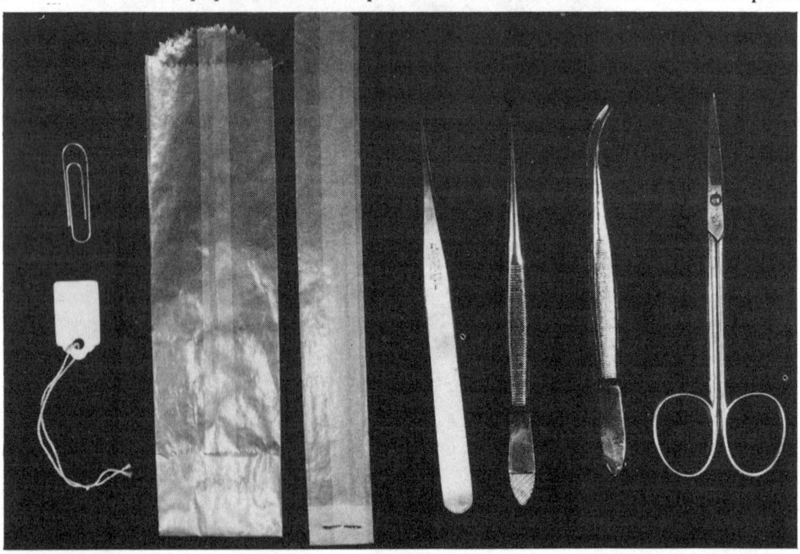

Fig. 4—Equipment for making oat hybridizations: tag and paper clip, alternative sizes of glassine bags, alternative types of forceps, and scissors.

straight ends (Fig. 4). Sometimes the tip 0.5 to 1 cm of each prong is ground or filed to a more or less flat edge to minimize damage to the glumes, lemma, and palea.

Glassine bags for protecting emasculated florets can be of various sizes depending to a large extent on preference of the breeder (Fig. 4). Many breeders use a bag that is just large enough to cover the emasculated florets and long enough to be secured to the culm of the female plant. A bag 3 × 15 cm or smaller will usually cover the emasculated florets on a single panicle.

B. Preparation of the Female

Frey and Caldwell (1961) suggested that the optimum time to emasculate oats is when the anthers have obtained full size, but are not yet ready to dehisce. Natural pollination begins in the uppermost florets and proceeds according to the order of emergence from the boot. Florets in individual spikelets will usually flower within 1 to 3 days after emerging from the boot, with the exact time being somewhat dependent on environmental conditions (Fig. 5). Anthers in florets near the tip of the panicle and at the ends of longer lateral branches are more mature than those below or, in the case of

Fig. 5—Panicle with spikelets near the top in approximate stage for emasculation.

lateral branches, those to the inside or on shorter branches. Anthers in the primary floret are more mature than those in the secondary floret of the same spikelet. Because of the above considerations, the hybridist can locate a primary floret containing anthers near maturity and be assured that those below have not dehisced and are good candidates for emasculation. Withered, empty anthers from natural anthesis the previous day are usually apparent the following morning, hanging from between the glumes of a closed spikelet, and these anthers provide an excellent benchmark below which florets can be selected for emasculation.

Emasculation can be started at any point on the panicle, but most hybridists emasculate florets near the top. Five to eight primary florets are usually emasculated on each panicle, but more can be emasculated on large panicles particularly if florets near the center of the panicle are ready for emasculation or if both primary and secondary florets are used. Hadden (1952) and Zillinsky (1955) found that secondary florets produce as many successful crosses as primary florets, but they are not as easy to use because of their smaller size and later stage of development. The most important consideration in making emasculations is to be sure that anthers are removed before they have dehisced any pollen. Although the more mature florets are easier to emasculate, they are more subject to accidental selfing during the emasculation process. A panicle with spikelets near the top in approximate stage for emasculation is shown in Fig. 5.

Although florets can be emasculated at any time of day that natural anthesis is not in progress, most hybridists emasculate in the morning to avoid the period of natural anthesis which occurs in the afternoon. The risk of accidental selfing during emasculation increases as the period of natural anthesis is approached; however, I have successfully emasculated and pollinated florets simultaneously. When using this method, emasculations must be made very near the period of natural anthesis to have usable pollen, and only a few (three to five) florets can be used on each panicle. While this method is convenient and useful, it should not be used by inexperienced workers or for genetic or other studies where an occasional self cannot be tolerated.

Each hybridist develops, through practice, an individual technique for manipulation of floral parts during emasculation. The spikelet is held near the base and usually between the thumb and index finger of one hand with the ventral (underlapping and longest) outer glume facing the hybridist (Fig. 6). The ventral glume is easy to identify because its tip extends slightly above the tip of the dorsal glume. Using the index finger instead of forefinger leaves the forefinger free to manipulate the floral parts. The outer glumes are separated with forceps held between the thumb, forefinger, and index finger of the other hand, resting the hand holding the forceps on the index finger of the hand holding the spikelet. The exposed secondary floret can be removed by snipping it with the forceps. The palea is separated from the lemma by inserting the tip of one prong of the forceps between the palea and lemma and pulling forward on the palea to expose the three anthers of the primary floret (Fig. 6). With practice, the hybridist will discover that the forefinger of the hand holding the spikelet will be extremely useful in separating first the glumes, lemma, and palea by applying light pressure on the dorsal glume. The forefinger is used to hold the floral parts open once they have been separated. The three anthers are extracted from the floret cavity with the forceps using care to avoid breaking any ripe anthers that are present and to avoid damage to the two stigma branches. The floral parts are replaced in their original positions. A slight rotation of the spikelet helps in tucking the edges of the ventral glume inside the edges of the dorsal glume.

The palea is usually subjected to less bending or creasing, especially by inexperienced workers, if the secondary floret is not removed until after the primary floret has been emasculated. In this case, the palea is pulled out and down and held by the forefinger against the secondary floret while the anthers in the primary floret are being removed. After the palea is pushed back into its original position, the secondary floret is snipped off with the forceps or pulled out and down breaking the rachilla attachment.

Several authors (Brown and Shands, 1956; Coffman, 1937; Zillinsky, 1955) have emphasized that care should be exercised to avoid mutilation of flower parts. However, my more recent experiences and the results reported by McDaniel et al. (1967, 1969) suggest that a reasonable amount of mutilation can be tolerated with little detrimental effect on seed set, particularly when other conditions are favorable.

The emasculated florets are covered with a glassine bag to exclude outside pollen. The bag should be just wide enough to cover the emasculated spikelets and long enough to be folded and secured with a paper clip or staple (Fig. 7). When securing the bag, the paper clip should not be extended across the culm because it will often cause the culm to break at that point. Coffman (1956) reported that bagging was detrimental to seed set during hot weather, but it is doubtful that this is very important under most conditions. When crossing in the field, panicles may need to be supported by a stake, but this is usually not necessary if a small bag is used. Bagging of emasculated florets is sometimes omitted in the greenhouse if sufficient isolation is available or if pollen drift is minimal.

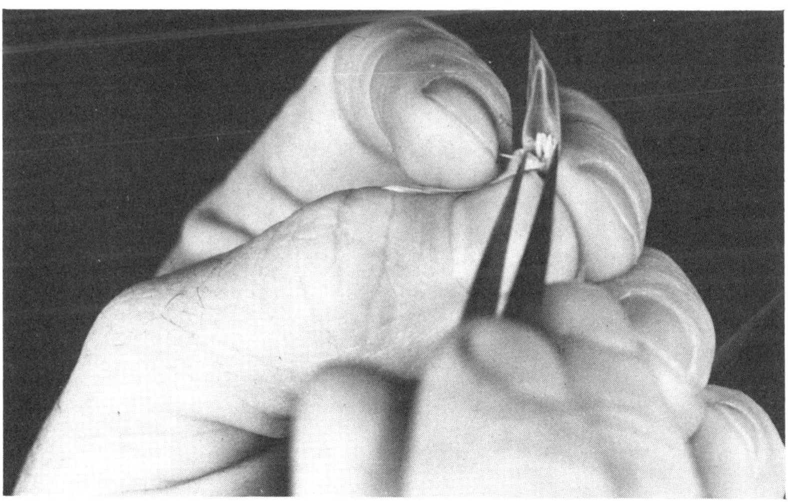

Fig. 6—Floret opened for emasculation. Note position of fingers on onehand holding spikelet and the other holding forceps.

The emasculated panicle should be labelled with a tag designating the female parent and date of emasculation (Fig. 7). A small tag attached on the main panicle axis just below the last emasculated floret can indicate that all florets above that point have been emasculated. A small length of yarn can also serve the same purpose. When this method is used, only the unemasculated spikelets that interfere with bagging need to be removed, but any spikelets above this point that had pollinated naturally prior to emasculation should be removed so that only emasculated florets are under the bag.

C. Pollination

Ideal environmental conditions for making successful pollinations are clear days and moderate to low temperatures. High temperatures (30 C and above) reduce the availability of usable pollen, and contribute to floral dessication and loss of stigma receptivity (Brown and Shands, 1956). High temperatures on days between emasculation and pollination cause reduced seed set, particularly in cases where 3 or 4 days elapse between emasculation and pollination.

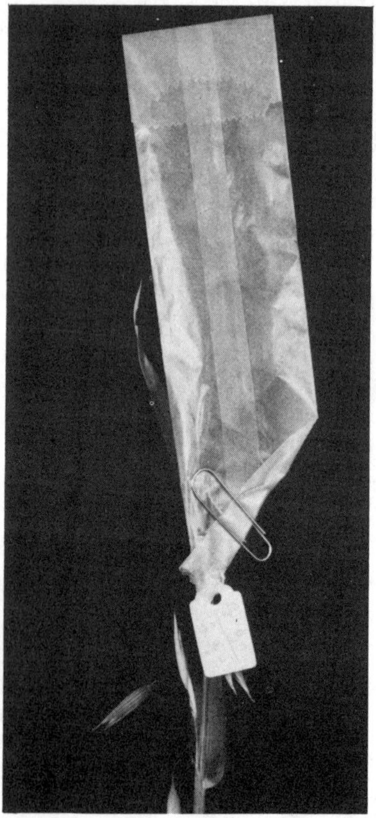

Fig. 7—Emasculated florets bagged to exclude outside pollen.

Brown and Shands (1956) reported that the optimum time between emasculation and pollination was 1 to 3 days. Hadden (1952) reported excellent results from emasculations made in the morning and pollinated on the afternoon of the same day. I also have experienced some excellent results when making pollinations immediately after emasculation. If emasculations and pollinations are to be made simultaneously or even on the same day, more mature and fewer florets need to be selected on each panicle, and it should be recognized that this will increase the probability of natural selfs. Brown and Shands (1956) suggested that selfing can be reduced to a minimum by emasculation 1 or more days previous to natural anthesis and then allowing 2 to 3 days between emasculation and pollination. Usually the more mature stigmas begin to dehydrate and lose receptivity if more than 3 days elapse between emasculation and pollination. If a large number of florets have been emasculated on an individual panicle, it may be desirable to pollinate them over a period of 2 to 3 days due to differences in maturity and receptivity of the females.

Most pollinations of oats are made between 1300 and 1600 hours, but the best time will vary from day to day. The important consideration is to make sure that pollinations are made when usable and viable pollen is available. This optimum period for good pollen usually begins 1 to 2 hours prior to natural anthesis and extends until natural anthesis has occurred. The problem of poor pollen quality on very hot days can sometimes be overcome by waiting until the temperature drops so that the time of hand pollination more nearly approximates that of natural flowering. However, the hybridist must be careful that pollination is not delayed to the point where natural anthesis has occurred such that all sources of viable pollen are eliminated for that day's crossing. Even the most experienced hybridist has to determine the optimum period each day through trial and error by examining anthers in the most mature florets that have not pollinated.

Anthers with usable pollen will be yellow, plump, and should dehisce within 1 min when removed from the floret and placed on the hand. Anthers can be collected from primary or secondary florets. Several florets with suitable pollen often will be found on the same panicle so that the hybridist can find a sufficient number of mature anthers on a single panicle for pollinating all the emasculated florets on one or more panicles. Within a panicle, those spikelets immediately below those with withered anthers from the previous day of natural anthesis are the most likely to be in the proper stage for use as pollen sources.

Anthers can be moved from one location to another by transporting entire panicles, individual spikelets, or anthers that have been removed from panicles on intact plants. Anthers that have been removed from florets should be used almost immediately. The anthers used for making pollinations are held in the crease of the palm, on the back of the hand between the thumb and forefinger, or on the wrist at the base of the thumb. When making pollinations, the lemma and palea of each emasculated floret are separated in the same way as for emasculation, and a mature anther held between the prongs of the forceps is gently tapped against the inside of the lemma to be sure the anther opens and spreads pollen on the stigma hairs.

The used anther may be left in the floret or in some cases it can be used to pollinate additional florets. The palea and glumes are returned to their original position. When all emasculated florets on a panicle have been pollinated, the tag is labeled with the male parent and date of pollination, and the glassine bag is replaced to prevent outcrossing.

The approach method is an alternative procedure for pollination of oats. It initially had good success in wheat and barley, but poor success in oats (Rosenquest, 1927). Curtis and Croy (1958) refined the method and found it quite satisfactory for making oat crosses in Oklahoma. McDaniel et al. (1967) further refined the technique and have used it with excellent success for making oat crosses in Texas. These workers found that approach crossing offers several important advantages over the conventional method. Crosses can be made at any time of day and many more florets can be pollinated than would be possible by hand. Even very young florets can be emasculated because the male parent sheds pollen over several days, providing pollen when the stigmas become receptive. More than one cross can be made under the same bag by bagging one or two panicles of the same male with several different female parents. Labor is reduced because it is not necessary to return to florets to apply pollen. Their seed set percentage was much higher with the approach method of crossing than with the conventional method.

With the approach method, each spikelet used as a female is prepared by removing the secondary floret and anthers of the primary floret. After emasculation, the upper portion of each spikelet is removed by clipping straight across the glumes, lemma, and palea just above the stigma (Fig. 8). The clipped spikelets remain erect so that pollen from the male parent falls directly onto the stigmas. Panicles used for male parents require little preparation except that heavy awns, if present, are removed to facilitate bagging.

In the greenhouse, pots containing the male and female parents are placed together so that the lowest floret on the male parent is just above the uppermost floret on the female parent. Blocks, pots, or other devices can be used to adjust the height of the parent panicles to the desired level. Panicles from both parents are enclosed in a bag. McDaniel et al. (1967) used a bag approximately 35 cm in length made from 2.9 cm cellulose dialyzing tubing.

Approach crossing can also be used to cross greenhouse-grown plants with field-grown plants (McDaniel et al., 1967). They cut culms of field-grown plants that were in the appropriate stage for use as male parents, and placed them in test tubes of water attached to stakes positioned near the female parents. The top of each test tube was then placed below the emasculated female panicle and the length of detached male culm was cut so that its panicle was just above the female. A pollinating bag was placed over both panicles and drawn over the open end of the test tube. The evaporation and transpiration were thought to add humidity to the atmosphere under the bag, providing a more favorable environment for seed set.

Even the experienced hybridist will encounter considerable variation in seed set from time to time. Although technique and environmental conditions influence success, pollen quality at time of making pollinations appears to be the major determining factor. An experienced hybridist working in near optimum conditions should expect 25% or more of the pollinated florets to set seeds in the field, 50% or more in the greenhouse.

D. Factors Affecting Efficiency

Field nurseries for crossing usually are made in rows, although they can be made by any method that will provide vigorous plants. Two to three plantings of one row 2 to 3 m long will provide plenty of material for crossing. Parental rows should be sufficiently far apart to allow space for working. Single rows 60 cm apart or paired rows with 30 cm between members of the pair and 60 cm between pairs provides sufficient space for most workers. Some breeders space plant or plant thin populations in the row to stimulate tillering and provide more crossing material for parents that are in small supply. Rows in the crossing nursery are sometimes arranged so that parents to be crossed with each other are adjacent.

Flowering date can be altered by varying planting dates. Low planting rates or even space planting will stimulate tillering and provide more usable male and female florets over a longer period of time. Clipping of plants in the late tillering or preboot stages to a height of 15 cm also will delay flowering and stimulate additional tillering.

Oats for crossing grown in the greenhouse can be handled most efficiently in pots. Two to three plantings of several pots of each parent made at 5- to 7-day intervals will usually provide ample material for crossing, and

Fig. 8—Spikelets clipped in preparation for approach crossing.

will extend the crossing period to accommodate the maturity ranges of most parents. More dates or more pots per date may be required where certain parents are to be used for a large number of crosses.

V. NATURAL HYBRIDIZATION

To date there has been no report of a reliable and efficient way of producing natural hybrids in oats. Cytoplasmic male sterility has not been reported. Although several cases of varying degrees of male sterility have been attributed to genetic causes or to pollen control chemicals, none has provided a suitable method for making hybrids. Lafever and Patterson (1964) suggested the possible use of a male-sterile nullisomic as the female parent to produce a fertile F_1 monosomic, but they obtained only 11.7% natural cross-pollination by using the method.

VI. SEED DEVELOPMENT, HARVEST, AND STORAGE

The reader is referred to Bonnett (1961) for an excellent treatment of all aspects of seed development in oats. Brown and Shands (1957) examined the time sequences involved in pollen tube growth, fertilization, and early development of the seed. Pollen germinated and tubes had entered the style 5 min after pollination. The male generative nuclei were near the egg nucleus and polar nuclei within 30 min, and male gametes had entered the egg after 4 hours. The fertilized egg was in prophase 13.5 hours after pollination and a two-celled proembryo was observed 19 hours after pollination.

Examination of the pistil of pollinated florets 2 or 3 days after pollination will distinguish between successful and unsuccessful hybridization. If fertilization has occurred, the floret will be closed, the stigma wilted, and the ovary noticeably enlarged.

Crossed seed harvested as soon as 12 to 14 days after fertilization will usually provide viable seed, but such seed will be small and poorly filled. After 25 to 30 days, seed will be fully mature, large, and plump. Oat genotypes exhibit varying degrees of seed dormancy, and chilling often is required to break this dormancy when crossed seed are to be planted soon after harvest. Wild oats usually exhibit a high level of seed dormancy.

Schwendiman and Shands (1943) found that dormancy in oats could be overcome by germinating at continuous low temperatures between 8 and 12 C for 10 to 14 days or by prechilling moistened seed for 4 days at 4 C and then placing them at 22 C for 6 days. Removing the hulls also increased the rate and total percentage germination. I have obtained excellent germination of freshly harvested seed by placing them between wet germination blotters or paper towels at 18 C for 7 days prior to planting in the field or greenhouse.

REFERENCES

Aung, Taing, and Hugh Thomas. 1976. Transfer of mildew resistance from the wild oat *Avena barbata* into the cultivated oat. Nature 260:603–604.

Baum, B. R. 1977. Oats: Wild and cultivated. A monograph of the genus *Avena* L. (Poaceae). Thorn Press Ltd.,

Bonnett, O. T. 1961. The Oat plant: Its histology and development. Illinois Agric. Exp. Stn. Bull. 672.

———. 1966. Inflorescences of maize, wheat, rye, barley, and oats: Their initiation and development. Illinois Agric. Exp. Stn. Bull. 721.

Brown, C. M., and H. L. Shands. 1956. Factors influencing seed set of oat crosses. Agron. J. 48:173–177.

———, and ———. 1957. Pollen tube growth, fertilization and early development in *Avena sativa.* Agron. J. 49:286–288.

Coffman, F. A. 1937. Factors influencing seed set in oat crossing. J. Hered. 28:296–303.

———. 1956. Bagging possibility reduces success in oat crossing. Agron. J. 48: 191.

———. 1961. Origin and history. *In* F. A. Coffman (ed.) Oats and oat improvement. Agronomy 8:15–40. Am. Soc. of Agron., Madison, Wis.

———, and H. Stevens. 1951. Relation of temperature and time of day of pollination to seed set in oat crossing. Agron. J. 43:498–499.

Curtis, B. C., and L. V. Croy. 1958. The approach method of making crosses in small grains. Agron. J. 50:49–51.

Frey, K. J., and R. M. Caldwell. 1961. Oat breeding and pathologic techniques. *In* F. A. Coffman (ed.) Oats and oat improvement. Agronomy 8:227–262. Am. Soc. of Agron., Madison, Wis.

Hadden, S. J. 1952. Unusual success in crossing oats. Agron. J. 44:452–453.

Lafever, H. N., and F. L. Patterson. 1964. Aneuploids for commercial hybrid oats. Crop Sci. 4:635–638.

McDaniel, M. E., B. R. Hathcock, and D. J. McDonald. 1969. Effect of stigma mutilation on oat seed set. Crop Sci. 9:384–385.

———, H. B. Kim, and B. R. Hathcock. 1967. Approach crossing of oats. Crop Sci. 7:538–540.

Misonoo, G. 1936. Ecological and physiological studies on the blooming of oat flowers. J. Fac. Agric., Hokkaido Imp. Univ. (Sapporo, Japan) 37:24–237.

Rajhathy, T., and H. Thomas. 1974. Cytogenetics of Oats, *Avena* L. Misc. Pub. Genet. Soc. of Canada—No. 2. Ottawa, Ontario, Canada.

Rosenquist, C. E. 1927. An improved method of producing F_1 hybrid seed of wheat and barley. J. Am. Soc. Agron. 19:968–971.

Schwendiman, A., and H. L. Shands. 1943. Delayed germination or seed dormancy in Vicland oats. J. Am. Soc. Agron. 35:681–688.

Sharma, D. C., and R. A. Forsberg. 1977. Spontaneous and induced interspecific gene transfer for crown rust resistance in *Avena.* Crop Sci. 17:855–860.

Zillinksy, F. J. 1955. Cross compatibility relationships among some *Avena* species and polyploids. Can. J. Agric. Sci. 36:107–113.

31 Peanut

A. J. Norden
University of Florida
Gainesville

The cultivated peanut (*Arachis hypogaea* L.) is an annual legume that probably originated in Bolivia, South America, along the eastern slopes of the Andes (Krapovickas, 1968). It now is widely grown and well adapted to tropical, subtropical, and warm temperate areas of the world. In the United States, peanuts are utilized primarily as whole seeds and for making peanut butter, while elsewhere in the world they are mainly crushed for oil. Seed of the commonly grown cultivars contain 42 to 52% oil and 25 to 32% protein.

I. PARENTAL MATERIAL

The genus has been classified into seven distinct sections (Gregory et al., 1973). The *Arachis* section includes four annual diploids, (2n = 20), at least five perennial diploid species, and two annual tetraploids (2n = 40), one of which is the cultivated peanut. Gregory and Gregory (1976) postulated, ". . .the cultivated peanut consists of two genomes, both of which came from species of the section *Arachis,* and . . .one of these genomes came from a perennial species and one from an annual."

The cultivated species has been taxonomically classified by Krapovickas (1968) into two subspecies and four interfertile varieties: (i) subspecies *hypogaea*; variety *hypogaea* (includes the U.S. market types Virginia and Runner), and variety *hirsuta*; (ii) subspecies *fastigiata*; variety *fastigiata* (Valencia market type) and variety *vulgaris* (Spanish market type). Although many interspecific crosses in the genus *Arachis* have failed, successful crosses have been reported. The hybrids from the successful crosses are largely sterile, but Smartt (1964) found that some sterility obstacles may be overcome by artificially doubling the chromosome number. Gregory et al. (1973) list several species of *Arachis,* a few of which are freely cross-compatible with the cultivated species. *A. monticola,* a wild annual species, has the same chromosome number, a similar karyotype, and complete interfertility with *A. hypogaea.*

Varnell and McCloud (1975) listed a total of 10,790 accessions of the cultivated peanut in nine countries of the world. The Southern Regional Plant Introduction Station at Experiment, Ga. maintains 3,990 accessions and is the most complete and accessible collection in the United States.

II. PLANT CULTURE

A. Field

Like most crop plants, peanuts do best with a balanced supply of available plant nutrients. The soil should be especially well supplied with calcium. Fruits that appear normal may be empty or contain only shriveled seeds due to a lack of readily available calcium in the fruiting zone.

Smith (1954) observed that changes in flower frequency may be influenced by climatic factors which determine relative humidity. In field plantings, flower frequencies reach low values under drought conditions, but high flower frequencies occur within 2 or 3 days after the soil has been saturated by rainfall or irrigation.

Ono et al. (1974) reported optimum soil moisture content in the podding zone to be about 40% of the total soil volume, regardless of the soil moisture content in the rooting zone. Soil moisture content lower than 40% reduced the podding percentage and pod enlargement, but moisture greater than 40% only reduced enlargement.

Gibbons (1977) had very low success rates in India when field hybridizations were performed during the dry season (December to May), presumably due to the low relative humidity.

Optimum plant population depends mainly on the purpose of the planting and secondarily on the subspecies involved. For example, F_1 plants should be spaced from 45 to 90 cm apart in rows 90 to 180 cm apart to produce large quantities of seed. Spacings from 30 to 45 cm apart in rows 91 cm apart are recommended for segregating populations to permit uniform opportunity for the expression of each plant's potential, and permit visual evaluation and manual separation by the breeder. The closer spacings are satisfactory with erect or semi-erect growing types.

Field nursery plantings of wild peanut species are in isolation type plots, i.e., planted rows have three or four non-planted 90 cm plots between them. Species that will cross are not planted in adjacent plots. If two species will cross, the hybrid should be easily recognized by a genetic marker, such as flower color (Gregory, 1977).

Umen (1933) has reported that *Arachis hypogaea* is photoperiod-insensitive under constant temperatures, but did show photoperiod sensitivity under variable temperatures. Fortanier (1957) demonstrated that the initiation of flower production was independent of photoperiodism, as well as thermoperiodism. These results were corroborated by Wynne et al. (1973). Peanuts require a high light level of not less than 45% full sunlight (Simpson, 1977). Fortanier (1954) stated that flower bud elongation and opening were light-wave-length-dependent, and that the time of flower expansion was regulated by a light stimulus received by the bud 3 days earlier.

Flower expansion was also shown to be dependent on white and blue wavelength.

Temperatures between 22 and 33 C are most ideal for flowering and fruiting of cultivated peanuts. Ono et al. (1974) noted that the beginning time and rate of pod development were significantly affected by soil temperature in the podding zone. Optimum soil temperature was 31 to 33 C; minimums were 15 to 17 C and maximums 37 to 39 C.

Pollen is sensitive to environmental changes such as light, temperature, and humidity. DeBeer (1963) noted that plants grown at temperatures of 33 C produced pollen of low viability, and that the temperature 36 to 96 hours preceeding flower opening also influenced viability.

Hybridization of the cultivated peanut is commonly done in pots or flats in greenhouses so the plants can be moved around. However, in some countries hybridization is done in field nurseries. Many wild species are not successfully hybridized when grown in pots, but must be grown in deep benches or beds. Several wild species do not set fruit inside a greenhouse and hybridization must be done outdoors (Gregory, 1977).

B. Growth Chamber and Greenhouse

The female plants of cultivated species that are to be used in crosses are generally grown in 15-liter, 30-cm diam pots on greenhouse benches. The containers may be rotated to provide convenient access to the flowers during emasculation and pollination. In Florida, hybridization has been more successful from plantings made in the greenhouse in early spring, probably because lower temperatures during the winter season, cloudy days, or both retard floral development and flowering (Norden and Rodriguez, 1971).

Banks and Jordan (1971) experimented with different sources of water (tap, lake, distilled) and different fertilizer treatments for growing peanuts under greenhouse conditions in a high pH soil. They found distilled water was best and acidified tap water second best. They obtained good results using acidifying soluble fertilizers of 21-7-7 and 20-20-20 NPK in combination with distilled water and steam-sterilized sandy loam soil in plastic pots.

New or sterilized soil to be used in hybridization studies should be inoculated with *Rhizobium* bacteria (Simpson, 1977). He found that 80 to 85% of soil capacity is the best moisture level for greenhouse plantings.

Apparently maintaining a high relative humidity is critical for successful artificial hybridization in greenhouses.

Peanut hybridization usually has been done by emasculating flowers in the evening and pollinating them the following morning. Two modifications that increase crossing efficiency recently have been reported. Profuse flowering can be induced with artificial lighting from 1700 to 0130 hours (Hildebrand, 1974). Following that treatment, buds can be emasculated between 0730 and 1130 hours. Pollen parents can be grown outdoors and will provide ample fresh pollen during this time enabling the breeder to make pollinations immediately after emasculating flowers of the female parent. The average time required for this crossing method is 4 min. Hammons (1973) estimated that the conventional method required 10 min per flower.

A second modification involves the use of a growth chamber to reverse the day-night cycle (Banks, 1976). A 12-hour, 29 C day begins at 1630 and a 12-hour, 21 C night at 0430 hours. Emasculations are made between 0800 and 1000 hours and pollinations are done immediately with fresh pollen coming from plants grown in a greenhouse. He found this crossing method yielded more than 50% viable hybrids per pollination and, because the growth environment was artificially controlled, permitted up to three crossing cycles per year. The crossing methods of Hildebrand (1974) and Banks (1976) share the obvious advantage of allowing hybridization to be done during normal working hours when labor can be more efficiently utilized.

III. FLORAL CHARACTERISTICS

The peanut is normally a self-pollinating species. Anthesis generally occurs prior to flower expansion. In mature flowers, the stigma lies buried among the dehisced anthers in the tightly closed keel petal, thus assuring self-fertilization (Smith, 1950; Capinpin and Guevara, 1951).

The occurrence of natural hybrids in the cultivated peanut was reported as early as 1910 by van der Stok (1910), and plants suspected of having originated by natural hybridization are not infrequent in most breeding nurseries. Hammons (1973) made numerous studies of natural hybridization using a dominant leaf marker called, krinkle. He found rates of outcrossing in breeding nurseries at Tifton, Ga. ranging from 0.25 ± 0.002% to 6.16 ± 0.007%. Culp et al. (1968) obtained much lower rates in Virginia, averaging from 0.09% to 0.27% which is similar to findings in Puerto Rico (Stone et al., 1973). Gibbons and Tattersfield (1969) reported from none to 1.67% outcrossing in Malawi, Rhodesia, and Zambia.

Various wild bee species are the principal vectors of natural cross-pollination in *A. hypogaea,* and the frequency is associated with the production of atypical flowers that vary considerably for different cultivars (Hammons, 1963; Hammons and Leuck, 1966; Leuck and Hammons, 1969).

The inflorescences consisting of three or more flowers are spikelike and always occur in the axils of cataphylls or of foliage leaves. The flowering axis never occur at the same nodes as vegetative branches, although they may appear to do so because of the shortness of the internode below the first cataphyll of a branch. Each flower is subtended by a bract and is borne on a minute branch of the inflorescence which arises in the axil of a second bract. Thus, two bracts are found below each flower, but the lower bract actually occurs on an axis of the infloresence (Fig. 1a). Mainstem inflorescences occur in the Spanish (*A. hypogaea fastigiata vulgaris*) and Valencia (*A. hypogaea fastigiata fastigiata*) cultivars, while inflorescences occur on the lateral branches of the Virginia (*A. hypogaea hypogaea hypogaea*) cultivars. The calyx and corolla are located at the top of the hypanthium. The staminal column is surrounded by and runs the entire length of the hypanthium (Fig. 1b). The calyx is five-lobed, one lobe being separate and opposite the keel and the other four fused, except for their tips. The keel closely embraces the stamens and style. Flower size varies with environment.

The staminal column is usually composed of 10 filaments, 8 of which normally bear anthers (Fig. 1b). The filaments are fused through one-half to two-thirds of their lengths. At the point of separation, the free ends of the filaments are sharply reflexed toward the standard forming acute angles with their fused bases. The pistil consists of a single ovary 1.5 mm long and 0.5 mm in basal diameter (Smith, 1950). It is surmounted by the long style which extends through the hypanthium, bends sharply through the staminal column, bends sharply again through the filaments, and ends in a club-shaped stigma (Fig. 1b). The style is clothed with upward slanting hairs near its summit on the surface facing the standard (Gregory et al., 1951).

Arachis hypogaea is indeterminate in growth, and flowering begins from 25 to 35 days after planting, depending on the cultivar. The most prolific period of flowering occurs between 6 and 11 weeks, depending on the cultivar, with a high proportion of the first flowers producing mature fruits.

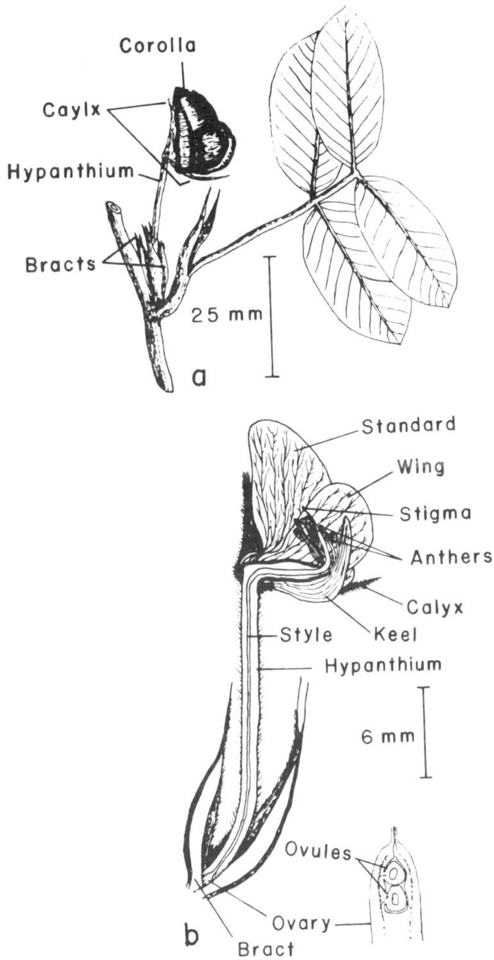

Fig. 1—Peanut inflorescence with flower in leaf axil (a) and longitudinal schematic drawing of the flower (b). (After Smith, 1950.)

Few of the flowers that open late produce mature fruits, unless something in the production environment prevents the early flowers from functioning (Bear and Bailey, 1973). peanuts produce many more flowers than the plants can support in fruit and seed production. Approximately two-fifths of the flowers fail to begin fruit development and an additional two-fifths produce immature pegs which abort before pod enlargement occurs (Smith, 1954). Removing all flowers every day can cause *A. hypogaea,* normally an annual, to live as long as 240 days or more and insures that flowers are available over a long period (Bear and Bailey, 1973).

Usually only one flower of an inflorescence reaches anthesis on a given day, occasionally two or more in Spanish types. Intervals between successive flowers on an inflorescence vary from one to several days. Twenty-four hours before anthesis, the flower bud is 6 to 10 mm long. During the day, the hypanthium elongates slowly and the buds attain lengths of 10 to 20 mm. At night, elongation accelerates, so that at anthesis flowers are usually 50 to 70 mm long (Smith, 1950). The bud usually opens at the beginning of the light period; however, this may be delayed by cool or wet weather. Pollen liberation may occur as early as 7 to 8 hours before flower opening, but even upon flower opening the anthers may not have dehisced in some floers (Bolhuis et al., 1965). The stigma is receptive from 24 hours before to 12 hours after the opening of the flower (Hassan and Strivastava, 1966).

On warm, sunny days the flowers often wither within 5 or 6 hours after expansion. In cooler weather, wilting is delayed. Only the ovary and the base of the style remain in a turgid condition after the day of anthesis.

IV. ARTIFICIAL HYBRIDIZATION AND SELF-POLLINATION

A. Equipment

The main instruments required for emasculation are a forceps and a sharp penknife, scalpel, or razer blade. Unless one has exceptional eyesight, a 2× to 3× magnifier attached to a head band is helpful.

B. Preparation of the Female

Usually only one flower bud per inflorescence is emasculated on the female parent. The flower bud selected should be as near to the main stem as possible. The percentage of successful hybridizations per emasculation is usually higher when the plants are young and in the early stage of flowering.

Stages in the emasculation of the peanut flower are illustrated in Fig. 2. Flower buds appear above the leaf axils during the afternoons of warm, bright days, and emasculations can be accomplished without difficulty as early as 1700 hour (Fig. 2-2). On cloudy or rainy days, fewer flower buds develop, and their growth is usually retarded to the extent that emasculations are easier to accomplish if delayed until 2100 or 2200 hours, or until the bud has emerged approximately 2 cm from the axil and is large enough to be manipulated.

Emasculation begins by grasping the bud with thumb and index finger of one hand (Fig. 2-3). The sepal in front of the keel is removed, and the sepal on the side of the standard is folded down (Fig. 2-4). The standard is opened with the forceps, and the wing petals pulled out and down (Fig. 2-5). As the standard is held back with thumb and index finger, the point of the forceps is used to break the keel and to move up the keel to pull it free of the stigma and anthers (Fig. 2-6). The keel is pulled down and held out of the way with thumb and index finger while all anthers and as much of the filaments as possible are removed with forceps (Fig. 2-7). Some operators find it difficult to hold the wing and the keel petals out of the way during the process; therefore, they remove them without noticeably reducing the percentage of successful hybridizations (Fig. 2-8). After the anthers are removed, the standard is returned to its original position over the stigma. No attempt usually is made to protect the emasculated flowers.

An alternative method was described by Reddy et al. (1970) who used a razor blade to cut around the flower bud two-thirds down from the tip for removal of the standard and a portion of the wing petals. The anthers were then removed and a folded piece of drinking straw inserted over the emasculated flower.

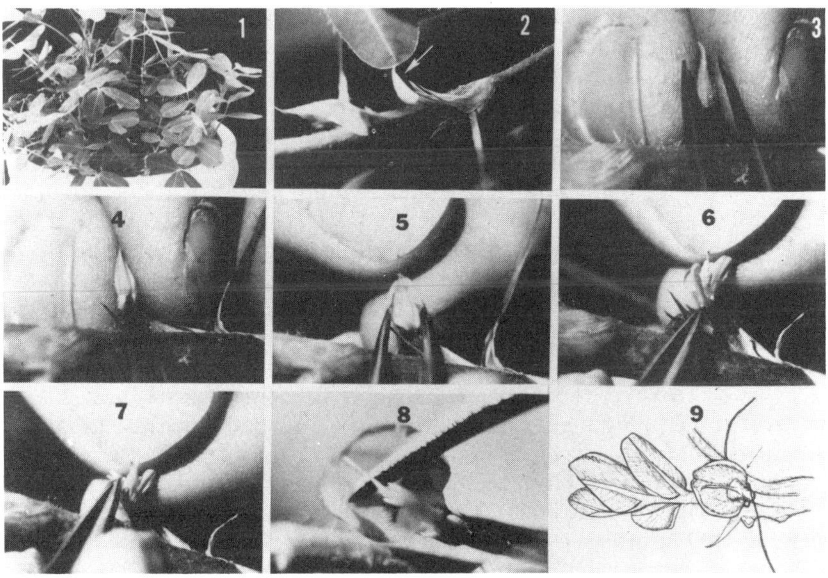

Fig. 2—Stages in the emasculation of a peanut flower. (1) Potted plant, (2) bud at proper stage for emasculation, (3) bud held in position for emasculation, (4) the sepal on the side of the standard is folded down, (5) the standard is opened with the forceps and wing petals pulled out and down, (6) the standard is held back with thumb and index finger and with the forceps the keel is pulled free of the stigma and anthers, (7) anthers and as much of the stamens as possible are removed, (8) and emasculated flower with the wing and keel petals removed, (9) emasculated flower identified with small colored string attached to hypanthium. (Photographs 3 through 7 provided through the courtesy of C. E. Simpson, Stephenville, Texas.)

The emasculated flowers can be identified in a number of ways. A common method is to attach a small thread to the hypanthium of the emasculated flower and drape the other end over the edge of the pot (Fig. 2-9). Different colored strings may be used to identify the flowers emasculated on different days. Mixon (1977) tags an emasculated flower with a strip of heavy duty aluminum foil that is cut 6 to 8 mm wide and 5 to 6 cm long and looped around the hypanthium. Gibbons (1977) marks the emasculated flower with a colored thread and uses a second thread of a different color to show that the flower has been pollinated.

C. Pollination

On the morning of the day following emasculation, the standards are usually expanded and the stigmas exposed, so that application of pollen is possible in many cases without handling the flower. Pollinations are accomplished between 0700 and 1000 hours by removing a healthy flower from the male parent (Fig. 3-1), squeezing its pollen onto a forceps, and transferring the pollen by means of the forceps to the stigma of the emasculated flower (Fig. 3-2). Some operators detach the keel from the male parent flower and squeeze the pollen from it directly onto the stigma; others apply pollen to the stigmas with a small camel hair brush. The plants and flowers must be handled very carefully following pollination because the pollen grains are easily dislodged from the stigma (Fig. 3-3). The operator should dip the forceps and rinse fingers in an alcohol solution before changing from one pollen source to another to reduce the possibility of pollen contamination.

To provide shade and a higher relative humidity for the germination and subsequent growth of pollen, a paper towel, approximately 12 × 12 cm in size, can be dampened with water and placed carefully over the flower without touching the stigma immediately after it is pollinated (Fig. 3-4a). Sharma (1964) found that placing a polyethylene tube cover, punched with a few holes to allow aeration, over the female plants after pollinating resulted in better pollen germination and ultimate fertilization without the necessity of covering the pollinated stigma. He reported little success when he covered the pollinated stigma with the keel, probably because the pollen was dislodged from the stigma in the process.

Peanut pollen is most viable when collected in the early morning (Bolhuis, 1959; DeBeer, 1963). Oakes (1958) found that pollen collected as early as 0300 hours was physiologically immature, but that collected after 0900 hours was overmature. He, as well as Hassan and Srivastava (1966), reported that the period of maximum physiological development of peanut pollen is between 0500 and 0700 hours.

Faucette and Emery (1974) found that pollen desiccated for 6 to 7 days and stored in dry containers for 9 days at 8 C was still viable. Hassan and Srivastava (1966) reported that pollen was viable up to 8 days when stored in a refrigerator at 6 C over calcium chloride in sealed desiccators, but for only 8.5 hours when stored at 28 C and a relative humidity of 56%.

Smith (1956) reported a time lapse of 12 to 16 hours between self-pollination and fertilization, and Oakes (1958) 8 to 9 hours.

After pollinations are completed, all unpollinated flowers on the plant are removed by breaking the hypanthium near the base with forceps by 1000 hours. If fertilization was successful, an aerial peg will usually become visible 7 to 10 days after pollination, rarely as long as 15 days. Smith (1954) and Stokes and Hull (1930) observed that the ovaries of pollinated peanut

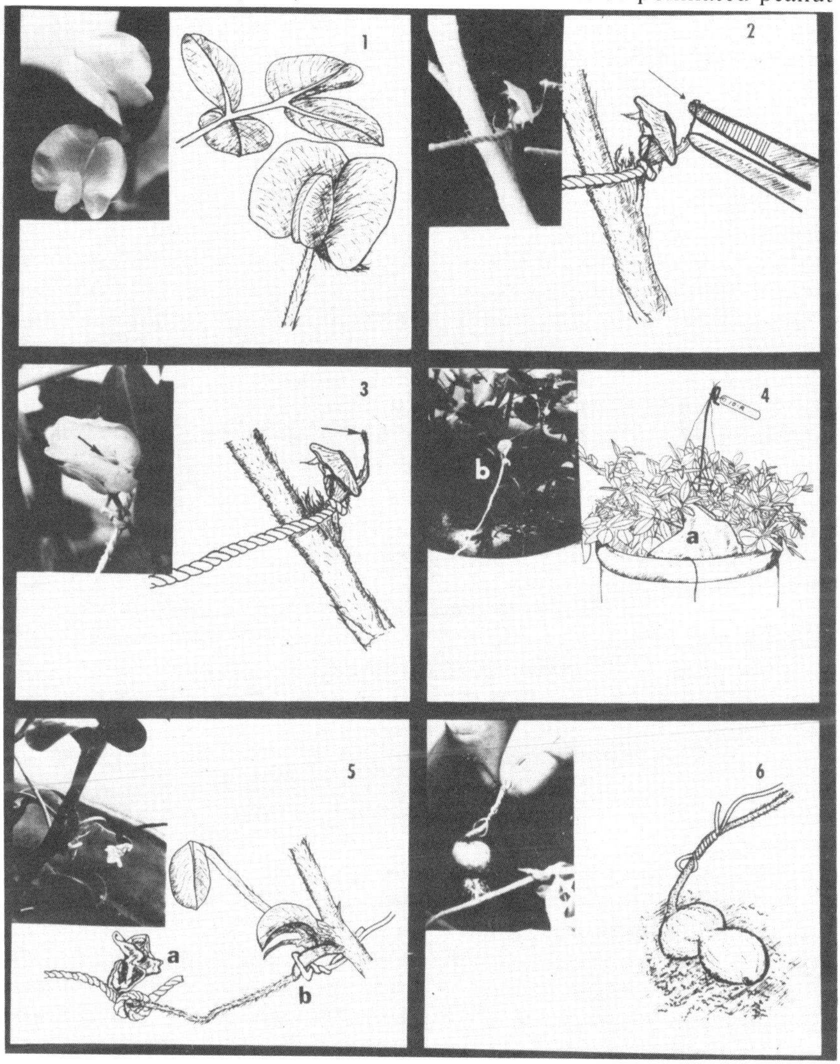

Fig. 3—Stages in the artificial pollination and subsequent development of hybrid peanut seed. (1) A flower removed from the male parent plant; (2) pollen that has been squeezed onto forceps and transferred to the stigma of the emasculated flower; (3) cluster of pollen grains on the stigma; (4) a dampened paper towel (a) placed over the pollinated flower to provide a higher relative humidity for germination, and the wilted flower (b) 5 to 6 hours after pollination; (5) a color-coded wire looped around the developing peg (b), which is still attached to the withered flower and its identifying string (a); (6) fruit which is generally harvested 55 to 65 days after the peg is wired. (Photographs by John Swearingen and line drawings by Ashley Wood, Univ. of Florida, Gainesville.)

flowers can remain dormant for several weeks without losing their ability to resume active fruit development.

After fertilization occurs, the ovary elongates because of intercalary meristem growth in its base. This peg grows down into the soil where the peanut develops. Figure 3-5a shows the withered flower attached to the peg and Fig. 3-5b the peg that has been wired for identification. Approximately 2 months later, the mature peanut fruit with the peg still attached can be harvested (Fig. 3-6).

The developing peg, with the withered flower and string still attached, can be identified by means of a color-coded wire which is looped around the peg before it penetrates the soil. The other end of the wire can be attached to the stake that supports the label of the female plant (Fig. 3-4 to 3-6). Discarded sections of telephone cable provide an excellent source of different colored wires for identifying crossed seed. The use of a different wire color each day provides the opportunity for allowing each hybrid seed a prescribed period of time in which to develop and mature. Simpson (1977) identifies the pollinated flowers with nylon string attached to a dated stake. The string is transferred to the peg when it emerges.

Immediately after pollinating a flower, Mixon (1977) records the number of the pollinator plant and the date of pollination on the foil strip that was attached to the hypanthium at the time of emasculation. When the resulting peg reaches 12 to 16 mm, the foil marker is removed from the hypanthium and secured around the elongating peg. This labeling procedure is particularly useful in making random crosses between several plants.

Artificially hybridizing peanuts in the greenhouse has consistently resulted in over 70% success. Gibbons (1977) reported a success rate of 44% in India from field hybridization starting at 0630 hours during the monsoon season (July to August), compared to 27% when pollination was started at 0830 hours.

D. Factors Affecting Efficiency

Flowers and developing pegs without attached strings or wires are removed daily for a period of 10 days after the last cross has been completed. The lateral vegetative growth and fruiting branches that do not contain hybrid pegs are pruned back. When more than 5 to 10 pegs are developing per plant, the efficiency of the operator's time is reduced in terms of the number of crossed seed harvested relative to the number of pollinations made.

To facilitate record keeping, it is recommended that the plant in any container donates or receives pollen from the plant in only one other container. Because plants involved in various crosses do not generally begin to flower at the same time, the flowers are removed daily until the majority of the plants are flowering.

Genetic markers are not commonly used in peanut breeding, but are used in peanut genetic studies and for detecting contamination from natural outcrossing. Hammons (1964) reported a monogenic dominant krinkleleaf marker that is easily differentiated from normal leaves from seedling leaf expansion through harvest maturity.

V. NATURAL HYBRIDIZATION

Van der Stok (1923) stated that natural hybridization had been used to develop new peanut cultivars in Buitenzorg, Java. Hammons (1973) discussed the utilization of natural crosses as a new genetic technique he termed pedigree natural crossing. This technique permits the production of larger numbers of F_1 hybrids than is possible with the same expenditure of time using manual procedures. The chief disadvantage of the technique is that the marker stock must have in its genetic makeup those characteristics desirable for cultivar improvement.

VI. SEED DEVELOPMENT, HARVEST, AND STORAGE

The appearance of the aerial peg with withered hypanthium and string adhering to it is evidence, but no proof that the fertilization was successful (Fig. 4-5). Seed harvest is usually 55 to 65 days after the peg was wired depending on environmental conditions and maturity of the parental lines.

Seed of the cultivated peanut generally becomes non-viable within 2 years when stored at room temperature. Seed storage studies indicate that shelled peanut seeds with not more than 6% moisture can be stored for at least 9 years without appreciable loss in germination when held at −4 to 5 C (Norden, 1975).

VII. TECHNIQUES FOR SPECIAL SITUATIONS

In hybridization programs, the number of F_1 seeds sometimes are insufficient to produce the large F_2 populations desired, in which case the F_1 hybrids may be multiplied by vegetative propagation.

Nuchowicz (1955) cultured excised embryos and sections of embryos of peanuts as a means of increasing the number of descendants, but found that the divided embryos grew poorly. Martin (1970) conducted research on methods of in vitro embryo culture in attempting to overcome some of the barriers to interspecific hybridization. He has been able to produce viable seedlings from ovules measuring 0.3 mm. Banks (1977) has had some success culturing embryos that were differentiated to the extent that cotyledons were visible, but as yet has not been successful with very young embryos.

ACKNOWLEDGEMENTS

In addition to the contributions of those persons whose names have already been mentioned in the text and figures, the author acknowledges his appreciation to William R. Carpenter, Paul E.Cassidy, and Miriam D. Brown for their assistance in preparing the chapter and to D. J. Banks, W. C. Gregory, R. O. Hammons, and C. E. Simpson for their critical review of the manuscript.

REFERENCES

Banks, D. J. 1976. Hybridization of peanuts in growth chambers. Peanut Sci. 3:66–69.

———. 1977. Personal communication. USDA, SEA, Dep. of Agronomy, Oklahoma State Univ., Stillwater, OK 74074.

———, and B. G. Jordan. 1971. Effects of water sources and fertilizers on peanut plants in the greenhouse. Proc. Okla. Acad. Sci. 51:47–50.

Bear, J. E., and W. K. Bailey. 1973. Earliness of flower opening and potential for pod development in peanuts *Arachis hypogaea.* J. Am. Peanut Res. Educ. Assoc. 5:26–31.

Bolhuis, G. G. 1959. Observations on the flowering and fructification of the groundnut, *Arachis hypogaea.* Neth. J. Agric. Sci. 7:138–140.

———, H. D. Frinking, I. Leeuwaugh, R. G. Rens, and G. Staritsky. 1965. Observations on the opening of flowers, dehiscense of anthers and growth of pollen tubes in *Arachis hypogaea* L. Neth. J. Agric. Sci. 13:361–365.

Capinpin, J. M., and V. F. Guevara. 1951. The floral biology and fructification of peanut, *Arachis hypogaea.* Philipp. Agric. 35:137–192.

Culp, T. W., W. K. Bailey, and R. O. Hammons. 1968. Natural hybridization of peanuts, *Arachis hypogaea* L., in Virginia. Crop Sci. 8:108–111.

DeBeer, J. F. 1963. Influences of temperature on *Arachis hypogaea* L. with special reference to its pollen viability. Versl. Landbouwkd. Onderz. (Wageningen) N.R. 69. 2:1–81.

Faucette, D. F., and D. A. Emery. 1974. Preliminary evaluations of techniques for collecting and culturing peanut (*Arachis hypogaea*) pollen. Peanut Sci. 1:14–18.

Fortanier, E. J. 1954. Some observations on the influence of spectral regions of light on stem elongation, flower bud elongation, flower bud opening and leaf movement in *Arachis hypogaea* L. Mededel. Landbouw. (Wageningen) 54:103–114.

———. 1957. De beinvloeding van de bloei bij *Arachis hypogaea* L. Mededel. Landbouw. (Wageningen) 57:1–116.

Gibbons, R. W. 1977. Personal communication. ICRISAT, Hyderabad-500 016, A.P., India.

———, and J. R. Tattersfield. 1969. Out-crossing trials with groundnuts (*Arachis hypogaea* L.). Rhod. J. Agric. Res. 7:71–85.

Gregory, W. C. 1977. Personal communication. North Carolina State Univ., Raleigh, NC 27607.

———, and M. P. Gregory. 1976. Groundnut, *Arachis hypogaea.* p. 151–154. *In* Evolution of Crop Plants. Longman, London.

———, ———, A. Krapovickas, B. W. Smith, and J. A. Yarbrough. 1973. Structures and genetic resources of peanuts. p. 47–133. *In* Peanuts-culture and uses. Am. Peanut Res. Educ. Assoc., Stillwater, Okla.

———, B. W. Smith, and J. A. Yarbrough. 1951. Morphology, genetics, and breeding. p. 28–88. *In* The peanut—the unpredictable legume. Natl. Fertilizer Assoc. Washington, D.C.

Hammons, R. O. 1963. Artificial cross-pollination of the peanut with bee-collected pollen. Crop Sci. 3:562–563.

———. 1964. Krinkle, a dominant leaf marker in the peanut, *Arachis hypogaea* L. Crop Sci. 4:22–24.

———. 1973. Genetics of *Arachis hypogaea.* p. 135–173. *In* Peanuts-culture and uses. Am. Peanut Res. Educ. Assoc., Stillwater, Okla.

———, and D. B. Leuck. 1966. Natural cross-pollination of the peanut, *Arachis hypogaea* L., in the presence of bees and thrips. Agron. J. 58:396.

Hassan, M. A., and D. P. Srivastava. 1966. Floral biology and pod development of peanut studied in India. J. Indian Bot. Sci. 45:92–102.

Hildebrand, G. L. 1974. A note on artificial hybridization of the groundnut (*Arachis hypogaea* L.). Rhod. J. Agric. Res. 12:195–196.

Krapovickas, A. 1968. The origin, variability and spread of the groundnut (*Arachis hypogaea*). p. 427–441. *In* P. J. Ucko and I. S. Falk (eds.) The domestication and exploitation of plants and animals. Gerald Duckworth Co., Ltd., London.

Leuck, D. B., and R. O. Hammons. 1969. Occurrence of atypical flowers and some associated bees (Apoidea) in the peanut, *Arachis hypogaea* L. Agron. J. 61: 958–960.

Martin, J. P. 1970. Culture in vitro d'ovules d'arachide. Oleagineux 25:155–156.

Mixon, A. C. 1977. Personal communication. USDA, SEA, Coastal Plain Station, Tifton, GA 31794.

Norden, A. J. 1973. Breeding of the cultivated peanut (*Arachis hypogaea* L.). p. 175–208. *In* Peanuts-culture and uses. Am. Peanut Res. Educ. Assoc., Stillwater, OK.

———. 1975. Effect of preparation and storage environment on lifespan of shelled peanut seed. Proc. Am. Peanut Res. Educ. Assoc. 7:64 (Abstr.).

———, and V. A. Rodriguez. 1971. Artificial hybridization of peanuts. Oleagineux 26:159–162.

Nuchowicz, A. 1955. Research on the culture of excised embryos and sections of embryos of *A. hypogaea.* Agric. Louvain 3:3–37.

Oakes, A. J. 1958. Pollen behavior in the peanut, *Arachis hypogaea* L. Agron. J. 50: 387–389.

Ono, Y., K. Nakayama, and M. Kubota. 1974. Effects of soil temperature and soil moisture in podding zone on pod development of peanut plants. Proc. Crop Sci. Soc. Japan 43:247–251.

Reddy, G. P., P. S. Reddy, and A. N. Murthi. 1970. An improved crossing technique in groundnut (*Arachis hypogaea*). Andhra Agric. J. 17:124–127.

Sharma, K. D. 1964. An improved crossing technique in groundnut. Indian Central Oilseeds Committee, Third Conf., Oilseed Res. Workers in India:36–37.

Simpson, C. E. 1977. Personal communication. Texas Agric. Exp. Stn., Texas A&M Univ., Stephenville, TX 76401.

Smartt, J. 1964. Interspecific hybridization in relation to peanut improvement. Proc. Third Natl. Peanut Res. Conf.:53–56.

Smith, B. W. 1950. *Arachis hypogaea,* Aerial flower and subterranean fruit. Am. J. Bot. 37:802–815.

———. 1954. *Arachis hypogaea.* Reproductive efficiency. Am. J. Bot. 41:607–616.

———. 1956. *Arachis hypogaea.* Normal megasporogenesis and syngamy with occasional single fertilization. Am. J. Bot. 43:81–89.

Stokes, W. E., and F. H. Hull. 1930. Peanut breeding. J. Am. Soc. Agron. 22: 1004–1019.

Stone, E. G., W. K. Bailey, and J. E. Bear. 1973. Natural outcrossing of peanuts *Arachis hypogaea* in Puerto Rico. J. Am. Peanut Res. Educ. Assoc. 5:134–140.

Umen, D. P. 1933. Biology of peanut flowering. Lenin Acad. Agric. Sci., Scientific-practical, issue No. 6. VNIIMK. (1976 Transl. from Russian by A. K. Dhote.) Amerind Publ. Co. Ltd., New Delhi.

Van der Stok, J. E. 1910. Onderzoekingen omtrent rijst entweedegewassen. (Katang-Tanah *Arachis hypogaea*). Meded. Vitg. v.h. Dept. Landbouw. 12:176–221.

———. 1923. Erdnuss, *Arachis hypogaea* L. *In* C. Fruwith, Handbuch der landwirthschaftenlichen pflanzen zuchrung 5:202–205. [Translated into English by Elke Humpries and R. O. Hammons.]

Varnell, R. J., and D. E. McCloud (eds.). 1975. Germplasm preservation and genotype evaluation in *Arachis* (peanuts). Report of a Workshop, Gainesville, Fla. 11–15 July.

Wynne, J. C., D. A. Emery, and R. J. Downs. 1973. Photoperiodic responses of peanuts. Crop Sci. 13:511–514.

32 Pearl Millet

Glenn W. Burton

USDA-SEA, Georgia Coastal Plain Station
Tifton

Pearl millet, *Pennisetum americanum* (L.) Leeke, is a robust, annual bunchgrass that is grown on more than 20 million ha in the world. It is the world's sixth most important cereal. In Africa and India, it is the principal grain crop in areas too hot and dry for other cereals. In the United States, it makes a high-yielding summer forage of excellent quality.

I. PARENTAL MATERIAL

Pearl millet is an extremely variable species. Plants may consist of a single culm less than 0.5 m in height or many culms 5 m tall. Culms may be simple or branched, slender or stout, smooth or hairy. Leaves and sheaths also may be smooth or hairy. Plant parts may be green, purple, red, or golden yellow.

According to Brunken et al. (1977), ". . .the greatest morphological diversity in pearl millet occurs today in West Africa south of the Sahara Desert and north of the forest zone. The wild progenitor also occurs in the drier, northern portion of this zone." Present-day distribution indicates that the sahel zone extending from western Sudan to Senegal is the original home of pearl millet. Other centers of origin that have been proposed include the Ethiopian highlands, Sierra Leone, and the headwaters of the Niger river in the central Sahara (Brunken et al., 1977). The world collection of pearl millet germplasm is small compared to the great variety of landraces in the crop. The International Crops Research Institute for the Semi-Arid Tropics (ICRISAT) in Hyderabad, India, is presently centralizing and enlarging the world collection.

Of the many species in the genus *Pennisetum,* only napiergrass, *P. purpureum* Schumach, is closely related to *P. americanum.* Both species occur in the *Penicillaria* section and have a basic chromosome number of seven. Attempts to hybridize pearl millet with species in other sections that have different basic chromosome numbers have had little success (Burton and Powell, 1968).

Krishnaswamy (1962) reported that *P. americanum* and *P. purpureum* have a common genome, A, and the latter has a second genome, designated B. The triploid interspecific hybrids produce the expected seven bivalents and seven univalents at meiosis.

Many of the sterile triploids are vigorous and have forage potential (Burton, 1944). When such triploids were used as females and backcrossed to each parent, one weak hybrid was found in an estimated 5,000 potential hybrids with each parent. The best of the sterile F_1 hybrids could be propagated vegetatively as a forage crop.

It has been suggested that pearl millet × napiergrass hybrids could be propagated by seed produced by planting cytoplasmic-genetic male-sterile (CMS) pearl millets between rows of napiergrass at a time that would bring both into flower simultaneously (Powell and Burton, 1966). Pearl millet × napiergrass F_1 hybrids established from seed start off slower than pearl millet hybrids, but produce about as much forage in a growing season because they make much more growth in the fall. The future of F_1 pearl millet × napiergrass hybrids depends on the development of a commercially feasible method of seed production. Napiergrass genotypes studied to date have been short-day sensitive and have not started to initiate floral organs until the photoperiod was less than 12 hours. As a consequence, it has been impossible to produce seed of CMS pearl millet × napiergrass before frost in most of the continental U.S.

Several workers have developed amphidiploids of pearl millet × napiergrass hybrids. Although somewhat fertile, these amphidiploids have been unstable and nothing of economic importance has resulted from the considerable effort expended (Burton and Powell, 1968).

II. PLANT CULTURE

A. Field

Most pearl millet genotypes stool a great deal when given adequate space. At Tifton, Ga., nurseries are planted with cone planters and measured volumes of seed that place 10 to 15 seeds per 1 m of row. Plants are thinned to one plant for every 0.3 to 0.4 m. Rows are alternately spaced 0.6 and 1.2 m apart to give more room to bag heads and make hybrids. Growth of the millet in this 0.6, 1.2, 0.6, 1.2 m spacing is similar to that in rows uniformly spaced 0.9 m apart. Propazine [2-chloro-4,6-bis(isopropylamino)-*s*-triazine], applied at a rate of 2 kg/ha on loamy sand soils immediately after planting, effectively controls weeds without apparent injury to pearl millet.

Photoperiods that bring pearl millets into flower permit hybridization and seed set. Photoperiods of less than 12 hours tend to bring all genotypes into flower at the same time and thus facilitate the hybridization of genotypes that differ in photoperiodic response.

The temperature extremes that limit flowering and seed set in pearl millet have not been established. It has been known to flower and set seed at temperatures between 25 and 45 C.

Pearl millet flowers in a relative humidity of 100%, such as that obtained when heads are enclosed in plastic bags, but it does not set seed in such bags. Under natural conditions, the relative humidity never seems to get too low for hybridization and seed production.

Pearl millet can grow, flower, and produce seed on soils that range from infertile sands to highly fertile loams. Because most pearl millets reach heights of 2 to 4 m on fertile, well-watered soils, plants can be more easily hybridized if the soil fertility is kept low.

Pearl millet tolerates little flooding, but can be hybridized and made to set seed under considerable moisture stress.

B. Growth Chamber and Greenhouse

Maintaining temperatures above 25 C seems to be the only special requirement for growing and hybridizing pearl millet in greenhouses, nurseries, and growth chambers during the winter months.

Supplemental light has not been required for normal flowering and seed set in pearl millet during winter months at Tifton, Ga. Pearl millet has flowered and set seed in growth chambers with only 8 hours of light per day at 43 to 48 klux, attained from 70% input wattage of cool-white florescent lighting and 30% input wattage from incandescent lighting (Hellmars and Burton, 1972).

III. FLORAL CHARACTERISTICS

The inflorescence (head) of pearl millet is a false spike ranging from 5 to more than 150 cm in length and from 1 to 5 cm in diam. The involucre, borne on a stalk up to 15 cm long, consists of a cluster of bristles that are usually inconspicuous in mature heads, but may extend several cm beyond the grain. The spikelets, 4 to 7 mm long in each involucre, usually occur in pairs with a sessile male floret and a short pedicelled bisexual floret. The latter has a single pistil with two feathery stylar branches and three anthers enclosed between the lemma and palea. The stylar branches are usually colorless, but in one genotype they develop a red color when exposed to the sun.

Pearl millet is a protogynous species, and the stylar branches are exserted from the florets several days before the anthers. The protogynous flowering habit facilitates cross pollination. Natural crossing is not complete because plants have several culms that flower in succession. This allows the head that reaches anthesis first to pollinate other heads on the same plant that are just exserting anthers. As a consequence, self-pollination as high as 31% has been observed (Burton, 1974).

Stylar branches are first exserted from florets in the upper half of the head, and by the third day nearly all styles emerge on heads less than 25 cm long. Longer heads usually require more days for complete exsertion of all stylar branches. Styles are usually exserted after the heads emerge from the boot, but in some genotypes they are exserted before head emergence. Seed set is usually very poor if anthesis occurs in the boot.

The first anthers generally emerge from the first florets to exsert styles at least 1 day after most styles on a head are exserted. Anthers in the sessile male florets are exserted 2 or 3 days later. Most heads shed pollen for 4 to 6 days. When temperatures exceed 25 C, anthesis occurs anytime during the day with the greatest flush of anthers appearing soon after sunrise. On cool fall nights, anthesis is delayed until one or more hours after sunrise and occurs first on the east side of the heads.

Stigmas in a floret remain receptive from emergence until 1 day after its anthers have been exserted. Heads of cytoplasmic male-sterile 'Tift 23A', pollinated 2 to 3 days before anthesis, set more seed than those pollinated after anthesis begins (Burton, 1966). Many florets that set seeds on heads pollinated 3 days before anthesis had not exserted stigmas when the head was pollinated. Thus, pollen either entered the opening at the apex of the floret to effect fertilization immediately or remained viable at the stylar opening of the glumes until the styles were exserted.

Pearl millet pollen collected in glassine bags and stored in sealed glass jars at 27 C for 1, 2, 3, and 4 days before application to receptive heads of cytoplasmic male sterile Tift 23A was 59, 10, 3, and 1% as effective as fresh pollen, respectively (Cooper and Burton, 1965). When stored at 4 C for 1, 2, 3, and 4 days, it was 96, 95, 82, and 71% as effective as fresh pollen, respectively. When stored at 4 C for 21 days and then applied to Tift 23A, it was still 8% as effective as fresh pollen.

Fertilization occurs within a few hours after pollination. Stylar branches wilt and usually are dry 24 hours after pollination.

IV. ARTIFICIAL HYBRIDIZATION AND SELF-POLLINATION

A. Equipment

The equipment for hybridization and self-pollination includes bags 7.5 × 35 cm in size made of 11.35 kg (26 lb) bleached waterproof glassine sealed with waterproof glue, similar bags made of 22.70 kg (50 lb) kraft paper, a stapling plier or standard paper clips to fasten the bags on the heads, and a coarse pen with waterproof ink.

If grain-destroying insects, such as the sorghum webworm, *Celama sorghiella*, Riley, and the corn earworm, *Heliothis zea* Boddie, are present, they must be controlled by spraying the millet with a suitable insecticide before bagging or by treating the bags with an insecticide. An insecticide such as aldrin (1, 2, 3, 4, 10, 10-hexachloro-1,4,4a,5,B,Ba-hexahydro-1,4-endo-exo-5,8,-dimethanonaphthalene) mixed with mineral spirits (1 liter to 0.908 kg/3.784 liter (2 lb/gal) aldrin emulsion per 40 liters of mineral spirits) can be impregnated in the kraft bags by dipping bundles of the bags in a wash tub of the chemical mix, draining off the excess, and drying them in the greenhouse.

B. Preparation of the Female

To prepare the female parent, the top one or two leaf blades are removed and the partially exserted head and culm are covered with a glassine bag before styles appear (Fig. 1 and 2). The bag is fastened by folding the open end at a 45° angle and stapling it close enough to the stem so that it does not blow off. It must be loose enough, however, to be moved up the culm by the exserting head.

The protogynous habit of pearl millet makes emasculation unnecessary. Instead of emasculation, the hybridizer examines the head by looking through the glassine bag for the presence of exserted styles. When styles are exserted, the head is ready to pollinate (Fig. 3). Some genotypes start ex-

Fig. 1—Pearl millet head with top two leaf blades removed ready to be bagged.

serting anthers before the florets at the top and the bottom of the head are exserted. On such genotypes, the florets without stigmas can be removed at the time of pollination without adversely affecting the pollination and seed set of those remaining on the head.

C. Pollination

Selfing in pearl millet is easily accomplished by removing the top one or two leaf blades of the culm and enclosing the head and culm in a kraft bag before any styles appear (Fig. 4). The bag is stapled together as described for the glassine bags. The bag is left on the head until it has been harvested, dried, and prepared for threshing.

Crosses are made by using as females, heads that have exserted their styles but not their anthers (Fig. 3). Pollen can be collected any time that it is dry. Glassine pollen-collecting bags allow the worker to see the amount of pollen available. Shaking the pollen-collecting bag vigorously and carefully

Fig. 2—Head enclosed in 7½ × 35 cm glassine bag.

removing it from the head maximizes the amount of pollen harvested. Although crossing can be accomplished at any time, the seed set from crosses made at midday is lower than that from crosses made earlier and later in the day (Cooper and Burton, 1965).

Pollen is usually applied by putting the female head in the bag in which the pollen was collected and shaking it. If the female millet is tall, the culms are bent over so that the head assumes a nearly horizontal position before the pollen-containing bag is pulled over it. If the female is 1 to 2 m tall, the opened pollen-containing bag can be rapidly inverted and pushed down over the head before pollen can fall from it. It is possible to pollinate two female heads at the same time in a good bag of pollen.

After pollination, the head or pair of heads are enclosed in a kraft bag on which is written the pedigree of the cross, the date, and any other pertinent information (Fig. 4). These bags are left on the heads until they are threshed. Insecticide-treated bags protect the seed from insects, and bags usually protect the seed from birds. If pollen is fresh and abundant, nearly 100% seed set can be expected.

Fig. 3—Head with styles exserted ready to be pollinated.

D. Factors Affecting Efficiency

Nurseries are arranged to minimize walking by grouping together lines to be hybridized, lines with similar maturities, and lines that otherwise form logical units.

Hybridization of lines with different flowering dates is accomplished by planting lines to be hybridized at 10-day intervals to cover the expected range in maturities. Most millet lines develop branches from the upper nodes that bear flowering heads 3 to 5 weeks after the primary head flowers. These secondary heads can often be used to make hybrids missed on primary heads.

Photoperiod, temperature, and drought alter flowering dates in pearl millet. Some genotypes fail to flower when day lengths exceed 12 hours (Burton, 1965). Other genotypes flower in long days (16-hour photoperiod), but flowering is earlier under short days (Begg and Burton, 1971; Burton,

Fig. 4—Head enclosed in 7½ × 35 cm kraft bag after pollination.

1965). When planted in the winter in Puerto Rico or in the greenhouse at Tifton, Ga., with temperatures above 30 C, both "day neutral" and short-day millets flowered at about the same time (Barnes and Burton, 1966). An automatic darkbox can be used in summer to subject 3-week-old potted plants of pearl millet to 10 hours of light for 4 weeks and induce near simultaneous flowering in genotypes with different day-length requirements (Burton and Stansell, 1971).

Increasing the growing temperatures up to 32 C and providing the appropriate photoperiod reduces the number of days from planting to flowering (Hellmars and Burton, 1972).

Observations suggest that drought may hasten flowering in some genotypes, like 'Tift 186', and may delay flowering in other genotypes, such as 'Tift 23DB' and 'Tift 239'.

Pearl millet hybrids are much more vigorous than inbred lines and can be easily identified on this basis if the parents are inbreds. The hybrid nature of offspring can be determined with a number of simply inherited plant characters in pearl millet such as dwarfness, trichomelessness, and plant colors.

V. NATURAL HYBRIDIZATION

Before CMS was discovered, 'Gahi 1' pearl millet gave growers of forage most of the increased production that results from hybrid vigor (Burton, 1958a). Gahi 1 seed is produced commercially by harvesting all seed from isolated fields planted to a mixture of equal numbers of pure live seeds of the four inbreds 13, 18, 23, and 26. These inbreds, selected from many that were tested in diallels, gave high-yielding hybrids in all combinations. Because Gahi 1's parental inbreds flower at the same time, complete intermating occurs and about 75% of the seed produced is a mixture of the six possible crosses among the four inbred parents. The other 25% of the seed consists of selfs and sibs of the four parents. When planted at recommended seeding rates, the Gahi 1 seed mixture yields as well as the double cross between the four lines because the more vigorous F_1 hybrid seedlings crowd out and eliminate the inbred seedlings (Burton, 1948).

Inbreds 13, 18, 23, and 26 are increased in isolation and are mixed together in equal numbers of pure live seeds by the National Foundation Seed project to produce foundation seed of Gahi 1. Seed producers must plant only Gahi 1 foundation seed to produce commercial Gahi 1 seed. Only seed harvested from fields planted to Gahi 1 foundation seed can bear the name Gahi 1.

CMS was first discovered in pearl millet in 1956 and was incorporated into inbred Tift 23, an excellent maintainer (Burton, 1958b; Burton, 1965). Two other cytoplasm sources (A_2 and A_3) were later discovered, and 'Tift 239' became an excellent maintainer for the A_2 source (Burton and Athwal, 1967, 1969). Introducing the d_2 gene that cuts plant height in half by a backcrossing program produced 'Tift 23DA', which differed fromTift 23A only by this recessive dwarfing gene (Burton, 1969a; Burton and Fortson, 1963). Tift 239 is also dwarfed by the d_2 gene. Maintainer lines carry the letter B in their name. Thus 'Tift 23B' is the maintainer for Tift 23A. Lines that carry the dominant fertility restorer gene, *Ms,* are called R lines (Burton and Athwal, 1967).

The first commercial F_1 grain-producing hybrid was released in India in 1965. Named 'HB1', this hybrid had outyielded many other experimental hybrids and had produced 88% more grain than the best open-pollinated check in numerous yield trials throughout India's millet belt (from 11 to 31°N Lat) prior to its release. HB1 is a cross between Tift 23A and the restorer R line Bil 3B (Burton and Powell, 1968). Indian farmers planted HB1 as rapidly as seed could be produced and increased pearl millet grain production in India from 3.5 million metric tons in 1965 to 8.0 million metric tons in 1970. A number of other grain-producing hybrids have been released and are currently being grown in India.

In the United States, several F_1 hybrids have been released for forage production. Most of these have Tift 23A or Tift 23DA as the female parent. Forage hybrids remain vegetative longer and do not become weeds for succeeding crops if they produce no seed. Using sterility maintaining B lines as male parents will achieve this end.

The d_2 gene that reduces plant height 50% and facilitates seed production also increases leafiness, forage quality, and animal performance. Tifleaf 1 is the first commercial forage hybrid dwarfed with the d_2 gene (Burton et al., 1969; Johnson et al., 1976).

Commercial F_1 hybrid seed is produced by planting 6- to 12-row strips of the CMS female parent between 2-row strips of the male parent. Only the CMS rows are harvested and sold as hybrid seed.

Seed-certifying agencies set minimum isolation requirements for certified seed. Pearl millet pollen, although largely wind-borne, can be carried up to 1.6 km by bees (*Apis mellifera* L.) (Leuck and Burton, 1966). Isolations for breeders seed should be 1.6 km or more, and fields planted to breeders seed should be checked for rogues. In the tropics and subtropics, the growing season is long enough to mature two or more crops per year in the same field. Time, in the form of staggered plantings, can provide isolation. Both A_1 and A_2 CMS cytoplasms mutate to normal pollen-shedding cytoplasm (Burton, 1977b). These pollen-shedding plants seem stable and act as maintainers. Although the frequency of pollen shedding mutants is low, pollen from them can fertilize florets on CMS plants in seed fields. Seeds fertilized by the pollen shedders give rise to weak CMS plants that may reduce forage yield of the hybrid. In India and Africa where ergot infects pearl millet florets and poisons the grain, CMS plants are more apt to develop ergots than fertile pollen-shedding hybrids. Pollen-shedding mutants should be rogued from CMS seed increase fields and limited generation increase should be practiced to reduce the frequency in hybrid seed production fields.

Pearl millet hybrids outyield cultivars under favorable conditions and show even greater superiority when grown under the stresses to which pearl millet is exposed. For these reasons, hybrids would be used in the developing countries if hybrid seed were easily obtained. Research at Tifton, Ga., suggests the following procedure to solve the problem of seed supply. A central government agency can increase the parents and produce a hybrid foundation mix containing 85 to 90% of the CMS parent and 10 to 15% of the fertility-restoring male parent. This foundation mix can be distributed to selected farmers for increase in isolation. These farmers can harvest all the seed produced (the small amount of male inbred seed harvested would be crowded out by the hybrids, as in Gahi 1) and exchange it with their neighbors for two or three times as much grain. The neighbors planting the hybrid seed can harvest enough additional grain (more than that from their old cultivars) to much more than cover the grain cost of hybrid seed.

VI. SEED DEVELOPMENT, HARVEST, AND STORAGE

Seed set can be checked visually by examining the heads about 10 days after pollination.

Pearl millet should be harvested as soon as the grain is mature. The peduncles remain green for several weeks after the grain is mature; thus the heads must be dried before they can be threshed. At Tifton, Ga., bulk lots of heads are dried on plastic sheets on the floor of empty greenhouses. A concrete floored greenhouse sprayed with a dilute white latex paint to reduce light and heat penetration is used to dry and store bundles of selfs and hybrids after harvest. After they have been dried in the greenhouse, single heads are threshed in small nursery threshers. Either tooth or bar cylinder threshers work well with pearl millet if the heads are dry. Nursery threshers work best if the heads are preconditioned for several hours in a forced-air electric oven set at 40 C.

Pearl millet seeds are difficult to differentiate other than by number and pedigree.

At Tifton, Ga., pearl millet seeds dried to less than 12% moisture and stored in airtight containers at 5 C have germinated well (70%) after 26 years (Burton, 1976). I believe that good quality millet seeds, dried to 5 to 7% moisture content in a forced-air oven at 40 C, sealed in airtight containers, and stored at −20 C, will retain their viability for at least 100 years.

VII. TECHNIQUES FOR SPECIAL SITUATIONS

Seeds of most pearl millet genotypes are dormant for several weeks after they mature. Seed dormancy usually breaks down in a few weeks at warm temperatures, but it may last for more than a year if the seeds are stored at 4 C soon after they mature. To grow several generations a year, dormancy must be broken as soon as seeds mature. Soaking seeds for 1 hour at 20 to 25 C in a water solution of 1% 2-chloroethanol plus 0.5% sodium hypochlorite has increased the germination of dormant pearl millet seeds more than a number of other treatments tested (Burton, 1969b).

Birds are fond of pearl millet seed and can quickly consume all the grain in small seed fields or yield plots. At Tifton, Ga., spraying the heads with the cotton insecticide, monocrotophos [dimethyl phosphate ester with *(E)*-3-hydroxy-*N*-methylcrotonamide], has controlled grain-destroying insects and has also served as an excellent repellent for sparrows and blackbirds. Rates of 375 g/ha (0.75 lb/A) of actual monocrotophos have given good results. Repeated applications have sometimes been necessary.

REFERENCES

Barnes, D. K., and G. W. Burton. 1966. Tropical environment of Puerto Rico useful for studying day length sensitivity in pearl millet. Crop Sci. 6:212.

Begg, J. E., and G. W. Burton. 1971. Comparative study of five genotypes of pearl millet under a range of photoperiods and temperatures. Crop Sci. 11:803–805.

Brunken, Jere, J. M. J. de Wet, and J. R. Harlan. 1977. The morphology and domestication of pearl millet. Econ. Bot. 31:163–174.

Burton, G. W. 1944. Hybrids between napiergrass and cattail millet. J. Heredity. 35: 726–732.

———. 1948. The performance of various mixtures of hybrid and parent inbred pearl millet, *Pennisetum glaucum* (L.) R. Br. J. Am. Soc. Agron. 40:908–915.

———. 1958a. New millet from Georgia, Gahi 1 outyields Starr. The Progressive Farmer, p. 138–139 (April).

———. 1958b. Cytoplasmic male sterility in pearl millet, *Pennisetum glaucum* (L.) R. Br. Agron. J. 40:230.

———. 1965. Pearl millet Tift 23A released. Crops & Soils 17:19 (April–May).

———. 1965. Photoperiodism in pearl millet, *Pennisetum typhoides.* Crop Sci. 5: 333–335.

———. 1966. Pearl millet breeding. African Soils 11:39–47.

———. 1969a. Registration of pearl millet inbreds Tift 23B, Tift $23A_1$, Tift 23DB, and Tift 23DA. Crop Sci. 9:397.

———. 1969b. Breaking dormancy in seeds of pearl millet, *Pennisetum typhoides.* Crop Sci. 9:659–664.

———. 1974. Factors affecting pollen movement and natural crossing in pearl millet. Crop Sci. 14:802–805.

———. 1976. Gene loss in pearl millet germplasm pools. Crop Sci. 16:251–255.

———. 1977a. Registration of Gahi 3 pearl millet (Reg. No. 40). Crop Sci. 17: 345–346.

———. 1977b. Fertile sterility maintainer mutants in cytoplasmic male sterile pearl millet. Crop Sci. 17:635–637.

———. 1978. Registration of Tifleaf 1 pearl millet (Reg. No.). Crop Sci. (In press).

———, and D. S. Athwal. 1967. Two additional sources of cytoplasmic male sterility in pearl millet and their relationships to Tift 23A. Crop Sci. 7:209–211.

———, and ———. 1969. Registration of pearl millet inbreds Tift $239DB_2$ and Tift $239DA_2$. Crop Sci. 9:398.

———, and J. C. Fortson. 1963. Inheritance and utilization of five dwarfs in pearl millet (*Pennisetum glaucum*) breeding. p. 230. *In* Genetics Today, Proc. XI Int. Genetic Congress, Vol. 1.

———, W. G. Monson, J. C. Johnson, Jr., R. S. Lowrey, H. D. Chapman, and W. H. Marchant. 1969. Effect of the d_2 gene on the forage yield and quality of pearl millet. Agron. J. 61:607–612.

———, and J. B. Powell. 1968. Pearl millet breeding and cytogenetics. Adv. Agron. 20:48–89.

———, and J. R. Stansell. 1971. An automatic darkbox to induce flowering in short-day plants in midsummer. Crop Sci. 11:595–596.

Cooper, R. B., and G. W. Burton. 1965. Effect of pollen storage and hour of pollination on seed set in pearl millet, *Pennisetum typhoides.* Crop Sci. 5:18–20.

Hellmars, Henry, and G. W. Burton. 1972. Photoperiod and temperature manipulation induces early anthesis in pearl millet. Crop Sci. 12:198–200.

Johnson, J. C., Jr., W. G. Monson, G. W. Burton, and W. C. McCormick. 1976. Performance of dairy heifers grazing pastures of either Gahi 1, Gahi 3, or Tifleaf 1 millet. J. Dairy Sci. 59:19.

Krishnaswamy, N. 1962. Bajra *Pennisetum typhoides* S. & H. Indian Council Agric. Res., Cereal Crop Ser. II., p. 94.

Leuck, D. B., and G. W. Burton. 1966. Pollination of pearl millet by insects. J. Econ. Entomol. 59:1308–1309.

Powell, J. B., and G. W. Burton. 1966. A suggested commercial method of producing an interspecific hybrid forage in *Pennisetum.* Crop Sci. 6:378–379.

33 Pigeonpea

D. Sharma

J. M. Green

International Crops Research Institute for the Semi-Arid Tropics
Hyderabad, India

Pigeonpea [*Cajanus cajan* (L.) Millsp.] is grown in most tropical countries and has many common names, including red gram, tur, arhar, gandul, and pois d'Angole. India has over 90% of the hectarage and maximum variability for almost all the characters in the germplasm of pigeonpea and the genus *Atylosia,* a probable progenitor. The African origin proposed by de Candolle is controversial (De, 1974). Other important pigeonpea-growing countries are Burma, Uganda, Kenya, Dominican Republic, Panama, Puerto Rico, and the West Indies. Dry grain of the commonly grown cultivars contains 21 to 24% protein, and is usually consumed as dhal, a type of split peas, and as green peas.

I. PARENTAL MATERIAL

Pigeonpea belongs to the genus *Cajanus*, tribe Phaseoleae, subtribe *Cajaninae,* subfamily Papilionoideae, and family Leguminosae. Early subdivisions, *C. bicolor* DC. and *C. flavus* DC., based on variation for plant form, maturity, inflorescence structure, flower size, and color, are now considered the same species. *Cajanus indicus* Spreng., *C. luteus* Bello, *Cytisus cajan* L., *Cytisus pseudo-cajan* Jacq., and *Cajan cajan* (L.) Huth are synonyms (Pathak, 1970; Westphal, 1974; Morton, 1976; Huth, 1893). The *Cajanus kerstingii* Harms, found in Senegal, Togo, Ghana, and Nigeria, also listed by Hutchinson and Dalziel (1958) would have been placed in the genus *Atylosia* if Harms had known the strophioled seed (van der Maesen, 1977).

The genus *Atylosia* W. and A. forms a secondary gene pool of 34 known species. *Atylosia* and *Cajanus* are mainly distinguished by a persistent aril or strophiole on the seeds of *Atylosia.* The character is simply inherited and does occur in some cultivars of *Cajanus.* The two genera, however, are quite distinct because *Cajanus* is found only under cultivation and the *Atylosia* spp. are all uncultivated, wild, weedy forms. Van der Maesen (1975) reviewed the distribution of *Atylosia* and reported the occurrence of 22 out of 34 species in India, Sri Lanka, and Burma, one in Mauritius, seven in Australia, one in Malaysia, two in China, and one in Thailand.

Collections of 10 species, *A. lineata* W. and A., *A. scarabaeoides* (L.) Benth., *A. sericea* Benth. ex Baker, *A. platycarpa* Benth., *A. volubilis* (Blanco) Gamble (*A. crassa* Prain ex King), *A. trinervia* (DC.) Gamble (*A. candollei* W. and A.), *A. albicans* (W and A.) Benth; *A. rugosa* W. and A., *A. grandifolia* (F. von Muell. Benth.) and *A. cajanifolia* Haines are being maintained at the International Crops Research Institute for the Semi-Arid Tropics (ICRISAT), Hyderabad, India.

Cytogenetic relationships of the *Atylosia* species have not yet been elucidated. Deodikar andThakar (1956) observed close affinity between *Cajanus* and some erect species of *Atylosia* and suggested the possibility of transferring wilt resistance to *Cajanus.* Three species *A. lineata, A. sericea,* and *A. scarabaeoides,* were successfully crossed with *Cajanus.* Cytological studies showed that all were close to *Cajanus,* but *A. lineata* was the closest (Kumar and Thombre, 1958; Kumar et al., 1966; Reddy, 1973; and De, 1974). Reddy (1973) and De (1974) concluded that *Cajanus* originated through selection of gene mutations without any change in chromosome number ($n = 11$) from an advanced *Atylosia* species, most likely *A. lineata.* However, it would appear that *A. cajanifolia,* only recently added to the ICRISAT collection, is the closest relative of *C. cajun*. Except for pod and seed characters, it is difficult to distinguish from *C. cajun.*

Recent crossing attempts were unsuccessful with *Cajanus* × *A. platycarpa* and *A. volubilis (A. crassa),* but apparently successful with *A. albicans* and *A. trinervia (A. candollei)*. Success in crossing with *Cajanus* as female has varied from 0.0 to 32.9% (De, 1974). Reciprocal crosses have not been successful, except with *A. sericea.*

The Regional Pulse Improvement Project of USAID in India and Iran made the first comprehensive collection of 4,000 accessions in the 1960's. The bulk of this collection is now at ICRISAT, which has the responsibility of collecting and maintaining pigeonpea germplasm at the international level. Approximately 5,000 accessions from India and 300 from other countries are being maintained. Several small collections are being maintained at the national and regional level.

II. PLANT CULTURE

A. Field

Pigeonpea is sensitive to photoperiod and is a short-day plant (Sen Gupta, 1955; Gooding, 1962; Spence and Williams, 1972). Cultivars differ widely in time from planting to flowering at different latitudes due to the varying degree of sensitivity to day length and temperature. Most cultivars flower in day length of 11 to 11.5 hours (Akinola et al., 1975). In Puerto Rico, Riollano (1964) was successful in shortening the time from planting to flowering by 4 months in medium maturing and 7 weeks in late maturing cultivars by reducing the day length to 8 hours, but it did not effect an early maturing, all season cultivar. This so called all-season cultivar in Puerto Rico (18°N Lat) is not necessarily photoperiod insensitive. In India, all extra early cultivars flowered within a reasonable period of time when planted every month of the year at 17°N Lat, but they took much longer to flower when grown at 29°N Lat.

No study has been done under controlled temperature and photoperiod conditions to separate photoperiod and temperature effects. Preliminary observations made under field conditions indicate that no pigeonpea genotype is truely photoperiod insensitive. The degree of sensitivity varies quantitatively, however, with the earliest maturing types being the least sensitive. Monthly plantings of 21 cultivars of varying maturity at Hyderabad, India, indicated at least 4 major response groups.

Pigeonpea is successfully grown between 30°N and 30°S Lat. In India and Java, productive plants were found at altitudes up to 1,830 m, but seed set was adversely affected at 1,070 to 1,525 m in Hawaii where minimum night temperature was 10 C (Akinola et al., 1975). In Kenya, production has been successful at 2,000 m. Pigeonpeas generally are highly susceptible to frost. Pollination and pod setting is adversely affected at high temperatures in dry weather and low soil moisture conditions. At Hyderabad, pollinations can be successfully done until the end of March with maximum temperatures of 38 to 40 C, but in April when temperature exceeds 40 C, pod set is adversely affected.

Inflorescence development and pod setting is influenced by the availability of light. No pods are set under the crop canopy. Dry, bright days are favorable for fertilization, while cloudy, damp weather results in excessive flower drop (Howard et al., 1919; Mahta and Dave, 1931). Precise information on the quality, level, and the duration of light required for fertilization and seed setting is lacking.

Pigeonpea can grow and produce relatively good yield on fertile soils with a soil pH of 5 to 8 (Akinola et al., 1975). It is highly susceptible to waterlogging, and genetic differences exist for tolerance to temporary waterlogged conditions. The deep tap root system sustains the crop on residual moisture during dry periods.

Pigeonpeas grow slowly during the first 45 to 60 days, but become large bushy plants by flowering time. For seed production purposes, a wide range of plant densities from 6,000 to over 300,000 plants/ha are being used in different regions, depending on the maturity, determinate or indeterminate plant type, soil type, and time of planting (Abrams and Julia, 1973; Ariyanayagam, 1975; Saxena and Yadav, 1975).

Common spacings used in experimental plots in India are 0.75 × 0.25 m or 0.50 × 0.25 m for early types, 0.75 × 0.30 m or 0.90 × 0.30 m for medium cultivars, and 0.90 × 0.30 m or 1.50 × 0.30 m for late cultivars. Spacings of 0.76 × 1.53 m to 1.53 × 1.53 m were recommended for pedigree breeding by Krauss (1932).

In the crossing block, spacings of 1.50 × 0.50 m for medium and late and 0.75 × 0.30 m for early types have been found convenient. Where land is a limiting factor, medium and late types can be planted in paired rows 5 m long at 0.75 × 0.30 m spacing with a one row gap between the pairs. This provides satisfactory access for emasculating and crossing.

Tall growing plants should be cut back for easy handling of flowers for emasculation and pollination and for allowing sufficient penetration of light in the plant canopy, which is essential for floral initiation and development.

Winter nurseries are in use with short and medium duration cultivars. The chief constraint is photoperiod reaction, but most cultivars which have a short life cycle will flower within 80 to 100 days when planted after the normal season harvest date, and two generations can be obtained each year.

B. Growth Chamber and Greenhouse

Only limited work has been done on production of pigeonpeas in growth chambers and greenhouses. Satisfactory growth in greenhouses has been observed where light was adequate and temperatures were kept between 38 and 40 C. Summerfield et al. (1977) reported successful growth of pigeonpea to maturity in growth chambers, greenhouses, and under a plastic shelter in the field. They provide details on rooting substrate, nutrient solution, irrigation and light requirements but they are not included herein because it seems unlikely breeders would use so expensive a facility for making crosses.

III. FLORAL CHARACTERISTICS

The complete flower is similar in structure to that of other legumes. The raceme inflorescences form a terminal panicle in indeterminate and somewhat corymb shape bunch in determinate plant types. These are grouped together at the ends of the branches in late types, and distributed along the branches in early and medium, indeterminate types.

Individual flowers consist of a calyx with five sepals, and a corolla with a standard, two wings, and a keel. There are 10 stamens, 9 fused in a column and 1 free. The ovary consists of two to nine ovules. The long, club-shaped style curves upward toward the standard. Flower size is positively correlated with seed size and is larger in vegetable types than in grain types.

Flower opening in an individual inflorescence and within a branch begins at the base and progresses upward. There are exceptions where the first flower opens in the middle of a flowering branch and successive flowering proceeds in either direction. The size of the inflorescence varies in different types and there may be as many as 10 flowers in each inflorescence. Usually two flowers open at a time on the same inflorescence, but the process of flowering in a plant continues until pod maturation. An inflorescence may hold from two to eight pods; however, the number in successive inflorescences decreases progressively towards the apical end of the branch. If old inflorescences and pods from the lower nodes are removed, pod setting on successive nodes improves and compensates for the loss. This compensatory mechanism helps in prolonging the growth of apical meristem and extends the flowering period.

Durga Prasad and Narsimha Murthy (1963) and Datta and Deb (1970) studied floral development in a few grain types at Hyderabad and a vegetable type at Calcutta and observed that a bud visible to the naked eye develops and blooms within 15 to 20 days time. In a fully developed bud, anthers surround the stigma and dehisce a day before the flower opens. This normally results in self pollination of all flowers, however, on an average 15 to 20% outcrossing is common.

Sen and Sur (1964) suggested that thrips (*Taeniothrips distalis*), which enter the mature buds in large numbers may be responsible for cross pollination. At ICRISAT, however, thrips were abundant on male-sterile plants bagged together with normal plants, but almost no pod set occurred. It seems that bees, *Megachile* spp. and *Apis* spp. are the main pollinating insects. Foreign pollen tube growth is probably faster than that of pollen from the same plant, which favors cross fertilization in the species. Datta and Deb (1970) studied pollen tube growth in a style pollinated with pollen from the same flower and observed it to be very slow, taking 54 hours to reach the base of the ovary.

Prasad et al. (1977) reported that receptivity of stigmas starts 68 hours before anthesis and continues for 20 hours after anthesis. The best time for stigma receptivity is a day before anthesis.

Flowers start opening early in the morning in the summer and by noon during winter, and continue opening throughout the day. The length of time flowers remain open is influenced by the weather. Mahta and Dave (1931) observed that the flowers remained open at Pusa (Bihar) in north east India as long as a day and a half, but at Nagpur in central India it was normally about 6 hours.

IV. ARTIFICIAL HYBRIDIZATION AND SELF-POLLINATION

A. Equipment

Emasculation is carried out with a fine forceps. Items used for pollination include a petri dish containing moist filter paper and threads of different colors. Muslin cloth bags, nylon stockings, or fine-mesh nylon bags are used to cover flowers for selfing.

B. Preparation of the Female

Flowering in pigeonpeas normally continues until 75 to 80% of the pods mature; however, only 10 to 20% of the buds develop into pods. At the onset and during peak flowering, 50% or more of the buds may set pods, while toward the end of the flowering period pod set diminishes to zero. Crossing success, therefore, is higher if early developing buds are chosen on the female parent. Usually two flower buds should be selected for crossing, and all smaller buds should be removed to prevent competition within the inflorescence. Only 2 to 10 buds should be emasculated on a branch.

Tightly closed buds, approximately two-thirds the size of mature buds, should be emasculated (Fig. 1). Buds at the correct stage should show a bright corolla color without any greenish hue. The selected bud should be firmly held between the thumb and the middle finger with the index finger used to support the flower (Fig. 2). The curved side of the standard is held toward the crosser and the sepal covering the keel is removed (Fig. 2A). The corolla is opened by inserting one of the tips of a fine pointed forceps at the base of the keel and moving upward to the tip of the standard (Fig. 2B). The bud will open with slight pressure of the supporting index finger, and the well-developed yellow anthers can be removed from the staminal column with forceps (Fig. 2C).

C. Pollination

The stigma is receptive before anthesis, and pollination can be done immediately after emasculation. Pollen source buds from male parent should be collected between 0800 and 1000 hours. These should be large, unopened buds in which the anthers will dehisce on the day collected (Fig. 1). They can be used throughout the day by keeping the buds in a covered petri dish on moist filter paper. When a flower has been emasculated, the staminal column of the pollen bud is used to brush pollen on the stigma of the female (Fig. 2D). If pollen buds are plentiful, the staminal column of the pollinator bud is left in contact with the pollinated stigma by trapping it between the edges of the corolla. Single flower buds can be used for two to three pollinations when necessary.

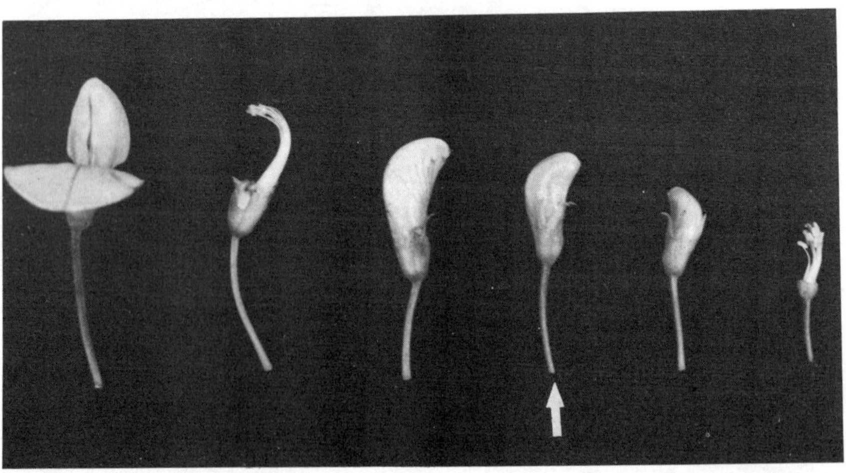

Fig. 1—Stages of flower development. Right to left. Flower before anthesis with calyx and corolla removed, an immature bud, bud at the stage for emasculation (arrow), bud at the stage for pollen collection, bud at the stage for pollen collection with calyx and corolla removed, and open flower.

Prasad et al. (1977) found that pollen in buds remained viable up to 42 hours at room temperature (25 to 28 C, 50.6% relative humidity) and up to 11 days in the refrigerator (10 C, 37.5% relative humidity). Successful pollinations have been made with *Cajanus* with detached mature flower buds of *Atylosia* spp. that had been stored up to 72 hours in polythene bags kept in a Thermos flask with ice.

For selfing, muslin cloth bags, reject nylon stockings, or fine-mesh nylon cloth bags have been found quite effective to exclude insect vectors. One can self the entire plant or an individual branch depending on the quantity of selfed seed required. At ICRISAT, bags 120 × 80 cm and 60 × 30 cm are used for entire plants and an individual branch, respectively. It is desirable to spray insecticide before bagging to control pod borers and avoid damage to the pods. Murthi (1977) obtained 15 to 18 selfed pods from a branch and over 200 pods from an entire plant. Kelkar and Pandya (1934) suggested smearing candle wax to seal individual flower buds in preference to muslin cloth bags and reported 15% success in obtaining selfed pods. The method has never been tried, however, on a large scale and does not seem practical.

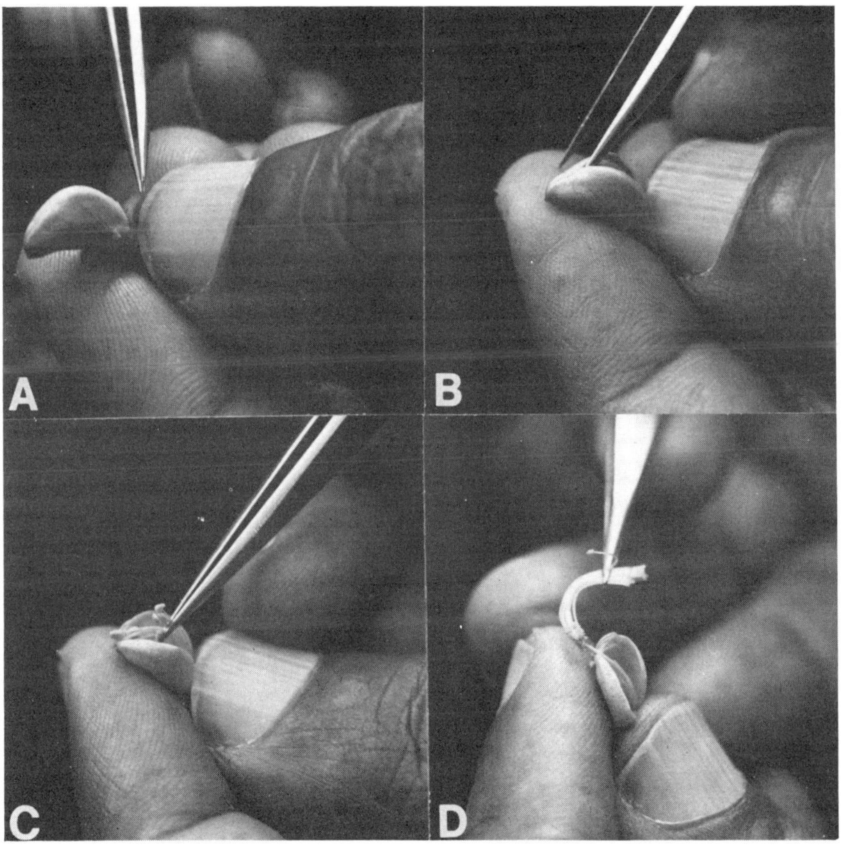

Fig. 2—Artificial hybridization of pigeonpea. A. Removing the sepal in front of the keel; B. Opening the flower; C. Removing the anthers; D. Brushing pollen on the stigma of the emasculated flower.

To identify cross-pollinated flowers, colored thread is better than a jeweller's tag commonly used for cereals and plants with bigger flowers. Different colored threads can be used to identify different crosses. Pollinated buds are generally not bagged because pod setting is greatly reduced under glassine bags. Our experience shows that outcrossing, if any, is negligible in unprotected pollinated buds. In a large crossing program, covering individual pollinated buds is not practical.

Success of crossing is influenced by the environment and skill of the pollinator. An average pod set of 20% was obtained at ICRISAT by skilled laborers who made 447,000 pollinations over a period of 8 months.

D. Factors Affecting Efficiency

The time of initiation and duration of flowering is influenced by the degree of perenniality of the cultivars, and by their photoperiod and temperature responses. Monthly plantings of early, medium, and late cultivars at Hyderabad (17°N Lat) have shown that by planting in September, the difference in days to flowering between the early and the late types is reduced to about 68 days compared with 85 days when planted in July. With plantings in September and October, there is overlap in the flowering period of early and medium, and medium and late. Successive plantings are needed to match flowering dates of early and late parents.

Some degree of perenniality is present in almost all pigeonpea cultivars and provides an opportunity to prolong the flowering period by 1 to 2 months in any cultivar if flowers are removed as they appear. Ratooning of early and medium maturing types and utilization of flowers on regenerated growth has been successfully used to cross divergent maturity types by Saxena et al. (1976).

Hybrids can be easily identified at flowering and pod formation by comparing various plant characteristics with the parents. For early detection of hybrids, stem pigmentation and obtuse leaf shape are good seedling markers. Deep purple pigmentation of the stem is partially dominant over green (F_1 is intermediate) and is monogenically inherited (Kumar and Haque, 1973). Normal lanceolate leaf with acute leaf apex is partially dominant over obtuse leaf (F_1 has broad leaf and acute apex).

V. NATURAL HYBRIDIZATION

Natural hybridization has been recognized as a major source of variability in released cultivars, as well as in landraces in farmers' fields. Both sources of variability have been extensively used by breeders for selection of improved types. In the absence of male sterility, Rachie and Gardner (1975) proposed a dual population system for effecting large-scale natural hybridization for population improvement. The system involves utilization of a simply inherited character like seedling coloration, plant growth habit (determinate or indeterminate), or leaf shape. The recessive condition of these characters is used in the female parent, so that the F_1 hybrids are quickly detected. Plantings of the female and male parent are made in alternate rows or hills to maximize outcrossing. A bulk hybrid population constituted by compositing equal quantities of seed from each female parent row is planted in isolation. In F_1 seed is harvested only from plants with the dominant marker (hybrid plants). In F_2 seed is harvested only from plants with the recessive marker and in succeeding generations alternate selection for the dominant and recessive marker ensures intermating. After 2 to 3 generations of intermating, S_1 testing can be used to evaluate lines and selected ones can be used to reconstitute two populations, one with the recessive and the other with the dominant marker. The system depends heavily on insect pollination which may vary from location to location and from one season to another, and the desired level of intermating may not be achieved in each generation.

Reddy et al. (1977) recently reported five types of male-sterile plants from 35 sources among the germplasm collections, and from F_4 derivatives of *Cajanus cajan* × *Atylosia* spp. crosses. Of the five types, translucent male-sterile seems to be most promising for use in the natural hybridization, either for population improvement or for the production of hybrid seed because male sterility is complete and stable, phenotypic identification by anther color is easy, and monogenic recessive inheritance permits relatively easy maintenance (Reddy et al., 1978). Preliminary observations have indicated that in a population of fertile and sterile sibs, seed set on sterile plants under natural conditions with insect cross pollination was 84% of the normal fertiles.

Isolation distance for maintaining cultivar purity will depend on the pollinating agents. Published recommendations vary widely. Sen and Sur (1964) suggested a distance of 5.5 m while FAO recommended an isolation distance of 126 to 360 m (Akinola et al., 1975). Bhatia (1980) observed only 3% outcrossing at 100 m isolation. More detailed work on isolation distance, field layouts, and proportion of male and female rows required in a hybrid seed production block is in progress at ICRISAT.

VI. SEED DEVELOPMENT, HARVEST, AND STORAGE

Fertilization occurs in 48 to 54 hours after pollination (Datta and Deb, 1970). Preliminary observations to check the success of pollination can be made after 1 week, when petals dry and the young developing ovary can be clearly seen. There is substantial abscission of young pods, particularly in windy wet weather. Pod development ordinarily is completed in about 20 days after blooming. Seeds attain physiological maturity in 30 days and are ready for normal harvesting in about 40 days with 18 to 20% moisture (Rao and Rao, 1974).

Seed can be harvested and germinated in petri dishes at room temperature of 30 C after 20 to 30 days, when they are pale green or pale white in color. Germination of about 35% has been increased to 60% when a small portion of the cotyledons was cut or the testa was ruptured. Dry pods generally are harvested by hand into cloth bags and are kept for further drying in the sun or in a drier before being shelled by hand.

Information on drying and storage of pigeonpea seeds is limited. Seeds with 10% moisture have been stored at room temperature (21 to 37 C) for about a year, and in a cool room at 14 to 18 C and 50 to 70% relative humidity for up to 3 years without affecting seed viability (Murthi, 1977).

Pigeonpeas are highly susceptible to stored grain pests and the primary infection of *Callosobruchus chinensis* and *C. maculatus* starts in the field. Napthalene balls placed in individual seed packets or bottles are highly effective against these pests and ensure safe storage.

REFERENCES

Abrams, R., and F. J. Julia. 1973. Effect of planting time, plant population and row spacing on yield and other characteristics of pigeonpeas, *Cajanus cajan* (L.) Millsp. Univ. of Puerto Rico J. Agric. 57:275–285.

Akinola, J. O., P. C. Whiteman, and E. S. Wallis. 1975. Agronomy of the pigeonpea (*Cajanus cajan*). Commonwealth Bureau of Pastures and Field Crops, Maidenhead, U.K. Review series No. 1/1975.

Ariyanayagam, R. P. 1975. Status of research in pigeonpea in Trinidad. p. 131–139. Int. Grain Legumes Workshop Proc., ICRISAT, Hyderabad, India.

Datta, P. C., and Arati Deb. 1970. Floral biology of *Cajanus cajan* (L.) Millsp. var. bicolor D.C. (Papilionaceae). Bull. Bot. Soc. Beng. 24(1/2):135–145.

De, D. N. 1974. Pigeonpea. p. 79–87. *In* J. Hutchinson (ed.) Evolutionary studies in world crops diversity and change in the Indian subcontinent. Cambridge Univ. Press, Cambridge, England.

Deodikar, G. B., and C. V. Thakar. 1956. Cyto-taxonomic evidence for affinity between *Cajanus indicus* spreng. and certain erect species of *Atylosia* W. and A. Proc. Indian acad. Sci. 43B:37–45.

Durga Prasad, M. M. K., and B. L. Narasimha Murthy. 1963. Some observations on anthesis and pollination in redgram (*Cajanus cajan* L.). Andhra Agric. J. 10: 161–167.

Gooding, M. J. 1962. The agronomic aspects of pigeonpeas. Field Crops Abstr. 15: 1–5.

Howard, Albert, G. L. C. Howard, and A. R. Khan. 1919. Studies of the pollination of Indian crops. I. Mem. of the Dep. of Agric. in India (Botanical series) Vol. III:195–225.

Hutchinson, J., and J. M. Dalziel. 1958. Flora of West Tropical Africa. 2nd ed., revised by R. W. J. Keay. Crown Agents for Oversea Governments and Administration, London.

Huth, E. 1893. Difen in der botanischen nomenclatur Genus—and speciesnamen gleich lauten? Helios 11:133.

Kelkar, S. G., and P. S. Pandya. 1934. New method of selfing tur (*Cajanus indicus*). Poona Agric. Coll. Mag. 26:108–111.

Krauss, F. G. 1932. The pigeonpea (*Cajanus indicus*), its improvement culture and utilization in Hawaii. Hawaii Agric. Exp. Stn. Bull. 64.

Kumar, A., and M. F. Haque. 1973. Inheritance of stem pigmentation and flower color in pigeonpea. Mysore J. Agric. Sci. 7:312–314.

Kumar, L. S. S., M. K. Shama Rao, and M. V. Thombre. 1966. Interspecific and intergeneric hybridization in the breeding of crop plants. Indian J. Genet. 26-A: 114–120.

————, and M. V. Thombre. 1958. An intergeneric hybrid of *Cajanus cajan* (L.) Millsp. × *Atylosia lineata* W. and A. J. Univ. Poona, 12:13–16.

Mahta, D. N., and B. B. Dave. 1931. Studies in *Cajanus indicus*. Mem. Dep. Agric. India. Bot. 19:1–25.

Morton, J. F. 1976. The pigeonpea (*Cajanus cajan* (L.) Millsp.) a high-protein, tropical bush legume. Hort. Sci. 11:11-19.

Murthi, A. N. 1977. Personal communication. International Crops Research Institute for the Semi-Arid Tropics, Hyderabad, India.

Pathak, G. N. 1970. Red gram. *In* Pulse crops of India. 14-53.

Prasad, S., R. Prakash, and M. F. Haque. 1977. Floral biology of pigeonpea. Tropical Grain Legume Bull. No. 7:12.

Rachie, K. O., and C. O. Gardner. 1975. Increasing efficiency in breeding partially out-crossing grain legumes. p. 285-300. *In* Int. Grain Legumes Workshop Proc., ICRISAT, Hyderabad, India.

Rao Madhava, K. V., and G. Rajeswara Rao. 1974. Gibberellin like substance in developing and germinating seeds of pigeonpea (*Cajanus indicus* spreng.). Indian J. Plant Physiol. 17(1/2):65-72.

Reddy, B. V. S., J. M. Green, and S. S. Bisen. 1978. Genetic male sterility in pigeon pea. Crop Sci. 18:362-364.

———, L. J. Reddy, and A. N. Murthi. 1977. Reproductive variants in *Cajanus cajan* (L.) Millsp. Tropical Grain Legume Bull. No. 7:11.

Reddy, L. J. 1973. Interrelationships of *Cajanus* and *Atylosia* species as revealed by hybridization and pachytene analysis. Ph.D. thesis. Indian Institute of Technology, Kharagpur, India.

Riollano, A. 1964. Effects of photoperiodism and other factors on the improvement of pigeonpea varieties. Univ. of Puerto Rico J. Agric. 48:232-235.

Saxena, K. B., D. Sharma, and J. M. Green. 1976. Pigeonpea ratooning—an aid to breeders. Tropical Grain Legume Bull. No. 4:21.

Saxena, M. C., and D. S. Yadav. 1975. Some agronomic considerations of pigeonpeas and chickpeas. p. 31-61. *In* Int. Grain Legumes Workshop Proc., ICRISAT, Hyderabad, India.

Sen Gupta, J. C. 1955. The physiology of growth and development of some crop plants. Bot. Soc. Bengal Bull. 9:62-80.

Sen, S. K., and S. C. Sur. 1964. A study on vicinism in pigeonpea (*Cajanus cajan* (L.) Millsp.). Agricultura (Louvain) 12:421-426.

Spence, J. A., and S. J. A. Williams. 1972. Use of photoperiod response to change plant design. Crop Sci. 12:121-122.

Summerfield, R. J., P. A. Huxley, and F. R. Minchin. 1977. Plant husbandry and management techniques for growing grain legumes under simulated tropical conditions in controlled environments. Exp. Agric. 13:81-92.

Van der Maesen, L. J. G. 1975. Germplasm collection and evaluation in *Cicer* and *Cajanus.* p. 229-237. *In* Int. Grain Legumes Workshop proc., ICRISAT, Hyderabad, India.

———. 1977. Persònal communication. International Crops Research Institute for the Semi-Arid Tropics, Hyderabad, India.

Westphal, E. 1974. Pulses in Ethiopia, their taxonomy and agricultural significance. Center for agricultural publishing and documentation, Wageningen, The Netherlands.

34 Potato

Robert L. Plaisted
Cornell University
Ithaca, New York

The potato of commerce in the United States, Europe, and other temperate and subtropical climates is *Solanum tuberosum,* subspecies *tuberosum.* This crop is often referred to as the Irish potato or white potato to differentiate it from the sweet potato. Both names are misleading in that the origin of potato is not Irish and they have colors other than white. Native cultivars of the same subspecies grow in Chile. The subspecies *andigena* is widely cultivated at high elevations from Argentina to Mexico with a center of concentration in Peru, Bolivia, and Ecuador. Some believe that North temperate clones were derived from subspecies *andigena* through selection for adaptation, but others believe that Chilean tuberosum was the source (Grun et al., 1977).

I. PARENTAL MATERIAL

Solanum contains about 2,000 species, of which about 150 are tuber-bearing. These occur in five cytological groups, with somatic chromosome numbers of 24, 36, 48, 60, and 72.

About 70% of all potato species are diploid (Hawkes, 1958). Most of these are self-incompatible and are able to produce seeds only when they are pollinated with pollen containing a different S allele. Diploid species which are self-compatible are *S. morelliforme, S. polyadenium,* and *S. verrucosum.* A prominent cultivated diploid is *S. phureja,* which is grown in mountain valleys throughout South America. It is distinguished by an absence of tuber dormancy, which permits immediate replanting in areas where continuous cropping is possible. This species has been used extensively as a pollinator to produce haploids of *S. tuberosum* cultivars, as a bridging parent between other diploid species and *S. tuberosum,* and as a base for genetic studies (Peloquin et al., 1966). Other cultivated diploids are *S. stenotomum,* which is cultivated in regions similar to where *S. phureja* is grown; *S. ajanhuire,* a hardy species grown in the highlands of Bolivia where its tubers are used for production of tunta, a freeze-dried produce; and *S. goniocalyx,* which is grown in the lower valleys in Peru.

Tetraploid, tuber-bearing *Solanums* may be self-fertile. Cipar et al. (1964) showed that haploids (diploids) of the tetraploids contain the same self-incompatibility system as the diploid species. The system conforms with the *Nicotiana* scheme of oppositional S alleles. Presumably the tetraploids are self-compatible when the pollen possesses two S alleles which induce mutual weakening. Even though the majority of diploid species are self-incompatible, variation exists within species. One cause for this variation may be the spontaneous production of unreduced gametes in plants which are heterozygous for S alleles. The other is the occurrence of an S-bearing translocation (Olsder and Hermsen, 1976), which would have the same mutual weakening effect as long as the translocation carried a different S allele.

Crosses between tetraploids and diploids rarely produce triploid offspring because of a "triploid block" arising from an imbalance of endosperm factors in the nucleus (den Nijs and Peloquin, 1977b). A few triploid clones are cultivated, but these are sterile and are preserved by asexual propagation. The non-bitter triploid cultivars are often classified as *Solanum* × *Chaucha.* The hybrids *Solanum* × *Juzepczukii* are assumed to be hybrids between the frost-resistant, tetraploid species, *S. acaule,* and cultivated diploids. Christiansen (1977) has identified more than 120 different clones and described the impact of these "bitter potatoes" upon the economy and nutrition of people living in the Altiplano region of Peru and Bolivia.

Tetraploids form about 15% of the tuber-bearing species. Clones of *S. tuberosum* subspecies *andigena* will cross readily with clones of subspecies *tuberosum,* but generally produce late maturing hybrids unless the *andigena* parent has been selected for earliness. Clones of this subspecies are valuable sources for many desirable characters.

The predominant form of chromosome pairing in tetraploids is as bivalents. Swaminathan and Howard (1954) reported an average quadrivalent and trivalent frequency of two to three per nucleus. Tetraploids may be hybridized with diploids directly or by first doubling the chromosome number of the diploid with colchicine. Tetraploid offspring are produced in crosses between diploids and tetraploids as a consequence of the unreduced gametes produced in small frequency by most diploids and in high frequency by a few. Hanneman and Peloquin (1968) obtained 91.9% tetraploid progeny in crosses between tetraploid × diploid parents and 94.8% in the reciprocal crosses. The diploid pollen appears to have a competitive advantage over the haploid pollen, and when crossed to a tetraploid species, the triploid block permits only the tetraploid ovules to develop.

Pollen fertility of interspecific hybrids frequently is dependent on the species used as female. Lamm (1953) reported ssp. *tuberosum* cytoplasm to be sensitive to *S. acaule* nuclear factors. Dionne (1961) reported male sterility in crosses between *S. demissum* and diploid species when *S. demissum* was used as a female. Grun et al. (1977) studied various manifestations of male sterility in *tuberosum* and its possible ancestors. They described eight cytoplasmic factors for male sterility which differentiated spp. *tuberosum* from ssp. *andigena, S. vernei,* and *S. stenotonum-S. phureja.* The ssp. *tuberosum* clones considered were from the northern hemisphere and from coastal Chile. In general, clones of ssp. *tuberosum* from both sets behaved the same. Sources were bilaterally fertile in crosses to each other, whereas ssp. *tuberosum* × ssp. *andigena* hybrids usually will display male sterility, although the reciprocals do not.

Solanum × *curtilobium* is a pentaploid, presumably a cross between *S.* × *juzepczukii* and subsp. *andigena. Solanum* × *curtilobium* and *S.* × *juzepczukii,* constitute the "bitter potatoes" grown on the Altiplano of the Andes.

Hexaploid species constitute about 8% of the tuber bearing *Solanum* species. These are generally self-fertile. They can be crossed directly to tetraploid cultivars with a small percentage of success or they may be crossed to a diploid to produce a tetraploid hybrid, which is then crossed to a tetraploid cultivar. The hexaploid species most widely used by potato breeders has been *S. demissum* which grows at high elevations in Mexico. This species served as the source of the major *R* genes for late blight resistance.

Major collections of tuber bearing Solanums are maintained at the International Potato Center in Peru, the IR-1 project at Sturgeon Bay, Wis., the Commonwealth Potato Collection at Pentlandfield, Scotland, the German-Dutch Potato Collection at Braunschweig, West Germany, the Central Colombian Collection at Bogota, Colombia, and the Potato Collection at Balcarce, Argentina. The International Potato Center has the best collection of cultivated species, particularly *andigena, phureja,* and the "bitter potatoes". The IR-1 collection is an excellent source of the wild species, as well as many accessions of the cultivated species. Because of quarantine restrictions on importation of tubers into the United States, the IR-1 project is an invaluable source of germplasm for interested users in the United States. Quarantine restrictions do not apply to the importation of true seeds, if material is desired from the other collections.

II. PLANT CULTURE

A. Field

Clones of temperate zone tuberosum will generally bloom 60 days after planting in the field. Pollination of the flowers on the plant in the field is generally unsuccessful for producing seeds. Many breeding programs overcome this difficulty by using the technique of cut stems described by McLean and Stevenson (1952) and developed further by Peloquin and Hougas (1959). This method involves cutting a stem that contains a large inflorescence with several unopened buds. The stem is cut 20 to 30 cm long with at least one fully expanded leaf. The cut stem is immediately placed in tap water and carried to an air-conditioned greenhouse or growth chamber. Any opened flowers are removed.

Where nurseries are grown as a source of cut stems for pollination, the plants should be at a spacing that will facilitate identification and access. Rows approximately 2 m apart and plants 1 m apart in the rows will accomplish this. The planting stock should be virus-free and the nursery managed to avoid infection by contact or aphids. If possible, the nursery should be planted where irrigation can be applied during the time of flower development, if rainfall is insufficient. The nursery should be a short distance from the place where the cut stems are maintained.

B. Growth Chamber and Greenhouse

When using cut stems, the greenhouse or growth chamber should have at least 16 hours of sunlight or 20 klux of artificial light. It is important that the temperature be maintained at about 19 C with a high relative humidity during the period of pollination. Usually an inflorescence will produce flowers for 5 to 10 days. After fertilization has been achieved, the stems can be placed in a warmer greenhouse if space is limiting. The most critical aspect of this technique is to keep the cut stems from rotting. This can be done if the water in the containers is replaced with fresh water two or three times a week during the first 2 weeks. Few stems will rot if 5 to 10 ppm of streptomycin sulfate are added to the water, and the water is changed once a week while at the 19 C temperature. If rot does begin, the stem can be saved by trimming it back. Stems treated in this fashion can develop mature fruits that can be harvested in about 6 weeks. Some breeders use milk bottles to hold individual stems. Another technique, more amenable to changing the water, is to use plastic paint pails with chicken wire over the top to support four stems.

Traditionally potato crosses have been made in the greenhouse during the winter when the temperature can be maintained at 19 C. The day length must be extended to 16 hours from the time of emergence to pollination. Lights with a high light level such as metal halide lamps, result in the best growth. Flowering will be benefitted if the fertility is somewhat restricted and tuberization is reduced. Dutch breeders have developed a system of placing the potato seed piece on a brick and covering it with sand (Thijn, 1954). As soon as the roots have penetrated the soil, the sand is washed away and tubers removed as they are formed. Another technique is to plant the tubers in 20 cm clay pots and set them on the greenhouse bench or ground bed. This constricts the root growth and promotes flowering. The plants need to be pruned to one or two main stems and tied to 2 m stakes.

III. FLORAL CHARACTERISTICS

It is difficult to classify the cultivated tetraploid potato as to extent of natural cross-pollination. It displays the inbreeding depression characteristic of cross-pollinated species, yet most of the seeds obtained from open-pollinated fruits are the result of self-fertilization. This is discussed in greater detail in section V.

The inflorescences are borne terminally on the stems and the number of flowers ranges from 1 to 30, with 7 to 15 being the most usual (Fig. 1A). The number of inflorescences per plant and the number of flowers per inflorescence are greatly influenced by the cultivar. About the time the first flowers are fully expanded, a new shoot develops at the proximal leaf axis and ultimately grows beyond the first flower cluster to produce a second inflorescence.

The flower is 3 to 4 cm in diam with five petals which give the flower the shape of a star (Fig. 1B). The size of the flowers and their distribution in the inflorescence make them easy to locate and to manipulate. The five anthers are 5 to 7 cm long, and openly displayed in a column surrounding the pistil (Fig. 1C). The stigma generally protrudes beyond the ring of anthers (Fig. 1D). The corolla may be white or a complex range of blue, red, and purple depending on the anthocyanins present. The anthers are a bright yellow, except for male-sterile clones in which they may be light yellow or

Fig. 1—Emasculation and anthesis of potato. (A) The four buds of the inflorescence marked by arrows are at proper stage for emasculation; (B) emasculated flower at the stage for pollination; (C) complete single flower prior to anthesis; (D) anthers ready to shed pollen.

yellow green. The stigmas are usually green, though some clones may have pigmented stigmas. The extrusion of the stigma beyond the anthers can range from essentially none to the style being half again as long as the anthers. The petals generally will conceal the anthers in the mature bud, though in some clones the petals of the bud will not be expanded fully enough to conceal the anthers. The stigma generally will not be extruded through the tip of the anther cone in the mature buds until a day before the flower opens.

The flowers on the branch closest to the base are the first to open with two or three opening each day. Flowers will remain open 2 to 4 days resulting in inflorescences with 5 to 10 flowers open at the time during the height of bloom. The receptivity of the stigma and duration of pollen production is about 2 days (Janssen and Hermsen, 1976). Clarke (1940) found that fertilization can occur as early as 36 hours after pollination. Hoopes (1977) found that pollen tubes of *tuberosum* × *andigena* hybrids grew the length of the styles within 20 hours. Viable embryos can be excised 14 days after pollination (Haynes, 1954). Six weeks after pollination is the minimum time for full development of viable seed (Simmonds, 1963b).

IV. ARTIFICIAL HYBRIDIZATION AND SELF-POLLINATION

A. Equipment

If many flowers are to be pollinated with one pollen source or a bulk collection of pollen is required, a mechanical vibrator will facilitate pollen collection. These vibrators may be constructed using a flashlight case with a modified door bell unit fastened to it (Fig. 1D) (Sanford and Hanneman, 1977). In operation, the looped wire fastened to the vibrating arm of the buzzer is placed against the base of the flower and the anthers vibrated to release the pollen (Fig. 1C). The pollen can be collected in small test tubes filled with paraffin to within 1 cm of the top, in gelatin capsules, or in modified plastic syringes where the plunger can regulate the depth of the pollen.

B. Preparation of the Female

Cut stems are best made in the field in the early morning. The opened flowers should be removed at that time and mature unopened buds can be emasculated. The buds at this stage are plump, the petals appear ready to separate, and display their final color (Fig. 1A).

When potatoes are grown in winter in closed greenhouses, and not subjected to vigorous shaking by ventilation air currents, opened flowers generally fail to set fruit unless they are hand pollinated. Consequently, emasculation is not essential in the greenhouse if the consequences of a chance selfing are not very serious. Where complete control is required, the emasculation can be accomplished easily as a consequence of the simple flower structure and large anthers. Either forceps or a scalpel reduced to about half its normal blade width can be used to remove the five anthers. Since pollinations are made in the morning after the petals have reopened, the emasculation of buds can be done in the afternoons. The petals can be forcibly opened to reveal the anthers (Fig. 1). Usually the anthers can be removed in groups of two or three without breaking the style. The emasculated flowers do not require protection under the conditions of crossing described, and generally are obvious enough not to require special identification at this stage (Fig. 1B).

C. Pollination

Pollination is most successful in the morning soon after the flowers are fully open. On dark days in the winter, it may be necessary to wait for about an hour after artificial lights are turned on. A pollination schedule of 3 days per week will generally catch most available flowers and reduce the extent of double pollination of individual flowers.

Flowers with plump, bright yellow anthers with a brownish tip are most apt to be good sources of pollen (Fig. 1D). Pollen is generally most abundant in the morning. Most flowers can be used as a pollen source for 2 days after the corolla first opens (Janssen and Hermsen, 1976). If flowers from the field are used as a source of pollen, stems of the male plants may be cut the day before pollination and held in water. If pollen is collected directly in the field, an early morning collection is most apt to be successful.

Potato pollen is robust, if it is kept cool and dry. It can be stored for a month at 2.5 C or a year at −24 C (Blomquist and Lauer, 1962). Longevity is improved if the storage is dry, such as a dessicator with silica gel (Howard 1958). Pollen stored at room temperature and humidity loses its viability after 1 day.

When only a few pollinations are to be made with a single pollen source, the pollen can be removed from mature anthers by inserting a blunted, narrowed scalpel at the base of the suture in the anther lobe and scraping it the length of the anther (Fig. 2A). If the clone is a good pollen source, this will produce sufficient pollen to pollinate three or four stigmas.

Mature, unopened buds and flowers which are newly opened are receptive to pollination. The pollen can be applied in many ways, including touching the stigma with the tip of the pollen-bearing scalpel (Fig. 2B) or inserting the stigma into the vial of collected pollen. A coating of germinating pollen is an effective barrier against further pollination.

When selfing is desired, the unemasculated flowers are allowed to open and the pollen is removed from an anther and applied to the stigmas of the opened flowers on that plant. Dipping the scalpel in alcohol will prevent contamination between pollinations. If volumes of selfed seed are needed, the vibrator can be used, inserting the stigma in the pollen as it is being collected among plants within that clone.

Generally all the flowers of a single inflorescence are assigned to the same pollen source. When the pollination is made, a string tag, which identifies the male or both the female and the male, can be looped about the stem. A color code for the tags can be used to simplify the search of flowers for pollination on subsequent days.

The success in cross-pollination is dependent upon the two parents. If parents of unknown value are used, success greater than 50% is good. If the female is known to produce good mature flowers and the male to produce good pollen, the percentage success is higher, but is not guaranteed. Pollination with bulk pollen from several clones can produce 90% success.

D. Factors Affecting Efficiency

In some breeding programs, potatoes are grafted onto tomato roots to promote flowering of difficult clones. The amount of flowering and duration of flowering will be improved and the abcision of newly pollinated flowers will be reduced. This technique is particularly useful for those clones which tend to abort their flowers prior to pollination.

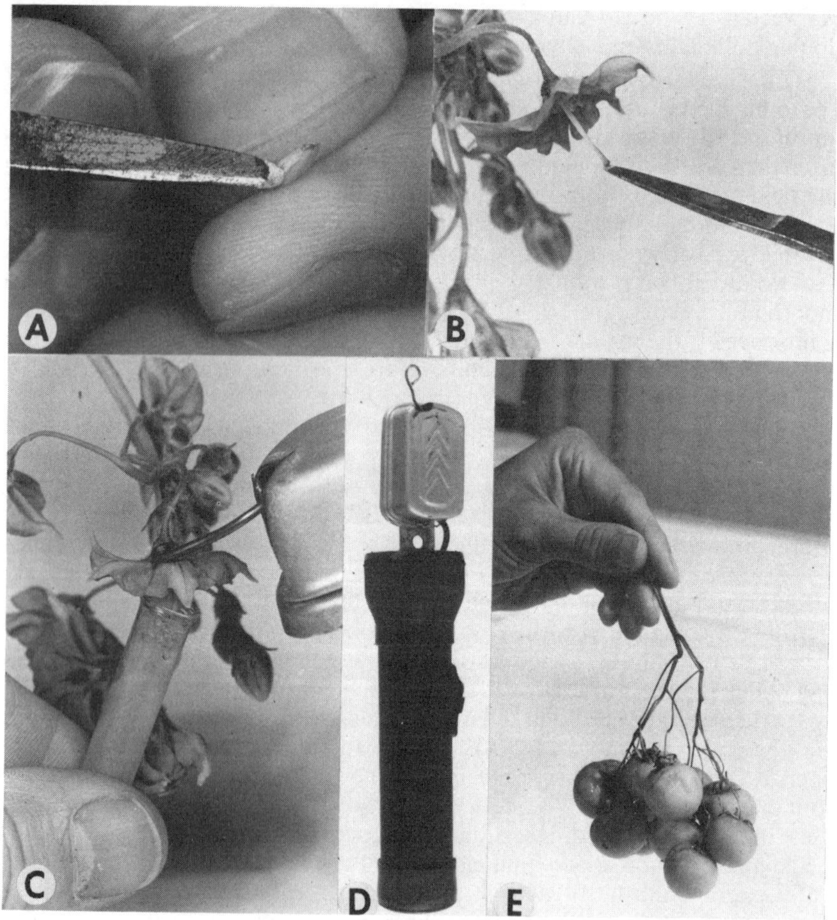

Fig. 2—Pollination and seed. (A) Collecting pollen from a single anther using a blunt scalpel; (B) applying pollen to a stigma with the scalpel; (C) collecting bulk pollen using a vibrator; (D) battery operated vibrator; (E) mature fruits ready for seed extraction.

The most efficient marker for identifying hybrid seed is an embryo spot described by Dodds and Long (1956) and used by Peloquin and Hougas (1959). This marker was found in the diploid species *S. phureja.* The dominant genes B^c and B^d in the presence of P or R condition a blue or red spot at the cotyledonary node of the embryo. The most efficient mature plant markers are also anthocyanin pigments, which are inherited as dominant genes. Simply inherited disease resistance alleles like the H_1 gene for Race A golden nematode resistance also can be used.

V. NATURAL HYBRIDIZATION

Many cultivars of potatoes, particularly those of recent vintage, will bear fruits in the field if the weather during blooming is cool and humid. The primary insect visiting the flowers is the bumblebee. Unopened flowers on fertile plants do not set fruits if they are protected by screened cages. Glendinning (1976) described bumblebee behavior: ". . .bee alights on the anther-cone and its weight causes the flower to hang down. It then vibrates the anthers, and pollen falls onto its abdomen and also onto the stigma. If it obtains much pollen it hangs from the flower while collecting it to its pollen baskets; occasionally, when pollen is abundant, it may renew its grip on the anther and repeat the process. Visits to less productive flowers are briefer and do not involve collection of pollen to the pollen baskets". Estimates of cross-pollination in the field range from zero to about 20% (Glendinning, 1976; Cunningham, 1977).

Since potatoes have such a high percentage of self-pollination in the field, crossing blocks are rarely used in most potato breeding. They are used however, when population development, involving a large number of clones, is the objective. This system is particularly appropriate for diploid species which are generally self-incompatible, so that all the seed that is produced is from cross-pollination. When field nurseries are used in these programs, they should be planted at latitudes or altitudes where prolonged high temperature and low humidity can be avoided. The field should not be subject to volunteer plants from previous crops and should be isolated from other plantings. The plants should be spaced approximately 1 m apart in and between rows to facilitate the harvest and identification of the fruits. Insect control should be done with systemic insecticides or with target specific chemicals during the period of bloom to avoid killing the bumblebees.

VI. SEED DEVELOPMENT, HARVEST, AND STORAGE

Enlargement of the ovary can be observed within a week after pollination and usually is a sign of a successful cross; however, parthenocarpic development can occur and be misleading. The fruits which develop are shaped like tomatoes about 2.5 cm in diam (Fig. 2E). As they reach maturity, they become soft and change from a dark green color to a grey-green. To avoid the loss of seed, about 4 weeks after pollination, each cluster of fruits should be inserted in a paper bag, and the bag stapled around the stem. When the fruits begin to soften, they should be removed from the plant or stem cutting and held in storage until the seed can be extracted. This will be 1.5 to 2 months after pollination (Simmonds, 1963b; Lam and Grenard, 1976).

The safest technique for threshing fruits is to half fill a liter beaker with tap water and squeeze the ripe fruits in the water. The seeds will sink and the pulp can be decanted off. The decanting is repeated once or twice until the beaker contains only seed and clear water. The water and seeds are poured onto a cheesecloth stretched over a pail and held by a rubber band. The seeds can be scooped from the cheesecloth onto a paper towel with a spoon. The tag from the bag holding the fruits is attached to the towel. The towels are laid upon a counter top for 2 or 3 days to dry. When the seeds are dry, the towels can be folded in half and the halves rubbed together to loosen the seeds. The seeds are stored in suitable envelopes.

An alternate technique which can be used more efficiently with large lots of seeds is to put the fruits with a small volume of water into a food blender. It is important that the blades be blunted and that the time of exposure to the blade action be only momentary or seed damage will occur. The contents of the blender can be poured through a small seive to remove most of the pulp. The rest of the procedure can be the same as for the hand extraction. Some workers digest the pulp by adding 1% volume of HCl and soaking the seeds for 1 hour (Slomnicki, 1961).

Seeds that have been dried and stored in coin envelopes at room temperature in a cabinet containing a room dehumidifier have maintained germinability more than 10 years. Storage life can be extended to 20 years by drying the seeds over silica gel and storing them in sealed glass vials or sealed envelopes of metalized polyester film with low water vapor permeability at 5 C (Ross, 1969; Howard, 1975).

Seeds of tetraploid potatoes will be dormant about 6 months, with some maintaining this dormancy as long as 2 years (Lam and Erickson, 1966; Simmonds, 1963a; Ross, 1969). Some breeding programs hold seeds for 1 year to overcome this dormancy. When it is necessary to overcome the dormancy, the seeds can be soaked for 24 hours with 2,000 ppm of gibberellic acid (Spicer and Dionne, 1961; Lam and Erickson, 1966).

VII. TECHNIQUES FOR SPECIAL SITUATIONS

Embryo culture has not been used extensively in potato breeding, probably because the frequency of unreduced gametes in diploids is sufficient to permit diploid × tetraploid crosses.

The special technique of using selected clones of *S. phureja* to derive haploids of tetraploid clones has created many opportunities in potato breeding. The haploids present a convenient means for producing hybrids with diploid species. They are also a means of conducting genetic analyses at the diploid level. The hybrids between haploids and *S. phureja* clones which produce a high percentage of 2n gametes by first division restitution may offer an opportunity for greater heterosis in tetraploid breeding programs (den Nijs and Peloquin, 1977a; Mendiburu and Peloquin, 1977; and

Mok and Peloquin, 1975). The basis of this concept is that much of the heterozygosity of the diploid hybrid will be transferred intact to the tetraploid. The selection of elite *S. phureja* × haploid hybrids could produce progenies with more predictable performance and uniformity.

REFERENCES

Blomquist, A. W., and F. I. Lauer. 1962. A simplified technique for handling and storing potato pollen. Am. Potato J. 39:340–343.

Christiansen, J. 1977. The utilization of bitter potatoes to improve food production in the high altitude of the tropics. Ph.D. thesis. Cornell Univ.

Cipar, M. S., S. J. Peloquin, and R. W. Hougas. 1964. Inheritance of incompatibility in hybrids between *Solanum tuberosum* haploids and diploid species. Euphytica 13:163–172.

Clarke, A. E. 1940. Fertilization and early embryo development in the potato. Am. Potato J. 17:20–25.

Cunningham, C. E. 1977. Personal communication. Campbell Institute for Agricultural Research, Cinnaminson, NJ 08077.

den Nijs, Ton P. M., and S. J. Peloquin. 1977a. Polyploid evolution via 2n gametes. Am. Potato J. 54:377–386.

————, and S. J. Peloquin. 1977b. The role of the endosperm in hybridization. Am. Potato J. 54:296. (Paper title).

Dionne, L. A. 1961. Cytoplasmic sterility in derivitives of *Solanum demissum*. Am. Potato J. 38:117–120.

Dodds, K. S., and D. H. Long. 1956. The inheritance of colour in diploid potatoes II. A three factor linkage group. J. Genet. 54:27–41.

Glendinning, D. R. 1976. Neo-tuberosum: new potato breeding material. 4. The breeding system of Neo-tuberosum, and the structure and composition of the Neo-tuberosum gene-pool. Potato Res. 19:27–36.

Grun, P., C. Ochoa, and D. Capage. 1977. Evolution of cytoplasmic factors in tetraploid cultivated potatoes (Solanaceae). Am. J. Bot. 64:412–420.

Hanneman, R. E., Jr., and S. J. Peloquin. 1968. Ploidy levels of progeny from diploid-tetraploid crosses in the potato. Am. Potato J.

Hawkes, J. G. 1958. Significance of wild species and primitive forms for potato breeding. Euphytica 7:257–270.

Haynes, F. L. 1954. Potato embryo culture. Am. Potato J. 31:282–288.

Hoopes, R. W. 1977. Personal communication. Dep. of Plant Breeding and Biometry, Cornell Univ., Ithaca, NY 14853.

Howard, H. W. 1958. The storage of potato pollen. Am. Potato J. 35:676–678.

————. 1975. The prolonged storage of true seeds of potatoes. Potato Res. 18: 320–321.

Jannssen, Aw. B., and J. G. Th. Hermsen. 1976. Estimating pollen fertility in *Solanum* species and haploids. Euphytica 25:577–586.

Lamm, R. 1953. Investigations on some tuber-bearing *Solanum* hybrids. Hereditas 39:97–112.

Lam, S. L., and H. T. Erickson. 1966. Interaction of light and gibberelin on potato seed germination. Am. Potato J. 43:443–449.

————, and R. Grenard. 1976. Potato selection for high dry-matter in seedling generation. Am. Potato J. 53:285–291.

Mendiburu, A. O., and S. J. Peloquin. 1977. The significance of 2n gametes in potato breeding. Theor. Appl. Genet. 49:53–61.

McLean, J. G., and F. J. Stevenson. 1952. Methods of obtaining seed on Russet Burbank and similar flowering varieties of potatoes. Am. Potato J. 29:206–211.

Mok, D., and S. J. Peloquin. 1972. Three mechanisms of 2n pollen formation in diploid potatoes. Am. Potato J. 49:362–363. (Abstr.).

———, and ———. 1975. Breeding value of 2n pollen (diplandroids) in tetraploid × diploid crosses in potatoes. Theor. Appl. Genet. 46:307–314.

Olsder, J., and J. G. Th. Hermsen. 1976. Genetics of self compatibility in dihaploids of *Solanum tuberosum* L. I. Breeding behaviour of two self-compatible dihaploids. Euphytica 25:597–607.

Peloquin, S. J., and R. W. Hougas. 1959. Decapitation and genetic markers as related to haploidy in *Solanum tuberosum.* Eur. Potato J. 2:176–181.

———, ———, and A. C. Gabent. 1966. Haploidy as a new approach to the cytogenetics and breeding of *Solanum tuberosum. In* Riley and Lewis (eds.) Chromosome manipulations in plant genetics. Oliver and Boyd, Edinburgh.

Ross, R. W. 1969. Seed dormancy and longevity in *Solanum* species. Am. Potato J. 46:438 (Abstr.).

Sanford, J. C., and R. E. Hanneman, Jr. 1977. The common potato pollen collector modified for bulk pollen collection. Am. Potato J. 54:573–574.

Simmonds, N. W. 1963a. Experiments on the germination of potato seeds. I. Eur. Potato J. 6:45–60.

———. 1963b. Experiments on the germination of potato seeds II. Eur. Potato J. 6:69–73.

Slomnicki, I. 1961. Problems of potato breeding in Israel. Eur. Potato J. 4:394–399.

Spicer, P. B., and L. A. Dionne. 1961. Use of gibberelin to hasten germination of *Solanum* seed. Nature 189:327–328.

Swaminathan, M. S., and H. W. Howard. 1954. The cytology and genetics of the potato (*Solanum tuberosum*) and related species. Bibliogr. Genet. 16:1–192.

Thijn, G. A. 1954. Observations on flower induction with potatoes. Euphytica 3: 28–34.

35 Rapeseed and Mustard

R. K. Downey
A. J. Klassen
G. R. Stringam
Agriculture Canada
Saskatoon, Saskatchewan

Rapeseed supplies about 8% of the world's edible oil. The seed yields over 40% oil on a dry weight basis, and the remaining meal is a high-quality feedstuff containing 38 to 41% protein. Although rapeseed is considered a single commodity by the edible oilseed industry, it is actually produced by plants of two distinct species, *Brassica napus* L. and *B. campestris* L. These crops are commonly referred to simply as "rape" and "turnip rape", respectively. Both species have summer and winter forms, but differ significantly from one another, both morphologically and agronomically. Major producing regions include the Indian Subcontinent, China, Canada, and Northern Europe.

The product commonly known as mustard seed also is produced by more than one species. At one time, *Brassica nigra* (L.) Koch, or black mustard, was the main source of hot mustard flour or paste, but today this species has been replaced largely by high-yielding, domestic cultivars of brown (brown seeded) and oriental (yellow seeded) mustard of the species *Brassica juncea* (L.) Coss. The hot dog or cream salad mustard is manufactured from yet another species, *Brassica hirta* Moench or *Sinapis alba* L., commonly called yellow or white mustard. Most of these *Brassica* species are believed to have originated in the Mediterranean region.

I. PARENTAL MATERIAL

The *Brassica* oilseed and condiment crops described above have evolved within closely related species which form part of a larger interspecific, intergeneric gene pool (Table 1). Cytological evidence strongly supports the thesis that all these crops originated from an extinct, common ancestor with a basic chromosome number of x = 5 or 6 (Heyn, 1977). The species with the lower chromosome numbers (n = 8 to 12) apparently developed into separate monogenomic species through a process of autoploidy followed by chromosomal loss and structural differentiation. The three species with the higher chromosome complements, *B. juncea, B. napus,* and *B. carinata* Braun., have resulted from natural crosses between the monogenomic species *B. campestris, B. nigra,* and *B. oleracea* L., with chromosome doubling occurring before, during, or after the interspecific cross to produce fertile amphidiploids (Singh, 1958; Heyn, 1977). Support for this evolutionary pathway is strong in that the three amphidiploids have been synthesized from their respective monogenomic species and analysis of meiotic pairing in synthetic amphihaploid hybrids has demonstrated partial homology between the A, B, C, E, S, and R genomes (Mizushima, 1968).

The probability of obtaining crosses within and between *Brassica* species and forms will vary with the species and cultivars chosen, as well as the condition of the plants and their environment. Extensive interspecific crossing experiments have been undertaken by many researchers to elucidate the relationships among the various species (Section VII).

Brassica seed collections are maintained at most of the major rape and mustard plant breeding stations. In addition, *Brassica* spp. introduced into the United States are available from the New York State Agricultural Experiment Station, Geneva, N.Y., while a cruciferous germplasm collection is maintained at the Instituto Nacional de Investigaciones Agrarias, Avad. Puerto de Herro, Madrid 3, Spain.

II. PLANT CULTURE

A. Field

Rapeseed and mustard crops are well adapted to a wide variety of soil types, although poor aeration of the root zone can permanently stunt growth. Crossing blocks should not be sown in areas where flooding or waterlogging of the soil is likely to occur. Crossing blocks should be sown at the recommended seeding dates for the commercial crop in that area. Such a procedure will normally provide the light, photoperiod, and temperature regime conducive to abundant flowering and good seed set. Plants to be crossed usually are space-planted to ensure vigorous growth, working space, and easy plant identification. Seed of *B. napus, B. campestris,* and *B. juncea* has been produced successfully in nearly all the important agricultural regions of the world, but *B. hirta* is highly photoperiod-sensitive. For floral initiation in *B. hirta* a 16 to 18-hour day is optimal. Plants tend to remain vegetative with a day length of less than 10 hours (Bernier, 1969).

Table 1—Nomenclature and genome relationships of some *Brassica* species and related genera.

Species name	Common synonym	Chromosome no. (n) & genomic formula	Common name
Brassica campestris L.			
ssp. *oleifera* (Metzg.) Sinsk. f. *annua*	*B. rapa* ssp. *oleifera* Metzg.	10 AA	Summer turnip rape
ssp. *oleifera (Metzg.)* Sinsk. f. *biennis*			Winter turnip rape
ssp. *eu-campestris* (L.) Olsson	*B. campestris* L.		Bird rape; wild turnip rape
ssp. *sarson* Prain	For list see Singh (1958)		Yellow and brown sarson
ssp. *dichotoma* (Roxb.) Olsson	*B. napus* ssp. *dichotoma* Prain		Toria
ssp. *chinensis* (L.) Makino	*B. chinensis* L.		Chinese mustard; Pe-tsai
ssp. *pekinensis* (Lour.) Olsson	*B. pekinensis* (Lour.) Rupr.		Celery cabbage
ssp. *nipposinica* (Bailey) Olsson	*B. nipposinica* Bailey		Curled mustard
ssp. *rapifera* (Metzg.) Sinsk.	*B. rapa* L.		Turnip
B. tournefortii Gouan			Wild turnip
B. nigra (L.) Koch		8 BB	Black mustard
B. oleracea L.			
ssp. *acephala* DC.		9 CC	Kale
ssp. *botrytis* L.			Cauliflower
ssp. *capitata* L.			Cabbage
ssp. *gongylodes* L.			Kohlrabi
ssp. *italica* Plenck			Broccoli
ssp. *sylvestris* L.			Wild cabbage
B. juncea (L.) Coss	*B. cernua* Forbes & Hemsl.	18 AABB	Brown, Oriental, leaf or Indian mustard
B. napus L.			
ssp. *oleifera* (Metzg.) Sinsk. f. *annua*	*B. napus* L. *annua* Koch	19 AACC	Summer or Argentine rape
ssp. *oleifera* (Metzg.) Sinsk. f. *biennis*	*B. napus* L. *biennis* (Schuebl. Mart.) Reichenb. *B. napella* Chaix.		Winter rape
ssp. *pabularis* (DC.) Reichenb.			Rape-kale
ssp. *rapifera* (Metzg.) Sinsk.	*B. napobrassica* (L.) Mill.		Rutabaga, swede
B. carinata Braun		17 BBCC	Abyssinian mustard
B. hirta Moench	*Sinapis alba* L.	12 DD	White or yellow mustard
B. kaber (DC.) L.C. Wheeler	*Sinapis arvensis* L.	9 SS	Charlock, wild mustard
Eruca sativa Mill.		11 EE	Rocket salad
Raphanus sativus L.		9 RR	Radish
Camelina sativa Crantz.		20	Camelina, false flax
Crambe abyssinica		45	Crambe, Abyssinian kale

B. Growth Chamber and Greenhouse

Emasculation and pollination can be facilitated by growing plants in individual pots rather than in beds because plants can be moved as required. Excellent growth is obtained in greenhouses and growth chambers by growing individual rape plants in 10 or 13-cm pots containing a soil-free mix. A good mix for rape and mustard plants developed by W. H. Leonard for the Agriculture Canada Research Station staff at Saskatoon consists of 2 parts peat moss, 2 parts fine vermiculite, and 1 part medium grade sand. To 36 liters of this mix the following fertilizers are added: 135 g ground limestone,

70 g superphosphate, 120 g control release fertilizer 18–6–12 (Osmocote, Sierra Chemical Co., Newark, Cal.), 1 g iron chelate (Geigy Sequestrene 330 Fe), 2 g fritted trace elements. This mix is stable and weed-free and provides excellent root aeration essential for healthy, vigorous growth. The most satisfactory method of watering is to place pots in galvanized metal trays which can be filled with water. After the potting medium is saturated, excess water is allowed to drain out of the trays.

Spring rape, turnip rape and the mustards can be grown under variable day length and temperature. In the greenhouse and growth chamber, 16 to 18 hours of high light level with 20 day/16 C night temperatures result in good plant development. High light level prevents excessive plant height and weak plants. Relative humidity of at least 40 to 50% is desirable, but should be reduced during the ripening phase to prevent premature germination of seeds in the pod. Increasing the air temperature at this stage will also hasten maturation.

Aphid infestation of racemes contained under pollination bags may be a problem, but good control can be obtained by applying a systemic insecticide such as Temik (Union Carbide, N.Y.) to the potting medium. Mildew can be controlled by dusting plants with sulfur once a week.

Vernalization of winter annuals can be accomplished by germinating seeds on moist filter paper for 1 or 2 days, or until the seed coats rupture. The moist seeds are then maintained at about 4 C with an 18-hour photoperiod for a minimum of 6 weeks, after which the resulting seedlings can be planted in pots or beds in the greenhouse or growth room (Rakow, 1973).

III. FLORAL CHARACTERISTICS

Brassica napus and *B. juncea* are self-fertile, with about 80% of the seed normally arising from self-pollination, although a greater degree of cross-pollination may occur when insect pollinators are abundant. Most *B. campestris* and *B. hirta* strains, on the other hand, are highly self-incompatible, and self-pollinated seed can be produced only with difficulty. Under large-scale commercial production, wind is the main pollinating agent, with insect pollinators of secondary importance.

Flowering is indeterminate and begins at the lowest part of the raceme and continues upward. Flowers begin opening very early in the morning and are fully open by 0800 or 0900 hours. The stigma is receptive from 3 days before to 3 days after the flower opens (Mohammad, 1935). As the flower opens, the anthers become extrorse and dehisce when the petals completely unfold. Pollen is shed through two longitudinal slits on the upper side of the anthers. If favorable warm, dry weather occurs, nearly all the pollen is shed the day the flower opens. In the evening the flower tends to close, approaching a funnel shape, and opens again the following morning. On the third day, the flower remains nearly closed and the petals and sepals begin to wilt. If necessary, *Brassica* pollen can be stored for up to 4 or 5 weeks without complete loss of viability (Bach, 1917; Chiang, 1974). Studies of pollen-tube growth in several *Brassica* species indicate that fertilization is effected within about 24 hours after pollination (Akhtar, 1932; Khanna and Chowdhury, 1974).

Both the onset of flowering and duration of the flowering period are variable and quite dependent on weather, particularly temperature. Low temperatures decrease the rate of plant development and hence delay the onset and rate of flowering. Low plant density allows secondary branching, thus extending the flowering period. Although rape and turnip rape are annual or biennial species, regrowth will occur in both forms if plants are pruned back while still green. Flowers which are produced on the regrowth are typically smaller and less productive than the first formed flowers.

The floral arrangement of the *Brassicas* is typically a corymbiform raceme. The flowers are regular, bisexual, and hypogynous with four free sepals in two whorls, median and transverse, and four free diagonally placed petals (Fig. 1). Flowers have one pair of lateral stamens with shorter

Fig. 1—Main raceme of *B. napus* L. Arrows indicate buds at correct stage for emasculation; inset details flower structure.

filaments and four median stamens with longer filaments. Sutures of the anthers are introrse in the bud stage, but the anthers of the four long stamens become extrorse after flower opening in all forms of rape, except an Indian subspecies of *B. campestris* known as Yellow Sarson (Table 1). Two functional nectaries are located at the base of the short stamens and two nonfunctional nectaries at the base of the pairs of long stamens. The ovary is syncarpous and normally consists of two carpels, but three or four also occur. After fertilization, the ovary develops into a bivalved silique with a longitudinal septum. Flowering plants of *B. napus* and *B. campestris* can be distinguished by the relative position of the buds to the open flowers on a given raceme. In *B. napus* the crown of developing buds is borne above the open flowers, while in *B. campestris* the bud cluster is more compact and the open flowers are borne at or above the level of the buds.

IV. ARTIFICIAL HYBRIDIZATION AND SELF-POLLINATION

A. Equipment

Medium or fine tweezers are the one essential tool needed for self and cross-pollination in *Brassica.* Emasculating scissors, as described by Nieuwhof (1969), also may be used. These have a sharp, semicircular opening cut into each blade so that when the scissors are closed a 1-mm diam opening remains in the center of the closed blades. The scissor opening is positioned around the bud about 3 mm below the bud tip. By closing and gently pulling the scissors upward toward the bud tip, sepals, petals, and some or all of the anthers are removed.

B. Preparation of the Female

For controlled pollination, buds which are about to open or will open the following day are selected for emasculation. Such flowers will not have been contaminated by self or foreign pollen and usually give the best fertilization rate. Some workers prefer to remove the petals and sepals while others leave them intact to protect the stigma. In the greenhouse or growth chamber with relatively high humidity, petals and sepals may be removed with no adverse effects. The six undehisced anthers are removed with tweezers and discarded. Emasculation can be accomplished any time during the day, but afternoons or evenings are the most favorable times.

If a large number of seeds from one cross are required, all open flowers on a raceme are removed and all buds of a suitable age are emasculated. The remaining young buds are discarded. Emasculated buds should be protected from foreign pollen for up to 1 week following pollination.

If a number of different cross combinations are to be made on the same raceme, or if all possible flowers on a raceme are to be individually fertilized, small plastic bags may be used to cover each flower (Kondra and Downey, 1968). When the same cross-pollination or self-pollination is practiced on all the flowers of a given raceme, then large conspicuous red marking tags (4 × 2.8 cm) containing parental details and pollination date are attached to the raceme immediately below the basal flower. However, smaller, lighter tags (1 × 2 cm) are needed for attachment to the flower pedicel when the individual flowers on a raceme are pollinated by different parents.

In the self-fertile species, *B. napus* and *B. juncea,* self-pollinated seed can be obtained by simply enclosing one or more racemes, free of open flowers, within a 16 kg (30 lb) weight glassine bag approximately 38.1 × 11.4 × 7 cm in size. The bag is closed and secured by stapling the flap around a cane stake and the raceme. To accommodate raceme elongation, the bags normally are moved up the raceme at least once during the flowering period. Bag agitation may assist pollen dispersal and increase seed set.

C. Pollination

In controlled crosses or bud pollinations, pollen transfer is accomplished by picking selected anthers with a pair of forceps and applying them to the stigma of an emasculated flower, or by collecting pollen on a small camel hair brush and dusting the stigma. Many European workers cross winter rape in field nurseries. Racemes of flowers from plants selected as male parents are gathered early in the morning and stored with their stems in water until the female parents are emasculated. Pollen is applied by dusting emasculated stigmas with racemes of male flowers. In all cases, care must be taken to clean equipment between pollinations. Washing equipment in 95% ethanol is effective in controlling pollen contamination. Intraspecific hybridizations normally are accomplished without difficulty and pollination of a single flower can yield 20 or more seeds.

D. Factors Affecting Efficiency

Genetic markers are useful tools in hybridization and plant breeding, but have not been studied extensively or utilized in the oilseed and condiment *Brassicas* (Stringam, 1973). Few genetic markers are known in the mustard crops, *B. juncea* and *B. hirta*, but a number are available in *B. napus* and *B. campestris.* Two biochemical mutants, low erucic acid and low glucosinolate, which have profound effects on the quality of rapeseed oil and meal, have been used as markers in both rapeseed species. Since erucic acid synthesis is controlled by additive genes in the developing embryo, its presence can be used to detect hybrid seed from crosses involving high and low erucic acid parents (Dorrell and Downey, 1964). The glucosinolate content is more difficult to use as a genetic marker because the presence and amount of these sulfur compounds in the seed are determined by the genotype of the maternal parent, and normally several dominant genes are involved (Kondra and Stefansson, 1970). In *B. campestris,* a monogenic, recessive, chlorophyll-deficient mutant has been utilized in a study of isolation distance requirements and a similar marker in *B. napus* is being utilized for the same purpose (Stringam and Downey, 1978; Stringam, 1978).

Flower color mutants are useful particularly in breeding programs because they appear to have fewer deleterious, pleiotropic effects than many other mutants so far isolated in the rapeseed and mustard species. Olsson (1955) used a flower color mutant to study the effect of isolation distances in *B. hirta,* and several recessive, pale flower color genes are available for the same purpose in *B. napus* and *B. campestris* (Alam and Aziz, 1954; Morice, 1960; Stringam, 1978). Additional genetic markers in *B. campestris* are now becoming available. These markers display a wide variety of effects on plant characteristics such as shape and size of petal and leaf, and plant height, anthocyanin pigmentation, sensitivity to light, pod shape, size and orientation to the raceme, and waxy bloom (Stringam, 1978; Williams, 1978).

V. NATURAL HYBRIDIZATION

Many species of the Cruciferae contain a self-incompatibility (SI) system which forces natural outcrossing. The compatibility of the pollen and stigma is controlled sporophytically by a series of alleles called S alleles or S factors (Frankel and Galun, 1977). In its most simplified form the presence of the same S allele in the pollen and the stigma will inhibit the pollen grain from germinating, or will prevent the pollen tubes of such grains from penetrating the stigmatic surface and effecting fertilization. This natural barrier to self-fertilization can be circumvented or partially broken in a number of different ways. In addition to high humidity after pollination (Carter and McNeilly, 1976), mutilation or removal of the stigma, pollen transplantation, chemical treatment of the stigma, washing the pollen, high temperature, application of an electric potential difference, high CO_2 concentration after pollination and bud pollination have been used with various degrees of success (Roggen, 1974). Bud pollination is undoubtedly the most widely employed method, but CO_2, high humidity, and electrical treatments also are used in many cruciferous vegetable and forage crops to produce and maintain inbred lines for hybrid seed production.

In the oilseed and condiment crops, this SI system is present in both *B. campestris* and *B. hirta,* although it has never been exploited in these crops for the production of hybrid cultivars. In general, the high cost of producing self or sib-seed to maintain the parental lines, plus the extensive rather than intensive cultivation practices employed in the commercial production of these crops, discouraged the development of single or multiple hybrid combinations. The amphidiploid *B. napus* and *B. juncea* species are largely self-fertile although some self-incompatible plants have been found in both species, just as some self-compatible plants can be selected in *B. campestris* and *B. hirta.*

The SI system in *B. campestris* and *B. hirta* makes possible the natural crossing between paired plants in the field by enclosing racemes of two or more parents, prior to flowering, within a single glassine bag. Shaking the bag each day during the flower period usually results in good seed set. Unfortunately, because of the presence of SI and the annual habit of these crops, there is no easy means of further propagating the genotypes of desirable parents.

Genetic male sterility (*ms*), determined by single genes, has been reported in *B. napus* (Takagi, 1970; Koch and Peters, 1953), and *B. campestris* (Anon., 1967). However, in these crops no practical means has been devised to utilize such *ms* plants in the production of hybrid cultivars. More recently, certain intraspecific crosses have yielded cytoplasmic male-sterile plants in *B. napus, B. campestris,* and *B. juncea* (Thompson, 1972; Shiga and Baba, 1973; Williams, 1978). Restorer genes have been found for the Shiga-Baba system, and rapid progress is being made toward producing hybrid *B. napus* cultivars. No cytoplasmic male sterility has yet been reported in *B. hirta.*

Complete isolation under field conditions is difficult to achieve without the use of selfing bags. It is important to keep in mind that both wind and insects are effective pollen vectors of the *Brassica* oilseed and condiment species. Since these vectors act additively, contamination levels may be higher than might be expected in crops pollinated only by wind or insects. The pollen size of these crops varies, but is generally smaller than most cultivated crops. Dry pollen of *B. campestris* measures approximately 15 × 30 μ, and *B. napus* 20 × 40 μ, about one-third the size of corn pollen and half that of forage grasses (Wodehouse, 1935). Thus the pollen can be carried long distances by either the wind or pollinating insects, and an isolation distance of at least 366 m is desirable between increases of advanced breeding material belonging to the same species. The opportunity for contamination from foreign pollen is less for the largely self-fertile *B. napus* and *B. juncea* species because self-pollination usually occurs as the flower opens or shortly thereafter. Since interspecific crosses between the *Brassica* species rarely occur in nature because of barriers to natural fertility or different times of flowering, breeding material of the various species can be increased in the same area, providing safeguards are taken to prevent mechanical mixtures.

Because of the small pollen size, cloth or mesh cages are not completely effective barriers to foreign pollen, although some breeders use cheesecloth cages for isolation and crossing of individual *B. oleracea* plants or racemes. (For details of cages, see Nieuwhof, 1969.) In the oilseed crops it is more common to enclose several racemes within a pollen-tight glassine bag (Section IV).

Small isolation compartments or rooms in heated or unheated greenhouses are used extensively for crossing and selfing. The compartments vary in area from a few square meters to a few dozen square meters. These compartments usually are equipped with an air-intake filter to exclude contaminant pollen. Pollination within compartments can be effected by manipulating the open flowers with a camel hair brush or by blasting a stream of air into the compartment to distribute the pollen. Some authors report effective pollination can be obtained in rooms and cages with the use of the meat-fly, *Phormia terra-novae* R. D., which remains active at lower temperatures. For those interested, Nieuwhof (1969) provides details on fly rearing and use for *B. oleracea.* However, experience at the Saskatoon Research Station with oilseed rape and mustard has been disappointing since the relatively high temperature required to stimulate fly activity tends to interfere with normal flowering.

Honeybees and bumblebees also may be used for greenhouse pollination, but here again it is somewhat difficult to regulate a greenhouse temperature which is both suitable for maximum bee activity and normal flowering. With honeybees, a miniature hive with two or three frames is used per compartment. Care must be taken to balance the number of bees to the available bloom. Too high a bee concentration will result in heavy overworking of the flowers, causing reduced seed set per pod from either floral injury or shortage of pollen. Captured bumblebee queens and males can be useful pollinators in isolation compartments. Such bees usually live for 10 to 25 days, and two bees are sufficient for a small compartment. Worker bumblebees, because they are short-lived, are less suitable for isolation chambers. Nieuwhof (1969) provides details of mini-hive construction and handling of bees. Of the bee pollinators available, leaf-cutter bees are the easiest to rear and manage, but the higher temperatures required to stimulate leaf cutters to forage makes them less suited to pollinate rape and mustard.

VI. SEED DEVELOPMENT, HARVEST, AND STORAGE

The first evidence of a successful cross is the rapid elongation of the pistil into a developing silique or pod. As the pod elongates, developing seeds can be observed by examining the pod against a background of strong light.

When individual pods or seed are being harvested, the seed should be completely ripe as indicated by the seed's firmness when rolled between the thumb and forefinger, and a seed color that is uniformly black, dark brown or yellow, depending upon the seed color of the maternal parent. If whole racemes are being harvested, at least two-thirds of the seeds should have begun to turn from green to dark brown or yellow. Pods or racemes should be allowed to dry naturally at room temperature until the pods become brittle. The pods thresh easily. Whole plants or racemes can be passed through a Vogel-head thresher or the Labour ear thresher (Hergert, 1968a). A single pod or pods of a particular cross can be placed in an envelope and crushed by hand or with a hand block. The seed can be separated from the chaff by a hand-held, round-holed sieve, by air with a Dakota or E.R.S. seed-cleaning blower (Hergert, 1968b), or by gravity on an inclined plane covered with velvet, having the nap facing against the flow of seed. Seed tags with cross and plant identification should accompany the seed through all operations.

Seed stored at 5 to 10 C with less than 9% moisture will remain highly viable for several years. In relatively dry climates, such as Western Canada (<5% relative humidity), seed stored at room temperature will remain viable for 5 to 8 years. In more humid climates, such as Japan and parts of Western Europe, seed of these crops can lose its viability within 1 to 2 years if not refrigerated. A high correlation exists between air humidity and seed-moisture content; so for long-term storage, seeds should be dried using temperatures of less than 60 C and immediately placed in moisture-proof containers for storage at or below freezing.

VII. TECHNIQUES FOR SPECIAL SITUATIONS

Interspecific crosses are important in extending the genetic variation of rape and mustard crop plants. Many interspecific crosses have been made, some easily, others only after thousands of pollinations (Table 2). Some species have never been successfully crossed.

Generally the interspecific cross is more successful if an allopolyploid species, such as *B. napus* or *B. juncea,* is used as the female parent, particularly if the allopolyploid has one genome in common with the pollen parent (Heyn, 1977). Hybrids between monogenomic species, however, are more difficult to obtain. Success tends to vary between 0.002 and 0.03 hybrids per bud-pollinated flower (Olsson, 1960; Robbelen, 1966).

Most investigators have tended to rely on making large numbers of crosses with the expectation that some foreign pollen grains would effect fertilization and by chance some hybrid embryos would survive natural barriers. In some crosses, a high failure rate has been attributed to an imbalance between the developing embryo and the endosperm (Neumann, 1973; Hakansson, 1956), which can sometimes be overcome through in vitro embryo culture using solid (Nishi et al., 1959) or liquid (Harberd, 1969, 1971) media. In other interspecific crosses the main barrier may lie in the failure of the foreign pollen tube to penetrate the stylar tissue and effect fertilization (Robbelen, 1966). Techniques which tend to overcome the self-incompatibility system of many *Brassica* could increase the success rate of some interspecific crosses. The time at which pollen is applied to the emasculated bud may influence the number of hybrids obtained. Olsson (1960) found pollen application immediately following emasculation resulted in a higher proportion of successful crosses than applying pollen 36 hours after emasculation. On the other hand, Eenink (1974) reported a delay in pollination of 25 to 96 hours increased the number of hybrid seeds set.

Table 2—Ease of crossing among self-compatible (SC) and self-incompatible (SI) oilseed and condiment *Brassicas* and some close relatives.

Species		*B. campestris*	*B. napus*	*B. juncea*	*B. hirta*	*B. oleracea*	*B. carinata*	*B. nigra*	*B. kaber*	*B. tournefortii*	*B. fruticulosa*	*Raphanus sativus*	*Eruca sativa*
B. campestris	♀	SI	3†	1	0	2	2	1	1	2	2	1	2
	♂	SI	5	4	0	1	2	0	1	1	--	1	--
B. napus	♀		SC	2	1	2	1	1	1	2	1	1	1
	♂		SC	3	0	2	0	0	--	--	--	--	--
B. juncea	♀			SC	0	1	2	3	1	--	--	1	--
	♂			SC	0	0	0	--	0	--	--	0	--
B. hirta	♀				SI	1	--	0	0	--	--	--	--
	♂				SI	0	--	0	0	--	--	--	--

† Ratings of 0 to 5 indicate approximately nil, 0.002, 0.02, 0.5, 1, and over 2 hybrids per flower pollinated, respectively; dashes indicate insufficient information to assign a rating.

Pod and seed set observations following interspecific *Brassica* crosses frequently can give a false sense of success. Pods will often develop without containing seeds, and if seeds are obtained they may not be hybrids. Both male and female parents frequently may produce functional unreduced gametes. Thus, interspecific or intergeneric cross-pollinations can produce several different types of offspring, including (a) hybrids from the reduced gametes from both parents, (b) amphidiploid hybrids from the union of unreduced gametes of both parents, (c) triploid hybrids from parental-reduced and unreduced gametes, and (d) matromorphic plants from either parthenogenetic development of unreduced gametes or the spontaneous chromosome doubling and embryogenic development of reduced female gametes (Heyn, 1977; Olsson, 1960). In crosses between the amphidiploid *B. napus* and one of its putative parental species, *B. campestris,* there is a tendency for gametes with high chromosome numbers to be more functional.

Embryo culture has not been used extensively in *Brassica* although the techniques are relatively simple. Cultures can be initiated on either solid media (Nishi et al., 1959) or liquid media (Harberd, 1969, 1971). Embryos can be cultured well on a range of different media, the main requirements being the elevated sucrose level (8%) and the addition of at least 400 mg/liter of casein hydrolysate.

One schedule which has proved successful utilizes liquid medium and is a modification of Harberd's (1969, 1971) technique. With this method, siliques at approximately 10 to 14 days following fertilization are removed and surface-sterilized in a solution of 10% bleach containing a small drop of Tween 80. The solution is agitated for 5 min. All subsequent operations are carried out in a laminar flow hood under aseptic conditions. The pods are rinsed four times in sterile, distilled water and placed in a sterile petri dish where they are slit down the suture with a scalpel, and the individual ovules removed with a fine forceps. With the aid of fine needles, the ovules are then ruptured, each with a slit extending about one-half its circumference. Care must be taken to ensure that the embryos are not removed or damaged at this stage of the process. The ruptured ovules containing the young embryos are then placed in 30 ml McCartney tubes charged with 10 cm^3 of sterile liquid Murashige and Skoog medium (MS) (Murashige and Skoog, 1962) containing the following modifications or additions: 8% instead of 3% sucrose, 400 mg/liter casein hydrolysate, and pH adjusted to 5.7 prior to autoclaving. The tubes are placed at a slant in a growth cabinet under 18-hour day length and a 20/16 C day-night temperature regime, with a light level of approximately 7 klux. The tubes are shaken for a few seconds daily but require no further agitation for normal embryo development. As the embryos grow they emerge from the ruptured ovules into the culture medium. After 2 to 3 weeks they are ready for transfer to a 0.8% agar medium containing one-half the mineral salts of MS and no sucrose. Within 10 days, embryos root sufficiently to allow transfer of the young plantlets into potting compost treated with a fungicide to control damping off. Embryos can be cultured as early as 1 week following fertilization; however, the survival rate is only about 25%. The success rate increases at successively later stages. Two weeks after fertilization, 80% survival is common. For interspecific crosses, the success rate is proportionately lower and one must allow for a high frequency of maternal embryos arising from functional unreduced egg cells (Heyn, 1977).

The newly potted plants are very susceptible to drying and should be covered with a polyethylene bag which is opened gradually over a period of several days until the plants are hardened sufficiently to growth cabinet conditions.

Intra and interspecific grafting within the *Brassica* is accomplished easily. The stalk and scion should be growing actively with stems filled with succulent green pith. The scion usually is small (4 cm or shorter) and the foliage trimmed back to reduce transpiration. The scion base is sometimes cut to form a wedge which is inserted into a similarly shaped cut made in the main stem of the stalk. The area of the graft is then wrapped firmly with thread and the air excluded with Vaseline or grafting compound, such as Braco. Some workers cut off the stalk, about 8 cm above the next node, leaving one leaf above the graft (Lein, 1972). Other workers have tended to leave the young stalk uncropped with 8 to 10 leaves (Namai, 1971). Pawlowski (1974) was successful in simply cutting a branch of the stalk and firmly securing one or more small scions to the cut surface with fine mounting pins. The stalk and scion were then maintained under high humidity by enclosing the moistened plant in a sealed plastic bag until a successful graft union was accomplished. It is normal to obtain inflorescences on both the stalk and scion.

REFERENCES

Akhtar, A. R. 1932. Studies in Indian Brassicae. I. Sterility and selective pollen tube growth. Indian J. Agric. Sci. 2:280–292.

Alam, Z., and M. A. Aziz. 1954. Inheritance of flower color in some self-fertile oleiferous Brassicae. Pakistan J. Sci. Res. 6:27–36.

Anonymous. 1967. Annual Report of the Ayub Agric. Res. Inst., Lyallpur for 1966/67. (Plant Breed. Abstr. 40:2292, 1970.)

Bach, S. 1971. Zur Pollenbiologie von Raps and Rubsen. Z. Pflanzenzucht. 5:337–345.

Bernier, G. 1969. Sinapis alba L. *In* L. T. Evans (ed.) Induction of flowering, some case histories. Cornell Univ. Press, Ithaca, N.Y.

Carter, A. L., and T. McNeilly. 1976. Increased atmospheric humidity post pollination: A possible aid to the production of inbred line seed from mature flowers in the Brussels sprout (*Brassica oleracea* var. *gemmifera*). Euphytica 25:531–538.

Chiang, M. S. 1974. Cabbage pollen germination and longevity. Euphytica 23:579–584.

Dorrell, D. G., and R. K. Downey. 1964. The inheritance of erucic acid content in rapeseed (*Brassica campestris*). Can. J. Plant Sci. 44:499–504.

Eenink, A. H. 1974. Matromorphy in *Brassica oleracea* L. III. The influence of temperature, delayed prickle pollination and growth regulators on the number of matromorphic seeds formed. Euphytica 23:711–718.

Frankel, R., and E. Galun. 1977. Pollination mechanisms, reproduction and plant breeding. Springer-Verlag, Berlin.

Hakansson, A. 1956. Seed development of *Brassica oleracea* and *B. rapa* after certain reciprocal pollinations. Hereditas 42:373–396.

Harberd, D. J. 1969. A simple effective embryo culture technique for *Brassica*. Euphytica 18:425–429.

———. 1971. Addendum to 'a simple effective embryo culture technique for *Brassica*'. Euphytica 20:138.

Hergert, G. B. 1968a. A list of manufactured equipment for experimental plot mechanization. Pt. 1. Field equipment. Can. Dep. Agric. Eng. Res. Serv., Eng. Spec. 6715-1.

———. 1968b. A list of manufactured equipment for experimental plot mechanization. Pt. 2. Associated laboratory equipment. Can. Dep. Agric. Eng. Res. Serv., Eng. Spec. 6715-2.

Heyn, F. W. 1977. Analysis of unreduced gametes in the Brassicae by crosses between species and ploidy levels. Z. Pflanzenzucht. 78:13–30.

Khanna, V. K., and J. B. Chowdhury. 1974. Studies on pollen tube growth and seed set in some interspecific crosses in genus *Brassica.* Haryana Agric. Univ. J. Res., Hissar 4:231–235.

Koch, H., and R. Peters. 1953. Neue Gesichtspunkte der Rapszuchtung. Wissensch. Zeitsch. der Martin-Luther Univ., Halle Wittenberg 2:363–367.

Kondra, Z. P., and R. K. Downey. 1968. Pollination bags for individual florets of *Brassica*. Can. J. Plant Sci. 48:108–109.

———, and B. R. Stefansson. 1970. Inheritance of the major glucosinolates of rapeseed (*Brassica napus*) meal. Can. J. Plant Sci. 50:643–647.

Lein, K. A. 1972. General and physiological studies on the formation of glucosinolates in rape seeds: Localization of the main site of glucosinolate biosynthesis by grafting experiments. Z. Pflanzenphysiol. 67:333–342.

Mizushima, U. 1968. Phylogenetic studies on some wild *Brassica* species. Tohoku J. Agric. Res. 19:83–99.

Mohammad, A. 1935. Pollination studies in toria (*Brassica napus* L. var. *dichotoma*) and sarson (*Brassica campestris* L. var. *sarson* Prain). Indian J. Agric. Sci. 5:125–154.

Morice, J. 1960. Heredite de la couleur des fleurs chez le colze *Brassica napus* L. var *oleifera* Metzger. Ann. Inst. Natl. Rech. Agron., Paris Ser. B. 10:155–168.

Murashige, T., and F. Skoog. 1962. A revised medium for rapid growth and bioassays with tobacco tissue cultures. Physiol. Plant 15:473–497.

Namai, H. 1971. Studies of the breeding of oil rape (*Brassica napus* var. *oleifera*) by means of interspecific crosses between *B. campestris* ssp. *oleifera* and *B. oleracea*. Jpn. J. Breed. 1:40–48.

Neumann, M. 1973. On the postgamic incompatibility of *Brassica*. Arch. Zuechtungsforsch 3:133–140.

Nieuwhof, M. 1969. Cole crops; botany, cultivation and utilization. Leonard Hill, London.

Nishi, S., J. Kawata, and M. Toda. 1959. On the breeding of interspecific hybrids between two genomes, C and A of *Brassica* through the application of embryo culture techniques. Jpn. J. Breed. 8:215–222.

Olsson, G. 1955. Vindpollinering hos Kosshblomstriga oljevaxter. Sv. Utadesf. Tidskr. 65:418–422.

———. 1960. Species crosses within the genus *Brassica.* II. Artificial *Brassica napus* L. Hereditas 46:351–386.

Pawlowski, S. H. 1974. Personal communication. St. Vincent, Alberta, Canada.

Rakow, G. 1973. Personal communication. Norddeutsche Pflanzenzucht, H. G. Lembke K. G. 2331 Hohenlieth, Eckernforde, West Germany.

Robbelen, G. 1966. Observations on interspecific *Brassica* crosses, in particular about the origin of the matromorphous form of F_1 plants. Angew. Bot. 39: 205–221.

Roggen, H. P. J. R. 1974. Pollen washing influences (in)compatibility in *Brassica oleracea* varieties. *In* H. F. Linskens (ed.) Fertilization in higher plants. North-Holland Publishing Co., Amsterdam.

Shiga, T., and S. Baba. 1973. Cytoplasmic male sterility in oilseed rape, *Brassica napus* L., and its utilization to breeding. Jpn. J. Breed. 23:187–197.

Singh, D. 1958. Rape and mustard. Indian Central Oilseeds Committee. Examiner Press, Bombay.

Stringam, G. R. 1973. Inheritance and allelic relationships of seven chlorophyll-deficient mutants in *Brassica campestris* L. Can. J. Genet. Cytol. 15:335–339.

————. 1978. Personal communication. Agriculture Canada Res. Stn., Saskatoon, Sask.

————, and R. K. Downey. 1978. Effectiveness of isolation distances in turnip rape. Can. J. Plant Sci. 58:427–434.

Takagi, Y. 1970. Monogenic recessive male sterility in oil rape (*Brassica napus* L.) induced by gamma radiation. Z. Pflanzenzucht. 64:242–247.

Thompson, K. F. 1972. Cytoplasmic male sterility in oil-seed rape. Heredity 29: 253–257.

Williams, P. H. 1978. Personal communication. Dep. of Plant Pathology, Univ. of Wisconsin, Madison, WI 53706.

Wodehouse, R. P. 1935. Pollen grains; their structure, identification and significance in Science and Medicine. McGraw-Hill, N.Y.

36 Rice

W. Ronnie Coffman
R. M. Herrera
International Rice Research Institute
Los Baños, Laguna, Philippines

Rice (*Oryza sativa* L.) is a semiaquatic, annual grass that probably evolved along the foothills of the Himalayas and associated mountain ranges in South and Southeast Asia. A second cultivated species, *Oryza glaberrima* Steud., of minor importance, probably originated in the Niger River delta of West Africa (Lu and Chang, 1978). Rice is now grown on about 137 million ha in more than 100 countries, mostly in the tropics and subtropics and predominantly in Asia. Several million hectares are cultivated in temperate zones up to a limit of 53° Lat. Rice is the primary staple food of more than 1.3 billion people and it is an important part of the diet of more than half of the world's population.

I. PARENTAL MATERIAL

The genus *Oryza* has about 23 well-defined species (Chatterjee, 1948; Sampath, 1962), but only two are cultivated. *Oryza glaberrima* is grown on limited areas in Africa. The major cultivated species, *Oryza sativa,* is grouped into three varietal types: indicas, which predominate in the tropics and subtropics; japonicas, grown only in the temperate and subtropic zones; and javanicas, found mostly in Indonesia.

Sampath (1966) divided the genus *Oryza* into four series. Cultivated rice belongs to the series *Sativae* which includes *O. perennis* Moench, *O. nivara* Sharma et Shastry, *O. breviligulata* A. Cheval. et Roehr, and *O. glaberrima,* as well as the *O. sativa* complex. Khush (1975) discusses inter-

---

relationships within the series. *Oryza sativa* has been crossed with all other members of the series *Sativae*, but the F_1's are often sterile. Nevertheless, Khush et al. (1971) successfully transferred grassy stunt virus resistance from the wild species *O. nivara* into cultivars that are now widely grown in the Philippines and elsewhere.

Hybrids between distantly related varieties of the species *sativa,* particularly between japonicas and indicas, are often sterile (Jennings, 1966).

The most complete genetic collection of rice is maintained at the International Rice Research Institute (IRRI) in the Philippines. Other important collections are maintained at the National Institute of Agricultural Science Seed Storage Laboratory in Japan, the Central Rice Research Institute in India, and the United States Department of Agriculture in Beltsville, Md.

II. PLANT CULTURE

A. Field

Rice can be grown in a wide range of soils. It grows in both flooded and upland conditions, but moisture should be at least near field capacity. Flooding the soil normally minimizes nutritional problems. Xuan and Ross (1976) discuss appropriate production practices.

Rice cultivars vary widely in sensitivity to photoperiod and temperature. The optimum temperature for growth is about 27 C, but the optimum temperature to promote flowering depends on the photoperiod. Rice is a short-day plant; day length of about 10 hours is usually optimal for flowering. Longer days may progressively delay or inhibit flowering, depending on the sensitivity of the cultivar. Highly sensitive types will never flower if grown continuously under long day length. The critical photoperiod for sensitive cultivars varies from 12 to 14 hours and depends on the latitude of origin (Vergara and Chang, 1976).

In the monsoon tropics, rice is not usually crossed in the field because of inclement weather. Even in fair weather, working in the flooded paddies is difficult, and wind may interfere with pollination. Crossing in greenhouses or screenhouses is much more convenient. At IRRI, most parent plants are grown in the field until the appropriate stage for crossing and then lifted from the mud and transferred to pots in the screenhouse.

B. Growth Chamber and Greenhouse

The best planting medium is soil taken directly from the rice field and kept flooded to minimize nutritional problems. Seed should be planted in moist soil. The seed will meet the nutritional requirements of the young seedling for about 11 days, after which flooding is advisable if deficiency symptoms develop.

A photoperiod of 10 to 12 hours is preferred. If the natural day length is longer, a darkroom treatment may be necessary for sensitive cultivars.

The basic vegetative phase, during which the plant will not respond to photoinductive cycles, ranges from 14 to 63 days after which 5 to 24 treatments are required to induce flowering, depending on the cultivar (Vergara and Chang, 1976).

The light level required for good growth is about 500 μE $sec^{-1}m^{-2}$ at the plant level. Light quality is not of crucial concern.

Rices vary widely in sensitivity to temperature, which affects anthesis and seed set. Adair (1934) found 27 to 31 C to be optimal for anthesis. Chandraratna (1964) summarized reports on air temperature limits for anthesis. He also reported that anthesis occurred over a relative humidity range of 40 to 95% with the optimum at 70 to 80%. Noguchi (1936) specified 30 C and 80% relative humidity as ideal for hybridization. Yoshida (1977) specified 22 C as the critical low temperature for anthesis, 35 to 36 C as the critical high temperature, and 30 to 33 C as the optimum.

III. FLORAL CHARACTERISTICS

Rice is a self-pollinated crop, although some outcrossing (usually less than 1%) does occur (Beachell et al., 1938). Meiosis occurs during the boot stage around the time when the auricles of the flag leaf are opposite the auricles of the next lower leaf. This stage is followed by emergence of the panicle, which bears perfect flowers in single-flowered spikelets. A flower consists of six stamens, each composed of a two-lobed, four-loculed anther borne on a slender filament, and a pistil containing one ovule. The short style bears a feathery stigma with two branches. The flower is fully developed and the stigma fully receptive at the time that pollen sheds. At this stage, the lodicules become turgid and force the lemma and palea apart. Anther dehiscence and extrusion occur more or less simultaneously after which the lemma and palea close. This stage is usually reached in the uppermost florets when the panicle is about 50 to 60% emerged from the boot, but may occur earlier in certain cultivars. This is the proper time to attempt artificial hybridization.

IV. ARTIFICIAL HYBRIDIZATION AND SELF-POLLINATION

A. Equipment

The equipment required varies with the method used, but may include a Thermos bottle or a hot water bath, scissors (small, sharp, pointed, and of good quality), fine forceps (not sharp), a vacuum emasculator, a stool, and an apron with pockets or a small box containing a pencil, glassine bags (approximately 5 × 15 cm), paper clips, felt pen or wax pencil, pot labels, and tags.

The vacuum emasculator was developed by Kirk (1930) and first used on rice by Van der Meulen (1933). The techniques and equipment were recently modified by Herrera and Coffman (1974). Simple to construct, the vacuum speeds emasculation and relieves much of its tedium (Chapter 6). The components can usually be purchased locally; if not, they are available through IRRI.

B. Preparation of the Female

Rice is ready for emasculation when the panicle is from 50 to 60% emerged from the boot (Fig. 1). Individual flowers are ready for emasculation after emergence from the boot and prior to anthesis. In the tropics, emasculation should not be attempted until anthesis has ceased for the day, usually by mid-afternoon. In temperate areas, emasculation can be performed either early in the morning or late in the afternoon. The optimal times for emasculation and pollination can best be determined by observation and experience at each location. Chandraratna (1964, p. 52) summarizes times of anthesis (beginning-peak-ending) reported at 18 locations in 10 countries.

The hot-water method of emasculation is widely practiced in Japan (Matsubayashi et al., 1965). The panicles are soaked for at least 5 min in water held at a constant temperature of 43 C and then pollinated within 30 to 60 min. Any unopened spikelets are cut off. The water temperature is critical. A Thermos bottle is usually the most convenient container, but it is not essential. This method is relatively slow and pollen must be immediately available to introduce into the florets before they close. If no pollen is available, the spikelets must be clipped or forcibly reopened later.

The most simple and efficient emasculation technique is to clip the spikelets and remove the remnant anthers with forceps or by vacuum. Some breeders prefer a combination of the above techniques. The procedure used at IRRI is outlined below.

Selection of the cultivars to be emasculated depends on the program objectives. A list is made that indicates the number of panicles to be emasculated. One or two panicles are alloted for a single cross, seven for a backcross or topcross, and 11 for a double cross.

Individual plants are selected that are healthy and representative of the cultivar or line. The mud is cut from around the base of the plant with a knife or sickle and the plant is removed from the plot. The plant is tagged with the designation and plot number. The plant is potted in soil and water and this must be completed at least 6 hours before emasculation to permit recovery from the slight wilting that may occur.

Individual panicles are selected, preferably those that are 50 to 60% emerged (Fig. 1). In some cultivars, especially dwarf types, the panicle may not attain 50% emergence before anthesis begins. Such panicles must be taken earlier, but the success rate will be lower due to dessication of the newly exposed, tender spikelets. The selected panicle is separated from surrounding ones to make it easy to emasculate. The flag leaf sheath is removed carefully to avoid breaking the stem.

All florets that have pollinated in the top of the panicle are cut off with scissors. Such spikelets appear translucent and often have anthers clinging to the outside. Young florets from the bottom of the panicle wherein height of the anthers is less than half that of the floret are also cut off. About one-third to one-half of each of the remaining florets is cut away obliquely to expose the anthers (Fig. 2). If the anthers are to be removed by forceps, some breeders prefer to cut directly across each floret and through most or all of the anthers. The stigma can be injured if the cut is made too low. If the cut is too high, emasculation is difficult and pollen may not reach the stigma.

Anthers are removed with either forceps or a vacuum. With forceps, the tip of one prong is used to gently press the anthers against the side of the floret and lift them out (Fig. 3). Extreme care must be used to avoid damage to the stigma, and to be certain that all six anthers are removed. If the vacuum is used, the suction is adjusted until there is about 500 mm Hg (Chapter 6). Excessive suction may damage the stigma. The extractor tip is touched to the floret and gently moved until the sound of the airflow is interrupted by the extracted anthers (Fig. 4). All six anthers must be removed.

After all the florets on the panicle are emasculated, the date of emasculation and one's initials are marked on the bag with a felt pen or wax pencil. A bag is placed over the panicle, the bottom edge is folded over, and a paper clip is placed on the fold against the stem to hold it securely in place.

Fig. 1—Rice panicle at the appropriate stage for emasculation. Fig. 2—Clipping a floret for emasculation.

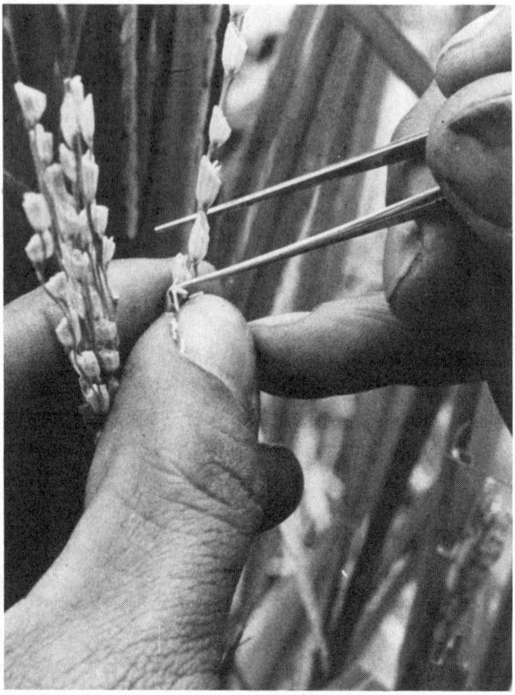

Fig. 3—Removal of the anthers by forceps. Note how the hands are positioned for maximum stability which allows for precise manipulation of the forceps.

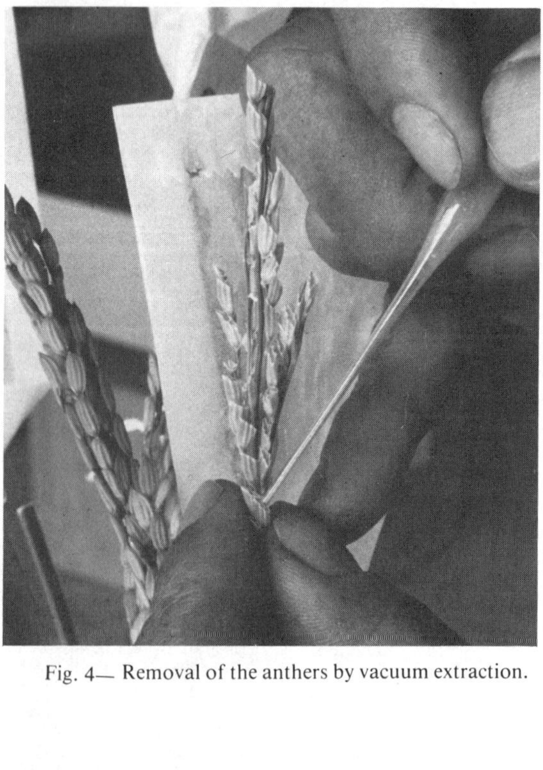

Fig. 4— Removal of the anthers by vacuum extraction.

C. Pollination

As anthesis commences the morning after emasculation, a blooming panicle of the male parent is cut and carried to the emasculated panicle. If not pollinated, the stigma remains receptive for 4 or 5 days, but receptivity drops at an exponential rate after the first day. Pollen normally remains viable for a very short time after anther dehiscence (Jodan, 1938), probably less than 5 min under most conditions. However, Chandraratna (1964) cites Noguchi (1931) and Noguti and Hamada (1927) who reported viability under nonspecified conditions for as long as 50 hours. Temperature and perhaps other environmental factors determine the length of time from pollination to fertilization, although Akemine (1914) and Terada (1928) reported that the time is generally about 12 hours. Cho (1955) found that fertilization was completed within 3 hours after pollination.

The time of anthesis varies among locations and, to a lesser extent, among cultivars. Pollination should only be attempted during the period of peak anther dehiscence, usually about 2 hours each day. This peak usually occurs in mid to late morning in the tropics and around noon in temperate areas.

Pollination techniques vary considerably among breeders and depend, to some extent, on local conditions and facilities. The steps and key points of the procedure used at IRRI are outlined below.

Pots containing female plants are arranged for easy access during pollination. A large circle or semicircle is usually best. The emasculated panicles are checked for any anthers that may have been left and any floret is reclipped where the opening is small. If two panicles on the same plant are to be pollinated by the same male, they are brought together by wrapping the two culms once with one of the glassine bags and fastened with a clip. The other bag is placed over both panicles.

Pollen is gathered just before anther dehiscence. The crossing tags should have already been prepared and grouped by male parent. There should be at least one male panicle for each tag. Panicles from healthy, representative plants are selected that will have a large number of florets undergoing anther dehiscence (indicated by several florets at the top with anthers showing). The culm is cut at a convenient length below the panicle and the flag leaf is removed. The panicles of each male parent are fastened together with a rubber band. Appropriate tags are attached to each group of panicles and the stems are clipped to a uniform length. The stems are placed in a water-filled container to transport to the site of pollination. The male panicles are placed in water-filled pots, with stems immersed, in the center of the circle of the female plants where air movement is minimal and access is easy. The panicles are arranged so that any one can be removed without disturbing the others.

Panicles are watched closely for anther extrusion and pollination begins as soon as possible. It is important to work quickly to take advantage of the period of maximum pollen production. To apply pollen, the glassine bag is removed from the female, the blooming panicle is lifted from the water, and it is shaken over the emasculated female (Fig. 5).

The glassine bag is placed over the pollinated panicle and the bottom edge is folded over. A paper clip is placed on the fold against the stem to keep the bag secure. The string of the identification tag is attached to the stem and the tag is placed under the paper clip at the bottom of the glassine

Fig. 5—Application of pollen.

bag with the side showing the parents of the cross inward and the pollination date outward. The pots with female plants are placed in an area protected from wind, rain, and pests but with good exposure to sunlight.

If pollen is not available or is insufficient due to unfavorable conditions, plants may be pollinated on subsequent mornings because the stigmas of emasculated spikelets usually remain receptive for 4 to 5 days. This ability to pollinate later is an advantage of emasculation by clipping.

If pollination must be done in the field or in other situations where there is appreciable air movement, bags can be fabricated from dialysis tubing of a diameter smaller than the glassine bags used on the female panicles, and they are placed over the male panicles before anthesis. At anthesis, the male panicles are cut from the plant with bags intact, placed in a box, and carried to the female. The top of the glassine bag, on the female panicle is clipped, and blowing into it forms a cylinder around the panicle. One of the male panicles is removed from the box, keeping it in a horizontal position. The top end of the panicle bag is held downwind, then it is clipped off and gently shaped into a circular form. The male panicle bag is placed into the bag containing the female panicle, where it is held by the stem above the female, and twirled vigorously for a few seconds to cause pollen shed.

Dialysis tubing is strong and light, and it is ideal for protecting the female panicles in the field while the hybrid seed matures.

D. Factors Affecting Efficiency

Flowering dates can be altered by manipulating temperature and photoperiod. The most common and practical way to have flowering at the desired time is to plant at four or five dates, except for highly photoperiod-sensitive rices which may require darkroom treatment if the natural day length exceeds 12 to 14 hours. At IRRI, a hybridization block is planted with five dates at 2-week intervals. A single seedling per hill is transplanted in plots of 1 $\times$ 2 m at 20 $\times$ 20 cm spacing. Valuable plants are sometimes grown directly from seed in pots.

Rice is unique among small grains in that it can easily be ratooned at any time from panicle initiation to maturity. New tillers develop from nodal primordia and will flower in 25 to 30 days after the original panicles are cut. Ratooning ability is a cultivar characteristic to some degree. Ratoon panicles, although adequate for crossing, are always smaller and less desirable.

Khush (1975) listed and described traits controlled by well-known rice genes. The most commonly used markers for identifying hybrids include plant height (tall is dominant), anthocyanin production (dominant), earliness (usually dominant), presence of awns (dominant), glabrous leaves (recessive), and waxy endosperm (recessive). Other markers that are sometimes useful are colored apiculi (dominant), clustered spikelets (dominant), gold hull (recessive), and shape of spikelet (broad is dominant). The fact that F_1 hybrids should tiller at least as well as the heavier-tillering parent can be used to confirm hybrids, especially when tall types (usually lower tillering) are used as females.

Consistent use of certain marker genes to identify hybrids may result in many commercial cultivars with the same cytoplasm. For example, the first dwarf rices developed by IRRI were used almost exclusively as females in subsequent crosses because tall height is dominant. As a result all IRRI cultivars, many IRRI breeding lines named as cultivars in rice-growing countries, and many locally developed cultivars from national breeding programs have the same cytoplasm, from the cultivar 'Cina.'

V. NATURAL HYBRIDIZATION

Natural cross-pollination in cultivated rice is rare (Beachell et al., 1938), normally less than 1% (Virmani and Athwal, 1973). Male sterility has been reported in rice (Shinjyo and Omura, 1966; Erickson, 1969; Athwal and Virmani, 1972; Li, 1977), but seed set on male-sterile plants is poor, even when they are surrounded by good pollinators (Stansel and Craigmiles, 1966; Athwal and Virmani, 1972; Carnahan et al., 1972). Virmani and Athwal (1973, 1974) investigated the floral characters that influence outcrossing in rice, but systems for natural hybridization remain unsatisfactory for practical purposes.

F_1 hybrid rice cultivars are said to be cultivated on a large scale in the Peoples Republic of China (Lin, 1977). Cytoplasmic-genetic male sterility is employed in hybrid seed production, but natural crossing is supplemented by hand pollination and augmented by cutting the flag leaves and tearing the leaf sheath that encloses the panicle.

VI. SEED DEVELOPMENT, HARVEST, AND STORAGE

If pollination is successful, the ovary should begin to swell in 3 or 4 days. Seed set should be at least 50% and often higher if the technique is good. The crossed seed is mature when it loses its green color about 25 days after pollination. Panicles are cut, the seeds are hand threshed, and the glume remnants that adhere to the base of the seeds are removed. The naked seeds are counted and placed in small envelopes marked with the designations of the two parents and the number of crossed seeds.

The incomplete covering of the developing seed by the clipped glumes results in viable, but abnormally shaped seeds that are sometimes more dormant than unclipped grains of the female parent. Dormancy is broken by placing the seeds in a dry-air oven for about 7 days at 50 to 55 C. The crossed seeds are stored in a cold room or refrigerator until they are germinated to produce F_1 plants.

Because they have no protective covering, the hybrid seeds may be attacked by fungi and other organisms during germination. It is best to surface-sterilize the seeds and then germinate them in sterile petri dishes. Soon after germination, before the roots entangle, the seedlings should be transferred to soil flats.

VII. TECHNIQUES FOR SPECIAL SITUATIONS

No references were found on the use of hormones or grafting techniques in rice. Such techniques have not been used at IRRI. Watanabe (1975) and Watanabe and Koga (1975) used embryo culture in interspecific hybridizations in connection with cytogenetic studies on rice and its wild relatives. Noguchi (1936) demonstrated what he termed post-harvest hybridization. He harvested panicles, along with the culm, 1 day before flowering, placed them in water at 30 C, and then crossed them in the usual manner. The culms developed roots at the lowest nodes and he harvested viable hybrid seed from 13 of the 17 emasculated panicles.

REFERENCES

Adair, C. R. 1934. Studies on blooming in rice. J. Am. Soc. Agron. 26:965-973.

Akemine, M. 1914. Ueber das bluken des reises und einige sich anknupfende erschernungen. Z. Pflanzenzucht 2:339-375.

Athwal, D. S., and S. S. Virmani. 1972. Cytoplasmic male sterility and hybrid breeding in rice. p. 615-620. *In* Rice breeding. Int. Rice Res. Inst., Los Baños, Philippines.

Beachell, H. M., C. R. Adair, N. E. Jodon, L. L. Davis, and J. W. Jones. 1938. Extent of natural crossing in rice. J. Am. Soc. Agron. 30:743-753.

Carnahan, H. L., J. R. Erickson, S. T. Tseng, and J. N. Rutger. 1972. Outlook for hybrid rice in the U.S.A. p. 603-607. *In* Rice breeding. Int. Rice Res. Inst., Los Baños, Philippines.

Chandraratna, M. F. 1964. Genetics and breeding of rice. Longman, Green and Co. Ltd., London.

Chatterjee, D. 1948. A modified key and enumeration of species of *Oryza sativa* L. Indian J. Agric. Sci. 18:185-192.

Cho, J. 1955. Cytological study of double fertilization in *Oryza sativa* L. Jpn. J. Breed. 5:131-148.

Erickson, J. R. 1969. Cytoplasmic male sterility in rice (*Oryza sativa* L.). Am. Soc. Agron. Abstr., p. 6.

Herrera, R. M., and W. R. Coffman. 1974. Emasculation of rice by vacuum extraction. Proc. Crop Sci. Soc. Philippines 5:12-14.

Jennings, P. R. 1966. Evaluation of partial sterility in indica × japonica rice hybrids. Int. Rice Res. Inst. Tech. Bull. 5. Los Baños, Philippines.

Jodon, N. E. 1938. Artificial hybridization of rice. J. Am. Soc. Agron. 30:295-305.

Khush, G. S. 1975. Rice. p. 31-58. *In* R. C. King (ed.) Handbook of genetics (plants, plant viruses, and protists). Plenum Press, N.Y.

————, E. Torres, and R. C. Aquino. 1971. Exploiting wild germplasm of *Oryza* for improving cultivated rice. Proc. Crop Sci. Soc. Philippines 2:311-320.

Kirk, L. E. 1930. Abnormal seed development in sweet clover species crosses—a new technique for emasculating sweet clover flowers. Sci. Agric. 10:321-327.

Li, P. 1977. How we studied hybrid rice. [In Chinese, English summary] Acta Bot. Sin. 19:7-10.

Lin, Shih-Cheng. 1977. Rice breeding in China. (Notes from a seminar given by Mr. Lin Shih-Cheng, Plant Breeder and member of the Rice Improvement Team of the Agricultural Association of the Peoples Republic of China, visiting IRRI, 17 Oct. 1977). Int. Rice Res. Newsl. 2(5):27–28.

Lu, J. J., and T. T. Chang. 1978. Rice in its temporal and spatial perspectives. *In* B. S. Luh (ed.) Rice: production and utilization. AVI, Westport, Conn.

Matsubayashi, M., R. Ito, T. Nomoto, T. Takase, and N. Yamada (eds.). 1965. Theory and practice of growing rice. Fuji Publishing Co., Tokyo.

Noguchi, Y. 1931. Environmental influences on flowering in rice with special reference to germination of pollen grains and development of pollen tube. (In Japanese). Proc. Crop Sci. Soc. Jpn. 3:22–23.

————. 1936. The application of post-harvest pollination in the hybridization of rice plant. Jpn. J. Genet. 12:324–325.

Noguti, Y., and N. Hamada. 1927. Uber die befruchlungsfahigkeit der narbe und pollen bei wasserreisplflangen. J. Sci. Agric. Soc. 300:515–524.

Sampath, S. 1962. The genus *Oryza*: its taxonomy and species interrelationships. Oryza 1:1–29.

————. 1966. The genus *Oryza*; an evolutionary perspective. Oryza 3:30–34.

Shinjyo, C., and T. Omura. 1966. Cytoplasmic-genetic male sterility in cultivated rice, *Oryza sativa* L. I. Fertilities of F_1, F_2 and off-spring obtained from their mutual reciprocal backcrosses and segregation of completely male sterile plants. [In Japanese]. Jpn. J. Breed. 16:179–180.

Stansel, J. W., and J. P. Craigmiles. 1966. Hybrid rice—Problems and potentials. Rice J. 69:14–15.

Terada, S. 1928. Embryological studies in *Oryza sativa* L. Jpn. J. Bot. 4.

Van der Meulen, J. G. 1933. Over het kunstmalig kruisen van Rijst op Java (English summary). Korte Medelingen van het algemeen Proefstation voor den landbouw 13:19–23.

Vergara, B. S., and T. T. Chang. 1976. The flowering response of the rice plant to photoperiod, 3rd ed. Int. Rice Res. Inst., Los Banos, Philippines.

Virmani, S. S., and D. S. Athwal. 1973. Genetic variability in floral characteristics influencing outcrossing in *Oryza sativa* L. Crop Sci. 13:66–67.

————, and ————. 1974. Inheritance of floral characteristics influencing outcrossing in rice. Crop Sci. 14:350–353.

Watanabe, Y. 1975. Cytogenetic studies on rice and its wild relatives. (1) Cytogenetic investigations of induced polyploids with special reference to the sterility of synthesized amphiploids. [In Japanese] Bull. Natl. Inst. Agric. Sci. (D) 26:1–90.

————, and Y. Koga. 1975. Cytogenetic studies on rice and its wild relatives. (II) Genetic and cytogenetic studies on the trisomic plants of rice, *Oryza sativa* L. [In Japanese]. Bull. Natl. Inst. Agric. Sci. (D) 26:91–138.

Xuan, Vo-Tong, and V. E. Ross. 1976. Training manual for rice production. Int. Rice Res. Inst., Los Banos, Philippines.

Yoshida, S. 1977. Ecophysiology of rice. *In* Ecophysiology of tropical crops. Academic Press, N.Y.

37 Rye

D. D. Morey
University of Georgia
Tifton

R. D. Barnett
University of Florida
Quincy

Rye (*Secale cereale* L.) is important as a food grain, especially in parts of northern Europe where soils are poor and winters severe. The leading rye grain producers are U.S.S.R., Poland, West Germany, East Germany, Czechoslovakia, Hungary, Canada, and the United States. Winter and spring ryes for all purposes were seeded in the United States on about 1.4 million ha annually during the 3-year period from 1975 to 1977 with nearly two-thirds of the crop grown in Minnesota, Nebraska, North Dakota, South Dakota, and Georgia. In the North Central States, rye is grown primarily for grain and for plow-down as a green manure crop, but is also used for hay and pasture. Only 27% of the planted hectarage is harvested for grain in the United States. The grain is used for bread making, feeding livestock, and producing alcoholic beverages. Grain yields in the United States are rather low, averaging only 1,380 kg/ha. In the southern U.S. where it is grown alone or in combination with ryegrass or clovers, rye is used primarily for temporary winter grazing and also as a cover or green manure crop.

I. PARENTAL MATERIAL

Rye is an erect annual or winter annual classified in the tribe *Triticeae* within the grass family *Gramineae.* The genus *Secale* contains several species but only one, *S. cereale* L., is widely cultivated. Other rye species are

---

 Hybridization of Crop Plants.

principally weedy or wild types which often occur as mixtures in other grains and grasses. Annual species, in addition to *S. cereale,* include *S. vavilovii* Grossh., *S. silvestre* Host., *S. ancestrale* Zhuk., and *S. segetale* Zhuk. Perennial species include *S. montanum* Guss., *S. africanum* Stapf., *S. anatolicum* Boiss., and *S. kuprijanovii* Grossh. Stutz (1972) discussed the origin of cultivated rye and gave a description of the various wild species. Evans and Scoles (1976) summarized the species situation in rye by stating that *S. montanum, S. africanum, S. silvestre,* and *S. vavilovii* differ from *S. cereale* by two and in the case of *S. silvestre* by perhaps three reciprocal translocations. Crosses between species have been successful in most instances. Price (1975) found in crosses between *S. cereale* × *S. africanum* that viable seeds were formed only when *S. cereale* was the female parent. He concluded the seed failure in the reciprocal cross was due to abnormal endosperm development.

Rye breeding has been enhanced over the past 50 years by induced polyploidy. Autotetraploid ryes with 28 instead of the normal 14 chromosome have been produced by colchicine. Muntzing (1951) reviewed the research related to the production and testing of tetraploid rye.

Some of the advantages of tetraploid rye are larger kernels, higher protein content, and better seedling emergence. Disadvantages include lower grain yield, reduced tillering, greater height, and later maturity. Another objection to tetraploid ryes is the necessity of isolation from diploid ryes to avoid formation of triploids. Triploid zygotes abort so yields are reduced in both diploids and tetraploids when grown together. Koo (1958) found that seed set in the diploid cultivars 'Emerald' and 'Imperial', when exposed to tetraploid pollen, was reduced two-thirds. In the reverse situation, the seed set of 'Tetra Petkus', when exposed to diploid pollen, was one-half to two-thirds of that when isolated.

The U.S. Department of Agriculture maintains and distributes rye seed for experimental purposes from a world collection of rye. This collection contains approximately 600 entries and is stored and catalogued in Beltsville, Md.

The Food and Agricultural Organization of the United Nations (FAO) has periodically issued lists of germplasm collections including rye (Saint, 1972). Mao (1961) lists stations where plant breeding is being carried out, including the breeders and their respective crops.

Triticale breeding has created new interest in rye germplasm because rye is a parent of primary triticales. Examples of possible sources of rye germplasm are CIMMYT in Mexico and experiment stations in Germany, Canada, the United States, and other countries doing research on rye and triticale.

II. PLANT CULTURE

A. Field

Rye is the most widely adapted of the cereals because of its extreme winterhardiness and ability to grow on marginal soils. The most drought-resistant of the cereals, rye has an extensive root system and adjusts maturity to moisture. Rye uses 20 to 30% less water per unit of dry matter formed than wheat (*Triticum* sp.) (Starzycki, 1976). Tetraploid rye cultivars are more sensitive to drought than diploids. Although rye is also able to tolerate a range of soil acidity, it does best at a pH of 6.0 to 6.5. Rye responds well to fertilizers, but care must be used to prevent excess nitrogen fertilization, which will cause the plants to lodge. The crop tillers well; normally for seed production the seeding rate is about 63 to 78 kg/ha. The culture of rye is similar to that of wheat and other cereal crops.

B. Growth Chamber and Greenhouse

Cultural requirements for growing rye in the greenhouse is similar to those for wheat and other small grains. Normally slightly larger pots are needed because of the extensive root system, and the plants have to be staked and tied up to prevent stem breakage. Rye is a long-day plant and extending the light period to 14 to 18 hours per day will hasten flowering. Photoperiod-insensitive types may be available; however, we are not aware of documented evidence. Temperatures in the greenhouse are not as critical as they are for the other small grains because rye will grow at lower temperatures, but it grows best at moderate temperatures. Winter, spring, and intermediate cultivars are available and differ considerably in vernalization period. The winter types require 40 to 60 days, whereas the spring types require only 10 to 12 days of cold temperatures to shift into the reproductive stage. As previously mentioned, rye tillers profusely, so individual plants can easily be split into several clones.

III. FLORAL CHARACTERISTICS

The rye inflorescence is a dense, terminal spike 10 to 15 cm long, more slender and nodding than wheat. The spikelets are solitary at each node and are alternately arranged on a zig-zag rachis. Each spikelet usually consists of only two fertile florets, with supernumerary florets vestigial or absent (Fig. 1). The florets are perfect, each containing three anthers and a pistil consisting of two feathery stigmas and an ovary.

Rye differs from other small grains in that it is cross-pollinated. Although pollen can and does fall into flowers of the same plant, little self-pollination takes place even when the spikes of individual plants are bagged. Leith (1925) studied over 84,000 rye flowers and concluded that self-sterility ranged from 96 to over 99% in rye. This and other evidence showed conclusively that commercially grown rye depends almost entirely upon cross-pollination for seed set and grain production. However, some inbreds selfed consecutively for 20 years or more have improved in self-fertility, perhaps as high as 20 to 40%.

The most extensive analyses of the genetics of self-incompatibility in rye have been performed by Lundquist (1954, 1956). He postulated that incompatibility is controlled by two loci, designated S and Z, each containing a multiple allelic series. A pollen grain is not inhibited unless both incompatibility loci in the stigma have alleles common to the two present in the pollen grain. If at least one of the two alleles in the pollen grain does not have a similar matching allele in the style, the pollen is compatible and will fertilize the ovary.

Landes (1939) postulated that self-incompatibility was partially the result of inhibition of pollen tube growth. Ruebenbauer (1973) proposed a general hypothesis aimed at unraveling the genetic phenomena of self-incompatibility.

Ovule or zygote abortion also reduces seed set in open-pollinated fields. On 185 random heads from a field in Georgia, we found only 80.9% seed set. Landes (1939) found a large proportion of self-pollinated ovaries which had aborted in the early stages.

Meiotic divisions begin in the middle part of the spike and gradually proceed upwards and downwards in the spike, usually a few days before the emergence of the spike from the leaf sheath. Bennett et al. (1973) have shown that male meiosis precedes female meiosis by about 15 hours in the same floret. They found only a single megaspore mother cell per floret in wheat and rye compared with about 2,100 microspore mother cells in 'Chinese Spring' wheat and 16,000 in 'Petkus' rye. Sapra and Hughes (1975) reported an average of 42,000 pollen grains per anther in rye and 11,000 in wheat, a further indication of large anther size in rye. Pederson et al. (1961) found pollen production per unit area was 2.25 to 2.50 times greater in diploid than in tetraploid rye.

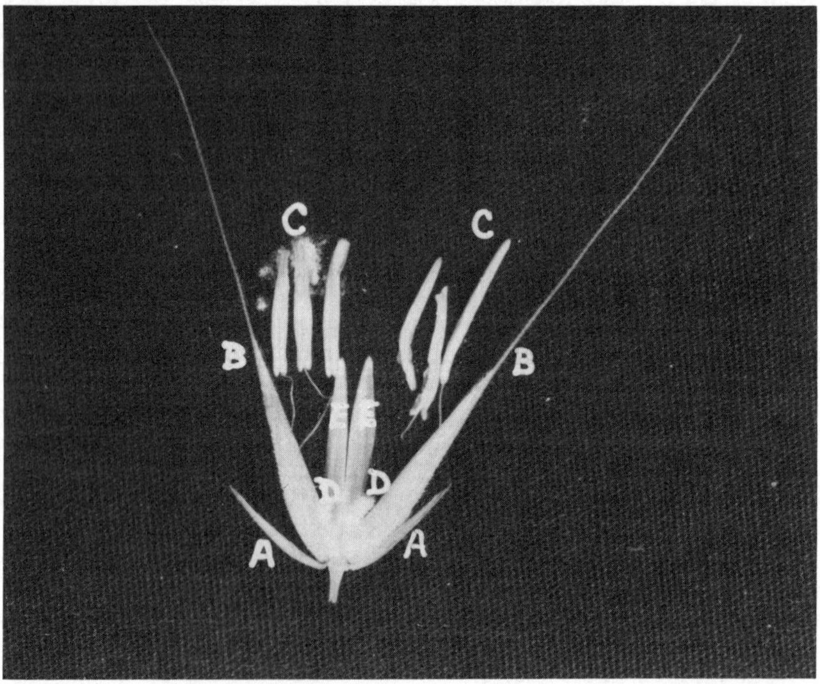

Fig. 1—Spikelet of rye, consisting of a pair of fertile florets. A) Outer or empty glume; B) Lemma with barbed awn; C) Three anthers at anthesis showing pollen grains above and threadlike, connecting filaments below; D) Feathery stigma (ovary below not visible); and E) Palea.

Wodehouse (1935) described rye pollen grains as being rather large (62 × 40 μ) and ellipsoidal with the germ pore on the side near one end. This character is rather unique among the grasses and serves to distinguish this species from others. The pollen grains of diploid rye (62.5 μ) are smaller than those of wheat (70.0 μ) (Sapra and Hughes, 1975).

The peak of rye pollen release comes at different times of day in different locations and is influenced by temperature, humidity, and sunlight. Jones and Brown (1951) observed that 'Balbo' rye in Oklahoma shed pollen from 0800 to 1200 hours with a peak between 1030 and 1100 hours. Under favorable weather, rye flowers have opened after sunrise with anthesis lasting until evening.

The first flowers to open are usually two-thirds of the way up the spike. They then open in both directions, up and down the spike, over a period of several days. The flowers usually remain open from 15 to 45 min. The lodicules swell, causing the lemma and palea of a floret to open exposing the three anthers. Brewbaker (1926) observed that rye florets open slowly at first and then more rapidly until the filaments and anthers hang out of the florets releasing large amounts of pollen carried by air currents to other plants.

Brewbaker (1926) found that unfertilized florets stayed open as long as 19 days in the greenhouse. Laube and Quadt (1959) stated that rye stigmas could be receptive for 7 days after opening of the flower. Heribert-Nilsson (1916) found rye stigmas receptive 14 days after anthesis.

D'Souza (1972) found rye stigmas (5.4 mm) to be longer than wheat stigmas (3.5 mm). He further showed that stigmas have two phases of receptivity which are dependent upon temperature and humidity. In the first phase, stigmas are receptive and seed set is good. The second phase is marked by a sudden drop in seed set. In rye at 20 C and 60% relative humidity, the first phase lasted 6 days and the second 5 days. At 30 C and 40% relative humidity, the two phases had a duration of only 3 days each. Temperature affects stigma receptivity more than humidity.

About 80% of pollen shed is viable for at least 10 hours according to D'Souza (1972) who stated that pollen tubes germinate about 3 min after coming in contact with the stigmas. High temperatures and low humidity are detrimental to pollen. Pollen storage is normally more successful at low temperatures and humidity. Gramineae pollen is an exception in that it requires storage in high humidities of 80 to 100%. Collins et al. (1973) reported that rye pollen stored in liquid air (−192 C) germinated in vitro and produced seed set in vivo.

Artificial germination of pollen from species within Gramineae has proven difficult because of the wide differences in species acceptance of the growth media. The addition of 3.5% bacto-agar to solidify a 25% sucrose and 0.02% boric acid solution was found essential for germinating rye pollen (Pfahler, 1965).

Viable pollen will travel in air currents and still be effective. Heribert-Nilsson (1916) used a wax-free (bloom-less) genetic marker and observed 54.5% natural crossing with plots separated by 50 m, 46.3% at 250 m, 29.7% at 350 m and 19.0% at 400 m.

IV. ARTIFICIAL HYBRIDIZATION AND SELF-POLLINATION

A. Equipment

Wells and Mahdawi (1953) found kraft corn tassel bags were the most convenient, dependable, and economical method for selfing single plants and for crossing paired plants. These waterproof pollen bags can be tied shut with string or stapled. It is usually necessary to support plants in the nursery with stakes to prevent plants from lodging and breaking. Rye breeders in the U.S.S.R., Poland, and other countries use cloth or heavy muslin cages to cover several plants or entire plots. Cages also can be made of wire covered with plastic film. Various sizes and weights of semitransparent paper sleeves are used for greenhouse rye breeding and are available from supply companies for corn and sorghum pollinators.

The equipment for making hand crosses includes a small pair of sharp-pointed scissors, narrow or sharp-pointed forceps, thin but high quality glassine bags, paper clips, and small tags with attached strings.

B. Preparation of the Female

Rye is cross-pollinated, and most genotypes are highly self-sterile so all that is required for hybridization is to place spikes from two different plants beneath one pollinating bag. Tapping the bags as they are raised on the supporting stake to accommodate culm elongation facilitates pollen dispersal within the bag. Normally good seed set is obtained and reciprocal crosses are obtained from one simple set-up. For critical crosses involving inheritance studies, species crosses, and self-fertile inbred lines, most workers completely emasculate rye and pollinate under controlled conditions.

At the onset of anthesis, the spikes most suitable for emasculation are those which have emerged from the leaf sheath by about one spike length. As the flowering period progresses, emerging spikes are suitable for emasculation (Bauer, 1965). The entire spike usually is emasculated by cutting across the tips of the spikelets with small, sharp scissors and removing three green anthers with forceps from each of the two primary florets (Fig. 2). Usually the left-hand column of spikes on one side of the spike is emasculated from the spike base to tip, followed by the right-hand column of florets. This method eliminates shifting the left forefinger from side to side as it is used to press outward from behind each floret tip to force the lemma and palea apart, providing easy entry for the forceps (Fig. 2). The left forefinger hooks around behind the spike for pushing on the tips of florets in the right column. Florets on the opposite side of the spike are emasculated in the same manner. After all anthers have been removed, each floret may be cut back further, just above the stigma branches. This insures stigma protrusion outside the lemma and palea, and facilitates pollination 2 to 4 days later. The florets should not be cut back if there is danger of dessication. This entire operation is tedious and time-consuming, especially if large numbers of crosses are needed.

During the 1930's Laube and Quadt (1959) described "schnittkastration," or scissor emasculation, and it is claimed to be 90% as effective as the classical method just described. The florets are cut across about one-half of the way down from the tip with scissors just as the head emerges from the boot, or about a week before anthesis (Fig. 3). All the anthers are cut in half in this operation but the portions remaining in the florets are not removed because they dry up and do not produce viable pollen. After emasculation, the spikes must be covered to exclude foreign pollen, and tagged for easy identification. About 5 days later stigmas will protrude above the edges of the cut lemma and palea, their accessibility allowing easy pollination by the desired male parent. If conditions are favorable, good seed set should be obtained (Fig. 4).

C. Pollination

Ripe anthers are collected from male plants and placed in a petri dish or watch glass. These anthers are then placed near a warm light bulb or in sunlight for 1 to 2 min to encourage natural dehiscence of the anthers. The pollen is applied with forceps to the stigmas of the female by rubbing them with a ruptured anther. Stripping the spike between thumb and forefinger encourages anther extrusion, which makes pollen easier to collect and aids in identifying mature anthers. Spikes gathered from male parents can be laid on paper to prevent anthers from whipping in the wind and losing pollen. Two or three anthers may be enough to pollinate an entire spike; one anther will certainly suffice for several florets. In making inbred-cultivar

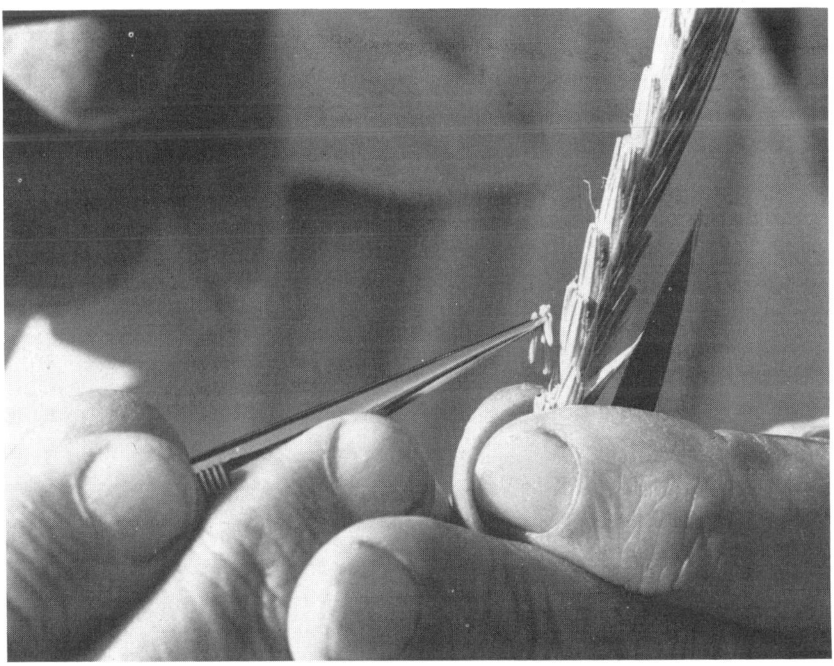

Fig. 2—Removing anthers from the floret during emasculation.

crosses, the flowers of the inbred lines may be emasculated in the usual manner, and when the florets open several days later a large amount of pollen may be collected by enclosing a large number of spikes of the male parent in a paper bag and applying the pollen to the emasculated flowers with a camel hair brush. Another method, which is similar to that used in barley (*Hordeum* sp.) and wheat loose smut inoculation is called the poof method. It involves collecting relatively large amounts of pollen, mixing the pollen with talc, and injecting the pollen-talc mixture into each female floret by means of a hypodermic needle attached to a small rubber bulblet. Preparing the mixture is time-consuming, but pollination is rapid (Shands, 1978).

Another method which works well for hand crossing is the approach method first described by Rosenquist (1927), especially with the improved techniques described by Curtis and Croy (1958) and McDaniel et al. (1967). The approach method involves placing an emasculated spike slightly beneath a nonemasculated or male spike and covering both spikes with a polli-

Fig. 3 (Left)—Scissor emasculation of rye. A) Rye spike at the proper stage, and B) emasculation of the spike. Fig. 4 (Right)—Seed set in rye spike after scissor emasculation and 10 days after pollination.

nation bag. Pollen from the male spike falls on the female spike, affording good opportunity for pollination. The approach method offers several advantages over hand pollination. The crosses can be set up when the emasculation is done so it will not be necessary to go back several days later. The crosses can be made up any time during the day, and many more florets can be pollinated than would be possible by hand. Important keys to successful approach crossing are the use of dialysis tubing for pollination bags and occasional thumping of the bags during the pollination period to get good pollen distribution. Hybrid seeds are often smaller with this method, but they are much more easily obtained. Rye is relatively easy to cross and, under most conditions, if self-sterility is not involved, 50 to 75% seed set can be expected.

D. Factors Affecting Efficiency

Temperatures are not critical for flowering, but 20 to 30 C are the temperatures which prevail in many rye fields at flowering time. Greenhouse temperatures in the spring may exceed 30 C, thus reducing the period of stigma receptivity and pollen effectiveness. Cooling the greenhouse below 25 C will improve seed set. Low humidity (40%) is more detrimental to stigmas than higher humidity (60%).

With many crops, altering planting dates will alter heading dates. In the United States, altering the planting date of winter rye in the field has a relatively small effect on heading date in the spring. A spread of 3 months in planting dates in Georgia changed heading dates only 11 days. Varying the planting date for material in the greenhouse generally is more effective. Also the vernalization period can be varied for different cultivars. Seed of the extreme winter-types can be vernalized by sprouting them on absorbent paper in petri dishes in a cold chamber or refrigerator at 1 to 2 C for 4 to 6 weeks before planting in the greenhouse, and thus can be matched to spring or intermediate types. The spring types can be planted about the time the winter types are removed from vernalization. Care must be taken after removing plants from vernalization because the effects can be reversed by high temperatures.

Cutting rye plants back once or twice in the vegetative stage will delay flowering. Care must be taken not to cut below the growing points and destroy or injure the reproductive meristems.

Spaced plants are best to work with in field crossing blocks, and room must be provided for staking and crossing. Planting two rows and skipping one will give some room to work around the plants. When bagging rye, space must be allowed for the continued upward growth of the spike because the rye stem continues to elongate for an extended period of time after heading.

V. NATURAL HYBRIDIZATION

Spatial isolation from other rye plants can be used for crossing blocks, increasing inbred lines, or developing synthetic populations. The disadvantage of this method is the difficulty of finding suitable isolation in frequency and distance.

Recently, rye breeders in Germany, U.S.S.R., and Poland have developed a possible way to utilize inbred lines or early generation material in the production of hybrid rye. Systems for controlled pollination based on cytoplasmic-genetic male sterility are now available (Geiger and Schnell, 1970; Gulyaeva, 1971; Klyuchko and Belousov, 1972; Kobylyanskii and Katerova, 1973; and Madej, 1976). Crossing schemes for hybrid rye are believed to be similar to those employed for the production of hybrid corn and hybrid sorghum. Several reports are available which indicate that a considerable amount of heterosis is available in rye (Pfahler, 1968a, 1968b; and Sechler and Chapman, 1967).

VI. SEED DEVELOPMENT, HARVEST, AND STORAGE

Rye seed development follows that of wheat very closely. Special care must be taken when threshing seed by machine because the pointed embryo of the seed protrudes and can easily suffer mechanical damage. Rye normally threshes free of the chaff easily and it is not necessary to use high speeds with threshing equipment. Seed viability is short-lived when rye is stored under ordinary laboratory conditions (Morey, 1976). For long-term storage, rye seed can be dried at 40 C to a moisture content of 5%, then dried at a temperature of 50 to 60 C. The seed may be safely dried to a moisture content of about 3%, but further dehydration may cause some somatic damage. For long-term storage, it is necessary to store the seed in airtight containers to retard senescence (Grzesik and Molski, 1975).

Rye does not offer the problem of a long period of afterripening dormancy as found in some crops. In Canada, Depauw (1976) studied seed dormancy or afterripening in rye and found no difference between cultivars, but did obtain a difference between years. In 1974, about 37 days of afterripening were required, but in 1975 only 1 to 7 days were required. At Tifton, Georgia, in 1977, 'Athens Abruzzi' required 21 days of afterripening to gain its full germinating capacity. Bekendam (1975) suggested the use of gibberellic acid to break dormancy of cereal seeds.

VII. TECHNIQUES FOR SPECIAL SITUATIONS

The development of homozygous lines from tissue culture of microspores of F_1 plants would be a great advancement in the development of hybrid rye. Thomas and Wenzel (1975) utilizing F_1 winter hybrids of *S. cereale* which included genes for short culm and self-fertility from *S. vavilovii* suggested that this would be feasible, and they cultured calluses bearing primordia as well as plantlets and perfect embryoids. Malepszy (1975) reported success in culturing rye microspores into haploid callus, as well as plants grown from the somatic callus. Despite the problems inherent in these procedures, the ability to obtain homozygous diploids immediately would be of great advantage to breeders.

REFERENCES

Bennett, M. D., R. A. Finch, J. B. Smith, and M. K. Rao. 1973. The time and duration of female meiosis in wheat, rye and barley. Proc. R. Soc. London (Ser. B) 183:301–319.

Bauer, F. 1965. A study of time and method of emasculation in rye. Novenytermeles 14:321–336.

Bekendam, J. 1975. Report of the working group on the application of gibberellic acid in routine germination testing to break dormancy of cereal seed. Seed Sci. Technol. 3:92–93.

Brewbaker, H. E. 1926. Studies of self-fertilization in rye. Minnesota Agric. Exp. Stn. Tech. Bull. 40.

Collins, F. C., V. Lertmongkol, and J. P. Jones. 1973. Pollen storage of certain agronomic species in liquid air. Crop Sci. 13:493–494.

Curtis, B. C., and L. I. Croy. 1958. The approach method of making crosses in small grains. Agron. J. 5:49–51.

Depauw, R. M. 1976. After-ripening in fall rye. Can. J. Plant Sci. 56:395–396.

D'Souza, L. 1972. A comparative study of the size and receptivity of the stigma in wheat, rye, triticale and secalotricum. Z. Pflanzenzucht. 68:73–82.

Evans, L. E., and G. J. Scoles. 1976. Cytogenetics, plant breeding and agronomy. p. 13–26. *In* Walter Bushuk (ed.) Rye: production, chemistry and technology. Am. Assoc. Cereal Chemists, Inc., St. Paul, Minn.

Geiger, H. H., and F. W. Schnell. 1970. Cytoplasmic male sterility in rye (*Secale cereale* L.). Crop Sci. 10:590–593.

Grzesik, M., and B. Molski. 1975. The effect of temperature and dehydration of the dried seeds on the biological properties of rye introductions. Hodowla Ros. Aklim. Nasienn. 19:449–458.

Gulyaeva, Z. B. 1971. A study of male sterility in rye. Genetika 8:18–24.

Heribert-Nilsson, H. 1916. Populationsanalysen und Erblickeitsversuche. Z. Pflanzenzucht. 4:1–44.

Jones, M. D., and J. G. Brown. 1951. Pollination cycles of some grasses in Oklahoma. Agron. J. 43:218–222.

Klyuchko, P. F., and A. A. Belousov. 1972. Genetic studies on cytoplasmic male sterility in winter rye. Genetika 8:9–15.

Kobylyanskii, V. D., and A. G. Katerova. 1973. Genetics of cytoplasmic and nuclear male sterility in diploid winter rye. Genetika 9:5–11.

Koo, F. K. S. 1958. Deleterious effects from interpollination of diploid and autotetraploid winter rye varieties. Agron. J. 50:171–172.

Landes, Margaret. 1939. The cause of self-sterility in rye. Am. J. Bot. 26:567–571.

Laube, W., and F. Quadt. 1959. Roggen (*Secale cereale*). p. 35–102. *In* H. Kappert and W. Rudorf (ed.) Handbuch Pflanzenzucht. Bd. II. Breeding of grain species. Paul Parey, Berlin.

Leith, D. B. 1925. Sterility of rye. J. Am. Soc. Agron. 17:129–132.

Lundquist, Arne. 1954. Studies on self sterility in rye. *Secale cereale* L. Hereditas 40:278–294.

———. 1956. Self incompatibility in rye: genetic control in the diploid. Hereditas 42:293–348.

Madej, L. 1976. The genetical characteristics of three sources of male sterility in rye (*Secale cereale* L.). Hodowla Ros. Aklim. Nasienn. 20:157–174.

Malepszy, S. 1975. A contribution to the production of haploids in rye. Bull. Acad. Pol. Sci., Ser. Sci. Biol. 23(3):167–172.

Mao, Y. T. 1961. World list of plant breeders. FAO. United Nations, Rome.

McDaniel, M. E., H. B. Kim, and B. R. Hathcock. 1967. Approach crossing of oats (*Avena* spp.). Crop Sci. 7:538–540.

Morey, D. D. 1976. The effect of storage conditions and containers on the viability of rye, wheat and oat seed. Seed Sci. Technol. 4:313–332.

Muntzing, A. 1951. Cytogenetic properties and practical value of tetraploid rye. Hereditas 37:17–84.

Pederson, P. N., H. B. Johansen, and J. Jorgenson. 1961. Pollen spreading in diploid and tetraploid rye. Arsskr. K. Vet. Landbohoejsk.:54–86.

Pfahler, P. L. 1965. *In vitro* germination of rye (*Secale cereale* L.) pollen. Crop Sci. 5:597–598.

———. 1968a. Heterosis and homeostasis in *Secale montanum* Guss. × *S. cereale* L. hybrids. I. Vegetative growth and development. Crop Sci. 8:663–666.

———. 1968b. Heterosis and homeostasis in *Secale montanum* Guss. × *S. cereale* L. hybrids. II. Reproductive characteristics. Crop Sci. 8:667–670.

Price, H. J. 1975. Unilateral seed failure after reciprocal crosses of *Secale cereale* with *Secale africanum*. Can. J. Bot. 53:929–932.

Rosenquist, C. E. 1927. An improved method of producing F_1 hybrid seeds of wheat and barley. J. Am. Soc. Agron. 19:968–971.

Ruebenbauer, T. 1973. Effect of self-incompatibility on the genetic system in rye. Genet. Pol. 14:157–171.

Saint, S. 1972. World list of germplasm collections—barley, oats and rye. Plant Genetic Res. Newsl. 28:24–33.

Sapra, V. T., and J. L. Hughes. 1975. Pollen production in hexaploid triticale. Euphytica 24:237–243.

Sechler, Dale, and W. H. Chapman. 1967. Hybrid vigor in intervarietal rye crosses. Proc. Soil Crop Sci. Soc. Fla. 27:81–87.

Shands, H. L. 1978. Personal communication. Dep. of Agronomy, Univ. of Wisconsin, Madison, WI 53706.

Starzycki, Stanislaw. 1976. Diseases, pests, and physiology of rye. p. 27–61. *In* Walter Bushuk (ed.) Rye: production, chemistry and technology. Am. Assoc. Cereal Chemists, Inc., St. Paul, Minn.

Stutz, H. C. 1972. On the origin of cultivated rye. Am. J. Bot. 59:59–70.

Thomas, E., and G. Wenzel. 1975. Embryogenesis from microspores of rye. Naturwissenschaften 62:40–41.

Wells, D. G., and Abdul-Ridha Alwan Mahdawi. 1953. Pollination control in rye. Agron. J. 45:379–380.

Wodehouse, R. P. 1935. Pollen grains: their structure, identification and significance in science and medicine. McGraw-Hill, N.Y. p. 318.

38 Safflower

P. F. Knowles
University of California
Davis

Safflower (*Carthamus tinctorius* L.) probably evolved as a domesticated crop in the Middle East. It was first grown for the pigment in its flowers that was used for coloring foods and dyeing cloth. Later it was grown for the oil in the seed.

The largest hectarage of safflower is in south-central India. Commercial production in the United States began in the 1940's in Nebraska and Colorado, and expanded to California in 1950. By states, the average hectarage for 1975, 1976, and 1977 was: California, 42,100; Arizona, 7,000; and the Northern Great Plains (mostly eastern Montana), 14,460. Average yields in those areas were respectively 2370, 2800, and 820 kg/ha. Other important producers of safflower are Mexico, Australia, Spain, and Portugal.

Safflower yields two types of oil. The standard polyunsaturated type is used in soft margarines, for cooking, and in surface coatings because of its high levels of linoleic acid. The oleic type, so named because of its high level of the monounsaturated oleic acid, is a high-quality frying oil and also has industrial uses. The meal, with or without the hulls, is fed to livestock.

I. PARENTAL MATERIAL

Safflower (n = 12) is the only species of *Carthamus* that is domesticated (Fig. 1). All the remaining species are successful weeds, collectively ranging from northwestern India westward to and around the Mediterranean Sea. Wild species fall into five chromosome groups with n = 10, 11, 12, 22, and 32.

Knowles (1969) identified seven centers of safflower culture in the Old World: Far East (China, Japan, and Korea); India-Pakistan (India, Pakistan, and Bangladesh); Middle East (Afghanistan to Turkey, southern U.S.S.R. to the Indian Ocean); Egypt (bordering the Nile River north of

Aswan); Sudan (bordering the Nile in southern Egypt and northern Sudan); Ethiopia; and Europe (southern Europe and northwest Africa). The types in each area are distinctive. Except for the Far East Center, each is well-represented in the USDA world collection.

Five species with $n = 12$ are closely related to cultivated safflower (Hanelt, 1963). *Carthamus palaestinus* Eig, a self-compatible species, occupies desert areas of western Iraq, Jordan, and southern Israel. *Carthamus persicus* Willd. (=*C. flavescens* Spreng), inhabiting areas of Turkey, Syria, and Lebanon, is entirely self-incompatible. *Carthamus oxyacanthus* M.B. (=*C. oxyacantha* M.B.), distributed from northwestern India to central Iraq, is a mixture of self-incompatible and self-compatible types. *Carthamus gypsicolus* Ilj., in the U.S.S.R., and *C. curdicus* Hanelt, in Iraq, have not been studied. When those species were crossed with cultivated safflower, the F_1's had normal meiosis and were fertile.

Another species with $n = 12$, *C. nitidus* Boiss., found in Syria, Israel, Lebanon, Jordan, and Nubia, will cross with *C. tinctorius* to give an F_1 that is completely sterile (Knowles and Schank, 1964). It is self-compatible.

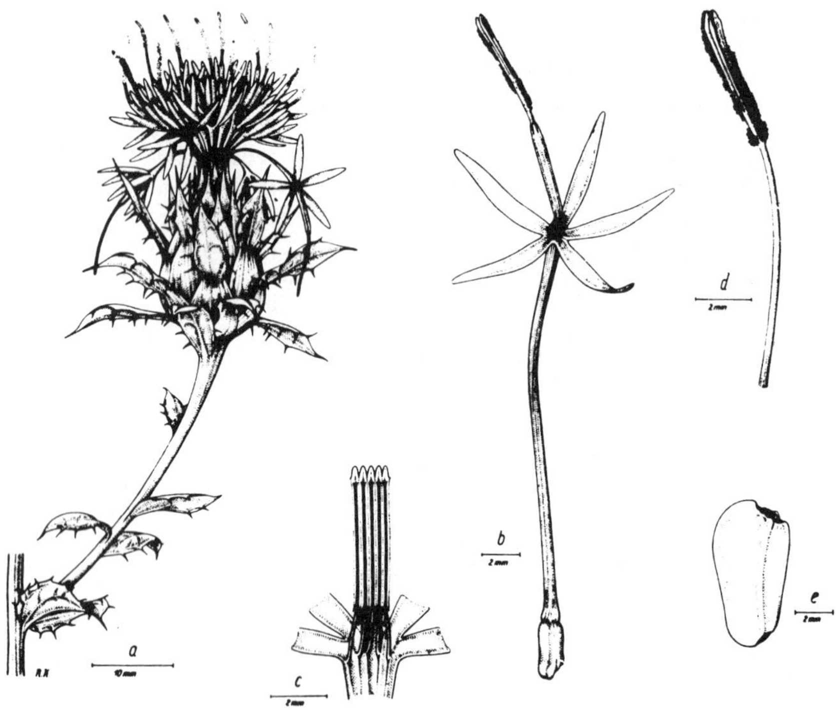

Fig. 1—Reproductive structures of safflower. a) Head showing spiny outer bracts and several flowers; b) disk, or tubular flower, showing ovary at base, tubular corolla terminating in five lobes, an anther tube with short filaments attached to the top of the corolla tube, and the stigma with attached pollen above the anther tube; c) anther tube slit on one side and opened up to show the five attached anthers; d) stigma with attached pollen and upper portion of the style; 3) achene (seed). From Hanelt (1963).

A large group of species with $n = 10$ is found in the Middle East, in southeastern Europe and northeastern Africa. They are closely related to one another and show variable self-incompatibility (Schank and Knowles, 1964). One exception, *C. leucocaulos*, is entirely self-compatible. Most will cross with cultivated safflower, but the F_1's are highly sterile.

One species, *C. divaricatus* (Beg. and Vacc.) Pamp., from Libya, has 11 pairs of chromosomes. It is closely related to species with $n = 10$, and is self-incompatible (Estilai and Knowles, 1976). It will also cross with cultivated safflower, but the F_1 is highly sterile.

All species with $n = 22$ and 32 are self-compatible. *Carthamus lanatus* L., with $n = 22$, and *C. lanatus* ssp. *creticus* (L.) Holmb. (= *C. baeticus* (Boiss et Reut.) Nym.), with $n = 32$, (Fig. 2), have extended the distribution of the genus to western Europe, northwestern Africa, and several other areas with a Mediterranean climate, including California and Australia. *Carthamus lanatus* ssp. *turkestanicus* (M. Popov) Hanelt (= *C. turkestanicus* Popov), with $n = 32$, is in the area occupied by the genus east of central Turkey and the Near East. Only *C. lanatus* has been crossed to cultivated safflower, giving a sterile F_1.

The USDA Western Regional Plant Introduction Station, Pullman, WA 99164, maintains the World Collection of domesticated safflower (Dietz et al., 1977). Plant introductions and cultivars are preserved at the USDA National Seed Storage Laboratory at Colorado State University, Fort Collins, CO 80521.

II. PLANT CULTURE

A. Field

Commercial production of safflower has been most successful where soils are deep, fine-textured, well supplied with moisture at planting, and well drained (Knowles and Miller, 1965). These requirements relate in part to the deep root system, often reaching to 4 m in California. It has about the same salt tolerance as barley when grown on dryland, but is less tolerant when irrigated. Where salinity is a problem, safflower is planted on the sides of sloping beds, avoiding the greater salt accumulations on ridges.

Fertilizer practices are like those for wheat and barley, except that nitrogen may be increased about 20%. Immediately after rice, where safflower has often developed slowly, it has responded dramatically to phosphorus, if it is applied with or immediately below the seed (Martin et al., 1971).

Water requirements are high, particularly just before and during maximum flowering, when daily consumptive use on a hot day can reach 1.2 cm. Under irrigation, yields above 4,000 kg/ha will require as much as 1 m of water, including rainfall. Dryland yields below 2,000 kg/ha consume 45 to 50 cm of water.

Safflower is very sensitive to excess water where root rot (*Phytophthora drechsleri* Tuck.) is usually severe. Irrigations should conform to the following guidelines. If the deep subsoil is not moist, preirrigate to the rooting depth expected at maturity. Use a root-rot-tolerant cultivar. Use beds and furrow irrigation. Irrigate before the lower leaves are fired by drought stress. Avoid prolonged irrigation and ponding of water, especially in warm weather.

Safflower is grown commercially in rows 15 to 30 cm apart if sowing is with a grain drill, or in rows 50 to 75 cm apart if a row planter is used. Bed plantings are preferred for furrow irrigation with row spacings of 70 to 75 cm. Seeding rates are 17 to 45 kg/ha, the lower rates being used with wide-row plantings, early seedings, and dryland culture.

Because most safflower genotypes are spiny, breeding nurseries commonly use paired rows 50 to 75 cm apart, with the space between pairs double that within the pair. This arrangement increases the number of plants available to the researcher and permits furrow irrigation between each pair of rows, while the spaces separating the pairs are left dry.

Cultivated safflower is grown under different day lengths as a winter crop in south central India (day lengths of less than 12 hours), as an early summer crop in California (about 14 hours during flowering), and as a mid-summer crop in the Northern Great Plains of the United States (16 hours). Variations in light level and quality in the field do not seem to affect safflower hybridizations.

Under field conditions, plant development differs markedly with dates of planting, presumably reflecting both temperature and day length effects. Sown in the late fall at Davis, Calif., it will remain more or less in a rosette stage until spring, when stems elongate and produce plants 2 m high with over 50 heads each, if conditions are optimum. Plantings in mid-May or early June advance very quickly through the rosette stage and produce plants about 60 cm high with 5 to 10 heads per plant. Winter types require a period of cold, the amount not yet determined, before advancing beyond the rosette stage (Ghanavati and Knowles, 1977). If sown in the spring, these types flower very late in the fall, about the time when spring types mature.

In the field, safflower cannot thrive under high humidity, primarily because of diseases. Leaf diseases caused by *Alternaria carthami* Chow., *Ramularia carthami* Z., *Cercospora carthami* Sun and Rao, and *Puccinia carthami* Corda have been very damaging. *Botrytis cinerea* Pers. causes severe damage in coastal areas of California.

B. Growth Chamber and Greenhouse

Safflower has been grown successfully in a variety of planting media that provide good drainage. The planting medium used at the University of California at Davis is: sand, 0.38 m^3; peat mix, 0.38 m^3; and UC Mix, 6.6 kg. The UC Mix is a mixture of 681 g 38–0–0 (NPK), 113 g 10–10–5, 170 g K_2SO_4, 1,135 g single superphosphate, 3,405 g dolomitic limestone, and 1,135 g of hydrated lime.

Greenhouse potting mixes are supplemented with nutrients as usual for other crops. Care must be taken to avoid wilting when the plants are in bloom. Water is best supplied by placing the pots in shallow trays, in which the water is maintained at a depth of 1 to 2 cm, with nutrients added

(Zimmerman, 1972). Hoagland's solution, the most satisfactory, is supplied full-strength twice weekly. Algae that grow in the water appear to benefit the safflower, so no attempt is made to control them with fungicides. A float-and-valve arrangement will regulate the water level.

When a plant is fully developed it will need about 0.16 to 0.20 m^2 of space. Heads of adjacent plants should not touch each other.

Flowering date can be advanced in winter in a greenhouse by providing long days which shortens the rosette period. The duration of the rosette period for six genotypes averaged 38.9 days under 10-hour days and 23.2 days for 14-hour days (Zimmerman, 1973). Horowitz and Beech (1974) grew two cultivars under 9-hour and 16-hour days. The periods between emergence and floral initiation for 9-hour days were 97 days for cultivar 2, and 128 days for cultivar 9042. With 16-hour days, the emergence to floral initiation period was 48 days for cultivar 2, and 77 days for cultivar 9042. Some wild species will remain indefinitely in a rosette stage when day lengths are less than 12 hours.

Under low humidity in growth chamber experiments, high temperatures (max 43 C and min 27 C) gave 35% lower yield than low temperatures (max 31 C and min 16 C) (Zimmerman, 1972). Medium temperatures (max 38 C and min 20 C) gave 7% lower yields than the low temperatures. Under high humidities, high temperatures decreased yield 77% and medium temperatures 27% compared with low temperatures. Heads were most sensitive to high temperature just before and at anthesis, but were sensitive at all stages if humidities were high. Most of the yield responses were due to differences in numbers of seeds in the mature heads. Cultivars varied in sensitivities to combinations of high temperature and high humidity. It is apparent that safflower hybridizations should be made where temperatures do not exceed 38 C and humidities are below 40%.

III. FLORAL CHARACTERISTICS

Safflower has a composite-type inflorescence with several flowering heads (capitula), each of 20 to 180 flowers (Fig. 1a). Surrounding the flowers are bracts, which are modified leaves. The tubular (disk) flowers are located on a flat or convex receptacle. Interspersed among the flowers are bristles or hairs.

At the base of a flower is an inferior ovary that develops into a single-seeded fruit called an achene, referred to herein as a seed (Fig. 1e). The corolla tube is 2 to 3 cm long, and the five petal lobes are 6.5 to 8.0 mm long. Inside the corolla lobes and surrounding the style are five fused anthers that form a tube about 4 mm long (Fig. 1c and 5F). The pollen of cultivated safflower and several other species is yellow and covered with sharp spines. The filaments of the anthers are short and separate, and rise from the top of the corolla tube (Fig. 1c). Inside the anther tube is the style, terminated by a stigma that projects beyond the top of the anther tube by 5 to 6 mm (Fig. 1b, d). The stigma, often divided at the tip for 1 to 2 mm, is covered with short hairs. At the base of the stigma is a collar with longer hairs than those on the rest of the stigmatic surface.

A tuft of bristles that adhere after harvest may develop on the top of seeds in the center of the head. Commercial cultivars are essentially free of bristles, whereas some experimental lines may have bristles on 30% of the seeds.

Flowers are classified for color when they are dry. Common colors are yellow, orange (golden yellow when fresh), red (brilliant orange when fresh), white, and cream. The color may vary in intensity with the genotype.

Developing flower buds are completely enclosed by the bracts (Fig. 2 and 5A). The first flowers to open are on the periphery, and flowering proceeds toward the center. Succeeding flowers appear in whorls about two flowers wide on each day for 3 to 7 days, depending on head size and environment. For all heads on a plant, flowering may last up to 30 days.

The first one to three rows of flowers around the periphery of the head usually have rudimentary ovaries, and are female-sterile. Otherwise they are normal in every respect, and produce pollen. Cultivars of domesticated safflower and species of wild safflower vary in the percentage of rudimentary flowers. Schank and Knowles (1964) found a range of 10.3 to 67.1% in wild species. Rudimentary ovaries are elongated and narrower than normal ovaries (Fig. 3). It is obvious that the first flowers to open should not be emasculated and used as a female parent.

Fig. 2—Left to right: *Carthamus tinctorius* (cultivated safflower), *C. lanatus*, and *C. baeticus*. 0.4 ×. A bract of each is in the top row, inflorescences (heads) in the center row, and mature seeds in the lower row.

The day before a flower opens it is about 15 mm long (Fig. 5C). At that stage, the petal lobes enclose the anther tube. Pollen is not released, and the style and stigma do not project beyond the tips of the corolla lobes.

The night before a flower opens, the corolla tube and style elongate, projecting them beyond the bracts by morning (Ebert and Knowles, 1968). Before the flower emerges through the bracts, the pollen dehisces. Corolla lobes open in the early morning when the flower is 3 to 4 cm long. About 2 hours later, the style elongates rapidly for a brief period, forcing the stigma through the anther tube. Short hairs on the stigma and longer hairs on the collar at the base of the stigma sweep out the pollen released by the anthers. Pollination occurs immediately.

The seeds are usually fully developed 30 days after pollination, but may require another 2 weeks to dry enough for harvest.

Some genotypes of safflower have fused corolla lobes (closed flowers) (Fig. 4). When closed flowers were first identified, it was hoped that they might provide a means of ensuring self-pollination (Dille and Knowles, 1975). Unfortunately seed set compared to 100% in normal flowers was 1% in the bar-type, 75% in the cage-type, and 90% in the arrowhead-type. Failure to set seed was apparently a consequence of the changed environment in the flower because the bar-type produced normal pollen and set the normal amount of seed, if corolla lobes were separated early in flower development.

Rubis (1967) identified a very thin-hull mutant in which the flowers produced no pollen or pollen release was delayed. Failure or delay in release of pollen by thin-hull types was considered to be a form of structural male

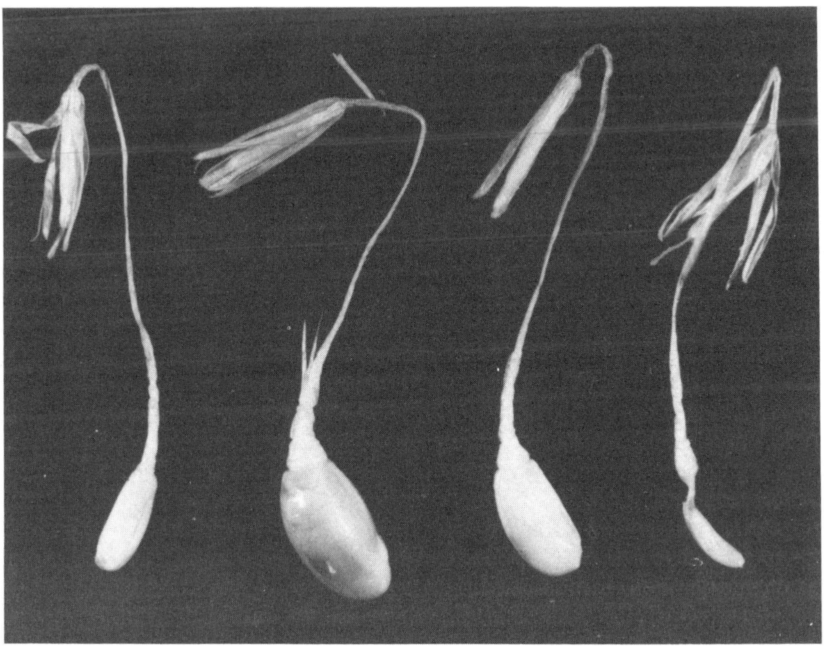

Fig. 3—Flowers about 15 days after pollination. The two in the center have normal ovaries that are fertile, and those on the left and right have rudimentary ovaries that are sterile. 2 ×.

sterility. It led to attempts to use them as female parents of hybrid cultivars. When the thin-hull female parent was grown under isolation, enough pollen was produced to permit seed production. When the structurally male-sterile plant was grown adjacent to a selected male parent producing normal amounts of pollen, most of the seed produced on the thin-hull parent was F_1 hybrid seed. The attempts failed because the frequency of selfed seed produced by the female parent was usually high enough to depress yield when F_1 plants were grown commercially (Urie and Zimmer, 1970). Because structurally male-sterile types always produced a variable amount of pollen, they have not been considered for use to facilitate hybridization in breeding programs. Heaton and Knowles (1980) reported the occurrence of genetic male sterility governed by the recessive gene *ms*.

The level of natural crossing in domesticated safflower depends on the genotype and insect activity. Reported levels range from 1.4 to 3.4% in Egypt to 8.3 to 100% in Nebraska (Claassen, 1950; Knowles, 1969).

Wild species are often self-incompatible. Self-incompatibility in *C. flavescens* (Imrie and Knowles, 1971) and in *C. oxyacantha* (Ramanamurthy, 1963) has been found to be of the sporophytic type, with alleles at one locus being involved. Where self-incompatibility was expressed, the pollen germinated but did not penetrate the stigmatic surface (Imrie and Knowles, 1971).

Safflower pollen is transferred by insects, not by wind. The most active are domesticated bees, although several other insects visit the flowers. Free (1970) reported the following: bees visit the flowers early in the day, mostly

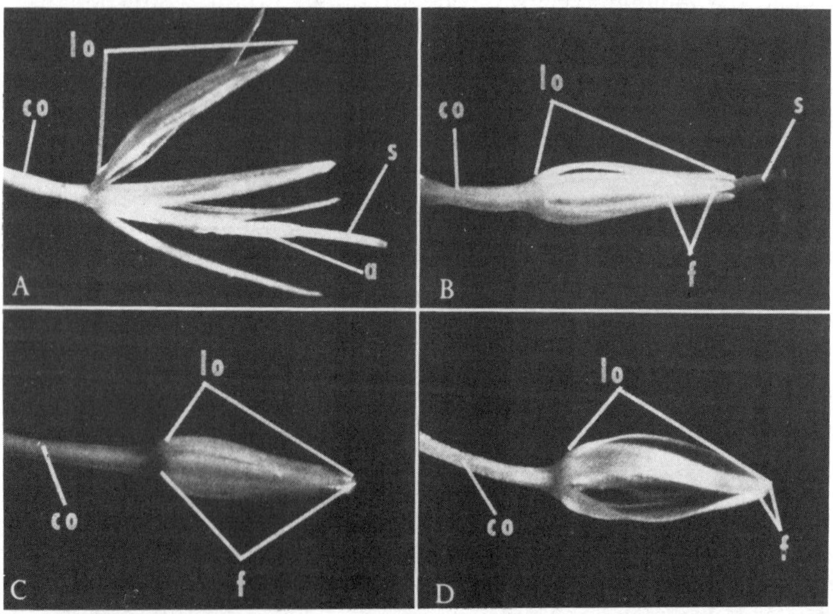

Fig. 4—Corolla lobes of normal and closed flowers of safflower. A) Normal; B) arrowhead, showing fused region extending from the tip to half the corolla lobe length; C) bar, with corolla lobes fused throughout their length; D) cage, with fusion at lobe tips only. 7 ×. Symbols: a = anther, co = corolla tube, f = fused portion of lobes, lo = corolla lobe, and s = stigma. From Dille and Knowles (1975).

for pollen; bee numbers decrease after midday; the proportion of bees collecting nectar increases as the day progresses; bees do not increase seed yields where cultivars are self-fertile and produce abundant pollen; and bees increase yields of genotypes with low levels of pollen production (e.g., types with structural male-sterility). Safflower pollen has a very short life. No experimental evidence has been published showing viability beyond the day pollen was released. Partial viability of pollen on the second day after release in some genotypes has been reported by commercial breeders. The stigma is receptive for about 2 days after its exsertion from the corolla tube.

IV. ARTIFICIAL HYBRIDIZATION AND SELF-POLLINATION

A. Equipment

Emasculation requires a pair of high-quality, sharp-pointed scissors, tweezers with narrow points and flat surfaces, labeling tags, and 0.5 kg paper bags that will allow moisture loss, but exclude insects. Emasculations of some wild species will require a sharp dissecting needle and both a 57% ethyl alcohol solution and water rinse.

B. Preparation of the Female

Emasculation of the female parent generally is done in the afternoon. Heads are selected after the first whorl of flowers appear above the bracts (Fig. 5A). Many, and perhaps all, of these will have rudimentary ovaries. The first heads produced on a plant are preferred because they produce both more flowers and flowers that are easier to manipulate.

The bracts are removed by cutting them with scissors at the base, just at the level of the tops of the ovaries (Fig. 5B). The elongated outer rings of flowers are removed with their ovaries because they have been pollinated (Fig. 5C). The inner rings of flowers have not elongated and have not shed any pollen.

Each individual flower bud is emasculated (Knowles, 1958). With the point of the tweezers, the corolla tube is gently squeezed just below the attachment of the filaments of the anther tube (Fig. 5D and F). The tissues of the corolla tube should be crushed, but not the style. The corolla tube is bent at the point where it was crushed until it breaks. With practice, this can be done without breaking the style. The tip of the closed corolla lobe is grasped with the tweezers, taking care to avoid the tip of the stigma (Fig. 5E). The top of the corolla and anther tube are pulled from the flower, leaving the base of the corolla and the projecting style and stigma (Fig. 5F). This technique completely removes the anthers without loss of pollen. If the style breaks, it is not necessary to remove the style and ovary below the break because the ovary will not develop further.

In some cases, all the unelongated buds of the head are emasculated. More often, 10 to 15 are done around the periphery of the head, and the buds in the center of the head are removed by cutting them with the tweezers. Immature buds should not be pulled off because that will damage

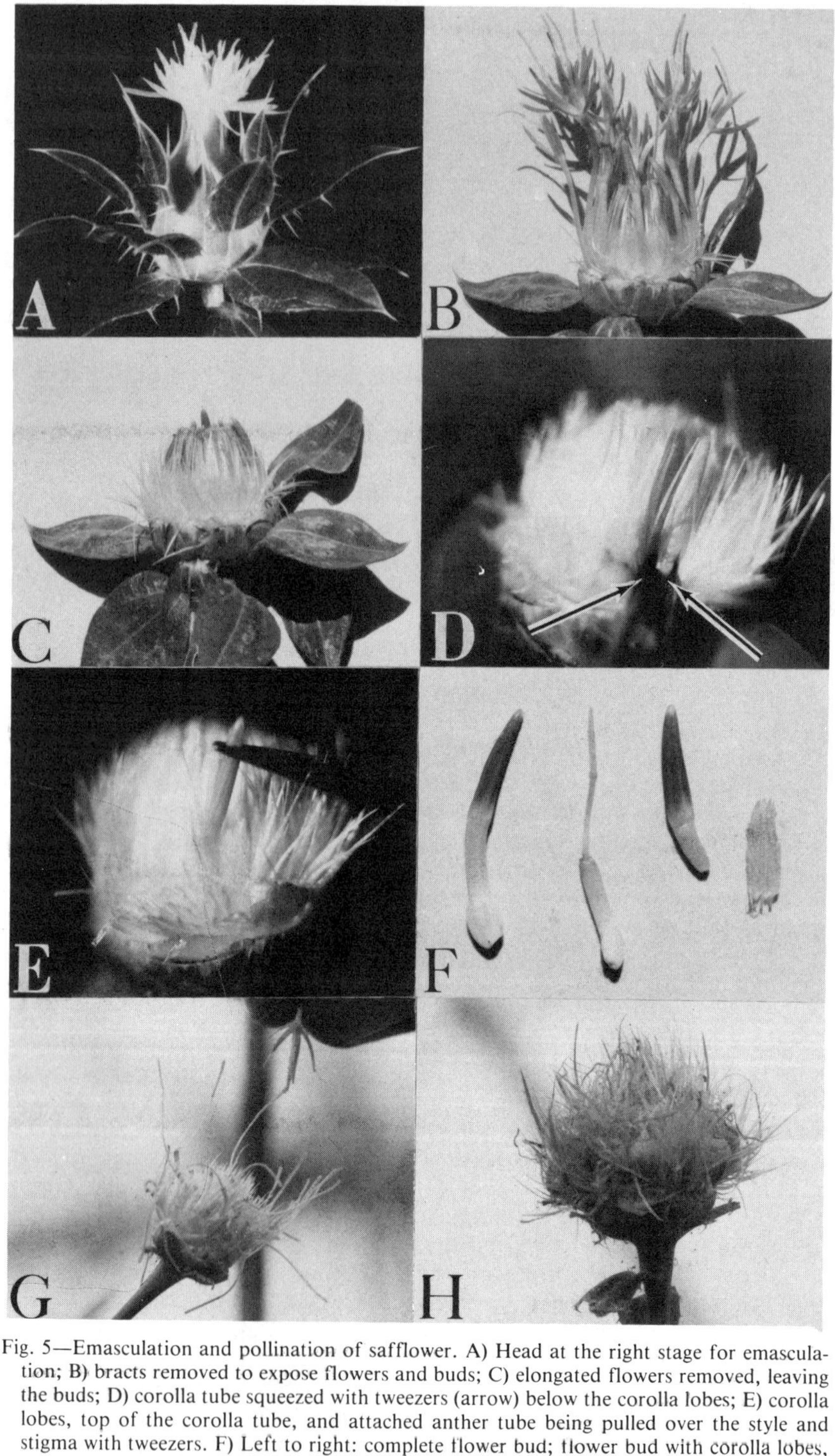

Fig. 5—Emasculation and pollination of safflower. A) Head at the right stage for emasculation; B) bracts removed to expose flowers and buds; C) elongated flowers removed, leaving the buds; D) corolla tube squeezed with tweezers (arrow) below the corolla lobes; E) corolla lobes, top of the corolla tube, and attached anther tube being pulled over the style and stigma with tweezers. F) Left to right: complete flower bud; flower bud with corolla lobes, anthers, and part of corolla tube removed; portion removed; and anther tube slit lengthwise on one side and opened up. G) Pollination using a single flower of the pollen parent; H) crossed seeds about 1 month after pollination.

the receptacle. Heads are covered with a paper bag in the field, but not in a greenhouse free of insects. The head is labeled with the cross number and other appropriate information.

Experience has shown that seed set will be higher if removal of plant parts is kept to a minimum, and if removed parts are cut off, not pulled off. Seed set is reduced even when hairs are pulled off the receptacle.

Before the above method was developed, Claassen (1950) used a technique that is still preferred by some researchers. Sharp-pointed tweezers or a dissecting needle are used to split the corolla and anther tube, and both are removed. After all flower buds are emasculated, the head is soaked for 10 seconds in 57% ethyl alcohol, and rinsed in water to destroy any pollen released when the anther tube is split. All other procedures are the same as described earlier. This technique must be used for wild species having a corolla tube so narrow that it will not pull over the stigma without breaking the style.

The styles will elongate during the night, and will have reached their full length by morning (Fig. 5G). Under field conditions stigmas are receptive early in the morning before pollen is available. They remain receptive at least for that day, and if temperatures are not too high they may be receptive on a second day. When they begin to wither they are no longer functional.

C. Pollination

In both field and greenhouse, heads of the male parent are identified the night before they are used as a source of pollen. All elongated flowers are cut from the head to remove all danger that foreign pollen will be present the next day. In the field, the heads are bagged to exclude insects, and identified. The next morning, all flowers that project above the bracts will serve as a source of pollen, as soon as the pollen is brushed out of the anther tube by the emerging stigma. The first heads on the plant are usually a better source of pollen than heads produced at the end of the flowering period.

Pollinations begin as soon as pollen appears and continue as long as pollen is available, even to the end of the day. To permit an early start, heads of the male parent may be removed the night before, kept at room temperature, and exposed to artificial light at an early hour.

Pollen can be transferred by brushing one or all flowers of a head on the stigmas of the female parent (Fig. 5G). That is usually done the day after emasculation but can be done 2 days after emasculation, or on both days. Pollinated heads are bagged in the field, but not in a greenhouse free of insects.

Self-pollinated seed for genetic studies and breeding programs is obtained in the field by bagging heads to exclude insects. Bagging is not required in a greenhouse where insects are excluded. Selfing bags should be porous enough to prevent moisture accumulation within the bag. Various sizes of bags are used, most researchers preferring small bags enclosing one to three heads, but others choosing larger bags covering 10 or more heads. Heads and branches should not be squeezed together in a bag because the increased humidity will lead to head rot and reduced seed set. Genotypes seem to vary in seed set when bagged.

D. Factors Affecting Efficiency

Success largely reflects researcher skill and favorable environment. Desiccating winds and high temperatures reduce seed set. Ample soil moisture increases seed set. A skilled researcher working in a favorable environment should obtain a seed set of 80% or better.

In the field a well-developed safflower plant will produce heads that flower over a period of 2 to 3 weeks, thus permitting the overlap of flowering of genotypes that differ in the time that they commence flowering. In the greenhouse, where the flowering period may be short because head number is reduced, it is advisable to plant parental materials at two dates spaced about 3 weeks apart.

Developing seeds have swollen ovaries and can be counted 10 days after pollination (Fig. 5H). Because no protective bracts are present, harvest in a greenhouse should begin just as the seeds mature; otherwise, they will fall off.

V. NATURAL HYBRIDIZATION

Natural hybridization is not used in safflower breeding because of the low levels of cross-pollination. When structurally male-sterile types were used as female parents to produce hybrids (Section III), caged bees were used to make crosses. The cages, variable in size and each containing a small hive of bees, were wide enough to include the male-sterile female and the male parent in adjacent rows.

VI. SEED DEVELOPMENT, HARVEST, AND STORAGE

Hill and Knowles (1968) found that the seed weight of safflower increased most rapidly during the first 15 days after flowering, and reached maximum dry weight in about 28 days. Ten to 15 additional days would be required for seed to reach maturity through loss of water.

Usually when safflower crosses are made, achenes that expand fully after pollination will contain a viable seed. However, under environmental stress, such as lack of water during or after flowering, safflower heads will produce some achenes with normally developed hulls, but containing no seeds.

When seeds reach maturity, they will germinate immediately. Seed germination in the head has been a problem in areas where prolonged rains occur at harvest time. The seeds of many wild species remain almost completely dormant for long periods of time after maturity. Such seeds can often be induced to germinate by acid or mechanical scarification.

For best viability, seed should be stored at moisture contents of 6% or less. When seeds are stored at room temperatures, germination levels will remain above 80% for up to 5 years. Thereafter, viability will decrease gradually with time, though precise measurements of the decline in viability have not been made. Some seed samples in storage at room temperatures for 16 years gave low levels of germination in the field at Davis, Calif. Viability is extended by low temperatures, but there are no reports of the duration of viability at different temperatures.

VII. TECHNIQUES FOR SPECIAL SITUATIONS

Emasculation is not necessary when crosses are made with self-incompatible plants of wild species. The self-incompatible plant used as female is pollinated in the same manner as emasculated flowers.

Embryo culture has not been used on safflower but probably would be useful in some interspecific crosses.

REFERENCES

Claassen, C. E. 1950. Natural and controlled crossing in safflower, *Carthamus tinctorius* L. Agron. J. 42:381–384.

Dietz, S. M., A. M. Davis, L. W. Hudson, and G. R. Pesho. 1977. Safflower inventory. USDA Western Regional Plant Introduction Station, Washington State Univ., Pullman, WA 99164.

Dille, J. E., and P. F. Knowles. 1975. Histology and inheritance of the closed flower in *Carthamus tinctorius* (Compositae). Am. J. Bot. 62:209–215.

Ebert, W. W., and P. F. Knowles. 1968. Developmental and anatomical characteristics of thin-hull mutants of *Carthamus tinctorius* (Compositae). Am. J. Bot. 5:421–430.

Estilai, Ali, and P. F. Knowles. 1976. Cytogenetic studies of *Carthamus divaricatus* with eleven pairs of chromosomes and its relationship to other *Carthamus* species (Compositae). Am. J. Bot. 63:771–782.

Free, J. B. 1970. Insect pollination of crops. Academic Press, N.Y.

Ghanavati, N. A., and P. F. Knowles. 1977. Variation among winter-type selections of safflower. Crop Sci. 17:44–46.

Hanelt, P. 1963. Monographische Ubersicht der Gattung *Carthamus* L. (Compositae) (Monograph on the genus *Carthamus* L. (Compositae)). Zeit. Feddes Repert. 67:41–180.

Heaton, T. C., and P. F. Knowles. 1980. Registration of UC-148 and UC-149 male-sterile safflower germplasm.

Hill, A. B., and P. F. Knowles. 1968. Fatty acid composition of the oil of developing seeds of different varieties of safflower. Crop Sci. 8:275–277.

Horowitz, B., and D. F. Beech. 1974. Photoperiodism in safflower (*Carthamus tinctorius* L.). J. Aust. Inst. Agric. Sci. 40:154–155.

Imrie, B. C., and P. F. Knowles. 1971. Genetic studies of self-incompatibility in *Carthamus flavescens* Spreng. Crop Sci. 11:6–9.

Knowles, P. F. 1958. Safflower. Adv. Agron. 10:289–323. Academic Press, N.Y.

————. 1969. Centers of plant diversity and conservation of crop germ plasm. Safflower. Econ. Bot. 23:324–329.

————, and M. D. Miller (eds.). 1965. Safflower. Univ. of Calif. Circ. 532.

————, and S. C. Schank. 1964. Artificial hybrids of *Carthamus nitidus* Boiss. and *C. tinctorius* L. (Compositae). Crop Sci. 4:596–599.

Martin, W. E., R. L. Sailsbery, D. M. Brandon, and R. T. Peterson. 1971. Answering the riddle of poor safflower after rice. Calif. Agric. 25(9):4–6.

Ramanamurthy, G. V. 1963. Relationships of cultivated safflower (*Carthamus tinctorius* L.) to the wild species *C. oxyacantha* M.B. Ph.D. thesis. Univ. California, Davis.

Rubis, D. D. 1967. Genetics of safflower seed characters related to utilization. Safflower Utilization Conference. USDA Publ. 74-93.

Schank, S. C., and P. F. Knowles. 1964. Cytogenetics of hybrids of *Carthamus* species (Compositae) with ten pairs of chromosomes. Am. J. Bot. 51:1093–1102.

Urie, A. L., and D. E. Zimmer. 1970. Yield reduction in safflower hybrids caused by female selfs. Crop Sci. 10:419–422.

Zimmerman, L. H. 1972. Effect of temperature and humidity stress during flowering on safflower (*Carthamus tinctorius* L.). Crop Sci. 12:637–640.

————. 1973. Effect of photoperiod and temperature on rosette habit in safflower. Crop Sci. 13:80–81.

39 Sesame

Demetrios M. Yermanos
University of California
Riverside

Sesame (*Sesamum indicum* L.) is one of the oldest oil crops cultivated by man. Several historical records indicate that sesame probably originated in the general area of Ethiopia, from there it was introduced into India and China, and it became a popular food in southern Europe, North and East Africa and Southern Asia as early as 2000 B.C. Its cultivation gradually spread around the world within a zone extending from 40° N to 40° S Lat. Sesame was introduced to the Americas around the 17th century, and its acreage and use increased rapidly in several Latin American countries including Mexico, Guatemala, Nicaragua, and Venezuela. Strong market demand for sasame in the United States lead to the cultivation of some 5,000 to 7,500 ha in the early 1950's, mostly in Texas. Heavy dependence on manual labor for harvesting, however, caused sesame acreage to decline rapidly and no commercial acreage of sesame has been recorded in the United States during the last 20 years. Major producers of sesame include mainland China, India, Sudan, and Mexico.

Sesame seed contains up to 58% oil. Though high-priced, the oil is in demand in the food industry because of its excellent cooking quality, flavor, and stability. The meal is a superior source of protein, high in methionine but somewhat low in lysine content. Blended in equal proportions with soybean meal, it constitutes an excellent protein source of plant origin and with a well-balanced amino acid profile. Other minor uses of sesame oil are in the preparation of antioxidants, cosmetics, and medicinal formulations.

I. PARENTAL MATERIAL

The genus *Sesamum* has been placed in the family Pedaliaceae which comprises 16 genera with about 60 species found in Africa, Asia, and Australia. The family Pedaliaceae has a superior ovary, usually two-celled, which is completely or partially divided by false septa, each compartment having one to several ovules attached to a central placenta.

The general name of *Sesamum* may have been taken from the Arabic word for sesame, "simsim." Linneaus described the genus in 1754 and suggestions were made by later workers to revise and subdivide the genus into subgenera and species on the basis of leaf and seed morphology (Nayar and Mehra, 1970). Index Kewensis includes 36 species of *Sesamum*; however, the validity of two of these has been questioned (Nayar and Mehra, 1970).

The cytogenetic work on *Sesamum* has been reviewed by Joshi (1961). The chromosome number of cultivated sesame first published as 2n = 26 has been confirmed by us and by several other workers. The chromosome number of 7 of the 36 species listed in the Index Kewensis has been reported as follows: *S. capense, S. malabaricum,* and *S. schenkii,* 2n = 26; *S. angolense, S. laciniatum, S. prostratum,* 2n = 64. The chromosome number of the other genera in the Pedaliaceae has not been determined, except for *Pedalium* and *Ceratotheca* in which 2n = 16.

Attempts to intercross different sesame species have not always been successful. Crosses between species having the same chromosome number were successful within the groups having n = 13 and n = 32. In the n = 16 group, crosses between *S. prostratum* and *S. laciniatum* were fertile. Crosses between these two species and *S. angolense* were not fertile. When crosses between species having unequal numbers were attempted, either no seed was set at all, or the seed set was not fertile (Nayar and Mehra, 1970).

II. PLANT CULTURE

Sesame can be grown readily in the field or in controlled environment. It reacts strongly to environmental stresses; thus, the same genetic materials may look and behave differently when grown under varying environmental conditions.

A. Field

Sesame does best on light or medium-textured soils with good water penetration and drainage. While seed yield and plant development are better at higher levels of soil fertility, reports on sesame response to fertilization lead to contradictory conclusions. Sesame is reported to respond strongly to fertilizer applications in some cases, and in others hardly at all. My experience at Riverside is that nitrogen applications make plants larger and higher in seed protein content, but they do not increase seed yields. Side-dressing nitrogen in excess of 75 kg/ha depresses yield, causes lateness and lodging, and prolongs the time interval within which pods attain maturity.

Sesame is damaged by excessive moisture at all stages of development. It may be planted on preirrigated beds or it may be irrigated after planting. In the absence of summer rainfall, three to four irrigations timed 10 to 15 days apart seem to be adequate for optimum development. Irrigations applied after the plants have completed three-fourths of their life cycle merely delay maturity without increasing yields significantly.

Sesame is planted in rows, preferrably on raised beds. Row spacings of 45 to 75 cm produce the highest seed yields. Where space between rows has to be 1 m or more to make mechanical cultural operations possible, optimum yields are obtained when two rows are planted per bed, 25 to 30 cm apart. Spacing within rows may range from 8 to 40 plants/m without altering yields substantially. In crossing blocks, spacing of 1.5 m between rows is favored to allow freedom of movement.

Germination, emergence, and development of sesame are optimized when soil temperature is over 22 C. Lower temperatures prolong the emergence period and lower the ability of sesame to compete with faster growing weeds. Coupled with high soil moisture, low temperatures promote seedling diseases, which often have devastating effects on sesame stands. These problems become more serious for indehiscent cultivars, which generally have lower seedling vigor.

Sesame performs best when temperature is high throughout the entire growing season. A hard freeze before maturity would kill the plants and reduce seed quality. With indehiscent cultivars, a freeze after maturity is helpful because defoliation and desiccation of the plants makes mechanical harvesting more efficient.

Most sesame cultivars respond strongly to photoperiod. Tropical cultivars grown in California have a prolonged vegetative cycle, become excessively large, enter the reproductive phase late in the fall, and produce few capsules. By contrast, temperature zone cultivars grown in tropical latitudes are dwarfed and mature earlier.

B. Growth Chamber and Greenhouse

Sesame may be grown under controlled conditions in containers ranging in size from 4 × 4 × 15 cm paper sleeves to 5 liter or larger plastic, metal, or clay pots. Plants grown in the greenhouse have longer than normal internodes and weak stems. The plants need to be supported with stakes when they are about 30 cm tall. A soil mix consisting of equal parts of peat moss, sand, and loamy soil is used extensively with good results. Frequency of watering of potted plants depends on size of containers, stage of development, and ambient temperature. With greenhouse temperature in the 22 to 30 C range and liter pots, watering every 2 days is adequate during the first 3 to 4 weeks of development. At full bloom, watering may need to be done as frequently as three times daily. Weekly use of a weak nutrient solution is recommended for watering potted plants in the greenhouse. Generally, pots should be kept relatively dry to avoid the development of pests and diseases. The most frequent pests are aphids and white flies; both can be controlled easily with systemic insecticides.

We use commercially available styrofoam speedling blocks with 4 × 4 × 15 cm perforations, or open-ended paper containers of similar size for growing breeding materials. When planted in these blocks or paper containers, sesame plants grow to about 50 cm in height and produce only a total of two to four capsules which can be used for making crosses. Because of reduced height, plants do not require staking. This system is helpful for growing large segregating populations indoors because about 2,000 plants can be accommodated on a 1 × 10 m greenhouse bench.

III. FLORAL CHARACTERISTICS

Sesame flowers appear in clusters borne on short peduncles at the base of leaf axils. Individual flowers have a small calyx divided into five segments 3 to 7 mm in length (Fig. 1 and 2). The corolla is tubular, with an entire upper lip and a three-lobed lower lip, the middle one of which is larger than the two side lobes. The corolla tube is open at the distal end, generally curved downward, and measures 15 to 35 mm in length. The calyx is usually green, but in rare cases is red-brown or slightly yellowish. The corolla tube is light green and folded closed at the distal end while the flower is developing. When the flower is fully extended, the corolla is usually pale white, but is sometimes rose or of greenish hue and in rare cases red-brown. The corolla tube opens in most cultivars; however, sesame selections are available in which the corolla tube never opens completely, thus blocking the entry of most pollinating insects. Some cultivars have a 'V'-shaped pigmented pit inside the corolla on the larger petal which forms the base of the floral tube.

In cultivars bearing a single flower per axil, each flower is flanked by a nectary on both sides. In cultivars where three flowers per axil may appear, the flowering pattern usually varies within the plant. In the middle half of the plant, one or both of the nectaries flanking the middle flower develop into side flowers and each flower produces a capsule in the leaf axil. In the lower and upper quarter of the plant, only one nectary may develop next to the middle flower, therefore, only one or two capsules per axil may appear.

Fig. 1—Sesame flowers. Center, open; side, closed.

Flowers open from the base to the top of the plant. Where more than one capsule forms per axil, the middle flower opens earlier than the side ones and usually develops into a considerably larger capsule than the side flowers. Instances have been observed where more than three flowers per axil develop, giving rise to as many capsules.

The sesame flower typically has four green-white stamens arranged in pairs, one pair shorter than the other, attached to the upper lip of the corolla tube. In some cases, a fifth stamen is present which is usually nonfunctional. The anthers are white or yellow and 1 to 2 mm in length. In male-sterile types of sesame they remain green, closed and underdeveloped in size throughout anthesis. Pollen grains are light yellow. In the common bicarpellate sesame, the style is cylindrical ending in a stigma with two hairy lobes (Fig. 3). In tetracarpellate sesame, the style is flattened and has a four-lobed stigma (Hanawa, 1953). Morphological variations of the sesame flower have been discussed by several workers (Joshi, 1961).

In field plantings of sesame in California, most flowers open within a 2-hour period in the early morning. On clear, sunny, dry, warm mornings this period lasts from 0500 to 0700 hours. On cloudy, foggy, damp, chilly mornings, this period occurs 1 to 2 hours later in the day.

Before the flower opens, the four anthers are closed and are below the level of the stigma. About 2 hours before flower opening, the filaments start to elongate rapidly and as the higher positioned two anthers reach the level of the stigma, they burst longitudinally and release their pollen at about the time the flower opens. The second pair of anthers then burst in quick succession. At about the same time, the two hairy lobes of the stigma separate and receive large quantities of pollen. Thus, pollination takes place shortly before or shortly after flower opening. The stigma is receptive 24 hours

Fig. 2—Flowers with corolla partly removed to reveal anthers, style, bifid stigma, and ovary.

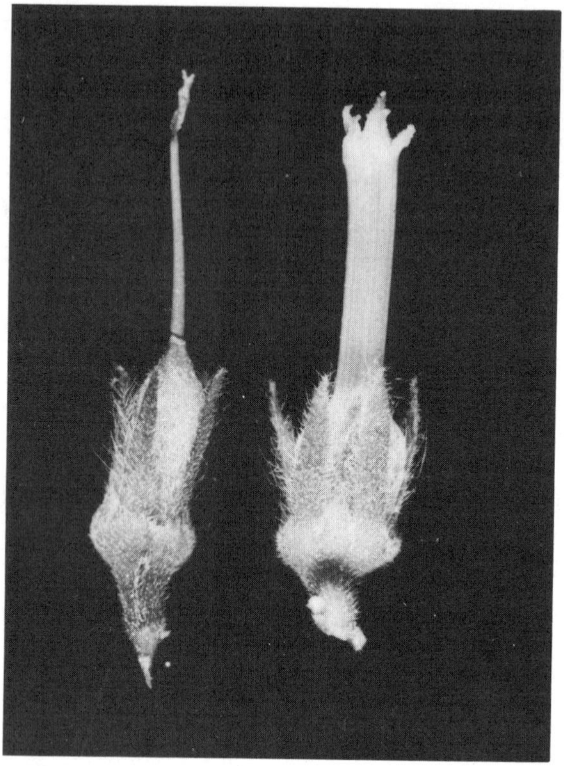

Fig. 3—Style of bicarpellate (left) and tetracarpellate (right) sesame capsules.

Fig. 4—Closed sesame flower with protruding style.

prior to flower opening; therefore, thrips or other insects carrying pollen could contribute to natural cross-pollination if they entered a flower prior to the time its own pollen becomes available for self-pollination. The probability for natural cross-pollination increases on cloudy, cool, damp mornings when pollen release is often delayed until 1 to 2 hours after flowers open.

While the timetable of anthesis is fairly well-defined in field-grown sesame within a given environment, it may undergo major modifications under controlled environmental conditions. In the greenhouse, flowers may open and may be pollinated at almost any time of the day. The sequence of events described above may be upset greatly by different schedules of pot watering and fertilizer application, and by fluctuations in temperature and light level. When light level is reduced, the style of a number of cultivars elongates at a faster rate and protrudes outside the corolla prior to flower opening (Fig. 4). This phenomenon has been observed in both the greenhouse and in the field, especially in irrigated sesame, which may be over 2 m tall with large leaves shading the flowers and maintaining high levels of humidity around them.

Few reports are available on the duration of stigma receptivity. Khidir and El Awad (1972) and Arzumanova (1950) found that the stigma remained receptive for about 24 hours after flowering. Aiyadurai et al. (1963) reported that receptivity was lost in about 14 hours. Arzumanova (1950) reported that pollen kept at 20 to 25 C remains viable for 4 days; however, other reports indicate that viability is lost in about 17 to 24 hours (Khidir and El Awad, 1972).

IV. ARTIFICIAL HYBRIDIZATION AND SELF-POLLINATION

A. Equipment

A pair of tweezers is used to detach anthers and brush them against stigmas for crossing or selfing sesame. In rare cases, a long sharp needle may be useful in splitting partially closed anthers and in straightening bent styles. Soda straws are inserted over crossed flowers to eliminate unwanted cross-pollinations and jeweller's tags are tied to those flowers to identify each cross. Grocery type, 12 × 35 cm brown paper bags secured to the plants by twistems, or staples are used for selfing sesame in the field.

B. Preparation of the Female

In its early stages of development, the sesame flower has a green corolla which is somewhat thick and rigid. As it enlarges and approaches the pollination stage, the corolla color changes progressively from green to white and its walls become thinner and softer. A flower at the proper stage of development for emasculation is shown in Fig. 5. The corolla tube is still closed, but its margins are not zipped together tightly. It may still have a greenish hue towards the base, but it is pale white towards the top. Its

Fig. 5—Sesame flower at proper stage for emasculation.

general appearance suggests that it is ready to unfurl and attain the stretched, open tube configuration within the next 24 hours. Inside the corolla, the anthers are at about the height of the stigma; they are pale white, turgid, with a moist surface, but have not ruptured yet.

To emasculate sesame flowers, breeders routinely remove the entire corolla while it is still greenish and closed, instead of removing each anther individually. In doing so, they also remove the stamens which are attached to the corolla. Because the stigma is receptive at this stage of development, it is advisable to check each corolla that is removed to verify that the anthers inside it are still closed. On rare occasions, some anthers may burst open and release pollen prematurely. This inspection is not time-consuming and removes any suspicion of accidental self-pollinations which could ruin an entire experiment, if they remained unnoticed. The style and stigma are left intact to complete their development until the next morning when the cross is actually made.

A brightly colored 8 × 16 mm jeweller's tag is tied loosely around each emasculated flower. Because plants in greenhouses are usually crowded, these tags help the breeder to identify the emasculated flowers without wasting time. If the female parent has three flowers per axil, the middle flower at each axil is emasculated while the two side flowers are removed with tweezers.

We have found it unnecessary to protect emasculated flowers against accidental pollination in an insect-free greenhouse. In the field, emasculated flowers are protected by inserting a 4 to 6 cm long piece of paper soda straw over the style. The distal end of this straw is either folded or stapled to block the entrance of pollinating insects.

C. Pollination

In field plantings of sesame, flowering generally is determinate, with most flowers opening in the early morning. We perform emasculations in the late afternoon at 1600 to 1800 hours on the day before flowers are expected to open, and pollinations early the next morning at 0700 to 0900 hours. Because flowering in the greenhouse is rather indeterminate, emasculations and pollinations may be made at almost any time that pollen is available and female parent flowers are at the right stage. The stigma is receptive at the time flowers are emasculated; thus, if pollen is available, crosses may be made at the same time as emasculations.

Open flowers of the predetermined male parent are cut off and placed in a plastic cup. Open anthers covered with pollen are pinched from these flowers with tweezers and are lightly brushed against the stigmas of the emasculated flowers. Each crossed female flower is identified by a white 8 × 16 mm jeweller's tag on which the parentage and the date of the cross are recorded. The tweezers are dipped a few times in a vial containing 95% ethanol before being used to make a cross involving different parents. We do not think it necessary to protect the pollinated stigmas in the greenhouse.

In fields having large insect populations, male parent flowers may be collected the afternoon before and the day they are expected to open, and stored overnight in a moist chamber or a plastic bag at room temperature. Abundant quantities of pollen are available from these flowers the following morning for pollinations. The soda straw protecting the emasculated female flower is removed briefly for the pollination and then reinserted over the style and left there until it is pushed off by the developing F_1 capsule. Some breeders find it helpful to secure the soda straw placed on the crossed flower by tying or attaching it with string or a rubber band to the leaf below the flower.

When healthy, insect-free plants and flowers are used, success in cross-pollination among cultivated strains approaches 100%. We avoid making crosses when the light level is drastically reduced, as it is on cloudy winter days, because low light level contributes to excessive flower drop and scant pollen. We avoid making pollinations in the field during the first 2 days after the field is irrigated because irrigation, like cloudy days, contributes to excessive flower abscission.

Self-pollinated seed may be produced by growing plants in a greenhouse or screen cage. In the field, selfed seed may be produced by the same basic techniques as in producing crossed seed, except that each flower is pollinated with its own pollen. The safest, fastest, and most efficient method we have used is to slip a 12 × 35 cm brown paper bag over the top of the plant while it is blooming profusely (Fig. 6, 7). The bag is slipped over the top of the plant, and its open end is folded tightly and stapled or tied with a twistem to keep insects out. Prior to placing the bag, all open and fertilized flowers of the inflorescence are removed by hand or with tweezers. We avoid using plastic or plasticized waterproof bags for selfing because exces-

Fig. 6—Selfing bag on indehiscent sesame.

sive amounts of moisture condense inside the bag, causing severe molding, poor seed set, and other problems. At maturity, bagged parts of the stem may be cut off and harvested individually. Other workers have produced selfed seed by covering each individual flower with a gelatine bag or capsule the day before it opens. The next morning, the corolla is removed from the covered flower and pollen from its anthers is applied to the stigma. Some investigators place a large, insect-proof sack or inverted cage-like structure over the entire plant and tie it at the base of the plant or over the flower-bearing part of the stem (Martinez and Quilantan, 1963-64). Any one of these techniques produces fairly large quantities of selfed seed.

D. Factors Affecting Efficiency

Sesame cultivars differ in the number of days required to flowering. Crossing between early and late flowering types can be accomplished by planting the late flowering types a few days before the early types, so that both types may be in bloom at the same time. Flowering dates may be altered by exposing sesame cultivars to differing photoperiods. In most cases, short photoperiods accelerate the onset of flowering (Rhind and Thein, 1933). Reducing the number of days to flowering is usually accompanied by reduced vegetative growth. Little effort has been devoted to altering the dates of flowering by growth regulator treatment.

Genetic markers are used to identify hybrid seed or plants. There is a strong association between capsule indehiscence and leaf enations, i.e., leaf-like outgrowths. Appearance of leaf enations on the leaves of 2 to 4-week-old seedlings has been used universally for the early identification of indehiscent plants in segregating generations of crosses between dehiscent and indehiscent strains (Fig. 8). It has not yet been resolved whether the two traits are due to pleiotropic gene effects or whether they are controlled by different loci with extremely tight linkage. Recombinant types of dehiscent sesame with leaf enations have not been reported. The lack of such potential recombinants has led many workers to hypothesize that leaf enations result from pleiotropic gene action. Capsule indehiscence and, by extension, presence of leaf enations have been reported as being controlled by a single re-

Fig. 7—F_1 hybrids between tropical and temperate zone strains growing at the University of California plots in Shafter.

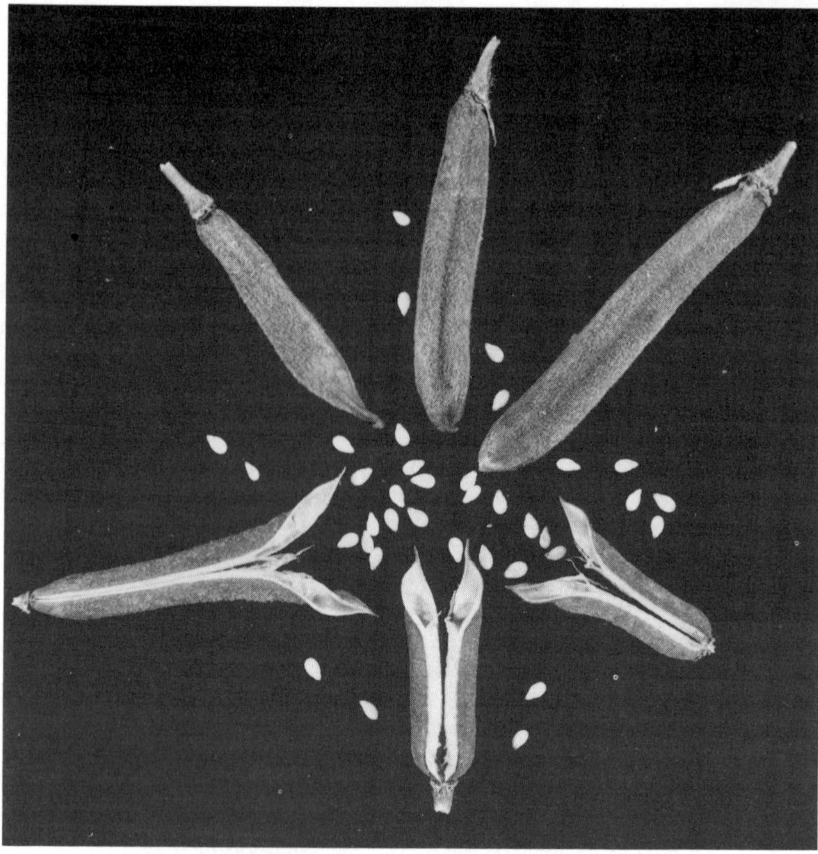

Fig. 8—Dehiscent (lower) and indehiscent (upper) sesame capsules.

cessive gene. Data from segregating populations collected by us do not conform to a 3:1 ratio, but rather to an 8:1 ratio. Thus, the inheritance of capsule indehiscence in sesame may not be as simple as it has been thought to be.

Practically all the cultivated sesame seed has a smooth seed coat which adheres to the seed very strongly. Removal of the seed coat requires soaking of the seed in water, heating, treatments with salt and lye, and mild rubbing action to detach the seed coat from the cotyledons. A dominant single gene gives rise to the rough seed coat, which is easily recognizable by its coarse appearance. In addition, the rough seed coat type of sesame can be recognized by rubbing the seed between the fingers or palms. Due to the mechanical weakness of the cell layers between the seed coat and the cotyledons, the seed coat crumbles and flakes off the cotyledons and the seed is instantly decorticated. To make use of this genetic marker, plants must be allowed to complete their life cycle to maturity.

All cultivated sesame has bicarpellate capsules. A single recessive gene gives rise to tetracarpellate capsules. This is, also, a dependable genetic marker that can be identified soon after the first flowers become pollinated.

Sesame cultivars may have one to three capsules per leaf axil. Cultivars with a single capsule per axil present no difficulty in their identification; however, cultivars with three capsules per leaf axil may sometimes present some ambiguity. Under conditions of stress (e.g., low soil moisture and fertility), the side capsules may not develop at each leaf axil and a three-capsule per axil type may be mistakenly classified as a one-capsule per axil type. While this genetic marker may not be as reliable as others, it has the advantage of being identifiable soon after anthesis.

Various botanical traits that could be used as genetic markers include: purple flower color dominant to purple-white and to white (Pal, 1934); red flower color dominant to white (Khidir, 1973); hairy leaves dominant to glabrous (Langham, 1945); colored seed dominant to white (Nohara, 1933); open corolla dominant to normal (Nohara, 1933); branching dominant to non-branching (Nohara, 1933); one capsule per leaf axil dominant to three (Nohara, 1933); bicarpellate capsule dominant over tetracarpellate capsule (Nohara, 1933); dehiscent capsule dominant to indehiscent, and color patterns of the corolla (Nohara, 1933; Pal, 1934; Sikka and Gupta, 1947; John, 1934; Khidir, 1973).

V. NATURAL HYBRIDIZATION

Cultivated sesame has been described by most workers as a self-pollinated species. However, several reports indicate that under particular environmental conditions, cross pollination levels of up to 65% may be reached. Values of 3 to 15% natural cross pollination were reported by Martinez and Quilantan (1963) in Mexico; 1 to 17% by Ali and Alam (1933) and Sikka and Gupta (1949) in India, and 3 to 6% by Khidir (1973) in the Sudan. My experiments in Shafter, Calif., indicated less than 1% of natural cross pollination when sesame plots were surrounded by large hectarages of cotton and other crops. There was 68% natural cross-pollination in Moreno, Calif., where sesame was the only blooming plant material in the midst of a semiarid area with little other vegetation.

Ali and Alam (1933) name *Apis florea* (the honeybee), *Andrena ilerda, Ceratina sexmaculata,* and *Trichometallae pollinosa* as the common visitors of sesame flowers in Lyallpur, India. I erected clear plastic cages (5 × 10 m base × 4 m height) over rows of cultivars bearing contrasting genetic markers in California. Four hives with honey bees were placed in each of several cages during anthesis to evaluate the effectiveness of domesticated bees in natural sesame cross pollination. The level of cross pollination was less than 2%, and it appeared that this resulted from the inability of the large bodied domesticated bees to enter the sesame flowers deep enough to be effective pollinators. The smaller-bodied wild bees would appear to be better suited for that function.

VI. SEED DEVELOPMENT, HARVEST, AND STORAGE

Once pollen lands on the stigma, the process of fertilization starts rapidly. Pollen grains germinate within 1 to 2 hours (Aiyadurai et al., 1963), within 4 hours pollen tubes enter the ovary cavity, and within 6 to 12 hours the process of fertilization has been completed (Joshi, 1961). An abscission layer forms at the base of the style 22 hours after pollination and 80% of the styles drop naturally 27 hours after pollination (Datta and Sikul, 1967).

Successful pollinations become obvious within 2 to 4 days because fertilized ovaries swell and their styles quickly wilt and break off. Within 4 to 6 weeks after pollination, the capsule and the seed reach their maximum size, but the seed remains white and milky as long as the capsule is green outside, As the capsules progressively turn brown, the seed hardens, dries up, and attains its final coloration. At the same time, the attachment of the seed to the placenta gradually weakens and is obliterated when the capsule is completely desiccated. In the dehiscent strains, the seed drops from the mature capsule if the latter is inverted or vigorously shaken.

To avoid losses of seed, crossed and selfed capsules are removed from the plant at physiologic maturity and are stored in bags or coin envelopes. The bags are placed in a vented oven to dry slowly at 25 to 30 C. Drying temperature may be raised to 40 C without affecting viability adversely. Dehiscent strains may be threshed after about 10 days in the drying oven by merely shaking the bags in which they were stored. Capsules of indehiscent cultivars appear to have much more moisture at physiologic maturity and must remain in the drying oven for at least 4 to 5 weeks; otherwise they are rubbery and difficult to thresh.

REFERENCES

Abe, A. 1919. Preliminary notes of the genetic studies on some characters of *Sesamum indicum* L. Agron. Rep. Formosa No. 153.

Aiyadurai, S. G., M. S. Dorairaj, and C. Murugarajendran. 1963. Anthesis and pollination studies in *Sesamum indicum* L. Madras Agric. J. 50:43–475.

Ali, Mohammad, and Z. Alam. 1933. Types of *Sesamum indicum* D. C. in the Punjub. Indian J. Agric. Sci. 3:897–911.

Arzumanova, A. M. 1950. Biological characteristics of flowering and fruit setting in sesame. [In Russian]. Bull. App. Bot. Gen. Plant Breed. 28(3):81–87.

Datta, P. C., and B. Sikul. 1967. A report on the floral biology of *Sesamum indicum* L. growing under Calcutta conditions. Sci. Cult. 33:531–533.

Hanawa, J. 1953. The embryosac formation and embryogeny of *Sesamum indicum* L. in the diploid and the tetraploid strains. Bot. Mag. Tokyo 66:98–102.

John, C. M. 1934. Inheritance studies in Gingely (*Sesamum indicum* L.). Proc. Assoc. Econ. Biol. Boimbatore, 2:33–34.

Joshi, A. B. 1961. *Sesamum.* Indian Central Oilseeds Committee, Hyderabad, India.

Kashi, Ram. 1930. Studies in Indian oilseeds. 4: The types of *Sesamum indicum* D. C. Mem. Dept. Agric. India (Bot. Ser.) 18(5):127–147.

Khidir, M. O. 1973. Genetic studies in sesame. II. Inheritance of flower colour and number of locules per pod. Exp. Agric. 9:361–364.

———, and S. H. El Awad. 1972. Studies on floral biology in sesame. Sudan Agric. J. 7:17–26.

Langham, D. G. 1945. Genetics of sesame. J. Hered. 36:135–142.

Martinez, H. D., and V. L. Quilantan. 1963–64. Percentage of natural cross pollination of sesame in Iguala. Agric. Tec. Mex. 2:175–177.

Nayar, N. M., and K. L. Mehra. 1970. Sesame: its uses, botany, cytogenetics and origin. Econ. Bot. 24(1):20–31.

Nohara, S. 1933. Genetical studies on *Sesamum indicum* L. J. Coll. Agric. Imp. Univ. Tokyo 12(2):227–386.

Pal, B. P. 1934. Recent progress in plant breeding at Pusa. Agric. Live-Stock. India 4:505–515.

Rhind, D., and U. B. Thein. 1933. The classification of Burmese sesamums (*Sesamum orientale* L.). Indian J. Agric. Sci. 3:478–495.

Sikka, A., and A. Gupta. 1947. Inheritance studies in *Sesamum indicum*. Ind. J. Genet. Plant Breed. 7(1–2):35–42.

————, and A. Gupta. 1949. Pollination studies in *Sesamum orientale* L. Ind. J. Agric. Sci. 9(1):33–41.

40 Sisal and Other Long Fiber Agaves

John F. Osborne
Department of Land Development
Bangkok, Thailand

D. P. Singh
Jute Agricultural Research Institute
West Bengal, India

The major source of hard vegetable fibers is the long, rigid, fleshy, fiber-containing leaves of species of the genus *Agave*. The genus originated in tropical and subtropical America, including the Caribbean countries (Lock, 1969; Wienk, 1976). The fiber of many species of the genus has been used by the indigenous inhabitants of these areas, but only a few species of the section Rigidae in the sub-genus *Euagave* Baker have been used for commercial long fiber production. Two species, *Agave sisalana* (Perrine ex Engelm.) Drummond and Prain (sisal) and *A. fourcroydes* Lemaire (henequen) are the source of 85% or more of the world production of hard, long vegetable fibers. *Agave cantala* Roxburgh (maguey), *A. letonae* F. W. Taylor (Salvador henequen), *A. zapupe* Trelease, *A. angustifolia* Haw. (*A. wightii* Drummond and Prain), *A. lurida* Ait. (*A. veracruz* Miller), and *A. amaniensis* Trelease and Nowell have contributed, or still do contribute, on a small scale to long, hard fiber production.

Cultivation of *Agave* spp. for fiber production occurs mainly in tropical countries. The two principal countries in which sisal is produced are Brazil and Tanzania. Smaller quantities are produced in Kenya, Mozambique, Angola, Malagasy Republic, Haiti, Taiwan, Venezuela, and South Africa. The principal country in which henequen is produced is Mexico, there being smaller outputs in Cuba and El Salvador. Some of the current production of sisal in Tanzania and Kenya is the fiber from an *Agave* hybrid.

The long fibers from *Agave* spp. are used mainly in the production of cordage, of which agricultural twines are the most important products. Some yarns are used in the production of carpets, mattings, and sacks. The shorter fibers are used in the production of padding. Some material has been used for reinforcement of plaster boards, and other possible uses as reinforcing materials are being studied.

I. PARENTAL MATERIAL

The taxonomy of the genus *Agave* is still obscure, and many of the species described may be synonymous. The basic chromosome number of the genus is x = 30, but polyploidy occurs. Somatic chromosomes have been found ranging from 2n = 58 to 2n = 240. The following list records species (with subgenus and section) which have produced viable seed in crosses or from self-pollination. Where known, the ploidy level and characteristics of interest in hybridization are given. Abbreviations used in the text are noted for each species.

A. amaniensis Trelease and Nowell (*Euagave,* Rigidae); 2n = 2x = 60; good leaf configuration and acceptable fiber (am).
A. americana L. (*Euagave,* Americanae); 2n = 2x to 8x = 60 to 240 (ar).
A. angustifolia Haw. (*Euagave,* Rigidae); 2n = 2x = 60; high leaf number; also *A. angustifolia* var. *marginata* (ag).
A. brandegeei Trelease (*Euagave,* Deserticolae) (br).
A. cantala Roxburgh (*Euagave,* Rigidae); 2n = 3x = 90; acceptable fiber (ca).
A. decipiens Baker (*Euagave,* Rigidae); 2n = 4x = 120; resistance to *Phytophthora* (de).
A. fourcroydes Lemaire (*Euagave,* Rigidae); 2n = 5x = ca. 140; good leaf configuration, although it has marginal spines; acceptable fiber (fc).
A. horrida Lemaire (*Littaea,* Pericamptagave) (ho).
A. ingens var. *picta* (Saim.) Berger (*Euagave,* Americanae); 2n = 4x = 120 (in).
A. lespinassei Trelease (*Euagave,* Rigidae); 2n = 2x = 60; resistance to *Phytophthora* (le).
A. potatorum var. *verschaffeltii* (Lemaire) Berger (*Euagave,* Scolymoides) (po).
A. sartorii C. Koch (*Littaea,* Anacamptagave) (sa).
A. sisalana (Perrine ex Engelm.) Drummond and Prain (*Euagave,* Rigidae); 2n = 5x = ca. 138 to 149; good leaf configuration; acceptable fiber (si).
A. zapupe Trelease (*Euagave,* Rigidae); 2n = 4x = ca. 110 (za).
Additionally two Agave species or clones with unpublished names were used in the East African breeding work—*A. fortuna* (fo) and *A. nirvana* (ni).

No record of successful intergeneric crossing has been located. Attempts to cross *Agave* with *Furcraea* (*F. cubensis* Haw., *F. gigantea* Vent., *F. selloa* C. Koch) were not successful (Doughty, 1963, 1965).

It is possible to achieve fertile crossings between many species of *Agave.* Apparent natural hybridization has been reported between *A. victoria-reginae* T. Moore (subgenus *Littaea*) and *A. asperimma* Jacob (sub-genus *Euagave*) (Gentry, 1967).

Table 1—Crossing relationships between commercial and other agave species.

	Male plant												
	Commercial			Other									
Female plant	si	fc	ca	am	ar	ag	br	ho	in	le	po	ni	za
A. sisalana (si)	( ) 5	--	†2,3										
A. fourcroydes (fc)	†14	( ) 5											
A. cantala (ca)	†3												
A. amaniensis (am)	†5,6		†5	( ) 5		†5,8	†5	†5	†5	†5,6	†6	†5	†6
A. americana (ar)				†5									
A. angustifolia (ag)	†5,6		†5	†5		( ) 5				†5			
A. brandegeei (br)				†5									
A. decipiens (de)				†9									
A. horrida (ho)				†5									
A. ingens var. *picta* (in)						†5			( ) 5				
A. lespinassei (le)				†5							†7		
A. potatorum (po)	†6												
A. sartorii (sa)				†5									
A. zapupe (za)	†6												
A. nirvana (ni)					†5					†5		( ) 5	

A † indicates a successful cross; -- indicates unsuccessful cross; ( ) indicates seed from selfing, natural or artificial; no symbol means no report of the combination. Figures refer to references as follows: 1—Vidal (1925); 2—Garrido (1939); 3—Kist and Friedrich (1947); 4—Medina (1951); 5—Doughty (1965); 6—Lerche (1965); 7—Burgwin (1971); 8—Ciaramello (1975); 9—Haki (1975).

The major polyploid species used for production of long agave fibers are not promising for use in hybridization. *Agave sisalana* was introduced into Florida from Yucatan during 1836 and from this base was distributed to many countries. *Agave fourcroydes* is grown in many tropical countries, but is used commercially only in Mexico and Cuba. *Agave cantala* was introduced to the Philippines from Mexico and later to Indonesia.

Table 1 shows crosses that have been attempted. Only the sica crossing has produced promising material from the commercial species (Garrido and Aguedan, 1965). The other crosses producing useful material were amag and agam. The progeny of these crosses have been widely used in East Africa by Doughty, Lock, and other workers in backcrossing and more complex combinations; these appear to retain the fertility of the original combinations. There are also reports of outcrossings of amag material to other species (de, le, po, si, and za). From the work carried out to date it seems clear that *A. amaniensis* and hybrids derived from it are the principal sources of germplasm carrying the characteristics of good leaf configuration and acceptable fiber.

Only one report has been noted of alteration of ploidy by chemical treatment in *Agave*. The use of 0.4% colchicine on seeds of *A. cantala* doubled its chromosome number giving plants with thicker broader leaves (Thombre and Desai, 1960). Most success in *Agave* breeding will be obtained by using species with the same chromosome numbers, e.g., *A. amaniensis* and *A. angustifolia*. Where it is essential to introduce a character from a species with a different chromosome number, a suitable technique will have to be worked out so that backcrossing can be effected as required. It seems probable that the resistance to *Phytophthora* present in *A. decipiens*, a tetraploid, will have to be combined with the high yields of various diploid hybrids. Direct crosses would produce triploid offspring likely to be sterile, and it has been suggested that it would be preferable to double the diploid chromosome complement of the hybrids before crossing so that work could be conducted at the tetraploid level rather than double the triploid offspring to hexaploids (Wienk, 1970).

In East Africa, there are collections of *Agave* spp. and hybrids at the Agricultural Research Institute, Mlingano, Tanzania and at the Horticultural Research Station, Thika, Kenya. In America, there is material at the Instituto Agronomico, Campinas, Brazil, and at the Campo Agricola Experimental de la Zona Henequenesa, Mochocha, Yucatan, Mexico. Many Botanical Gardens throughout the world maintain large selections of *Agave* material.

II. PLANT CULTURE

A. Field

Agave spp. will grow in a wide range of soils, but fertile, well-drained soils should be chosen for breeding work. An adequate supply of calcium is essential.

Although agaves grow naturally under low rainfall regimes, breeding material is best grown in areas with well-distributed rainfalls up to 1,800 mm per annum. Long periods of high humidity are not ideal.

Shade has an adverse effect and 4 or more hours of sunshine per day are preferred. The effects of day length have not been studied. The principal breeding work has been carried out in the tropics. Although much work has been done at low altitudes, some species flower more easily at higher altitudes. Frost causes damage to the plants and temperatures above 10 C provide for more rapid growth.

Four to 5 months after harvest, seed should be sown in shaded nurseries. The soil should be lightly mulched with sisal waste or other organic material to prevent capping of the soil. The young plants should be left in this nursery until they reach a height of 15 cm. This normally takes about 9 to 10 months. In dry periods the soil should be kept moist, but not wet.

Seedlings are planted in open field nurseries when they are 15 cm high using a spacing of 50 cm between the rows and 30 cm between plants in the rows. An extra space in the row is normally left between families and a marker tag inserted in the soil. At the first transplanting, seedlings with excessive spines are discarded and often those lacking in vigor also are eliminated.

Field nurseries should be mulched, if possible. Old sisal waste can be spread over the ground in a layer 2 to 3 cm thick. A black polythene mulch is also effective. If polythene is not used, the nursery must be weeded frequently. Hand weeding with a hoe is adequate, but the use of a herbicide is better. The most effective chemical is bromacil, normally applied at a rate of 3 kg/ha active ingredient. If the area has the sisal weevil (*Scyphophorus interstitialis* Gylh.) as a pest, the plants should be protected by applications of aldrin as a dust or dieldrin as a spray.

When the nursery plants are about 60 cm tall, they are transplanted into field plantations, normally about 3 years from sowing. Although double rows are generally preferred in commercial practice, the material from crosses is best planted in single rows. A reasonable spacing would be 3 m between rows and 1 m between plants in the rows. Such plantings need to be recorded in maps and should be arranged in families. Additional plants whose leaves have spiny margins and slow-growing plants can be discarded at this second transplanting. The plants in the field are allocated numbers (e.g., a plant numbered 77052/16 would be the 16th plant of the family produced by the 52nd crossing during 1977) noted by metal tags on metal sticks alongside the plants.

A 1 m wide strip of soil along the plant rows should be kept free of weeds for 2 years. The strip of vegetation between the rows should be kept short. If the area has the sisal weevil as a pest, aldrin dust should be applied to the planting holes. Follow-up treatments with dieldrin or carbicron sprays will be needed annually because the higher yielding plants provide a more favorable habitat for the weevil. Considerable losses of promising plants were sustained in the East African breeding program before the advent of dieldrin and aldrin (G. W. Lock, personal communication, 1978).

A breeding register must be maintained which, for any cross, would record both parents, the number of flowers pollinated, the number of capsules obtained, the number of seeds obtained, the number of plants in the seedling nursery with a record of their characteristics, the number of seedlings transplanted to the field nursery, and the number of seedlings transplanted to the field plantation. In the field plantations separate records are kept for each plant of the number of leaves unfurled and the length of the last unfurled leaf. At each harvest, the number and weight of leaves cut per plant and the weight of fiber obtained from these leaves is recorded. In the early stages, the time between each cut for high-yielding plants will be about 6 months and leaves at an angle of more than 45° to the vertical are harvested. After three or four cuts, it may be possible to identify promising clones, and suckers can be collected for multiplication. Care needs to be taken to ensure the suckers are at the ends of the rhizomes of the plants selected. A final selection is only possible when each clone flowers. Those plants selected for further study on the basis of high fiber yields per plant, at rates of yield per annum and per 1,000 leaves as good or better than *A. sisalana,* require further study in observation plots under various growing conditions and finally comparative trials.

When a large quantity of planting material of a selected clone is required and no bulbils are available, the rapid propagation method of cutting rhizomes into pieces with at least one lateral bud and planting these in mulched nurseries can be used (May and Diekmahns, 1960). Another method is to place outer leaves, stripped off bulbils or suckers, in water (Medina and Inforzato, 1950; Wienk, 1969). If an axilliary bud accompanies the leaf, it develops into a new plant after the leaf has developed roots.

From the foregoing discussion, it is apparent that one of the main requirements of a vigorous *Agave* breeding program is a large area of land. In 1969, the breeding program at the Sisal Research Station in Tanzania carried out 44 crosses (5,677 pollinations) of which 23 set fruit (837 capsules). From these, 37,813 seeds were obtained from 727 capsules in 21 crosses. Germination that year of 116,719 seeds from 84 earlier crosses gave only 710 seedlings from 26 crosses, of which 472 seedlings were selected for planting in the open nursery.

B. Growth Chamber and Greenhouse

The use of growth chambers or greenhouses in breeding agaves has not been noted in the literature.

III. FLORAL CHARACTERISTICS

The inflorescence of *Agave* species in the section Rigidae is essentially a very large panicle. In both *A. sisalana* and *A. fourcroydes,* the normal size of the inflorescence is 6 to 7 m in height (Fig. 1). Somewhat smaller panicles of 3 to 6 m in height are produced by *A. cantala, A. lurida,* and *A. zapupe*, while *A. angustifolia* has a panicle less than 3 m tall. The shape of the panicle varies from oblong through oblong-pyramidal to elongate-pyramidal.

The erect, hermaphroditic flowers are borne in clusters at the ends of each main branch of the panicle, these having branched trichotomously five or six times to provide the pedicels. The sisal panicle normally has from 25

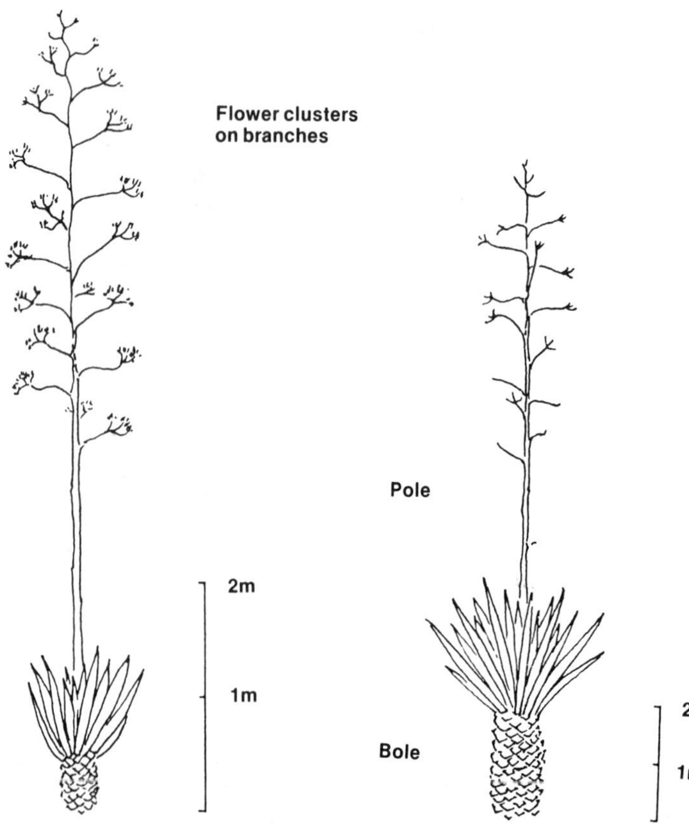

Fig. 1—Inflorescences of *A. sisalana* (left) and *A. fourcroydes* (right).

to 40 main branches, and there are 40 flowers in each cluster.

The epigynous flowers are large, ranging from 2.5 cm in *A. zapupe* to 8.5 cm in *A. lurida*, and are green to greenish-yellow in color. The perianth is composed of six petals of nearly equal length united at the base to form a tube. There are six stamens which, when fully developed, are much longer than the perianth. Each stamen consists of a two-lobed, speckled anther on a long, thin, coppery filament; the filaments being inserted in the base of the perianth. The slender, awl-shaped style bears a three-lobed stigma. The inferior ovary is three-celled, each cell containing two rows of superposed axile ovules which can total over 100 per cell. Flowers of *A. sisalana* from pre- to post-anthesis are illustrated in Fig. 2. After fertilization, the ovary develops into a large green capsule, which turns black on ripening. The fertile seeds are black, thin, triangular-rounded, and papery. Capsules of several species are illustrated in Fig. 3.

The onset of flowering is indicated by the production of a tight rosette of narrow, tapering, sword-like leaves around the growing apex. These leaves rapidly become shorter until the tip of the reproductive shoot (pole) emerges. The pole is fast-growing, with an elongation of 10 to 12 cm per day in the early stages. It has 50 to 60 purplish-brown bracts 10 to 15 cm in length that protect axilliary buds. As the pole approaches its full height, flowering branches develop from these buds in the upper portion of the pole. The flowering branches are about 2 cm thick and oval in cross-section. They protrude horizontally for about 30 cm before branching to provide the flower clusters.

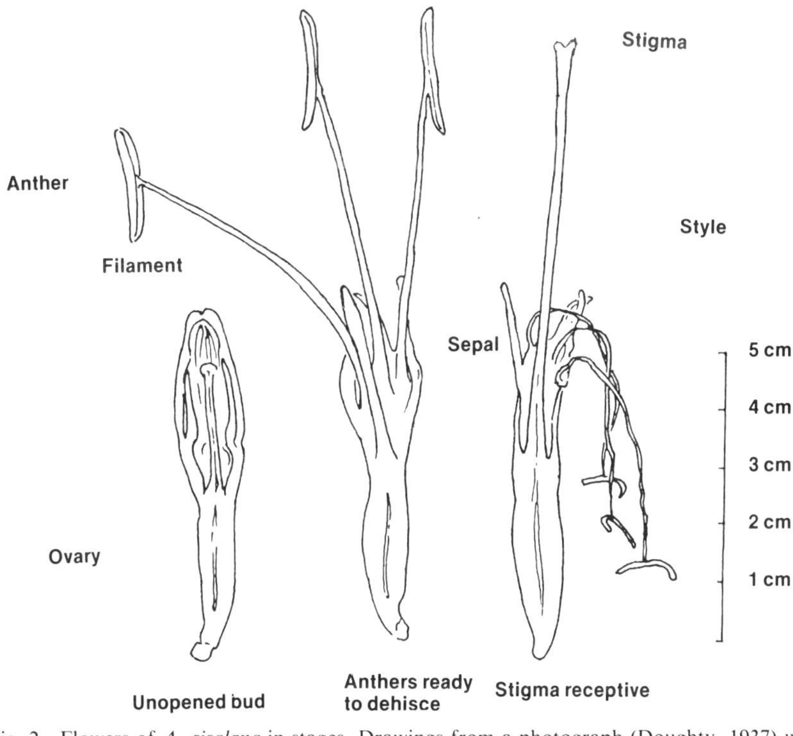

Fig. 2—Flowers of *A. sisalana* in stages. Drawings from a photograph (Doughty, 1937) with grateful acknowledgement to the Editor, *East African Agriculture and Forestry Journal.*

Flowering begins on the lowest branch of the panicle and progresses up the pole. Flowers in all stages of development are found on any one branch. It takes about 1 week for flowering to be completed on each branch, and the whole panicle requires several weeks for complete blossoming.

Pollen from upper branches can fall by gravity to receptive stigmata on lower branches, but it is probable that the bulk of natural pollination is done by insects, especially bees, which are attracted by the large quantities of viscous liquid exuded in the perianth tube during anthesis. Bats also may be pollinators.

The anthers are extruded about 36 to 48 hours before they dehisce (Fig. 2). In East Africa, anther dehiscence occurs early in the evening and by next morning most of the pollen has been shed.

The style begins to elongate once the anthers are fully extruded, but the stamens have withered by the time it reaches its full length (Fig. 2). The stigma becomes receptive (indicated by a sticky exudate) 3 to 4 days after anther extrusion, and remains so for up to 3 to 4 days. The style withers 1 or 2 days after the stigmas has been receptive, whether fertilization has occurred or not. The literature contains no record of the time taken from pollination to ovule fertilization. Capsules take about 6 months to mature.

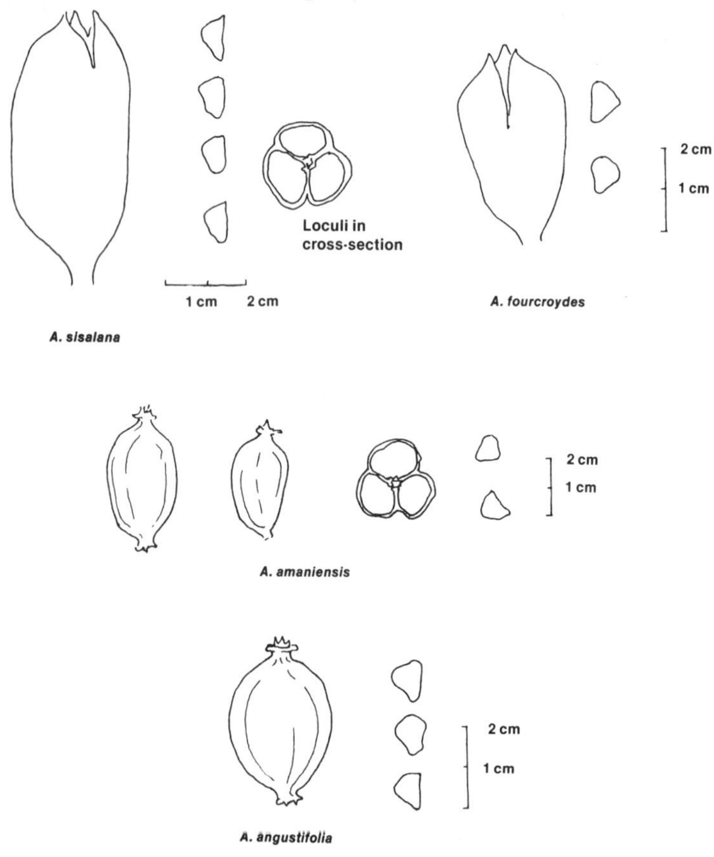

Fig. 3—Capsules and seeds of several *Agave* species.

In some species, fertilized ovaries fail to mature into ripe capsules because an abscission layer develops between the bases of the ovaries and the pedicels shortly after the flowers have withered. This layer fractures easily so that the ovaries and withered flowers fall to the ground.

IV. ARTIFICIAL HYBRIDIZATION OR SELF-POLLINATION

A. Equipment

Pollen collection requires some small flasks or beakers, sheets of shiny black paper, large jars, specimen tubes, and a desiccator. The female parents need forked supports for the flower stalks. Cloth bags are used to protect the selected female flowers, marker tags to label, and camel hair brushes to apply pollen.

B. Preparation of the Female

Although the large size of the inflorescences makes the flowers inaccessible in a normal plant, this drawback can be overcome. Scaffolds have been used (Doughty, 1937; Garrido, 1939; Vidal, 1925), but an easier procedure has been used in East Africa (Lock, 1969). The plants to be used in hybridization are watered thoroughly once the flowers begin to develop, the roots are chopped on one side of each plant close to the bole, and the plants are tilted so that the poles are nearly horizontal.

A branch of unopened buds is selected and covered with a finely-woven cloth bag. Neighboring branches are removed, but several branches can be chosen on one inflorescence. The large flowers are easily emasculated. The six anthers are pulled off soon after extrusion and before dehiscence, which does not take place for 1 or 2 days after extrusion. Because branches on the inflorescence have been removed, there is no danger of their pollen falling on selected female flowers during emasculation or artificial pollination. Each selected cluster of flowers is inspected each morning, and those with anthers extruded are emasculated.

C. Pollination

The branches of flowers to be used as male parents are cut off and put in conical flasks or beakers containing water. These flasks or beakers are placed on shiny black paper and covered with large jars. The flowers steadily open inside the jars and the shed pollen falls onto the black paper. This pollen and any remaining on dehisced anthers can be easily collected in small specimen tubes, which are then stored in desiccators until needed, but not for more than 7 days.

Pollen is applied in the morning to receptive stigmata (indicated by the sticky exudate) with a camel hair brush. After pollination, metal or card tags bearing the cross number are tied to the branches or flowers which are then covered with a cloth bag to avoid contamination from natural pollination.

D. Factors Affecting Efficiency

The long life cycle in agaves of 7 years or more is a drawback. Even with annual plantings of species or clones selected as germplasm, it is not possible to guarantee a regular supply of flowering plants to enable the desired crosses to be made. Although the flowering stimulus can affect suckers on rhizomes still attached to the parent plant, no records have been found of hormones being effective in altering flowering dates.

Self-pollination can be done easily between male and female flowers from the same plant or different plants of the same clone or species, but no record has been found of the use of pollen from one flower to its own stigma. Cross-pollination is determined by the availability of parents of various species. In commercial plantations, open self-pollination within a clone can occur naturally, but seed setting is rare, except in the East African hybrid 11648. Because seeds are not used commercially for crop propagation, control of natural pollination has not been attempted.

Artificial pollination usually results in better fruit set than open pollination, but even so, normally less than half the ovules become fertilized. In 1969, the crosses at the Sisal Research Station, Tanzania, produced capsules containing an average of 52 seeds each. The degree of success will be much less with triploid and pentaploid species, some of which exhibit some pollen sterility. Irregular pairing at meiosis, which could result in pollen sterility, has been reported for *A. sisalana* (Ewusie and Ghatak, 1972).

Those species which develop abscission layers are far from ideal as female parents, but this problem can be partially overcome by decapitating the pole before its branches develop. The branches which develop low on the pole sometimes retain capsules to maturity.

V. NATURAL HYBRIDIZATION

Natural hybridization does not fulfil any useful purpose in a breeding program for long fiber agaves.

VI. SEED DEVELOPMENT, HARVEST, AND STORAGE

Successful fertilization is indicated by swelling of the green ovaries. The capsules turn black as they mature. At 5 months from pollination, each cluster with maturing fruits should be bagged in muslin or mosquito gauze to prevent loss of seed when the capsules dehisce. Harvesting is best done manually, after which the capsules can be fully split open and the black, papery seeds separated from the unfertilized, white ovules. The seeds should be stored for 4 to 5 months before being sown. The optimum conditions for this storage have not been evaluated.

REFERENCES

Burgwin, W. A. 1971. High-level Sisal Research Station, Thika. p. 197–210. *In* Annu. Rep. Dep. Agric. Kenya, 1968. II. Rec. Invest.

Ciaramello, D. 1975. Production de nuevas lineas de Agave. Paper presented at Primer Simposia del Fique, el Sisal y Otras Fibras Duras, Medellin, Colombia.

Doughty, L. R. 1937. Experimental breeding of fibre agaves in East Africa. I. The material and methods of the breeder. E. Afr. Agric. J. 3:222–228.

———. 1963. Sisal breeding. Kenya Sisal Board Bull. 45:28–33.

———. 1965. A review of *Agave* breeding in Tanganyika. Res. Bull. T.S.G.A. Sisal Res. Stn. Mlingano, Ngomeni 39:22–24.

Ewusie, J. Y., and J. Ghatak. 1972. Studies on reproduction and cytology of sisal (*Agave sisalana* Perrine ex Engelm.). Ghana J. Sci. 12:42–50.

Garrido, T. G. 1939. Results of sisal and maguey hybridization. Philipp. J. Agric. 10:233–246.

———, and H. B. Aguedan. 1965. Comparative study of different varieties and strains of *Agave.* Philipp. J. Plant Ind. 30:63–82.

Gentry, H. S. 1967. Putative hybrids in *Agave*. J. Hered. 58:32–36.

Haki, J. M. 1975. Sisal research. Sisal breeding. p. 22–24. *In* Annu. Rep. Agric. Res. Inst. Mlingano, 1974–75, Crop Dev. Div. Min. Agric., Tanzania.

Kist, J. M., and J. C. Friedrich. 1947. De cultuur van enkele bast- en blad-vezeleverende gewassen. Landbouwk. Tijdschr. 59:337–345.

Lerche, K. 1965. Annual report of the officer-in-charge, sisal research, 1963. p. 102–115. *In* Annu. Rep. Dep. Agric., Kenya, 1963. II. Rec. Invest.

Lock, G. W. 1969. Sisal. Thirty years' sisal research in Tanzania. Chap. 2, 12. 2nd ed. Longmans, Green and Co. Ltd., London.

May, W. B., and E. C. Diekmahns. 1960. Rapid propagation of sisal hybrid No. 11648. E. Afr. Agric. For. J. 26:68.

Medina, J. C. 1951. L' experimentation sisaliere au Bresil. La selection du sisal *Agave sisalana* Perrine. Coton Fibr. Trop. 6:133–140.

———, and R. Inforzato. 1950. Propagacao de sisal a partir de folhas de bulbilhos. Bragantia 10:33–34.

Thombre, M. V., and M. C. Desai. 1960. A preliminary note on autopolyploidy in *Agave cantala* L. Sci. Cult. 26:276–277.

Vidal, R. 1925. Breeding work with henequen and sisal. J. Hered. 13:9–12.

Wienk, J. F. 1969. Long fibre Agaves. *Agave sisalana* Perr. and *A. fourcroydes* Lem. p. 1–21. *In* F. P. Ferwerda and F. Wit (eds.) Outlines of perennial crop breeding in the tropics. Misc. Pap. No. 4, Univ. Agric., Wageningen, the Netherlands.

———. 1970. The long fibre agaves and their improvement through breeding in East Africa. Chap. 8. p. 209–230. *In* C. L. A. Leakey (ed.) Crop improvement in East Africa. Tech. Commun. 19, Commonw. Bur. Pl. Breed. Genet. Commonw. Agric. Bur., Farnham Royal, England.

———. 1976. Sisal and relatives. *Agave* (Agavaceae-Agaveae). Chap. 1, p. 1–4. *In* N. W. Simmonds (ed.) Evolution of crop plants. Longmans, London and New York.

41 Sorghum

K. F. Schertz
USDA-SEA and Texas A&M University
College Station

L. G. Dalton
Pioneer Hi-Bred International, Inc.
Plainview, Texas

Sorghum, *Sorghum bicolor* (L.) Moench, is a major food and feed grain crop, and its vegetative parts are used as forage, syrup, and shelter. Of tropical origin, sorghum is grown extensively in marginal rainfall areas of the tropics and subtropics, and has been adapted to and is widely grown in temperate climates.

I. PARENTAL MATERIAL

Cultivated sorghum consists of several morphologically distinct races which can be readily crossed. Although most scientists now consider all of these types as a single species, many older reports treat the types within *S. bicolor* (formerly called *S. vulgare* Pers.) as distinct species, and one must interpret the literature with discretion.

Several grassy types, e.g., *S. arundinaceum, verticilliflorum,* and *aethiopicum,* with the same chromosome number as *S. bicolor* (2N = 20), can be crossed with *S. bicolor* with only slight barriers to gene exchange. These types usually are considered to be species different from *S. bicolor.* Sudangrass, sometimes called *S. sudanense,* is fully compatible with and is usually considered to be a variety of *S. bicolor.*

The subsection Halepense, typified by Johnsongrass, *S. halepense,* contains species which are related to *S. bicolor* and have a chromosome number of 2N = 40, except for *S. propinquum* which has 2N = 20. Other groups (Parasorghum, Stiposorghum, Chaetosorghum, and Heterosorghum) have chromosomes that are morphologically distinct from those of the species previously discussed and have never been successfully crossed with those species. Classifications and species relations are discussed in several reports (Harlan and de Wet, 1972; Celarier, 1958). Sorghum has been crossed with sugarcane (*Saccharum*) (Thomas and Venkatraman, 1930), and it has also been indicated that sorghum and *Zea mays* have been successfully crossed (James, 1976).

The collection, assembly, and classification of sorghum entries into a World Collection by K. Rachie and associates provided the material for expanded germplasm introduction (Murty et al., 1967). A sorghum conversion program was initiated by the U.S. Department of Agriculture and the Texas Agricultural Experiment Station using the assembled lines of the World Collection (Stephens et al., 1967). Many photoperiod-sensitive (short day) lines have been converted to temperate adaptation by this and other similar backcrossing programs and will flower and set seed in long days. As a result, many diverse types are available for breeding programs throughout the world. The World Collection is being maintained at the International Crops Research Institute for Semi-Arid Tropics (ICRISAT) at Hyderabad, India and a large part of it is in long-term storage at the USDA National Seed Storage Laboratory, Fort Collins, Colo.

II. PLANT CULTURE

A. Field

Sorghum grows well in soils of many different types when other environmental conditions are suitable. Recommended fertilization practices depend on soil, climatic, and past cropping variables (Ross and Webster, 1964).

Adequate moisture is needed for maximum production and should always be available at heading and flowering. Sorghum can, however, tolerate rather dry conditions. It has the ability to slow its growth during a dry period and resume development when conditions improve. Cultivars differ considerably in their reaction to moisture stress, and those that would ordinarily flower at the same time may differ in flowering date after an extended dry period. High humidity and early morning dews will delay flowering on specific days, and little or no flowering occurs during a cool, rainy day.

The efficiency of a sorghum hybridization program is enhanced by winter nurseries which can be used effectively for making crosses or growing progeny. Plants, however, are usually smaller than normal, with fewer seed per panicle, and selection for productivity potential may not be reliable. Unusual disease or insect problems may affect unadapted material in a winter nursery; one should avoid eliminating germplasm simply because of its response in that nursery. If the weather is too cool, plant development might be delayed and might cause "blasting" (incomplete development of some spikelets). A distinct advantage of winter nurseries, in addition to the opportunity to obtain another generation, is that photoperiod-sensitive and nonsensitive types will flower at the same time and crosses can be made between them. Such crosses are difficult to make in long summer days because of different flowering responses. To accomplish such crosses in the summer, it is frequently necessary to induce flowering in the photoperiod-sensitive types by excluding light from them for all but 10 hours on 14 consecutive days.

B. Growth Chamber and Greenhouse

Greenhouses are useful for sorghum hybridization. Careful attention to watering, nutrition, and pest control is required. Various potting media may be used, including a light soil or a peat and perlite mixture. Care should be taken not to allow the potting medium to become alkaline or too acidic because micronutrient deficiencies will likely result. Constant temperatures of about 26 C are effective, but day temperatures of 26 to 32 C and night temperatures of 17 to 22 C are especially conducive to good plant growth and development of large panicles. Supplemental light is usually not needed to grow sorghum in greenhouses, but requirements depend on the locality and climate. Although growth chambers have been used for growing sorghum, they are rarely used for hybridization purposes. This is because inconsistent, and frequently discouraging, results have been obtained in attempts to grow sorghum to maturity in growth chambers where light quality and level and temperature requirements are more difficult to satisfy than in greenhouses. Sorghum has been induced to flower in growth chambers at 27 C day/22 C night with about one-half the light level of full sunlight (Eastin, 1976).

III. FLORAL CHARACTERISTICS

Sorghum is mainly self-fertilized with cross-pollination ranging from 2 to 35% and averaging about 6%. Wind and convection currents are the chief agents for pollen movement.

The inflorescence of sorghum is a panicle that varies morphologically from compact to open (Fig. 1). The spikelets are usually in pairs on the branches, one being sessile and fertile and the other being pedicelled and male or sterile (Fig. 2, 3). The terminal sessile spikelet of each branch has two pedicelled spikelets associated with it. The glumes of the fertile sessile spikelet enclose two florets, the upper being perfect and fertile, and the lower being sterile. The fertile floret has a membranous lemma, a palea, two lodicules, three stamens, and an ovary with two long styles with plumose stigmas (Doggett, 1970; Freeman, 1970).

The process of blooming, stigma receptivity, dehiscence of pollen, and fertilization have been reviewed in detail (Stephens and Quinby, 1934; Doggett, 1970). The essential features are that flowering begins 0 to 3 days after panicle emerges from the boot. It begins at or near the panicle apex and proceeds downward for a period of 4 to 7 days. Stigmas are receptive 0 to 2 days before blooming and if unpollinated remain receptive for 5 to 16

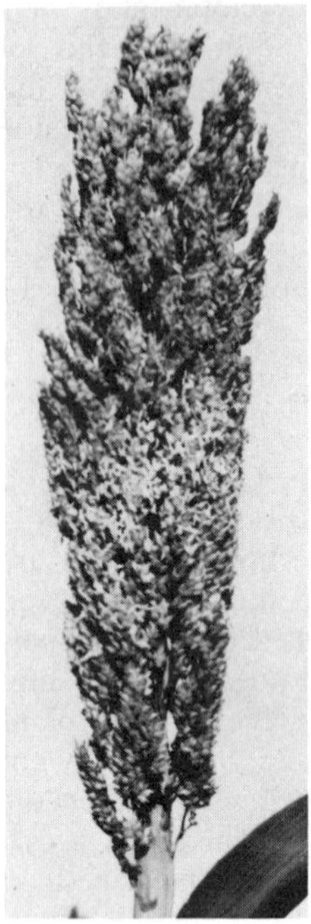

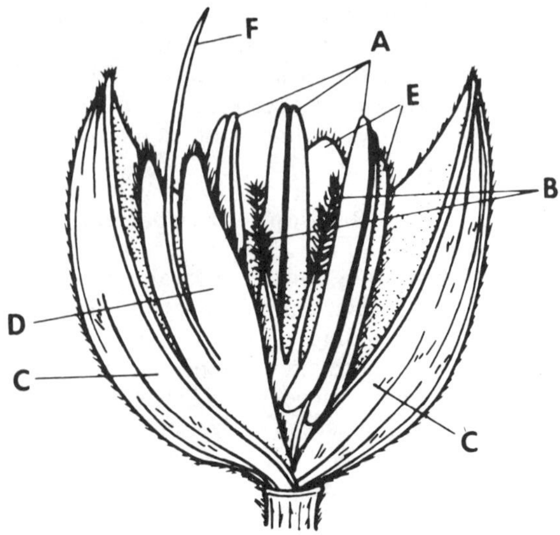

Fig. 1—Panicle of sorghum in mid flowering. *Fig. 2*—Spikelet: (A) Anthers, (B) Stigmas, (C) Glumes, (D) Lemma, (E) Palea, (F) Awn. Adapted from Poehlman (1977).

days after anthesis, depending on the environment. Anthesis generally occurs in the morning, depending on the cultivar and environment, but has been noted as early as 2200 hours (Stephens and Quinby, 1934). Cool, wet conditions delay flowering. Viable pollen is shed from the anthers until about noon of calm days, but pollen may be all dislodged earlier on windy days. Pollen is generally highly functional for about 30 min after dehiscence and decreases to negligible viability at about 4 hours (Stephens and Quinby, 1934). Environmental conditions such as temperature and humidity affect the duration of viability. Pollen kept under refrigeration is capable of fertilization for 3 to 4 days (Ayyangar and Rao, 1936; Sanchez and Smeltzer, 1965). Sorghum pollen is reported to require light and to germinate only after daybreak (Artschwager and McGuire, 1949). Other requirements being satisfied, pollen germinates when in contact with a receptive stigma, and fertilization usually takes place in about 2 hours. Organ differentiation occurs during the following 12 days, and the embryo continues to grow until the seed is mature. Cultivar differences have been found for stigma receptivity, order of stigma and anther emergence, pollen dehiscence, and flowering response to wet, cool conditions.

Sorghum is a short-day species sensitive to changes of day length. It is a warm season crop with a minimum of 7 to 10 C for germination of seed and 21 C for plant growth (Leonard and Martin, 1963). Optimum temperatures

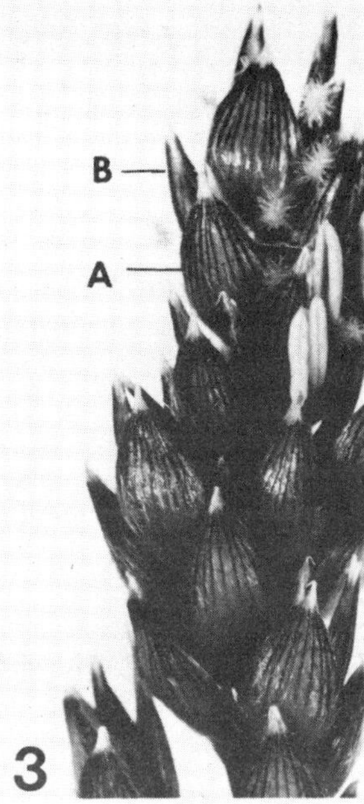

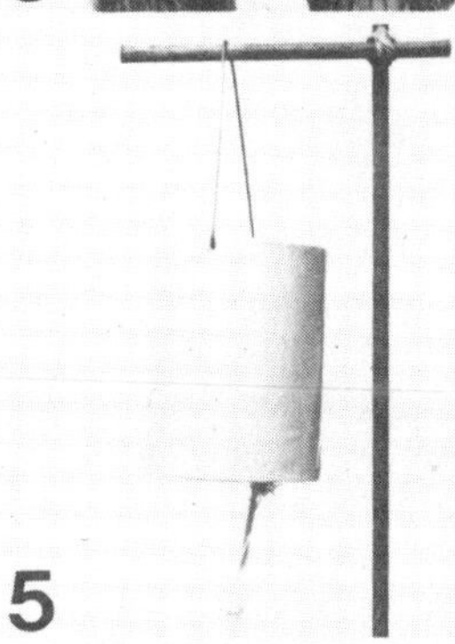

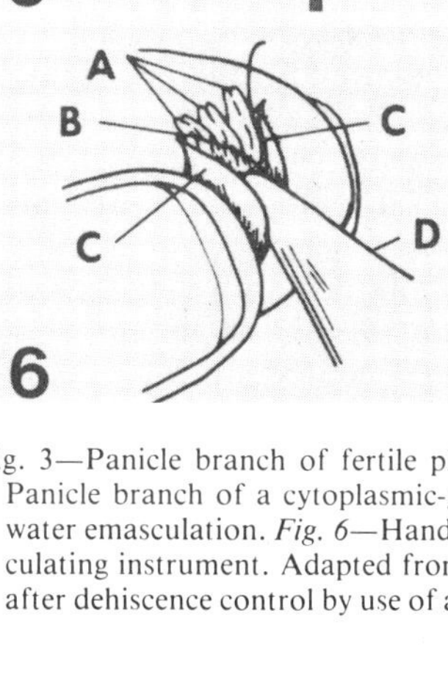

Fig. 3—Panicle branch of fertile plant. A) Sessile spikelet, B) pedicelled spikelet. *Fig. 4*—Panicle branch of a cytoplasmic-genetic male-sterile plant. *Fig. 5*—Plastic sleeve for hot-water emasculation. *Fig. 6*—Hand emasculation. A) Anthers, B) palea, C) glumes, D) emasculating instrument. Adapted from Poehlman (1977). *Fig. 7*—Panicle ready for pollination after dehiscence control by use of a polyethylene bag.

for sorghum flowering seem to be between 21 and 35 C, and flowering is delayed by temperatures outside of this range. When temperatures drop below 12 C at night, flowering is limited the following day, even at optimum temperatures. Warm nights have a tendency to shorten the growing period of cultivars that are late maturing under other conditions. Each cultivar has a characteristic flowering response at specific planting dates and locations.

IV. ARTIFICIAL HYBRIDIZATION AND SELF-POLLINATION

A. Equipment

A diversity of procedures is used to prepare the female parent for crossing, depending on the pollen control method employed. Equipment needed includes paper pollinating bags of approximately 6 × 12 × 40 cm, smaller bags made of paper, parchment, or dialyzing tubing of approximately 5 × 10 cm dimensions, pollinating aprons, stapler, knives, marking pencils, forceps, clips, small scissors, hot water storage container, thermometer, plastic or rubber sleeve, support for sleeve, polyethylene bags, and string. The first four items can be purchased from suppliers of pollinating or scientific equipment, whereas the others are available at local grocery, stationery, or hardware stores.

B. Preparation of the Female

Pollen control in this perfect-flowered species may be obtained by any of four general methods: male sterility, hot-water emasculation, hand emasculation, and control of anther dehiscence. Preparation of the female varies, depending on the pollen control method used.

Male sterility, genetic or cytoplasmic-genetic (Fig. 4), is the preferred method of pollen control because it takes little time to prepare the female parent and an abundance of hybrid seed can be obtained from a single pollination. Male-sterile plants are selected just before anthesis, and the entire panicle is covered with a paper bag which is stapled or clipped at its base. If anthesis has begun, the opened spikelets are removed with a knife or scissors before the panicle is bagged. Panicles of similar maturity are bagged on any one day so that the bagging date written on the bag can be used to determine when the panicle is ready to pollinate. Covering panicles of unpollinated male-sterile plants with distinctly colored bags facilitates distinguishing them from pollinated panicles of the same plot.

Hot-water emasculation is effective when a large number of seed of a specific hybrid are desired, but male-steriles are not available in the intended female and a few selfs would not be objectionable. This method developed by Stephens and Quinby (1933) is employed today with only minor modifications. A panicle that has just begun to flower is chosen and opened spikelets are removed. The panicle is then enclosed in a sleeve of rubber or plastic which is tied tightly around the peduncle and open at the top (Fig. 5). The panicle is covered for 10 min with water at 47 to 48 C. Some cultivars require slightly different maximum temperatures to kill the pollen, but not the ovaries. It is quite simple in the greenhouse to immerse a panicle in hot water by inverting the potted plant. After treatment with hot water, the panicle is covered with a paper pollinating bag, usually after the panicle has dried somewhat. The identity of the emasculated panicle is provided by a note on the bag or by a bag of distinctive color. A check on the effectiveness of the hot water treatment can be determined by covering and not pollinating one branch of the panicle.

Hand emasculation is used when male sterility is either unavailable or undesirable, when a small quantity of seed is needed, and when selfing cannot be tolerated because it would be impossible to distinguish hybrids from selfs. Hand emasculation, however, takes special skill and is time consuming, so it is used only when more rapid methods do not suffice.

Florets are emasculated the day before anthesis. Florets below and within about 3 cm of opened florets usually will open the next day (Fig. 3). Hand emasculations in the field are usually made in the afternoon to avoid contamination by viable pollen from other plants. All open spikelets are removed with scissors, and viable pollen is rinsed from the panicle with water. If there is viable pollen on the hands, scissors, or emasculating tool, they too are rinsed. In most greenhouses, emasculations may be made at any time of the day because air movement is minimal, but rinsing should always be practiced because pollen may remain viable for extended periods in the greenhouse.

Any of three basic instruments are used for hand emasculations. Some workers remove the anthers with a pointed instrument such as a sharpened pencil or a stick of approximately the same dimensions. The point is entered between the outer glumes of the sessile spikelet and the anthers are teased out (Fig. 6). Others use a forceps, usually with flat, broad points, to enter the spikelets and to lightly tease out the anthers. It helps when using either of these instruments to pull one of the glumes partially away from the other to provide room for manipulation of the instrument. Extreme care must be taken to avoid damage to anthers because of selfing which may occur from the released pollen. If an anther is broken or punctured, that spikelet or its ovary is removed and the pollen is removed from the emasculating instrument. The third instrument used is a scissors with which the top one-third of the spikelet and a portion of the anthers are removed (Crook et al., 1972).

Male-fertile, pedicelled spikelets are removed with a forceps or scissors to prevent selfing and to mark emasculated spikelets. Unemasculated spikelets in the region of those emasculated are removed, and a small bag made of paper, parchment, or dialyzing tubing is placed over the emasculated florets, folded, and clipped into place.

Control of anther dehiscence by humidity is used to limit self-fertilization of intended female plants when 1) many seeds are desired, and 2) plants derived from selfing can be distinguished from hybrids. In areas of very high humidity, a paper bag placed over a panicle will delay anther dehiscence about 30 min the next morning, which is sufficient time to pollinate it by the "pour" method (Quinby, 1965).

Polyethylene bags may be similarly used to control dehiscence by high humidity for an extended period each day, so many plants can be pollinated (Schertz and Clark, 1967). Panicles that have just begun to flower are chosen and opened spikelets are removed. A polyethylene bag may be placed over the panicle at any time of day and tightly secured around the peduncle with a string or rubber band. Panicles prepared for crossing are easily identified by the attached polyethylene bag which is left on the panicle for successive days until the desired number of spikelets have opened (Fig. 7). If conditions are hot and calm, however, temperature under the bag may be high enough to injure the flowers. In such cases, one must bag the panicle late in the day, cover the polyethylene bag with a paper bag, pollinate the next day, and remove the unopened spikelets at the base of the panicle. By using polyethylene bags to control anther dehiscence, even an inexperienced person can prepare numerous females in a day.

C. Pollination

Sorghum is self-pollinated simply by placing a paper pollinating bag over the panicle before flowering. Selfs can be made on panicles that have commenced flowering if the flowered portion of the panicle is removed, usually with a knife, before it is covered by a bag. A plant bagged as a pollen source also can be selfed if open-pollinated spikelets are removed and the remainder of the panicle is bagged.

Pollen for crossing should be collected in the morning when anthers dehisce, usually between 0700 and 1200 hours. Any panicle that is flowering can be used as a source of pollen, but pollen is most abundant when taken from the center of the panicle (Fig. 1). Stray pollen on the panicle of the pollen parent usually is not a problem and pollen can be collected from open heads. The male can be covered the day before crossing if one desires absolute protection from pollen contamination or if the wind may agitate the pollen from the anthers of unbagged panicles before all pollinations can be completed. Pollen from bagged heads is shed later and is available for a longer time than is pollen from open heads. To collect pollen, the bag covering the panicle is tilted toward the horizontal and the bag with enclosed panicle is rapped sharply.

On *male-sterile and hot-water emasculated* panicles, crosses are made after the panicles have completed flowering and the receptive stigmas are extruded, usually in 5 to 8 days after bagging. The bag containing pollen is inverted over the panicle of the female and shaken to promote movement of the pollen onto the stigmas. Alternative methods are to remove the male panicle or a portion of it and brush it on the stigmas of the female or to bag together the panicles of the male and female plants. The panicle of the male plant may be placed in a vial of water attached to the peduncle of the female.

If all of the *hand emasculated* spikelets on a panicle are intended for the same cross, the previously discussed methods of pollen transfer, or modifications of them are used. Frequently, a single female plant may be crossed by several males or pollen from a single male may be needed for crossing several females. To pollinate under these circumstances, more care is needed in pollen transfer, and an artist's brush is useful for transferring the pollen to the intended stigmas from the bag in which it was collected.

After pollination is completed, the panicles or the individually pollinated panicle branches are covered and the bag is secured with a staple or clip. The identification of the cross (usually a pedigree or plot number) is written on the bag or a tag attached to the bag or peduncle. Stigmas of flowers properly pollinated will lose their freshness in a few hours.

When *anther dehiscence is controlled*, the polyethylene bag is removed in 1 to 5 days, whenever the desired number of spikelets has opened. The panicle is rapped to remove excess moisture and anthers extruding from the florets (Fig. 7). The panicle is pollinated as described for male-steriles. If it is not desired to pollinate an entire panicle, the unopened spikelets at the base of the panicle should be removed, unless they were removed at the time of female preparation.

The success of pollination varies with personal skill, environment, age of stigmas, abundance of pollen, interference from insects, and amount of injury sustained by the flower parts during emasculation and pollination. Optimally, 80 to 90% of the spikelets will set seed. Pollinations of hand emasculations are consistently good in the greenhouse, but are variable in the field. Also seed set is sometimes low because of heat damage to panicles emasculated by hot water and those covered with polyethylene bags. Excellent success is ordinarily obtained when crossing onto genetic or cytoplasmic-genetic male-steriles.

D. Factors Affecting Efficiency

Field nursery arrangements and procedures vary with the types of material to be handled, the tasks to be performed, and the nature of the pollinating crew. Plots are single to multiple rows about 5.5 to 7.6 m long, but may be much longer for certain materials, such as segregating populations. Parallel and adjacent plots constitute a range or tier which is separated from adjacent ranges by alleys. Rows are from 10 to 100 cm apart and plants within rows are from 5 to 30 cm apart, with about 10 cm spacing being common where hand crossing is intended. For some purposes, the plant density within rows should be that used by farmers in production fields. The nursery usually is overplanted and hand-thinned to the desired plant population.

Materials are grouped in blocks within the nursery according to the study and the intended pollinating procedure, e.g., selfing, crossing, population breeding, observations, forage breeding, and tropical conversion. This arrangement permits efficient management of labor because it allows simplified work assignments to be made within clearly identifiable boundaries. A field map of the nursery is useful for orientation. Plot numbers are written on stakes or tags to be placed at the front of each plot, or at least every 10th plot. Different colored tags help identify special materials, such as steriles. Field books contain corresponding plot numbers, pedigrees, and special notes. Notes are recorded in the field book identifying all operations performed on the plot.

Ranges of 100 or fewer plots are most convenient and lines to be crossed are planted near each other, adjacent if possible, to reduce walking distances. Frequently, male-sterile (A) lines are paired with maintainer (B) or restorer (R) lines with which they are to be crossed. Sometimes specific females or males are planted in rows extending along one side of the block, thus placing them within easy reach from anywhere within the block. It is advisable to put restorer and nonrestorer lines in separate blocks when pollen is to be collected from open-pollinated heads, or in population work in which natural pollination occurs.

Lines to be crossed with several other lines are best planted two or three times at 7 to 10-day intervals to assure that flowering dates are matched. Flowering dates of plants within a plot can be varied by cutting off the panicles of some plants before they flower, thus promoting the formation of tiller panicles to flower at a later date.

At the time of flowering, different-colored pollinating bags can be effectively used to identify panicles handled differently, e.g. plain bags for selfing, red striped for steriles, black striped for crosses. Bagging or pollinating dates are written or stamped on the pollinating bag. Some workers put the cross identification on the pollinating bag, while others prefer to attach to the bag or the peduncle a tag with such information. The nursery is harvested by hand according to a plan prepared by the breeder.

Knowledge of genetic characters permits distinguishing hybrids from selfs in a progeny. Some of the more useful genes as dominant markers are *A, B, D, Dw*, Ma_1, *P, Pc, R, Rs, Wx,* and *Y* (Schertz and Stephens, 1966). Of the seedling characters, red seedling stems (*Rs*) is usually reliable. Normal endosperm (*Wx*) has a unique advantage as a marker when a waxy endosperm (*wx*) line can be used as the female because the hybrid seed can be identified. Of the dominant mature plant characters, dry midrib (*D*), plant color (*P*), seed color (red or yellow, *Y, R*) and tallness (*Dw*) are especially useful. Even without the intentional use of a specific dominant genetic marker, it is not too difficult to distinguish hybrids from selfs. If general characteristics rather than specific markers are used, the best times to make observations are at flowering and grain maturity. Sorghum genetic information, including descriptions and linkages are provided in several reviews (Ayyangar, 1942; Quinby and Martin, 1954; Doggett, 1970).

V. NATURAL HYBRIDIZATION

Natural hybridization is used effectively in sorghum. Genetic steriles are used for natural hybridization to promote crossing in random mating populations (Eberhart, 1970). Genes for male sterility, such as ms_3 (Webster, 1965) are introduced into a population, and hybridization and recombination are continued by selection of male-sterile plants in subsequent segregating generations.

Cytoplasmic-genetic male-steriles can be used effectively to increase hybrid seed for testing. A normal procedure is to plant a fertility-restoring pollinator in an isolation block with several male-sterile females. The wind disperses the pollen under natural conditions and hybrid seed is harvested from each female. This method is an inexpensive and efficient way in which to produce seed for evaluation of a breeder's selections. Information on the use of natural hybridization for production of commercial seed of hybrid sorghum is provided in Chapter 8.

VI. SEED DEVELOPMENT, HARVEST, AND STORAGE

Seed development can be observed by 7 days after successful pollination and fertilization by removal of opaque bags or viewing through those bags made of transparent material. Physiological maturity, as indicated by the formation of a black layer at the base of the seed, occurs from 35 to 45 days after pollination. Seed is sufficiently mature for hand harvesting at about 30 days after pollination. Successive rains or heavy dews can rapidly deteriorate seed; therefore, under such conditions it is best to harvest seed produced by hand crossing about 30 days after pollination. For harvest, the peduncle is cut below the panicle. One procedure for transporting and drying the panicles is to place a tag identifying the plot or plant on a wire, and to thread either the peduncles or the bags covering the panicles onto the wire. The wire can be hung in a shelter to dry the panicles. Another procedure is to place the harvested panicles in cloth bags and place them in a drying chamber.

Seed will usually be dry enough to thresh in a few days. Seed produced from pollinations of hand-emasculated florets are usually few and may be rubbed out on a corrugated board. Seed from other crosses are usually more numerous, and a small head thresher is the proper choice. Care must be taken to clean the machine between threshing of seed with different parentages. Seed from an individual cross is placed in an envelope on which the identity is written, and the identifying tag used at the time of crossing is inserted into the envelope.

Sorghum seed may be damaged by grain insects. Seed should be fumigated if it was produced in the warmer climates and is already infested with insects when harvested. Stored seed can be protected by placing napthalene crystals in the envelopes or in the box containing seed-bearing envelopes. Storage in metal boxes or in sealed jars adequately protects seed from rodents. Environment of storage is also important. A temperature of 10 C and relative humidity of 50% or lower will increase the storage life at most locations and are absolutely essential in areas of high temperature and high humidity. In drier areas, temperature control alone is sometimes reasonably effective. In some of the driest areas, sorghum seed has been stored for many years with neither temperature nor humidity control.

REFERENCES

Artschwager, E., and R. C. McGuire. 1949. Cytology of reproduction in *Sorghum vulgare*. J. Agric. Res. 78:659–673.

Ayyangar, G. N. R. 1942. The description of crop-plant characters and their ranges of variation. IV. The variability of Indian sorghum (*Jowar*). Indian J. Agric. Sci. 12:529–563.

————, and V. P. Rao. 1936. Studies in sorghum: The great millet. III. Anther, pollen, and stigma. Indian J. Agric. Sci. 6:1299–1322.

Celarier, R. P. 1958. Cytotaxonomy of the Andropogoneae. III. Subtribe Sorgheae, genus *Sorghum.* Cytologia 23:395–418.

Crook, W. J., A. J. Casady, and L. G. Campbell. 1972. Scissor emasculation of sorghum. Crop Sci. 12:709–710.

Doggett, H. 1970. Sorghum. Longmans, Green and Co. Ltd. London.

Eastin, J. D. 1976. Temperature influence on sorghum yield. p. 19–23. *In* 31st Corn and Sorghum Res. Conf. Am. Seed Trade Assoc.

Eberhart, S. A. 1970. Techniques and methods for more efficient population improvement in sorghum. p. 197–213. *In* N. G. P. Rao and L. R. House (eds.) Sorghum in seventies. Oxford and IBH Publishing Co., New Delhi, India.

Freeman, J. 1970. Development and structure of the sorghum plant and its fruit. p. 28–72. *In* J. S. Wall and W. M. Ross (eds.) Sorghum production and utilization. AVI Publishing Co., Inc., Westport, Conn.

Harlan, J. R., and J. M. J. de Wet. 1972. A simplified classification of cultivated sorghum. Crop Sci. 12:172–176.

James, J. 1976. Personal communication. International Maize and Wheat Improvement Center, Londres 40, Mexico 6, D.F., Apdo. Postal 6-641.

Leonard, W. H., and J. H. Martin. 1963. Cereal crops. Macmillan Co., N.Y.

Murty, B. R., V. Arunachalam, and M. B. L. Saxena. 1967. Classification and catalogue of a world collection of *Sorghum.* Indian J. Genet. Plant Breed. 27:1–312. (Special number.)

Poehlman, J. M. 1977. Breeding field crops. AVI Publishing Co., Inc., Westport, Conn.

Quinby, J. R. 1965. Personal communication. Pioneer Hi-Bred International, Inc., Dep. of Sorghum Breeding, Box 1506, Plainview, TX 79072.

———, and J. H. Martin. 1954. Sorghum improvement. Adv. Agron. 6:305–359.

Ross, W. M., and O. J. Webster. 1964. Culture and utilization of grain sorghum. USDA, Agric. Inform. Bull. 218.

Sanchez, R. L. and D. G. Smeltzer. 1965. Sorghum pollen viability. Crop Sci. 5: 111–113.

Schertz, K. F., and L. E. Clark. 1967. Controlling dehiscence with plastic bags for hand crosses in sorghum. Crop Sci. 7:540–542.

———, and J. C. Stephens. 1966. Compilation of gene symbols, recommended revision and summary of linkages for inherited characters of *Sorghum vulgare* Pers. Tech. Monograph 3. Texas Agric. Exp. Stn., College Station.

Stephens, J. C., F. R. Miller, and D. T. Rosenow. 1967. Conversion of alien sorghums to early combine genotypes. Crop Sci. 7:396.

———, and J. R. Quinby. 1933. Bulk emasculation of sorghum flowers. J. Am. Soc. Agron. 25:233–234.

———, and ———. 1934. Anthesis, pollination, and fertilization in sorghum. J. Agric. Res. 49:123–136.

Thomas, R., and T. S. Venkatraman. 1930. Sugarcane-sorghum hybrids. Agric. J. India 25:164.

Webster, O. J. 1965. Genetic studies in *Sorghum vulgare* (Pers.). Crop Sci. 5:207–210.

42 Soybean

Walter R. Fehr
Iowa State University
Ames

The soybean [*Glycine max* (L.) Merrill] is an annual legume that originated in China and has been cultivated for centuries. The crop is grown throughout much of the world with the largest production in the United States, Brazil, People's Republic of China, Mexico, Indonesia, and Argentina. Soybean seed contains about 40% protein and 20% oil and is used for human consumption, livestock feed, and industrial purposes.

I. PARENTAL MATERIAL

Cultivated soybean belongs to the family *Leguminosae*, subfamily *Papilionoideae,* and genus *Glycine.* The genus presently includes six perennial species in the subgenus *Glycine* Willd. and two annual species in the subgenus *Soja* (Moench) F. J. Herm. The annual species, consisting of the cultivated type *max* and the wild type *soja* Sieb. and Zucc., have a chromosome number of $2x=40$ and can be readily intercrossed. The F_1 hybrids from some *G. max* × *G. soja* crosses are semisterile due to chromosome translocations (Hadley and Hymowitz, 1973). No successful crosses have been reported between *G. max* and the six species of the subgenus *Glycine* (Hadley and Hymowitz, 1973).

More than 5,000 strains of *G. max* and a smaller number of strains of the other *Glycine* species are maintained by the U.S. Department of Agriculture at Urbana, Ill., and Stoneville, Miss. Germplasm collections also are maintained by the Applied Genetics Laboratory of the Korea Atomic Energy Research Institute in Seoul, Korea, the National Institute of Agricultural Sciences at Hiratsuka, Japan, and the Asian Vegetable Research and Development Center in Taiwan.

II. PLANT CULTURE

A. Field

Soybeans are adapted to a wide range of soil types, but grow best in well-drained soils with a pH of 6.5 to 7.0. Mineral nutrients required to produce 2,600 kg/ha of seed include 164 kg/ha of N, 16 kg/ha of P, and 45 kg/ha of K (deMooy et al., 1973). Soybeans can fix symbiotically much of their own N if adequate numbers of rhizobia [*Rhizobium japonicum* (Kirchner) Buchanan] are applied at planting or are present in the soil. Soybeans respond equally well to direct application of P and K or residual from the previous crop. Secondary elements and micronutrients are adequate in most areas, but deficiencies can occur on light textured, acid, or calcareous soils.

With adequate mineral nutrition, vegetative and reproductive development usually are proportional to the available moisture supply (Howell, 1963). Moisture stress, one of the principal reasons for flower and seed abortion, is associated with inadequate soil moisture and high evapotranspiration (Shibles et al., 1975). Plants may wilt when evapotranspiration is high, even if the soil moisture supply is adequate. The interrelationship between evapotranspiration and flower and seed abortion may explain the common observation of soybean breeders that successful hybridization may be difficult on sunny, windy days when the temperature exceeds 35 C and relative humidity is low. Irrigation can be valuable for successful hybridization and for high yields of self-pollinated seed.

The soybean is a short-day plant, but there is considerable genetic variation for sensitivity to photoperiod (Hamner, 1969; Criswell and Hume, 1972). The critical day length for flowering ranges from about 13 hours for genotypes adapted to tropical latitudes to 24 hours for photoperiod-insensitive genotypes grown at higher latitudes (Shibles et al., 1975). Soybeans seem to be insensitive to day length for 9 days after emergence. Photoperiods shorter than the critical day length are required for 7 to 26 days to complete flower induction (Borthwick and Parker, 1938; Shanmugasundaram and Tsou, 1978).

Sensitivity to day length is an important consideration when genotypes are grown outside of their area of adaptation. When genotypes adapted to tropical latitudes are grown in the field at higher latitudes, they may not mature before frost occurs. They can be induced to flower and mature earlier by creating artificially short days or by grafting (Section IV D). Soybeans frequently are grown in winter nurseries located at sea level in tropical latitudes where day lengths are much shorter than their critical photoperiod. The short day lengths and warm temperatures encourage early flowering and seed maturation, and genotypes can produce a seed crop in 90 days or fewer after planting. Early flowering is useful for generation advance when only a few self-pollinated seeds per plant are needed, but not for artificial hybridization because the flowers self-pollinate before they are large enough to manipulate for hybridization. Artificial lighting is used to extend the natural day length to about 14.5 hours to obtain flowers suitable for hybridization and to increase yields of self-pollinated seed.

The effect of short photoperiod on flowering and seed yield can be partly offset by altitude, probably due to the effects of cool temperature (Major et al., 1975). At tropical latitudes, cultivars adapted to the northern U.S. perform more like those adapted to the southern U.S. at high altitudes than they do at sea level.

The light level required to delay flowering is dependent on the quality of light emitted from the source and the genotype being grown. Blue light with a wavelength of about 480 nm requires more than 30 times the energy to inhibit flowering as red light with a wavelength of about 640 nm (Parker et al., 1946). Flowering of genotypes that possess the e_3 gene can be delayed by incandescent light, but not by cool-white fluorescent light (Buzzell, 1971). The level of light required to control flowering was estimated as 5.4 lux by Borthwick and Parker (1938) with the cultivar 'Biloxi' adapted to the southern U.S. My experience at Isabela, Puerto Rico, using quartz-iodine lamps to extend the day length is that genotypes adapted to the northern U.S. frequently self-pollinate before they are large enough to manipulate for hybridization if the light level is less than 32 lux. However, a level of 5.4 lux in Puerto Rico will prolong vegetative growth so that plants are taller and produce more seed than those grown without an extended day length.

Temperature can also play a significant role in the flowering and development of soybean (Major et al., 1975). It can influence the time of flowering and suitability of flowers for hybridization. Temperatures below 21 C or above 32 C can reduce floral initiation or seed set (Hamner, 1969; van Schaik and Probst, 1958). Artificial hybridization is most successful between 26 and 32 C because cooler temperatures reduce pollen shed and result in flowers that self-pollinate before they are large enough to manipulate. Warmer temperatures frequently are associated with increased flower abortion caused by moisture stress; however, successful crosses are possible at about 35 C if soil moisture is adequate (Hartwig, 1978).

B. Growth Chamber and Greenhouse

Soybeans can be grown in an assortment of media ranging from a mixture of 1 soil:1 sand:1 compost to hydroponic solutions. Mineral nutrients can be provided as fertilizer in soil media or by watering with a solution containing 150 ppm N, 20 ppm P, and 60 ppm K (Summerfield, 1976). Seeds commonly are inoculated with rhizobia to encourage N fixation.

Plants for hybridization or for production of a large amount of self-pollinated seed commonly are grown in 25-cm diam pots. The genotype may be overplanted and thinned to two plants per pot, or seedlings may be transferred from a germination medium. When only a few self-pollinated seeds per plant are needed, a density of nine plants or more is common in a 15-cm pot and growth is controlled by a short photoperiod and an anti-gibberellin, such as Amo-1618 (Wilcox, 1974).

Artificial lighting in growth chambers and in winter greenhouses is needed for adequate photosynthesis and control of flowering. Adequate photosynthesis and control of flowering occurs with a light level of about 21 klux provided by a mixture of cool-white fluorescent and incandescent lights or wide-spectrum fluorescent lights (Tanner and Hume, 1976). For hybridization, a photoperiod of adequate length is required. A 12-hour photoperiod is used for early flowering and seed production (Wilcox, 1974). Temperature should be maintained at not less than 21 C, with 26 to 32 C being most desirable. High relative humidity seems to improve the success of artificial hybridization (Byth, 1966). Automatic humidifiers may be used or floors may be kept wet.

Soybeans often have a thin stem and long internodes and grow abnormally tall in growth chambers and greenhouses. Excessive plant height is encouraged by long photoperiods, high temperature, excessive N, low light levels, and inappropriate light quality (Tanner and Hume, 1976). When photoperiod is reduced, the first flowers to develop may be satisfactory for hybridization, but the later ones usually self-pollinate before hybridization is possible.

Successful hybridization in growth chambers or greenhouses is difficult, even when all environmental requirements seem to have been met. Some genotypes are more difficult to use as parents than others. 'Corsoy' is an example of a highly productive cultivar that is used readily as a parent for artificial hybridization in the field, but may not have normal seed set in the growth chamber or greenhouse. Sensitivity to light level and quality seems to be involved.

III. FLORAL CHARACTERISTICS

Soybean flowers typically are self-pollinated on the day when the corolla opens. The amount of natural crossing is approximately 1% for adjacent plants within a row and 0.5% between plants in adjacent rows (Weber and Hanson, 1961). Natural crossing is associated with insect vectors, primarily honeybees (*Apis mellifera* L.) (McGregor, 1976).

The structure of soybean flowers is similar to that of other legume species and consists of a calyx with five sepals, a corolla with five petals, 10 stamens, and a pistil (Carlson, 1973). The calyx encloses the corolla until the day before anthesis (Fig. 1A). The corolla emerges and unfolds to expose a standard, two wing petals, and two keel petals (Fig. 1D). An open flower is about 7 mm long from the base of the calyx to the tip of the standard and 6 mm wide across the standard.

The pistil consists of a single ovary that contains one to five ovules, a style that curves toward the standard, and a club-shaped stigma (Fig. 1C). The stigma is receptive to pollen about 1 day before anthesis and remains receptive for 2 days after anthesis, if the flower petals are not removed (Kuehl, 1961). Filaments of nine stamens are fused, and the one nearest the standard is free. The stamens form a ring below the stigma until about 1 day before anthesis (Fig. 1C), then their filaments begin to elongate rapidly and elevate the anthers around the stigma. The anthers dehisce on the day of anthesis, pollen grains fall on the stigma, and within 10 hours the pollen tubes reach the ovary and fertilization is completed (Johnson and Bernard, 1963).

Soybeans have been classified as indeterminate, semi-determinate, and determinate based on the abruptness of stem termination after flowering begins (Bernard and Weiss, 1973). When grown at their latitude of adapta-

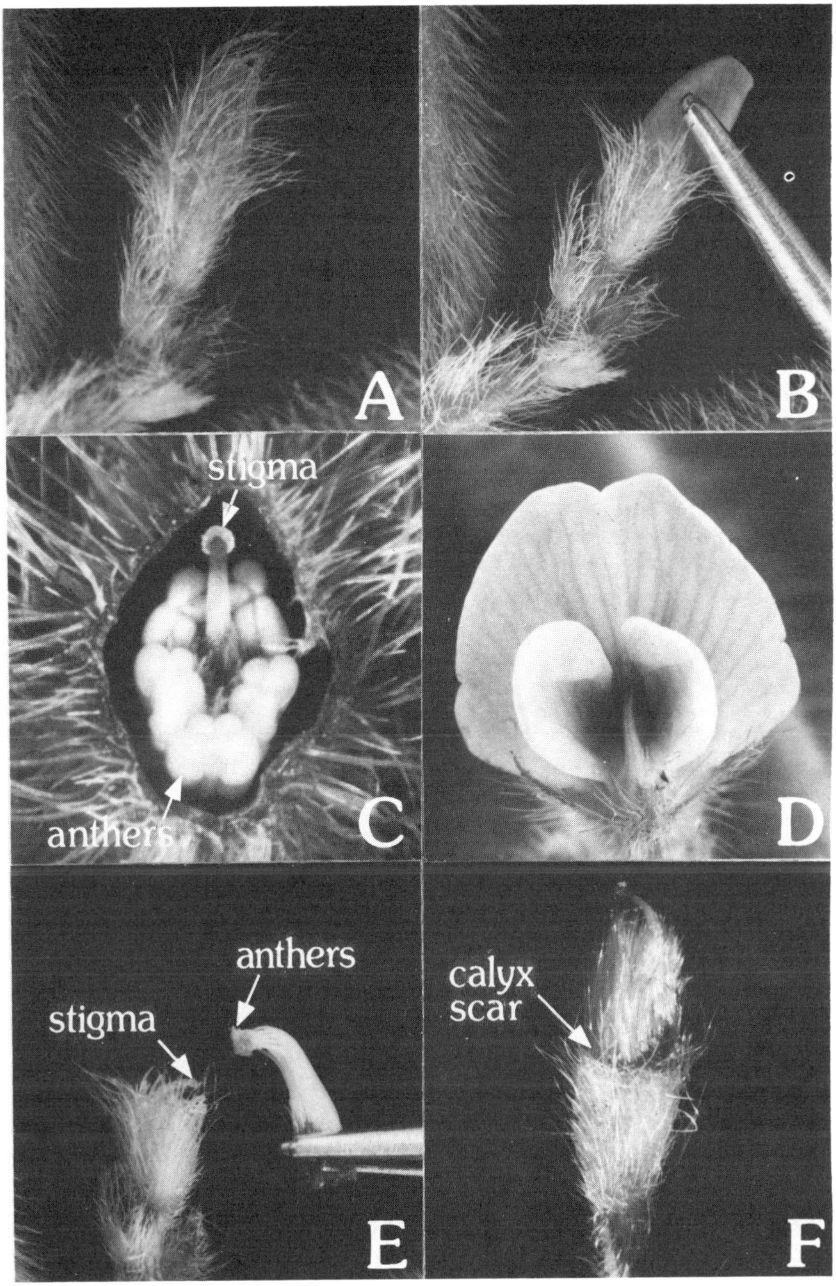

Fig. 1—Preparation and pollination of a soybean flower. (A) Flower at the stage for preparation and pollination, (B) removal of the corolla from the female flower after the calyx has been removed, (C) ring of 10 anthers and the stigma, (D) flower with pollen available, (E) pollination of the stigma, and (F) a pod 7 days after pollination with the calyx scar differentiating it from self-pollinated pods. (Photographs A, B, D, E, and F by C. J. Deutsch and P. A. Krumhardt, Iowa State Univ., Ames, Iowa; photograph C by G. L. Berkey, Ohio Agric. Res. and Devel. Ctr., Wooster, Ohio).

tion, indeterminate genotypes flower when about one-half of the nodes on the main stem have developed. They have short racemes with few flowers, and their terminal node has only a few flowers. Semi-determinate genotypes also flower when about one-half of the nodes on the main stem have developed, but node development and flowering on the main stem stop more abruptly than on indeterminates. Their racemes are short and have few flowers, except for the terminal one, which may have several times more flowers than those lower on the plant. Determinate cultivars begin flowering when all or most of the nodes on the main stem have developed. They usually have elongated racemes that may be several centimeters in length and may have a large number of flowers. Stem termination and flowering habit are reported to be controlled by two major genes (Bernard and Weiss, 1973).

IV. ARTIFICIAL HYBRIDIZATION AND SELF-POLLINATION

Self-pollination occurs naturally in soybean with no manipulation of the flowers. The following discussion pertains only to artificial hybridization.

A. Equipment

Forceps with fine tips and a smooth interior surface are required to manipulate the small flower. Plastic tags about 7 × 15 mm and wired with a flexible copper strand often are used to identify pollinated flowers. Plastic tags with a snap-on design can be attached quickly to the plant, but can be knocked or blown off more readily than those with a wire attachment (Ivers, 1977). Paper tags have been used successfully, but they are more susceptible to weather and insect damage than plastic tags.

Some breeders use magnifiers mounted on a headband or on a pair of glass frames. A magnification of 2.5 × provides satisfactory enlargement of the flower. Petri dishes or envelopes are used to collect male flowers. Desiccators containing calcium chloride crystals are used in some environments to dry male flowers to obtain adequate pollen shed.

B. Preparation of the Female

Flowers that are expected to open the following day are selected on the female parent. The buds are swollen and the corolla is just visible through the calyx or has begun to emerge (Fig. 1A). Usually no more than two buds on a raceme are prepared, and all self-pollinated flowers or immature buds are removed with forceps. Special care is required to remove immature buds that are hidden under the stipules at the leaf axil, and could develop into flowers at a later date.

The flower is grasped between the thumb and index finger and the location of the stigma determined by examining the sepals. A long, curved sepal covers the keel, and the stigma is on the opposite side of the flower (Fig. 1A). The calyx is removed by grasping a sepal with the forceps, pulling it down and around the flower, and repeating the procedure until the five sepals are removed. The exposed corolla is removed by grasping it just above the calyx scar, then lifting and wiggling the forceps simultaneously (Fig. 1B). Practice is required to grasp the corolla low enough to remove the keel petals without injuring the stigma. The ring of anthers is visible after the corolla is removed, unless the anthers were removed with the petals (Fig. 1C).

Recent evidence indicates that emasculation is unnecessary to prevent self-pollination of a flower when artificial pollination is unsuccessful (Walker et al., 1979). When emasculation is not used, the anthers near the stigma frequently are removed to make it clearly visible for pollination. I recommend that inexperienced persons use complete emasculation initially to familiarize themselves with a good female flower. They learn that the anthers are difficult to remove if the flower is immature or that the anthers surround the stigma if self-pollination has already occurred.

C. Pollination

The female flower usually is hand-pollinated immediately after it is prepared; although a delay of several hours does not seem to reduce seed set. The time of pollen shed is primarily a function of temperature. Pollen shed may begin at 0700 hours and end by 0900 hours when early morning temperatures are above 30 C, or may begin at 1000 hours and continue throughout much of the day with more moderate temperatures.

Pollen is available from a flower with a recently opened corolla, but the degree of corolla opening associated with pollen shed may vary during the day (Fig. 1D). In many environments, it is possible to collect male flowers and use them immediately without storage. In the southern U.S. and other humid climates, pollen shed occurs in the morning when female flowers are more immature and difficult to manipulate than in the afternoon, and the flowers may be damp from heavy dew. In those circumstances, male flowers are collected into envelopes or petri dishes at about 0800 hours and the open container is placed in a desiccator for about 4 hours at a room temperature of about 25 C. The desiccator is taken to the field in the afternoon and kept in the shade to prevent excessive temperatures from developing within it.

Pollen viability can be maintained in flowers for up to 2 days when stored at about 5 C. Kuehl (1961) demonstrated that flowers can be stored successfully for several weeks in a desiccator at 3 C; however, cultivars differed in the percentage of pollen that germinated after long-term storage. He found that flowers stored over calcium chloride for several weeks did not dehisce, but when they were placed over water in a closed container for about 30 min, the anthers would dehisce.

Hand pollination is carried out by removing the stamens and pistil with a forceps from a flower of the male parent and gently brushing the anthers against the stigma of the female flower (Fig. 1E). Access to the stamens can be achieved by removing the front sepal and keel petals, or piercing the keel with the closed forceps and allowing them to open to push the petals away. Brushing the anthers on the stigma causes them to rupture, and the highest percentage of successful crosses is obtained when pollen is clearly visible on the stigma. Some breeders prefer to tap the anthers on their thumbnail to check pollen shed before brushing the stigma. Several male flowers may have to be used to obtain suitable pollen shed when conditions are unfavorable, or the same male may be used to pollinate several flowers when pollen shed is excellent. Some breeders prefer to leave the stamens of the male flower hooked over the style of the female, but there are no data to suggest that it improves the percentage of successful pollinations. The forceps are cleaned after pollination by placing the tips in one's mouth or dipping them in 95% ethanol.

Pollinated flowers are identified with a tag attached to the internode or to the raceme below the pollinated flower. Information recorded may include the number of crosses, number of consecutive nodes above the tag where crosses were made, date, breeder, and parentage. The pollinated flowers are not protected.

There seems to be a considerable difference in the efficiency of hybridization for indeterminate and determinate cultivars. Indeterminate cultivars usually have short racemes in the leaf axil and the flowers are difficult to reach and manipulate. Determinate cultivars usually have long racemes at the top of the plant that are well above the leaf axil. An experienced breeder can hybridize about 20 flowers per hour with an indeterminate cultivar compared with about 50 flowers per hour with a determinate (Hartwig, 1978).

An experienced breeder has an expected success of near 50%, which is equivalent to one hybrid seed per cross, assuming two seeds per pod. Inexperienced individuals may have no success. The most common causes of failure include use of female flowers that are too immature, damage to the stigma, and inadequate pollination. Some breeders have indicated a tendency for success to be less when the first flowers on a plant are used as female.

D. Factors Affecting Efficiency

Efficiency of artificial hybridization can be increased when pollen transport and information written on the tag after pollination are minimized. When male flowers do not have to be collected and dried in a desiccator, it is useful to plant the parents of a cross adjacent to each other (Chapter 5).

Plants usually are grown in rows 65 to 100 cm apart to facilitate movement of personnel within the field nursery. Yield of self-pollinated seed from an individual plant may range from a few seeds to more than 1,000 as a function of plant density. A density of 30 plants/m of row can be used when 30 or fewer seeds per plant is adequate, 10 plants/m can be used to obtain about 100 seeds/plant, and 3 plants/m usually results in maximum seed production per plant. Densities of 12 plants/m or less commonly are used for artificial hybridization.

Multiple planting dates about 7 to 14 days apart usually are used to match parents of different flowering dates. When differences in flowering dates are extreme between parents, flowering of the later parent can be hastened by creating an artificially short day or flowering of the earlier parent can be delayed by use of artificially long days or delayed planting. For example, crosses with genotypes adapted to the southern U.S. are made in northern U.S. locations by covering the late genotype with a box, large can, or similar container to create an artificially short photoperiod of about 12 hours for about 15 days beginning when there are three nodes with trifoliolate leaves on the main stem. Plants induced to flower early tend to have flowers that self-pollinate when they are small and can be difficult to prepare for hybridization.

Grafting can be used to hasten the flowering of late flowering genotypes. A scion from a late genotype grafted on a stock that has begun to flower will begin to bloom up to 42 days earlier than normal (Kiihl et al., 1977). First flowers on the scion appear from 21 to 50 days after the graft.

Many qualitative characters have potential for use as genetic markers in soybeans; however, few are different among cultivars commonly used as parents (Bernard and Weiss, 1973). The most widely used genetic markers are flower color (purple dominant to white), pubescence color (brown dominant to gray), and pod color (brown dominant to tan). The association of purple hypocotyl color with purple flowers and green hyocotyl color with white flowers is commonly used to identify hybrids in the seedling stage. Differences in maturity, height, hilum color, and pest resistance between parents also are used to verify hybrid plants.

V. NATURAL HYBRIDIZATION

Genetic male sterility is available in soybeans and may be useful to facilitate hybridization, particularly for recurrent selection programs (Brim and Stuber, 1973). The distance required for complete isolation of a crossing block is not known; however, outcrossing is less than 0.5% when male-sterile plants are 12 m or more from a foreign pollen source (Boerma and Moradshahi, 1975). Plants on the boundaries of the crossing block probably sustain the most outcrossing with foreign pollen and can be eliminated at harvest to minimize contamination (Brim, 1977).

Little is known about the optimum plant density for natural hybridization. Cross-pollination is more common within rows than between adjacent rows; therefore, Brim (1977) prefers to grow populations with genetic male sterility on a square grid to create rows in all directions. He uses single-plant hills on 50-cm centers, subdivides the area into blocks of an equal number of hills for harvest, and bulks an equal amount of seed from male-sterile plants in each block to enhance random pollination.

VI. SEED DEVELOPMENT, HARVEST, AND STORAGE

Observing pod development 7 days after pollination generally is adequate to identify a successful cross (Fig. 1F). Abortion of pods and seeds can occur several weeks after pollination, but the percentage of abortion usually is low if plant stress is minimized (Shibles et al., 1975). During the check for successful pollinations, any new buds that may have developed at a node should be removed.

Pods that develop from artificial hybridization can be distinguished from self-pollinated pods by the presence of the calyx scar, caused by removal of the sepals (Fig. 1F). The sepals begin to fall off as the pods mature; therefore, harvest should be completed at or immediately before the time the pods reach their mature color. Harvesting pods early also avoids any loss by shattering. Marking the calyx or young developing pod of pollinated flowers with a felt-tip pen containing colored, nontoxic ink minimizes the chance of mistaking a selfed pod for a hybrid one (Ivers, 1977).

Pods are air-dried at not more than 38 C until the seeds contain 13% moisture or less, then the seeds are removed by hand. Each cross may be numbered for future reference using a system similar to that described in Chapter 5.

Seed can be stored satisfactorily at about 25 C for up to a year if relative humidity is 50% or less. In the humid tropics, germination percentage declines rapidly unless the seed is dried to 7% moisture and stored in an air-tight container at room temperature. Long-term storage in any climate is best accomplished by drying seed to 7% moisture and storing it at 10 C or less in a room maintained at 50% relative humidity or in an air-tight container.

REFERENCES

Bernard, R. L., and M. G. Weiss. 1973. Qualitative genetics. p. 117–154. *In* B. E. Caldwell (ed.) Soybeans: improvement, production, and uses. Am. Soc. of Agron., Madison, Wis.

Boerma, H. R., and A. Moradshahi. 1975. Pollen movement within and between rows to male-sterile soybeans. Crop Sci. 15:858–861.

Borthwick, H. A., and M. W. Parker. 1938. Photoperiodic perception in Biloxi soybeans. Bot. Gaz. 100:374–387.

Brim, C. A. 1977. Personal communication. Dep. of Crop Science, North Carolina State Univ., Raleigh, NC 27600.

———, and C. W. Stuber. 1973. Application of genetic male sterility to recurrent selection schemes in soybeans. Crop Sci. 13:528–530.

Buzzell, R. I. 1971. Inheritance of a soybean flowering response to fluorescent-daylength conditions. Can. J. Genet. Cytol. 13:703–707.

Byth, D. E. 1966. Hybridization and pollen germination in soybeans. Aust. J. Exp. Agric. Animal Husb. 6:371–373.

Carlson, J. B. 1973. Morphology, p. 17–95. *In* B. E. Caldwell (ed.) Soybeans: improvement, production, and uses. Am. Soc. of Agron., Madison, Wis.

Criswell, J. G., and D. J. Hume. 1972. Variation in sensitivity to photoperiod among early maturing soybean strains. Crop Sci. 12:657–660.

deMooy, C. J., J. Pesek, and E.Spaldon. 1973. Mineral nutrition. p. 267–352. *In* B. E. Caldwell (ed.) Soybeans: improvement, production, and uses. Am. Soc. of Agron., Madison, Wis.

Hadley, H. H., and T. Hymowitz. 1973. Speciation and cytogenetics. p. 97–116. *In* B. E. Caldwell (ed.) Soybeans: improvement, production, and uses. Am. Soc. of Agron., Madison, Wis.

Hamner, K. C. 1969. *Glycine max* (L.) Merrill. p. 62–89. *In* L. T. Evans (ed.) The induction of flowering: some case histories. Cornell Univ. Press, Ithaca, N.Y.

Hartwig, E. E. 1978. Personal communication. Delta Branch Experiment Station, Stoneville, MS 38776.

Howell, R. H. 1963. Physiology of the soybean. p. 75–124. *In* A. G. Norman (ed.) The soybean. Academic Press, N.Y.

Ivers, D. R. 1977. Personal communication. Land O'Lakes, Inc. Research Farm, Webster City, IA 50595.

Johnson, H. W., and R. L. Bernard. 1963. Soybean genetics and breeding. p. 1–73. *In* A. G. Norman (ed.) The soybean. Academic Press, N.Y.

Kuehl, R. O. 1961. Pollen viability and stigma receptivity of *Glycine max* (L.) Merrill. M.S. Thesis, North Carolina State College, Raleigh, N.C.

Kiihl, R. A. S., E. E. Hartwig, and T. C. Kilen. 1977. Grafting as a tool in soybean breeding. Crop Sci. 17:181–182.

Major, D. J., D. R. Johnson, J. W. Tanner, and I. C. Anderson. 1975. Effects of daylength and temperature on soybean development. Crop Sci. 15:174–179.

McGregor, S. E. 1976. Insect pollination of cultivated crop plants. USDA Handb. 496. U.S. Government Printing Office, Washington, D.C.

Parker, M. W., S. B. Hendricks, H. A. Borthwick, and N. J. Scully. 1946. Action spectrum for the photoperiodic control of floral initiation of short-day plants. Bot. Gaz. 108:1-26.

Shanmugasundaram, S., and S. C. W. Tsou. 1978. Photoperiod and critical duration for flower induction in soybean. Crop Sci. 18:598–601.

Shibles, R. M., I. C. Anderson, and A. H. Gibson. 1975. Soybean. p. 151–189. *In* L. T. Evans (ed.) Crop physiology, some case histories. Cambridge Univ. Press, Cambridge, England.

Summerfield, R. J. 1976. Use of glasshouses as an adjunct to field research on soybeans. p. 358–370. *In* L. D. Hill (ed.) World soybean research. The Interstate Printers and Publishers, Inc., Danville, Ill.

Tanner, J. W., and D. J. Hume. 1976. The use of growth chambers in soybean research. p. 342–351. *In* L. D. Hill (ed.) World soybean research. The Interstate Printers and Publishers, Inc., Danville, Ill.

van Schaik, P. H., and A. H. Probst. 1958. Effects of some environmental factors on flower production and reproductive efficiency in soybeans. Agron. J. 50: 192–197.

Walker, A. K., S. R. Cianzio, J. A. Bravo, and W. R. Fehr. 1979. Comparison of emasculation and nonemasculation for artificial hybridization of soybeans. Crop Sci. 19:285–286.

Weber, C. R., and W. D. Hanson. 1961. Natural hybridization with and without ionizing radiation in soybeans. Crop Sci. 1:389–392.

Wilcox, J. R. 1974. Response of soybeans to Amo-1618 and photoperiod. Crop Sci. 14:700–702.

43 Sugarbeet

Garry A. Smith
USDA-SEA and Colorado State University
Fort Collins

The beet root was a common part of the diet in Egypt during the building of the pyramids, but the potential of the sugarbeet (*Beta vulgaris* L.) as a source of sugar was not discovered until the middle of the 18th century. Beginning in the second half of the 19th century, plant breeders transformed the fodder beet into an efficient producer of sugar (sucrose). Sugarbeet now provides 45% of the world's sugar. The beet sugar industry is concentrated in the temperate climates of Europe, the United States, Canada, and the USSR. Sugarbeets are also grown in parts of South America, North Africa, Japan, China, and the Near East.

I. PARENTAL MATERIAL

Sugarbeet (*Beta vulgaris* L.) has several wild relatives, some of which may be hybridized with sugarbeet to obtain genes for disease resistance or other desired characters. Coons (1975) has extensively studied the genus *Beta* and the origin of species therein and classified the species into four sections (Table 1).

Table 1—Sections and species of *Beta* and normal chromosome numbers.†

Section Vulgares	*Section Corollinae*
Beta vulgaris L. (2n = 18)	*Beta macrorhiza* Stev. (2n = 18)
B. maritima L. (2n = 18)	*B. trigyna* Wald. et Kit (2n = 36,45,54)
B. macrocarpa Guss. (2n = 18)	*B. foliosa* (sensu Haussk.) (2n = 18)
B. patula Ait. (2n = 18)	*B. lomatogona* Fisch. et Mey. (2n = 18,36)
B. atriplicifolia Rouy (2n = 18)	*B. corolliflora* (2n = 36)
Section Patellares	*Section Nanae*
Beta patellaris Moq. (2n = 36)	*Beta nana* Bois. et Held. (2n = 18)
B. procumbens Chr. Sm. (2n = 18)	
B. webbiana Moq. (2n = 18)	

† Table adapted from Coons (1975). According to Coons, *Beta trigyna*, as collected in Hungary and the Crimea and distributed to various herbaria, is a hexaploid. The plants collected in the Caucasus and having the 2n number 36 were named *corolliflora* by Zossimovitch, but considered by Coons to be *B. trigyna*. A pentaploid (5 ×) form of *B. trigyna* found to be apomictic has also been identified.

The section *Vulgares* contains the common cultivated sugarbeet. All five species of this section are diploids (2n = 18 chromosomes) and all produce fertile hybrids when crossed with the cultivated beet. All cultivated beets may have originated from *Beta maritima* which is a prominent member of this section (McFarlane, 1971). Considerable genetic diversity is suggested by the wide variation for nearly all characters in the segregating generations from hybridizations among the species within the *Vulgares* section. Several characters including high *Cercospora* resistance have been transferred from *B. maritima* to *Beta vulgaris.*

The section *Patellares* includes three species that are characterized by hard-coated monogerm seed and a viny growth habit. Plants of the section offer the breeder a valuable source of genes for disease and nematode resistance. Successful crossing and gene transfer is difficult from species in the *Patellares* section to species in the *Vulgares* section. Intensive hybridization has been conducted between *B. vulgaris* L. and the species of the section *Patellares* because these are the only *Beta* species that are immune or highly resistant to the sugarbeet nematode (*Heterodera schachtii* Schm.). As a rule, these hybrids failed to grow because F_1 seedlings do not form secondary root systems and most of the few F_1 seedlings that have survived were sterile. Two methods have been used for growing hybrid plants from crosses with species of the section *Patellares.* Coe (1954) and Johnson (1956) developed a successful technique for grafting small interspecific F_1 seedlings on to the root system of young sugarbeet seedlings. A second method of overcoming the lethal condition of interspecific F_1 seedlings is the use of intermediate or bridge hybrids. For example, Swiss Chard (*Beta vulgaris* L.) is closely related to sugarbeet, crosses readily with it, and the hybrids are vigorous and fertile. Crosses between wild species of the section *Patellares* and Swiss Chard have been hybridized with sugarbeet. Forage and red garden beet (*Beta vulgaris* L.) also have been used as successful bridging hosts.

Oldemeyer (1970) obtained about 10% viable seed from crossing *B. maritima* × *B. promcumbens* hybrids to diploid sugarbeet. This fertility was attributed to unreduced egg cells, which gave rise to viable triploid (2n = 27) embryos. Several of these plants seemed to be nematode resistant, but progeny of a second backcross to sugarbeet were all susceptible. Savitsky (1960, 1975) reported that F_1 plants from tetraploid matings had higher viability and less sterility than diploid matings. Some viable seedlings were obtained from tetraploid matings without bridge hybrids or grafting.

The section *Corollinae* includes five species that are characterized by a corolla-like perianth. They are only distantly related to cultivated sugarbeet. In general, hybrids between each of the species of *Corollinae* and cultivated sugarbeet display little homology between chromosomes and are sterile. Savitsky (1969), however, crossed the curly top resistant tetraploid species *B. corolliflora* with sugarbeet and obtained highly resistant or immune aneuploid plants in the second backcross to sugarbeet. If the problem of chromosome nonhomology can be successfully solved, there are several characters in the *Corollinae* section of potential use to breeders including monogerm seed, apomixis, and curly top resistance.

The section *Nanae* is represented by a single species, *B. nana,* which is characterized by small plants with rosettes of leaves generally not more than 10 cm across. Plants have single flowers and hard nut-like monogerm seeds. Hybridizations with sugarbeet have not been reported.

Germplasm collections available to breeders are maintained in the United States at the U.S. Department of Agriculture sugarbeet research stations at Fort Collins, Colo.; Logan, Utah; Salinas, Calif.; East Lansing, Mich.; Beltsville, Md.; and at the Plant Introduction Station at Ames, Iowa.

II. PLANT CULTURE

A. Field

Cultivated sugarbeet is a herbaceous dicotyledon which normally completes its life cycle in 2 years. The biennial habit is variable and under certain conditions the plant may function as an annual or even as a perennial. When growing as a biennial, a fleshy tap root is produced the first year. After a mild winter or low temperature exposure in storage, the plant in the field or greenhouse produces a rosette of leaves like those of the first year. After 2 to 6 weeks, the apical portion begins to elongate to form an aerial stem and produces its flowering branches. Initiation of this aerial stem and resulting inflorescence is commonly referred to as bolting. Manipulation of the bolting process is important in the hybridization of sugarbeet.

Seed stalk formation varies widely among sugarbeet cultivars. To insure complete reproductive development, the plant must be exposed to a period of moderately low temperatures (4 to 7 C is ideal), followed by a period in which the daylight hours are long. With cultivars that bolt easily, comparatively short exposure to cool temperatures is adequate to induce plants to flower. Cultivars that do not bolt readily require longer periods of cold temperatures to insure floral induction. A wide range of climatic conditions exist in the present seed growing areas of the United States. Because of the mild winter climate in areas in Arizona and southern Utah (lowest winter temperatures range −4 to −1 C), only cultivars of easy to moderate bolting tendency are reproduced in those areas. The seed produced in these areas is used to plant fields for sugar production in areas which practice spring seeding and fall harvesting. Longer and colder winters, such as in the Oregon coastal valleys (lowest winter temperature range −9 to −6 C), favor the induction of bolting-resistant types. Much of this seed is used in the sugarbeet areas of California where the crop is planted during the fall and winter months.

Owen et al. (1940) used the term photothermal to describe the initiation of seedstalks and flowering of biennial beets brought about mainly by the cumulative effect of low-temperature exposure followed or accompanied by the effect of long photoperiod. The main difference found between biennial and annual beets for photothermal induction is the longer period of low temperature exposure required for flowering in biennials. Photothermal induction is frequently reversed in biennial beets, either by increasing the temperatures or by decreasing the length of the photoperiods. The temperatures that favor reproductive development in sugarbeets are fairly well established. Temperatures above 21 C favor vegetative growth, and temperatures between 4 and 13 C favor initiation of the reproductive phase. Most commercial cultivars of sugarbeet require 90 to 110 days of exposure to inductive temperatures for initiation of reproductive development.

Plants intended for hybridization are grown from seed the first year. In breeding programs located in regions which have extensive periods of freezing temperatures in the winter, sugarbeet roots grown in the summer are dug and stored indoors during the winter. The beets are prepared for storage by digging the roots and trimming the leaves and petioles without injuring the crown buds. Roots are placed in cold storage without light at a temperature of 5 to 7 C and a relative humidity approaching 100%. The length of storage time required for successful induction depends upon the genetic constitution of the line, but may vary from as few as 75 to over 140 days. Roots can be stored up to 230 days.

Soil types utilized for seed production include loam, sandy loam, and silty clay loam. Good drainage is always an important consideration in selecting seed fields especially in areas, such as Oregon, where heavy fall and winter rains prevail. Irrigation at planting time is necessary in all commercial areas of seed production in the United States. Furrow irrigation is practiced in Arizona and Utah and sprinkler irrigation in Oregon. Seeding rates for commercial seed production range from 11 to 20 kg per ha. Seed is planted 1.9 to 2.5 cm deep and plants are left unthinned to suppress weed growth and maintain cooler soil and beet crown temperatures in Arizona and Utah.

B. Growth Chamber and Greenhouse

The biennial nature of the sugarbeet is one of the chief obstacles to improvement of the crop through breeding. One of the most effective methods of reducing the length of the sugarbeet life cycles is to expose young seedlings to prolonged cool temperature and continuous light. Gaskill (1952) outlined a method in which two generations of sugarbeet seed can be produced in 12 months.

For the preinduction treatment, seedlings are grown in a warm greenhouse (20 to 23 C) in individual pots or tubes for 2 to 8 weeks after planting the seed. The longer period favors greater seed production. Continuous illumination is provided from time of emergence with light furnished by incandescent 150 W lamps equipped with reflectors. The lamps are spaced 1.5 m apart, and suspended about 76 cm above the plants.

To induce flowering, seedlings are transferred in their original containers to a chamber with constant temperature of 4 to 7 C and continuous artificial illumination. The light source may be incandescent lamps, fluorescent lamps, or a combination of both. Fluorescent lamps are more commonly used because they produce less heat per unit of light output than incandescent lamps. Reflector-type light units, equipped with two 20-W, 61 cm, standard cool-white fluorescent lamps suspended approximately 48 cm above the plants will provide light of sufficient intensity for an area of about 0.4 m^2. Duration of induction required depends upon the genetic resistance to bolting of the material being prepared for crossing. For cultivars having average bolting tendencies, 10 weeks of the induction treatment is sufficient.

The postinduction treatment consists of transplanting seedlings to the crossing location and providing continuous illumination as described for the preinduction period. Extensive study by Gaskill (1963) clearly demonstrated the need for incandescent, not fluorescent lamps, in the postinduction phase. Day/night temperatures during postinduction should be relatively cool (about 26/16 C) because high temperatures tend to reverse induction (Owen et al., 1940). Improper temperature during anthesis and seed maturation may adversely affect seed yield and quality.

III. FLORAL CHARACTERISTICS

The sugarbeet has a perfect flower that commonly consists of a tricarpellate pistil surrounded by five stamens and a perianth of five narrow sepals. Petals are absent. Flowers are located on the terminal portions of the main axes and on the lateral branches (Fig. 1). Flowers are sessile and occur singly

Fig. 1—Sugarbeet inflorescence and flowers. A) Complete inflorescence of sugarbeet plant. B) Branch of monogerm sugarbeet with opened flowers; note bracts, sepals and dehiscing anthers. C) Branch with ripening fruits. D) Open male-fertile flower with anthers dehiscing. E) Open male-sterile flower; note shrunken anthers.

or in clusters of two to seven. The ovary forms a fruit which is embedded in the base of the perianth of the flower. The ovaries are enclosed by the common receptacle of the flower cluster; therefore, a multiple fruit is formed by the aggregation of two or more flowers. This aggregation of two or more enclosed fruits forms the multigerm beet seed. Where flowers occur singly, a monogerm seed is formed (Fig. 2). Although, the terms monogerm and multigerm are used commonly to describe sugarbeet seed, botanically speaking they refer to the fruits. Each fruit contains a single seed whose shape varies from round to kidney-shaped.

Sugarbeet is naturally cross pollinated with pollen movement achieved by wind and occasionally by insects. Depending on the cultivar and environmental conditions, mature flowers begin anthesis about 5 to 6 weeks after initiation of reproductive growth and continue for several weeks. Flower openings begin at the base of each stem and continue upward as the stem elongates. Flower openings occur mainly in the morning, but continue throughout the day. The stigma matures about 6 days before anther dehiscence and remains receptive for up to 12 days after anthesis. Thus, the pistils of the beet become functional much before the flowers are opened

Fig. 2—Male-fertile sugarbeet floral buds and just opening flower. Note plump anthers (usually lemon-yellow color).

and the anthers are ripe. The embryo sac is fully formed when the stigma becomes functional, and the egg cell is capable of being fertilized 5 days before anthesis occurs (Kharetchko-Savitzkaya, 1931). The pollen tube commonly reaches the embryo sac and effects fertilization in 20 hours or less (Artschwager and Starrett, 1933).

IV. ARTIFICIAL HYBRIDIZATION AND SELF-POLLINATION

A. Equipment

Hybridization and self-pollination of sugarbeet in the greenhouse is an integral part of most sugarbeet breeding programs. Greenhouse hybridization and seed increases facilitate the low cost production of small quantities of seed with the assurance of no pollen contamination. Facilities used for greenhouse seed production range from large compartmentalized glass houses to small plastic isolation chambers. The essential features of these indoor isolators include a filtered air system to exclude foreign pollen, a heating and cooling system for year around use, and a supplemental incandescent light source. Detailed descriptions of isolators have been given by Wood et al., 1960; McFarlane and Skoyen, 1963; and Hecker and Gaskill, 1975. Although rarely used as a hybridization technique, manual emasculations of sugarbeet flowers are possible.

B. Preparation of the Female

In those cases where male sterility or genetic markers are not available to insure or recognize hybridization, a fine pointed jewelers' forceps can be used to remove the anthers of the sugarbeet flower. Anthers must be removed from flowers just ready to open and before anther dehiscence (Fig. 2). Sepals may be teased back to expose the anthers which may be removed from the flower with the forceps. Extreme care must be taken not to damage the stigma during anther removal. Any unemasculated flowers should be removed from the branch and the branch bagged to prevent any contamination of emasculated flowers with foreign pollen.

C. Pollination

Pollinations can be made immediately following emasculation or up to 12 days after normal anthesis would have occurred (Kharetchko-Savitzkaya, 1931). Pollen is collected by shaking flowering branches in a paper bag or by use of a vacuum aspirator. Pollen collected from desired male-fertile plants (Fig. 1D) may be dusted or brushed over the stigmas of the emasculated flowers.

Success of the pollination can be seen about 1 week after pollination by the swelling of the ovary. Fertilization failure may be indicated by the lobes of the stigma remaining open and by a pronounced hairy stigma compared with less pronounced stigmatic hairs of the successfully pollinated flowers.

Pollen viability may be zero if collected 3 to 4 days before anthesis, 75 to 84% if collected 1 day before dehiscence, and 94 to 99% if collected from dehiscing anthers (Karetchko-Savitzkaya, 1931). Pollen may remain viable for several hours or up to 20 days at room temperature. Pollen stored for 5 years in closed vials in a desiccator jar at 0 to −10 C has been used to achieve normal seed set (Hogaboam, 1977).

Inbred lines are used extensively by sugarbeet breeders to combine desired traits into hybrids. The sugarbeet normally exhibits a high degree of inherent self-incompatibility and consequently is not readily selfed. This self-incompatibility system seems to include four gametophytic *S*-loci with complementary interaction. This system requires that the four *S*-genes in the pollen be matched in the pistil to cause incompatibility (Larsen, 1977). An allele for self-fertility, however, enables production of highly self-compatible inbreds (Owen, 1942). Induction of self-pollination of otherwise self-incompatible plants also has been achieved by bagging plants at high altitudes. Selfing is achieved by bagging branches of plants just before flowering using white paper bags of suitable porosity and wet strength. A wooden stake or bamboo cane 1 to 2 m long is driven into the ground next to the root to provide a support for the bags placed on the plant. Larger leaves are removed from the branch before a bag is placed over it. Bags are fastened to the branch and to the wooden stake by soft wire ties or by string. Less severe inbreeding may be achieved by interpollination of plants of the same line (sib-mating). Advantages of sib pollination are a reduction of inbreeding depression and the production of sufficient quantities of seed for yield trials and disease nurseries.

Muslin tents capable of holding about 16 plants are used for sib matings and crossing to prevent entry or exit of pollen. Tents are used to obtain quantities of seed from sib matings that would not otherwise be obtainable by selfing because of self sterility. Tents also are used for indexing type O genotypes (Section V) by placing the tent over pollen fertile plants along with cytoplasmic-genetic male-sterile (CMS) plants. Seed harvested from the CMS plants is tested for pollen fertility. Crossing of CMS and pollen-fertile plants under the tents also facilitates seed production for yield tests used to estimate general and specific combining ability.

The problem of maintaining selected genotypes of self-sterile material may be overcome by the use of paired-plant crosses which can be used to maintain a high frequency of favorable genes. Plants from a heterogeneous population are pair-crossed under bags or tents and the progeny evaluated in a replicated yield trial. Progeny rows also are planted in a nursery to produce mother beets which can be induced to flower for purposes of obtaining seed. The best performing paired crosses are identified from the yield trials. Selections are made from the mother-beets representing the superior performing crosses. Seed from the selected mother beets is bulked to form the next population. The best paired-plant crosses identified from yield trials are bulked to form a pollinator line, which can then be used as a male parent for commercial hybrid production.

D. Factors Affecting Efficiency

Coordination of flowering between plants is achieved by placing plants in more advanced stages of flowering into chambers where a temperature of 5 C inhibits further flowering. These plants are placed back in the crossing isolation when the plants to which they are to be crossed are at the proper stage of flowering.

When hybridizing sugarbeet without the use of CMS, genetic markers have been identified which can be utilized to identify crosses. The most commonly used genetic markers in sugarbeet are hypocotyl color, genetic male sterility, monogerm-multigerm seed, and annual growth habit.

In the absence of cytoplasmic-genetic male sterility, hypocotyl color, which is readily identifiable in the seedling stage of growth, is used by breeders to identify hybrids. The gene, *R*, first reported by Kajanus (1917) and verified by Keller (1936), consists of a multiple allelic series ($R, R^t, R^P, R^h\ r$) and governs hypocotyl, root, and foliage color. The *R* allele produces a pink to reddish hypocotyl in the young beet seedling and usually red color in crown buds of older plants. R^t produces intense red stripes extending into the petioles and R^P produces pink petioles and hypocotyl and rose or light pink tinted roots (Pedersen, 1944, Theurer, 1968b). R^h conditions only red hypocotyl even when dominant genes are present at the *Y* locus (Pedersen, 1944). The recessive combination, *rr*, produces a plant with green hypocotyl and crown buds. Crossing of a plant carrying the *rr* combination to a plant carrying the *RR* combination, produces progeny seed harvested from the *rr* plant that will give red hypocotyl (*Rr*) seedlings. These plants will be easily identified from the *rr* green hypocotyl plants which may have resulted from either selfing or sibbing. Associated with the *R* gene is the *Y* gene also identified by Kajanus (1917) and verified by Keller (1936). The *Y* gene controls yellow pigmentation, but its expression is modified by the presence of the hypocotyl color *R* gene. The *rrY__* combination results in a yellow hypocotyl, yellow rooted plant with the yellow color extending to the petioles and larger leaf veins. The *R__yy* combination produces a red hypocotyl, white rooted plant. However, the *R__Y__* combination produces a red hypocotyl, red rooted plant. The presence of modifier genes along with the *R__Y__* combination may produce entirely red foliage and intense red root epidermis. Recently, Linde-Laursen (1972) identified a recessive gene, *p*, which when present causes the absence of red and yellow color, regardless of the *R* and *Y* color genes. The dominant allele *P* which is indispensable for red and yellow color formation is carried by the majority of beet strains.

Genetic male sterility (Owen, 1952) is useful for emasculation of individual plants in breeding programs and in genetic investigations. Genetic male sterility is especially useful for hybridization between self-fertile lines.

It also has been used along with cytoplasmic-genetic male sterility to produce double-cross hybrids (Owen, 1954). The dominant gene A_1, conditions male fertility, whereas the homozygous recessive condition a_1a_1 conditions male sterility.

Annual growth habit, designated *B* (Munerati, 1931, Abegg, 1936), can be used to speed generation cycles. Plants with the dominant *B* allele bolt and produce seedstalks in an environment of 18 to 24-hour photoperiod and 24 to 27 C. Homozygous *bb* plants remain vegetative under the same conditions. Additional modifying factors sometimes influence bolting and make it difficult to classify plants for annual habit.

Monogerm seed revolutionized the sugarbeet seed industry and is conditioned by the homozygous recessive condition of *mm* which produces plants with single germ fruits and seeds. A single fruit or a lateral branch can be born in the axil of a leaf, but never both together. Four multigerm alleles, designated M, M^1, M^{Br}, and M^2, show different degrees of dominance over *m*. Homozygous *mm* plants, normally monogerm, are sometimes observed to have double-germ fruits on the central floral axis and are thought to be the result of non-allelic genes.

Table 2 summarizes the gene symbols and the tentative linkage groups for all reported genes of sugarbeet.

V. NATURAL HYBRIDIZATION

Cytoplasmic genetic male sterility provides a practical method of emasculation and has facilitated the large-scale production of commercial hybrid seed. In sugarbeet, CMS is most correctly described as cytoplasmic-genetic due to the interaction of sterile cytoplasm (S) and normal cytoplasm (N) with two nuclear genes designated by Owen (1945) as *X* and *Z*. Plants with the *Sxxzz* genotype are completely male sterile; *SXxzz* and *SxxZz* are semi-male-sterile; and *SXxZz* are also semi-male-sterile, but may be visually indistinguishable from normal pollen fertile plants. Bliss and Gabelman (1965) described the dominant allele, gene *X*, as the restorer of complete male fertility and the dominant allele, gene *Z*, as a restorer of partial fertility to plants having *S* cytoplasm. Gene *Z* also was reported to be independent of and hypostatic to *X* and plants containing N cytoplasm were male-fertile regardless of the particular alleles of the *X* and *Z* genes. Crosses between plants with genotypes *Nxxzz* (normal cytoplasm and pollen fertile) and *Sxxzz* produce all male-sterile progeny. Maintainer lines of the *Nxxzz* genotype were designated type O by Owen (1945). The term B lines is used to describe equivalent maintainer lines in other crops, such as corn (*Zea mays* L.) and sorghum [*Sorghum bicolor* (L.) Moench]. Introduction of CMS into a type O breeding line is accomplished by crossing a female CMS plant with a type O male. Several backcrosses to the type O parent will produce a male-sterile equivalent of the original type O.

Flowers of male-sterile plants are characterized initially by white shrunken anthers which progressively change to dark brown or black (Fig. 1E). Flowers of male-fertile plants are recognized by their lemon yellow color anthers, that may range from very plump to mildly shrunken depending on the genetic constitution of the plant. Plants of female lines are checked before anthesis and any yellow anthered plants are rogued. Unopened flowers of each female plant are examined by opening the sepals with a pointed instrument. A hand lens is frequently used to closely examine questionable anthers.

Table 2—Characters and gene symbols in sugarbeet *Beta vulgaris* L.

Linkage group	Genetic symbols	Authority or verifier	Character
I	Y, Y^r, y	Kajanus (1917) Keller (1936)	Yellow pigment
	R, R^l, R^P, R^h, r	Kajanus (1917) Keller (1936)	Hypocotyl color
	Cl, cl	Owen & Ryser (1942)	Colored leaf
	Tr, tr	Owen & Ryser (1942)	Trout or spotted leaf
	Cv, cv	Deming† Owen & Ryser (1942)	Colored vein
	B, b	Munerati (1931) Abegg (1936)	Annual growth habit
	B, b	Owen et al. (1940)	Annual growth habit but less effect than B, considered allelic to B
	v_1, V_1	Owen & Ryser (1942)	Variegated foliage
	C, c	Abegg & Owen (1936)	Partially dominant curly top resistance
	cr, Cr‡	Stewart†	Crinkled foliage, reduced plant size
	p, P	Linde-Laursen (1972)	Non-production of color
II	m, M, M^{Br}, M^1, M^2	Savitsky (1950, 1952)	Monogerm seed
	lb	Savitsky (1952)	Late bolting
	C_2	Savitsky & Murphy (1954)	Partial curly top resistance
III§	a_1, A_1	Owen (1952)	Genetic male sterility
IV§	lu_2, Lu_2	Theurer (1968a)	Lutescens, green cotyledons followed by death
V§	ru, Ru	Owen & Ryser (1942)	Russet root
No linkage group established	ch_2, Ch_2	Savitsky (1957)	Chlorina cotyledons and all leaves are yellow green
	ch_1, Ch_1	Savitsky (1940)	Chlorina, more reduction in root yield than ch_2
	vi_4, Vi_4	Stander & Theurer (1970)	Virescens, pronounced delay in chlorophyl production in first true leaves, white to light green leaves
	X, x	Owen (1945)	Restores male fertility in sterile cytoplasm
	Z, z	Owen (1942, 1945)	Produces partial fertility in sterile cytoplasm
	Rf_1	Theurer & Ryser (1969)	Pollen restorer
	Rf_2	Theurer (1971)	Annual pollen restorer
	Rf_3	Roundy & Theurer (1974)	Weak pollen restorer from a yellow leaf mutant
	Sh, sh	Hogaboam (1957)	Enhances pollen production
	yl, Yl	Roundy & Theurer (1974)	Yellow leaf
	bl, Bl	Munerati & Costa (1930)	Black root
	d, D	Theurer (1968b)	Dwarf mature plants. Seedlings have thick hypocotyls
	pl, Pl	Abegg (1936)	Plantain leaf; semi-parallel veination
	S^s, S^f	Owen (1942)	Self fertility
	S^a, S^b	Owen (1942)	Self sterility
	$S_a, S_b, S_c S_d$	Larsen (1977)	Self-incompatibility, 4 gametophytic loci
	w, W	Stewart†	Albino lethal
	v_2, V_2	Abegg (1940)	Variegated cotyledons
	v_3, V_3	Abegg (1940)	Variegated foliage and root flesh
	f, F	Stewart†	Flaccid leaf
	n, N	Stewart†	Nana plants, leaves thick and leathery
	m, M	Abegg (1940)	Minature, lethal, conflicts with monogerm
	re, Re	Savitsky†	New leaves of mature plants reduced in size and lanceolate shaped
	w_1, W_1	Savitsky (1957)	Albino seedlings
	Au, au	Savitsky (1957)	Golden-yellow cotyledons, semilethal heterozygote, lethal homozygote
	lu, Lu	Savitsky (1957)	Lutescens, leaves produced after second pair become progressively yellow, some survival
	vi_1, vi_2, vi_3	Savitsky (1957)	Virescens, golden-yellow seedlings, varying degrees of delay of chlorophyll production

† As cited by Abegg (1940).
‡ Butterfass (1968) classes this as linkage group II.
§ Tentative linkage group assignments based on the fact that they were good marker genes not associated genes in group I and II.

In the production of hybrid seed, the primary concern is to produce seed free of contamination by foreign pollen. Wind-borne pollen can be distributed horizontally at least 4,500 m and has been observed at vertical distances of 5,000 m (Archimowitsch, 1949). In commercial practice, plantings of related seed types are separated by distances of 1.6 km or more. Fields of unrelated seed types usually are separated by 3.2 km or more, depending on terrain and prevailing winds (Campbell and Mast, 1971).

Two methods of planting are used to produce commercial hybrid seed. One method is to plant a seed mixture of 85 to 90% monogerm male-sterile and 5 to 10% multigerm pollinator. Care must be exercised to assure a uniform distribution of pollinator seed in the row. Seed is bulk harvested and the larger multigerm seed is mechanically separated, leaving only the hybrid seed from the monogerm male-sterile plants. In the second method, alternate strips of a male-sterile line and a pollinator line are planted. A distance of 1.83 m (or two blank rows) is left between the strips so that seed harvest can be accomplished without mixing the two lines. The ratio of pollinator rows to male-sterile rows is commonly 4 to 12 or 4 to 16, depending on the pollen-producing ability of the pollinator. If the pollinator seed is not used, it is tilled out about 3 weeks before harvest of the male-sterile line. This strip planting method is used when the pollinator line is a weak pollen producer, or when its growth habit will not permit effective competition in a mixed planting with more vigorous male-sterile lines.

Roots taken from storage are commonly planted 0.6 to 1.2 m apart in rows spaced 0.6 to 1.2 m apart to allow for cultivation and to prevent excessive tangling of male and female seed branches. Alternate rows of CMS and male-fertile rows are planted. Female lines are checked for male sterility just before flowering and any pollen fertile or partially fertile plants are rogued.

Field crossing nurseries often consist entirely of male-fertile plants which are planted in alternate rows of red hypocotyl (with or without red roots and foliage) and green or yellow hypocotyl plants. In such arrangements, the non-red lines are considered as females and resulting hybrids are identified by hypocotyl and root color. A field crossing nursery also can consist of a male-fertile pollinator and one or more lines segregating for genetic male sterility (a_1a_1). In such an arrangement, the segregating male-sterile lines are rogued before anthesis.

For recurrent selection and polycross breeding systems, some branches of male-fertile plants in field crossing nurseries are bagged to obtain selfed seed in addition to the open-pollinated seed obtained from the unbagged branches on the rest of the plant. Amount of seed obtained under the bags will depend on the degree of self sterility of the plants. With cool temperatures of 20 to 29 C, it is possible to obtain some selfed seed even from plants which are carrying genes for self sterility.

Roots from mass-selected populations are planted in field crossing nurseries to obtain open-pollinated seed for increase or for use in further mass selection cycles. To achieve equal and random pollination, roots frequently are cut into two to four sections with at least one crown bud and interplanted among the other roots of the nursery. In locations free of severe freezing, seed may be harvested in midsummer from plantings made the previous fall.

VI. SEED DEVELOPMENT, HARVEST, AND STORAGE

In commercial seed production areas of the United States, first blooms appear 5 to 6 weeks from the time seedstalks begin to form. The full-bloom stage of development is reached approximately 2 weeks later. The crop begins to mature and is ready for harvest about 6 weeks after the full bloom stage. Seed maturity is best determined by observing the stage of development of the endosperm. Fruits (seed balls) from representative branches are dissected and the endosperm observed. Fruits may be very green and high in moisture content, but if the true seed is in a moderate to hard dough stage, the seed crop is ready to harvest. Definite changes in color of the seedstalks and foliage, and shattering of early maturing seed, are indicators of proper time for harvesting the crop. As the seed crop approaches maturity, it usually takes on a straw color. The maturity of cultivars resistant to bolting, especially monogerm lines, can be difficult to determine because the seedstalks and foliage remain green when the seed is ready for harvest.

In breeding programs with small seed plots, entire plants are hand-cut, placed in burlap bags, dried naturally or artificially, and threshed with a Vogel thresher or similar device.

Commercial seed harvest is accomplished by windrowers with vertical sickle attachments which permit cutting through tangled masses of branches. Threshing is accomplished with a combine after drying in windrows for 7 to 14 days. Threshed seed is cleaned by passing through air separators and other equipment to remove trash and to size the seed to contract specifications. Seed is processed by passing it through equipment designed to polish the seed by grinding off a portion of the rough, corky material (pericarp) surrounding the true seed. This operation reduces weight of the seed and makes the product more uniform in size.

Processed seed is stored in bags under relatively cool temperatures 13 C and low humidity of 15%. Under low temperature and low humidity, seed viability may be maintained for extended lengths of time. Pack and Owen (1950) reported 75% seed germination of seed lots which had been stored in a commercial cold-storage house for 22 years at temperatures of -12 to -23 C. This was only 10% less than the original germination.

VII. TECHNIQUES FOR SPECIAL SITUATIONS

The normal diploid chromosome number of sugarbeet ($2n = 18$) may be doubled by colchicine treatment of dry, pregerminated, or germinated seed. Colchicine concentrations of 0.3 to 0.5% for dry and pregerminated seed and 0.1 to 0.5% for germinated seed are effective in inducing high percentages of tetraploid plants (Savitsky, 1966). Treatment of pregerminated seed for 16 hours is most effective for obtaining the highest number of tetraploids.

Production of clones by cuttings enables perpetuation and increase of plants from any individual genotype. Clonal propagation is accomplished by stem cuttings from bolting plants and by longitudinal splitting of roots in such a way that each root section contains at least one crown bud. Owen (1941) summarized effective methods for asexual propagation and Hogaboam (1962) described inexpensive plastic humidity chambers which facilitate the successful rooting of cuttings taken from seed-producing plants. Clonal propagation can be especially useful in polycross and recurrent selection breeding systems where self-sterile genotypes inhibit or prevent self-fertilized seed.

ACKNOWLEDGMENTS

The author wishes to acknowledge the constructive reviews and suggestions of Mr. J. O. Gaskill, Dr. R. T. Lewellen, Dr. R. J. Hecker, Dr. J. S. McFarlane, Dr. J. C. Theurer, Dr. E. G. Ruppel, and Mr. J. B. Bostick.

REFERENCES

Abegg, F. A. 1936. A genetic factor for the annual habit in beets and linkage relationship. J. Agric. Res. 53:493–511.

————. 1940. List of characters and gene symbols reported for the species *Beta vulgaris* L. Proc. Am. Soc. Sugar Beet Technol. 2(1):109–113.

————, and F. V. Owen. 1936. A genetic factor for curly top disease resistance in beets (*Beta vulgaris* L.) and linkage relationships. Am. Nat. 70:36.

Archimowitsch. A. 1949. Control of pollination in sugar-beet. Bot. Rev. 15:613–628.

Artschwager, E., and R. C. Starrett. 1933. The time factor in fertilization and embryo development in the sugar beet. J. Agric. Res. 47:823–843.

Bliss, F. A., and W. H. Gabelman. 1965. Inheritance of male sterility in beets, *Beta vulgaris* L. Crop Sci. 5:403–406.

Butterfass, Th. 1968. The position of the R locus of sugarbeet on chromosome II. [In German]. Theoret. Appl. Genet. 38:348–350.

Campbell, S.C., and A. A. Mast. 1971. Seed production. p. 437–450. *In* R. T. Johnson, J. T. Alexander, G. E. Rush, and G. R. Hawkes (eds.) Advances in sugarbeet production: principles and practices. Iowa State Univ. Press, Ames.

Coe, G. E. 1954. A grafting technique enabling an unthrifty interspecific hybrid of *Beta* to survive. Proc. Am. Soc. Sugar Beet Technol. 8:157–160.

Coons, G. H. 1975. Interspecific hybrids between *Beta vulgaris* L. and the wild species of *Beta*. J. Am. Soc. Sugar Beet Technol. 18:281–306.

Gaskill, J. O. 1952. A new sugar-beet breeding tool—two seed generations in one year. Agron. J. 44:338.

————. 1963. Comparison of fluorescent and incandescent lamps for promotion of flowering in sugar beet seedlings. J. Am. Soc. Sugar Beet Technol. 12:623–634.

Hecker, R. J., and J. O. Gaskill. 1975. Plastic isolation chambers for sugarbeet seed production. J. Am. Soc. Sugar Beet Technol. 18:264–268.

Hogaboam, G. J. 1957. Factors influencing phenotypic expression of cytoplasmic male sterility in the sugar beet (*Beta vulgaris* L.). J. Am. Soc. Sugar Beet Technol. 9:457–465.

———. 1962. Plastic chambers for humidity and temperature control in vegetative propagation and growth of sugar beets. J. Am. Soc. Sugar Beet Technol. 11: 661-667.

———. 1977. Personal communication. USDA-ARS, P.O. Box 1633, Michigan State Univ., East Lansing, MI 48823.

Johnson, R. T. 1956. A grafting method to increase survival of seedlings of interspecific hybrids within the genus, *Beta.* J. Am. Soc. Sugar Beet Technol. 9:25-31.

Kajanus, B. 1917. On the color variation of *Beta.* [In German]. Z. Pflanzenzucht. 5:357-372.

Keller, Wesley. 1936. Inheritance of some major color types in beets. J. Agric. Res. 52:27-38.

Kharetchko-Savitzkaya, E. I. 1931. Methods and technique of artificial hybridization of the sugar beet. [In Russian; German summary]. Works Sci. Res. Inst. Sugar Ind. 13:613-632.

Larsen, K. 1977. Self-incompatibility in *Beta vulgaris* L. I. Four gametophytic complementary S-loci in sugar beet. Hereditas 85:227-248.

Linde-Laursen, Ib. 1972. A new locus for colour formation in beet, *Beta vulgaris* L. Hereditas 70:105-112.

McFarlane, J. S. 1971. Variety development. p. 401-435. *In* R. T. Johnson, J. T. Alexander, G. E. Rush, and G. R. Hawkes (eds.) Advances in sugarbeet production: principles and practices. Iowa State Univ. Press, Ames.

———, and I. O. Skoyen. 1963. Greenhouse chambers for small seed increases. J. Am. Soc. Sugar Beet Technol. 12:323-325.

Munerati, O. 1931. Inheritance of the annual tendency in the common cultivated beet. [In Italian]. Z. Zuchtung, Reihe, A., Pflanzenzuchtung 17:84-89.

———, and T. Costa. 1930. Observations on the transmission of the "black skin" character in beet. [In Italian]. Z. Induk. Abstamm. Vererbungsl. 54:458-465.

Oldemeyer, R. K. 1970. A sugarbeet × *Beta procumbens* hybrid and its backcross derivatives. J. Am. Soc. Sugar Beet Technol. 15:641-646.

Owen, F.V. 1941. Asexual propagation of sugar beets. J. Hered. 32:187-192.

———. 1942. Inheritance of cross- and self-sterility and self-fertility in *Beta vulgaris.* J. Agric. Res. 64:679-698.

———. 1945. Cytoplasmically inherited male-sterility in sugar beets. J. Agric. Res. 71:423-440.

———. 1952. Mendelian male sterility in sugar beets. Proc. Am. Soc. Sugar Beet Technol. 7:371-376.

———. 1954. Hybrid sugar beets made by utilizing both cytoplasmic and mendelian male sterility. Proc. Am. Soc. Sugar Beet Technol. 8(2):64.

———, E.Carsner, and M. Stout. 1940. Photothermal induction of flowering in sugar beets. J. Agric. Res. 61:101-124.

———, and G. K. Ryser. 1942. Some mendelian characters in *Beta vulgaris* and linkages observed in the Y-R-B group. J. Agric. Res. 65:155-171.

Pack, D.A., and F. V. Owen. 1950. Viability of sugar beet seed held in cold storage for 22 years. Proc. Am. Soc. Sugar Beet Technol. 6:127-129.

Pedersen, A. 1944. The colors of beets. [In German]. K. Vet. Landbohoejsk. Arsskr. 60-111.

Roundy, T. E., and J. C. Theurer. 1974. Inheritance of a yellow-leaf mutant and a pollen fertility restorer in sugarbeet. Crop Sci. 14:62-63.

Savitsky, H. 1960. Viable diploid, triploid and tetraploid hybrids between *Beta vulgaris* and species of the section *patellares*. J. Am. Soc. Sugar Beet Technol. 11:215–235.

———. 1966. Effectiveness of selection for tetraploid plants in C_0 generation on the basis of the number of chloroplasts in stomata. J. Am. Soc. Sugar Beet Technol. 13:655–661.

———. 1969. Meiosis in hybrids between *Beta vulgaris* and *Beta corolliflora* and transmission of resistance to curly top virus. Can. J. Genet. Cytol. 11:514–521.

———. 1975. Hybridization between *Beta vulgaris* and *B. procumbens* and transmission of nematode (*Heterodera schachtii*) resistance to sugarbeet. Can. J. Genet. Cytol. 17:197–209.

———, and A. M. Murphy. 1954. Study of inheritance for curly top resistance in hybrids between mono- and multigerm beets. Proc. Am. Soc. Sugar Beet Technol. 8(2):34–44.

Savitsky, V. F. 1940. Genetics of the sugar beet. [In Russian]. All-Union Scientific Research Institute of the Sugar Industry. Sugar Beet Cult. 1:549–684.

———. 1950. Monogerm sugar beets in the United States. Proc. Am. Soc. Sugar Beet technol. 6:156–159.

———. 1952. A genetic study of monogerm and multigerm characters in beets. Proc. Am. Soc. Sugar Beet Technol. 7:331–338.

———. 1957. Inheritance of chlorophyll deficiencies in *Beta vulgaris* L. J. Am. Soc. of Sugar Beet Technol. 9:321–336.

Stander, J. R., and J. C. Theurer. 1970. Inheritance and linkage of A Virescens and A Chlorina in *Beta vulgaris* L. Crop Sci. 10:548–549.

Theurer, J. C. 1968a. Inheritance of a lutescens mutant in sugarbeets, *Beta vulgaris* L. Crop Sci. 8:422–423.

———. 1968b. Linkage tests of mendelian male sterility and other genetic characters in sugarbeets, *Beta vulgaris* L. Crop Sci. 8:698–701.

———. 1971. Inheritance studies of a pollen restorer from ruby queen table beet. J. Am. Soc. Sugar Beet Technol. 16(4):354–358.

———, and G. K. Ryser. 1969. Inheritance studies with a pollen fertility restorer sugarbeet inbred. J. Am. Soc. Sugar Beet Technol. 15(6):538–545.

Wood, R. R., D. E. Conwell, C. W. Impey, and P. B. Smith. 1960. Development of air conditioned, compartmented greenhouse. J. Am. Soc. Sugar Beet Technol. 11:44–48.

44 Sugarcane

Norman I. James
USDA-SEA
Beltsville, Maryland

Cultivated sugarcane (*Saccharum* sp.) is a vegetatively propagated, perennial grass that originated in Northern India and New Guinea (Grassl, 1977). Most cultivars are complex tri-species hybrids adapted to tropical and subtropical climates. The soft, juicy stems contain 9 to 15% sucrose on a fresh weight basis. The juice is extracted by milling, evaporated, and the resulting crystalline sucrose is termed raw sugar. The familiar white sugar is obtained by further refining of raw sugar. In some countries, crude sugar products are prepared by boiling juice in open pans.

I. PARENTAL MATERIAL

Grassl (1977) considers that the genus *Saccharum* consists of four species: *S. officinarum* L., *S. spontaneum* L., *S. robustum* Brandes and Jeswiet ex Grassl, and *S. sanguineum* (Grassl) Grassl. He suggested referring to the formerly recognized species *S. sinense, S. barberi,* and *S. edule* as horticultural groups derived from natural intergeneric hybridization. Controversy exists about the taxonomy of the genus and additional data are needed before agreement is achieved (Daniels et al., 1975).

Saccharum officinarum (2n = 80) is characterized by large, juicy stalks with high sugar content, disease susceptibility, and low vigor. These clones and their natural hybrids were grown commercially for centuries until hybridization work began in the late 19th century. *Saccharum spontaneum* (2n = 40 to 128) and groups Sinense (2n = 118) and Barberi (2n = 82 to 124) are characterized by thin stalks with high fiber and low sugar content, disease resistance, vigor, and cold tolerance. *Saccharum robustum* (2n = 60 to 194) clones often have very large stalks, low sugar content, leaves that shed at maturity, and adaptability to wet conditions. *Saccharum sanguineum* (2n = 60), formerly included in *S. robustum*, has red interior stalk color.

Group Edule produces an edible inflorescence, often with abortive flower parts, that does not emerge from the enclosing leaf sheath and, therefore, cannot be hybridized.

Early sugarcane breeders achieved great success through interspecific hybridization, particularly in obtaining resistance to diseases, cold tolerance, and increased vigor. Because sugarcane is vegetatively propagated, the genetic constitution of the derived hybrids is fixed, and these hybrids could be grown commercially without developing pure lines. Most commercial cultivars today can be traced to the early hybridization work in Java and Coimbatore, India (Jeswiet, 1927a, Dutt et al., 1953). The species involved were *S. officinarum, S. spontaneum,* and horticultural group Barberi. Later, horticultural group Sinense and then *S. robustum* were included in breeding schemes.

Clones of the genus *Saccharum* were hybridized with at least the following seven genera: *Eccoilopus, Miscanthidium, Miscanthus, Ripidium, Sclerostachya, Sorghum* and *Zea.* Grassl (1977) speculates that this number may be increased to 20 as other genera become available.

A World Germplasm Collection of sugarcane is maintained by the Science and Education Administration-U.S. Department of Agriculture (SEA-USDA) at Beltsville, Md., and Canal Point, Fla., and by the Sugarcane Breeding Institute at Cannanore, India (Anonymous, 1971). These duplicate collections are maintained under the auspices of the International Society of Sugar Cane technologists and contain 1,200 to 1,800 representatives of *Saccharum,* horticultural groups, and related genera.

II. PLANT CULTURE

A. Field

Sugarcane is usually grown for hybridization in one to three-row plots 3 to 20 m in length with row spacings of 3 to 6 m to permit accessibility to flowering plants by people or vehicles. With few exceptions, flowering stalks are moved from the field to a central location for hybridization. Plantings generally are made in the fall of the year for use during the next flowering season. Stalks are cut into three-bud pieces, dropped in furrows at the rate of about 14 buds/m of row, and covered with 5 to 20 cm of soil. In some areas, fungicide treatment is applied to seed pieces at or before planting to control diseases. Fertilization is often less than that recommended for commercial culture to reduce lodging and assure low nutritional status during the period of floral initiation. The usual practice is to harvest the stalks after the flowering season and allow the planting to ratoon for an additional 1 to 5 years (Heinz, 1977; Miller, 1977).

At some locations where lodging is a serious problem, such as Canal Point, Fla., plots are about 3 m long, so that rope or nylon strapping can be interlaced among stalks and tied to posts placed at both ends of plots to prevent lodging (Dunckelman, 1959). In some hybridization programs at high latitudes, natural flowering in the field rarely occurs because of a limited number of daily photoinductive cycles or low temperature. In such cases, field plantings may serve only as a source of cuttings to plant in pots or cans in a greenhouse.

Clements and Awada (1965) and Coleman (1968) discussed factors that influence flowering of sugarcane and outlined the following optimum conditions for floral induction: a plant that has two to four elongated internodes exposed; photoperiod of 12 hours 16 min to 12 hours 26 min with high light level; day temperature of 26.7 to 30.6 C; night temperature of 21.7 to 23.2 C; vigorous plant cane (first year crop), but changing to a low nutritional status; and low soil moisture tension. In the Northern Hemisphere, floral initiation occurs from mid-August to mid-October and flowers emerge from mid-November to mid-January or later (James and Miller, 1971). Some *S. spontaneum* clones have different photoperiod requirements and as a group flower from early spring until late winter.

B. Growth Chamber and Greenhouse

Sugarcane is grown in pot cultures at all latitudes for photoinduction experiments and for all hybridization work at high latitudes where natural flowering rarely occurs in the field because of low temperatures. When temperatures can be less than 21 C, plants in pot cultures are left outside during the day and moved into heated greenhouses or photoinduction chambers at night (Fig. 1 and 2) (James, 1968; Paliatseas, 1962).

At Canal Point, Fla., single-bud seed pieces of chosen parental clones are planted upright in greenhouse flats filled with muck soil in February each year (James, 1968). In early April, the young plants are transplanted to 38-liter cans in a mixture of sand and muck soil. Cans of female parents are placed adjacent to outside racks equipped with tie bars to which developing stalks are tied to prevent lodging. A watering system constructed of plastic pipe fitted with nozzles provides an efficient means of watering plants during the growing season. Cans of male parents are placed on rail carts that can be moved into a heated crossing house at night to protect plants when temperature drops below 21 C during fall and winter (Fig. 1). Plants are fertilized at 3-week intervals from April to July, then fertilization is discontinued until October to assure low nutritional status during the floral initiation period in September. Various schemes were devised for managing plants in pot cultures, but the essential features are a potting medium that permits adequate drainage, proper fertilization, adequate water, and a method to prevent lodging as stalks elongate.

III. FLORAL CHARACTERISTICS

The inflorescence of sugarcane, referred to as an arrow or tassel, is an open panicle with main axis and lateral axes of first, second, and third orders (Artschwager et al., 1929). Branching is greatest at the lower end of the panicle and decreases in an apical direction (Fig. 3). Four to 100,000 or more spikelets occur in pairs alternately at rachis nodes. One spikelet of each pair is sessile and the other pedicellate (Fig. 4). Long, silky callus hairs are attached to the base of each spikelet. At maturity, the lateral axes of the inflorescence break below the spikelets at rachis nodes. The pedicellate spikelet breaks free, leaving the sessile spikelet attached to the rachis segment and the stalk of the pedicellate spikelet (Fig. 4).

Fig. 1—Sugarcane crossing house at Canal Point, Fla., showing rail carts for transporting male parents to the heated crossing house. The lighting system on the left is used to delay flowering. Tie bars between posts are used to support sugarcane stalks as they elongate during the growing season.

Fig. 2—Interior of crossing house at Canal Point, Fla., showing isolation cubicles. Sugarcane panicles that were bagged to save the first seeds to ripen are shown in the upper left-hand corner.

Fig. 3—Inflorescence of sugarcane.

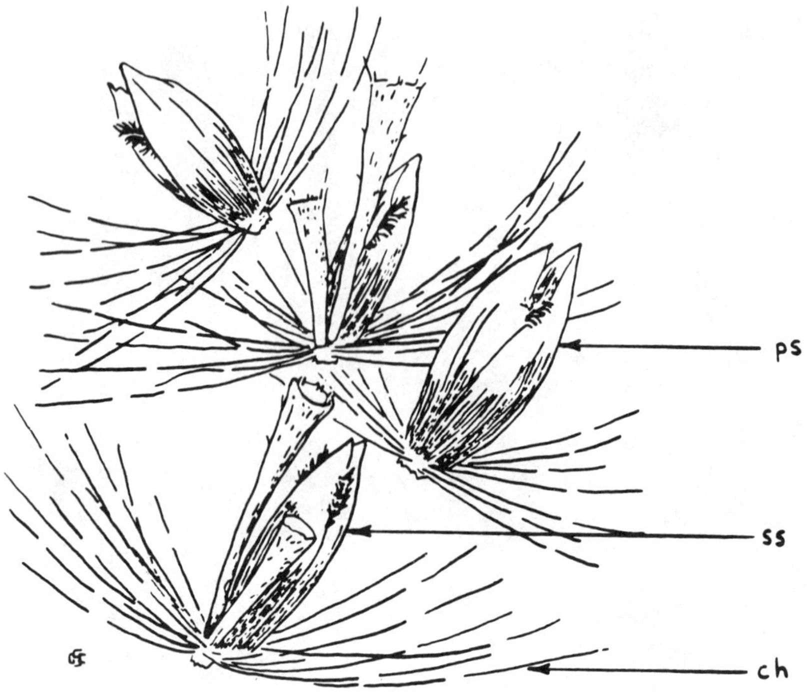

Fig. 4—Ripe seed (fuzz) of sugarcane showing breakage of the rachis, callus hairs (ch), sessile spikelets (ss), and pedicellate spikelets (ps).

Each spikelet contains a single flower consisting of three or four glumes, two lodicules, a whorl of three stamens, and a single ovary with two feathery, purple stigmas. Flowers that are about to open can be recognized by stigmas protruding from the closed glumes. As anthesis approaches, the filaments lengthen and the anthers are extruded. Often the flowers are imperfect with only rudimentary development of the ovary or with anthers devoid of pollen grains.

The sugarcane caryopsis is typical of grasses and remains enclosed in the glumes at maturity with stigmas attached (Fig. 4).

As the plant develops, internodes and leaf sheaths below the apical region become successively longer while leaf blades become successively shorter. The uppermost leaf consists of a sheath about 1 m long, fully enclosing the young panicle, with a blade only about 15 cm in length. This is known as the flag or boenting stage. Elongation of the peduncle causes emergence of the inflorescence. After emergence is complete, the panicle is well above the leaf canopy at a height of 3 to 6 m from the ground and in position for wind pollination to occur.

Anthesis may begin as the inflorescence starts to emerge from the flag leaf or as much as several days after full emergence. In all clones, anthesis takes place from the top of the panicle downward and from the ends of panicle branches inward. Some flowers usually open before sunrise each day, and the process is completed in 3 to 14 days. Following the opening of the flower, anther dehiscence usually takes place after some drying has occurred, if the humidity is high. Fertile anthers split along the antherior walls one-quarter to one-half the length of the anther from top to bottom to release pollen.

Stigmas are receptive when flowers open (Artschwager et al., 1929). Wind and humidity probably affect length of the receptive period because stigmas protrude from glumes. There has been little need to determine the period of receptivity because of the hybridization techniques that are employed throughout the world.

Degree of anther dehiscence and pollen viability ranges from 0 to 100% among sugarcane clones. Few complex species hybrids are completely male-sterile, and breeders must classify intended parental clones for use as male parents or female parents when making biparental crosses. This is usually based on percentage of anthers that dehisce and the percentage of released pollen that is viable as determined by the starch test (Van-Breemen et al., 1962), vital stains, or occasionally by germination of pollen (Moore and Jung, 1974).

Pollen germinates immediately after falling on the stigmatic papilla. The pollen tube grows down the style and enters the ovary (Artschwager et al., 1929). The fertilization process was not observed, but fertilization is considered to take place within 8 hours after pollination because fertilized egg sacs were detected within about 7 hours (Bremer, 1959; Dutt and Rao, 1933).

IV. ARTIFICIAL HYBRIDIZATION AND SELF-POLLINATION

A. Equipment

Facilities and methods required for a successful hybridization program with sugarcane are determined largely by the latitude where the work is undertaken. In more tropical climates between approximately 20° North and South latitudes, temperature is generally favorable for flowering and pollen fertility. Hybridization can be conducted in the open, usually among trees or other windbreaks, and sometimes under temporary or permanent shelters to protect inflorescences from wind and rain. The inflorescences are susceptible to breakage from high winds, particularly when laden with moisture.

At high latitudes, low temperature during the fall and winter adversely affects pollen fertility and depresses flowering (Brett, 1950). Heated greenhouses are required to protect at least the male parents during flower development and hybridization. Photoperiod houses may be required to induce flowering and thereby extend the range of parental clones available for hybridization.

At even higher latitudes where natural flowering rarely occurs, both male and female parents must be protected from low temperature by placing them in heated greenhouses. The restricted natural flowering may necessitate extensive photoperiod facilities for induction of most, if not all, parental clones. Because sugarcane is vegetatively propagated, all plants subjected to selection are derived directly from seed obtained by hybridization, and demands on hybridization programs range from several thousand to 2 million or more seeds each year. Other equipment depends on the method selected for maintaining the male and female parents during pollination and seed maturation.

B. Preparation of the Female

Sugarcane breeders in Barbados initially emasculated the small sugarcane flowers with the aid of a dissecting microscope while working in the field on platforms about 3 m above the ground. This practice was soon abandoned in favor of selecting female parents that were highly male-sterile. Various emasculation techniques were tried such as the use of suction, hot water, chemicals, and high humidity to prevent anther dehiscence (Stevenson, 1965). Low temperature during flower development at least partially destroys pollen viability and has been used to emasculate male-fertile clones by growing them outside in subtropical climates (Dunckelman, 1959) or placing them in cold rooms at 15 C during late stages of flower development (Paliatseas, 1977). A satisfactory emasculation technique is not yet available and, with few exceptions, sugarcane breeders select clones for female parents that are partially or highly male-sterile. In Barbados and other locations, portable enclosures (lanterns) were developed that could be placed over female inflorescences in the field to provide isolation from foreign pollen (Stevenson, 1965).

A marcotting (air-layering) method was developed in India that permitted isolation of field-grown flowering stalks of both male and female parents (Venkatraman and Thomas, 1926). At least several weeks before flowers emerged, two halves of a clay pot were tied in place around a stalk so that one or two root bands near the nodes of the stalk were covered when the pot was filled with moist soil or other rooting medium. After a few weeks, a root system developed from the root primordia in the root band. The stalks were cut below the pot either before or after flower emergence, transported to an isolated area, and hybridized under temporary crossing sheds (Sahib et al., 1927).

The marcotting technique was modified later by replacing pots with stove pipes (Paliatseas et al., 1953). Finally, plastic film was used to enclose sphagnum moss or other rooting medium (Dutt and Hussainy, 1956).

In Hawaii, Verret (1925) discovered that cut stalks would live for several days or weeks and many would develop mature seed when the cut ends were immersed in a 500 ppm sulfurous acid solution. The acid solution was modified later to include phosphoric acid (Mangelsdorf, 1953) and still later to include sulfuric and nitric acids (Warner, 1961). The present recommended solution consists of 150 ppm sulfurous acid + 75 ppm phosphoric acid + 37.5 ppm sulfuric acid + 37.5 ppm nitric acid (Heinz, 1977). Solutions are usually changed occasionally and replenished with fresh solutions or acid stock. In some programs, an internode is cut from the base of stalks one or two times during the period stalks are in solution. Rubber or plastic containers are used in preference to metal containers to prevent neutralization of the acid.

Over the years, the solution method for maintaining female parents was adopted by many sugarcane breeders. At some locations, it was necessary to use rainwater (Stevenson, 1965) or deionized water (James, 1968) to make the solution, particularly when water from the normal source was highly alkaline. The advantage of the solution method is that neither male nor female parents require prior preparation. Labor costs are reduced and all inflorescences that emerge can be hybridized. The flowering stalks are simply cut at the proper time after flower emergence and moved to the hybridization area.

C. Pollination

In the early years of hybridization, male parents had to be replaced each day because a method to keep them alive was not available. Much of this work was done at night to have the male parents in place when female flowers opened about sunrise (Jeswiet, 1927b; Stevenson, 1965).

With development of the marcotting and solution techniques, it was no longer necessary to replace male parents each day. Typical pollination procedures now consist of cutting both male and female flowering stalks as anthesis begins and moving them to a crossing shelter where they are placed together in a crossing lantern or cubicle (Fig. 2 and 5). Male flowering stalks are shaken each morning at about 0800 hours to dislodge pollen. After 10 to 14 days, male parents are discarded and female parents are retained an additional 10 to 20 days until seed is ripe.

In Hawaii, the Experiment Station of the Hawaiian Sugar Planters Association conducts an extensive hybridization program under tropical conditions (Heinz, 1977). Two types of crosses are made: biparental, in

Fig. 5—Crossing shelter at Maunawili, Hawaii, showing lanterns for biparental sugarcane crosses. (Photo courtesy D. J. Heinz, Hawaiian Sugar Planters Association).

which one male parent and one highly male-sterile female parent are placed together in a lantern for isolation (Fig. 5), and melting pot (modified polycross) in which several elite male parents and a large number of female parents are placed together under a crossing shelter. Stalks are held upright by inserting them through large mesh wire suspended in a horizontal position about 1 m above the ground. The cut ends are placed in solution containers. Solutions are changed twice weekly and replenished on intervening days. Male parents in biparental crosses are discarded after 10 to 12 days, but male parents are retained in melting pot crosses and all seed is harvested when ripe.

At Canal Point, Fla., the SEA-USDA conducts a hybridization program in a large plastic-covered crossing house (ca. 61 m long × 9 m wide × 7 m high) in which temperature is maintained above 21 C (Fig. 1). Cubicles on each side of the house provide isolation from foreign pollen (Fig. 2). Either the marcotting or solution technique is used to maintain both male and female parents during hybridization. When stalks are marcotted, male parents are cut below the rooted section as anthesis begins, leaf blades are removed, and the stalks are carried to the crossing house. The enclosing leaf sheath is removed to completely expose the panicle. The stalks are tied to support bars with the rooted section suspended in the air. Plastic bags are placed over the rooted sections, tied to stalks, and filled with water to keep the stalks alive. Female parents are cut and prepared for crossing in the same manner as male parents. Rooted sections are placed in troughs of tap water that is circulated and aerated by continuous pumping.

For the solution technique, cut ends of stalks from the male parents are placed in 1-liter jars containing a weak acid solution. The jars are tied to the stalks and the stalks are tied to the support bars in cubicles with the jars suspended in the air. Cut ends of stalks from the female parents are placed in rubber tubs containing a weak acid solution.

Height of male panicles for both the marcotting and solution techniques is adjusted so that they will be about 0.3 m above female panicles. All female stalks are tied to the same support bars as the male parents to keep them upright. Usually one male parent and several female parents are placed in a cubicle. At 0800 hours each morning, the support bars are shaken to disperse pollen. Male parents are discarded after a 10 to 14-day pollination period, and female parents are retained until seed is ripe.

D. Factors Affecting Efficiency

Several approaches have been used to alter flowering dates, such as time of planting (Stevenson, 1965), growing parental clones at different elevations (Van-Breemen et al., 1962) and on different soil types (Grassl and Belcher, 1953), and leaf trimming and chemical sprays (Buzacott, 1959).

Most research to alter time of flowering has dealt with manipulation of photoperiod. Researchers generally have attempted to induce earlier flowering by decreasing the natural photoperiod during the summer (Chilton and Paliatseas, 1956; Paliatseas, 1962; James and Miller, 1971). Others have attempted to delay flowering of early-flowering clones by interrupting the dark period with light to delay floral initiation and subsequent flowering

(Coleman, 1962; Moore and Heinz, 1971; James, 1972). More recently, extension of the photoperiod after floral initiation occurred naturally was shown to delay flowering by as much as 3 months (James and Miller, 1971; Moore and Heinz, 1971).

With proper application of current technology, sugarcane breeders can synchronize flowering of most clones, except those of *S. officinarum*. There is great variation in specific photoperiod requirements among *Saccharum* clones, and sugarcane breeders must determine the exact requirements for asynchronous and non-flowering clones that are chosen for hybridization.

The method selected for maintaining female stalks influences seed set, at least in some years. For example, at Canal Point, Fla., seed set for the weak acid solution method was only 60 to 92% as high as for the marcotting method (Miller, 1977). Selection of strong pollen producers as male parents increases seed set, but in most programs it is necessary to use weak pollen producers as male parents to utilize a wide range of germplasm. Seed set is extremely variable and ranges from 0 to 5,000 or more viable seed from one inflorescence. Average seed set per inflorescence was 520 for a 9-year period at Canal Point and ranged from 214 to 774 (James and Miller, 1973). Average seed set above 1,000 per inflorescence is unusual (Paliatseas, 1977).

V. NATURAL HYBRIDIZATION

Natural hybridization is not an important technique in sugarcane. Polycross blocks occasionally are established in the field, but asynchronous flowering within as well as among clones reduces the effectiveness of the method.

VI. SEED DEVELOPMENT, HARVEST, AND STORAGE

Seed maturity in sugarcane follows the same pattern as anthesis. Ripening of seeds begins at the top of the panicle and proceeds downward to the base. The minimum length of time from the beginning of the pollination period to seed maturity is about 22 days (Heinz, 1977). As the inflorescence begins to disintegrate, the ripe seeds at the top are lost before seeds at the base are mature. In some programs, panicles are harvested 22 to 36 days from the beginning of the pollination period after one-third of the seed has fallen (Heinz, 1977).

Stevenson (1965) described a method used in Barbados of taping paper cones around and below panicles to catch the first seeds to ripen and fall. At Canal Point, Dunckelman (1963) enclosed panicles in lightweight paper clothing bags and increased seed production by 50% when panicles were harvested at 30 to 36 days after the beginning of pollination compared to harvesting when the first seeds began to fall. Currently, panicles are bagged about 20 days after pollination begins. Panicles of flowering stalks maintained in weak acid solution are harvested 10 days after they are bagged and those of marcotted stalks 14 to 16 days after they are bagged.

At harvest, peduncles are cut below the panicles and the bagged panicles are dried in a forced air oven at 29 to 40 C until spikelets can be hand-stripped from panicle branches, usually 2 or 3 days. Seeds are placed in kraft paper or plastic bags. Bags are weighed and identified by year, cross number, and parentage. Seeds are either planted immediately or stored at temperatures ranging from −10 to −30 C (Heinz, 1977; Miller, 1977). At Canal Point, a 1-g random sample is obtained from each cross for germination tests before planting or storage to estimate seed production and to determine the proper planting rate in greenhouse flats.

Sugarcane seed (fuzz), being a mass of spikelets with callus hairs attached, is difficult to plant evenly in flats. Various methods were tried to remove callus hairs, such as use of sulfuric acid and burning and rubbing. A procedure utilizing a rice scarifier has now been developed that seems to be both effective and efficient (Breaux, 1977).

REFERENCES

Anonymous. 1971. Report of Standing Committee on Germplasm and Breeding. Int. Soc. Sugar Cane Technol. Proc. 14:1651–1654.

Artschwager, E., E. W. Brandes, and R. C. Starrett. 1929. Development of flower and seed of some varieties of sugarcane. J. Agric. Res. 39(1):1–30.

Breaux, R. D. 1977. De-fuzzing sugarcane seed with a small seed scarifier. Sugar y Azucar 72(6):34 (Abstr.).

Bremer, G. 1959. Increase of chromosome number in species hybrids of *Saccharum* in relation to embryo-sac development. Bibl. Genet. 18(1).

Brett, P. G. C. 1950. Flowering and pollen fertility in relation to sugarcane breeding in Natal. Int. Soc. Sugar Cane Technol. Proc. 7:43–56.

Buzacott, J. H. 1959. Methods of cane breeding used by the Queensland Bureau of Sugar Experiment Stations. Int. Soc. Sugar Cane Technol. Proc. 10:690–694.

Chilton, S. J. P., and E. D. Paliatseas. 1956. Studies on the flowering of sugarcane. Int. Soc. Sugar Cane Technol. Proc. 9:652–656.

Clements, H. F., and M. Awada. 1965. Experiments on the artificial induction of flowering in sugarcane. Int. Soc. Sugar Cane Technol. Proc. 12:795–812.

Coleman, R. E. 1962. Control of flowering and the use of pollen storage as techniques in a sugarcane breeding programme. Int. Soc. Sugar Cane Technol. Proc. 11:533–540.

———. 1968. Physiology of flowering in sugarcane. Int. Soc. Sugar Cane Technol. Proc. 13:992–1000.

Daniels, J., P. Smith, and N. Paton. 1975. The origin of the genus *Saccharum.* Int. Sugar Cane Technol. Sugarcane Breed. Newsl. 36:24–31.

Dunckelman, P. H. 1959. Transition: Effectiveness and implications of breeding sugarcane in an indoor environment at Canal Point, Florida. Sugar Bull. 37(19): 229–252.

———. 1963. Production crossing and technical improvements in sugarcane breeding at Canal Point, Florida—1961–1962 season. Sugar J. 25(10):6–20, 36.

Dutt, N. L., and S. A. Hussainy. 1956. Some recent features of cane breeding technique at Coimbatore. Int. Soc. Sugar Cane Technol. Proc. 9:607–617.

———, and S. Rao. 1933. Observations on the cytology of the sugarcane. Indian J. Agric. Sci. 3:1.

———, J. T. Rao, and S. A. Hussainy. 1953. On certain recent aspects of the breeding programme at Coimbatore. Int. Soc. Sugar Cane Technol. Proc. 8: 402–410.

Grassl, C. O. 1977. The origin of the sugar producing cultivars of *Saccharum*. Int. Soc.Sugar Cane Technol. Sugarcane Breed. Newsl. 39:8–24.

———, and B. A. Belcher. 1953. Sugarcane breeding techniques practiced at Canal Point, Florida. Int. Soc. Sugar Cane Technol. Proc. 8:443–451.

Heinz, D. J. 1977. Personal communication. Experiment Station, Hawaiian Sugar Planters Association, P.O. Box 1057, Aiea, HI 96701.

James, N. I. 1968. Improved methods and facilities in the sugarcane crossing program at Canal Point, Florida, during 1967–68. Sugar Bull. 46(21):7–11.

———. 1972. Delayed flowering of *Saccharum spontaneum* L. and related hybrids. Crop Sci. 12:425–427.

———, and J. D. Miller. 1971. Photoperiod control in the USDA sugarcane crossing program. Int. Soc. Sugar Cane Technol. Proc. 14:341–347.

———, and J. D. Miller. 1973. Sugarcane crossing program at Canal Point, 1972–73. Sugar Bull. 51(23):10–13.

Jeswiet, J. 1927a. The history of sugarcane selection work in Java. Int. Soc. Sugar Cane Technol. Proc. 2:115–122.

———. 1927b. The flowering of sugarcane. Int. Soc. Sugar Cane Technol. Proc. 2:133–137.

Mangelsdorf, A. J. 1953. Sugarcane breeding in Hawaii, Part II. Hawaiian Planters Rec. 54:101–137.

Miller, J. D. 1977. Personal communications. Sugarcane Field Station, Star Route Box 8, Canal Point, FL 33438.

Moore, P. H., and D. J. Heinz. 1971. Increased post-inductive photoperiods for delayed flowering in *Saccharum* sp. hybrids. Crop Sci. 11:118–121.

———, and W. L. Jung. 1974. Studies in sugarcane pollen I. In vitro germination of pollen. Phyton 32(2):147–157.

Paliatseas, E. D. 1962. Further studies on flowering of sugarcane in Louisiana. Int. Soc. Sugar cane Technol. Proc. 11:504–515.

———. 1977. Personal communications. Dep. of Plant Pathology, Louisiana State Univ., University Station, Baton Rouge, LA 70893.

———, R. J. Steib, P. H. Dunckelman, and S. J. P. Chilton. 1953. The production of true seed of sugarcane in Louisiana. Int. Soc. Sugar Cane Technol. Proc. 8:361–364.

Sahib, R., T. S. Venkatraman, and R. Thomas. 1927. Sugar cane hybridization sheds. Int. Soc. Sugar Cane Technol. Proc. 2:124–131.

Stevenson, G. C. 1965. Genetics and breeding of sugarcane. Longmans, Green and Co. Ltd., London.

Van-Breeman, J. F., Lii-Jang Liu, T. O. Ellis, and G. Arceneaux. 1962. Effect of elevation on arrowing and pollen fertility in sugarcane. Int. Soc. Sugar Cane Technol. Proc. 11:540–545.

Venkatraman, T. S., and R. Thomas. 1926. Sugarcane breeding technique. Isolation of live arrows from undesired pollen through artificial rooting of canes. Agric. J. India 21(3):203–209.

Verret, J. A. 1925. A method of handling cane tassels for breeding work. Hawaiian Planters Rec. 29:84–94.

Warner, J. N. 1961. Int. Soc. Sugar Cane Technol. Sugarcane Breed. Newsl. 7:6.

45 Sunflower

Walter Dedio
Eric D. Putt
Agriculture Canada
Morden, Manitoba

The cultivated sunflower [*Helianthus annuus* var. *macrocarpa* (DC.) Ck11.] is one of the four most important annual crops in the world grown for edible oil. There is archaelogical evidence that the present sunflower with a single head was grown in the prehistoric North American Indian culture and may have been domesticated before the introduction of corn (*Zea mays* L.). The Indians used the crop as a source of food. Early Spanish explorers introduced the plant to Europe where it was largely used as an ornamental.

Extensive use of sunflower as a source of edible oil began in 1830 after the crop reached Russia. The sunflower has become the main source of edible oil in the USSR, other eastern European countries, and Argentina. The crop was re-introduced to North America from Europe in the late 19th century with initial interest being its use as silage. Commercial extraction of oil began in 1946 in Manitoba when Canada promoted it as an oilseed crop. Some of the crop grown in North America is for confectionery and birdfeed purposes.

In the last 10 to 15 years, production increased greatly in some other countries, including the United States where plantings in 1977 were approximately 1,000,000 ha. Much of the stimulus in the United States has been due to the importation of Russian germplasm with oil content consistently above 40%, and the discovery of a cytoplasmic male sterile and fertility restorer system which is used to produce hybrid seed. The introduction of hybrid seed has improved the yield and uniformity of the crop, as well as disease resistance.

I. PARENTAL MATERIAL

The genus *Helianthus* comprises over 100 species but about 50 species found in North America have been studied in more detail. Considerable differences of opinion exist as to the classification and grouping of the *Helianthus* species. Heiser (1965) pointed out three distinct groups in the genus: (1) North American annuals and tap-rooted perennials; (2) North American herbaceous perennials, mostly rhizomatous; and (3) South American, more or less shrubby perennials. More recently, Heiser et al. (1969) delineated the second group into western perennials or Ciliares and eastern perennials or Divaricati.

Not all possible hybrid combinations have been made among the species in the genus. Most of the species in the first group can be crossed with *H. annuus,* a member of the same group, but the seed set is usually low. The percentage seed set and germination may depend on the species used as female.

H. annuus itself is a highly variable species. Wild *H. annuus* is abundant, particularly in north central U.S. It crosses readily with the cultivated *H. annuus* and creates a problem of contamination for hybrid seed producers. Wild *H. annuus* is usually taller than the cultivated sunflower with profuse branching. It has been a source of genes for fertility restoration, and for rust and Verticillium resistance. The haploid chromosome number of *Helianthus* species is 17 or multiples of it. Both the cultivated and wild *H. annuus* are diploids with chromosome number (n) of 17.

A widely exploited cross has been *H. annuus* with *H. tuberosus* L. (Jerusalem artichoke). The latter is a hexaploid, perennial species which is grown occasionally for its tubers. The initial cross can be achieved easily and sometimes occurs naturally, but backcrosses using *H. annuus* as the recurrent parent are more difficult to obtain. Pustovoit and Pustovoit (1972) reported successful backcrosses by exposing the F_1 hybrids to thermal shock during meiosis.

H. petiolaris Nutt., an annual, diploid species, has been crossed by *H. annuus* (LeClercq, 1969). Progeny of the cross revealed a cytoplasmic male sterility system which is presently exploited on a commercial scale to produce sunflower hybrids. *H. petiolaris* closely resembles wild *H. annuus,* but can be distinguished by a white deposit on its face. Mature heads of *H. petiolaris* which have dispersed their seed have lighter colored chaff than those of *H. annuus.*

Other annuals species that cross readily with cultivated *H. annuus* are *H. argophyllus* and *H. paradoxus.* These species are being investigated by American breeders as a possible source of insect resistance (Fick et al., 1976). Two diploid species occurring frequently in the North American plains, *H. maximiliana* Schrad. and *H. giganteus* L., have been crossed to cultivated *H. annuus* by using wild *H. annuus* as a bridge (Whelan, 1976). Other species reported by Soviet breeders that can be crossed with cultivated *H. annuus* include *H. grosseserratus, H. rigidus, H. decapetalus* and *H. hirsutus* (Table 1).

Table 1—List of *Helianthus* species which have been crossed with cultivated *H. annuus*.

Species	Haploid chromosome no.	Fertility of hybrids	Plant habit
Wild *H. annuus*	17	High	Annual
H. petiolaris	17	High	Annual
H. argophyllus	17	High	Annual
H. paradoxus	17	High	Annual
H. tuberosus	51	Medium	Perennial
H. rigidus	51	Medium	Perennial
H. maximiliani	17	Low	Perennial
H. giganteus	17	Low	Perennial
H. grosseserratus	17	Low	Perennial
H. decapetalus	17,34	Low	Perennial
H. hirsutus	34	Low	Perennial

The largest collection of *Helianthus* species in North America is being maintained by the United States Department of Agriculture at Bushland, Tex. Smaller collections are kept by Government Research Stations at Fargo, North Dakota and Morden, Manitoba. The USDA world sunflower collection consisting of about 500 entries, mostly cultivated types, from over 30 countries is maintained at the North Central Regional Plant Introduction Station at Ames, Iowa. In Europe, large collections of wild and cultivated *Helianthus* cultivars are maintained at the Station d'Amelioration des Plantes, Clermont-Ferrand, France, at the Research Institute for Cereals and Industrial Crops in Fundulea, Romania, the N.I. Vavilov All-Union Research Institute of Plant Industry (VIR) at Leningrad, USSR, and by the All-Union Scientific Research Institute of Oil-Bearing Crops (VNIIMK) at Krasnodar, USSR.

II. PLANT CULTURE

A. Field

Sunflower will grow in a wide range of soils, but prefer those that warm up quickly, such as well-drained sandy soils. Sunflower can tolerate drought better than many annual crops because they are more efficient in extracting soil moisture. Photosynthesis has been found to continue in sunflower at high levels of moisture stress (Boyer, 1970). Soil moisture is most critical at flowering time for proper seed development. Soil fertility requirements are similar to those for most other crops with the best growth being obtained where levels of N are adequate (Zubriski and Zimmerman, 1974). Sunflowers are adapted to high light level, and some studies in France have shown that sunflowers grown in shade produce low seed yields. Commercially, sunflowers may be planted at populations of between 4 to 7 plants per m^2 without significantly affecting the yield. It may be desirable to have a lower plant density in breeding nurseries.

Many sunflower lines and cultivars are short-day plants, although some of the more popular lines are not sensitive to photoperiod. Breeders have observed that the shorter photoperiod in Florida and Hawaii hastens the initiation of flower primordia of some genotypes, while others are not affected. Seed yield often is reduced under these growing conditions.

Sunflower requires warm temperatures for fertilization and seed development. Cool temperatures can delay or prevent the pollination and fertilization processes by affecting the activity of pollinators and the metabolism of the plant. Temperatures in excess of 30 C also may prevent normal pollination and fertilization. This particularly applies to plants with erect heads where the upper part of the head is exposed to direct sunlight which can raise the temperatures of the pollen and stigma to injurious levels. Smith (1978), for example, has noted pale pollen following periods of high temperature and bright light in California. Even though fertilization does not occur, the ovary wall will usually develop almost normally to produce empty hulls.

Many companies and public agencies have breeding programs in Hawaii, Florida, or South America where seed is increased and crosses are made in the winter. Seed set usually is lower in winter nurseries, but the crop matures and can be harvested considerably earlier than in the temperate zones.

B. Growth Chamber and Greenhouse

Sunflower can be grown successfully to maturity in the greenhouse or growth chamber. The plants may be grown in 15 or 20 cm pots provided soil that drains easily is used. Because sunflower is a large user of water, daily watering is necessary. Nutrients should be supplied frequently either in soluble form with water or in the slow release form.

An important requirement in the greenhouse is sufficient light. Low light level, particularly in the blue and ultra-violet spectra, produces tall plants with weak stems that break easily. Plants produced in the greenhouse generally have smaller heads, smaller seed, and lower seed yields than those grown under normal field conditions. The senior author and other breeders have found that greenhouse-produced seed tends to be more dormant and requires a longer rest period than field-produced seed.

Sunflower may be dwarfed to facilitate crossing and moving in the greenhouse. A reduction in height of up to 40% has been obtained with (2-chloroethyl) trimethylammonium chloride (CCC) and succinic acid, 2,2-dimethylhydrazide (SADH) (Dorrell, 1973).

III. FLORAL CHARACTERISTICS

Sunflower is a highly cross-pollinated plant. The individual disc flowers are effectively protandrous and the positioning of the stigma above the anthers make self-pollination difficult. A genetically controlled system of self-incompatibility in certain lines prevents the pollen from penetrating the styles and carrying out fertilization (Putt, 1941; Schuster, 1961).

Sunflower is pollinated mostly by insects. Bees are frequent visitors to flowers on warm, sunny days. Little pollination is accomplished by wind (Putt, 1940). Sunflower pollen is rather heavy and sticky and most of it drops on the leaves or on the ground in clumps of five or more grains.

Fig. 1—Sunflower head showing development of rings of flowers proceeding from the outside to the center.

The head of the sunflower (*Helianthus annuus* L.) is a compound inflorescence composed of many individual flowers in a large disc subtended by large ray flowers (Fig. 1). The ray flowers are normally asexual, but some may produce pollen. The disc flowers are perfect with petals and five anthers that are united in separate tubes. The disc flowers are arranged in concentric circles radiating from the center of the head. The ray flowers open first and flowering then proceeds from the periphery to the center of the head at the rate of one to four rows per day.

The development of individual disc flowers has been described by Plotnikov (1940) and Putt (1940). Early in the morning the staminal filaments rapidly elongate and exert the anther tube from the corolla (Fig. 2A). This occurs about 0700 hours on warm, sunny days, but later on cool, wet days. Immediately after this stage is reached, the anther locules dehisce, releasing their pollen inside the anther tube (Fig. 2B). An elongation of the lower portion of the style follows, which forces the two-lobed pubescent stigma up the anther tube (Fig. 2C). During its upward movement the stigma pushes the dehisced pollen through the upper end of the anther tube. The stigma is not receptive at this stage because the two lobes are held together covering the inner receptive surface. At the same time, the staminal filaments lose turgidity and the anther tube begins to recede into the corolla. The next day, the stigma lobes are fully emerged and have separated to expose the receptive surfaces to pollination (Fig. 2D). Pollination and fertilization soon occur, after which the stigma withers and recedes. By the morning of the second day, after the flower opens, both the anther and the stigma have receded almost completely into the corolla (Fig. 2E).

IV. ARTIFICIAL HYBRIDIZATION AND SELF-POLLINATION

A. Equipment

Emasculation of the female parent can be performed with forceps. Workers have individual preferences for the type of forceps. Some prefer a fine, sharp-pointed surgical type; others use a somewhat larger, coarser type with points about 3 mm wide.

Putt (1941) has described a knife for removing unwanted flowers that is made by splitting a pair of cover slip forceps and mounting each portion in a small wooden handle. The tip is sharpened to obtain a knife with a small blade set at an angle to the handle (Fig. 3). A disinfectant, such as ethanol, is used for sterilizing the forceps between emasculations to prevent unwanted cross-pollination.

B. Preparation of the Female

The ray flowers and bracts usually are removed before emasculation to make the flowers on the disc more accessible and to eliminate a large surface area on which pollen could lodge. Unless a large quantity of seed is desired, it is prudent to emasculate only those flowers which open on a single day and remove all the others. This precaution reduces the hazards of uncon-

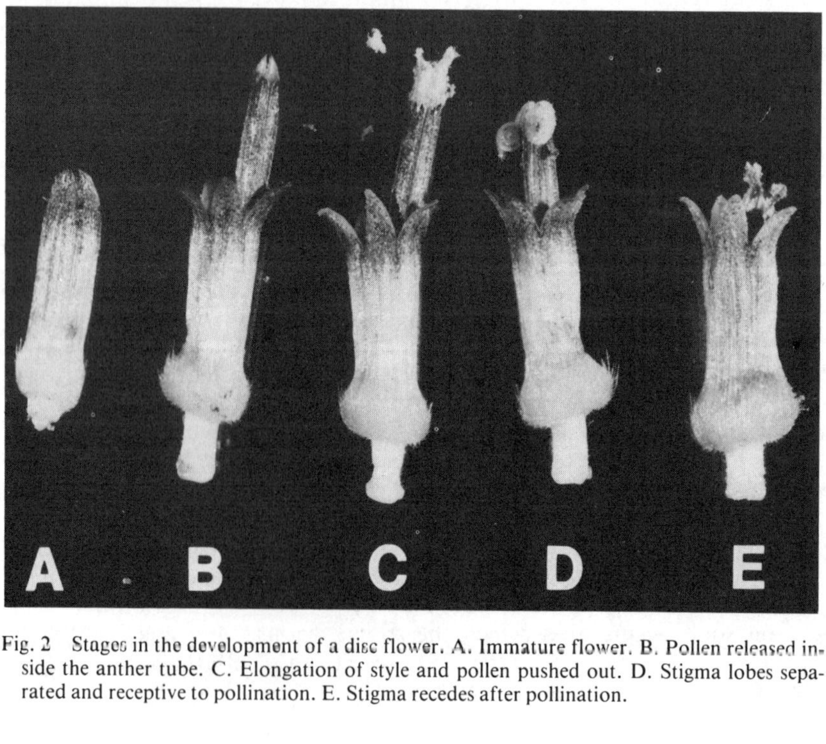

Fig. 2 Stages in the development of a disc flower. A. Immature flower. B. Pollen released inside the anther tube. C. Elongation of style and pollen pushed out. D. Stigma lobes separated and receptive to pollination. E. Stigma recedes after pollination.

trolled cross-pollination, as well as self-pollination. The flowers opening prior to the day of emasculation can be removed from their ovaries by a simple sideways pulling motion with the thumb and forefinger.

In the field, emasculation should be carried out very early in the morning. On sunny, warm days this would be prior to 0700 hours, but may be later on cool, cloudy days and in the greenhouse in winter. The ideal time to emasculate is in the period when the anther tube is extended sufficiently to be grasped with the forceps but the pollen has not yet dehisced. In practice, however, because of the short duration of this period, the anther tube often is removed after dehiscence (Fig. 2B and 4A), but before the stigma has grown far enough into it to be injured or for the stigmatic lobes to separate when the tube is removed. Free pollen on the outside of the stigma lobes must be blown off.

Undeveloped central florets are removed, usually by cutting them off with a knife at a point just above the ovaries (Fig. 4B and C). A few flowers closely adjacent to those emasculated cannot be cut off with a knife without danger of damaging the emasculated flowers. These can be removed with forceps.

Fig. 3—A knife used for cutting off the undeveloped central florets.

The stigma normally remains receptive for 3 to 5 days, although its viability can be preserved up to 17 days under field conditions (Plotnikov, 1940). Stigmas of flowers which are not fertilized continue growth for several days to form a coil which will allow the stigmatic surface to contact pollen adhering to the outer surface of the stigma lobes (Fig. 2D).

Emasculated flowers may be protected from being pollinated with unwanted pollen by covering the head with paper, cotton, or perforated plastic bags.

Emasculation is not necessary in some cases. Heads of male sterile or highly self-incompatible plants, for example, can be pollinated directly provided they have been protected to prevent undesirable outcrossing.

C. Pollination

Self-pollination is achieved by placing a bag over the head before anthesis or anther dehiscence. The amount of seed set depends on the degree of self-compatibility of the lines and type of bag used. Cloth bags have been found to be the most efficient among several kinds tested for effecting self-fertilization of sunflower (Kalton, 1951; Putt, 1940). Seed set was increased considerably when heads were manipulated by brushing with cotton batting (Putt, 1940).

Pollen for crossing is collected by placing bags over the heads of the male parents a day or two before flowering begins (Fig. 5). Paper bags are most convenient and adequate pollen will be found in the bags 1 or 2 days after flowering commences. It is best to use the pollen shortly after collection. According to Arnoldova (1926), pollen can be stored for a year, but other workers have found that viability is lost within a month whether the pollen is stored at room temperatures or is refrigerated (Plotnikov, 1940; Putt, 1941).

Pollen should be applied when the stigmatic lobes have separated and have exposed their receptive surfaces (Fig. 4C). Receptive stigmas are brushed with a pollen laden cotton wad or camel's hair brush usually a day after emasculation. Some breeders prefer the cotton wad because pollen adheres to it better and it can be discarded after one pollination, thus reducing the hazard of contamination and removing the need to sterilize with ethanol.

Fig. 4—Stages in the emasculation of a sunflower head. A. Before emasculation. B. Several rings emasculated with part of the outside rings removed. C. A head 1 day after emasculation showing the stigma lobes ready for pollination with ray and central florets removed.

Pollination in field nurseries can be effected simply by rubbing heads from the two parents against each other. If one of the heads is emasculated or male sterile, the cross is one-way. Because of the high degree of self-incompatibility in the sunflower, rubbing unemasculated heads of two normal lines will result in reciprocal crosses.

The pollinated head is maintained in isolation by rebagging. The bags also serve as a protection against birds. Identity of the cross may be lettered on the bag or indicated on a tag attached to the plant, depending on the preference of the breeder and the procedures used from harvest to storage of seed.

Fertilization of the flowers is indicated by withering and receding of the stigmas. This condition may be observed the morning after pollination. The success rate of pollination should be almost 100%. Lower seed sets usually result from non-viable pollen or stigmas.

Fig. 5—Sunflower head just before anthesis.

D. Factors Affecting Efficiency

The most common field plan to facilitate crossing places the female and male parents of planned hybrids adjacent to each other to facilitate the transfer of pollen. This may require planting by hand rather than using the spacings and seed arrangements feasible with a commercial planter. Plant population normally is not critical in field nurseries and the population often is lower than that used commercially. A lower population results in larger seeds and easier maneuverability among plants.

The cultivated sunflower normally has only one head per plant which requires the plants to bloom at the same time for cross-pollination to take place. Inbred lines with a branching character are available which have numerous heads which enables the plant to produce pollen over a longer period than the single-head type. The branching trait is conditioned by recessive genes in all commercial lines used for hybrid seed production, thus branching does not appear in commercial hybrids (Fick and Zimmer, 1974). The branching character evident in wild *Helianthus* species generally is controlled by dominant genes.

Branching can be induced in normal sunflower plants with a single head by removing the apical floral bud to stimulate the formation of new floral buds in the leaf axils. The time of flowering has been hastened in *H. tuberosus* by shortening the photoperiod (Shchibrya, 1938). We have observed more vigorous initial growth and earlier flowering in *H. annuus* following application of high phosphate fertilizer.

Hybrid sunflower seed cannot be distinguished from self-pollinated seed on the female parent because the hull characteristics are controlled by the female genotype.

Self-pollination and fertilization might inadvertently occur in making controlled crosses because it is difficult to achieve absolute emasculation. This problem is more common when self-compatible lines are used as the female parent. Hence, some method of distinguishing hybrids from maternal plants is desirable. Usually hybrid plants are more vigorous than either of the parents and this vigor may be observed at the seedling stage. Other characters controlled by dominant genes that may aid in differentiating hybrids from the female parent include anthocyanin pigment in the stalks, leaves, corolla and anthers; yellow coloration and ligulate shape of petals; and some types of branching.

V. NATURAL HYBRIDIZATION

Natural hybridization in sunflower is made possible by using male sterility and self-incompatibility systems. Both systems have been exploited in commercial production of hybrids.

The first hybrids used for commercial planting were developed in Canada using a type of self-incompatibility system or conversely a feature of high cross compatibility (Unrau and White, 1944). The female parent was not highly self-compatible or was a sparse pollen producer. It was inter planted with an abundant pollen producer, and a proportion of its seed was from hybridization. The proportion of fertilization occurring as a result of cross-pollination depended on the ratio of male to female plants, degree of isolation, and weather conditions which would favor insect activity. In commercial practice the percentage of hybrid seeds averaged only from 39 to 45%, with some samples being as low as 19% (Putt, 1962). The full potential of heterosis was not realized and other means of exploiting it were sought.

Both genetic and cytoplasmic male sterility occur in the sunflower. Genetic male sterility is found frequently in nature and has been used for commercial hybrid seed production in Romania. This type of sterility is controlled by any one of at least five recessive genes (ms_1, ms_2, ... ms_5). Only ms_1 has been exploited commercially because it is linked closely with the locus for a dominant gene controlling anthocyanin production (Vranceanu and Stoenescu, 1970). In the coupling phase, this linkage permits the producer to remove purple seedlings which will produce male-fertile heads. The male-sterile plants will lack the purple color characteristic of anthocyanin. This system theoretically should result in 100% hybridization, but the purple color is not always expressed and some male-fertile plants will escape removal.

More recently, LeClercq (1969) discovered cytoplasmic male sterility (CMS) which is used extensively now for hybrid seed production. Shortly after its discovery, fertility restoring genes were discovered (Anonymous, 1970). In hybrid seed production, rows of the CMS line are interplanted with rows of a fertility restorer line in ratios from 2:1 to 6:1. If the time from planting to flowering differs for the parents, they must be planted at different dates to foster coincidence of bloom. Other practices to overcome this difference include planting the earlier blooming parent deeper to delay emergence and blooming or using a fertilizer practice which will hasten bloom of the later parent.

VI. SEED DEVELOPMENT, HARVEST, AND STORAGE

Seed set can be detected by the plumpness of the seed and the uniform dense color of the pigmented layer when it is a character of the female parent. These characteristics are usually visible at about 20 days after fertilization. Sunflower heads should not be harvested before the onset of yellow color in the head and browning of the bracts. This stage usually occurs at about 45 days after fertilization. The junior author has observed that plantings in greenhouses can be hastened to maturity by ceasing to apply water 20 days after blooming terminates. If not completely dry when harvested, the heads can be dried in an oven at about 40 C for several days.

Individual heads can be threshed easily, without hazard of mixing, with a circular, wooden disc mounted on a stand above a chute (Fig. 6). Several wires are mounted, equally spaced across the face of the disc. The disc is rotated by a motor and the rubbing action of the wires dislodges the seed when the sunflower head is pushed against the disc. If only a few heads are involved in a program other simple devices can be used in removing seed

Fig. 6—Machine for threshing individual heads.

from the head. Basically these can be any rough surface against which the heads can be rubbed, such as pieces of coarse screen, wood with staples or nails protruding 3 to 5 mm above the surface, or wood wrapped with coarse wire at 10 to 15 mm intervals. After cleaning, the seed should be kept in cool storage free of insects and mice.

VII. TECHNIQUES FOR SPECIAL SITUATIONS

A technique often used to obtain male-sterile plants is the application of gibberellic acid at the bud stage (Anaschenko, 1967; Seetharam and Kusumakumari, 1975). Gibberellic acid, in a 50 to 100 ppm solution, is applied shortly after floral bud initiation. Seetharam and Kusumakumari (1975) emphasized the importance of this stage to obtain maximum male sterility.

Dormancy is a problem often encountered in efforts to germinate sunflower seeds shortly after harvesting. Udayakumar and Krishnasastry (1974) reported that much higher germination was obtained in 20-day developing, dehusked seeds than in the 30- and 40-day-old seeds. The same workers were able to obtain better germination in dehusked seeds with applications of different growth regulators. Most effective was ethrel at 25 ppm concentration.

REFERENCES

Anaschenko, A. V. 1967. Chemical castration of sunflower. Dokl. Vsesojuz. Akad. Sel'skokhoz. Nauk. V.I. Lenina SS SR, 2:1718.

Anonymous. 1970. Key to hybrid sunflowers found by U.S. Dept. Agric. scientist. Crops Soils 3:21.

Arnoldova, O. N. 1926. To the biology of sunflower blooming in connection with the technics of its crossing. J. Expt. Landev. Sudost. Eur. Russ. 3:131–143.

Boyer, J. S. 1970. Leaf enlargement and metabolic rates on corn, soybean, and sunflower at various leaf water potentials. Plant Physiol. 46:233–235.

Dorrell, D. G. 1973. Controlling plant height in sunflowers with growth retardants. Can. J. Plant Sci. 53:417–418.

Fick, G. N., and D. E. Zimmer. 1974. RHA 271, RHA 273, and RHA 274 sunflower parental lines for producing downy mildew resistant hybrids. Farm Res. 32:7–9.

———, ———, and T. E. Thompson. 1976. Wild species of Helianthus as a source of variability in sunflower breeding. p. 4–5. *In* Proc. Sunflower Forum (Fargo, N.D.).

Heiser, C. B. 1965. Species crosses in *Helianthus*: III. Delimitation of "sections". Ann. Missouri Bot. Bard. 52(3):364–370.

———, D. M. Smith, S. B. Clevenger, and W. C. Martin. 1969. The North American sunflowers (Helianthus). Memoirs of the Torrey Botanical Club 22(3):4–6.

Kalton, R. R. 1951. Efficiency of various bagging materials for effecting self-fertilization of sunflowers. Agron. J. 43:328–331.

LeClercq, P. 1969. Un sterilite male cytoplasmique chez le tournesol. Ann. Amelior. Plantes 19(2):99–106.

Plotnikov, A. I. 1940. *In* Sunflower. Krasnodar Kray book publishers.

Pustovoit, V. S., and G. V. Pustovoit. 1972. Results of sunflower breeding and seed production. Genetika 8(12):37–46.

Putt, E. D. 1940. Observations on morphological characters and flowering processes in the sunflower (*Helianthus annuus* L.). Sci. Agric. 21(4):167–169.

———. 1941. Investigations of breeding technique for the sunflower (*Helianthus annuus* L.). Sci. Agric. 21(11):689–702.

———. 1962. The value of hybrids and synthetics in sunflower seed production. Can. J. Plant Sci. 42:488–500.

Schuster, W. 1961. Investigations on artificially induced pollen sterility in sunflowers, *Helianthus annuus* L. Z. Pflanzenz. 46:389–404.

Seetharam, A., and P. Kusumakumari. 1975. Induction of male sterility by gibberellic acid in sunflower. Indian J. Genet. Plant Breeding 35:136–138.

Shchibrya, N. A. 1938. Reports of the Academy of Sciences USSR, Biological Series No. 3.

Smith, D. L. 1978. Planting seed production. *In* Jack F. Carter (ed.) Sunflower science and technology. Agronomy 11:371–386. Am. Soc. of Agron., Madison, Wis.

Udayakumar, M., and K. S. Krishnasastry. 1974. Effect of exogenous application of growth regulators on germinating ability of developing sunflower seeds. Ind. J. Exper.Biol. 12(6):543–545.

Unrau, J., and W. J. White. 1944. The yield and other characters of inbred lines and single crosses of sunflowers. Sci. Agr. 24:516–525.

Vranceanu, V., and F. Stoenescu. 1970. "Obtinerea de hybrizi de floare asoarelul pe basa de androlterilitate genica", An. I.C.C.P.T., 36, Ser. C, 281.

Whelan, E. D. P. 1976. Sterility problems in interspecific hybridization of Helianthus species. p. 5–6. *In* Proc. Sunflower Forum (Fargo, N.D.).

Zubriski, J. C., and D. C. Zimmerman. 1974. Effects of nitrogen, phosphorus, and plant density on sunflower. Agron. J. 66:798–801.

46 Sweet Potato

A. Jones
USDA-SEA
Charleston, South Carolina

The sweet potato [*Ipomoea batatas* (L.) Lam.) is an asexually propagated vegetable that is grown mainly in tropical and subtropical regions. Its exact center of origin is not known but is probably just south of the Isthmus of Panama. In the temperate United States it provides variety to the diet. In Japan it is used as a source of industrial starch and in the tropics it is an important staple of subsistence farmers. The vines, as well as the roots, may be eaten or fed to animals (Yen, 1974, 1976).

I. PARENTAL MATERIAL

The sweet potato is a member of the *Convolvulaceae* and the only known natural hexaploid morning glory (6n = 90) (Jones, 1965). Most other species of the genus are diploids (2n = 30), although there are some wild tetraploids (4n=60) (Jones, 1968; Martin et al., 1974; Martin and Jones, 1973). A 90-chromosome wild plant collected in Mexico and reported as *I. trifida* (H.B.K.) G. Don has received considerable attention. The author and others do not accept that classification as correct but believe it to be *I. batatas* (Jones, 1967; Martin et al., 1974). The most recent taxonomic report on *Ipomoea* was prepared by Austin (1978).

To cross other species with the sweet potato, it is necessary to develop bridging types to overcome the ploidy differences (Austin, 1977). Japanese workers have succeeded in synthesizing 90 chromosome plants from diploid and tetraploid sweet potato relatives (Nishiyama et al., 1975; Kobayashi and Miyazaki, 1976). These results show that hybridization with other species of the genus is possible. Unfortunately, because of taxonomic difficulties, considerable uncertainties still exist relative to the exact species involved (Martin and Jones, 1973; Martin et al., 1974).

Because the sweet potato is the only natural hexaploid species presently known in the genus, hybridization is largely restricted to crosses within the species. Variability within the species is extensive and we have only begun to utilize what is available. Every trait I have studied from wide genetic backgrounds have varied over an extensive range. I firmly believe that with proper screening procedures one can find desirable plant types for almost any need within the species. Each seedling plant is heterozygous and genetically different from any other. Each cultivar represents vegetative multiplication by roots or stems from one seedling. Inheritance patterns are better described in quantitative genetic terms rather than in qualitative terms because of the continuous range of plant types that occur in any lot of seedlings (Jones, 1966; Yen, 1974).

The quantitative mode of inheritance influences breeding strategies (Jones et al., 1976). Although the following discussion will deal in some detail with techniques of crossing two selected parental plants, the best breeding strategy for some objectives may be to use open-pollination or polycross breeding methods.

The exchange of breeding materials is rather informal. There is no one main center for maintenance of genetic collections. I maintain a considerable collection of named cultivars, breeding lines from discontinued programs, plant introductions, and true seed from various open pollinated populations. Most sweet potato breeders maintain similar collections and exchange material freely within the constraints of quarantine regulations and national policies.

II. PLANT CULTURE

A. Field

There are genetic differences in flowering incidence, as well as strong environmental influences (Yen, 1974; Martin and Jones, 1971; Jones, 1965). The seasonal flowering response of cultivars observed in temperate climates may be lost in a tropical climate (Warmke and Cruzado, 1949). More recent work shows that sweet potatoes actually vary widely in photoperiod responses. Some flower in the spring, others in the fall, some in the spring and fall, some in the summer, and others flower all season (Jones, 1966; Martin and Jones, 1971). The lack of flowering of U.S. cultivars and breeding lines may be due to chance associations in the original plants used and a history of forcing normally non-flowering plants into flower, thus perpetuating the non-flowering trait in subsequent generations. In population studies where selection for good flowering has been very effective, no adverse associations of good flowering with agronomic characters have been found (Jones and Dukes, 1976; Jones et al., 1976; Martin et al., 1974). It may be best to use sparse flowering plants only as a last resort or to make a special cross. Excessive dependence on plants forced into flower can only enforce the sparse flowering problem.

Because of the variability in sweet potato, it is difficult to specify optimum environmental conditions that will hold for all plants. In Argentina, Folquer (1974) obtained best seed yields during a period of minimum daily temperatures from 12.8 to 18.6 C, maximum daily temperatures of 23 to 34 C, relative humidity from 62 to 75%, and a photoperiod of about 14.5 hours. The high seed set during this period was related to good flowering and may not represent ideal conditions for seed set. In the United States, a number of studies on improved flowering have indicated optimum photoperiods to be about 12 hours (Edmond and Ammerman, 1971). Personal observations would indicate that seed do not set well in very low humidities, even when flowering is abundant. Therefore, environmental factors do affect seed set in ways other than flowering rate, but few controlled studies have been conducted (Montelaro and Miller, 1951). Personal observations indicate that sweet potatoes set more seed when under environmental stress. Seed set is reduced when vine growth is rapid and vigorous. Low nitrogen fertility and low soil moisture probably improve set.

It is known that insects and diseases can reduce seed set and that chemical treatments can alleviate the problem (Yen, 1974; Jones and Jackson, 1968; Jones, 1964; Jones et al., 1977). I have seen corn earworm (*Heliothis zea* (Boddie)) and fall armyworm (*Spodoptera frugiperda* (Smith)) infestations so severe that virtually every flower contained at least one worm. When using chemical sprays, one should treat in the evenings when bee activity is less to avoid killing insect pollinators. Contact sprays or baits should be favored over sprays with residual activity for the same reason.

B. Growth Chamber and Greenhouse

Many sweet potatoes rarely flower under normal temperate conditions and special efforts are necessary if they are to be used for breeding (Miller, 1937; Edmond and Martin, 1946). Some respond to short day lengths and can be crossed in the greenhouse during winter months. Edmond and Ammerman (1971) and Wang (1975) have reviewed ways to induce flowering. Various physiological shocks, including grafting, girdling and chemical treatment, seem to encourage flowering. The primary stem can be cut about half way through in a slanting direction and prevented from healing back by inserting a wooden wedge or a piece of cardboard. Terminal portions of stems with 6 to 8 nodes can be grafted to other species which flower freely.

Greenhouse crossing is generally done during the winter months when days of about 12 hours are favorable and temperatures are not excessively high. Montelaro and Miller (1951) found about 24 C to be optimum. Edmond and Martin (1946) considered 16 to 17 C night temperatures and 24 to 30 C day temperatures to be optimum. Some breeders use ground beds while others use benches, a choice which is probably more a factor of availability than of plant growth. Plants are generally tied on stakes, canes, wires, or strings. Care must be taken not to over fertilize so as not to produce excessive vine growth. Transplanting vine cuttings from flowering plants may promote more flowering in the greenhouse. Sometimes restricting root growth by planting in a pot or by restricting nitrogen fertilization will improve flowering. In general, a large plant is not necessary for good seed set and rapid vine growth is not desirable.

III. FLORAL CHARACTERISTICS

Flowers of sweet potatoes are similar to those of other morning glories with cultivar differences occurring in size, shape, and color (Fig. 1). They occur in axillary inflorescences of 1 to 22 buds which open singularly or in groups of two or more (Jones, 1966). The corolla is funnel shaped, generally with white limbs and lavender to light purple throats. The depth of the tube varies from 28 to 63 mm and the width of the limb from 26 to 56 mm. Color varies from pure white through degrees of lavender of various patterns to complete lavender (Yen, 1974; Jones, 1966). Each of the five fused petals has a stamen attached at the base with anthers that are usually white, but may be light to dark lavender. Filaments vary in length from 5 to 21 mm among cultivars and within the flower of one plant filaments may vary in length and placement in relation to the stigma, such that any number of anthers may be below, equal to, or above the stigma. The stigma is bilobed and generally white, but sometimes light or dark lavender. Styles may be from 8 to 29 mm long. The pistil has two ovaries, each containing two ovules. Sepals are leaf-like and persistent and may be glabrous or pubescent. Conspicuous yellow glands that contain nectar which attracts insects are at the base of the corolla (Martin and Jones, 1973). Mature capsules may be glabrous or pubescent and contain one to four seeds.

Sweet potato flowers open soon after daybreak and generally are fading by noon depending on environmental conditions. On cloudy days, flowers may remain open longer than on hot, sunny days. Each flower opens for only a short time, then withers, and the corolla drops off. Charles et al. (1974) found that within 3 hours of opening the anthers showed signs

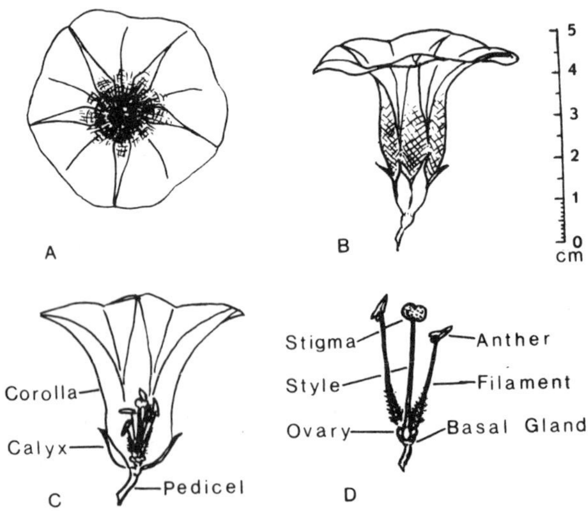

Fig. 1—Views of an intact sweet potato flower from above (A) and the side (B). Parts of the flower (C) with calyx and corolla removed on one side, and (D) with calyx and corolla completely removed.

of withering, and that by evening the styles began to shrivel. Not much is known about sweet potato pollen longevity and viability because no one has successfully germinated it in culture. Studies with fresh pollen indicate 80 to 85% viability (Togari and Kawahara, 1942; Jones, 1965; Jones, 1966). Even though some pollen grains germinate within minutes after pollination, germination may continue for 3 or 4 hours (Martin and Cabanillas, 1966). Pollen tubes may be detected in the ovary 4 hours after pollination and by 8 hours pollen tube growth is essentially complete.

IV. ARTIFICIAL HYBRIDIZATION AND SELF-POLLINATION

A. Equipment

Equipment needed for controlled hybridization is a pair of tweezers, small marking tags (about 1.3 to 1.9 cm), paper clips, and some means of protecting the emasculated flowers, such as soda straws or small glassine bags (Wang and Burnham, 1968).

B. Preparation of the Female

Buds that will open the next day are easily recognized because they are much larger than other buds (Fig. 3). Emasculation can be done after about 1400 hours by slitting the corolla into two parts from the sepals through the tip with tweezers or some other sharp instrument, and pulling each half of the corolla down and out with the forefinger and thumb (Edmond and Ammerman, 1971; Edmond and Martin, 1946). The attached stamens will be removed with the corolla and the pistil will be exposed (Fig. 2B). The pistil can be protected with a glassine bag or a short piece of soda straw pinched over at one end to prevent insect entry (Fig. 2C). The soda straw or bag also serves to identify emasculated buds.

C. Pollination

During the same time of day that emasculations are made, buds to be used in self-pollinations and as pollen sources are prevented from opening the next morning by placing a paper clip or piece of soda straw over their tips (Fig. 3D and E). The clips or straws can be knocked off easily by wind movement, people, or even bumblebees, so the breeder should prepare more than one expects to use. The next morning, pollinations can be made as soon as anthers begin dehiscing, about 0500 to 1000 hours (Edmond and Ammerman, 1971; Hernandez et al., 1967). The anthers are removed from the protected flowers with tweezers and the pollen applied to stigmas of desired female parents. The same flowers used as male parents may be selfed by removing the corolla and protecting the corolla against outcrossing. If the stigmas and anthers are the same height, self-pollination may be accomplished by shaking the corolla and leaving the straw or paper clip in place. The straw or bag is replaced over the pistil after pollination, and a tag attached giving the male parent's designation and any other information de-

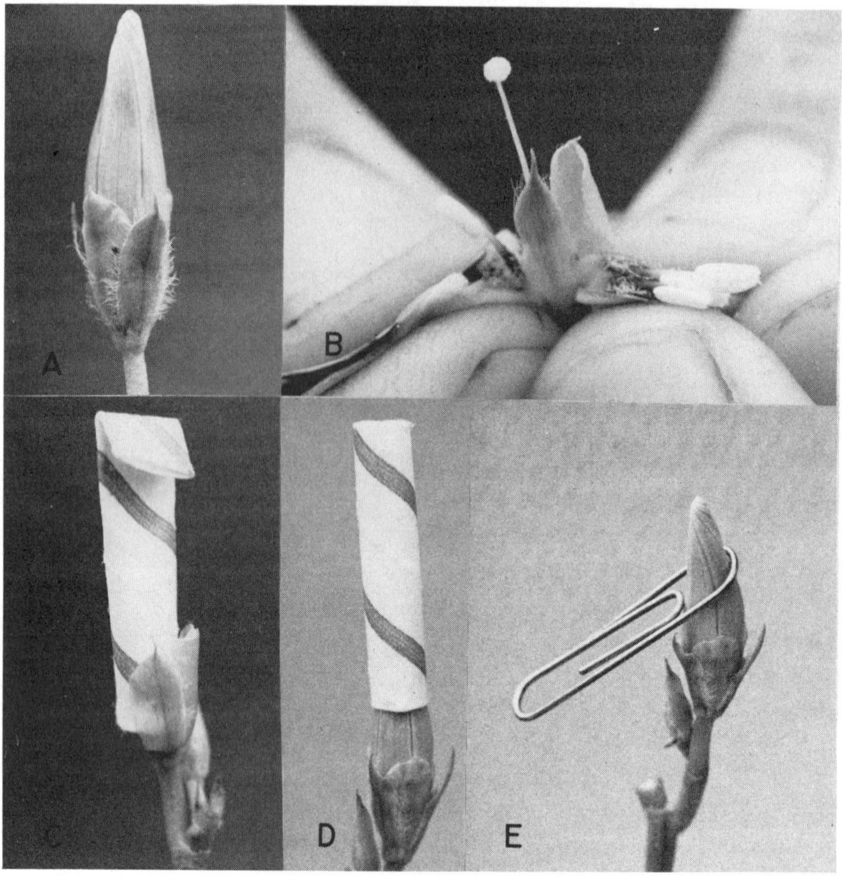

Fig. 2—Emasculation and self-pollination of a sweet potato flower: (A) flower at the stage for emasculation; (B) emasculated flower; (C) emasculated flower with soda straw protecting the stigma; (D) and (E) closing pollinator flowers with a soda straw or paper clip.

Fig. 3—View of the seed capsule from the side (A). The seeds (B) are usually dark brown to black, sometimes speckled or tan.

sired. Tags of different colors may be used to designate different parental lines or types of crosses. After 2 days, the protective covering should be removed. Crossed buds should be observed closely to prevent younger buds from growing through the string of the tag, which could result in misidentification of the cross.

D. Factors Affecting Efficiency

Field nurseries are used during the frost free months by southeastern U.S. breeders to make controlled crosses. If prolific early flowering types are selected as parents, field nurseries probably can be used anywhere sweet potato breeding is conducted. Procedures similar to those established by Miller (1937) generally are followed. This involves spacing plants about 3 m apart and training them up hog-wire trellises in a fan shape. If open-pollinated seed as well as those from controlled crosses are desired, the nursery should be nearly square rather than a few long rows to better insure random crossing. Nutrition must be carefully controlled to prevent excessive vine growth. Good air movement probably contributes to disease control and higher seed set. Therefore, weed control is important and one may consider using herbicides or mulches of black plastic or pine straw.

Sterilities, incompatibilities and environmental conditions all affect capsule set percentages (Folquer, 1974; Fujise et al., 1955; Martin and Cabanillas, 1966; Martin and Jones, 1971; Jones et al., 1977; Montelaro and Miller, 1951; Edmond and Martin, 1946; Wang and Burnham, 1968). The success rate in the same crossing block normally varies from 1 to 47% for different female parents and an average success of 35% should be considered good. Success from selfing seldom exceeds 3% capsule set (Yunoue and Hirosaki, 1975; Edem, 1975). The average number of seed per capsule seldom exceeds 1.7.

Complex compatibility and sterility systems complicate crossing of sweet potatoes. The literature has been reviewed by Martin (1965) and Williams and Cope (1967). Different theories have been developed to explain the systems and to put certain plants into compatibility groups, but none has held up when applied to plants from different origins. Jones (1965) studied the cytology and fertility of sweet potato and concluded that although chromosome behavior in some plants is cytologically abnormal, pairing generally is highly regular. Burnham (1967) and Wang and Burnham (1968) found that some plants with female sterility have ovaries with less than the normal four ovules. The hexaploid nature of the species also obviously could contribute to the general low fertility observed. Apparently a heavy load of deleterious genetic factors can be tolerated by this vegetatively propagated crop. These factors are expressed at various stages in the sexual process as low seed germination, low seedling vigor, abnormal plant types, reduced flowering, ovule abortion, ineffective pollen tube growth or embryo abortion, and poor seed set.

Several techniques have been developed to observe pollen tube growth in vivo which are useful for determining whether a particular cross is likely to be successful and which direction of a cross is more likely to be obtained. Buds that are to open the next day can be collected and held in vials of water in the laboratory (Fig. 2A). Reciprocal pollinations can be made as they open the next morning and the relative rates of pollen tube growth measured with the use of various stains and rapid squash techniques. With each procedure of staining used, the stigmas and styles of about five flowers are excised 4 to 5 hours after pollination. One staining procedure is to place the stigmas and styles on a glass slide, stain with 0.5% cotton blue in lacto-phenol, place a cover glass over the stained organs, and press with the finger. The prepared slides are kept at room temperature for microscopic obsevations the next day (Togari and Kawahara, 1942). Another procedure is to fix the material in formalin-alcohol-acetic acid, stain with aniline blue, and examine the material under UV light of 365 mu wavelength (Martin, 1959). The pollen tubes fluoresce a dark yellow while the stigma and stylar tissue fluoresce a light yellow (Charles et al., 1974; Martin and Ortiz, 1966). With compatible crosses, pollen grains will adhere to the stigma and pollen tubes can be seen penetrating the stylar tissue.

Genetic markers are not used in sweet potato breeding as a rule, however, parental characteristics often can be recognized in progeny.

V. NATURAL HYBRIDIZATION

Much of the intensive labor required by controlled crossing procedures can be avoided through the use of open-pollination by naturally occurring insects. Less elaborate trellises than used in controlled crossing can be constructed by use of metal reinforcing rods, stakes, or bamboo canes placed by each plant. These are tied together by a single cord near the top of the supports and anchored at the ends of the rows. These trellises are more easily constructed and more easily removed at the end of the season than hog-wire trellises commonly used for controlled crossing. Each plant is secured to its support with a string tied at the base, twined around the plant and support, and again tied higher near the top of the support. Each plant should be identified with a wooden pot label rather than a tag because the latter usually does not last for the whole season.

Rows can be spaced sufficiently far apart to allow tractor cultivation between them and plants can be spaced about 1.5 m in the row. No special field layout is necessary, except to avoid long narrow nurseries where cross pollinations would not likely be random. This kind of nursery has worked well for mass selection (Jones et al., 1976). Isolation requirements have not been determined specifically, but one probably should avoid other flowering sweet potatoes by a few hundred meters.

Natural pollination is accomplished during the early morning hours, chiefly by *Hymenoptera.* Very little information has been published on this subject, but many insects have been observed working the flowers and undoubtedly affect some pollination. There are hive differences in honey bee preference for sweet potato flowers. In some locations, large numbers of honey bees work these flowers while in other locations very few can be found working them even though hives are close by. Large numbers of bumble bees can be found working sweet potato flowers.

Bee hives may be placed near the end of the trellis if natural populations of insect pollinators are low. Two self-incompatible but cross-compatible parents can be planted in isolation from other flowering sweet potatoes and crossed by naturally occurring insects or honey bees.

VI. SEED DEVELOPMENT, HARVEST, AND STORAGE

Seed mature in about 1 month, a little sooner under hot, summer temperatures and later in cool, fall temperatures. When the capsules are completely dry and brown, the seeds are mature (Fig. 3). Each capsule contains a maximum of four seeds, but the average is much less, about 1.1 to 1.7 (Wang and Burnham, 1968; Fujise et al., 1955; Jones, 1965; Jones, 1966). Mature seed are about half their maximum green size. They have diameters of 3 to 5 mm and are flat on two sides and round on the other (Martin and Cabanillas, 1966). The usual color is dark brown or black, but some are tan and others speckled. The seed weigh about 2 g per 100 seeds, but this may vary with parental types from about 1.3 to 3.0 g (Fujise et al., 1955; Martin and Cabanillas, 1966; Jones and Dukes, 1976).

Since flowering is over an extended time, seed harvest also is extended. Some plants hold mature seed well and do not need frequent harvesting while others, if not harvested promptly, drop the capsules to the ground where they rot and allow the seed to sprout. In some plants, the capsules open soon after drying and the seed shatter. Seed are usually threshed by hand and stored in envelopes properly identified with their origin.

The seed are hard and retain viability for long periods. Under favorable conditions of 50% relative humidity and 18 C, they may maintain viability for 20 years or more. Because of the hard seed coat, germination is very irregular unless some means of seed scarification is used (Miller, 1937; Montelaro and Miller, 1951). The method of Steinbauer (1937) or some modification of it is still most commonly used (Martin, 1946). Seed are soaked in concentrated sulfuric acid for 20 to 60 min, immediately washed in water or neutralized with a solution containing baking soda, and rinsed in clear water. Seed also may be scarified by mechanical means, such as rubbing with emery paper or pricking each seed with a sharp instrument (Martin, 1946). Percent germination can be increased by putting seed in water and discarding those that float. A high proportion of those that sink are viable.

A seed weevil, *Megacerus impiger* Horn., can infest seed under field conditions and reinfest other seed during storage (Jones and Cuthbert, 1971). Satisfactory control has been obtained by enclosing a small segment of a household pest strip (20%, 2,2-Dichloroxinyl dimethyl phosphate) in a plastic bag with the seed for about 1 week.

VII. TECHNIQUES FOR SPECIAL SITUATIONS

The frequency of somatic mutations in sweet potato is quite high and some cultivars tend to have higher mutation rates than others (Hernandez et al., 1964). Many of the 'Jersey' types are mutations from older cultivars. Orange flesh types may mutate to lighter color and vine lengths and coloration may change. Some mutations may improve market value, such as occurred with 'Goldmar', a mutation of 'Redmar' (Anon., 1972). Because of this tendency to mutate, hill selection is necessary to preserve cultivars and is an integral part of certification programs (Edmond and Ammerman, 1971).

REFERENCES

Anonymous. 1972. About our cover. Maryland horticulture. HortScience 7:538.

Austin, D. F. 1977. Hybrid polyploids in *Ipomoea* section *batatas.* J. Hered. 68: 259–260.

———. 1978. *Ipomoea batatas* complex. I. Taxonomy. Bull. Torrey Bot. Club 105:114–129.

Burnham, M. 1967. Ovule number as a factor in low seed set of certain sweet potato clones. Proc. Am. Soc. Hort. Sci. 90:313–315.

Charles, W. B., D. G. Hoskin, and P. J. Cave. 1974. Overcoming cross- and self-incompatibility in *Ipomoea batatas* (L.) Lam. and *Ipomoea trichocarpa* (Elliot). J. Hort. Sci. 49:113–121.

Edem, U. E. 1975. Results of sweet potato (*Ipomoea batatas*) research 1957–1966. Memo Niger Dep. Agric. Res. 113.

Edmond, J. B., and G. R. Ammerman. 1971. Sweet Potatoes: Production, Processing, Marketing. AVI Pub. Co., Westport, Conn.

———, and J. A. Martin. 1946. The flowering and fruiting of the sweet potato under greenhouse conditions. Proc. Am. Soc. Hort. Sci. 47:391–399.

Folquer, F. 1974. Eficiencia varietal en la produccion primaveral de semillas de batata (*Ipomoea batatas* (L.) Lam.). Rev. Agron. N. O. Argent. 9:193–225.

Fujise, K., T. Yunoue, and T. Chishike. 1955. Studies on the habits of flowering and seed-setting in varieties of sweetpotatoes (*Ipomoea batatas* Lam.). Bull. Kyushu Agric. Extp. Sta. 3:109–142.

Hernandez, Teme P., Travis Hernandez, and Julian C. Miller. 1964. Frequency of somatic mutations in sweet potato varieties. Proc. Am. Soc. Hort. Sci. 85:430–433.

———, ———, R.J. Constantin, and R. S. Kakar. 1967. Improved techniques in breeding and inheritance of some of the characters in the sweet potato, *Ipomoea batatas* (L.). Proc. Int. Symp. Tropical Root Crops 1:31–44.

Jones, Alfred. 1964. A disease of pollen mother cells of sweet potato associated with *Fusarium moniliforme.* Phytopath. 54:1494–1495.

———. 1965. Cytological observations and fertility measurements of sweet potato (*Ipomoea batatas* (L.) Lam.). Proc. Am. Soc. Hort. Sci. 86:527–537.

———. 1966. Morphological variability in early generations of a randomly intermating population of sweet potatoes (*Ipomoea batatas* (L.) Lam.). Univ. Ga. Agric. Expt. sta. Tech. Bull. N.S. 56:1–31.

———. 1967. Should Nishiyama's K123 (*Ipomoea trifida*) be designated *I. batatas*? Econ. Bot. 21:163–166.

———. 1968. Chromosome numbers in *Ipomoea* and related genera. J. Hered. 59:99–102.

———, and F. P. Cuthbert, Jr. 1971. Unpublished data. U.S. Vegetable Laboratory, U.S. Department of Agriculture, Charleston, SC 29407.

———, and P. D. Dukes. 1976. Some seed, seedling and maternal characters as estimates of commercial performance in sweet potato breeding. J. Am. Soc. Hort. Sci. 101:385–388.

———, ———, and F. P. Cuthbert, Jr. 1976. Mass selection in sweet potato: Breeding for resistance to insects and diseases and for horticultural characteristics. J. Am. Soc. Hort. Sci. 101:701–704.

———, ———, and ———. 1977. Pesticides increase true seed production of sweet potato. HortScience 12:165–167.

———, and C. R. Jackson. 1968. Fungi from floral parts of sweet potato. HortScience 3:76–77.

Kobayashi, M., and T. Miyazaki. 1977. Sweet potato breeding using wild related species. Proc. Fourth Symp. Int. Soc. Tropical Root Crops 1:53–57.

Martin, F. W. 1959. Staining and observing pollen tubes in the style by means of fluorescence. Stain Technol. 34:125–128.

———. 1965. Incompatibility in the sweet potato. A review. Econ. Bot. 19:406–415.

———, and E. Cabanillas. 1966. Post-pollen-germination barriers to seed set in sweet potato. Euphytica 15:404–411.

———, and A. Jones. 1971. Flowering and fertility changes in six generations of open-pollinated sweetpotatoes. J. Am. Soc. Hort. Sci. 96:493–495.

———, and ———. 1973. The species of *Ipomoea* closely related to the sweet potato. Econ. Bot. 26:201–215.

———, ———, and R. M. Ruberte. 1974. A wild *Ipomoea* species closely related to the sweetpotato. Econ. Bot. 28:287–292.

———, and S. Ortiz. 1966. Germination of sweet potato pollen in relation to incompatibility and sterility. Proc. Am. Soc. Hort. Sci. 88:491–497.

Martin, J. A., Jr. 1946. Germination of sweet potato seed as affected by different methods of scarification. Proc. Am. Soc. Hort. Sci. 47:387–390.

Miller, J. C. 1937. Inducing the sweet potato to bloom and set seed. J. Hered. 28: 347–349.

Montelaro, J., and J. C. Miller. 1951. A study of some factors affecting seed setting in the sweet potato. Proc. Am. Soc. Hort. Sci. 57:329–334.

Nishiyama, I., T. Miyazaki, and S. Sakamoto. 1975. Evolutionary autoploidy in the sweet potato (*Ipomoea batatas* (L.) Lam.) and its progenitors. Euphytica 24: 197–208.

Steinbauer, C. E. 1937. Methods of scarifying sweet potato seed. Proc. Am. Soc. Hort. Sci. 35:606–608.

Togari, Y., and U. Kawahara. 1942. Studies on the different grades of self and cross incompatibilities in the sweet potato. II. Pollen behavior. Imperial Agric. Exp. Stn. Tokyo Bull. 52:21–30.

Wang, H. 1975. The breeding and cultivation of sweet potatoes. Tech. Bull. No. 26, ASPAC Food and Fertilizer Technology Center (Chiayi Agric. Exp. Stn., Taiwan).

———, and M. Burnham. 1968. A study of fruit and seed setting ability and female sterility in the sweet potato (*Ipomoea batatas* (L.) Lam.). Bot. Bull. Acad. Sinica 9:139–153.

Warmke, H. E., and H. J. Cruzado. 1949. The flowering and seed-setting of sweet potatoes in Puerto Rico. Science 109:62–63.

Williams, D. B., and F. W. Cope. 1967. Notes of self-incompatibility in the genus *Ipomoea* L. Proc. Int. Symposium on Tropical Root Crops 1:16–30.

Yen, D. E. 1974. The sweet potato and oceania. B. P. Bishop Museum Bull. 236:1–389.

———. 1976. Sweet potato. p. 42–45. *In* N. W. Simmonds (ed.) Evolution of crop plants. Longman, London.

Yunoue, T., and S. Hirosaki. 1975. Effects of the selection of fertility of inbred parent lines in sweet potato breeding. Bull. Kyushu Agric. Exp. Stn. 18:1–41.

47 Tobacco

E. A. Wernsman
D. F. Matzinger
North Carolina State University
Raleigh

Tobacco (*Nicotiana tabacum* L.) is a perennial allotetraploid that arose by the natural hybridization of *N. sylvestris* Speg. and Comes with *N. tomentosiformis* Goodsp. Although tobacco is considered to be of tropical origin, it is cultivated worldwide in areas with a frost-free period sufficiently long to permit crop maturity. The species is commercially cultivated for its cured leaf. Its ability to produce large quantities of seed and ease of culture frequently make it the organism of choice for research in biological sciences.

Tobaccos of commerce are classified according to their method of leaf curing. Tobaccos cured in heated barns are of the flue-cured (bright) and fire-cured types. Air-cured tobaccos include light air-cured (burley and Maryland), dark air-cured, and cigar tobaccos, filler, binder, and wrapper. Flue-cured and burley are the principal tobaccos used in cigarette manufacture and occupy predominant hectarages.

I. PARENTAL MATERIAL

The genus *Nicotiana* contains 65 species of which two, *N. tabacum* and to a lesser degree *N. rustica* L., are cultivated. All cultivars in the U.S. classification system are *N. tabacum*, but their cured leaf is not interchangeable in manufactured products and the utilization of germplasm from one type for the improvement of another is usually limited. Cured leaf quality of lines developed from inter-type crosses is generally considered to be inferior to those from intra-type hybridizations. However, qualitatively inherited genes for disease resistance have been transferred between types.

The germplasm base of flue-cured tobacco is currently narrow and most modern cultivars can be traced directly or indirectly to two cultivars, 'Hicks Broadleaf' and 'Coker 139'. Acceptability of diverse germplasm is restricted by minimum quality standards for physical, chemical, and smoking properties of cured leaf which new breeding lines must precisely meet before their adoption into the trade. Leaf quality requirements for acceptable burley tobacco may be less restrictive, thereby allowing greater diversity within this type (Matzinger et al., 1971).

The large number of species in the genus, their ability to produce large quantities of seed (up to 3500 seed per capsule), and the desire to transfer qualitatively inherited genes from wild species to *N. tabacum* have fostered the production of many interspecific hybrids. Over 300 interspecific hybrids in this genus have been reported (East, 1928; Goodspeed, 1954; Smith, 1968), and genes conditioning resistance to most major tobacco diseases have been transferred from wild *Nicotiana* species to *N. tabacum.* Examples include resistance to tobacco mosaic virus from *N. glutinosa* (Holmes, 1938); black shank, *Phytophthora parasitica* var. *nicotianae* (Breda de Haan) Tucker from *N. plumbaginifolia* Viv. (Apple, 1962) and *N. longiflora* Cav. (Valleau et al., 1960); blue mold, *Peronospora tabacinia* Adam, and black root rot, *Thielaviopsis basicola* Berk and Br., from *N. debneyi* Domin (Clayton, 1968, 1969); and wildfire, *Pseudomonas tabaci* (Wold and Foster) Stevens from *N. longiflora* (Clayton, 1969). In addition, the introgression of germplasm from *N. sylvestris* and *N. otophora* Gries. into *N. tabacum* has been used for the improvement of quantitatively inherited traits in the cultivated allotetraploid (Wernsman et al., 1976; Oupadissakoon and Wernsman, 1977).

Germplasm collections of tobacco cultivars and *Nicotiana* species are maintained primarily by individual scientists working at agriculture experiment stations in tobacco growing states, by private plant breeders, and by personnel of the U.S. Department of Agriculture. Coordinated efforts to maintain germplasm within the genus are limited.

II. PLANT CULTURE

A. Field

Tobacco can be grown to maturity and abundant seed produced on a wide array of soil types that are adequately drained. The plant flowers and produces seed over a range of soil fertilities, although flower number per plant and seed quality are significantly influenced by fertility. Fertilization recommendations depend on the type of tobacco being grown and the relative productivity of the site. Phosphorus and K fertilization rates normally employed in flue-cured leaf production (39 kg/ha P; 83 kg/ha K) are adequate for seed production, but total seed yields may benefit from N applications two times the normal rate of 75 kg/ha. The additional N has no major effect on seed quality.

Tobacco is very sensitive to saturated or water logged soils, and at high temperatures the plant may be severely damaged or killed in 24 to 48 hours. In light, sandy soils whose moisture content at field capacity is approximately 10% by weight, successful hybridization becomes difficult as soil moisture drops to 2 to 3%. Under these conditions, pollinated flowers frequently abscise from the inflorescence 24 to 36 hours following pollination.

Most tobacco cultivars are photoperiod insensitive, although genetic stocks of the mammoth type and other breeding lines have been developed which flower under short photoperiods only. In addition, certain wild species of *Nicotiana* are temperature and photoperiod sensitive (Burk, 1967a). Tobacco can be successfully hybridized at a wide range of temperatures, but the proportion of pollinated flowers which develop capsules and set seed seems to be reduced when daily maximum temperatures exceed 32 to 35 C.

Effects of light intensity and relative humidity on tobacco hybridization are largely unknown. Daily and seasonal fluctuations of these factors occur, but their influence on seed production is not obvious. Nevertheless, conditions of persistent rainfall and high relative humidities at the time of capsule and seed maturity are conducive to growth of fungi within the capsules which reduces seed quality.

B. Growth Chamber and Greenhouse

Growth chamber or greenhouse plantings may be successfully employed for advancing generations and obtaining self- and cross-pollinated seed. Growth rate and morphology of plants produced in artificial environments more closely resemble field grown materials when day/night temperatures of 26/22 C and 12-hour photoperiods are followed (Raper et al., 1971). Successful hybridization or self-pollination on greenhouse plants are obtained following similar procedures employed on field grown plants, although capsule size and seed number per capsule are generally reduced.

Chemical insect control, particularly for species of *Aphididae* and whitefly (*Trialeurodes vaporariorum* Westwood), is frequently necessary in greenhouse plantings and conveniently achieved by fumigation with insecticide generators. Pollinations should not be made 24 to 36 hours prior to fumigation, as flowers frequently abscise following these treatments.

Supplemental lighting to lengthen photoperiods in greenhouse plantings during winter months may shorten the transplanting to flowering period, but is not necessary for successful hybridization once flowering has been initiated.

III. FLORAL CHARACTERISTICS

Tobacco is a highly self-pollinated species and natural cross-pollination is usually low. Pollen is not windborne and cross-pollinations which occur are attributed to insect and bird vectors, such as honey bees (*Apis mellifera* L.), sweat flies (*Didea fasciata* Macq.), bumblebees (*Bombus* species), hawk moths (*Theretra tersa* L.), and hummingbirds (*Archilochus colubris* L.) (McMurtrey et al., 1960). The amount of cross-pollination varies with location, season, and activity of pollen vectors (McMurtrey et al., 1960). They observed seasonal ranges of 2.5 to 11.3% cross-pollination. Cross-pollination frequency also is influenced by plant genotype according to Litton and Stokes (1964). They recorded 9.1% outcrossing for 'KY 16', but only 1.3% for 'Burley 37'. When an individual capsule contained cross-pollinated seed, the proportion of crossed to selfed seed was usually high in their study.

The inflorescence of tobacco is a terminal panicle but smaller, well-developed inflorescences can develop from axillary buds in the leaf nodes. The upper leaf axils each possess primary, secondary, and tertiary meristems or suckers. With the elimination of apical dominance by decapitation (topping) or by complete flowering of the terminal inflorescence, a sucker at each axil is capable of developing an inflorescence. Each branch of the inflorescence closes to a flower or a fork. Anthesis commences with the first single terminal flower of the inflorescence and proceeds laterally and downward (Goodspeed, 1954). Individual flowers are perfect and sufficiently large prior to anthesis to be easily manipulated. Each flower possesses a five-lobed calyx and an elongated cylindrical corolla consisting of a tube, throat, and a five-cleft limb (Fig. 1A). Anthocyan pigmentation is usually limited to the distal 8 to 10 mm of the corolla. Of the five stamens inserted at the base of the tube, two have long filaments, two medium, and one short. The long and medium filaments, approximately 40 mm, project the anthers above the stigma on many genotypes, while the anther of the short stamen is usually below the stigma. The ovary is truncated at the base, bearing a single, white or light green style, 35 mm long, supporting the stigma. Mature capsules are approximately 20 mm in length and on most cultivated forms are generally indehiscent.

The stigma becomes receptive 24 to 48 hours before anthesis, and pollination of flowers at earlier stages of development is much less successful. Stigma receptivity can be estimated by flower length measurements, and is closely correlated with pigment development in the corolla tip. The period of stigma receptivity after anthesis has not been investigated because dehiscence usually occurs as the flower opens, making such flowers unsuitable for pollination.

IV. ARTIFICIAL HYBRIDIZATION AND SELF-POLLINATION

A. Equipment

Equipment used in obtaining self- and cross-pollination of tobacco includes a pollinating tray of 6.35 mm plywood construction 23 × 60 × 4.5 cm (Fig. 1B). This tray possesses four 4.5 cm holes, one for an alcohol sterilization bottle and three for 50 ml beakers used to carry male flowers. It has 8 mm holes for pencil and forcep storage, two 9 × 12 cm compartments for soda straws and pollinating tag storage, a 4 × 8 cm compartment for pollinating cards, and a 3.5 × 3.5 cm compartment for pollinating wire. Supplies needed include a 65 ml bottle containing 95% ethanol for sterilization, 50 ml beakers for male flower storage, a pair of 12 cm forceps, paper soda straws cut to 5 to 6 cm lengths, 0.25 to 0.30 mm diam copper wire cut in 6 cm lengths, 17 × 26 mm strung marking tags, and 7.5 × 12.5 cm index cards.

B. Preparation of the Female

Flowers selected for emasculation and pollination should be at a stage of development at which anthesis is expected in 12 to 24 hours (Fig. 1A). Corollas of flowers at this stage of development will be approximately 55

Fig. 1—Tobacco inflorescence, equipment, and procedures for hybridization. (A) Flowers of an appropriate size for pollen collection, emasculation, and pollination. (B) Equipment used in making pollinations including an alcohol bottle, 50 ml beaker with male flowers, forceps, pencil, identification tags, soda straws, and wire. (C) Emasculation of a flower. (D) Application of pollen to the stigma. (E) Protection of a pollinated flower with a soda straw, and cross identification. (F) Mature capsule from a hybridization.

mm long, tightly closed, and inaccessible to insects. They can be recognized by significant development of anthocyan pigmentation in the distal tip of the corolla. Corolla pigmentation in most genotypes is not highly evident until approximately 36 hours before anthesis. Hybridization can be achieved on more immature flowers up to 48 hours prior to anthesis, but the frequency of successful pollination is reduced.

Forceps are placed in ethanol for 3 to 4 min before emasculation for sterilization. The breeder also may use a small quantity of alcohol to clean the hands. Pollinations can be made at any time during the day, but flowers of the appropriate size for emasculation are more abundant in the morning hours. Points of the forceps are inserted into the tube of the corolla 20 to 25 mm above the pedicel and an incision is made to the tip of the corolla. Filaments of the stamens are cut with the forceps and the anthers removed (Fig. 1C). The emasculated flower should be pollinated immediately.

C. Pollination

Twenty-four hours before a pollination is to be made, flowers of the male parent are removed from the plant, placed in a 50 ml beaker to which 10 ml of water has been added, and stored in an insect-proof room at room temperature and relative humidity. Stage of development of selected flowers should be similar to that previously described for emasculation i.e., the corollas are tightly closed, inaccessible to insects, and anthers are expected to dehisce in 12 to 24 hours (Fig. 1A).

Pollen normally is used on the day of flower anthesis, but remains viable for a number of days. Good capsules containing large quantities of seed have been obtained from pollen stored at room temperature and relative humidity for up to 10 days. Dean (1965) demonstrated that viability, as measured by pollen tube growth, of pollen stored at room temperature and relative humidity decreased from 64% at anthesis to 2% after 3 weeks. Nevertheless, capsules and seed were obtained with 3-week old pollen provided that the stigma of a pollinated flower was heavily covered with pollen.

Pollen can be stored, and viability retained for long periods of time under low temperature and relative humidity. Dean (1965) found fair viability of pollen frozen over calcium chloride, and stored at 0 C and less than 5% relative humidity for 136 weeks. Gwynn (1972) developed a simplified pollen storage technique in which anthers are collected at anthesis, placed in No. 00 gelatin capsules, and stored over silica gel desiccant in tightly sealed bottles at −17 C. Pollen treated in this manner exhibited 33% germination after 24 months in storage.

A flower from the male parent is selected from the storage beaker and the corolla removed with the forceps. The corolla of the emasculated flower is opened with the index finger by pressing down on the throat on the side opposite the incision. The stigma is thoroughly covered with pollen by brushing with an anther of the male flower (Fig. 1D). A soda straw is folded at one end and the open end placed over the stigma and style (Fig. 1E). The corolla is closed around the soda straw and wrapped with a segment of copper wire, thereby protecting the pollinated flower from insect visitations and possible contamination with pollen from other flowers on the same plant. Parentage of the male and female parent is recorded on a marking tag and the tag gently tied to the pedicel of the pollinated flower. All pollinations are recorded on a 7.5 × 12.5 cm index card where parentage, number of flowers pollinated, and the date are noted. Cards are later employed in recording pollination success and in harvesting seed.

When large quantities of selfed seed are needed from individual plants, all open flowers and capsules are removed from the inflorescence, it is thoroughly dusted with an insecticide, and covered with a 7.3 kg (16 lb) wet-strength paper bag or a fine-mesh, porous polyethylene bag stapled around the base of the inflorescence. Protective bags prevent pollinators from transporting pollen from plant to plant. Capsule numbers and seeds per capsule are much greater for inflorescences covered with plastic than with paper bags, and is probably due to the reduced temperatures inside the porous bag.

D. Factors Affecting Efficiency

Tobacco nurseries and crossing blocks are usually planted at plant populations and row spacings similar to those employed in commercial leaf production. Plots may consist of 22 to 25 plants and multiple plots of the particular entry planted when large quantities of seed are to be produced, genetic segregation within plots is to be observed, or selected plants are to be pollinated. For genetically stable entries for which limited seed quantities are to be produced, a five-plant plot is frequently employed and all pollinations from a large number of male parents can be completed on a single female plant.

The primary inflorescence of a field-grown plant will produce 200 to 350 flowers over a 4 to 6-week period. Additional inflorescences can be developed from suckers to extend the flowering period of an individual genotype. If necessary, pollen from one flower can be collected and stored until the female parent of a desired cross has flowers.

Nursery plantings are frequently subjected to serious insect infestations because egg laying adults of a number of insect species are attracted to flowers of untopped plants. The most serious pests include species of *Aphididae,* tobacco hornworm (*Manduca sexta* L.), tobacco budworm (*Heliothus virescens* F.) and other species of this genus, and flea beetles (*Epitrix hirtipennis* Melsheimer). *Heliothus* species are particularly destructive because larvae feed readily on developing capsules and immature seed. A tobacco nursery should include provisions for insecticide applications to control these pests. In our flue-cured tobacco nurseries, every eighth row is left unplanted to provide alleys throughout the nursery. These alleys allow access for spraying and overhead irrigation equipment. Where mechanical spraying equipment cannot be employed (e.g., excessive plant height or severe plant lodging), capsules on plants used for hybridization can be protected by insecticide applications with hand dusting equipment.

Numerous qualitatively inherited, dominant and recessive genetic markers are available for the identification of hybrid plants in tobacco crosses. Examples include chlorophyll markers, genes affecting leaf shape, number, and texture, disease reactions, photoperiod sensitivity, and flower pigmentation. The reader is referred to Smith's compendium (1968) listing many of these markers, their chromosomal location, and a description of the complete monosomic series for this species. Availability of genetic markers is usually restricted to specific breeding lines.

Qualitative genetic markers for identifying hybrids in crosses of cultivars within a tobacco class are very limited. Flue-cured cultivars frequently differ for qualitatively-inherited genes affecting leaf shape, and these markers can be beneficial. Pollen control in tobacco is easily obtained and contamination in controlled crosses is very low provided that the breeder exercises adequate care in pollination and seed processing. With experience, subtle differences between cultivars are readily apparent to the breeder and marker stocks for hybrid identification are usually unnecessary.

V. NATURAL HYBRIDIZATION

Numerous sources of cytoplasmic-genetic male sterility (CMS) are available in tobacco and have been produced by the transfer of the *N. tabacum* nucleus into cytoplasms of alien *Nicotiana* species (Clayton, 1950; Burk, 1960; Chaplin, 1964). Nuclear genes for fertility restoration for most CMS forms have not been concomitantly transferred from cytoplasmic donor species to tobacco because the product of commercial interest is leaf rather than seed. CMS is employed in the production of F_1 hybrid seed of burley tobacco where hybrids are grown on an estimated 60 to 65% of the commercial acreage. Natural cross-pollination is so low that flowers of the male-sterile female parent must be hand-pollinated with pollen from a normal, fertile line usually grown in another field. Planting arrangements consist of four rows of the female parent followed by an unplanted row used to transport pollinating personnel and equipment through the field. Female plants are hand-pollinated four to five times per season to fertilize sequentially developing flowers.

Production of certified seed of pure line cultivars or hybrids requires 46 m isolation from fertile genotypes of the same tobacco class. This isolation distance may be waivered if four border rows of each cultivar are allowed to flower, but are not harvested for seed. If a second class of tobacco is grown in the same area, 402 m isolation from unprotected females is required (AOSCA, 1971).

VI. SEED DEVELOPMENT, HARVEST, AND STORAGE

Flowers of unsuccessful pollinations usually abscise from the plant 24 to 48 hours following pollination. In 3 to 4 days, the corolla of a fertilized flower withers and dries, and the corolla, straw, and wire fall from the flower (Fig. 1F). These events are followed by pedicel thickening, capsule enlargement, and a successful pollination is evident at the end of a week. With the aid of the pollinating cards previously described, numbers of successful crosses on a plant should be checked to determine if adequate numbers of capsules are developing. Gradual changes in capsule color from green to brown serve as a guide to capsule and seed development, and seed are expected to mature in 3 to 4 weeks following pollination. Gwynn (1973) showed that seed are sufficiently mature to germinate 14 days after pollination, but germination percentage continued to improve up to 26 days.

Capsules should be harvested as soon after maturity as possible to maintain maximum seed quality. Although capsules of cultivated forms of tobacco are generally indehiscent, capsules damaged by insect feeding readily disperse their seed. Care must be exercised in harvesting pollinated capsules so that open-pollinated seed does not adhere to the technician's hands or to harvested capsules.

Harvested capsules can be placed in small self-sealing envelopes and air-dried. If artificial heat is used to speed drying, temperatures should not exceed 32 C (Akehurst, 1968).

Threshing small numbers of capsules is accomplished by removing the tip of the capsule and pouring the seed into a small seed envelope. If large numbers of capsules are to be threshed, capsules may be hand-crushed and passed over a 20-mesh screen to remove course trash. Blower-type seed cleaners can be calibrated and utilized to remove fine chaff and light seed.

Seed can be placed in 5.5 × 9 cm sift-proof envelopes possessing the appropriate identification of the seed, and stored in a household food freezer at −17 C. Under these conditions good viability is maintained for 3 to 5 years. Kincaid (1958) demonstrated that excellent viability can be maintained for up to 25 years if seed are stored over desiccants in airtight bottles at either ambient temperatures or in a refrigerator.

VII. TECHNIQUES FOR SPECIAL SITUATIONS

Interspecific hybridization has extended the range of genetic variability in tobacco and three general approaches for the production of F_1 interspecific hybrids have been followed. These include direct hybridization of a diploid *Nicotiana* species female with a normal tetraploid (2n = 48) *N. tabacum* male; crosses of diploid, *Nicotiana* species female with a colchicine-doubled (2n = 96) *N. tabacum* male; and hybridization of autotetraploid *Nicotiana* species female with an octoploid *N. tabacum* male. Reciprocal hybridization (*N. tabacum* as female) is frequently less successful (Chaplin and Mann, 1961), but should be considered where viable F_1's can be produced because problems with alien cytoplasms (male sterility) in advanced backcrosses to tobacco would be diminished.

The philosophical basis for the choice of alternate ploidy levels in the hybridization attempt is to circumvent, where possible, sterility barriers in the F_1 interspecific hybrid. In direct hybridization (diploid *Nicotiana* species × tetraploid tobacco), the F_1 hybrids are triploids and highly sterile. Chromosome doubling of F_1's with colchicine usually results in greater fertility, but a significant proportion of the seedlings may be lost in colchicine treatment. Fertility of F_1 interspecific hybrids produced through the use of octoploid tobacco in the initial cross is usually enhanced, and F_1 hybrids do not have to be subjected to colchicine treatment, but can be directly backcrossed to the tobacco parent.

In interspecific hybrids where viable F_1 seeds are rarely produced, applications of naphthaleneacetic acid sprays (1 ppm) to flowers at fertilization followed by additional applications during capsule development may decrease capsule abscission (Gerstel, 1977). Apparao and Gopinath (1968) obtained enhanced capsule and seed set in a highly sterile trispecies interspecific hybrid by spraying flowers with 0.05% boric acid solutions before and after pollination. Kuenzel (1977), however, observed that a high proportion of pollinated flowers developed mature capsules in a (*N. tabacum* × *N. otophora*) × *N tabacum* backcross without hormone applications, but viable BC_1 seed were rare and many capsules were completely devoid of seed. Number of viable seed to prevent abscission of a developing capsule was very low.

Burk (1972) produced viable interspecific hybrids of *N. tabacum* × *N. langsdorffii* Weinm. using monosomic stocks of tobacco, where similar hybrids using normal disomic tobacco could not be produced. Bridgecrosses, i.e., 4x (*N. sylvestris* × *N repanda* Willd.) × 8x *N. tabacum,* have been employed in transferring alien chromosomes to tobacco where direct hybridization of *N. repanda* × *N. tabacum* was not possible (Burk, 1967b).

Tobacco is very amenable to tissue culture techniques. F_1 interspecific hybrids, normally lethal in seedling stages, have been grown in culture and viable hybrid sporophytes regenerated from callus (Lloyd, 1975). Large numbers of haploid sporophytes are readily obtained from anther cultures (Bourgin and Nitsch, 1967; Collins and Legg, 1974), and their chromosomes doubled to produce true breeding tetraploid sporophytes. Mature sporophytes can be regenerated from plant protoplasts (Nagata and Takebe, 1971), and parasexual interspecific hybrids have been regenerated from protoplast fusions (Carlson et al., 1972). As techniques are developed and refined for the selection of protoplast fusion products in culture, new interspecific and intergenetic hybrids will be feasible. It is emphasized, however, that interspecific gene transfer in *Nicotiana* is seldom limited by the failure to produce viable F_1 interspecific hybrids, but more commonly, by the failure to obtain sufficient recombination between alien chromosomes and those of tobacco in advanced generations.

REFERENCES

Akehurst, B. C. 1968. Tobacco. Humanities Press, N.Y.

Apparao, K., and D. M. Gopinath. 1968. Colchicine-induced mutations in a trispecific *Nicotiana* hybrid. Can. J. Genet. Cytol. 10:782–787.

Apple, J. L. 1962. Transfer of resistance to black shank *Phytophthora parasitica* var *nicotianae* from *Nicotiana plumbaginifolia* to *N. tabacum.* Phytopathology 52:1. (Abstr.)

Association of Official Seed Certifying agencies (AOSCA). 1971. Certification handbook. Pub. No. 23. Clemson, SC.

Bourgin, J. P., and J. P. Nitsch. 1967. Obtention de *Nicotiana* haploides a partir de'etamines cultivees in vitro. Ann. Physiol. Veg. 9:377–382.

Burk, L. G. 1960. Male sterile flower anomalies in interspecific tobacco hybrids. J. Hered. 51:27–31.

———. 1967a. Preservation of a *Nicotiana* species collection. Tob. Sci. 11:120–123.

———. 1967b. An interspecific bridge-cross: *Nicotiana repanda* through *N. sylvestris* to *N. tabacum*. J. Hered. 58:215–218.

———. 1972. Viable hybrids from monosomics of *Nicotiana tabacum* by *N. langsdorffii*. Tob.Sci. 16:43–45.

Carlson, P. S., H. H. Smith, and R. D. Dearing. 1972. Parasexual interspecific plant hybridization. Proc. Natl. Acad. Sci. USA 69:2292–2294.

Chaplin, J. F. 1964. Use of male-sterile tobaccos in the production of hybrid seed. Tob. Sci. 8:105–109.

———, and T. J. Mann. 1961. Interspecific hybridization, gene transfer and chromosomal substitution in *Nicotiana*. North Carolina Agric. Exp. Stn. Bull. 145.

Clayton, E. E. 1950. Male-sterile tobacco. J. Hered. 41:171–175.

———. 1968. The transfer of blue mold resistance to tobacco from *Nicotiana debneyi* Part IV. Breeding progress 1957–1967. Tob. Sci. 12:112–124.

———. 1969. The study of resistance to the black root rot disease of tobacco. Tob. Sci. 13:30–37.

Collins, G. B., and P. D. Legg. 1974. The use of haploids in breeding allopolyploid species. p. 231–247. *In* K. J. Kasha (ed.) Haploids in higher plants. Advances and potential. Proc. First Int. Symp. Guelph, Ontario, Canada.

Dean, C. E. 1965. Effect of temperature and humidity on the longevity of tobacco (*Nicotiana tabacum* L.) pollen in storage. Crop Sci. 5:148–150.

East, E. M. 1928. The genetics of the genus *Nicotiana*. Bibliogr. Genet. 4:243–320.

Gerstel, D. U. 1977. Personal communication. Dep. of Crop Science. North Carolina State Univ., Raleigh, NC 27650.

Goodspeed, T. H. 1954. The genus *Nicotiana*. Chronica Botanica Co., Waltham, MA.

Gwynn, G. R. 1972. Simplified methods of pollen storage in tobacco. Tob. Sci. 16: 84–85.

———. 1973. Effect of maturity on germination of seed of six tobacco cultivars. Tob. Sci. 17:108–109.

Holmes, F. O. 1938. Inheritance of resistance to tobacco-mosaic disease in tobacco. Phytopathology 28:553–561.

Kincaid, R.R. 1958. Tobacco seed storage for 25 years. Florida Agric. Exp. Stn. Bull. 593.

Kuenzel, K. A. 1977. Progenitor species germplasm as a source of genetic variation for the improvement of *Nicotiana tabacum* L. M. S.Thesis. North Carolina State Univ., Raleigh, NC.

Litton, C. C., and G. W. Stokes. 1964. Outcrossing in burley tobacco. Tob. Sci. 8: 113–115.

Lloyd, R. 1975. Tissue culture as a means of circumventing lethality in an interspecific *Nicotiana* hybrid. Tob. Sci. 19:4–6.

Matzinger, D. F., E. A. Wernsman, and H. F. Ross. 1971. Diallel crosses among burley varieties of *Nicotiana tabacum* L. in the F_1 and F_2 generations. Crop Sci. 11:275–279.

McMurtrey, J. E., Jr., D. B. Wilson, and J. P. Pointer. 1960. Natural crossing of tobacco under Maryland conditions. Tob. Sci. 4:243–247.

Nagata, T., and I. Takebe. 1971. Plating isolated tobacco mesophyll protoplasts on agar medium. Planta 99:12–20.

Oupadissakoon, S., and E. A. Wernsman. 1977. Agronomic performance and nature of gene effects in progenitor species-derived genotypes of *Nicotiana tabacum* L. Crop Sci. 17:843–847.

Raper, C. D., Jr., W. H. Johnson, and R. J. Downs. 1971. Factors affecting the development of flue-cured tobacco grown in artificial environments. I. Effects of light duration and temperature on physical properties of fresh leaves. Agron. J. 63:283–286.

Smith, H. H. 1968. Recent cytogenetic studies in the genus *Nicotiana.* Adv. Genet. 14:1–54.

Valleau, W. D., G. W. Stokes, and E. M. Johnson. 1960. Nine years' experience with the *Nicotiana longiflora* factor for resistance to *Phytophthora parasitica* var *nicotianae* in the control of black shank. Tob. Sci. 4:92–94.

Wernsman, E. A., D. F. Matzinger, and T. J. Mann. 1976. Use of progenitor species germplasm for the improvement of a cultivated allotetraploid. Crop Sci. 16: 800–803.

48 Tomato

Charles M. Rick
University of California
Davis

The tomato (*Lycopersicon esculentum* Mill.) is a herbaceous perennial that is usually grown as an annual crop. Wild forms of this species are known, as well as eight other exclusively wild species of the genus. Although the natural distribution of the wild species is restricted to the Andean region, the site or sites of domestication are far less certain; most evidence favors southern Mexico (Rick, 1976). The tomato is the second most important vegetable in dollar value in the United States and is highly popular elsewhere. It ranks at the top of all fruits and vegetables as a source of vitamins and minerals in the U.S. (Stevens, 1974). Because of its ease of culture, high reproductive rate, and genetic uniformity resulting from autogamy, the tomato is amenable for physiological and cytogenetic investigations.

I. PARENTAL MATERIAL

A great wealth of tomato germplasm is available, consisting of modern and old-fashioned cultivars, primitive cultivars, breeding lines, wild forms of *L. esculentum,* and the exclusively wild species. Large collections are maintained by the North Central Plant Introduction Station of the U.S. Department of Agriculture at Ames, Iowa, and by the Tomato Genetics Cooperative at the University of California, Davis, Calif. The specialty of the collection at Ames is cultivars and breeding lines of *L. esculentum* and that at Davis is tomato species and accessions of genetic and cytogenetic interest.

The wild tomato species play an important role in tomato improvement. Scores of cultivars with desired characters derived from these species have already been introduced, and further progress is anticipated from current breeding projects utilizing such germplasm.

All known wild species of *Lycopersicon—cheesmanii, chilense, chmielewskii, hirsutum, parviflorum, peruvianum,* and *pimpinellifolium*—can be hybridized with *L. esculentum.* Crosses also succeed with *Solanum lycopersicoides* and *S. pennellii*, although hybrid sterility in combinations with the former has precluded its exploitation in breeding programs. The wild forms (var. *cerasiforme*) and primitive cultivars of *L. esculentum* also provide important germplasm.

Most interspecific combinations produce viable seeds without special aids. Crosses between *L. esculentum* and the wild species fare better if *L. esculentum* is used as the female parent. In combinations with the self-incompatible species, Hogenboom (1973) demonstrated that the barrier is genetically independent of the self-incompatibility reaction and proposed the term incongruity. Crosses with typical *L. peruvianum* stimulate normal fruit development, but have subnormal or no embryo development. Smith (1944) demonstrated that sterile embryo culture in a standard medium will permit development of hybrid seedlings, after which special aids are not needed for further growth. Crosses with *L. chilense* also are facilitated by embryo culture, although occasional normal seeds develop (Rick and Lamm, 1955). Hogenboom (1972) succeeded in selecting genotypes in inbred *L. peruvianum* that accept *esculentum* pollen and permit normal embryo and seed development. Gunther (1964) accomplished the same objective by synthesizing an *esculentum-peruvianum* chimera that has *peruvianum* sporogenous tissue and will cross as female parent with *L. esculentum.*

II. PLANT CULTURE

A. Field

Tomato plants can be grown successfully to flowering and fruiting under a wide range of environmental conditions. It must be remembered, however, that, as a native of the dry tropics of western South America, the tomato is a warm season crop that does not tolerate excessively wet situations or prolonged exposure to temperatures below 12 C. Plants intended for hybridization should be given optimum cultural conditions. Field plantings can be direct seeded or transplanted, but plants should be spaced at least 1 m apart. Although standard fertilization practices applied to most crops are satisfactory, excessive applications of N can reduce fruit set if plants are stressed by reduced light level or other environmental restrictions.

B. Growth Chamber and Greenhouse

Tomatoes are generally hybridized with greater success in greenhouses or growth chambers than in the field because the former afford greater protection from environmental extremes to the plants and workers. No special precautions are necessary for growing tomatoes in such enclosures, except that the sprawling habit of the plant requires some type of staking or trellising.

Flowering is relatively independent of day length. Wittwer (1963) observed earlier emergence of inflorescences with a photoperiod of 9 hours than one of 18 hours in a test of 15 cultivars, but the mean difference in first flowering date was only 3 days. In contrast, Aung and Austin (1971) detected somewhat earlier flowering in long-day treatments, but admit that the effects might have been confounded with light level. In neither experiment did any tested photoperiod completely inhibit flowering. Short days are mandatory, however, for the flowering of certain wild tomato species—a factor that must be respected in interspecific hybridization. Once started, flowering continues indefinitely unless the plant bears a heavy fruit load, hence the flowering of contemporary plantings of male and female parents are naturally synchronized. The tolerances of fruiting are more exacting than those of flowering. Common causes of flower drop are low light level, mean night temperatures below 13 C or above 21 C, and daytime maxima above 40 C, particularly if reached suddenly after a cool period.

III. FLORAL CHARACTERISTICS

The statements throughout this presentation apply to the normal or standard genotype. It must be remembered, however, that genetically regulated deviations from normal in many aspects of floral morphology and physiology are known.

Tomato flowers develop on cymes of various degrees of complexity, appearing regularly on the stem at intervals of one per every three leaf nodes. Flowering is centrifugal with one or two new flowers opening per day (Fig. 1). Flowering starts in a given inflorescence before it ceases in the

Fig. 1—Inflorescence of small-fruited tomato cultivar (2 ×). The most mature flower is at the base of the inflorescence and successively earlier stages are arrayed in a counter clockwise fashion. The first two buds in the array are too advanced for emasculation, the next two buds are at the ideal stage, and the last two are too immature. The center two buds are unemasculated in the left photo and emasculated in the right photo. The open flower at the top is in ideal stage for pollen collection.

next oldest inflorescence, so that flowers and buds of all stages are usually present on a single branch.

The flower is perfect with the pistil enveloped by the solid tube formed by the stamens (Fig. 1). Pollen is released by lateral slits in the anthers into the central cavity, and, since the flower is pendant, drifts by gravity to the mouth of the tube, where the stigma is situated. Automatic self-pollination, thereby, is accomplished. Position of the stigma relative to the mouth of the anther tube markedly affects success of pollination, the optimum position being slightly depressed within the tube (Rick and Dempsey, 1969).

When grown in temperate zones, tomato cultivars are usually self-pollinated to the extent of 95 to 100%. In contrast, certain species, particularly the self-incompatible obligate outcrossers, are cross-pollinated by various bee species—solitary, bumble, carpenter, and stingless (Rick, 1950).

In the North Temperate Zone, activity of solitary and bumble bees has been observed, but usually at greatly reduced rates. Honeybees very rarely visit tomato flowers, and, despite claims to the contrary, the activity of such other agents as wind, thrips, mites, and other tiny flower-inhabiting arthropods has not been proven. In many years of observation in the field and greenhouse at Davis, Calif., we have never witnessed fruit setting of emasculated, male-sterile, or self-incompatible plants isolated from male-fertile, compatible plants by cages of coarse screen mesh, which do not exclude pollen borne by wind or tiny insects.

Flowers are observed to open throughout the daylight hours, although anthesis varies with temperature and reaches a maximum rate in the late morning. The life of the flower varies with the environment from 1 day under hot, dry field conditions to as long as 1 week in the cool, humid greenhouse. The receptivity of the stigma 1 day prior to anthesis facilitates controlled hybridization by permitting both emasculation and pollination at

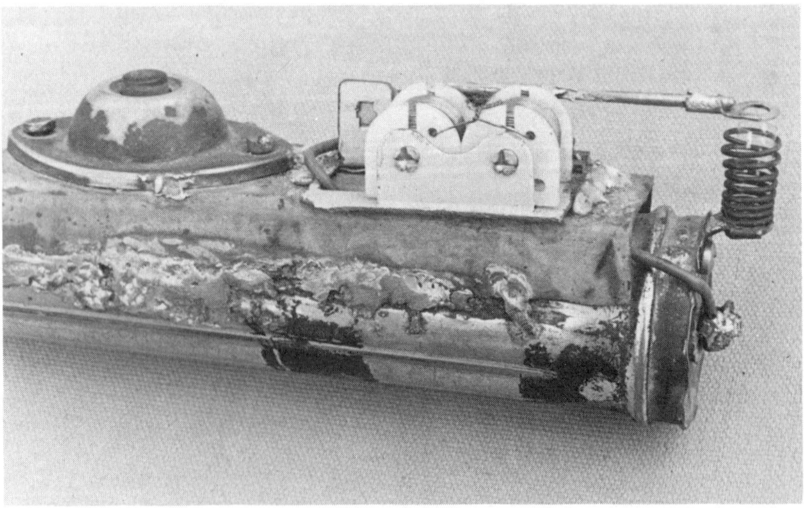

Fig. 2—Front assembly of a vibratory collector for tomato pollen designed by Martin (1960). An anther tube of a tomato flower is inserted in the metal loop at the upper right. The button switch is pushed to complete the electric circuit which vibrates the anther tube. Pollen is collected in the half capsule situated in the spring holder.

that time; the stigma remains receptive until shortly before the flower withers. Pollen is released, starting at anthesis or shortly thereafter, and, except for situations of excessive activity of pollinating insects (Rick, 1950), continues for the life of the flower.

Typical of the *Solanaceae*, tomato pollen is relatively long lived, retaining good viability under dry room conditions for several weeks. If stored at 0 to 5 C in a dessicator, it will remain viable as long as 6 months (McGuire, 1952). If lyophilized and stored at 4 C and 50% relative humidity, it retains viability for at least 2 years (King, 1965).

Fink's (1896) report of 12 hours required for pollen tubes to traverse the style was disputed by Smith (1935), who claimed that fertilization did not occur in less than 50 hours. Utilizing Martin's (1959) improved technique for observing pollen tubes, Dempsey (1970) noted tubes at the base of styles in 10 hours at 25 C, the time varying inversely with temperature. By removing styles of male-sterile flowers at various times after pollination, we have confirmed the observations of Fink and Dempsey.

IV. ARTIFICIAL HYBRIDIZATION AND SELF-POLLINATION

A. Equipment

Emasculation is usually accomplished with the use of a forceps. Pollination equipment includes a flattened or spear-point needle, electric pollen collector (Fig. 2), pollen carrier, and identification tags. More elaborate and sophisticated devices are used for large-scale hybridization.

B. Preparation of the Female

Flowers of most genotypes automatically self-pollinate; for those that fail to do so, pollen can be moved to stigmas of flowers on the same plant by procedures described in Section C.

The methodology of tomato hybridization has been described by Barrons and Lucas (1942). Flowers of the female parent must be emasculated to avoid self-contamination. Any floral bud can be emasculated provided it is large enough to manipulate and its corolla tips have not separated (Fig. 1). The risk of pollen release and consequent self-pollination is too great at later stages. The fusion of corolla and androecium to form a cap surrounding the pistil greatly facilitates emasculation. This cap is grasped with a pair of forceps, one tip inserted between the style and anther and the other between the anther and calyx. By pulling steadily, directly away from the floral axis, one can remove the entire cap in a single step without damaging the slender style (Fig. 1). If the cap is not detached as a unit, the stamens can be removed individually. According to extensive experience of many investigators, emasculated flowers do not need covering to avert contamination.

C. Pollination

For small-scale operations, pollen is collected by stripping the anther cone with a flattened or spearpoint needle. The same instrument is used to transfer pollen to the stigma of the desired female parent. Improvised or commercially manufactured devices are used for large-scale pollen collection. Cottrell-Dormer (1945) designed the first portable vibratory pollen collector with separate power source. Martin (1960) and others improved the device by combining the vibrator and batteries into a single unit. The usual design consists of a two-cell flashlight housing, whose bulb assembly is replaced by a metal plate and insulated terminal. A doorbell vibrator is mounted in front of the switch, the bell removed, and a small metal loop soldered to the end of the vibrator arm (Fig. 2). The batteries, vibrator, and switch are connected in a series circuit. On the front plate below the loop is mounted a small spring, as a holder for the inside portion of a #0 gelatin capsule, which serves as the pollen receptacle. Pollen is collected by placing the anther tube through the loop into the capsule and completing the circuit. The closed capsules serve as satisfactory containers for pollen and can be conveniently stored in a small board with holes of appropriate size and identified by notations on the board. With this device, a worker can normally harvest sufficient pollen in 30 min to supply him for a day of pollination. Massive operations require the use of more complex and stationary vibrators, examples of which are described by Dorossiev (1962) and Frankel and Galun (1977).

Pollen can be transferred from the capsules by means of the aforementioned needles. Various pollen carriers have been devised for larger scale operations. A simple, efficient device consists of a 12 to 15 cm piece of glass tubing cut transversely at one end and the other end blown to a lateral orifice of 2 to 3 mm diam, which is sufficiently large to admit a tomato stigma. The tube is filled with pollen and plugged with cotton at the transversely cut end. The tube is held upright, like a pencil, as the stigma is inserted into the lateral orifice. Pollen drifts by gravity onto the stigma. Munger (1969) described a modified plastic syringe of 1 ml capacity to serve the same purpose. Ethanol is used to sterilize all pollination instruments.

Pollinations are equally successful throughout the daylight hours, except for a decline in effectiveness during the late afternoon (Dempsey and Boynton, 1962). Best results are obtained by pollinating the largest buds immediately after emasculation or during the normal length of flower life. All naturally set fruit are removed to ensure maximum flowering and set of controlled matings. Barrons and Lucas (1942) and others report that fruit set is improved by repeatedly pollinating the same flower. A fruit can set if it contains only one seed (Dempsey and Boynton, 1962).

Applications of growth substances may improve fruit set 1) of partly sterile parents, 2) under unfavorable conditions, or 3) of interspecific crosses. In such situations, we systematically put 0.05% p-chlorophenoxyacetic acid in lanolin on the pedicel or calyx of pollinated flowers. This treatment approximately doubles the percentage of fruit set, but does not improve the seed count per fruit (Rick and Hunt, 1953).

The hybridized flowers can be marked in various ways. Small string tags with the parentage and other data marked thereon are usually used for limited numbers of matings. For larger operations, flowers can be marked with yarn or short sections of plant ties which are either color coded or stamped to identify parents (Rick, 1947a). Even when only a single male parent is used, the controlled crosses must be marked, if only by removal of calyx segments, to distinguish them from spontaneous selfs.

With fertile parents and good growing conditions, a mean fruit set of 50% can be achieved readily. Basal flowers of an inflorescence set better than terminal ones; therefore, Barrons and Lucas (1942) recommend pollinating only the first three flowers of each inflorescence.

D. Factors Affecting Efficiency

Various kinds of male-sterile mutants are widely used to circumvent the need for emasculation and to eliminate the risk of self-contamination. All of the reported mutants are determined by recessive genes and no satisfactory cytoplasmic control of male sterility has been reported.

With male sterility, in a strict sense, anthers fail to produce functional pollen. At least 44 mutants are known in many different cultivars with varying degrees of anther modification, flower size, and female fertility (Clayberg et al., 1971). Of these, several are variable in expression, notably *vms,* which is fertile at lower temperatures, but maintains effective male sterility under field conditions in central California (Rick and Boynton, 1967).

In stamenless mutants, stamens fail to develop in most environments. Examples include stamenless (*sl*—Bishop, 1954), corollaless (*cs*) and stamenless-2 (*sl*-2) (Hafen and Stevenson, 1955), and pistillate (*pi*—Rick and Robinson, 1951).

With functional male sterility, anthers contain viable pollen, but structural defects prevent release of pollen and obstruct self-pollination (*ps*—Larson and Paur, 1948). Pollen can be extracted manually and self-pollination effected. An experimentally induced, potentially functional sterility is the excessively exserted stigma condition resulting from gibberellin treatment of normal plants (Bukovac and Honma, 1967).

The totally male-sterile mutants must be maintained in segregating lines—a procedure that does not ordinarily pose serious problems if well-distinguished mutants are used. The density of planting is increased proportionately and the stand is thinned to sterile plants after flowering has started. The *ps* and *vms* mutants can be maintained homozygous by selfing under the appropriate conditions. Fertility also has been restored by gibberellin applications in several male-sterile mutants (Schmidt and Schmidt, 1975) and in *sl*-2 (Phatak et al., 1966).

V. NATURAL HYBRIDIZATION

Transferring pollen from male to female parents remains the most formidable obstacle to large-scale production of hybrid tomato seed. It is the costliest operation and one that has not yet been satisfactorily solved by use of mechanical devices. Bees, the natural tomato pollinators, have received the most attention in research.

Studies have revealed that the rates of natural cross-pollination (NCP) in most of the production areas of the world are too low to provide an economical means of accomplishing the collection and transfer of pollen (Lapushner and Frankel, 1967; Reeves, 1973). The highest rates of NCP have been recorded in various regions of Latin America, including the native region of the tomato species. By detecting hybrids in the progeny of tomato leaf (recessive—*c*) lines when interplanted with normal (dominant—*t*) lines, Richardson and Alvarez (1957) detected NCP rates of 1.5 to 15% in various localities in Mexico, and Rick (1958) reported 15 to 26% in southern Peru. By measuring the seed yields of male-sterile mutants interplanted with fertile plants, Lona (1965) observed yields approximately 30% as high as the fertile controls in Mendoza, Argentina, while Rick (1950) found 41 and 43% for valid comparisons at two stations in the valley of Rio Rimac in Central Peru. North temperate regions should not be dismissed from consideration as potential sites for exploiting NCP. In tests at various localities in California, rates measured by the male-sterile method ranged from 1.9 to 47% and were consistently high in regions where the natural ecology favored high activity of solitary bees (Rick, 1949).

Studies of other factors influencing the rates of NCP suggest methods for improving pollen transfer by insects. Seasonal differences in rates have been observed (Richardson and Alvarez, 1957; Rick, 1949; Rick et al., 1978; Soressi and Lorenzoni, 1970). Planting design can also influence the rate of NCP (Rick, 1947b). Differences between various male and female parents have been reported (Rick, 1948; Rick et al., 1978; Soost and Rick, 1957). In *L. pimpinellifolium,* a species very closely allied to *L. esculentum,* high positive correlations were detected between NCP and size of corolla of the female parent and between NCP and degree of stigma exsertion. Analysis by partial correlation indicated that flower size was probably the more influential factor (Rick et al., 1977, 1978).

A scheme to expedite NCP by incorporation of self-incompatibility was proposed by Denna (1971). It was suggested that the strict self-incompatibility of *L. peruvianum* be transferred by backcrossing and selection to *L. esculentum.* Simultaneous selection for exserted stigma and floral attractivity to bees would be mandatory. The difficulties of such transfers, particularly of the multigenic traits, are formidable. In transmitting self-incompatibility from *L. chilense* to *L. esculentum,* Martin (1961) encountered incompatibility between *esculentum* derivatives with differing *S* alleles.

Another approach to hybrid seed production by natural pollination was taken by Soressi and Lorenzoni (1970). The brown seed character results from an aleurone pigment determined by the recessive *bs* gene (Soressi, 1967). Homozygous normal (+/+) and hybrid (+/*bs*) seeds are white. By

Table 1—Comparative labor requirements of various schemes for hybrid tomato seed production (Amplified from Lapushner and Frankel, 1967).†

Operation	Standard method fertile plants	Male-sterile artificial pollination	Male-sterile natural pollination	GA-induced exs. stigma artificial pollination Bukovac & Honma (1967)	bs × bs-2 natural pollination Soressi (1970)	Self-incompatibility natural pollination Denna (1971)
Maintenance of male-sterile lines	−	+	+	−	−	−
Identification and removal of heterozygotes	−	+	+	−	−	−
Emasculation	+	−	−	−	−	−
Pollen collection and storage	+	+	−	+	−	−
Pollination	+	+	−	+	−	−
Marking of flowers	+	−	−	−?	−	−
Removal of fruits from uncontrolled pollination	+	−	−	−?	−	−
Selective harvesting of fruits	+	−	−	−?	−	−
Selection of hybrid seeds	−	−	−	−	+	−

† −, indicates a non-applicable operation; +, an applicable operation; ?, an operation of uncertain applicability.

interplanting + and *bs* homozygotes and harvesting seeds from *bs*, hybrid seeds can be identified and sorted from selfs. If *bs* were interplanted with non-allelic *bs*-2, white hybrid seeds could be identified in the progeny of both parents (Soressi, 1970). Soressi and Lorenzoni (1970) also have ascertained that massive hand application of + pollen to stigmas of *bs* immediately after anthesis results in very high proportions of hybrid progeny.

The aforementioned schemes for producing hybrid tomato seeds are compared with the requirement of hand labor in Table 1. The last three schemes in Table 1 are hypothetical because they have not been utilized on a commercial scale.

VI. SEED DEVELOPMENT, HARVEST, AND STORAGE

The first signs of ovary enlargement can be detected as early as 4 days after pollination. Barring extremely unfavorable conditions, the fruits do not abscise. Fruits ripen and seeds mature 6 to 9 weeks after pollination according to genotype and environment. Applying ethephon to intact or harvested fruit in late developmental stages can hasten the process by as much as 10 days (Sims and Kasmire, 1972).

The tomato fruit is a fleshy berry containing 50 to 300 seeds, each enveloped in a mucilaginous sheath. Small seed lots can be separated from the fruit, washed, and dried on paper. Fermentation, acid hydrolysis, or use of a blender serve as alternate methods for small lots. For fermentation, the contents of a fruit are squeezed into a cup or other container and allowed to stand at room temperature. Natural retting frees seeds from the placenta and sheaths in 1 to 5 days to permit separation of seeds by flotation and screening. Germination inhibitors usually prevent damage to seeds for periods of fermentation as long as a week, provided the normal level of liquid is maintained. Treatment with 1% HCl for 20 to 30 min can substitute for fermentation. As an alternate method, whole fruits can be blended until seeds are separated from sheaths. The blender speed must be sufficiently slow to prevent damaging seed coats (Boynton, 1962). After each of these treatments, seeds are extracted, treated with a solution of trisodium phosphate to kill virus particles adhering to seed coats, washed thoroughly, and allowed to dry. Large commercial lots generally are extracted by fermentation. Seeds of *L. esculentum* can be germinated as soon as mature, but dormancy is encountered in certain other species.

Tomato seeds retain viability for at least 10 years if stored under dry room conditions, and for much longer periods at 5 C and 40% relative humidity. They lose viability rapidly in the hot, humid tropics.

REFERENCES

Aung, L. H., and M. E. Austin. 1971. Negative and reproductive responses of *Lycopersicon esculentum* Mill. to photoperiods. J. Exp. Bot. 22:906–914.

Barrons, K. C., and H. E. Lucas. 1942. Production of first-generation in hybrid tomato seed for commercial planting. Proc. Am. Soc. Hort. Sci. 40:385–404.

Bishop, C. J. 1954. A stamenless male-sterile tomato. Am. J. Bot. 41:540–542.

Boynton, J. E. 1962. The use of blenders to facilitate seed extraction. Rep. Tomato Genet. Coop. 12:16–17.

Bukovac, M. J., and S. Honma. 1967. Gibberellin-induced heterostyly in the tomato and its implication on hybridization. Proc. Am. Soc. Hort. Sci. 91:514–520.

Clayberg, C. D., L. Butler, E. A. Kerr, C. M. Rick, and R. W. Robinson. 1971. Report of the Gene List Committee. Rep. Tomato Genet. Coop. 21:2–10.

Cottrell-Dormer, W. 1945. An electric pollinator for tomatoes. Queensland J. Agric. Sci. 2:157–169.

Dempsey, W. H. 1970. Effects of temperature on pollen germination and tube growth. Rep. Tomato Genet. Coop. 20:15–16.

———, and J. E. Boynton. 1962. Effect of time of day on controlled pollinations. Rep. Tomato Genet. Coop. 12:23–24.

———, and ———. 1965. Effect of seed number on tomato fruit size and maturity. Proc. Am. Soc. Hort. Sci. 86:575–581.

Denna, D. W. 1971. The potential use of self-incompatibility for breeding F_1 hybrids of naturally self-pollinated vegetable crops. Euphytica 20:542–548.

Dorossiev, L. 1962. A new technique for pollination of the flowers of tomatoes, eggplants, and other crops in hybrid seed production (in Bulgarian with English summary). Izv. Nauk. izled. Inst. Rasteniev, Sofia 14:203–212.

Fink, B. 1896. Pollination and reproduction of *Lycopersicum esculentum.* Minnesota Bot. Studies 1:636–643.

Frankel, R., and E. Galun. 1977. Pollination mechanisms, reproduction, and plant breeding. Springer Verlag, Berlin.

Gunther, E. 1964. Reciproke Bastarde zwischen *Lycopersicon esculentum* Mill. und *Lycopersicon peruvianum* (L.) Mill. Naturwiss. 18:443–444.

Hafen, L., and E. C. Stevenson. 1955. New male-sterile and stamenless mutants. Rep. Tomato Genet. Coop. 5:17.

Hogenboom, N. G. 1972. Breaking breeding barriers in Lycopersicon. 4. Breakdown of unilateral incompatibility between *L. peruvianum* (L.) Mill. and *L. esculentum* Mill. Euphytica 21:397–404.

———. 1973. A model for incongruity in intimate partner relationships. Euphytica 22:219–233.

King, J. R. 1965. The storage of pollen—particularly by the freeze-drying method. Bul. Tor. Bot. Club 92:270–287.

Lapushner, D., and R. Frankel. 1967. Practical aspects and the use of male sterility in the production of hybrid tomato seed. Euphytica 16:300–310.

Larson, R. E., and S. Paur. 1948. The description and inheritance of a functionally sterile flower mutant in tomato and its probable value in hybrid tomato seed production. Proc. Am. Soc. Hort. Sci. 52:355–364.

Lona, J. L. 1965. Tests of natural cross-pollination and hybrid seed production. Rep. Tomato Genet. Coop. 15:42–43.

Martin, F. W. 1959. Staining and observing pollen tubes in the style by means of fluorescence. Stain Technol. 34:125–128.

———. 1960. Some improvements in pollen collection and storage techniques. Rep. Tomato Genet. Coop. 10:23.

———. 1961. The inheritance of self-incompatibility in hybrids of *Lycopersicon esculentum* Mill. × *L. chilense* Dun. Genetics 46:1443–1454.

McGuire, D. C. 1952. Storage of tomato pollen. Proc. Am. Soc. Hort. Sci. 60:419–424.

Munger, H. M. 1969. Another device for the collection and application of tomato pollen. Veg. Imp. Newsl. 11:7.

Phatak, S. C., S. W.Wittwer, S. Honma, and M. J. Bukovac. 1966. Gibberellin-induced anther and pollen development in a stamenless tomato mutant. Nature 209:635–636.

Reeves, A. F. 1973. An observation on natural outcrossing in the tomato (*Lycopersicon esculentum* Mill.) in northwest Arkansas. Proc. Ark. Acad. Sci. 27:24–25.

Richardson, Jr., R. W., and E. Alvarez L. 1957. Pollination relationships among vegetable crops in Mexico. I. Natural cross-pollination in cultivated tomatoes. Proc. Am. Soc. Hort. Sci. 69:366–371.

Rick, C. M. 1947a. A flower marker for plant-breeding operations. Science 106:645.

———. 1947b. The effect of planting design upon the amount of seed produced by male-sterile tomato plants as a result of natural cross-pollination. Proc. Am. Soc. Hort. Sci. 50:273–284.

———. 1948. Genetics and development of nine male-sterile tomato mutants. Hilgardia 18:599–633.

———. 1949. Rates of natural cross-pollination of tomatoes in various localities in California as measured by the fruits and seeds set on male-sterile plants. Proc. Am. Soc. Hort. Sci. 54:237–252.

———. 1950. Pollination relations of *Lycopersicon esculentum* in native and foreign regions. Evolution 4:110–122.

———. 1958. The role of natural hybridization in the derivation of cultivated tomatoes of western South America. Econ. Bot. 12:346–367.

———. 1976. Tomato, *Lycopersicon esculentum*. p. 268–273. *In* N. W. Simmons (ed.) Evolution of crop plants. Longman, London.

———, and J.E. Boynton. 1967. A temperature-sensitive male-sterile mutant of the tomato. Am. J. Bot. 54:601–611.

———, and W. H. Dempsey. 1969. Position of the stigma in relation to fruit setting of the tomato. Bot. Gaz. 130:180–186.

———, J. F. Fobes, and M. Holle. 1977. Genetic variation in *Lycopersicon pimpinellifolium*: evidence of evolutionary change in mating systems. Plant Syst. Evol. 127:139–170.

———, M. Holle, and R. W. Thorp. 1978. Rates of cross-pollination in *Lycopersicon pimpinellifolium.* Plant Syst. Evol. 129:31–44.

———, and D. G. Hunt. 1953. The use of growth-promoting substances as an aid in tomato breeding. Rep. Tomato Genet. Coop. 3:22–23.

———, and R. Lamm. 1955. Biosystematic studies on the status of *Lycopersicon chilense.* Am. J. Bot. 42:663–675.

———, and J. Robinson. 1951. Inherited defects of floral structure, affecting fruitfulness in *Lycopersicon esculentum.* Am. J. Bot. 38:639–652.

Schmidt, V., and H. Schmidt. 1975. Anatomical and cytological investigations of male-sterile mutants of tomato with tendency to the stamenless character. Proc. XII Int. Bot. Cong. Leningrad 1:233.

Sims, W. L., and R. F. Kasmire. 1972. Ethephon response favorable on fresh-market tomatoes. California Agric. 26(5):3–4.

Smith, O. 1935. Pollination and life-history studies of the tomato (*Lycopersicon esculentum* Mill.). Cornell Univ. A.E.S. Memoir 184.

Smith, P. G. 1944. Embryo culture of a tomato species hybrid. Proc. Am. Soc. Hort. Sci. 44:413–416.

Soost, R. K., and C. M. Rick. 1957. Effect of varieties of pollen and ovule parents on natural cross-pollination of tomatoes. Proc. Am. Soc. Hort. Sci. 70:357–365.

Soressi, G. P. 1967. Brown seed (*bs*), a tomato seed character which behaves as an endosperm trait. Rep. Tomato Genetics Coop. 17:50.

———. 1970. Tomato hybrid seed production: a further scheme. Rep. Tomato Genet. Coop. 20:58–59.

———, and C. Lorenzoni. 1970. Crossing rate after natural or hand pollination with and without flower emasculation. Rep. Tomato Genet. Coop. 20:61–63.

Stevens, M. A. 1974. Varietal influence on nutritional value. p. 87–110. *In* P. L. White and N. Selvey (eds.) Nutritional qualities of fresh fruits and vegetables. Futura Publ. Co., Mt. Kisco, N.Y.

Witter, S. H. 1963. Photoperiod and flowering in the tomato (*Lycopersicon esculentum* Mill.). Proc. Am. Soc. Hort. Sci. 83:688–694.

49 Triticale

E. N. Larter
J. P. Gustafson
University of Manitoba
Winnipeg, Manitoba, Canada

Triticale (X *Triticosecale* Wittmack) was first recognized as an intergeneric hybrid between wheat (*Triticum* ssp.) and rye (*Secale* ssp.) just over a century ago (Wilson, 1876). These early hybrids were highly sterile, however, and could not be perpetuated. With the discovery of the chromosome doubling properties of colchicine in the 1930's and with the improvement of techniques for culturing hybrid embryos, the synthesis of fertile hybrids from wheat-rye crosses became relatively routine. Today, a number of triticale breeding programs are underway in North America, and also in less-developed countries supported by international funding agencies in Canada and the United States. Current yields of triticale in Canada, the United States, and Mexico are equal to those of the highest yielding wheats and the protein content of triticale is higher.

Because of its recent development as a crop species, commercial production of triticale is limited. In Canada and the United States, small amounts are used in the livestock feeding industry and for the manufacture of certain bread products and breakfast cereals. As plant breeders continue to improve the yields and nutritional quality of triticale, it is expected that an increase in use will be found for this species both as a feed and as a food.

I. PARENTAL MATERIAL

Two major classes of triticale based on their ploidy level have been developed. The hexaploid type (2n = 42) is the most advanced in its development and is recognized by its scientific name X *Triticosecale* Wittmack. It combines the A and B genomes of tetraploid wheat with the R genome of diploid rye. Of less agronomic potential as a crop species is the octoploid form of triticale (2n = 56) which combines the three genomes of hexaploid wheat (A, B, and D) with the R genome of rye.

The hexaploid and octoploid forms of triticale are cross-compatible and hybridization of the two has been used extensively in triticale breeding programs (Muntzing, 1974). Experience has shown that the best results are achieved from such a cross if the octoploid form is used as a female. The greatest proportion of the fertile progeny derived from octoploid × hexaploid crosses are themselves hexaploids (2n = 42) and are termed secondary hexaploids (Fig. 1). Secondary hexaploids in triticale have also been produced from crosses between primary hexaploids × hexaploid wheat (*T. aestivum* L. em Thell.). For best results, these crosses are made using triticale as the female parent because the reciprocal cross results in very poorly-developed, shrunken seeds.

Secondary hexaploid triticales may either contain the full complement of rye chromosomes, i.e.—genomically AABBRR, or one or more of the R-genome chromosomes may be replaced by D-genome wheat chromosomes introduced by the hexaploid wheat parent (Gustafson and Zillinsky, 1978). The latter group, called substitutional polyploids (Gustafson and Qualset, 1974), has been shown to possess beneficial characteristics depending upon the R/D substitutions involved (Merker, 1975; Gustafson, 1976).

Although the frequency of seed set is low, hybrids from hexaploid triticale × rye crosses have also been produced. Krolow (1973) was successful in developing a series of tetraploid triticale (2n = 28) in this way (Fig. 2). and more recently, Tarkowski (1978) has also produced 2n = 28 chromosome triticales.

II. PLANT CULTURE

A. Field

Many Canadian and United States triticales are classed as long-day plants. Short photoperiod, e.g.—less than 12 hours, delays flowering by several days relative to a day length of 16 hours or more. Hybrid seed development is also retarded by short photoperiods; however, no information is available on the effect of photoperiod on percentage seed set. All of the triticales developed in Mexico (CIMMYT) are photoperiod-insensitive.

The influence of temperature on seed development is more clearly understood. Working with wheat × rye hybrids grown under controlled conditions, Taira and Larter (1977a) found that a day/night temperature of 17/17 C was optimum for hybrid embryo growth in vivo (Fig. 3). At 17 C, the highest frequency of normal embryos obtained was 33.4%. At a stage at which embryo culturing is necessary for wheat × rye embryos (12 to 14 days post-pollination), only those visually classified as normal were sufficiently differentiated to sustain growth and development in vitro (Fig. 4). For intercrosses involving various triticale strains where embryo culture techniques are unnecessary, it would be logical to assume that the optimum temperature for growth of wheat × rye hybrid embryos would also apply.

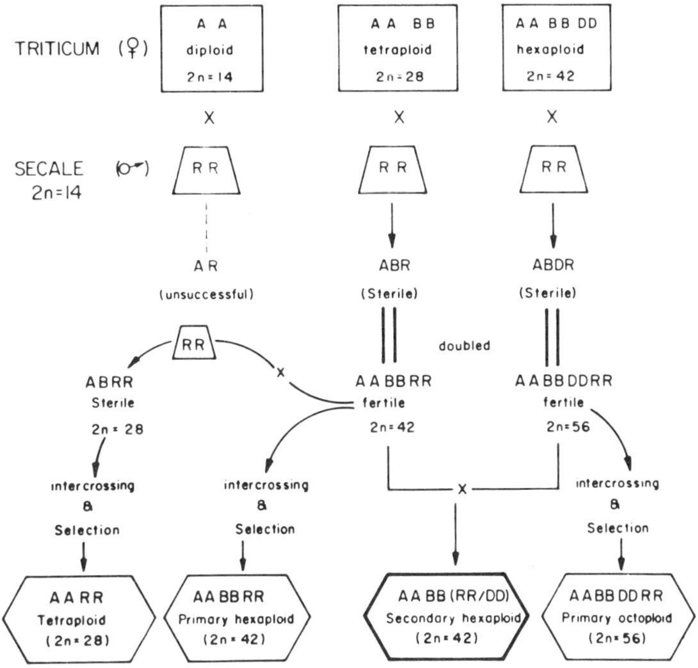

Fig. 1—Methods of synthesizing the various types of triticale.

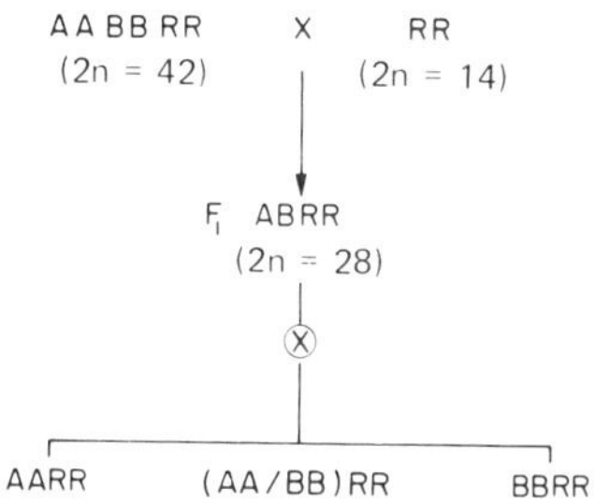

Fig. 2—Krolow's method of synthesizing tetraploid triticale (4n = 28).

Although other environmental factors such as relative humidity, soil fertility, and moisture requirements have a significant effect on the success of hybridization of triticale, no factual data regarding the optimum levels of each are available. It is known, however, that low humidity coupled with moisture stress have an adverse effect on the frequency of seed set and on the rate and degree of development of the hybrid seed. Such conditions also shorten the flowering period of the plant so that for a given number of technical workers, fewer crosses can be made. Either poor soil fertility or low light level would have a similar effect by reducing the number of vigorous tillers available for hybridization.

B. Growth Chamber and Greenhouse

During the winter months in the northern hemisphere, plant breeders carry out their hybridization programs in greenhouses and in controlled environment growth chambers. Under special circumstances, they may also choose to arrange with colleagues in a more southerly latitude to perform specific crosses in their field nurseries. When grown under greenhouse and growth chamber conditions, triticale exhibits vigorous growth and normal floral initiation at a day temperature of 22 to 24 C and a night temperature of 10 C. As indicated previously, the 17 C day/night temperature is optimum for seed development. One of the most common limiting factors of satisfactory growth of triticale under artificial conditions is sub-optimum light level. A level of 76 W/m^2 as provided by Gro-lux wide-spectrum florescent lamps produces excellent growth of triticale under greenhouse conditions during the winter months at the University of Manitoba.

III. FLORAL CHARACTERISTICS

The inflorescence (spike) of triticale resembles that of wheat rather than rye. The shape of the spike varies according to the strain or cultivar but is usually oblong, i.e.—uniform in width and thickness over its entire

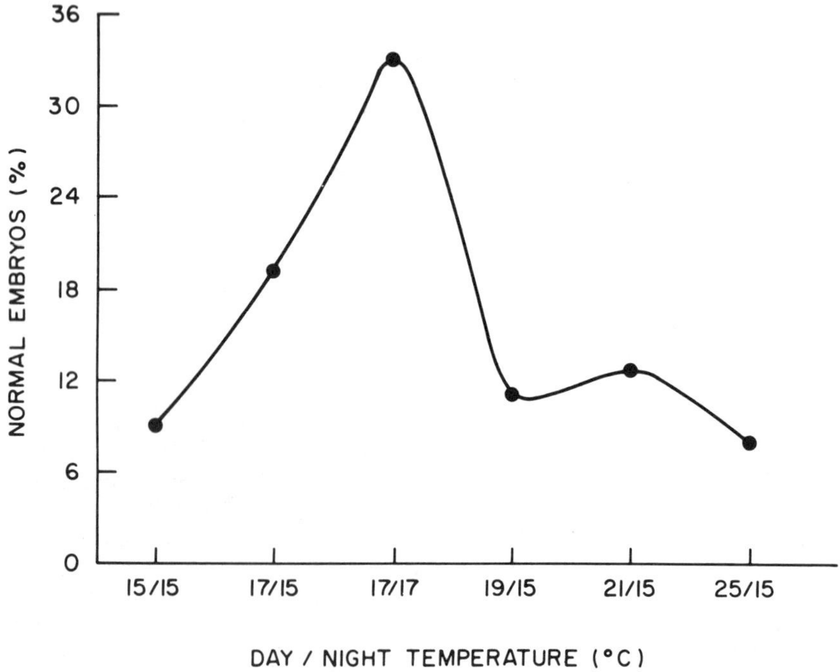

Fig. 3—Effect of various temperature regimes on wheat-rye embryo development.

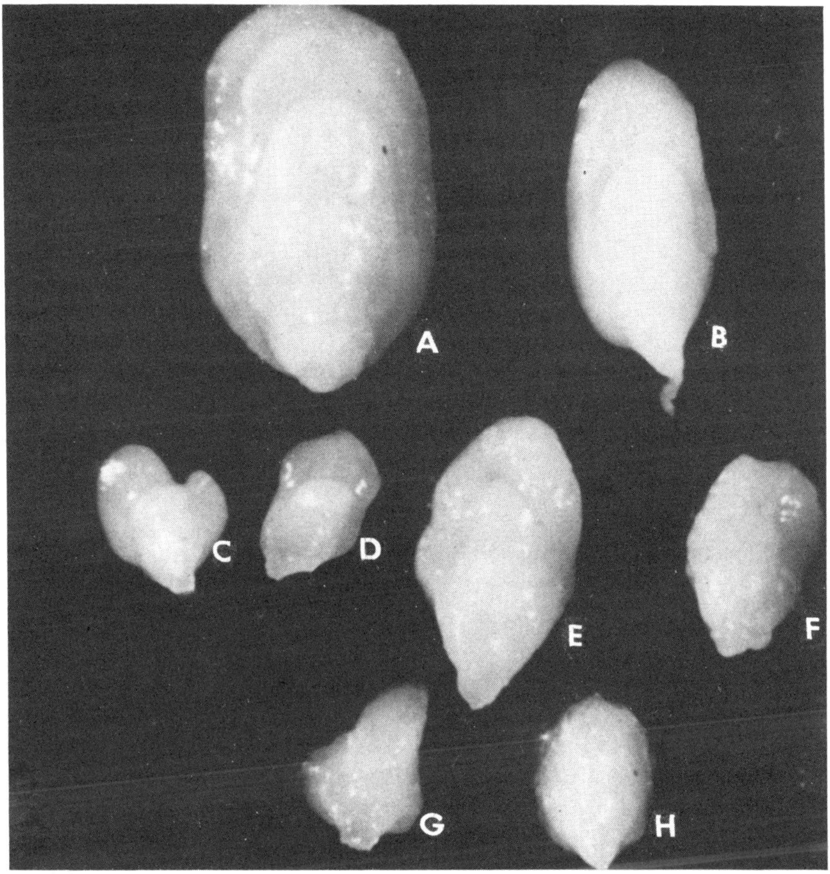

Fig. 4—Morphology of 15-day-old embryos: (A) tetraploid wheat; (B) rye; (C–E) normally differentiated hybrid embryos; (F–H) undifferentiated hybrid embryos.

length, or fusiform (tapering from base to apex). Compared to wheat, however, the spike of triticale is considerably larger (Fig. 5). The spike is composed of a series of spikelets alternately arranged on each side of a central axis, the rachis. Each spikelet (Fig. 6) is subtended by a pair of modified leaves (glumes) and consists of four to eight florets, usually only three to four of which are fertile. The reproductive parts of each floret are protected by another pair of modified leaves, the palea (innermost) and lemma. The lemma terminates in the form of an awn (beard) the length of which averages 7 to 10 cm depending upon the cultivar.

Triticale possesses a perfect flower and its typically monocotyledonous, containing three anthers and a bilobed stigma. Like wheat, each flower is cleistogamous, although it tends to open slightly 1 or 2 days after the flowering period.

Triticale is a self-pollinating species, notwithstanding the fact that one of its parents, rye, is open-pollinated. Some outcrossing can occur, the extent of which varies according to the climatic conditions at time of flower-

Fig. 5 to 8—Spikes and emasculation. Fig. 5—Spikes of triticale (left) and hexaploid wheat. Fig. 6—Spikelet of triticale. Fig. 7—Emasculation of triticale without removal of the lemma and palea. Fig. 8—Emasculation of triticale by clipping the floret to facilitate removal of the anthers.

ing. Yeung and Larter (1972) reported that in advanced strains of triticale at Winnipeg, Canada, approximately 5% outcrossing occurred. In new amphiploids and early generation hybrids, a much higher percentage may exist.

Anthesis is usually initiated when the base of the spike has emerged 6 to 10 cm from the leaf sheath. The first florets to undergo anthesis are located in the central portion of the spike, with successive pollinations occurring both basipetally and apically as the individual flowers reach maturity. Dehisced anthers are usually extruded from the florets soon after anthesis.

The period of receptivity of the stigmas of triticale is of relatively short duration. Larter (1977) conducted daily pollinations from 0 to 13 days after emasculation, and found that percentage seed set reached a maximum on the second and third day after emasculation. Virtually no success was achieved after the seventh day. The temperature regime at which these results were obtained was controlled at 20 C; higher temperatures would shorten and lower temperatures lengthen the period of stigma receptivity. As generally true of the grasses, the longevity and viability of triticale pollen is very short. Preliminary results from controlled crossing experiments indicated that freshly extracted pollen underwent a marked reduction in viability within 5 min at 20 C and for all intents and purposes was inviable after 1 hour. Attempts to store pollen of cereals at low temperatures have proved relatively unsuccessful (Harrington, 1970).

Environmental and genotypic differences profoundly affect all biological phenomena in plants, including the fertilization process. Kaltsikes (1973) demonstrated under greenhouse conditions and with a 16-hour photoperiod that the fertilization process in triticale takes place 10 to 15 hours after pollination. On the other hand, Bennett et al. (1975), using controlled temperatures of 20 C and continuous light, found fertilization to be completed within 1 hour after pollination in 'Rosner' triticale. Both studies demonstrated that the first zygotic division took place 20 to 24 hours after pollination, while the primary endosperm nucleus underwent its first division 6 to 8 hours after pollination.

IV. ARTIFICIAL HYBRIDIZATION AND SELF-POLLINATION

A. Equipment

The equipment required for the hybridization of triticale includes a pair of small scissors (approximately 15 cm in length) and a pair of inexpensive forceps (about 10 cm in length). Most workers secure the two instruments to either end of a cord which is worn around their neck during the emasculation and crossing operation.

B. Preparation of the Female

The first step in hybridization is the removal of all anthers from selected spikes of the female plant. Spikes of the female parent are selected visually on the basis of their morphological development on the particular day of emasculation. Those spikes that are at the optimum stage for emasculation can be quickly determined by examining the anthers and stigmas of the centrally-located (oldest) florets on the spike. If the stigma is perceptively feathery, yet the anthers are still green, the spike is in the proper stage for emasculation.

The florets comprising the most basal and apical spikelets of the spike are immature and are removed. Similarly, all but the primary and secondary florets of each spikelet are also removed by pulling them out and downward with the forceps. It is sometimes helpful to position a thumb on the base of each spikelet to provide support and to act as a fulcrum point during the removal of the floret. The three anthers of each of the remaining primary and secondary florets are extracted. Care must be exercised to avoid injury to the stigmas during the emasculation procedure.

There are two methods commonly used in the removal of anthers during the emasculation of triticale. One method involves entering the floret by separating the palea and lemma with the point of the forceps and simultaneously removing the three anthers (Fig. 7). This method leaves the floret intact and provides more protection for the stigma against desiccation.
Sometimes with gentle pressure from a forefinger on the rachis side of each floret, the breeder can separate the lemma and palea for each entry of the forceps. A second method involves removing the upper half of each floret with scissors, thereby exposing the tips of the three anthers for each extraction. All three anthers of each floret usually can be removed simultaneously using this method (Fig. 8). Its main disadvantage is that the stigma is left exposed and subject to desiccation should high temperatures prevail at the time of crossing.

Emasculated spikes must be protected against fertilization by foreign pollen. Immediately after each spike is emasculated, it is enclosed in a glassine envelope which is held in place by a paper clip. A marking tag is attached to the emasculated spike showing the identity of the female and male parents used in the cross and the day of emasculation.

C. Pollination

Triticale is usually pollinated the second day after emasculation. A quick examination of the stigma of one of the younger florets (located basally or apically) should be carried out, however, to ensure that it is feathery and receptive. Scissors and forceps are again used for the crossing process. For best results under field conditions, pollinations should be done in the morning. This is particularly important if hot, dry conditions prevail at the time of pollination. Under these conditions, the anthers do not dehisce normally and the pollen fails to flow freely. Warm sunny conditions favor rapid pollen development immediately before pollination. Since triticale pollen loses its viability within a few moments after dehiscence, it is necessary to pollinate within a short time after the pollen reaches maturity. For any particular spike, the pollen maturation period between earliest (central) and latest florets is about 2 days. When searching for mature pollen just prior to making the cross, spikes are selected that exhibit a few freshly-dehisced anthers extruding from the centrally located florets. An examination of the anthers of neighboring florets of such spikes will reveal anthers that are swollen and of yellowish coloration. These anthers may be either plucked out individually with forceps and collected in a shallow container, or the entire spike may be removed and carried to the female plant. The choice of collection methods is dependent upon the volume of material available as a pollen source. If the source is limited, the collection of individual anthers, although more time consuming, is more efficient in terms of maximum utilization of available pollen supply.

Two standard pollination techniques are used in triticale hybridization. If the anthers are collected individually, or if the entire pollen spike is used, the usual method involves removing the glassine envelope from the emasculated spike and by using a pair of forceps, placing a mature anther into each of the emasculated florets. In so doing, the anther is ruptured, spilling pollen onto the surface of the stigma (Fig. 9). One anther can be used to pollinate three to four florets. Those breeders who prefer to clip the top half of the floret at the time of emasculation may choose to dust pollen onto the exposed stigmas using a small camel's hair brush. This method is particularly useful when synthesizing new triticales from wheat × rye crosses. Rye is always used as the male parent, and being an open-pollinated species, it produces an abundance of pollen. The glassine envelope which was used to cover the emasculated spike is fastened back in place immediately after the spike is pollinated.

The second method of artificial pollination of triticale is by the swirl technique, also referred to as the twirl or go-go technique (Zillinsky and Borlaug, 1971). The male spike is removed from the plant and its florets clipped immediately above the top of the mature anthers. Both the action of clipping, together with the heat generated by the hand while holding the spike, causes the filament supporting the anther to elongate, thereby forcing the anther to protrude beyond the tip of the clipped florets (Fig. 10). At this stage the anthers are easily ruptured. For the pollination process, the top of the envelope is clipped and the pollen spike is inverted and lowered into the envelope. Holding the stem between the thumb and forefinger, the spike is twirled several times (Fig. 11). This creates a shower of pollen within the envelope, the top of which is then immediately folded over and clipped closed. A modified swirl method involves removing the glassine bag, clipping the tips of the forets to ensure all are open and receptive, then dusting with pollen from an inverted pollen spike held over the female spike and twirled. This method can lead to some outcrossing unless care is taken. The swirl method is rapid; however, a large volume of pollen must be available.

The final step in the hybridization process is the recording of the date of pollination on the tag already attached to the female plant at the time of emasculation. Sometimes, repeated pollinations for 3 to 6 days increase the seed set in wheat × rye crosses.

D. Factors Affecting Efficiency

The climatic conditions that prevail at the time of pollination plus proper choice of anthers dictate the success achieved. Under optimum conditions (temperature of 20 to 25 C and bright sunshine), a seed set of 70 to 90% can be expected in triticale using any of the methods described above.

Under controlled environment, it is possible to manipulate the overall flowering dates of triticale with changes in temperature, photoperiod, or both. For example, if one of two proposed parents is found to be inherently earlier than the other, its development can be delayed and the two made to coincide in their flowering periods by either lowering the temperature in which it is grown or by reducing the photoperiod.

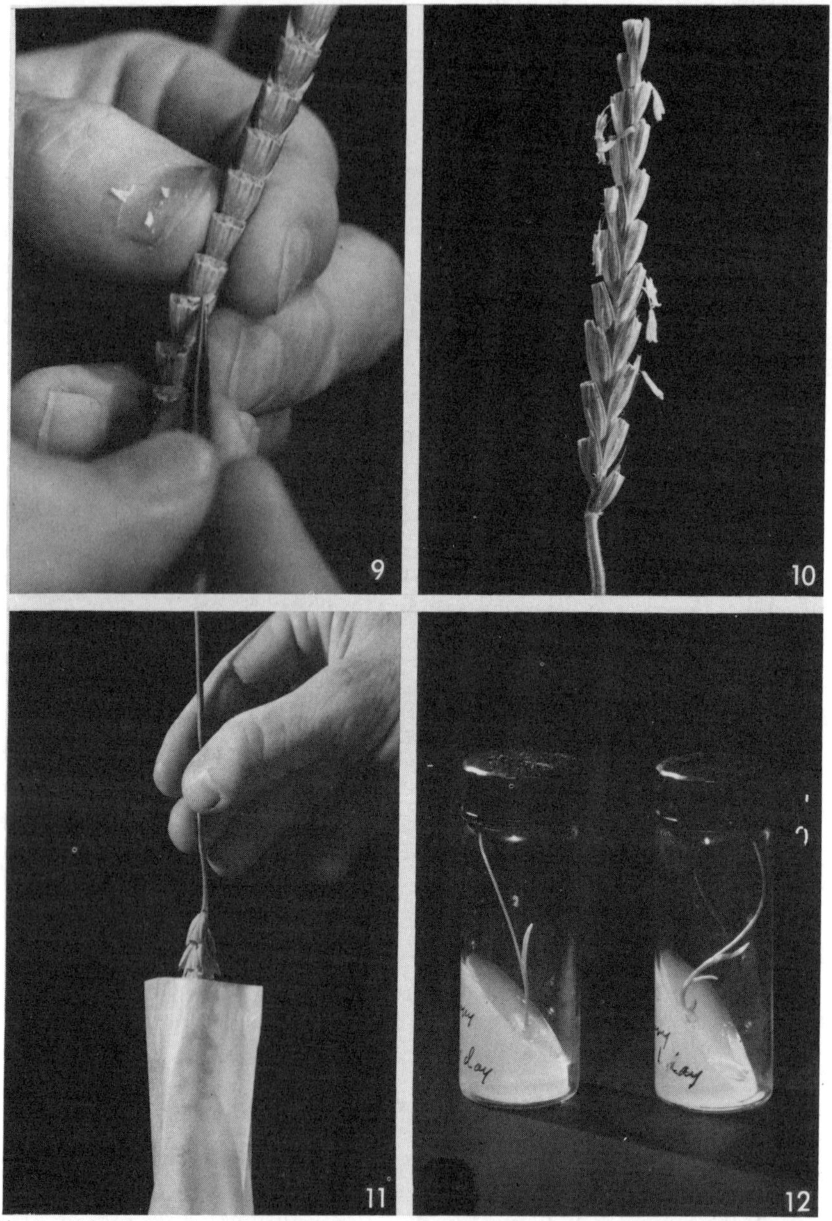

Fig. 9 to 12—Pollination. Fig. 9—Pollinating triticale using the single anther technique. Fig. 10—Spike showing extruding mature anthers. Fig. 11—Pollination using the swirl method. Fig. 12—Wheat-rye hybrids grown from embryos cultured in vitro.

A breeder with some experience can emasculate 20 to 25 florets on 8 to 10 spikes of triticale per hour and pollinate from 10 to 15 spikes per hour depending upon the procedure used.

All parental material to be used for crossing in any particular season is grown in a crossing nursery. Care in selection of the parental material to go into the crossing nursery is essential. Furthermore, the seeding arrangement of various parental stocks is important from the standpoint of saving time and effort during the hybridization period. As far as possible, the prospective parents of a cross should be sown adjacently, or at least within a short walking distance apart. It has been found convenient to sow one or two rows of each parent in 3 m rows, with inter-row distance alternating between 0.3 and 0.6 m. This row spacing is achieved by planting two rows and skipping one, which allows sufficient space for a breeder to move freely between the rows at the time of hybridization. Since the parental stocks might not coincide in their flowering periods, it is useful to sow each stock at two dates, with 10 to 14 days between seeding dates. This ensures a continuous supply of fresh pollen throughout the hybridization period. If the two parent populations are out of phase with respect to their flowering periods, one parent, usually the male, can be cut back and allowed to re-tiller.

A very limited number of genetic studies have been conducted with triticale because of its tendency to exhibit some degree of chromosomal instability. However, with continued improvement through breeding and selection, some strains which possess the full complement of the A, B and R-genome chromosomes have exhibited sufficiently high stability to enable them to be studied genetically. Three genetic seedling markers have been identified which are useful in the identification of hybrid triticale. Waxless (*Wx*) is a dominant gene which prevents the formation of the normal bluish waxy coat on all external structures of the plant. The *Wx* plants have a glossy, olive green coloration (Yeung and Larter, 1972). Another marker in triticale involves purple coleoptile, controlled by a pair of complementary genes (Kaltsikes, 1977). A third marker gene, hairy neck, causes pubescence on the peduncle and is controlled by a gene located on chromosome 5R of rye, although not all ryes have a hairy neck. Any of these three genetic factors when present in the pollen parent will be expressed in the F_1 progeny, substantiating their true hybrid nature.

V. NATURAL HYBRIDIZATION

Since triticale is predominantly a self-pollinating species, there has been very limited attention devoted to the possible use of open-pollination systems in this crop. However, one technique used with some success takes advantage of the fact that in triticale some outcrossing does occur and that pollen can be disseminated in the air over considerable distances (Yeung and Larter, 1972). In the synthesis of triticale, the F_1 hybrid of the wheat × rye cross in self-sterile. Rather than creating fertile amphiploids by the conventional process of colchicine-induced doubling of chromosomes, the self-sterile hybrids are grown in paired rows, each pair separated by a row of normal self-fertile triticale. Sufficient pollen is disseminated by wind to bring about some seed set (Zillinksy, 1974). Because of the aneuploid progeny, considerable sterility may persist in succeeding populations for a number of generations until chromosome stability is restored through either continuous self-pollination or controlled hybridization. A system of cross-pollination has the advantage of circumventing the artificial process of chromosome doubling, which at best is successful for only a small percentage of the treated population. The cross-pollinating technique is of value, however, only under optimum growth and an abundance of pollen.

VI. SEED DEVELOPMENT, HARVEST, AND STORAGE

It is possible to determine crossing success within 3 to 5 days from the time of making a triticale × triticale cross. In those florets in which fertilization occurred, the young hybrid seeds will have initiated growth and present a marked contrast to the pistils in unfertilized florets. It is important to conduct periodic checks on seed set during the crossing program so that if a particular parental combination is more difficult to hybridize than another, additional spikes of these parents can be crossed before the pollen supply becomes limiting.

Hybrid seeds usually are allowed to complete their development and to reach maturity while still on the maternal parent. Depending on the temperature during the maturation period, the time required to reach maturity is 40 to 50 days from the time of pollination. At maturity, spikes still enclosed within the glassine envelope and carrying identification tags are placed intact into individual seed envelopes. Material harvested in this way can remain stored under cool dry conditions for an indefinite period of time awaiting threshing. At the time of threshing, each spike is removed from its envelope and the parentage of the cross is copied from the tag onto the face of the envelope. The generation number should also appear on each envelope and be designated as F_1 seed. Finally, before threshing, it may be desirable to record the percentage seed set for each spike. The spike is threshed by hand by gently rubbing the seed into a clean tray. The seeds are returned to the seed envelopes and the total number of seed obtained is recorded both in a record book and on the face of the seed envelope. It is important that the identification tag also be placed in the envelope with the seed as this serves as a double-check on the parentage of each cross. The F_1 seed can be stored in this form indefinitely, with cool, dry conditions preferred.

VII. TECHNIQUES FOR SPECIAL SITUATIONS

Special procedures are required in the synthesis of primary triticale from crosses between wheat and rye. Hybrid embryos usually begin to degenerate in vivo within 12 to 16 days after fertilization. By culturing the embryos on a special nutrient media (Taira and Larter, 1978), they may be induced to differentiate into plantlets directly from the embryo phase (Fig. 12). Taira and Larter (1977b) found that certain growth substances (specifically ϵ-amino N-caproic acid and L-lysine) promoted embryo development in vivo, thereby enhancing the frequency with which they could be sustained in vitro.

As indicated previously, the F_1 hybrid plant cultured in vitro is sterile. To restore fertility, it is necessary to double its chromosome number. Colchicine, a natural alkaloid, is the most commonly used treatment for this purpose today. Although several methods of application have been described in the literature (Kaltsikes, 1974), one of the most effective procedures is the root immersion technique. Seedlings that have developed three or more tillers are removed from their growing media, the roots are washed and immersed in a 0.5% solution of colchicine for 5 hours. During this time, the solution is continuously aerated. Upon termination of treatment, the plants are removed from the solution, washed thoroughly, and repotted in soil. In the triticale program at the University of Manitoba, approximately 50% doubling is achieved using this method. Obviously, there is the need for further research to improve the techniques for amphiploid development. In this regard, studies recently conducted by Beatty (1976) indicate that a relatively higher frequency of doubling is achieved if the treated plants are retained at a cool temperature for a period of time immediately after colchicine treatment.

REFERENCES

Beatty, K. D. 1976. Triticale. II. Synthesis and description of new hexaploid and octoploid germplasm. Ph.D. Thesis. Univ. of California, Davis, California.

Bennett, M. D., J. B. Smith, and I. Barclay. 1975. Early seed development in the *Triticeae*. Phil. Trans. Royal Soc. Lond. 272:199–227.

Gustafson, J. P. 1976. The evolutionary development of triticale: The wheat-rye hybrid. Evol. Biol. 9:107–135.

———, and C. O. Qualset. 1974. Genetics and breeding of 42-chromosome triticale. I. Evidence for substantional polyploidy in secondary triticale populations. Crop Sci. 14:248–251.

———, and F. J. Zillinsky. 1978. Influence of natural selection on the chromosome complement of hexaploid triticale. Int. Wheat Genet. Symp., Proc. 5th (New Delhi, India).

Harrington, J. F. 1970. Seed and pollen storage for conservation of plant gene resources. p. 501–521. *In* O. H. Frankel and E. Bennett (eds.) Blackwell, Oxford.

Kaltsikes, P. J. 1973. Early seed development in triticale. Can. J. Biol. 51:2291–2300.

———. 1974. Methods for triticale production. Z. Pflanzenzucht. 71:264–286.

———. 1977. Personal communication. Dep. of Plant Science, Univ. of Manitoba, Winnipeg, Manitoba, R3T 2N2.

Krolow, K. D. 1973. 4x *Triticale* production and use in Triticale breeding. Int. Wheat Genet. Symp., Proc. 4th (Columbia, Mo.).

Larter, E. N. 1977. Unpublished data. Dep. of Plant Science, Univ. of Manitoba, Winnipeg, Manitoba, Canada, R2T 2N2.

Merker, A. 1975. Chromosome composition of hexaploid triticale. Hereditas 80:41–52.

Muntzing, A. 1974. Historical review of the development of triticale. *In* Triticale. Proc. Int. Symposium, El Batan, Mexico, 1–3 October 1973. Int. Devel. Res. Centre Monograph IDRC-024e:13–30.

Taira, Tomoaki, and E. N. Larter. 1977a. Effects of ϵ-amino-n-caproic acid and L-lysine on the development of hybrid embryos of triticale (X *Triticosecale* Wittmack). Can. J. Bot. 55:2330–2334.

———, and ———. 1977b. The effects of variations in ambient temperature alone and in combination with ϵ-amino-n-caproic acid on development of embryos from wheat-rye crosses (*T. turgidum* var. durum cv. Jori × *S. cereale*). Can. J. Bot. 55:2335–2337.

———, and ———. 1978. Factors influencing development of wheat-rye hybrid embryos in vitro. Crop Sci. 18:348–350.

Tarkowski, . 1978.

Wilson, A. S. 1876. On wheat and rye hybrids. Trans. a. Proc. Bot. Sco. Edinburgh 12:286–288.

Yeung, K. C., and E. N. Larter. 1972. Pollen production and dissemination properties of triticale relative to wheat. Can. J. Plant Sci. 52:569–574.

———, and ———. 1972. Inheritance of waxless in triticale. J. Hered. 63:384.

Zillinsky, F. J. 1974. The development of triticale. Adv. Agron. 26:315–348.

———, and N. E. Borlaug. 1971. Progress in developing triticale as an economic crop. CIMMYT Res. Bull. 17:1–27.

50 Warm-Season Grasses

Byron L. Burson
USDA-SEA
Temple, Texas

Warm-season grasses consist of many genera and species. In the United States, their adaptation ranges from the humid, semitropical conditions of the Southeast, to the temperate central states, and the arid Southwest. Many originated in tropical and semitropical regions of the world and were introduced into the United States. They are used as forage, turf, and ornamentals and differ in method of pollination, mode of reproduction, chromosome number, ploidy level, time of flowering, and numerous other characteristics.

I. PARENTAL MATERIAL

All grasses are members of the Gramineae family and the warm-season grasses are in the subfamilies Panicoideae, Eragrostoideae, Bambusoideae, Arundinoideae, and Oryzoideae. Most of the grasses discussed in this chapter are members of the Panicoideae and Eragrostoideae subfamilies. They are separated into tribes and numerous genera, many of which are of economic importance. The number of species within a genus may be quite limited or almost unlimited. For example, the genus *Eremochloa* has one species whereas *Paspalum* contains from 300 to 400 species.

The success of interspecific and intergeneric hybridization varies. Generally, it is easier to produce hybrids in cross-pollinated species than in self-pollinated species. Numerous interspecific hybrids have been produced among cross-pollinated *Cynodon* species (Burton, 1965; Harlan et al., 1969). Although most *Paspalum* species are self-pollinated and crossability in this genus is low, a number of interspecific *Paspalum* hybrids have been produced (Bennett and Bashaw, 1966; Burson and Bennett, 1976; Burson et al., 1973).

Crossability among *Sorghum* species is high if male sterility or perfected emasculation techniques are used. Interspecific hybrids have been produced between *S. almum* Parodi and sudangrass [*S. sudanense* (Piper) Stapf] (Pritchard, 1965a); *S. almum* and johnsongrass [*Sorghum halepense* (L.) Pers.] (Pritchard, 1965b); and johnsongrass and sudangrass (Casady and Anderson, 1952).

Since apomixis is prevalent in warm-season grasses and can affect the success of hybridization, the mode of reproduction of the species to be crossed should be known prior to making the cross (see Chap. 3). Interspecific hybrids have been produced between sexual buffelgrass, *Cenchrus ciliaris* L., and apomictic birdwoodgrass, *C. setigerus* Vahl (Read and Bashaw, 1969). An intergeneric hybrid has also been made between male-sterile pearl millet, *Pennisetum americanum* (L.) Leeke, and apomictic buffelgrass (Read and Bashaw, 1974). Napiergrass, *P. purpureum* Schumach, and pearl millet cross readily (Burton, 1944).

A number of interspecific and intergeneric hybrids have been made among the Old World bluestems, Bothriochloininae (Harlan et al., 1962) with a meticulous emasculation technique developed by Richardson (1958). Hybrids have also been produced between big bluestem, *Andropogon gerardii* Vitm., and sand bluestem, *A. hallii* Hack. (Peters and Newell, 1961).

Interspecific hybrids have been made in *Digitaria* (van Heemert and Schank, 1968), *Eragrostis* (Busey, 1976), *Setaria* (Hacker, 1968), and *Zoysia* (Forbes, 1952).

Maximum diversity of a species or genus exists at its center of origin. Centers of origin of many warm-season grasses are outlined by Vinall and Hein (1937). Most germplasm has been obtained from these areas by plant collection trips. After the material is collected, it is assigned a plant introduction number and is maintained at a USDA Regional Plant Introduction Station. Most accessions of warm-season grasses are kept at the Regional Introduction Station, Experiment, GA 30212. Seed is available upon request.

Germplasm is also available through exchange with other scientists or organizations. Plant breeders working with a particular genus or species generally maintain a large collection of plants. The Soil Conservation Service, USDA, maintains collections of native species at their regional plant materials centers. This material is also available upon request.

II. PLANT CULTURE

Plant culture for hybridization of warm-season grasses is almost exclusively limited to the greenhouse because of the instability of field environmental conditions and the floral characteristics.

A. Field

Grasses hybridized under field conditions should be grown on soils to which they are adapted. No one soil type is appropriate for all warm-season grasses. Species within the same genus vary in their soil requirements. For example, dallisgrass, *Paspalum dilatatum* Poir., is adapted to heavy clay soils; whereas bahiagrass, *P. notatum* Flugge, is well-adapted to sandy soils.

Most warm-season grasses respond to fertilization more readily with nitrogen than with phosphorus or potassium. A fertilizer ratio of 4–1–2 is generally used in the South (Wilkinson and Langdale, 1974). The rate applied depends on the inherent fertility of the soil and the response of the particular grass. High rates are usually applied to grasses to be crossed.

Moisture availability is very important. Not only is it critical for plant growth and development, but it is essential for the formation of viable gametes and subsequent seed formation. Provisions should be made for additional sources of water if supplemental moisture is needed during the growing season.

Large nurseries are not required for crossing warm-season grasses. In crossing blocks utilizing cross-pollinated species, plants are normally on 1 m centers.

Other cultural conditions promoting plant growth for hybridization under field conditions are very similar to those found in the greenhouse.

B. Growth Chamber and Greenhouse

Most grasses hybridized in the greenhouse are grown in pots. A variety of planting media are used. Many media consist of differing proportions of peat moss and pearlite or vermiculite. Sand or soil may be included in one of the above mixtures or may even be used alone if the plant is adapted to the particular soil. The soil is usually sterilized before potting to eliminate seed contamination. Occasionally, grasses are planted in greenhouses with soil floors.

Plants are kept moist and fertilized regularly with a low-concentration commercial fertilizer. It is generally accepted that hot, dry conditions or low soil fertility are not conducive to good pollen formation, pollination, or seed set. Soil moisture has little effect upon flowering rate; however, grasses grown under high-moisture regimes have larger inflorescences than those grown under dry conditions (Bennett, 1959b).

Most warm-season grasses flower during the summer months. They have indeterminate flowering, with new inflorescences initiated throughout the summer.

Warm-season grasses differ in their response to day length. Many are day-neutral, and others are either long-day or short-day. Within a genus, species commonly differ in their flowering response to photoperiod (Bogdan, 1977; Knight and Bennett, 1953; Evers et al., 1969). The latitude of origin of a species can be extremely important. Different strains of buffalograss, *Buchloe dactyloides* (Nutt.) Engelm., collected from Texas to North Dakota varied in their photoperiodic requirements (Olmsted, 1943).

To cross species or types that differ in their time of flowering, it is necessary to adjust the day length. Plants are usually grown in a greenhouse and day length is adjusted by covering plants or by providing supplemental lighting.

A high light level is necessary for floral initiation and the development of normal inflorescences. Plants grown under reduced light level produce elongated stems and inferior inflorescences.

Temperature can override photoperiodic effects. Night temperatures below 13 C inhibited flowering in common dallisgrass, bahiagrass, johnsongrass, carpetgrass (*Axonopus affinis* Chase), and bermudagrass (*Cynodon dactylon* (L.) Pers.) (Knight and Bennett, 1953). A night temperature of 18 to 21 C and a 14-hour photoperiod resulted in maximum flowering in common dallisgrass. However, night temperatures between 7 to 13 C inhibited seed head formation at the same photoperiod (Knight, 1955). In three *Zoysia* species under their correct day length, flowering was inhibited at a minimum temperature of 24 C, but was initiated at 29 C (Forbes, 1952).

Temperature also can be used to synchronize time of flowering between different species. Time of initial floret opening differs among common dallisgrass, vaseygrass (*Paspalum urvillei* Steud.), and *P. malacophyllum* Trin. (Table 1). They initiated flowering within an hour of one another when dallisgrass was subjected to a night temperature of 18 C and the other species were subjected to 13 to 16 C (Bennett, 1959c).

Warm temperatures are conducive to rapid floret opening. This principle can be used to open florets on a cool morning. Often floret opening is enhanced by carefully clasping hands around the inflorescence. Gently blowing into the clasped hands provides heat. Artificial heating of unopened florets accomplishes the same in rhodesgrass, *Chloris gayana* Kunth (Bogdan, 1959). Jones and Brown (1951) found that low temperatures delayed or inhibited blooming in several grasses. In Pensacola bahiagrass, maximum flowering occurred between 18 and 21 C and decreased as temperatures fluctuated in either direction (Hodgson, 1949).

Temperature also influences the time of anther dehiscence. Bennett (1959c) reported that in dallisgrass, vaseygrass, and *P. malacophyllum* anther extrusion and dehiscence occurred within 5 min after the florets were fully open at 21 to 24 C. At 29 C and above, the anthers dehisced before the florets open. Reducing the temperature below 21 C for dallisgrass and below 18 C for vaseygrass and *P. malacophyllum* increased the time the anthers remained pendent without dehiscing. Dehiscence was delayed for 4 hours at 11 C, for 2 hours at 13 C, and for 30 min at 16 C.

Generally, the optimum temperature for hybridization of warm-season grasses ranges from 26 to 32 C (day) and 17 to 21 C (night). A temperature of 21 C or lower is essential for proper emasculation when the anthers are removed from the florets before dehiscence.

Relative humidity is extremely important in hybridization because it delays anther dehiscence. Hodgson (1949) reported that high relative humidity prevented or delayed dehiscence in Pensacola bahiagrass. Heavy dew during the night delayed pollen shedding in weeping lovegrass, *Eragrostis curvula* (Schrad.) Nees. (Jones and Brown, 1951). Because high humidity delays anther dehiscence, this phenomenon has been used in different crossing techniques discussed in Section IV.

III. FLORAL CHARACTERISTICS

Most warm-season grasses have perfect flowers; however, there are exceptions. Buffalograss and pampasgrass, *Cortaderia selloana* (Schult.) Aschers. and Graebn., are dioecious species, and Eastern gamagrass,

Table 1—Duration of pollen dispersal and time of blooming in warm-season grasses.

Species: Scientific name	Species: Common name	Pollen dispersal period per inflorescence	Time of blooming (Daylight Standard Time)	Reference
		days	hours	
Andropogon gerardii Vitm.	Big Bluestem	7	0400–0700	Jones and Newell, 1946
A. hallii Hack.	Sand Bluestem	8	0400–0900	Jones and Newell, 1946
Bouteloua curtipendula (Michx.) Torr.	Sideoats Grama	8	0400–0900	Jones and Newell, 1946
B. gracilis (H.B.K.) Lag. ex Steud.	Blue Grama	6	0300–0900	Jones and Newell, 1946
Buchloe dactyloides (Nutt.) Engelm.	Buffalograss	Intermediate	0600–1100	Jones and Newell, 1946
Chloris gayana Kunth.	Rhodesgrass	--	1145–1230	Davidson, 1941
Chloris gayana Kunth.	Rhodesgrass	8	1245–1400	Bogdan, 1959
Eragrostis curvula (Schrad.) Nees.	Weeping Lovegrass	10.5	2400–0800	Jones and Brown, 1951
E. trichoides (Nutt.) Wood	Sand Lovegrass	7	0700–1100	Jones and Newell, 1946
Panicum maximum Jacq.	Guineagrass			
Cultivars				
Fine leaf		--	1600–2330	Warmke, 1951
Broad leaf		--	2000–2400	Warmke, 1951
Borinquen		--	2100–0100	Warmke, 1951
Regular guinea		--	2300–0400	Warmke, 1951
Gramalote		--	1330–1600	Warmke, 1951
Colonial		--	2130–2300	Javier, 1970
King Ranch #2		--	2130–2300	Javier, 1970
Common		--	2130–2300	Javier, 1970
Sabi		--	1730–1900	Javier, 1970
Green Panic		--	2130–2300	Javier, 1970
Sabi		--	1930–2100	Javier, 1970
P. purpurascens Raddi.	Paragrass	--	0700–1200	Warmke, 1951
P. virgatum L.	Switchgrass	12	1000–1400	Jones and Newell, 1946; Jones and Brown, 1951
Paspalum dilatatum Poir.	Dallisgrass	9	0430–0830	Bennett, 1959b
P. malacophyllum Trin.		11	0200–0830	Bennett, 1959b
P. notatum Flugge	Bahiagrass	5	2000–0900	Hodgson, 1949
P. urvillei Steud.	Vaseygrass	13	0300–0830	Bennett, 1959b
Phragmites communis Trin.	Common Reed	10	0600–1000	Jones and Newell, 1946
Sorghastrum nutans (L.) Nash	Indiangrass	8	0600–1000	Jones and Newell, 1946
Sorghum halepense (L.) Pers.	Johnsongrass	9.5	0900–1500	Jones and Brown, 1951
S. vulgare Pers.	Sorghum	10	0400–1000	Jones and Newell, 1946
S. sudanense (Piper) Stapf.	Sudangrass	10	0600–2400	Jones and Newell, 1946
Spartina pectinata Link	Prairie Cordgrass	10	0400–1000	Jones and Newell, 1946

Tripsacum dactyloides (L.) L., is a monoecious species.

A majority of the grasses are wind-pollinated, but insects are sometimes instrumental in pollination (Bogdan, 1977). Most warm-season grasses are naturally cross-pollinated; however, many are capable of producing some self-pollinated seed (Hanson and Carnahan, 1956). In addition to dioecious and monoecious plants, cross-pollination results from self-incompatibility, male sterility, and protogyny.

To produce hybrids, the plant breeder must know the flowering behavior of a species (Table 1). Flowering is a broad term that includes floret opening, anther and stigma extrusion, and anther dehiscence. Because locations differ within time zones, flowering times are general indicators and are not absolute.

Although flowering of most grasses starts slowly, increases to a peak, and gradually declines, some grasses have two flowering peaks. Peak pollen shedding in weeping lovegrass occurs shortly after 0100 hours and also near dawn; in Pensacola bahiagrass, a major peak occurs between 2130 and 2330 hours and a minor peak at 0730 hours (Jones and Brown, 1951; Hodgson, 1949).

The time of flowering is not uniform for a species. The variability among guineagrass, *Panicum maximum* Jacq., cultivars readily illustrates this fact (Table 1). Environmental factors also influence the time of flowering. Researchers have observed that flowering may be delayed or may not occur on cool, cloudy, humid days.

The length of time from floret opening or anther extrusion to pollen shedding varies between species and also is influenced by the environment. The whole process may take 6 to 12 min (Bogdan, 1959), 30 min (Jones and Newell, 1946), or 1 hour (Hodgson, 1949). Grasses that flower late in the day, when the humidity is low and the temperature is high, generally shed pollen quickly (Table 1).

Stigmas of warm-season grasses usually are considered receptive when they emerge from the floret. In dallisgrass, stigmas were receptive when the anthers dehisced, but they appeared dry and dead 2 hours after anthesis. In a humid environment some stigmas of dallisgrass remained receptive for 1 day after anthesis (Bennett, 1944). Stigmas of *Bothriochloa* and *Dichanthium* species remained receptive 18 to 24 hours after emasculation (Richardson, 1958) compared to 96 hours in bahiagrass (Burton, 1948c). In buffalograss, the stigmas remained receptive 20 days after anthesis, with maximum receptivity 5 days after stigma emergence and continued high receptivity until the 13th day (Jones and Newell, 1948).

Floral characteristics are sometimes conducive to natural hybridization. In dioecious and monoecious species, hybridization is necessary for seed production. In protogynous species, the stigmas are exerted from the florets before the anthers, which is also conducive to hybridization. Buffelgrass and birdwoodgrass are protogynous and their stigmas are exerted 2 to 3 days before anthesis (Fisher et al., 1954). Kikuyugrass, *Pennisetum clandestinum* Hochst. ex Chiov., is also protogynous and the presence of withered stigmas before the anthers are exerted indicate that cross-pollination has already occurred (Edwards, 1936). Species of *Zoysia* are protogynous (Forbes, 1952).

Under natural conditions, grass pollen retains its viability for only a short period after anthesis. The thin walls of pollen grains desiccate very rapidly under conditions of high temperature and low humidity (Hodgson, 1949). Dallisgrass pollen ceased to germinate in vitro 30 min after dehiscence (Bennett, 1959a), and sudangrass pollen viability was negligible 5 hours after shedding (Hogg and Ahlgren, 1943).

Although grass pollen does not generally lend itself well to storage, longevity of *Eragrostis superba* Peyr. pollen has been extended from less than 2 hours at room temperature to 4 days when stored at 2 C (Busey, 1976). Carmichael (1970) stored spikelets of *Digitaria pentzii* Stent with half-exerted anthers in a warm, humid environment. The pollen from spikelets stored for 8 hours germinated better in vitro than pollen from spikelets stored for 2, 4, and 6 hours. The addition of 50 ppm gibberellic acid to the medium increased germination.

Buffalograss pollen stored at 45 C and 30% relative humidity or at 25 C and 40% relative humidity remained viable for 24 hours, but was non-viable after 48 hours (Jones and Newell, 1946). When stored either as free pollen or in the spikelets at 4 C and 90% relative humidity, it remained viable from 6 to 8 days (Jones and Newell, 1948).

The time from pollination to fertilization varies for most warm-season grasses. The reported range is a minimum of 30 min to a maximum of 18 hours (Dewald and Harlan, 1961; Streetman and Wright, 1960).

IV. ARTIFICIAL HYBRIDIZATION AND SELF-POLLINATION

A. Equipment

In predominately self-pollinated species, it is necessary to emasculate the florets of the female parent to produce hybrids. The basic items include forceps, scissors, pollen collectors, pollination bags or plastic bags, stapler, paper clips, pencil, tags, artist brushes, string, stakes, and magnifying glasses.

B. Preparation of the Female

A successful, but tedious method used for self-fertile species involves emasculating the florets before anthesis (Richardson, 1958). Inflorescences are selected when a few anthers on the terminal rachis have dehisced. These florets are removed with small cuticle scissors, leaving only unopened florets. The hands and scissors are then rinsed in dilute alcohol or water bath to eliminate possible pollen contamination. Spikelet arrangement in many grasses are paired, and the spikelets differ in age. With the aid of a magnifying glass, the pedicellate spikelets are removed with scissors or forceps, leaving only the sessile spikelets.

Emasculation is initiated with the basal florets on the lowermost branch and gradually proceeds to the top floret. The emasculation procedure involves grasping the awn of the floret with the thumbnail and index finger of one hand. With forceps in the other hand, the tip of the lemma is grasped, and the lemma and palea are gently pulled apart. After the stigmas and anthers are exposed, the lemma is released and the bottom of the palea is pushed up. This dislodges the anthers, which can then be removed with forceps. The anthers, usually three, should be counted as they are removed to avoid leaving any in the floret. This process is repeated until the entire inflorescence is emasculated. The tools should be rinsed after each floret is emasculated.

The emasculated inflorescence is covered with a bag and the pertinent information (i.e., genus, species, date) is recorded on the bag. The sides of the bag are folded just below the inflorescence and secured to the stem with a paper clip. The bag is fastened to an overhead support so that the excess weight of the bag does not break the culm.

Other hand emasculation techniques are less laborious, but less effective. Burton (1948a) has used a humidifier to aid in the emasculation of *Paspalum* species. The night before flowering, plants are placed in a greenhouse where a continuous fog is maintained. All exerted anthers are removed with forceps before dehiscence, and the plants remain in the fog until flowering ceases. The plants are placed in a separate dry room to dry the stigmas for future pollination. After the stigmas are dry, the emasculated inflorescences are covered with an isolation bag until they are pollinated. This process is repeated daily until all the florets have flowered or the unemasculated florets are clipped from the inflorescence with scissors. In a modification of this method, the humidifier is placed inside a walk-in chamber in the greenhouse, and the plants are placed inside the chamber. This permits the plants to be in a heavier fog during emasculation (Read and Bashaw, 1969; 1974). Disadvantages of this technique are expense of a humidifier, discomfort of working in an environment at 100% relative humidity, and fogging of magnifying glasses used to remove the anthers. Species which shatter their seed readily lose their spikelets within 2 to 3 days after emasculation under these conditions.

Another technique that involves the principle of humidity is to spray water in the greenhouse prior to floret opening to increase the humidity at the time of anther exertion (Bennett and Bashaw, 1966). This method provides less time between anther extrusion and dehiscence than does the use of a humidifier, and the anthers must be removed quickly with forceps and the aid of a magnifier. This technique works best in locations where the relative humidity is 90% or higher early in the morning and the temperature is 21 C or lower.

Polyethylene bags also have been used to increase humidity as a means of preventing dehiscence. Hanna et al. (1973) placed the inflorescence of the female guineagrass parent into a plastic bag the evening before crosses were made. The next morning at about 0700 hours, high humidity inside the bag permitted removal of nondehiscent anthers. Stigmas were allowed to dry before pollination. The procedure was repeated for 2 to 3 days, after which unopened florets were removed.

Hot-water emasculation has been successful with forage sorghums, but has been less successful with other warm-season grasses. Induced male sterility by use of gametocides was attempted in weeping lovegrass, but the crossing frequency was not increased (Busey, 1976).

Selfed seed of self-compatible genotypes can be obtained by enclosing an inflorescence in a pollinating bag before flowering has begun. Seed set is usually higher if the bagged inflorescences are agitated daily after anthesis. Spatial isolation is another method for obtaining selfed seed. This usually results in a seed set higher than that from bagging because environmental conditions inside the bag can be adverse. However, complete spatial isolation from related species can be a problem if adequate plot land is unavailable.

C. Pollination

Pollen is collected when the anthers dehisce. This usually occurs early in the morning, near or shortly after sunrise (Table 1). However, dehiscence varies under different environmental conditions.

Crosses made in the greenhouse have the advantage of keeping the male parent in close proximity to the female and thus facilitating pollination. This can be done by growing the male parent in pots inside the greenhouse. Another method is to cut stems bearing inflorescences from plants in the field the evening before pollination and place them into a container of water in the greenhouse. Only those inflorescences which have begun to flower should be collected. The florets flower the next morning and continue to flower for several mornings. A large number of inflorescences can thus be maintained in a small area. The greenhouse should be closed with fans turned off because excessive air movement disperses the pollen at dehiscence. Pollination should be initiated immediately after collection because most grass pollen is short-lived.

Several pollen collection methods are used. With one method, a piece of slick-surfaced paper or a petri dish is held about 2 to 3 cm below the inflorescence, and the inflorescence is gently tapped. The pollen falls onto the paper or into the dish. A second method uses a small vacuum pump and a pollen collector. The pollen is pulled into the glass pollen collector and kept until needed. With a third method the inflorescence of the male parent is enclosed in a pollination bag. After dehiscence the bag is used to collect the pollen. This method is less desirable, because it can be used only on those grasses with many florets.

Pollen is applied to the stigmas of emasculated florets. If emasculations have been made the day before or within the hour, receptive stigmas will appear fresh and turgid. The manner of pollen application depends on the collection method. If the pollen is collected on slick paper, in a dish, or a pollen collector, the female inflorescence can be bent and placed in contact with the pollen or the pollen can be applied to the stigmas with an artist's brush. When pollen is collected in a bag, the bag is carefully transferred to the emasculated inflorescence; the inflorescence is inserted and agitated. Sometimes the bag can be torn apart and the pollen can be applied with a brush or the inflorescence can be brought in contact with the pollen on the bag.

After pollination, the inflorescence is enclosed in a labeled pollinating bag and stapled. An identification tag which was attached to the culm supporting the inflorescence should be attached to the base of the bag.

The expected success varies for each species. Success is usually higher in naturally cross-pollinated species than in self-pollinated species. The relationship between species being hybridized also affects success as do environmental conditions existing at the time of the cross.

D. Factors Affecting Efficiency

Limited pollen often reduces crossing efficiency. Some grass species produce very little pollen, and a large number of plants are required as a pollen source. These are often grown in the field and cuttings are the primary source of pollen.

The day length of species with different photoperiodic responses can be altered by the use of a growth chamber or lights or dark curtains in the greenhouse. If species undergo anthesis at different times of the day, their 24-hour rhythms can be altered by placing one plant in a growth chamber or a dark room to delay flowering. Proper management can provide synchronized flowering and anthesis in two species.

Although genetic markers are extremely valuable in identifying hybrids when one parent is carrying the marker, very few are reported in the warm-season grasses. These include a recessive white stigma color in common bahiagrass (Burton, 1948c), albino seedlings in Pensacola bahiagrass (Hodgson, 1949), and two chlorophyll deficient seedling characters in *Sorghum almum* (Pritchard, 1965b). Yellow-colored panicles in rhodesgrass (Bogdan, 1963), purple stolons in *Zoysia japonica* Steud. (Forbes, 1952), and hairiness in *Setaria trinervia* Stapf. (Hacker, 1968) also have been useful markers.

V. NATURAL HYBRIDIZATION

Many warm-season grasses are highly cross-pollinated because of self-incompatibility which facilitates natural hybridization. Male sterility and protogyny also allows natural hybridization.

Male-sterile lines are available in only a few warm-season grasses, probably because plant breeders have directed little attention to this effort. Male-sterile plants have been found or developed and have been used in Pensacola bahiagrass (Burton, 1948b), *Bothriochloa intermedia* (R.Br.) A. Camus (Harlan et al., 1962), sudangrass (Craigmiles, 1961), and johnsongrass (Casady, 1961).

In naturally cross-pollinated species, artificial hybrids are often produced by enclosing inflorescences of each parent at similar stages of anthesis in the same bag. Van Heemert and Schank (1968) used Terylene pollinating bags to produce *Digitaria* hybrids. They placed detached culms of the male parent in a test tube containing water and attached the tube to a stake below the bag. The culms extended inside the bag to a point where the inflorescence of the male parent was located above the female inflorescence to facilitate pollen transfer. Placing the inflorescences of the male parent in test tubes eliminated the possibility of seed forming in the male parent because the water supply in the test tube was gone by the end of anthesis. Mutual pollination has been accomplished with other cross-pollinated species by enclosing the inflorescences of both parents in the same bag (Hacker, 1968; Burton, 1948b, 1965).

In the usual design for crossing blocks in the field, one female plant is surrounded by pollen parents at a distance of 0.5 to 1 m between plants. In grasses that shatter their seed prematurely, the inflorescence of the female parent may be enclosed in a nylon cage, which permits the pollen to pass and yet retains the seed shattered from the female plant.

Each crossing block should be isolated from potential sources of contamination. This has been accomplished by spatial isolation, barriers to pollen movement, and a knowledge of the direction of the prevailing wind.

In kleingrass, *Panicum coloratum* L., crossing blocks were located at least 90 m apart, in an orchard where the trees served as a barrier (Burson and Bashaw, 1969). In sideoats grama, *Bouteloua curtipendula* (Michx.) Torr., isolation blocks were about 7.6 m^2 in size and 8.5 m apart in each direction, with a thick stand of sorghum planted between blocks (Harlan, 1950). Adequate isolation in Pensacola bahiagrass was 176 m for foundation seed production (Hodgson, 1949). Jones and Newell (1946) concluded that isolation distances of 302 m were necessary to avoid contamination in cross-pollinated grasses.

Culms of some cross-pollinated species can be removed shortly before anthesis and placed in water, which allows anthesis, pollination, fertilization, and seed formation. This method eliminates the necessity of planting specific field designs for particular crossing schemes. It also reduces problems related to contamination because the crosses are made inside a laboratory or greenhouse. This technique has been used to obtain seed for recurrent restricted phenotypic selection, single crosses, and polycrosses (Burton, 1948b, 1974).

VI. SEED DEVELOPMENT, HARVEST, AND STORAGE

Grass seed normally develop rapidly. Mature embryos developed in 14 to 18 days after pollination in dallisgrass (Bennett, 1944), 12 to 14 days in blue panicgrass, *Panicum antidotale* Retz. (Wright and Hall, 1965), and 8.5 days in weeping lovegrass (Busey, 1976). Seed development is commonly determined by removing several spikelets from the inflorescence and freeing the seed from the lemma and palea by rubbing them in the palm of the hand. This method is undesirable for hand-emasculated florets because hybrid seed may be lost. Seed development is best determined by germinating the seed after proper drying and storage.

In most grasses, the seed are harvested when the culm and inflorescence turn brown. Inflorescences from crosses are usually handled individually. Each culm is clipped from the plant by hand, and it is kept in the pollinating bag during drying. The bag should be folded and stapled at the opening where the culm enters to avoid the loss of seed inside the bag. The bags are placed in a warm dry environment to provide proper drying. In dallisgrass, a temperature from 38 to 60 C provided adequate drying without reducing seed viability (Bennett and Marchbanks, 1969). The desired final moisture level in the seed is about 10%. After the seed is dry, the inflorescences are removed from the bags, threshed by rubbing them between corrugated rubber blocks, and cleaned with a small forced-air blower or separated by screening. Hanson and Carnahan (1956) have discussed grass seed harvesting and processing in detail.

The cleaned seed are placed into coin envelopes with the essential information written on the outside. The environmental conditions are very critical if the seed is to be stored for an extended period. High temperature and humidity result in rapid seed deterioration. As a general rule, the combined sum of the percentage of relative humidity and temperature (°F) should not exceed 100. Storage conditions that do not exceed 10 C and 40% relative humidity or lower would be desirable.

Proper storage conditions include pest control. Napthalene crystals often are placed in the envelopes to protect against insects. Seed stored in paper envelopes should be placed in metal containers to protect against rodents. Sometimes the seed are stored in small glass jars for this purpose. Seed storage areas should also be rodent-proof; however, poison bait should be in the room as a precaution.

REFERENCES

Bennett, H. W. 1944. Embryology of *Paspalum dilatatum.* Bot. Gaz. 106:39–45.

———. 1959a. Artificial pollen germination for selection of improved seed production in *Paspalum dilatatum,* Poir. Agron. J. 51:109–111.

———. 1959b. The effect of soil moisture and light on flowering in *Paspalum* spp. Agron. J. 51:169–171.

———. 1959c. The effect of temperature on flowering in *Paspalum.* Agron. J. 51: 191–193.

———, and E. C. Bashaw. 1966. Interspecific hybridization with *Paspalum* spp. Crop Sci. 6:52–54.

———, and W. W. Marchbanks. 1969. Seed drying and viability in dallisgrass. Agron. J. 61:175–177.

Bogdan, A. V. 1959. Flowering habits of *Chloris gayana.* Proc. Linn. Soc. 170:Pt 2, 154–158.

———. 1963. *Chloris gayana* without anthocyanin coloration. Heredity 18:364–368.

———. 1977. Tropical pasture and fodder plants. Longman, London.

Burson, B. L., and E. C. Bashaw. 1969. Behavior, transmission, and morphological effects of accessory chromosomes in *Panicum coloratum* L. Crop Sci. 9:176–179.

———, and H. W. Bennett. 1976. Cytogenetics of *Paspalum conspersum* and its genomic relationship with yellow-anthered *P. dilatatum* and *P. malacophyllum.* Can. J. Genet. Cytol. 18:701–708.

———, H. Lee, and H. W. Bennett. 1973. Genome relations between tetraploid *P. dilatatum* and four diploid *Paspalum* species. Crop Sci. 13:739–743.

Burton, G. W. 1944. Hybrids between napiergrass and cattail millet. J. Hered. 35: 227–232.

———. 1948a. Artificial fog facilitates *Paspalum* emasculation. J. Am. Soc. Agron. 40:281–282.

———. 1948b. A method for producing chance crosses and polycrosses of Pensacola bahiagrass, *Paspalum notatum.* J. Am. Soc. Agron. 40:470–471.

———. 1948c. The method of reproduction in common bahiagrass, *Paspalum notatum.* J. Am. Soc. Agron. 40:443–452.

———. 1965. Breeding better bermudagrasses. Int. Grassland Congr., Proc. 9th. p. 93–96.

———. 1974. Recurrent restricted phenotypic selection increases forage yields of Pensacola bahiagrass. Crop Sci. 14:831–835.

Busey, P. 1976. Breeding and cytogenetics of lovegrasses (*Eragrostis* spp.). Ph.D. Thesis. Univ. Arizona. (Libr. Congr. Card No. Mic. 76-28,200). Univ. Microfilms. Ann Arbor, Mich. (Diss. Abstr. 37:2608B).

Carmichael, J. W. 1970. The effect of gibberellic acid on *in vitro* pollen germination in *Digitaria pentzii* Stent. Proc. Soil Crop Sci. Soc. Fla. 30:255–258.

Casady, A. J. 1961. Self sterility in johnsongrass, *Sorghum halepense.* Crop Sci. 1:154.

———, and K. L. Anderson. 1952. Hybridization, cytological, and inheritance studies of a sorghum cross—autotetraploid sudangrass × (Johnsongrass × 4n sudangrass). Agron. J. 44:189–194.

Craigmiles, J. P. 1961. The development, maintenance, and utilization of cytoplasmic male-sterility for hybrid sudangrass seed production. Crop Sci. 1:150–152.

Davidson, R. L. 1941. A note on anthesis in some common grasses near Johannesburg. S. Afr. Bot. 7:145.

Dewald, C. L., and J. R. Harlan. 1961. Stigma removal studies on certain accessions of *Bothriochloa intermedia* and *Dichanthium annulatum.* Crop Sci. 1:15–17.

Edwards, D. C. 1936. Three ecotypes of *Pennisetum clandestinum* Hocyst. (kikuyugrass). Emp. J. Exp. Agric. 5:371–376.

Evers, G. W., E. C. Holt, and E. C. Bashaw. 1969. Seed production characteristics and photoperiodic responses in buffelgrass, *Cenchrus ciliaris* L. Crop Sci. 9: 309–310.

Fisher, W. D., E. C. Bashaw, and E. C. Holt. 1954. Evidence for apomixis in *Pennisetum ciliare* and *Cenchrus setigerus.* Agron. J. 46:401–404.

Forbes, I., Jr. 1952. Chromosome numbers and hybrids in *Zoysia.* Agron. J. 44: 194–199.

Hacker, J. B. 1968. Cytology of species hybrids in the *Setaria sphacelata* complex. Aust. J. Bot. 16:533–540.

Hanna, W. W., J. B. Powell, J. C. Millot, and G. W. Burton. 1973. Cytology of obligate sexual plants in *Panicum maximum* Jacq. and their use in controlled hybrids. Crop Sci. 13:695–697.

Hanson, A. A., and H. L. Carnahan. 1956. Breeding perennial forage grasses. USDA Tech. Bull. No. 1145.

Harlan, J. R. 1950. The breeding behavior of side-oats grama in partially isolated populations. Agron. J. 42:20–24.

———, H. R. Chheda, and W. L. Richardson. 1962. Range of hybridization with *Bothriochloa intermedia* (R.Br.) A. Camus. Crop Sci. 2:480–483.

———, J. M. J. de Wet, and W. L. Richardson. 1969. Hybridization studies with species of *Cynodon* from East Africa and Malagasy. Am. J. Bot. 56:944–950.

Hodgson, H. J. 1949. Flowering habits and pollen dispersal in Pensacola bahiagrass, *Paspalum notatum* Flugge. Agron. J. 41:337–343.

Hogg, P. G., and H. L. Ahlgren. 1943. Environmental, breeding, and inheritance studies of hydrocyanic acid in *Sorghum vulgare* var. *sudanense.* J. Agric. Res. 67:197–210.

Javier, E. Q. 1970. The flowering habits and mode of reproduction of guineagrass (*Panicum maximum* Jacq.). Int. Grassland Congr., Proc. 11th. p. 284–289.

Jones, M. D., and J. G. Brown. 1951. Pollination cycles of some grasses in Oklahoma. Agron. J. 43:218–222.

———, and L. C. Newell. 1946. Pollination cycles and pollen dispersal in relation to grass improvement. Nebraska Agric. Exp. Stn. Res. Bull. 148.

———, and ———. 1948. Longevity of pollen and stigmas of grasses: Buffalograss, *Buchloe dactyloides* (Nutt.) Engelm., and corn, *Zea mays* L. J. Am. Soc. Agron. 40:195–204.

Knight, W. E. 1955. The influence of photoperiod and temperature on growth, flowering, and seed production of dallisgrass, *Paspalum dilatatum* Poir. Agron. J. 47:555–559.

———, and H. W. Bennett. 1953. Preliminary report on the effect of photoperiod and temperature on the flowering and growth of seven southern grasses. Agron. J. 45:268–269.

Olmsted, C. E. 1943. Growth and development in range grasses. IV. Photoperiodic responses in 12 geographic strains of side-oats grama. Bot. Gaz. 106:46–74.

Peters, L. V., and L. C. Newell. 1961. Hybridization between divergent types of big bluestem, *Andropogon gerardii* Vitm., and sand bluestem, *Andropogon hallii* Hack. Crop Sci. 1:359–363.

Pritchard, A. J. 1965a. Inheritance patterns in hybrids between *Sorghum almum* Parodi and perennial sweet sudangrass. Aust. J. Agric. Res. 16:525–532.

———. 1965b. Inheritance of two seedling characters in *Sorghum almum* Parodi, and cross pollination within *S. almum* and between *S. almum* and *S. halepense* (L.) Pers. Aust. J. Agric. Res. 16:533–540.

Read, J. C., and E. C. Bashaw. 1969. Cytotaxonomic relationships and the role of apomixis in speciation in buffelgrass and birdwoodgrass. Crop Sci. 9:805–806.

———, and ———. 1974. Intergeneric hybrid between pearl millet and buffelgrass. Crop Sci. 14:401–403.

Richardson, W. L. 1958. A technique of emasculating small grass florets. Indian J. Genet. Plant Breed. 18:69–73.

Streetman, L. J., and N. Wright. 1960. A cytological study of black gramagrass *Bouteloua eripoda.* Am. J. Bot. 47:786–793.

van Heemert, C., and S. C. Schank. 1968. New breeding techniques: Terylene bags for *Digitaria* breeding. Proc. Soil Crop Sci. Soc. Fla. 28:150–155.

Vinall, H. N., and M. A. Hein. 1937. Breeding miscellaneous grasses. USDA Yearb. 1937:1032–1102.

Warmke, H. E. 1951. Cytotaxonomic investigations of some varieties of *Panicum maximum* and of *P. purpurascens* in Puerto Rico. Agron. J. 43:143–149.

Wilkinson, S. R., and G. W. Langdale. 1974. Fertility needs in warmseason grasses. p. 119–145. *In* D. A. Mays (ed.) Forage fertilization. Am. Soc. of Agronomy, Madison, Wis.

Wright, L. N., and R. L. Hall. 1965. Cytological investigations of blue panicgrass, *Panicum antidotale* Retz. J. Arizona Acad. Sci. 3:136–141.

51 Wheat

R. E. Allan
USDA-SEA and Washington State University
Pullman

Cultivated wheats include einkorn (2x), emmer (4x), durum (4x), and common (6x) wheats, but einkorn and emmer wheats are seldom grown today. Wild einkorn was probably domesticated in southeastern Turkey and wild emmer in the upper Jordan Valley. Evidence suggests that hexaploid common wheat (*Triticum aestivum* L. em Thell.) was domesticated outside of these regions (Harlan, 1975).

The origin of wheat is speculative, but a leading theory suggests that hexaploid wheat arose from natural hybridization involving three diploid strains. The emmer and durum wheats probably originated from natural hybridization of wild *T. monococcum* L., the A-genome donor, and a currently unknown B-genome donor, probably of the *Sitopsis* section. Hexaploid common wheat probably arose when a tetraploid type [*T. turgidum* L. var. *dicoccoides* (Körn in litt. in Schweinf.) Bowden] with the A and B genomes hybridized with *T. tauschii* (Coss.) Schmal., the D-genome donor.

Wheat is grown worldwide and is the most widely adapted of the cereals. It exceeds all other cereals in total area and total production. Wheat grows best between 30 and 50° N lat and 25 and 40° S lat. The crop is primarily grown under dryland or rainfed conditions with about 85% of it produced in areas where rainfall is below 900 mm annually.

Winter wheat is more extensively grown than spring wheat and is favored anywhere that fall seedings survive. The main winter wheat regions include the central and southern plains of the United States, western Europe, the Balkans, southern USSR, and China.

Spring wheat is produced in colder, drier areas that are unfavorable for winter wheat. These regions include the northern plains of the United States, the Canadian prairies, Argentina, and northern and central USSR. Spring wheat is also grown as a fall-sown crop in countries with mild winters, such as Mexico, Brazil, India, Australia, and the southwestern U.S.

Wheat is used for food, feed, and industrial purposes. It is second only to rice as a food crop and serves as the chief food for about one-third of the world's population. It supplies about one-fifth of human caloric intake. It is used in a variety of products, including breads, tortillas, chapaties, crackers, cookies, cakes, crusts, confections, noodles, macaroni, bulgar, rolled flakes, many hot and cold breakfast foods, and other specialty products (Reitz, 1967). Wheat is an important livestock feed and accounts for about 10% of the grain fed to livestock. In many areas, it is used as pasture, hay, and silage. Rather small amounts of wheat are used industrially for starch, alcohol, oil, gluten, and paper products.

I. PARENTAL MATERIAL

Wheat breeders can draw from a large gene pool notable for its diversity. There are at least 23 species of *Triticum* with ploidy levels of 14, 28, and 42 chromosomes (Morris and Sears, 1967). Harlan (1975) described three subgene pools. The primary gene pool includes types that cross readily with one another. A very large secondary gene pool of about 35 species includes all species of *Aegilops, Secale, Haynaldia,* and some *Agropyrons.* Gene transfer is possible among members of the first and second pools, but special techniques may be needed. A tertiary gene pool comprised of *Hordeum vulgare* and several species of *Agropyron* and *Elymus* represents potential rather than presently useful variability.

Numerous types of wheat amphidiploids have been artificially synthesized (Morris and Sears, 1967). Species of *Triticum* hybridize with *Secale, Agropyron,* and *Haynaldia.* Many of these synthesized hybrids have been given names, such as Triticale, Agrotricum, and Haynaltricum. Such artificial species have potential for expanding the genetic diversity of cultivated wheat.

Several large collections of wheat germplasm exist, yet any single collection contains only a portion of the total genetic variability. An extensive and widely used collection is the USDA Wheat Collection, made up of over 34,000 accessions. Its accessions can be obtained from the Plant Genetics and Germplasm Institute, USDA, Beltsville, MD 20705. Other countries that hold large wheat stocks are the USSR, Turkey, Italy, Iran, India, Hungary, Switzerland, and East and West Germany. The Food and Agriculture Organization and the International Atomic Energy Agency have jointly listed researchers who hold stocks of wheat and its relatives. The Laboratory for Informational Sciences in Agriculture at Colorado State Univ., Ft. Collins, Colo., has compiled a wheat directory which lists several major data banks of wheat germplasm.

II. PLANT CULTURE

A. Field

Crossing nurseries are usually planted in easily accessible locations on fertile, well-drained soils where water can be provided. Manipulation of temperature and photoperiod may be possible by use of shading and supplemental light. These nurseries also facilitate manipulation of pest damage, nutrient imbalance, and other useful selective treatments. Use of wide-row spacing and within-row spacing simplifies the handling of the material during hybridization, and facilitates plant growth and development.

Environment has a greater influence on pollen production than on female receptivity. Generally, conditions that are conducive to maximum grain production benefit hybridization. Optimum pollen production of wheat occurs in a temperature range of 16 to 20 C and relative humidity of 70 to 75% (DeVries, 1971). Extremes in temperature and humidity inhibit pollen shedding and pollen viability. Temperature has more effect on pollen release than does humidity, but rainy, cloudy, windy weather and drought retard flowering. Solar radiation affects wheat pollen production, but less so than do temperature and humidity. Wheat stigmas remain most receptive at moderate temperature and humidity levels (DeVries, 1971).

Adequate supplies of water and soil nutrients during growth of the wheat plant are conducive to normal floral development and successful hybridization. Nitrogen deficiency prevents normal development of wheat flowers and kernels, and water stress subsequent to floral initiation reduces seed set and size. Rill and flood irrigation are more desirable methods than sprinkler irrigation for watering crossing blocks because they interfere less with normal pollen production and cause less lodging.

B. Growth Chamber and Greenhouse

When wheat plants are grown in greenhouses or controlled facilities, maximum production of seeds per spike is generally obtained with photoperiods of 14 to 16 hours and with mean daily radiation values above 6.3×10^6 J/m^2. Proper light level is particularly important 5 to 35 days before anthesis (Fischer, 1975). Temperatures ranging from 15 to 20 C during the day and 10 to 16 C at night are considered to be optimal for fertilization and subsequent kernel development (Thorne et al., 1968; Wardlaw, 1970).

Winter wheats need vernalization, but cultivars differ in length of vernalization period required. If temperatures range from 1 to 10 C, exposure for 6 to 8 weeks outdoors will vernalize most cultivars. Several methods can be used to vernalize wheat artificially, such as seeds in moist-rolled paper towels or on moist filter paper in Petri dishes in refrigerators, or in soil-filled wooden flats in cold rooms. Exposure of young wheat seedlings to 5 to 7 C for 6 to 10 weeks under 8 to 10 hours photoperiod is an effective method.

III. FLORAL CHARACTERISTICS

The inflorescence of wheat is a determinate, composite spike (Fig. 1A). Sessile spikelets are alternately arranged on opposite sides of the rachis of the main axis of the spike (Fig. 1A). Each spikelet has two bract-like empty glumes that enclose two to nine florets that are arranged alternately on a sub-axis or rachilla (Fig. 1C). The two outer parts of each floret are a flowering glume (lemma) and a thin-walled two-keeled glume (palea) (Fig. 1B). The lemma may have an awn or awnlet or may be awnless. The lemma and palea enclose the sex organs which are three stamens, a pistil, and two lodicules that regulate opening of the flower at anthesis (Fig. 1D) (Percival, 1921). Each stamen consists of a filament and an anther (Fig. 1D). The pistil consists of an ovary with two short styles and a branched feathery stigma (Fig. 1D and E). One or more of the upper florets of a spikelet usually is imperfect with either or both male and female organs nonfunctional. The basal spikelets are often entirely sterile (DeVries, 1971). Wheat spikes differ noticeably in shape, size, and density, although differences in wheat spike morphology do not usually complicate hybridization.

Although male and female floral development are closely synchronized, differentiation of the pistil generally follows that of the stamen. The stigma can remain receptive up to 13 days after anthesis, but it is most receptive from 0 to 3 days after anthesis (DeVries, 1971).

Flowering can occur anytime during daylight, but its intensity depends on variations of temperature, light, and humidity throughout the day. Flowering often is diurnal, with a large peak in the morning and a smaller one in the afternoon. Flowering generally begins midway up the spike and proceeds upward and downwards with opposite spikelets on the rachis generally undergoing synchronized flowering. The primary floret of a spikelet flowers first and is followed by the secondary, tertiary, etc. Successive florets often open on successive days (Percival, 1921). Anthers usually begin

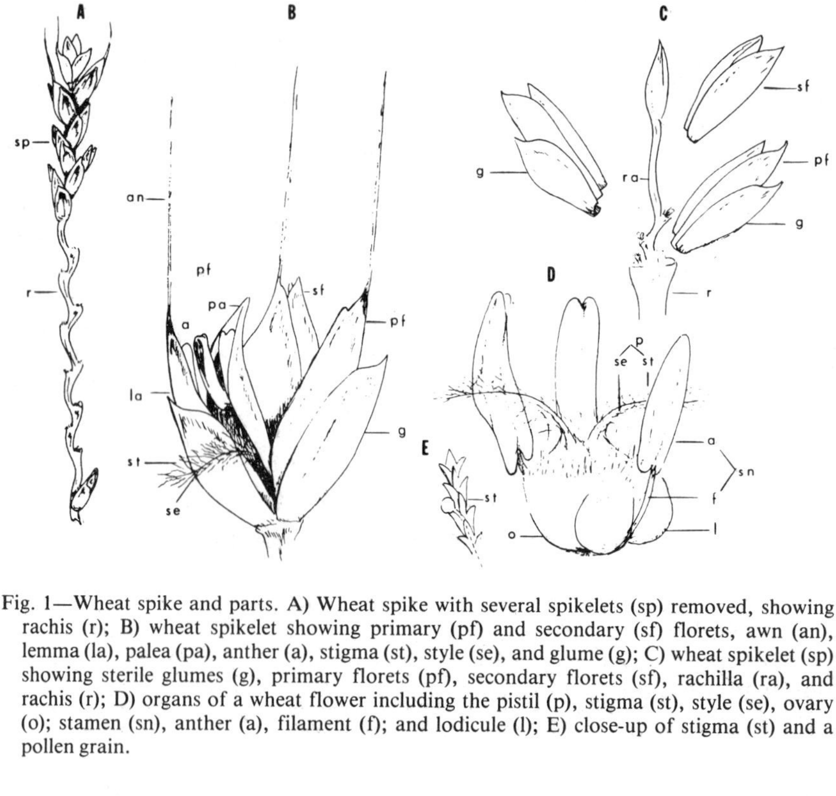

Fig. 1—Wheat spike and parts. A) Wheat spike with several spikelets (sp) removed, showing rachis (r); B) wheat spikelet showing primary (pf) and secondary (sf) florets, awn (an), lemma (la), palea (pa), anther (a), stigma (st), style (se), and glume (g); C) wheat spikelet (sp) showing sterile glumes (g), primary florets (pf), secondary florets (sf), rachilla (ra), and rachis (r); D) organs of a wheat flower including the pistil (p), stigma (st), style (se), ovary (o); stamen (sn), anther (a), filament (f); and lodicule (l); E) close-up of stigma (st) and a pollen grain.

to dehisce inside the floret, but cultivars differ in extent of anther extrusion. D'Souza (1970) found that about 5 to 7% of wheat pollen is shed on the stigma and 80% is shed outside the florets.

Wheat pollen is viable for a short time and rarely remains viable beyond 30 min in the field. The primary and secondary florets of a spikelet produce larger, and more viable pollen grains than do other florets (Ovcinnikov, 1952). Pollen on the stigma germinates readily, with pollen tube growth beginning within 15 to 120 min (DeVries, 1971).

IV. ARTIFICIAL HYBRIDIZATION OR SELF-POLLINATION

A. Equipment

Although chemical and cold treatments have been used to render stamens nonfunctional, the most reliable methods are mechanical. Tools for emasculation and pollination include a pair of smooth-jawed forceps that are 10 to 12 cm in length and a pair of scissors with sharp, close-fitting blades 30 to 40 mm long (Fig. 2A). The points of the forceps may be straight or curved depending on the user's preference, but they should fit closely together and be smoothly blunted so as not to pierce plant tissue.

Glassine bags (5 × 17 cm) are normally used to cover the female spikes. They may be fastened by paper clips, crimp-type fasteners, or staples (Fig. 2A). Dialysis tubing, closed at one end by a staple, has proved effective. Female spikes can be wrapped in sheets of bond typing paper of at least 25% rag that have been cut diagonally and secured with clips. Such coverings stay on securely in high winds and prevent rapid desiccation of

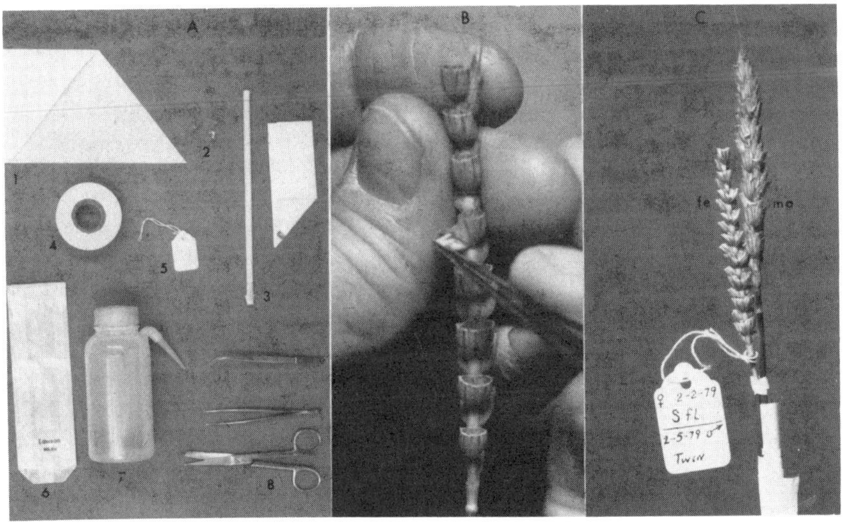

Fig. 2—Materials and methods of emasculation. A) Items used in hybridization—1) typing paper cut for spike wrapping, 2) crimp-type fasteners, 3) soda straw, 4) tape, 5) paper tag, 6) glassine bag, 7) squeeze-type wash bottle, 8) scissors, curved and straight type forceps; B) spike being emasculated by removing anthers from top of floret; C) wheat spikes prepared for the approach method; male spike shedding pollen (ma), female spike (fe) with receptive stigmas.

female organs. Paper tags (25 × 45 mm) are used for recording cross number, pedigrees, date, and worker's initials (Fig. 2A). Color-coded tags allow rapid identification of crosses.

B. Preparation of the Female

Compared to most cereals, wheat is rather simple to hybridize. Choice of spikes for emasculation requires experience and careful observation. Emasculation should be timed 1 to 3 days before normal anthesis. This may be done by observing flowering spikes and noting the extent they have emerged beyond the flag leaf. This stage may vary due to genotype and environment. Anthers should be well developed and light green, but not yellow or cream colored. The feathery stigmas should be clearly visible and extend to about one-fourth the length of the floret. By checking spikelets about midway up the spike, the most advanced stage of floral development can be ascertained. One to three of the basal and upper spikelets may have nonfunctional flowers, therefore, they are removed with scissors or forceps. Awns are excised with scissors. All but the primary and secondary florets of the remaining spikelets are removed, as follows: the spike is grasped gently but firmly between the thumb and fingers of one hand; the thumb is lightly pressed below the tertiary floret, and the center florets are pulled downward and outward with the forceps.

Anthers can be removed from each floret by carefully inserting the forceps between the lemma and palea and spreading them. The three anthers are removed with the forceps, but care must be used not to crush them or unduly injure the feathery stigmas. With experience, the breeder can usually grasp and remove all three anthers at once. Emasculated spikes should be covered immediately by bagging or wrapping.

In another procedure, the upper one-third of each floret is cut off with scissors to expose the sex organs. Anthers are then removed with forceps from the top of the floret. This method is useful in areas where cool temperatures or high humidities minimize floral desiccation (Fig. 2B).

Scissor emasculation also can be used (Wells and Caffey, 1956). Emasculation consists of merely cutting the primary and secondary florets at an earlier stage (5 to 7 days before anthesis). The anther remnants degenerate and most do not produce viable pollen. The method is fast, but requires considerable skill and experience. Selfing may be a problem, so some way to identify the hybrid seed or plants is important.

C. Pollination

Wheat flowers usually can be pollinated 2 to 4 days after emasculation. The time varies with climatic factors and stage of stigma development when the spike was emasculated. Florets should have well-developed feathery stigmas when they are pollinated (Fig. 2C).

Spikes with suitable pollen exhibit a few freshly extruded anthers (Fig. 2C). Mature pollen can be obtained from florets of spikelets in close proximity to these because alternating florets usually shed pollen on subsequent days. Anthers are removed with the forceps by entering between the lemma and palea of florets just before anthesis. If male spikes must be saved, anthers are removed from florets and quickly transported to the female in creases of the breeder's hand or in a small receptacle, such as a 35 mm plastic Petri dish. If there is no need to save the male spike, it can be removed and transported to the female.

Pollen can be applied several ways. The standard method is to remove the bag from the female, grasp with the forceps an anther which has just begun to shed pollen, and carefully brush it on the stigmas of the female plant. Usually one good anther can be used on three or four florets. If pollen is abundant, the breeder can repeat this process to insure better seed set.

Some breeders use a small brush to apply pollen, particularly if several spikes are to be fertilized with a single male. Several anthers are allowed to dehisce in the hand or a receptacle and the pollen is brushed on the stigmas. Residual pollen can be destroyed by placing the brush in the mouth or in alcohol.

The twirl pollination method can be used when the pollen source is plentiful (CIMMYT, 1976). This method is rapid and requires less skill than those described above. The upper one-third of each primary and secondary floret of the male spikes is removed to facilitate the extrusion of mature anthers and the release of pollen. During this operation, filaments of mature anthers usually elongate. The top of the bag covering the female flower is cut off and the male spike is inverted into the bag parallel to the female spike. The male spike is vigorously rotated by twirling its peduncle between the thumb and forefinger and the bag is resealed. The pollinating spike is often left in the inverted adjacent position for a day to provide additional pollen.

A third pollination system is the approach method (Rosenquist, 1927). The female spike is positioned slightly below the male spike and both spikes are covered with a bag. This method works well with moveable greenhouse plants. It can be used in the field by using detached male spikes or by growing male and females in closely-sown rows. Detached male spikes can be kept alive to shed pollen for 2 or more days by placing their culms in water. Such a system is used at Pullman, Wash., where one end of a plastic soda straw (Fig. 2A) is sealed with a plastic-bag sealing iron. The straw is taped to the female culm about 2 to 5 cm below the spike and is filled with water. The male spike is placed in the straw so that its midpoint is at the apex of the female spike, and their peduncles are taped together (Fig. 2C). Both spikes are covered with a glassine bag. On hot days, straws are refilled with water daily with a squeeze-type plastic wash bottle (Fig. 2A). The bags are shaken during peak anthesis periods to facilitate pollen dispersal.

The approach method has several advantages. Experienced people can be used to select pollen, the task requiring the most expertise in wheat hybridization. Lesser trained people can be used for emasculation, attaching straws, watering, etc. Because this method supplies pollen over a longer period than any other method, higher seed set can be achieved. Adverse weather affects the results less with it than with either the anther or twirl methods. The method is relatively fast, but slower than the twirl method. At Pullman, Wash., comparative studies based on 1,000 pollinations each indicated that 22% more F_1 seeds were produced by the approach method than the anther application method.

D. Factors Affecting Efficiency

Use of special nurseries or crossing blocks facilitates hybridization. Such nurseries contain parental material with genes for pest resistance, plant types, improved quality, and yield; elite lines; sterility sources; and special genetic stocks. Seeding breeding lines in wide rows 60 cm apart and spacing plants within rows 15 cm apart simplifies working with the material during hybridization and aids plant growth and development. Space planting can be done by hand or with specially designed seeders. Crossing blocks should be grown in close proximity to F_1 hybrid nurseries to facilitate backcrossing, top-crossing, and hybrid-parent comparisons.

Synchronization of flowering may be achieved by several methods. These include different seeding dates, space planting, clipping plants, and differential temperatures, photoperiods, light levels, and vernalization periods. Varying the seeding date is most effective in regions located within 30° N or S lat of the equator. It is less useful at Pullman, Wash. (46° N lat), where fall seedings made 75 days apart may flower within 5 days of one another.

Widely spaced seedings are often effective in synchronizing flowering particularly with winter wheats. Plants spaced 30 to 60 cm apart with optimum water and nutrients, have produced 50 to 100 tillers at Pullman. Single plants from these spacings have a flowering period of from 14 to 21 days. Clipping plants may be effective, especially for winter wheat. At Pullman, hybridization has been the most successful when plants that were cut back served as females. Plants should not be cut below the spike primordia or the primordia will be killed.

Growth-control facilities, including greenhouses, simplify floral synchronization. Such facilities make it possible to alter photoperiod, light level, vernalization, and temperature, all of which act individually and in various combinations to influence flowering. Increases in photoperiod (4.6 W/m^2 at plant height) at 2-hour intervals from 9 to 17 hours consistently reduced days to heading among several wheat cultivars (Levy and Peterson, 1972). Spring wheats even responded to vernalization treatment by heading 1.7 to 8.9 days earlier when vernalized 28 days. Staggered planting dates of spring wheats at transplanting time of winter wheats enables one to cross springs and winters in the greenhouse.

Temperatures 5 to 20 C above 10 C shorten heading date by 4 to 24 days. Floral initiation and heading date were shortened when light level was increased from 33 to 83 W/m^2 (Friend et al., 1963).

Several genetic markers available in wheat may aid in identifying hybrid plants, but very few facilitate detection of hybrid seed. In some environments, the blue-aleurone trait expresses xenia when parental selections with this gene are used as males (Suneson, 1962).

The F_1 hybrid plants often can be identified by comparing their gross morphology to their parents. When the female parent has the recessive allele for a certain morphological trait, an F_1 plant that expresses the male's dominant phenotype confirms hybridization. Useful morphological traits include awn expression, spike type, pubescence (on glumes, ligules, nodes) and nondwarfing (coleoptile and culms). Color of anther, glume, awn, coleoptile, and culm may be useful. Generally red, purple, and black pigments are dominant to the lighter colors and usually are under single gene control. Plant disease reaction may help verify hybridization.

Because wheat is a polyploid, many traits are actually influenced by several genes rather than one or two. Phenotypic expression for certain traits in F_1 plants may vary from the expected. An experienced breeder should be able to determine which traits are reliable for his material and environment.

Three independent complementary genetic systems may act to produce lethals or partial lethals that either kill or greatly reduce the productivity of F_1 plants. These traits are hybrid necrosis, hybrid chlorosis, and grass-clump dwarfness. Wheat breeders should be aware of the limitations that these genetic systems impose on wheat hybridization, and should be alert for these undesirable genotypes among their breeding lines.

Hybrid necrosis (progressive lethal necrosis) is controlled by a complementary, two-gene, multiallelic system. Hybrid chlorosis, governed by a complementary two-gene system, usually has a less drastic effect on F_1 plants than does hybrid necrosis and occurs less frequently. Listings of references giving genotypes of cultivars with genes for hybrid necrosis and chlorosis have been made (McIntosh, 1973).

The grass-clump dwarf trait is regulated by a rather complex genetic system involving at least four genes, possibly with multiple alleles. Frequently grass-clump plants fail to head or produce seed. Several genotypes produce dwarf plants and references with listings of the presumed grass-clump genotypes are available (McIntosh, 1973). This malady can be partially alleviated by growing plants at high temperatures (McVetty et al., 1976). Seed production of dwarfed plants of certain crosses can be enhanced by using gibberellic acid in combination with high temperature (Metzger, 1978).

V. NATURAL HYBRIDIZATION

Methods for increasing natural hybridization of wheat have been developed to exploit heterosis. Breeders can use such schemes to accomplish mass hybridizations needed for recurrent selection, composite cross breeding, and production of synthetics.

The *T. timopheevii* (Zhuk.) Zhuk. var. *timopheevii* cytoplasmic-sterility, genetic-fertility system is currently used in hybrid wheat programs. Adapted cultivars are converted into female cytoplasmic male steriles (A lines) by repeated backcrossing. The restorer lines (R lines) serve as pollinators possessing traits that complement those of the male-sterile females.

Stable high levels of male-fertility restoration have been difficult to achieve in hybrid wheat because several genes are involved. The effectiveness of some restorer genes can usually be increased and stabilized by selfing and reselection. Certain restorer genes function less effectively in some genetic backgrounds than in others. Because of these and other restrictions, full utilization of the cytoplasmic-sterility, genetic-fertility system to achieve mass wheat hybridization has not been realized.

Genic sterility also occurs in wheat (Suneson, 1962). Various schemes have been proposed to use this form of sterility (Briggle, 1970; Gill and Anand, 1970; Driscoll, 1972).

Use of chemical hybridizing agents may eventually offer a rapid and flexible way to induce male sterility in wheat. Some reports cite positive results. Using granular ethephon (2-chloroethylphosphonic acid), Fairey and Stoskopf (1975) reported sterility levels up to 100% with no impairment of female fertility or apparent morphological or physiological abnormalities. Several chemical companies are actively researching potential chemical hybridizing agents of wheat. If a reliable chemical agent is found, wheat hybridization will be greatly simplified, compared to hybridizations by the rather cumbersome cytoplasmic-genetic and genetic schemes.

Although wheat has a perfect flower, conducive to self-pollination, cross-pollination does occur, sometimes at a rather high frequency. The extent of cross-pollination is influenced by genotypic and environmental factors. Cross-pollination levels of about 3 to 4% are common, but much higher levels have been detected for some cultivars (Heyne and Smith, 1967).

Factors that affect natural cross-pollination of wheat include distance from pollen source, pollen concentration, weather, wind direction, and floral characters. Under North Dakota conditions, Miller and Lucken (1976) found that a 1:1 (3.1:3.1 m strips) ratio of male-sterile lines to restorer lines had several advantages over a 2:1 (6.2:3.1 m strips) ratio. The 1:1 ratio produced higher quality hybrid seed and buffered the effects of adverse environment on hybrid seed set. Both ratios were equally efficient in producing hybrid seed per unit area. Any agronomic practice that improved normal wheat yields also enhanced seed production of hybrid spring wheat.

VI. SEED DEVELOPMENT, HARVEST, AND STORAGE

The success of hybridization usually can be verified within 3 to 5 days after pollination. The lemmas and paleas of fertilized florets are no longer open, and kernel development (1.0 to 2.5 mm) is discernible. Hybrid seeds usually are allowed to ripen normally and are harvested when the female plant matures. For rapid reproduction, culms with hybridized spikes can be removed and dried 10 to 14 days after pollination. Post-harvest dormancy of immature seeds may be broken with low concentrations (10 to 50 ppm) of gibberellic acid (Metzger, 1978).

Seed may be threshed several ways. They can be removed from spikes by hand, threshed out with rubbing boards, or with impeller-type head threshers.

Hybrid seed packets must be properly identified. A convenient method is to fasten the crossing tag to the packet and arrange the packets in an order that catagorizes the purpose for each cross. Permanent numbers then can be assigned to indicate when and where each cross was made and to facilitate computerization for tracing the cross in subsequent generations.

VII. TECHNIQUES FOR SPECIAL SITUATIONS

Common wheat with its three genomes is genetically well buffered, which allows it to tolerate the addition and substitution of chromosomes from other species and genera. The existence of complete sets of aneuploid stocks greatly facilitates wheat chromosome engineering (Sears, 1972). Alien-addition lines are available with each of the seven chromosomes of rye and with various chromosomes of wild *Triticum (Aegilops), Agropyron,* and *Haynaldia.* Such lines have provided useful genetic information about the individual donor chromosomes and more importantly they have proven their value in gene transfers from the related species or genera to wheat, principally for their incorporation of disease and insect resistance.

The chromosomes of many of the wheat relatives also can be substituted for individual wheat chromosomes. Alien substitutions are produced by crossing alien-addition lines to a particular monosomic stock and identifying F_1 plants of the cross that have 20 bivalent wheat chromosomes plus a monosomic alien chromosome. Some selfed progeny of these individuals will have 20 bivalent wheat chromosomes and one bivalent alien chromosome. Such alien substitution lines are generally stable, but they almost always lack agronomic promise. Their main value has been to facilitate the exchange of a piece of alien chromosome for a piece of wheat chromosome.

Hybrid wheat plants can be successfully cloned to maximize F_2 seed return from a limited number of F_1 seeds (White, 1962). Usually four or more clones per plant may be obtained. Clones usually tiller less (about 30%) than seed stock plants and suffer slightly higher mortality rates (3 to 18%).

REFERENCES

Briggle, L. W. 1970. A recessive gene for male sterility in hexaploid wheat. Crop Sci. 10:693–696.

CIMMYT. 1976. The CIMMYT wheat training manual (Rev. February, 1976). Int. Maize and Wheat Improvement Ctr., Mexico 6, D.F.

DeVries, A. Ph. 1971. Flowering biology of wheat, particularly in view of hybrid seed production—A review. Euphytica 20:152–170.

Driscoll, C. J. 1972. XYZ system of producing hybrid wheat. Crop Sci. 12:516–517.

D'Souza, L. 1970. Untersuchungen uber die eignung des weizens als pollenspender bei der fremdbefruchtung, verglichen mit Roggen, Triticale und Secalotricum. Z. Pflanzenzucht. 63:246–269.

Fairey, D. T., and N. C. Stoskopf. 1975. Effects of granular ethephon on male sterility in wheat. Crop Sci. 15:29–32.

Fischer, R. A. 1975. Yield potential in a dwarf spring wheat and the effect of shading. Crop Sci. 15:607–613.

Friend, D. J. C., J. E. Fischer, and V. A. Helson. 1963. The effect of light intensity and temperature on floral initiation and inforescence development of Marquis wheat. Can. J. Bot. 41:1663–1674.

Gill, B. S., and S. C. Anand. 1970. Genetic male sterility for hybrid seed production in wheat. Crop Sci. 10:385–386.

Harlan, J. R. 1975. Crops and Man. Am. Soc. of Agron., Madison, Wis. 295 p.

Heyne, E. G., and G. S. Smith. 1967. Wheat breeding. *In* K. S. Quisenberry (ed.) Wheat and wheat improvement. Agronomy 13:269–306. Am. Soc. of Agron., Madison, Wis.

Levy, Jose, and M. L. Peterson. 1972. Responses of spring wheats to vernalization and photoperiod. Crop Sci. 12:487–490.

McIntosh, R. A. 1973. A catalogue of gene symbols for wheat. p. 893–937. *In* E. R. Sears and L. M. S. Sears (ed.) Proc. Fourth Int. Wheat Genet. Symp., Columbia, Mo.

McVetty, P. B. E., D. T. Canvin, and C. H. Hood. 1976. Temperature requirements for growth of grass-clump dwarf wheats and the inheritance of the trait. Crop Sci. 16:643–647.

Metzger, R. J. 1978. Personal communication. Science and Education Administration, USDA, and Dep. of Farm Crops, Oregon State Univ., Corvallis, OR 97331.

Miller, J. F., and K. A. Lucken. 1976. Hybrid wheat seed production methods for North Dakota. Crop Sci. 16:217–221.

Morris, Rosalind, and E. R. Sears. 1967. The cytogenetics of wheat and its relatives. *In* K. S. Quisenberry (ed.) Wheat and wheat improvement. Agronomy 13:19–87. Am. Soc. of Agron., Madison, Wis.

Ovcinnikov, N. N. 1952. The effect of pollen from different portions of the ear upon crosses. Sel. Semenovod. (Moscow) 19(1):76–77. [Plant Breed Abstr. 22:2604–1952.]

Percival, John. 1921. The wheat plant. Duckworth, London.

Reitz, L. P. 1967. World distribution and importance of wheat. *In* K. S. Quisenberry (ed.) Wheat and wheat improvement. Agronomy 13:1–18. Am. Soc. of Agron., Madison, Wis.

Rosenquist, C. E. 1927. An improved method of producing F_1 hybrid seeds of wheat and barley. J. Am. Soc. Agron. 19:968–971.

Sears, E. R. 1972. Chromosome engineering in wheat. Stadler Symp. 4:23–38. Univ. of Missouri, Columbia, Mo.

Suneson, C. A. 1962. Use of Pugsley's sterile wheat in cross breeding. Crop Sci. 2: 534–535.

Thorne, G. N., M. A. Ford, and D. J. Watson. 1968. Growth, development, and yield of spring wheat in artificial climates. Ann. Bot. (N.S.) 32:425–446.

Wardlaw, I. F. 1970. The early stages of grain development in wheat: response to light and temperature in a single variety. Aust. J. Biol. Sci. 23:765–774.

Wells, D. G., and H. R. Caffey. 1956. Scissor emasculation of wheat and barley. Agron. J. 48:496–499.

White, W. J. 1962. Clonal propagation of wheat plants for the evaluation of yield. Can. J. Plant Sci. 42:571–581.

52 Wild Rice

W. Anson Elliott
University of Minnesota
St. Paul

Wild rice (*Zizania aquatica* L.), an annual aquatic cereal, is native to the shallow lakes and streams in the upper Great Lakes region of North America. After processing, the nutritious wild rice grain has a unique texture and nutlike flavor. French explorers observed the Chippewa and Sioux Indians using wild rice as a staple food grain. Today, it is a delicacy often served with wild game dishes. Native stands continue to be harvested; however, over half of the production in Minnesota is from commercially grown wild rice in man-made paddies (Oelke et al., 1973; Elliott and Oelke, 1977).

I. PARENTAL MATERIAL

Most published reports identify wild rice as *Zizania aquatica* L. Fassett (1924) first used *Z. aquatica* L. to encompass all wild rice found in the United States, with the exception of that growing in a small area of Texas which he classed as *Z. texana* (Hitchc.). He also recognized *Z. latifolia* (Griseb.) as the type growing in China and northeast Asia. All North American wild rice species have the same chromosome number, 2n = 30. In *Z. latifolia,* however, the chromosome number is 2n = 34. Dore (1969) recently classified the genus into four species. He split *Z. aquatica* into two species. He designated the ecotypes growing in the Great Lakes region which have large kernels and provide the wild rice for commerce as *Zizania palustris* L. The small seeded ecotypes growing native along the St. Lawrence River estuary and along the eastern and southeastern United States coast retained the *Z. aquatica* nomenclature. Morphological characteristics have suggested the distinctness of *Z. aquatica, Z. palustris* and *Z. texana.* Emery (1977), however, has made interspecific crosses and reciprocal crosses among these three North American species. It seems probable that wild rice is a polymorphic species and that useful germplasm can be integrated as needed.

Eighty-nine ecotypes of *Z. palustris* from Minnesota and Canada have been collected and are being maintained by the University of Minnesota as a germplasm pool (Elliott, 1976).

II. PLANT CULTURE

A. Field

Wild rice cultivars adapted to Minnesota have successfully produced seed in many of the United States during the summer months. Snyder (1977), growing Minnesota-adapted wild rice in the warmer environment of Florida, found less tillering and lower yield per plant than are obtained in Minnesota, but there was no apparent blockage to pollination or seed set.

Most commercial hectares of wild rice in Minnesota are produced on organic soils, however, mineral soils can be as productive. Soil fertility programs recommended for wild rice are based upon soil tests and are similar to recommendations for small grains (Fenster et al., 1976). Oelke (1977) determined that the growth, pollination, and seed set of plants growing in peat and mineral soils occur normally when soil moisture is maintained at or above field capacity. Soils flooded with 25 cm of water have given maximum yields (Oelke, 1975), and flooding is the common practice in breeding nurseries.

When controlled pollinations are to be made in the field, planting may be done either in the fall or spring. Seeds are planted by hand or with a belt planter at 10 cm intervals in paired rows which are 1.5 m long and 0.3 m apart. Spacings of approximately 1 m between plots and 1.5 m alleys allow adequate access to plants. Immediate flooding or moistening of the fields after planting is necessary to avoid loss of seed viability from desiccation.

Number of culms per plant is greatly influenced by stand density. Oelke (1973) found 1 tiller per plant at stands of 19.3 plants per m^2 and 10 tillers at 4.6 plants per m^2. Low density planting will produce more tillers per plant, making it possible for more crosses to be made per plant. Another advantage of tillering is that it will extend the time of flowering for as long as 2 weeks.

Flowering in wild rice is affected by photoperiod and temperature. Elliott et al. (1974) reported that wild rice was a quantitative long-day plant. Under field conditions, I found that flowering occurred in fewer days in southern locations than in northern ones. This would suggest that temperature was more important than photoperiod in the control of flowering. Poor seed set has been observed in production fields and research plots when high temperatures exceeding 35 C and dry winds occurred during pollination.

Seed shatters before and during the harvest of wild rice nurseries. Because the seeds are dormant and can remain viable in the soil for several years, they will result in volunteer plants in succeeding years. To eliminate this source of contamination, we routinely fumigate nurseries before planting with 350 kg/ha of methyl bromide.

B. Growth Chamber and Greenhouse

Obtaining more than one generation of wild rice per year has been complicated by seed dormancy that exists for 3 to 5 months after seed harvest (Cardwell et al., 1973). A procedure used for germinating and growing wild rice presented by Elliott (1974) is presently being used with slight modifications for greenhouse and small field experiments.

The lemma and palea are removed from dormant seed with forceps and the pericarp over the embryo is removed by scraping. This allows germination to occur when seed is submerged in water (Cardwell et al., 1973; Woods and Gutek, 1974). Instead of scraping seeds, dehulled seeds may be tumbled (Oelke, 1975). Seeds are germinated in Petri dishes filled with water until secondary roots are 1.25 to 2.5 cm long. Optimum water temperature for germination has been shown to be 17 to 20 C (Cardwell et al., 1973). Water covering the germinating seed should be changed every other day to prevent build up of fungi and bacteria. Seedlings are transplanted into 15-cm pots containing sterile soil covered by approximately 0.5 cm of sand. The water level is maintained 2 to 4 cm over the soil top to enhance emergence and simultaneously promote faster plant growth than occurs in deeper water. Soil fertility level is that recommended by a soil test and is supplemented with iron at a rate of 48 kg/ha. The soil is in an oxidized state because it is sterilized by heat and is only slightly submerged in water. The oxidized state of the soil reduces the availability of iron. Iron chelate applied as a foliar spray will also alleviate the iron deficiency.

In a test of 43 genetic lines using the above procedure, approximately 30% of the scraped seeds became established. The success in plant establishment was greatly different among genetic lines (Elliott, 1974). Obtaining plants from only 30% of the seeds suggested that other factors need altering to obtain better results. Various types of artificial lighting have been used, but no type has been shown superior. The light level is routinely maintained in excess of one-tenth full sunlight and with a 14- to 16-hour photoperiod.

Temperature has been observed to have an effect on growth and pollination. Only minimal growth occurs when the temperature is approximately 10 C, but growth is improved when temperature is approximately 26 C. When temperatures are maintained below 21 C in efforts to promote more tillering, poor pollen production has been observed. This can be remedied by increasing the temperature to 26 C.

III. FLORAL CHARACTERISTICS

Wild rice plants range from 1.0 to 1.5 m tall. The plants are monoecious with the pistillate florets being above the staminate ones on the panicle (Fig. 1). The pistillate florets are borne erect on short stiff pedicels and held closely to the panicle branches, the staminate florets hang pendously on threadlike pedicels from the spreading panicle branches. Both types of florets have a lemma and a palea with glumes being rudimentary or absent.

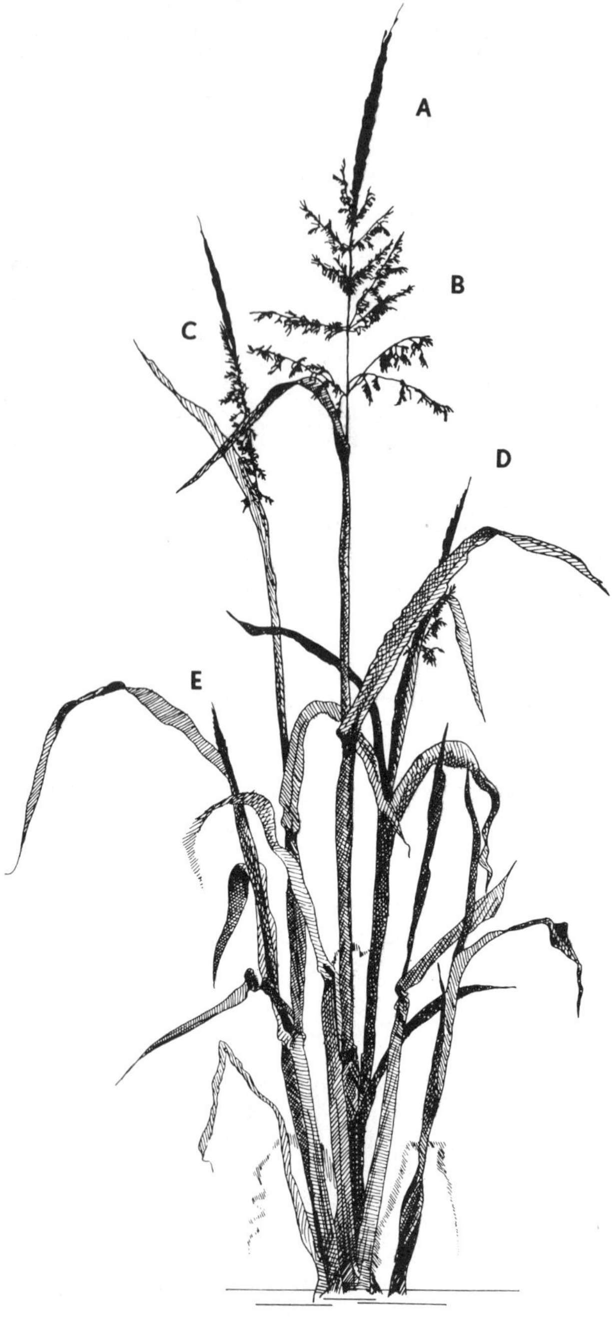

Fig. 1—A wild rice plant with the pistillate florets (A) above the staminate ones (B). Flowering begins in the primary culm followed by the first (C), second (D), and third (E) tillers.

As heading occurs, the two lodicules in each pistillate floret force the lemma, which is normally awned, to separate slightly from the palea (Weir and Dale, 1960). This allows the bilobed stigma to protrude and become pollinated. Factors affecting stigma receptivity in wild rice have not been investigated. However, scientists have assumed that stigmas are receptive when they are moist and feathery. Seeds have been obtained from crosses made during 0 to 4 days after stigma protrusion.

There are six stamens per staminate floret. The pollen mother cells undergo meiosis during early boot stage and anthesis occurs as the florets emerge from the boot. Observations made by Bush and Elliott (1977) on greenhouse grown material indicate that the greatest amounts of pollen are produced during the night or early morning period. Plants were bagged for pollen collection at 1600 hours for use the following day. Pollen viability was negatively correlated (−0.34) with temperature and positively correlated (0.58) with humidity. Pollen viability was determined by germination counts after placement on nutrient agar plates. Storage methods for maintaining high pollen viability have not been identified.

Weir and Dale (1960) found that fertilization occurs approximately 2 hours after pollination. The lodicules become flacid after fertilization and the lemma and palea closely surround, but do not adhere to the developing cylindrical grain (Fig. 2). Approximately 18 days after pollination, mature

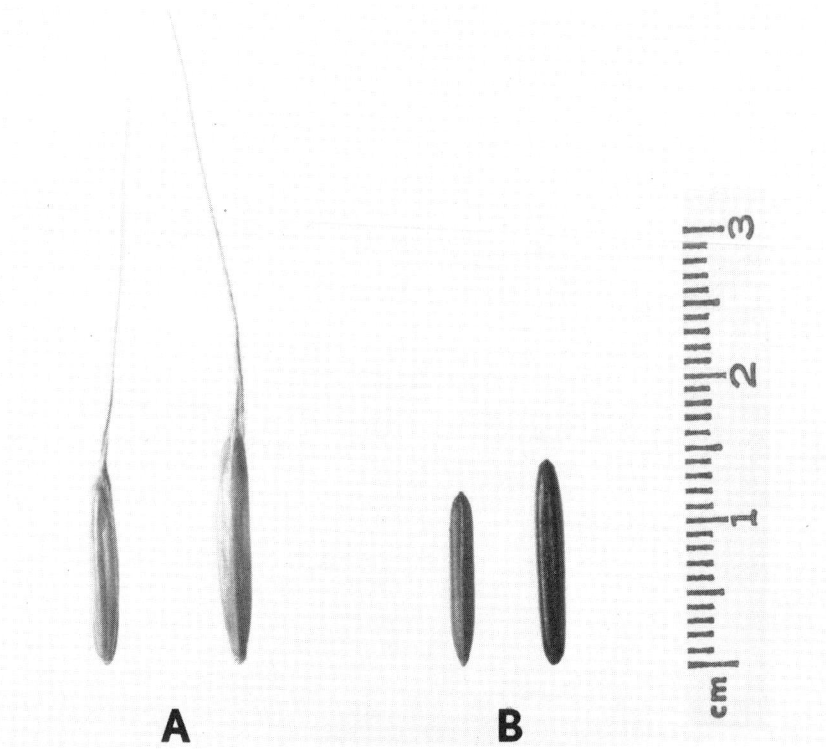

Fig. 2—(A) Grain of wild rice closely surrounded by the palea and the awned lemma and (B) the cylindrical grain that is light green in the dough stage, but turns dark brown or black at the hard kernel stage.

Fig. 3—(A) Wild rice tiller in the boot stage, (B) Pistillate florets emerging from the boot, and (C) Staminate florets at anthesis.

kernels of 1 to 2 cm length are formed. While much variation in seed maturity exists among and within the panicle branches, there seems to be a trend for maturity to progress from the tip toward the base of the panicle.

Wild rice is normally cross pollinated because the pistillate florets emerge from the boot and are receptive to wind-borne pollen before the staminate florets on the same panicle have yet emerged (Fig. 3). Pollination between tillers on the same plant is possible and results in some self-pollination. Elliott (1977) showed 94, 92, 92, and 88% cross-pollination in plants encircled by other plants at distances of 30, 60, 120, and 240 cm, respectively. Cross pollination was slightly higher on primary culms than on later-maturing culms. No evidence exists for self-incompatability systems in wild rice.

IV. ARTIFICIAL HYBRIDIZATION AND SELF-POLLINATION

A. Equipment

Because of the monoecious nature of wild rice, only minimum equipment is required for emasculation and pollination. The equipment needed includes scissors, paper clips, bamboo poles, plant ties, forceps, and pollinating bags. There are various sources and bags, however, we have been using Lawson pollinating bags #1812 (5.0 × 2.5 × 47 cm), Lawson Labels, Inc., 1206 N.E. Adams St., Peoria, IL 61603.

B. Preparation of female

The crossing process is analygous to that used for corn in that the female is first protected, while the pollen is collected in a panicle bag and then transferred to and dusted on the female.

Plants to be used for pollination should have extruded stigmas soon after emergence and should be male fertile. These desirable characteristics are lacking in many plants. The primary culm is useful in selecting desirable plants because it flowers as much as a week ahead of the other tillers (Fig. 1). Later tillers can be used for producing the crosses or selfs.

Plants of the female parent are bagged when pistillate florets are in boot stage (Fig. 3). After pistillate florets are emerged 2 to 3 cm, the flag leaf sheath is unrolled to remove the staminate florets. A small pair of scissors can be used to cut the primary lateral panicle branches which give rise to the staminate florets. The prepared panicle is rebagged and the ♀ sign and parents' name or number are written on the bag.

C. Pollination

Collection of pollen is normally done from panicles that are predominately emerged from the boot (Fig. 3). Panicles are bagged 12 to 24 hours before pollen is collected. This should be adequate time for the viability of any foreign pollen on the stamen to be lost and to allow time for newly developed pollen to mature in the anthers. Under conditions when pollen is scarce and high temperature and low humidity are expected, the most successful pollinations have been those made during the early morning hours. While factors such as temperature, light, and humidity affect pollen production and viability, no method is known for long-term storage of pollen. Therefore, synchronization of flowering in plants to be crossed is manditory.

Bagging for pollen collection is done by slipping the bag over the panicle, folding the open end tightly around the culm, paper clipping the fold, and tying the bagged panicle to a bamboo pole. Collection of pollen is accomplished by turning the bagged panicle of the pollen parent horizontally and tapping the side of the bag. The tapping causes a rupture of the anthers and collection of the yellow pollen in the bag. The bag is removed while it is held horizontally. The protective bag is removed from the prepared female panicle, and the panicle is inserted into the bag containing the pollen. With the open end of the bag folded around the culm, the bag is shaken to distribute pollen over the pistillate florets. After removal of the pollinating bag, pollen can be seen adhering to the stigmas. Both the pollinating bag and the bag used for protecting the female should be replaced on their respective parental plants. An examination of the female parent the following day will reveal withered and dried stigmas, if fertilization occurred. Because stigma extrusion from the florets on one panicle occurs over 2 to 3-day period, the pollination process is normally repeated to increase the number of possible F_1 seeds.

In producing selfed progeny of wild rice, pollen from the lower portion of a panicle must reach the pistillate florets at the top of a panicle. Four methods of obtaining selfs were investigated under natural conditions by Bush and Elliott (1977). The methods were a) bagging a single panicle when it was in the boot stage, keeping the panicle tied to a bamboo pole, and making necessary adjustments as growth demanded; b) identical to a), except that during 2 to 3 consecutive days at anthesis, the panicles were bent so that by tapping the bag, pollen fell over the pistillate florets; c) identical to a), except that two tillers from the same plant, differing slightly in maturity, were bagged together. This would allow pollen from the older tiller to fall downward onto the pistillate florets of younger tillers; d) identical to a), except carbon dioxide was injected under the bag for a few seconds during 2 to 3 days of anthesis. The carbon dioxide was used to blow the pollen upward over the pistillate florets.

Treatment a) resulted in 7% seed set, b) 34%, c) 13%, and d) 15%. Although treatment b) had the highest seed set, it also was most labor demanding. Approximately 50 seeds per panicle were obtained in this study which was conducted outdoors. Approximately 20 seeds per panicle have been more commonly obtained when using treatment b) for selfing in the greenhouse.

D. Factors Affecting Efficiency

Synchronization of flowering between parental lines is possible through altering planting dates, planting rates, photoperiods, and temperatures. Planting an early parent on weekly intervals after planting of the later parent often is practiced. The earlier maturing parent also may be planted at a lower plant population to allow development of a maximum number of tillers that result in a range of 2 weeks in maturity. Because wild rice has been found to be a quantitative long-day plant that is very temperature sensitive, hastening of flowering appears possible by increasing the day length, interrupting the night period, and increasing the temperature.

Shattering and plant pigmentation traits can be useful as genetic markers. Shattering was found to be controlled by two complementary dominant genes and has been used in studies of pollen movement (Elliott and Perlinger, 1977). Red pigmentation of the leaf sheath also was found to be due to two complementary dominant genes (Perlinger, 1976). The pigmentation trait is especially useful because green plants are rare in natural populations and the leaf sheath coloration can be detected both in the seedling and the mature-plant stage. Male floret coloration has been observed in F_2 populations to be completely linked with coloration in the sheath (Woods and Gutek, 1975; Perlinger and Elliott, 1975). Because the male floret coloration can be more easily observed at maturity than sheath color, it is usually used when classifying plants.

V. NATURAL HYBRIDIZATION

Male sterility systems, incompatibility systems, and isolation methods are pertinent considerations because wild rice is largely cross pollinated. Male sterility is commonly observed in wild rice, but it has not been studied. A completely pistillate mutant has been found and inheritance studies have been initiated (Elliott and Wood, 1976). No evidence exists for incompatibility systems in wild rice.

Because space in paddies is often limited and controlled pollinations are difficult, studies were conducted on the effect of plot size and design in reducing crossing with unwanted foreign pollen (Elliott, 1977). The number of border rows needed to reduce cross pollination also is being studied. An isolation distance of at least 225 m is required to prevent cross pollination between unique populations. This information is based upon pollen trapping from a 1 ha paddy of wild rice (Elliott, 1973).

Isolation in time may be accomplished by utilizing populations having different maturities and by staggering planting. There can be some difficulty in accomplishing this because tillers on the same plant mature over about a 2-week period.

Studies were conducted on the effectiveness of cages for providing isolation (Elliott, 1977). Cages were constructed and covered with muslin (50 threads/cm). While outcrossing was reduced under the cages, it was not prevented. Modification of the type of muslin is apparently needed. The necessity for very large cages to cover even a few plants makes cages expensive and time consuming to construct, install, and store.

VI. SEED DEVELOPMENT, HARVEST, AND STORAGE

If the stigma has withered 24 hours after pollination, fertilization probably has occurred. Two weeks after fertilization, the seed should be approximately 5 mm long, and after about 4 weeks the seed should be ready for harvest. Seeds may be harvested while they are green, as long as they are firm, or they may be harvested after the seed becomes darkened (mature). Harvesting of crosses is commonly done with clippers. The panicles are left in or placed into pollinating head bags to prevent loss of seed due to shattering. Seed identification is maintained on the bag. Selfed and crossed panicles are threshed by hand stripping the seed.

Viability of wild rice becomes reduced when allowed to become dryer than approximately 30% moisture (Brun, 1968). Seed can be held for as many as 16 weeks at 2 C and high humidity without lowering germination substantially (Elliott, 1974). Seed has routinely been submerged in water for storage and held at approximately 2 C (Muenscher, 1936). Other temperatures, e.g. −7, −2, and 13 C, and storage methods have been investigated, but no treatment has resulted in higher germination than that described by Muenscher (Elliott, 1973). No storage method presently known allows for the storage of seed for more than 1 year. Petri dishes are presently used for storing self and crossed seeds. The pedigree is written on the lids, rubber bands are used to keep the lids in place and to group dishes, and rubberized trays holding 45 dishes are used to hold the water for submerging the seed. Seed storage fungi can be reduced if the contaminated water is changed at monthly intervals with fresh, prechilled water.

VII. TECHNIQUES FOR SPECIAL SITUATIONS

Techniques to asexually maintain or multiply unique genotypes are needed. Stem cuttings have not been successful for propagation. Subdividing plants to promote more tillering for the purpose of producing more seed has been only partially successful. Transplanting the plants from one location to another in the tillering through the late boot stage is routinely done to facilitate crossing.

Tissue derived from the embryo of wild rice seed has been cultured for more than 1 year. Regeneration of plants, however, has not been achieved (Gengenbach, 1977).

REFERENCES

Brun, W. A. 1968. Effects of moisture percentage and alternating temperature on germination of wild rice (*Zizania aquatica* L.). Am. Soc. Agron. Abstr.:29.

Bush, T. C., and W. A. Elliott. 1977. Unpublished data. Dep. of Agron., Univ. of Minnesota, St. Paul, MN 55108.

Cardwell, V. B., E. A. Oelke, and W. A. Elliott. 1973. Seed dormancy studies. Progress Rep. of Wild Rice Res., Minnesota Agric. Exp. Stn., St. Paul. p. 14–16.

Dore, W. G. 1969. Wild-Rice. Can. Dep. Agric. Publ. No. 1393.

Elliott, W. A. 1973. Wild rice breeding research. Progress Rep. of Wild Rice Res., Minnesota Agric. Exp. Stn. St. Paul. p. 17–22.

———. 1974. Wild rice breeding research. Progress Rep. of Wild Rice Res., Minnesota Agric. Exp. Stn., St. Paul. p. 18–31.

———. 1976. Wild rice breeding research. Progress Rep. of Wild Rice Res., Minnesota Agric. Exp. Stn., St. Paul. p. 51–64.

———. 1977. Unpublished data. Dep. of Agronomy, Univ. of Minnesota, St. Paul, MN 55108.

———, and E. A. Oelke. 1977. New era for wild rice. Crops Soils 29:8–11.

———, ———, and W. A. Brun. 1974. The effect of photoperiod on heading date in wild rice (*Zizania aquatica* L.). Am. Soc. Agron. Abstr.:52.

———, and G. J. Perlinger. 1977. Inheritance of shattering in wild rice (*Zizania aquatica* L.). Crop Sci. 17:851–853.

———, and D. L. Woods. 1976. Unpublished data. Dep. of Agronomy, Univ. of Minnesota, St. Paul, MN 55108.

Emery, W. H. P. 1977. Personal communication. Southwest Texas State Univ., San Marcos, TX 78666.

Fassett, N. C. 1924. A study of the genus *Zizania.* Rhodora 25:153–160.

Fenster, W. E., C. J. Overdahl, C. A. Simpkins, and John Grava. 1976. Guide to computer programmed soil test recommendation in Minnesota. Agric. Ext. Serv. Spec. Rep. 1. Univ. of Minnesota, St. Paul.

Gengenbach, B. G. 1977. Unpublished data. Dep. of Agronomy, Univ. of Minnesota, St. Paul, MN 55108.

Muenscher, W. C. 1936. Storage and germination of seeds of aquatic plants. Cornell Univ. Agric. Exp. Stn. Bull. 652. p. 1–17.

Oelke, E. A. 1973. Wild rice cultural research. Progress Rep. of Wild Rice Res., Minnesota Agric. Exp. Stn., St. Paul. p. 1–14.

———. 1975. Wild rice cultural research. Progress Rep. of Wild Rice Res., Minnesota Agric. Exp. Stn., St. Paul, p. 1–12.

———. 1977. Unpublished data. Dep. of Agronomy, Univ. of Minnesota, St. Paul, MN 55108.

———, W. A. Elliott, M. F. Kernkamp, and D. M. Noetzel. 1973. Commercial production of wild rice. Univ. of Minnesota Agric. Ext. Folder 284.

Perlinger, G. J. 1976. The inheritance of sheath color, collar color, and shattering resistance in wild rice. M.S. thesis. Dep. of Agronomy, Univ. of Minnesota, St. Paul.

———, and W. A. Elliott. 1975. Unpublished data. Dep. of Agronomy, Univ. of Minnesota, St. Paul, MN 55108.

Snyder, G. H. 1977. Personal communication. Inst. of Food and Agric. Sci., Belle Grade, FL 33430.

Weir, C., and H. M. Dale. 1960. A developmental study of wild rice, *Zizania aquatica* L. Can. J. Bot. 38:719–739.

Woods, D. L., and L. H. Gutek. 1974. Germinating wild rice. Can. J. Plant Sci. 54: 423–424.

———, and ———. 1975. Unpublished data. Dep. of Plant Sci., Univ. of Manitoba, Winnepeg, Manitoba, R3T 2N2, Canada.

SUBJECT INDEX

A

B

K

L

M

P

S